Hybrid and Mixed Finite Element Methods

WILEY SERIES IN
NUMERICAL METHODS IN ENGINEERING

Consulting Editors
R. H. Gallagher, *College of Engineering, University of Arizona*
and
O. C. Zienkiewicz, *Department of Civil Engineering, University of College of Swansea*

Rock Mechanics in Engineering Practice
Edited by K. G. Stagg and O. C. Zienkiewicz

Optimum Structural Design: Theory and Applications
Edited by R. H. Gallagher and O. C. Zienkiewicz

Finite Elements in Fluids
Vol. 1 Viscous Flow and Hydrodynamics
Vol. 2 Mathematical Foundations, Aerodynamics and Lubrication
Edited by R. H. Gallagher, J. T. Oden, C. Taylor, and O. C. Zienkiewicz

Finite Elements for Thin Shells and Curved Members
Edited by D. G. Ashwell and R. H. Gallagher

Finite Elements in Geomechanics
Edited by G. Gudehus

Numerical Methods in Offshore Engineering
Edited by O. C. Zienkiewicz, R. W. Lewis, and K. G. Stagg

Finite Elements in Fluids
Vol. 3
Edited by R. H. Gallagher, O. C. Zienkiewicz, J. T. Oden, M. Morandi Cecchi, and C. Taylor

Energy Methods in Finite Element Analysis
Edited by R. Glowinski, E. Rodin, and O. C. Zienkiewicz

Finite Elements in Electrical and Magnetic Field Problems
Edited by M. V. K. Chari and P. Silvester

Numerical Methods in Heat Transfer
Edited by R. W. Lewis, K. Morgan, and O. C. Zienkiewicz

Finite Elements in Biomechanics
Edited by R. H. Gallagher, B. R. Simon, P. C. Johnson, and J. F. Gross

Soil Mechanics—Transient and Cyclic Loads
Edited by G. N. Pande and O. C. Zienkiewicz

Finite Elements in Fluids
Vol. 4
Edited by R. H. Gallagher, D. Norrie, J. T. Oden, and O. C. Zienkiewicz

Foundations of Structural Optimization: A Unified Approach
Edited by A. J. Morris

Creep and Shrinkage in Concrete Structures
Edited by Z. Bazant and F. Wittmann

Hybrid and Mixed Finite Element Methods
Edited by S. N. Atluri, R. H. Gallagher and O. C. Zienkiewicz

Hybrid and Mixed Finite Element Methods

Edited by

S. N. Atluri

Regents' Professor of Mechanics
Center for the Advancement of Computational Mechanics
Georgia Institute of Technology, Atlanta, Georgia

R. H. Gallagher

Dean of Engineering College
University of Arizona, Tucson, Arizona

and

O. C. Zienkiewicz, F.R.S.

Professor and Head, Civil Engineering Department
University College of Swansea, U.K.

A Wiley–Interscience Publication

JOHN WILEY & SONS
Chichester · New York · Brisbane · Toronto · Singapore

Library of Congress Cataloging in Publication Data:
Main entry under title:

Hybrid and mixed finite element methods.

(Wiley series in numerical methods in engineering)
' A Wiley–Interscience Publication.'
'List of publications by Theodore H. H. Pian'; p.
Includes indexes.
1. Finite element method. 2. Pian, H. H. Theodore
I. Atluri, Satya N. II. Gallagher, Richard H.
III. Zienkiewicz, O. C. IV. Series.
TA347.F5H93 1983 620′.0042 82-8615
ISBN 0 471 10486 8 AACR2

British Library Cataloguing in Publication Data:
Hybrid and mixed finite element methods.—
(Wiley series in numerical methods in engineering)
1. Pian, Theodore 2. Finite element method
—Addresses, essays, lectures
I. Atluri, S. N. II. Gallagher, R. H.
III. Zienkiewicz, O. C. IV. Pian, Theodore
515.3′53 TA347.F5

ISBN 0 471 10486 8

Printed in Northern Ireland at The Universities Press (Belfast) Ltd.
and bound by The Pitman Press, Bath.

Preface

This volume is dedicated to Professor Theodore H. H. Pian for his sixtieth birthday anniversary. It includes twenty-eight contributions by his colleagues, associates, and former students from across the world. The final chapter contains some reflections of Theodore Pian. Most of these contributions have been orally presented at an International Symposium to honour Professor Pian which was held during 8–10 April 1981 at the Georgia Institute of Technology, Atlanta, Georgia, under the sponsorship of the U.S. National Science Foundation, Office of Naval Research, and the Army Research Office. The symposium was organized by a committee which included:

S. N. Atluri, Center for the Advancement of Computational Mechanics, Georgia Institute of Technology, Chairman
R. H. Gallagher, University of Arizona, Tucson
P. Tong, Transportation Systems Center, Department of Transportation, Cambridge, Mass.
O. C. Zienkiewicz F.R.S., University College, Swansea, U.K.
K. Washizu, Univeristy of Tokyo, Japan

All the authors were given sufficient opportunity to update and revise their original oral presentations. Most of the chapters of this book are thus completely rewritten and are representative of the state of the subject at the time this volume is going to press.

Since this is the first and only comprehensive book on the subject of hybrid and mixed finite element methods, the editors hope that the book will be of permanent value to researchers and students in the area of computational mechanics.

In a book of this nature, it is almost unavoidable to have some non-uniformity of notation and presentation. The individual chapters are nevertheless reasonably self-contained, so that little difficulty will be encountered by the readers.

We wish to thank all the contributors to this volume and Ms. M. Eiteman for her assistance in the editorial tasks.

January 1982

S. N. Atluri
R. H. Gallagher
O. C. Zienkiewicz

Dedication

Theodore Pian was born in Shangai, China, on 18 January 1919. He earned the B.S. degree (1940) from Tsing Hua University in China and the M.S. (1944) and the Doctor of Science (1948) degrees from the Massachusetts Institute of Technology. He has spent all of his professional career to date at M.I.T., where he has been a Professor of Aeronautics and Astronautics since 1966. He has been a visiting professor at the University of Tokyo (1974) and at the Technical University, Berlin (1975).

Professor Pian has been the recipient of the Von Karman Memorial Prize from the TRE Corporation, Beverly Hills, California, in 1974, and the Structures, Structural Dynamics and Materials award from the American Institute of Aeronautics and Astronautics (AIAA) in 1975. He has served as an Associate Editor of the *AIAA Journal*, and continues to serve on the editorial boards of several international journals in the areas of structural mechanics and finite element methods.

Pian's contributions to the literature on structural mechanics and finite element methods have been truly inspiring. He is the author, coauthor, or editor of two books and more than ninety papers, a list of which is given at the end of this volume. Most of these articles deal with structural mechanics, and since the early 1960s his work has dealt primarily with finite element methods in solid mechanics. In 1964 he published a paper in the *AIAA Journal* which turned out to be the foundation of hybrid and mixed finite element methods, the subject dealt with in this volume.

His colleagues, associates, and former students who contributed to this volume did so not only as a tribute to his distinguished scientific career but also out of a genuine affection and admiration for him as a person. Many a graduate student has obtained his degrees at M.I.T. under his guidance, and each came away not only with a technical education but also as a mature human being with a strong sense of values such as humility, patience, and perseverence.

It is with a deep sense of affection and admiration that this volume is dedicated in honor of the sixtieth birthday anniversary of Theodore Pian.

Contributors

T. Aizawa — *Department of Nuclear Engineering, University of Tokyo, 7-3-1 Hongo Bunkyo-ku, Tokyo, Japan.*

D. N. Arnold — *University of Maryland, Institute for Physical Science and Technology, College Park, Maryland 20742, U.S.A.*

S. N. Atluri — *School of Civil Engineering, Georgia Institute of Technology, Atlanta, Georgia 30332, U.S.A.*

I. Babuska — *University of Maryland, Institute for Physical Science and Technology, College Park, Maryland 20742, U.S.A.*

K. J. Bathe — *Mechanical Engineering Department, 3-354 M.I.T., Cambridge, Massachusetts 02139, U.S.A.*

J. M. W. Baynham — *Department of Civil Engineering, University College of Swansea, Swansea SA2 8PP, U.K.*

T. Belytschko — *Department of Civil and Nuclear Engineering, Northwestern University, Evanston, Illinois 60201, U.S.A.*

M. Bercovier — *School of Applied Science and Technology, Hebrew University of Jerusalem, Jerusalem, Israel*

J. F. Besseling — *Laboratory of Engineering Mechanics, Mekelweg 2, Technische Hogeschool Delft, 2628 CD Delft, Holland*

F. Brezzi — *Instituto di Analisi Numerica del CNR, University of Pavia, Pavia, Italy*

Wei-Zang Chien — *Department of Mechanics, Tsinghua University, Peking, People's Republic of China*

B. Fransson — *Department of Structural Mechanics, Chalmen University of Technology, Goteborg 5, Sweden*

R. H. Gallagher — *Office of Dean of Engineering, University of Arizona, Tucson, Arizona 85721, U.S.A.*

E. Giencke — *Technische Universitat Berlin, Institut, fur Luft-und Raumfahrt, Gebaude 4.1, Sekr. SG 9 Salzufer 17–19, D-1000 Berlin 10, West Germany*

Y. Gilon — *School of Applied Science and Technology, Hebrew University of Jerusalem, Jerusalem, Israel*

Y. Hasbani — *School of Applied Science and Technology, Hebrew University of Jerusalem, Jerusalem, Israel*

J. C. Heinrich — *Office of Dean of Engineering, University of Arizona, Tucson, Arizona 85721, U.S.A.*

R. D. Henshell — *PAFEC Limited, Strelley Hall, Nottingham, U.K.*

L. R. Herrmann — *Department of Civil Engineering, University of California, Davis, California 95616, U.S.A.*

Dah-Wei Hsueh — *Peking Institute of Technology, Peking, People's Republic of China*

T. J. R. Hughes — *Department of Mechanical Engineering, Stanford University, Stanford, California 04305, U.S.A.*

T. Kawai — *University of Tokyo, Institute of Industrial Science, 22-1 Roppongi 7 Chome, Minato-ku, Tokyo 106, Japan*

F. Kikuchi — *Department of Mathematics, College of General Education, University of Tokyo, 3-8-1 Komaba, Meguro-ku, Tokyo 153, Japan*

R. Kohn — *Courant Institute, New York University, New York, U.S.A.*

Tang Limin — *Department of Engineering Mechanics, Dalian Institute of Technology, Dalian, Liaoing, People's Republic of China*

D. S. Malkus — *Department of Mathematics, Illinois Institute of Technology, Chicago, Illinois 60616, U.S.A.*

H. Murakawa — *Engineering Information Processing Department, 1-1 Saiwai-cho, 3-chome, Hitachi, Ibaraki 317, Japan*

S. Nemat-Nasser — *Department of Civil Engineering, Northwestern University, Evanston, Illinois 60201, U.S.A.*

A. K. Noor — *George Washington University, JIAFS, NASA Langley Research Center, Mail Stop 246, Hampton, Virginia 23665, U.S.A.*

J. T. Oden — *TICOM, University of Texas at Austin, Austin, Texas 78712, U.S.A.*

H. Okumura — *Institute of Industrial Science, University of Tokyo, 22-1 Roppongi 7-chome, Minato-ku, Tokyo 106, Japan*

M. D. Olson — *Department of Civil Engineering, University of British Columbia, Vancouver, B.C., Canada V6T 1W5*

J. Osborn — *University of Maryland, Institute for Physical Science and Technology, College Park, Maryland 20742, U.S.A.*

J. M. Peters — *George Washington University, JIAFS, NASA Langley Research Center, Mail Stop 246, Hampton, Virginia 23665, U.S.A.*

T. H. H. Pian — *Department of Aeronautics and Astronautics, Rm 33-311, Cambridge, Massachusetts 02139, U.S.A.*

F. N. Rigby — *PAFEC Limited, Strelley Hall, Nottingham, U.K.*

A. Samuelsson — *Department of Structural Mechanics, Chalmen University of Technology, Goteborg 5, Sweden*

N. Sarigul — *Office of Dean of Engineering, University of Arizona, Tucson, Arizona 85721, U.S.A.*

R. L. Spilker — *Department of Materials Engineering, University of Illinois at Chicago, Chicago, Illinois 60680, U.S.A.*

G. Strang — *Department of Mathematics, M.I.T., Cambridge, Massachusetts 20139, U.S.A.*

Y. Sumi — *Department of Naval Architecture and Ocean Engineering, Yokohama National University, Hodogaya-ku, Yokohama 240, Japan*

R. L. Taylor — *Department of Civil Engineering, University of California, Berkeley, California, U.S.A.*

N. Tokuda — *Research Institute, Ishikawajima-Harima Heavy Industries Limited, Koto-ku, Tokyo 135, Japan*

P. Tong — *Department of Transportation, Research and Special Programs Administration, Transportation Systems Center, Kendall Square, Cambridge, Massachusetts 02142, U.S.A.*

R. Valid — *Office National d'Etudes et de Recherches Aerospatials, 92320 Chatillon, France*

Chen Wanji — *Department of Engineering Mechanics, Dalian Institute of Technology, Dalian, Liaoing, People's Republic of China*

J. J. Webster — *Department of Mechanical Engineering, Nottingham University, Nottingham, U.K.*

W. Wunderlich — *Rurh-Universitat Bochum, Institut fur Konstruktiven Ingenieurbau, Lehrsuhl IV, 463 Bochum-Querenburg, Buscheystrasse, Postfach 2148, West Germany*

G. Yagawa — *Department of Nuclear Engineering, University of Tokyo, 7-3-1 Hongo-Bunkyo-ku, Tokyo, Japan*

Y. Yamada — Institute of Industrial Science, University of Tokyo, 22-1 Roppongi 7-chome, Minato-Ku, Tokyo 106, Japan

Y. Yamamoto — *Department of Naval Architecture, University of Tokyo, Bunkyo-ku, Tokyo 113, Japan*

Liu Yingxi — *Department of Engineering Mechanics, Dalian Institute of Technology, Dalian, Liaoing, People's Republic of China.*

O. C. Zienkiewicz — *Department of Civil Engineering, University College of Swansea, Swansea SA2 8PP, U.K.*

Contents

Preface

1 **Mixed Finite Elements for Couple-Stress Analysis** 1
L. R. Herrmann

2 **The Mixed Finite Element Method in Elasticity and Elastic Contact Problems** 19
Mervyn D. Olson

3 **Recent Studies of Hybrid and Mixed Finite Element Methods in Mechanics** 51
S. N. Atluri, P. Tong, and H. Murakawa

4 **Hybrid and Hellinger–Reissner Plate and Shell Finite Elements** 73
F. N. Rigby, J. J. Webster, and R. D. Henshell

5 **New Hybrid Stress Models in the Limit Analysis of Solids and Structures** 93
Tadahiko Kawai

6 **Bilinear Mindlin Plate Elements** 117
Robert L. Spilker and Ted Belytschko

7 **On a Mixed Method Related to the Discrete Kirchhoff Assumption** 137
Fumio Kikuchi

8 **A Variational Finite Strip Method with Mixed Variables** . . 155
Bo Fransson and Alf Samuelsson

9 **String Net Function Approximation and Quasi-Conforming Technique** 173
Tang Limin, Chen Wanji, and Liu Yingxi

10 **A Mixed Method for Non-Linear Plate and Shallow Shell Problems** 189
E. Giencke

11 Mixed Models for Plates and Shells: Principles—Elements—Examples 215
W. Wunderlich

12 Estimate of Dynamic Properties of Composites by Mixed Finite Element Methods 243
S. Nemat-Nasser

13 Finite Element Properties, Based Upon Elastic Potential Interpolation 253
J. F. Besseling

14 Hybrid Approximations of Non-Linear Plate-Bending Problems 267
F. Brezzi

15 Optimal Design for Torsional Rigidity 281
Robert Kohn and Gibert Strang

16 The Structural Stability Criterion for Mixed Principles . . . 289
Roger Valid

17 Finite Element Analysis of Stress and Strain Singularity Eigenstate in Inhomogeneous Media or Composite Materials 325
Y. Yamada and H. Okumura

18 Some Fracture Mechanics Analyses Using Finite Element Method with Penalty Function 345
Genki Yagawa and Tatsuhiko Aizawa

19 Accuracy Considerations for Finite Element Calculations of the Stress Intensity Factor by the Method of Superposition 361
Yoshijuki Yamamoto, Nasaki Tokuda, and Yoichi Sumi

20 Incompatible Plate Elements Based upon Generalized Variational Principles 381
Wei-Zang Chien

21 Mixed and Irreducible Formulations in Finite Element Analysis 405
O. C. Zienkiewicz, R. L. Taylor, and J. M. W. Baynham

22 Selection of Finite Element Methods 433
D. N. Arnold, I. Babuška, and J. Osborn

23 Complementary Energy Revisited 453
R. H. Gallagher, J. C. Heinrich, and N. Sarigul

24 Mixed Finite Element Approximations Via Interior and Exterior Penalties for Contact Problems in Elasticity . . 467
J. T. Oden

25 A General Penalty/Mixed Equivalence Theorem for Anisotropic, Incompressible Finite Elements 487
Thomas J. R. Hughes and David S. Malkus

26 On a Finite Element Procedure for Non-linear Incompressible Elasticity 497
M. Bercovier, Y. Hasbani, Y. Gilon, and K. J. Bathe

27 Four Theorems on the Limit analysis in Solid Mechanics . . 519
Hsueh Dah-Wei

28 Mixed Models and Reduced/Selective Integration Displacement Models for Vibration Analysis of Shells 537
Ahmed K. Noor and Jeanne M. Peters

29 Reflections and Remarks on Hybrid and Mixed Finite Element Methods 565
Theodore H. H. Pian

30 List of Publications by Theodore H. H. Pian 571

Author Index 577

Hybrid and Mixed Finite Element Methods
Edited by S. N. Atluri, R. H. Gallagher, and O. C. Zienkiewicz

Chapter 1

Mixed Finite Elements for Couple-Stress Analysis

Leonard R. Herrmann

1.1 INTRODUCTION

1.1.1 Objective

The objective of the paper is to investigate and compare four 'mixed' finite elements for plane couple-stress elasticity. The purpose of the paper is not to describe or justify couple-stress theory, but some preliminary remarks on this subject are given in the next two subsections for completeness. The contents of the paper are limited to linear elastic considerations, but they can be extended to non-linear and/or inelastic conditions.

1.1.2 Couple-stress theory

Couple-stress theory was originated by Cosserat [1] and often bears his name. More recent developments and extensions may be found in works on 'micropolar' elasticity and 'multimoment' plate and shell theories (e.g. see the papers by Eringen [2] and Gruzdev [3]).

The brief description of couple-stress theory that follows is taken from the work of Mindlin [4, 5] and is restricted to plane elasticity. Couple-stress theory admits the existence on a differential element of not only normal and shear but also couple-stress components (see Figure 1.1). While the usual stress components are functions of the strains, couple-stresses are functions of the strain gradients. These latter relations introduce (for an isotropic material) another elastic constant called the 'modulus of curvature'. The ratio of the modulus of curvature and the shear modulus (or Young's modulus) has the dimension of length squared and is a property of the material (denote this length by l).

In classical elasticity it is assumed that the couple stresses are zero [6]. This assumption has as its justification the fact that for most applications the predictions of the resulting theory are verified by experiments. A theoretical

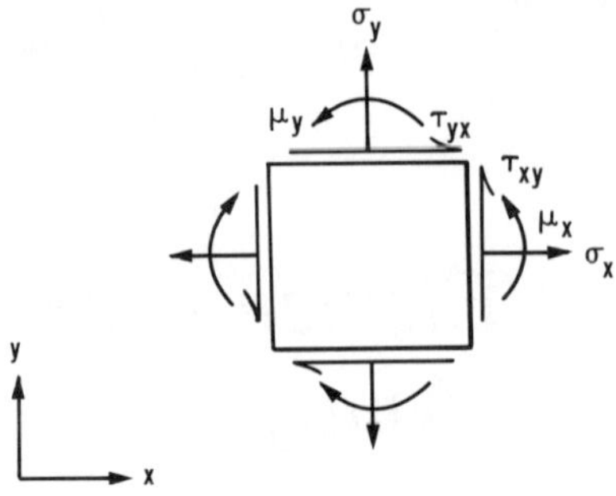

Figure 1.1 Stress and couple-stress components in a rectangular coordinate system

argument is sometimes advanced that the existence of a couple-stress on an infinitesimal element would require a discontinuous force distribution (a couple) at the microstructure level and that this is incompatible with the continuum concept. Considering the fact that no real material is truly a continuum at the microstructure level this argument is questionable.

The argument that the usual elasticity theory agrees with experimental results suffices for those cases where it is true. There is, however, some evidence that in certain situations classical elasticity theory is not completely adequate and that it may be necessary to include couple-stresses. The importance of the couple-stresses depends heavily on the value of the fundamental length quantity l; the larger the value of l the larger the effect. The length l is a property of the material's microstructure (for a true continuum, i.e. a hypothetical material without microstructure, even at the atomic level, presumedly $l = 0$). Mindlin [4] suggests that: 'In perfect crystals and amorphous materials like glass or plastics, l is probably submicroscopic; but might be of the order of the radius of the root of a crack. In polycrystalline metals or granular materials, l may be considerably larger. For example, in the idealized case of a simple cubic array of contiguous elastic spheres, l is about three quarters of the radius of a sphere.' For composite materials for which a 'representative volume' exists, the value of l appears to be of the order of its largest dimension (usually such materials are orthotropic and there is more than one characteristic length) [7, 8, 9]. Mindlin [4] concludes: 'Thus, dimensions of structural elements and machine parts, or dimensions of holes, fillets, notches or cracks in them, may approach l for a variety of materials.'

1.1.3 Analysis of couple-stress problems

Couple-stress elasticity is considerably more complicated than conventional theory and hence leads to more difficult boundary value problems. Analytical solutions have been found for a few basic couple-stress problems (e.g. see [4]). However, as is the case for conventional theory, one must resort to numerical analysis methods for the solution of all but the simplest problems. One approach is to use Galerkin's method or the theorem of minimum potential energy to develop a finite formulation in terms of displacement quantities. The resulting analysis would be considerably more complicated than for conventional elasticity. The added complexities result from the higher order continuity requirements imposed on the approximate displacement field. While a finite element formulation in terms of displacements can be found, it is of interest to determine whether or not a mixed approach would yield a simpler formulation (this possibility is suggested by related work for bending problems, that is [10, 11]).

1.2 GOVERNING EQUATIONS FOR PLANE COUPLE-STRESS ELASTICITY

1.2.1 Differential equations

The investigation is restricted to orthotropic materials. In addition, body forces and couples, and temperaure effects are not included. The notation and the governing differential equations are taken from [4] (with the exception that the stress–strain law has been generalized to include a class of orthotropic materials).

Linear and rotational equilibrium requires:

$$\frac{\partial \sigma_x}{\partial x} + \frac{\partial \tau_{yx}}{\partial y} = 0 \tag{1.1}$$

$$\frac{\partial \tau_{xy}}{\partial x} + \frac{\partial \sigma_y}{\partial y} = 0 \tag{1.2}$$

$$\frac{\partial \mu_x}{\partial x} + \frac{\partial \mu_y}{\partial y} + \tau_{xy} - \tau_{yx} = 0 \tag{1.3}$$

The presence of the couple-stresses μ_x and μ_y render the stress tensor unsymmetric and thus care must be taken in distinguishing betwen τ_{xy} and τ_{yx}. It is convenient to express the shear stress components in terms of symmetrical τ_s and antisymmetrical τ_a parts (see Figure 1.2), i.e.

$$\tau_s = \frac{\tau_{xy} + \tau_{yx}}{2} \tag{1.4}$$

$$\tau_a = \frac{\tau_{xy} - \tau_{yx}}{2} \tag{1.5}$$

Introducing these expressions into equations (1.1), (1.2), and (1.3) yields an

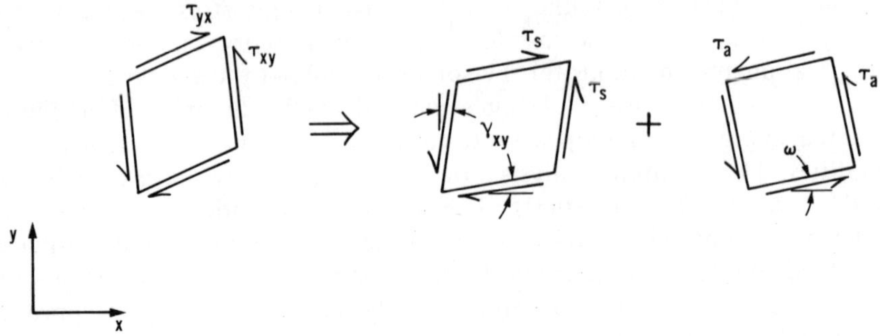

Figure 1.2 Shear strain and local rotation produced by the symmetric and antisymmetric components of the shear stress (note that the small deformation and rotation are greatly exaggerated)

alternative form for the equilibrium equations, i.e.

$$\frac{\partial \sigma_x}{\partial x}+\frac{\partial \tau_s}{\partial y}-\frac{\partial \tau_a}{\partial y}=0 \tag{1.6}$$

$$\frac{\partial \tau_s}{\partial x}+\frac{\partial \tau_a}{\partial x}+\frac{\partial \sigma_y}{\partial y}=0 \tag{1.7}$$

$$\frac{\partial \mu_x}{\partial x}+\frac{\partial \mu_y}{\partial y}+2\tau_a=0 \tag{1.8}$$

The above three equations can be reduced to two by using the last one to eliminate τ_a, i.e.

$$\frac{\partial \sigma_x}{\partial x}+\frac{\partial \tau_s}{\partial y}+\frac{1}{2}\frac{\partial^2 \mu_x}{\partial x\,\partial y}+\frac{1}{2}\frac{\partial^2 \mu_y}{\partial y^2}=0 \tag{1.9}$$

$$\frac{\partial \tau_s}{\partial x}+\frac{\partial \sigma_y}{\partial y}-\frac{1}{2}\frac{\partial^2 \mu_x}{\partial x^2}-\frac{1}{2}\frac{\partial^2 \mu_y}{\partial x\,\partial y}=0 \tag{1.10}$$

The deformation quantities of interest are the usual normal and shear strains (u and v are the displacement components in the x and y directions respectively):

$$\varepsilon_x=\frac{\partial u}{\partial x} \tag{1.11}$$

$$\varepsilon_y=\frac{\partial v}{\partial y} \tag{1.12}$$

$$\gamma_{xy}=\frac{\partial v}{\partial x}+\frac{\partial u}{\partial y} \tag{1.13}$$

the local rotation (see Figure 1.2; for simplicity the notation ω_z of [4] is

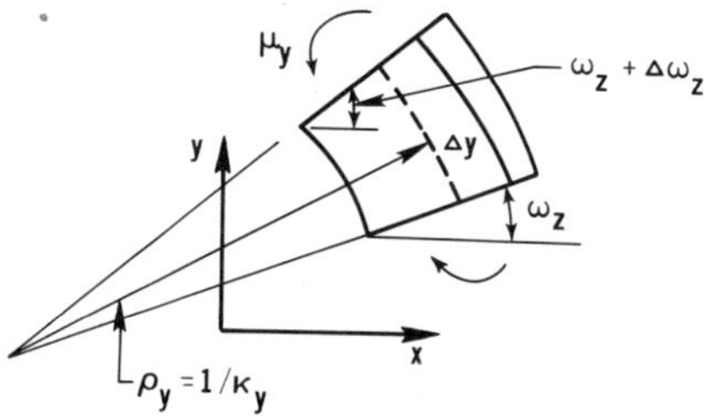

Figure 1.3 Curvature produced by the couple-stress μ_y

replaced by ω:

$$\omega = \frac{1}{2}\left(\frac{\partial v}{\partial x} - \frac{\partial u}{\partial y}\right) \tag{1.14}$$

and the curvatures* (see Figure 1.3):

$$\kappa_x = \frac{\partial \omega}{\partial x} = \frac{1}{2}\left(\frac{\partial^2 v}{\partial x^2} - \frac{\partial^2 u}{\partial x\,\partial y}\right) \tag{1.15}$$

$$\kappa_y = \frac{\partial \omega}{\partial y} = \frac{1}{2}\left(\frac{\partial^2 v}{\partial x\,\partial y} - \frac{\partial^2 u}{\partial y^2}\right) \tag{1.16}$$

The linear elastic constitutive equation is written in matrix notation:

$$[\sigma] = [D][\varepsilon] \tag{1.17}$$

where

$$[\sigma]^{\mathrm{T}} = [\sigma_x, \sigma_y, \tau_s, \mu_x, \mu_y] \tag{1.18}$$

$$[\varepsilon]^{\mathrm{T}} = [\varepsilon_x, \varepsilon_y, \gamma_{xy}, \kappa_x, \kappa_y] \tag{1.19}$$

The orthotropic behaviour is limited to that required for a study of structural grillages, i.e. the case when $D_{13} = D_{14} = D_{15} = D_{23} = D_{24} = D_{34} = D_{35} = D_{45} = 0$. For an isotropic material

$$D_{33} = \frac{E}{2(1+\nu)} = G \quad \text{and} \quad D_{44} = D_{55} = 4B \tag{1.20}$$

for plane strain

$$D_{11} = D_{22} = \frac{E(1-\nu)}{(1+\nu)(1-2\nu)} \quad \text{and} \quad D_{12} = \frac{E\nu}{(1-2\nu)(1+\nu)} \tag{1.21}$$

* While the advisability of using these particular 'curvature' definitions may be questioned on both theoretical and practical grounds (under certain circumstances they lead to somewhat ill-defined descriptions of boundary conditions and external work) they will adequately serve to illustrate the points of interest in this study.

while for plane stress

$$D_{11}=D_{22}=\frac{E}{1-\nu^2} \quad \text{and} \quad D_{12}=\frac{E\nu}{1-\nu^2} \tag{1.22}$$

Young's modulus and Poisson's ratio are denoted by E and ν, while B is the modulus of curvature (or bending) and has the dimensions of force. The value of B, in general, differs for plane stress and plane strain conditions; one would expect their ratio to be $(1-\nu^2)$.

The fact that the antisymmetric shear τ_a does not appear in the stress–strain equation does not pose a problem because it can be eliminated using the rotational equilibrium equation (1.8). Equations (1.6) to (1.8) and (1.11) to (1.17) are fourteen differential equations involving the fourteen unknowns σ_x, σ_y, τ_a, τ_s; μ_x, μ_y; ε_x, ε_y, γ_{xy}; κ_x, κ_y, ω; u and v.

Boundary conditions are discussed in the following sections.

1.2.2 Variational equations

1.2.2.1 General

The expression of the couple-stress elasticity problem in terms of minimum potential energy (or its least squared stress error interpretation [12]) is straightforward. The admissibility conditions require continuity of some of the first derivatives of the displacements, thus leading to quite complicated displacement-based finite element formulations. In order to avoid this problem, two alternative variational statements are considered in this section.

1.2.2.2 First mixed variational equation

The couple-stress problem can be expressed in terms of 'mixed' variational equations which utilize both displacement and force quantities as primary dependent variables. One such equation whose development was motivated by the work in [10] utilizes the displacement and couple-stress components as primary dependent variables, that is $\delta F_1=0$, where

$$\begin{aligned}F_1=\iint_B \frac{1}{2}\bigg\{ & D_{11}\left(\frac{\partial u}{\partial x}\right)^2+2D_{12}\frac{\partial u}{\partial x}\frac{\partial v}{\partial y}+D_{22}\left(\frac{\partial v}{\partial y}\right)^2+D_{33}\left(\frac{\partial u}{\partial y}+\frac{\partial v}{\partial x}\right)^2\\ & -\frac{1}{D_{44}}(\mu_x)^2-\left(\frac{\partial v}{\partial x}-\frac{\partial u}{\partial y}\right)\frac{\partial \mu_x}{\partial x}-\frac{1}{D_{55}}(\mu_y)^2-\left(\frac{\partial v}{\partial x}-\frac{\partial u}{\partial y}\right)\frac{\partial \mu_y}{\partial y}\bigg\}\,dx\,dy\\ & -\int_{S_1}\bar{\sigma}_n u_n\,ds-\int_{S_2}\bar{\tau}_{ns}u_s\,ds+\int_{S_3}\bar{\omega}\mu_n\,ds\end{aligned} \tag{1.23}$$

On the boundary S of the two-dimensional body B the outward normal and tangential coordinates are denoted by n and s, the displacement components by u_n and u_s, the normal and shear stress components of the surface traction by σ_n and τ_{ns}, and the couple-stress by μ_n. The overbars indicate specified quantities (for example $\bar{\sigma}_n$ is the specified value of σ_n on S_1). On the boundary segments S_1, S_2, and S_3 the quantities σ_n, τ_{ns}, and μ_n are respectively specified; on the remainder of the boundary S_1', S_2', and S_3' $(S_i+S_i'=S)$ the quantities u_n, u_s, and $\omega=(\partial u_s/\partial n-\partial u_n/\partial s)$ are specified.

The variations of F_1 with respect to the primary dependent variables u, v, μ_x, and μ_y respectively yield equations (1.9), (1.10), (1.15), and (1.16) as Euler equations where equations (1.11), (1.12), (1.13), and (1.17) are used to eliminate σ_x, σ_y, and τ_a from equations (1.9) and (1.10). The natural boundary conditions express $\bar{\sigma}_n$ in terms of the strains, $\bar{\tau}_{ns}$ in terms of the strains and the derivatives of the couple stresses,* and $\bar{\omega}$ in terms of the displacements (equation 1.14). The essential boundary conditions are the specifications of u_n, u_s, and μ_n. The admissibility conditions on the primary dependent variables are that u, v, and μ_n be continuous across element interfaces and satisfy specified boundary conditions.† The Euler equations and associated natural and essential boundary conditions are a proper (mixed) differential equation statement of the couple-stress problem.

1.2.2.3 *Second mixed variational equation*

A second mixed variational equation utilizing as primary dependent variables the displacement components (u, v), the rotation ω, and the antisymmetric component of the shear (τ_a) has the form $\delta F_2=0$, where‡

$$F_2=\iint_B \frac{1}{2}\left\{D_{11}\left(\frac{\partial u}{\partial x}\right)^2+2D_{12}\frac{\partial u}{\partial x}\frac{\partial v}{\partial y}+D_{22}\left(\frac{\partial v}{\partial y}\right)^2+D_{33}\left(\frac{\partial u}{\partial y}+\frac{\partial v}{\partial x}\right)^2\right.$$
$$\left.+D_{44}\left(\frac{\partial \omega}{\partial x}\right)^2+D_{55}\left(\frac{\partial \omega}{\partial y}\right)^2-4\tau_a\omega+2\left(\frac{\partial v}{\partial x}-\frac{\partial u}{\partial y}\right)\tau_a\right\}\,dx\,dy$$
$$-\int_{S_1}\bar{\sigma}_n u_n\,ds-\int_{S_2}\bar{\tau}_{ns}u_s\,ds-\int_{S_3}\bar{\mu}_n\omega\,ds \qquad (1.24)$$

* For example, using equations (1.4), (1.5), (1.8), (1.13), and (1.17) it is easy to show that $\tau_{yx}=\tau_s-\tau_a=D_{33}\gamma_{xy}-\frac{1}{2}(\partial\mu_x/\partial x+\partial\mu_y/\partial y)$, etc.

† Note that μ_x and μ_y need only satisfy continuity conditions to the extent that they render μ_n continuous.

‡ This expression can be regarded as an alternative statement for the potential energy where ω is treated as an independent variable with its actual dependence on u and v specified by the introduction of the Lagrange multiplier τ_a. Note the similarities of this variational equation to the one developed in [13].

The variations of F_2 with respect to u, v, ω, and τ_a give equations (1.6), (1.7), (1.8), and (1.14) as Euler equations, where equations (1.11) to (1.17) have been used to eliminate σ_x, σ_y, τ_s, μ_x, and μ_y. The natural boundary conditions express $\bar{\sigma}_n$ in terms of the strains, $\bar{\tau}_{ns}$ in terms of the strains and τ_a, and $\bar{\mu}_n$ in terms of ω (see equations 1.15, 1.16, and 1.17). The admissibility conditions are that u, v, and ω be continuous across element interfaces and satisfy specified boundary conditions; note that τ_a need not be continuous across element interfaces. The Euler equations and associated natural and essential boundary conditions are a proper (mixed) differential equation statement of the couple-stress problem.

1.3 MIXED FINITE ELEMENT ANALYSIS OF COUPLE-STRESS PROBLEMS

1.3.1 General

In the following subsections four easily implemented mixed finite elements for two-dimensional couple-stress elasticity are described. Other mixed elements are of course possible; these particular four were investigated because of their simplicity and the ease with which they can be adapted into

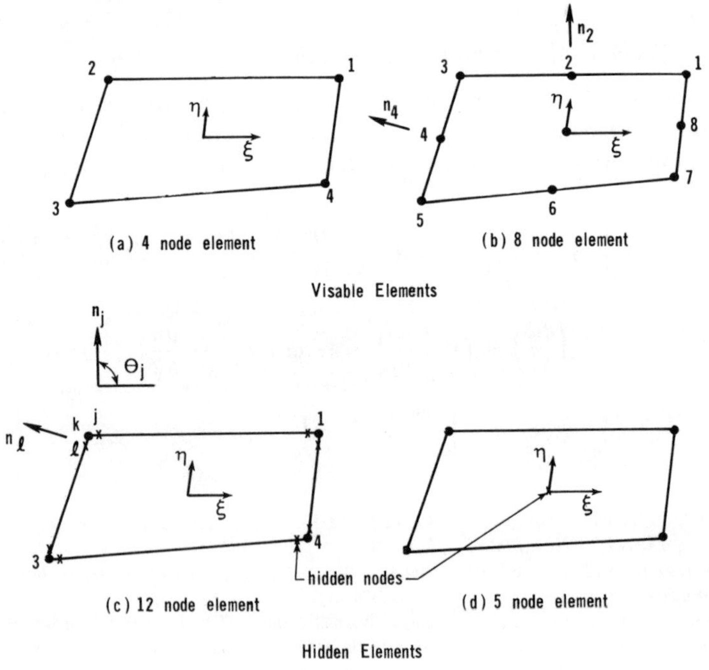

Figure 1.4 Element types

existing programmes. All four are straight-sided isoparametric elements with either four or eight nodes (see Figure 1.4). The first-order shape functions are denoted by N_i. All numerical integration is done with a four-point formula. For the sake of brevity standard details concerning the isoparametric elements are omitted; the reader is referred to a standard text such as [14]. The ranges of free and dummy indices are not stated when they are clear from the context; repeated dummy indices are summed over their ranges. A sample of numerical results obtained with these elements is given in the example at the end of this section.

1.3.2 Element Q4U16

The Q4U16 element is derived from the first mixed variational equation. A four-node isoparametric element (Figure 1.4a) with first-order approximations for all four primary dependent variables u, v, μ_x, and μ_y is used (the resulting continuity of μ_x and μ_y is more than is needed by the admissibility conditions which require continuity of μ_n but not μ_s across element interfaces). The four-node point unknowns* are denoted by U_i, V_i, M_{x_i}, and M_{y_i}, i.e.

$$u = u_iN_i, \quad v = V_iN_i, \qquad \mu_x = M_{x_i}N_i, \qquad \text{and} \qquad \mu_y = M_{y_i}N_i \tag{1.25}$$

Substituting these expressions into equation (1.23) and differentiating with respect to the node point unknowns yields (where $F_i \equiv \partial N_i/\partial x$, $G_i \equiv \partial N_i/\partial y$, and $|J|$ is the Jacobian)

$$\frac{\partial F_1}{\partial U_i} = \iint \{[D_{11}F_iF_j + D_{33}G_iG_j]U_j + [D_{12}F_iF_j + D_{33}G_iF_j]V_j + [\tfrac{1}{2}G_iF_j]M_{x_j} + [\tfrac{1}{2}G_iG_j]M_{y_i}\} |J| \, d\xi \, d\eta \tag{1.26}$$

$$\frac{\partial F_1}{\partial V_i} = \iint \{[\;\;]U_j + [D_{22}G_iG_j + D_{33}F_iF_j]V_j + [-\tfrac{1}{2}F_iF_j]M_{x_j} + [-\tfrac{1}{2}F_iG_j]M_{y_j}\} |J| \, d\xi \, d\eta \tag{1.27}$$

$$\frac{\partial F_1}{\partial M_{x_i}} = \iint \left\{[\;\;]U_j + [\;\;]V_j + \left[-\frac{1}{D_{44}} N_iN_j\right]M_{x_j} + [\mathrm{O}]M_{y_j}\right\} |J| \, d\xi \, d\eta \tag{1.28}$$

$$\frac{\partial F_1}{\partial M_{y_i}} = \iint \left\{[\;\;]U_j + [\;\;]V_j + [\;\;]M_{x_j} + \left[-\frac{1}{D_{55}} N_iN_j\right]M_{y_j}\right\} |J| \, d\xi \, d\eta \tag{1.29}$$

* In actual implementation the couple-stress variables are scaled by representative values of D_{44} and D_{55} in order to improve the numerical characteristics of the algebraic equations [10].

For the sake of brevity the symmetrical terms have not been repeated. The element matrix is formed from the above equations and with one exception the remainder of the finite element analysis proceeds as usual. At boundary nodes where either μ_n or a non-zero value of ω is specified the variables M_{x_i} and M_{y_i} (and the corresponding expressions $\partial F_1/\partial M_{x_j}$ and $\partial F_1/\partial M_{y_i}$) must be transformed to M_{n_i} and M_{s_i} with provision for either specifying M_{n_i} or subtracting the given value of ω_i (where 'distributed' specifications of ω are appropriately 'lumped' at the nodes) from the right-hand side of the system equation associated with M_{n_i}.* Most four-node isoparametric element, two-dimensional elasticity programmes can be easily modified to accept the Q4U16 couple-stress element.

1.3.3 Element Q8U16

When using the Q4U16 element (previous section), the specification of rotation boundary conditions at non-right-angled corners may be ambiguous. The problem is due to the ill-defined nature, in couple-stress theory, of boundary rotations. The element discussed in this section was developed to overcome this problem.

The Q8U16 element (Figure 1.4b) has two unknowns at each of its eight nodes. The disadvantages of this element when compared to the previous one are added complexities of data preparation resulting from the presence of the mid-side nodes, and the increased (for a given grid size) number of unknowns and bandwidth.

In the formulation of this element first consider the twelve-node element shown in Figure 1.4(c); the eight 'hidden' nodes* are located on the *sides* of the element at infinitesimal distances from the vertices. The hidden nodes on either side of vertex k (Figure 1.4c) are denoted j and l. The primary unknowns are again approximated using the first-order shape functions (equation 1.25, where i refers to the four vertex nodes in Figure 1.4c). The unknown displacements are assigned to the four vertex nodes of the eight-node element of Figure 1.4(b). At node k (Figure 1.4c) the values of M_{x_k} and M_{y_k} are expressed in terms of the normal couple-stresses (M_{n_j} and M_{n_l}) at the two adjacent hidden nodes (it is required that $\theta_j \neq \theta_l$); i.e.

$$\begin{bmatrix} M_{n_j} \\ M_{n_l} \end{bmatrix} = \begin{bmatrix} \cos\theta_j & \sin\theta_j \\ \cos\theta_l & \sin\theta_l \end{bmatrix} \begin{bmatrix} M_{x_k} \\ M_{y_k} \end{bmatrix} = [T] \begin{bmatrix} M_{x_k} \\ M_{y_k} \end{bmatrix} \tag{1.30}$$

Equation (1.30) is now used to transform the element matrix, found in the previous section for the Q4U16 element, so as to replace the M_x and M_y

* At corners provisions for specifying two couple-stress boundary conditions must be provided (e.g. at a free corner $\mu_x = \mu_y = 0$).

* A similar system of hidden nodes is used in the complementary energy element discussed in [15].

unknowns at the four vertex nodes by the M_n unknowns at the eight hidden nodes. Of course, both the rows and columns of the matrix are transformed in the process. For bookkeeping purposes, the two normal couple-stress unknowns at the two hidden nodes, for each of the four sides of the element of Figure 1.4(c), are associated with the corresponding mid-side node of the element of Figure 1.4(b); the result is the Q8U16 element. At each mid-side node the two unknowns are ordered such that the first refers to the normal couple-stress for the hidden node adjacent to the vertex node of the lower number. The resulting couple-stress approximation for the Q8U16 element does not exceed the admissibility requirement, as does the approximation for the Q4U16 element, that is μ_s is not forced to be continuous across element interfaces for the Q8U16 element as it is for the Q4U16 element.

Thus, the element of Figure 1.4(c) is treated as an eight-node element (Figure 1.4b) with two unknown displacements at each vertex node and two unknown normal couple-stresses at each mid-side node. The operation of transforming the vertex couple-stresses to normal components at the hidden nodes makes the form time for the Q8U16 element somewhat greater than for the Q4U16 element. Two boundary specifications are possible at each node (displacements or forces at the vertex nodes and couple-stresses or rotations at the mid-side nodes). In the specification of boundary conditions for a mid-side node it must be remembered that in reality the specifications are for the adjacent hidden nodes. Because the couple-stress boundary conditions are in terms of normal components on the *sides* of the elements (at the hidden nodes), no problems arise at corners of the body.

Once thc global system of equations are solved, values of μ_x and μ_y are calculated at the element level for the vertex nodes (equation 1.30); the values from two or more elements surrounding a common vertex node are averaged to give global node values (the Q4U16 element leads directly to such values).

1.3.4 Element Q8U12

The excessive number of unknowns and bandwidth of the Q8U16 element can be reduced, without a substantial loss in accuracy, by replacing the two couple-stress unknowns at each mid-side node by a single unknown. This reduction is accomplished by equating the two unknowns at each mid-side node, i.e. by approximating the normal component of the couple-stress to be constant along each element side. The construction of the Q8U12 element matrix from the Q8U16 matrix merely involves adding the two rows and columns associated with each mid-side node. The specification of a boundary condition at a mid-side node now involves the value of couple-stress or 'lumped' rotation at the actual point in question.

The disadvantages of the Q8U12 elements as compared to the Q8U16 element is some loss in accuracy and having to deal with a different number of unknowns at the vertex nodes than at the mid-side nodes.

1.3.5 Element Q4U13

The element considered in this section is actually a five-node element (Figure 1.4d) with three unknowns at each vertex node and one at the central node. The equation associated with the central node has a zero in the diagonal position, and thus the unknown cannot be eliminated at the element level by 'condensation' (condensation would be possible if a 'penalty' formulation were used). For the sake of simplicity of grid preparation, the author prefers to assign the unknown at the central 'hidden' node to the vertex node of highest number [16], and thus have a four-node element (Figure 1.4a). The resulting element has, of course, a varying number of unknowns per node.

The element is based on the second variational equation. Three of the primary dependent variables, u, v, and ω, are approximated using the first-order shape functions. The values of u, v, and ω at the vertex nodes are written as U_i, V_i, and Ω_i respectively. The fourth variable, τ_a, is approximated as a constant (T_{a_m}) within each element (recall that the admissibility condition for the antisymmetrical component of the shear τ_a does not require interelement continuity).* The largest of the four node numbers describing the element (Figure 1.4a) is denoted by m (if instead a five-node element is used then m is the node number of the central node).

Substituting these approximations into equation (1.24) and differentiating with respect to the node point unknowns yields

$$\frac{\partial F_2}{\partial U_i}=\iint\{[D_{11}F_iF_j+D_{33}G_iG_j]U_j+[D_{12}F_iG_j+D_{33}G_iF_j]V_j$$
$$+[O]\Omega_j+[-G_i]T_{a_m}\}\,|J|\,\mathrm{d}\xi\,\mathrm{d}\eta \tag{1.31}$$

$$\frac{\partial F_2}{\partial V_i}=\iint\{[\quad]U_j+[D_{22}G_iG_j+D_{33}F_iF_j]V_j$$
$$+[O]\Omega_j+[F_i]T_{a_m}\}\,|J|\,\mathrm{d}\xi\,\mathrm{d}\eta \tag{1.32}$$

$$\frac{\partial F_2}{\partial \Omega_i}=\iint\{[\quad]U_j+[\quad]V_j+[D_{44}F_iF_j+D_{55}G_iG_j]\Omega_j$$
$$+[-2N_i]T_{a_m}\}\,|J|\,\mathrm{d}\xi\,\mathrm{d}\eta \tag{1.33}$$

$$\frac{\partial F_2}{\partial T_{a_m}}=\iint\{[\quad]U_j+[\quad]V_j+[\quad]\Omega_j+[O]T_{a_m}\}\,|J|\,\mathrm{d}\xi\,\mathrm{d}\eta \tag{1.34}$$

* Because of the similarity between this element and ones used for 'incompressible' materials one should expect many of the related difficulties, e.g. 'checkerboarding'. Remedies for these problems are beyond the scope of this discussion; the reader is referred to other papers in this volume.

The element matrix is formed from the above equations in the usual way. At boundary nodes, values of displacements or 'lumped' forces are specified along with the rotation or 'lumped' couple-stress. As is the case with the Q4U16 element, at fixed–free corners the ambiguity in the rotation term raises a question concerning the proper boundary condition specification.

Most conventional two-dimensional elasticity programmes which utilize four-node elements and permit a variable number of unknowns at the nodes can be easily modified to accept the Q4U13 element. As an alternative the element can be installed in a programme that accepts five-node elements (with a different number of unknowns at the central nodes than at the vertex nodes).

1.3.6 Relative merits of the mixed elements

In this section the relative merits of the four mixed elements are discussed. The conclusions are based in part on the results of analyses of example problems. The comparisons of computational efficiency (see the example at the end of this section) assume the use of a 'skyline' band equation solver; the conclusions drawn from these comparisons might be somewhat different if either a 'constant band' or a 'frontal' solver were used.

On the basis of the limited number of numerical comparisons made to date, the Q8U12 element was found to be slightly more computationally efficient in the calculation of displacements and normal stresses, the Q8U16 element was more efficient in the calculation of the couple-stresses, and the Q4U13 element in the calculation of the shear stress. These differences in computational efficiencies are rather small and may be of less importance than other considerations.

As previously noted from a standpoint of having non-ambiguous rotation and couple-stress boundary condition specifications at corners, the eight-node elements are preferable to the four-node ones.* Thus, assuming that the slight added complexity of having differing numbers of unknowns at the mid-side and vertex nodes is of no concern, it appears that the Q8U12 element is preferable; otherwise the Q8U16 element appears to be the best choice. If, however, one wishes to avoid the inconveniences associated with data preparation for eight-node element grids then the Q4U16 element is probably best.

Example Consider an isotropic plate subjected to a uniform shear stress field that is disturbed by a very small hole (Figure 1.5). The exact solution of the couple-stress elasticity equations for this problem is given in [4].

For isotropic conditions the elastic coefficients are given by equations (1.20) and (1.22). Following [4] the 'modulus of curvature' is expressed in

* However, in the author's limited experience the Q4U13 and the Q8U16 elements gave essentially identical results for problems with fixed-free corners.

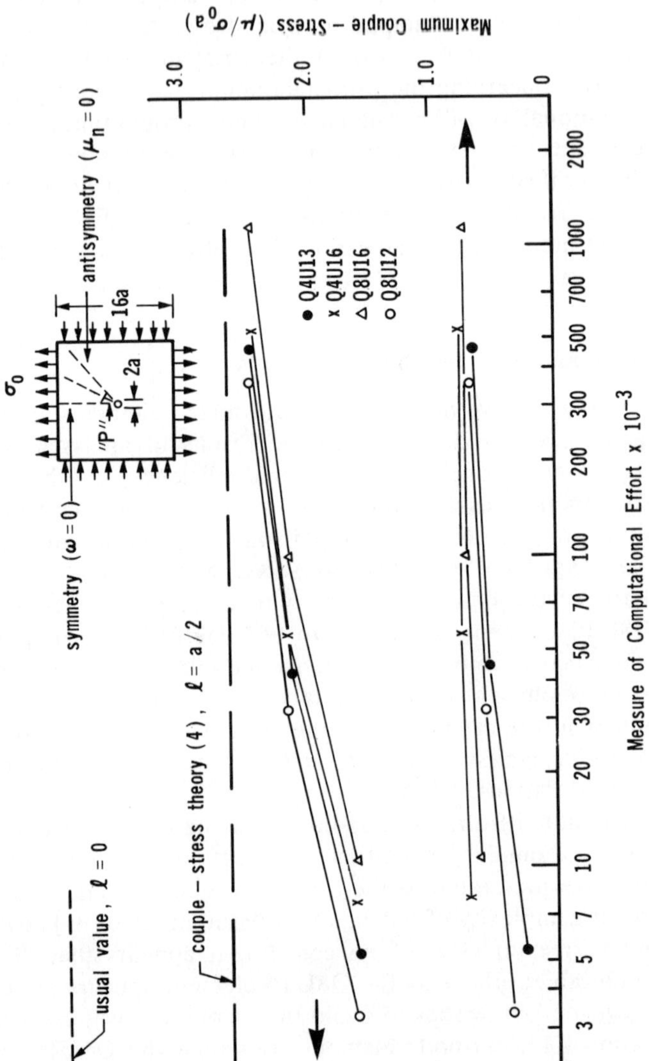

Figure 1.5 Stress concentration example

terms of a characteristic length l of the microstructure and the shear modulus G, that is $B = l^2 G$. For the purpose of comparing finite element results with the exact solution the particular case of $\nu = 0.3$ and $a = 2l$ is considered, i.e. the hole diameter is only four times the characteristic length. For this case the exact couple-stress elasticity solution [4] yields a stress concentration factor (ratio of the maximum hoop stress at the surface of the hole to the applied stress) of 2.67; in the absence of couple-stress effects ($l = 0$) the value is 4. Take advantage of the lines of symmetry and antisymmetry, only one-eighth of the plate is analysed. The stresses are applied at a distance of four times the hole diameter from its centre. The basic four-element grid is indicated in Figure 1.5; the refinement to sixteen elements was accomplished by dividing each element into four, etc. For the four-element grid the ratios of the lengths of successive element sides on the grid lines perpendicular and tangential to the hole are respectively 5 and 1, for the sixteen-element grid they are $5^{1/2}$ and 1, and for the sixty-four-element grid they are $5^{1/4}$ and 1. Finite element analyses were performed with each of the mixed elements discussed in Section 1.3; the resulting predictions for the stress concentration and the maximum couple-stress (point P) are given in the figure. The factor a is introduced into the couple-stress ordinate to non-dimensionalize the plot; it does not suggest that the couple-stresses are linear functions of a. The values are plotted versus a measure of the computational effort required to solve the simultaneous equations [13]. On the basis of simple examples of this type it is concluded that all four mixed elements are quite effective in solving couple-stress elasticity problems and that there is little difference in their computational efficiencies.

Space does not permit the reporting of the results of a study using couple-stress theory for capturing strain gradient effects in a continuum representation of structural grillages; these results will be contained in a later publication.

1.4 CONCLUSIONS

Mixed formulation finite elements for two-dimensional couple-stress elasticity are simple to derive, programme, and install into existing finite element programmes. All four elements investigated herein appear to be about equally computationally efficient and to give convergent results. However, because of some practial and theoretical questions for the four-node elements regarding the specification of rotation and couple-stress boundary conditions at corners, the eight-node elements appear to be superior. Of the eight-node elements, the twelve degrees of freedom element gives nearly the same accuracy with fewer degrees of freedom than the sixteen degrees of freedom one.

Because only relatively few analyses have been performed by the author to date, conclusions drawn concerning the importance of including couple-stresses must be considered as speculative. The need for considering couple-stresses in analyses of traditional homogeneous materials remains to be demonstrated. There appears to be good evidence that the inclusion of couple-stresses in analyses of equivalent homogeneous, orthotropic representations of composite materials increases the accuracy of the predictions for regions of severe strain gradients and/or edge effects. However, to be able to completely capture these effects the inclusion of additional micropolar variables would appear to be necessary for many composites.

Because of the substantial increase in cost of a couple-stress elasticity analysis as compared to a conventional analysis an investigation of ways of reducing computational cost is in order. One possible approach is to solve the system of simultaneous equations by block iteration with the coupling terms between the usual displacement unknowns and the added couple-stress unknowns moved to the right-hand side and approximated by iteration. A second possibility is to include couple-stress unknowns only in regions of severe strain gradient and edge effects; such regions could neither be *a priori* identified by the analyst nor established iteratively within the programme.

REFERENCES

1. E. Cosserat and F. Cosserat, *Theorie des Corps Deformables*, A. Hermann et Fils, Paris, 1909.
2. A. C. Eringen, 'Linear theory of micropolar elasticity', *J. Math. and Mech.*, **15,** No. 2, 909–923 (1966).
3. Iu. A. Gruzdev and V. K. Prokopov, 'Multi-moment theory of equilibrium of thick plates', *Applied Mathematics and Mechanics*, **32,** No. 1, 342–351 (1968).
4. R. D. Mindlin, 'Influence of couple-stresses on stress concentrations', *Experimental Mechanics*, **3,** 1–7 (1963).
5. R. D. Mindlin and H. F. Tiersten, 'Effects of couple-stresses in linear elasticity', *Archive for Rational Mech. and Anal.*, **11,** 413–448 (1962).
6. I. H. Shames, *Introduction to Solid Mechanics*, Prentice-Hall, Englewood Cliffs, New Jersey, 1975.
7. L. R. Herrmann, 'Finite element modeling of composite edge effects', *ASCE Proc. Seventh Conf. on Electronic Computations*, 1979, 593–607 (1979).
8. L. R. Herrmann and Z. Al-Yassin, 'Numerical analysis of reinforced soil systems', *ASCE. Proc. Symp. Earth Reinforcement*, **1979,** 428–457 (1979).
9. L. R. Herrmann and R. A. Schamber, 'Finite element analysis of layered systems with edge effects', *Proc. Int. Conf. Numerical Methods for Coupled Problems*, Swansea, September 1981.
10. L. R. Herrmann, 'Finite element bending analysis for plates', *J. Eng. Mech. Div., ASCE*, **93,** No. EM5, 13–26 (1967).
11. L. R. Herrmann and D. M. Campbell, 'A finite-element analysis for thin shells', *AIAA J.*, **6,** No. 10, 1842–19847 (1968).

12. L. R. Herrmann, 'Interpretation of finite element procedures as stress error minimization procedures', *J. Eng. Mech. Div., ASCE,* **1972,** EM5, 1330–1336 (1972).
13. L. R. Herrmann, 'Elasticity equations for incompressible and nearly incompressible materials by a variational theorem', *AIAA. J.*, **3,** No. 10, 1896–1900 (1965).
14. O. C. Zienkiewicz, *The Finite Element Method*, 3rd ed., McGraw-Hill, London, 1977.
15. L. R. Herrmann, 'Efficiency evaluation of a two-dimensional incompatible finite element', *J. Computers and Structures,* **3,** 1377–1395 (1973).
16. R. J. Farris, L. R. Herrmann, J. R. Hutchinson, and R. A. Schapery, 'Development of a solid rocket propellant nonlinear viscoelastic constitutive theory', *Final Report to Air Force Rocket Propulsion Laboratory*, AFRPL-TR-75-20, 1975.

Hybrid and Mixed Finite Element Methods
Edited by S. N. Atluri, R. H. Gallagher, and O. C. Zienkiewicz

Chapter 2

The Mixed Finite Element Method in Elasticity and Elastic Contact Problems

Mervyn D. Olson

2.1 INTRODUCTION

The earliest applications of mixed finite elements date back to the mid-1960s when the use of mixed finite element models for plate bending were proposed, independently, by Herrmann [1] and Hellan [2]. Dunham and Pister [3] used the Hellinger–Reissner variational principle to develop mixed finite elements for plane elasticity and plate-bending problems but few results were given.

The mixed methods involve indefinite systems and, despite their wide-spread use, their mathematical properties were not as well understood as those of the displacement and equilibrium methods. It is only recently that a mathematical framework has been built up for the mixed methods and their application to the finite element method by Oden [4], Oden and Reddy [5, 6], and Babuska, Oden, and Lee [7]. See also Raviart and Thomas [8] and Oden and Lee [9] for analyses of the mixed method for Poisson's equation and plane elasticity problems respectively. Concurrently, Mirza [10] tried to establish the convergence of mixed methods in the energy sense by defining an energy product and the associated energy norm directly in a restricted space on which the mixed operator was positive definite. His restriction imposed on the order of polynomials used for approximating displacements and stresses turned out to be essentially the same as the consistency condition in [6] and [7]. Mirza also carried out extensive numerical tests to illustrate how the method worked in plane elasticity, and some of these are discussed herein.

Subsequently, Tseng [11] extended the method to elastic contact problems. He added boundary integrals for the contact regions in order to incorporate the contact conditions directly into the variational principle. A significant novelty is that in some cases the displacement constraints become natural boundary conditions. Several applications are discussed to illustrate the methods.

The present paper represents a very brief overview of work carried out by the author and his former students, Mirza and Tseng (mainly by the latter). The interested reader is referred to the theses [10, 11] and other papers [12, 13, 14] for more details.

2.2 THEORY FOR MIXED FINITE ELEMENT FORMULATION

In linear elasticity, the linear, positive definite, and self-adjoint operators can be decomposed into two or more lower order operators of the form

$$\mathbf{A}=\begin{bmatrix}\mathbf{0} & \mathbf{T}^* \\ \mathbf{T} & -\mathbf{C}\end{bmatrix} \tag{2.1}$$

where the operator $\mathbf{T}^*$ is the formal adjoint of $\mathbf{T}$ and $\mathbf{C}$ is a compliance matrix. In plane stress linear elasticity, the lower order differential equations are

$$-\tau_{xx,x}-\tau_{xy,y}=f_x \tag{2.2}$$

$$-\tau_{xy,x}-\tau_{yy,y}=f_y \tag{2.3}$$

$$u_{,x}-\frac{1}{E}(\tau_{xx}-\nu\tau_{yy})=0 \tag{2.4}$$

$$u_{,y}-\frac{1}{E}(-\nu\tau_{xx}+\tau_{yy})=0 \tag{2.5}$$

$$u_{,y}+v_{,x}-\frac{2(1+\nu)}{E}\tau_{xy}=0 \tag{2.6}$$

where u,v are displacements in the x,y directions respectively, f_x, f_y are body forces, and $\tau_{xx}, \tau_{xy}, \tau_{yy}$ are the stress components. Equations (2.2) and (2.3) are the usual equilibrium equations, while equations (2.4) to (2.6) are the constitutive equations.

Using the vector notations:

$$\mathbf{u}=(uv)^T, \qquad \boldsymbol{\tau}=(\tau_{xx}\tau_{yy}\tau_{xy})^T, \qquad \mathbf{f}=(f_xf_y)^T \tag{2.7}$$

equations (2.2) to (2.6) may be put in the form

$$\mathbf{A}\boldsymbol{\Lambda}=\mathbf{p} \quad \text{on} \quad \Omega \tag{2.8}$$

where $\boldsymbol{\Lambda}^T=(\mathbf{u}^T\boldsymbol{\tau}^T)$, $\mathbf{p}^T=(\mathbf{f}^T\mathbf{0}^T)$, and

$$\mathbf{T}^*=-\mathbf{T}^T=\begin{bmatrix}-\dfrac{\partial}{\partial x} & 0 & -\dfrac{\partial}{\partial y} \\ 0 & -\dfrac{\partial}{\partial y} & -\dfrac{\partial}{\partial x}\end{bmatrix}, \qquad \mathbf{C}=\frac{1}{E}\begin{bmatrix}1 & -\nu & 0 \\ -\nu & 1 & 0 \\ 0 & 0 & 2(1+\nu)\end{bmatrix} \tag{2.9}$$

The simplest homogeneous boundary conditions for the equations represented by equations (2.8) are

$$\begin{aligned} \mathbf{u} &= 0 \quad \text{on} \quad S_u \\ \boldsymbol{\tau} \cdot \mathbf{n} &= 0 \quad \text{on} \quad S_\tau \end{aligned} \tag{2.10}$$

where S_u and S_τ are the displacement and stress portions respectively of the boundary S of Ω and $\mathbf{n}$ is an outward normal to the boundary.

A mixed functional for equations (2.8) may be written as

$$F(\mathbf{u}, \boldsymbol{\tau}) = F(\boldsymbol{\Lambda}) = \int_\Omega [\boldsymbol{\tau}^T \mathbf{T} \mathbf{u} - \tfrac{1}{2} \boldsymbol{\tau}^T \mathbf{C} \boldsymbol{\tau} - \mathbf{u}^T \mathbf{f}] \, d\Omega \tag{2.11}$$

with kinematic and natural boundary conditions on $\mathbf{u}$ and $\boldsymbol{\tau}$ respectively. This equation is merely a particular example of the general mixed Hellinger–Reissner variational principle. That is, equation (2.11) is an extremum and when its first variation is zeroed, it yields equations (2.8) as Euler equations and equations (2.10) as boundary conditions. Non-homogeneous boundary conditions can easily be included by adding to equations (2.11) the appropriate boundary integrals representing negative virtual work.

The first two terms in the integral of equation (2.11) represent strain energy U, i.e.

$$U = \int_\Omega [\boldsymbol{\tau}^T \mathbf{T} \mathbf{u} - \tfrac{1}{2} \boldsymbol{\tau}^T \mathbf{C} \boldsymbol{\tau}] \, d\Omega \tag{2.12}$$

It was shown in [10] that, when the $\mathbf{u}$ and $\boldsymbol{\tau}$ approximations are chosen independently, the expression in equation (2.12) will be definite (i.e. non-zero), provided the following is satisfied:

The strains from the stress approximations should possess at least all the strain modes that are present in the strains derived from the displacement approximations.

Of course, rigid body modes should also be precluded. The foregoing is the completeness criterion as defined in [10]. If this criterion is violated, the element matrix will contain 'mechanisms' and the matrix equations for an application will be singular. This is illustrated in the next section.

Lastly, the convergence estimate carried out in [10] indicated that the error in strain energy would be governed by the mean square error in the stress approximation $\boldsymbol{\tau}$, and all the numerical studies tended to verify this prediction. These studies included beam-bending applications (using four first-order equations and two second-order ones) as well as extensive plane elasticity applications.

Consider, as an example, the choice of complete linear interpolations for stresses and displacements within a triangular element. In this case, the error in stresses is $O(l_e^2)$, where l_e is the largest diameter within an element. Further, since this approximation satisfies the completeness criterion, and displacements and stresses are continuous across interelement boundaries, according to the above prediction, the error in strain energy is expected to be the same as the mean square error in the stresses, that is $O(l_e^4)$, which is $O(N^{-4})$ for a uniform grid.

2.2.1 Eigenvalue analysis of the element matrix

The mixed functional of equation (2.11) for homogeneous boundary conditions and zero body forces can be written as

$$F(\mathbf{\Lambda}) = \int_{\Omega} \left\{ \tau_{xx} u_{,x} + \tau_{yy}(u_{,y} + v_{,x}) - \frac{1}{2E}[\tau_{xx}^2 + \tau_{yy}^2 - 2\nu\tau_{xx}\tau_{yy} + 2(1+\nu)\tau_{xy}^2] \right\} \mathrm{d}\Omega \tag{2.13}$$

Since $F(\mathbf{\Lambda})$ in (2.13) represents strain energy, inspection of the right-hand side suggests the following three rigid body modes will yield zero strain energy:

$$\begin{array}{lll} \text{(a)} & u = \text{constant},\ v = 0; & \tau_{xx} = \tau_{yy} = \tau_{xy} = 0 \\ \text{(b)} & u = 0,\ v = \text{constant}; & \tau_{xx} = \tau_{yy} = \tau_{xy} = 0 \\ \text{(c)} & u = -cy,\ v = cx; & \tau_{xx} = \tau_{yy} = \tau_{xy} = 0 \end{array} \tag{2.14}$$

These rigid body modes are expected to be removed by the specified kinematic boundary conditions. Furthermore, it is required by the functional of (2.13) that the displacements satisfy the kinematic boundary conditions while the stresses emerge as natural boundary conditions. Therefore, the finite element approximations should be in compliance with these requirements, i.e. the discrete matrix equivalent of (2.13) should exhibit the rigid body modes of (2.14).

Now this discrete matrix is of the form (see [13])

$$[S] = \begin{bmatrix} \mathbf{0}_T & \mathbf{a} \\ \mathbf{a} & -\mathbf{b} \end{bmatrix} \tag{2.15}$$

The submatrix **b** is a symmetric, positive definite matrix and **a** can be either a rectangular or a square matrix. Then the matrix $[S]$ in (2.15) is symmetric but in general indefinite.

For the purpose of eigenvalue analysis, various combinations of interpolations for the displacements and the stresses over a triangular and a rectangular element are considered. The eigenvalue routine, which uses Householder transformations, is then used to solve the eigenvalue problem

$$[S]\begin{Bmatrix} \boldsymbol{\delta} \\ \boldsymbol{\tau} \end{Bmatrix} - \lambda[I]\begin{Bmatrix} \boldsymbol{\delta} \\ \boldsymbol{\tau} \end{Bmatrix} = 0 \tag{2.16}$$

where $\boldsymbol{\delta}^T = (\mathbf{u}^T\mathbf{v}^T)$, $\boldsymbol{\tau}^T = (\boldsymbol{\tau}_{xx}^T\boldsymbol{\tau}_{yy}^T\boldsymbol{\tau}_{xy}^T)$ and $[I]$ is the identity matrix. Since the matrix $[S]$ is symmetric, all eigenvalues are expected to be real but can be negative, zero, or positive.

The qualitative description of the eigenvalues and the composition of the eigenvectors for all the combinations of interpolations used for the displacements and stresses over a triangular element appear in Table 2.1 and those for a rectangular element are listed in Table 2.2. For both triangular and rectangular elements, the number of negative eigenvalues always corresponded to the number of stress degrees of freedom, i.e. they arise from the **b** matrix in equation (2.15). The correct three zero eigenvalues for the expected rigid body modes are obtained only for the displacement–stress combinations which satisfy the completeness criterion in the previous section. In the cases where more than three zero eigenvalues were obtained, the number of extra zeros corresponded to the number of modes present in the strains derived from the assumed displacements that were not contained in the strains from the assumed stresses. Further, the stresses in the eigenvectors for these extra zero eigenvalues were always zero. These extra modes are called mechanisms, since the element exhibits an apparent strain without any stress.

Table 2.1 Eigenvalue results for different interpolations in a triangular element

Interpolation		Degrees of freedom			Sign of eigen-	No. of eigen-	
u,v	τ's*	u,v	τ's	Total	values	values	Composition of eigenvectors.
Linear	Constant	6	3	9	(−)	3	τ's constant; u,v linear
					(0)	3‡	τ's $= u_{,x} = v_{,y} = u_{,y} + v_{,x} = 0$‡
					(+)	3	τ's constant; u,v linear
Linear	Linear	6	9	15	(−)	9	τ's,u,v linear
					(0)	3	τ's $= u_{,x} = v_{,y} = u_{,y} + v_{,x} = 0$
					(+)	3	τ's,u,v linear
Quadratic	Constant	12	3	15	(−)	3	τ's constant; u,v quadratic
					(0)	9†	$\tau_s' = 0; \{u_{,x} = v_{,y} = u_{,y} + v_{,x} = 0; u,v$ quadratic
					(+)	3	τ's constant; u,v quadratic
Quadratic	Linear	12	9	21	(−)	9	τ's linear; u,v quadratic
					(0)	3	τ's $= u_{,x} = v_{,y} = u_{,y} + v_{,x} = 0$
					(+)	9	τ's linear; u,v quadratic
Quadratic	Quadratic	12	18	30	(−)	18	τ's u,v quadratic
					(0)	3	τ's $= u_{,x} = v_{,y} = u_{,y} + v_{,x} = 0$
					(+)	9	τ's u,v quadratic

* τ's: All stresses τ_{xx}, τ_{yy}, and τ_{xy} have the same type of interpolation.
† Extra zero eigenvalues are associated with mechanisms which have the same u,v distributions as the approximating polynomials.
‡ Rigid body modes ($u_{,x} = 0$; $v_{,y} = 0$; $u_{,y} + v_{,x} = 0$).

Table 2.2 Eigenvalue results for different interpolations in a rectangular element

Interpolation		Degrees of freedom			Sign of eigen-values	No. of eigen-values	Composition of eigenvectors
u,v	$\tau_{xx},\tau_{yy},\tau_{xy}$	u,v	τ's	Total			
					(−)	3	τ's constant; u,v bilinear
Bilinear	Constant	8	3	11	(0)	5†	τ's $=0$; $\{u_{,x}=v_{,y}=u_{,y}+v_{,x}=0; u,v$ bilinear$\}$
					(+)	3	τ's constant; u,v bilinear
					(−)	12	τ's,u,v bilinear
Bilinear	Bilinear	8	12	20	(0)	3‡	τ's $=u_{,x}=v_{,y}=u_{,y}+v_{,x}=0$‡
					(+)	5	τ's, u,v bilinear
					(−)	12	τ's bilinear; u,v biquadratic
Biquadratic*	Bilinear	16	12	28	(0)	4†	τ's $=0$; $\{u_{,x}=v_{,y}=u_{,y}+v_{,x}=0; u,v$ biquadratic$\}$
					(+)	12	τ's bilinear; u,v biquadratic
					(−)	24	τ's, u,v biquadratic
Biquadratic	Biquadratic	16	24	40	(0)	3	τ's $=u_{,x}=v_{,y}=u_{,y}+v_{,x}=0$
					(+)	13	τ's, u,v biquadratic

* Full quadratic in x and y plus x^2y and xy^2.
† Extra zero eigenvalues are associated with mechanisms which have the same u,v distributions as the approximating polynomials.
‡ Rigid body modes ($u_{,x}=0$; $v_{,y}=0$; $u_{,y}+v_{,x}=0$).

Clearly, in these cases, the strain energy representation is not definite and, therefore, convergence in the energy sense is precluded. It is anticipated that, in general, these mechanisms cannot be removed by the kinematic boundary conditions.

2.2.2 Application to contact problems

In the studies of the contact phenomenon, the problem may be classified into one of three categories: unbonded contact, bonded contact, and stick-slip contact. Contact between well-lubricated metallic surfaces may be modelled as unbonded. For perfectly rough surfaces coming into contact, bonded conditions may govern. The stick-slip type of contact is the most general and is characterized by contact friction stresses varying from zero to the allowable maximum.

For a given pair of surfaces, the magnitudes of these frictional stresses are proportional to the normal stresses pressing the two surfaces together and may be expressed in the simplified relation:

$$\tau_{nt} \leqslant \mu\tau_{nn} \tag{2.17}$$

where $\tau_{nt} \neq$ surface tangential shear stress
$\tau_{nn} =$ surface compressive normal stress
$\mu =$ Coulomb's coefficient of friction

The inequality denotes adhering conditions and the equality denotes sliding conditions.

In the contact problem, the boundary conditions over the contact region are both displacement (the points on opposing surfaces in contact occupy the same space) and stress (the tangential shear stress is bounded by a fraction of the compressive normal stress). It is desirable to be able to express the stress conditions explicitly in the finite element formulation. In accomplishing this, as will be shown in the following, the stress variables in the contact region are associated with forced boundary conditions and the displacement variables with natural boundary conditions.

An addition of a line integral to the function of equation (2.13).

$$-\int_{S_C} (\delta\tau_{nn}\mathbf{u}\cdot\mathbf{n}+\delta\tau_{nt}\mathbf{u}\cdot\mathbf{t}+\tau_{nn}\boldsymbol{\delta}\mathbf{u}\cdot\mathbf{n}+\tau_{nt}\boldsymbol{\delta}\mathbf{u}\cdot\mathbf{t})\,\mathrm{d}s \tag{2.18}$$

where S_C is the part of the boundary in contact, will fulfil these requirements. Thus a functional defined as

$$F_1 = F - \int_{S_C} (p_x u + p_y v)\,\mathrm{d}s \tag{2.19}$$

where

$$\begin{aligned} p_x &= l\tau_{xx} + m\tau_{xy} \\ p_y &= l\tau_{xy} + m\tau_{yy} \\ u &= lu_n + mu_{nt} \\ v &= mu_n - lu_{nt} \end{aligned} \tag{2.20}$$

and

with l, m as the direction cosines of the outward normal on the boundary, will require displacements to be natural boundary conditions over S_C and forced elsewhere and stresses to be forced boundary conditions over S_C and natural elsewhere.

The Reissner principle with non-homogeneous boundary conditions may now be written as

$$\begin{aligned} F_1 = F &- \int_{S_C} (p_x u + p_y v)\,\mathrm{d}s \\ &- \int_{S_{FT}} (u\bar{p}_x + v\bar{p}_y)\,\mathrm{d}s - \int_{S_{FU}} [(u-\bar{u})p_x + (v-\bar{v})p_y]\,\mathrm{d}s \\ &+ \int_{S_{CT}} [u(p_x-\bar{p}_x)+v(p_y-\bar{p}_y)]\,\mathrm{d}s + \int_{S_{CU}} (\bar{u}p_x + \bar{v}p_y)\,\mathrm{d}s \end{aligned} \tag{2.21}$$

where S_F is that part of the boundary not in contact and

$$S_{CT} \cup S_{CU} = S_C$$
$$S_{FT} \cup S_{FU} = S_F$$
$$S_C \cup S_F = S$$

Subscript T denotes that part of the boundary where surface traction is prescribed and subscript U denotes that part of the boundary where surface displacements are prescribed, the barred variables being the prescribed values on the boundary.

In forming the finite element matrix equations, an element lying on S_C has an additional contribution from the boundary integral over the 'contact' side of the element. Along S_F, the surface tractions are consistent loads on appropriate displacement variables as indicated by the boundary integral of S_{FT}. Let f_U be the consistent load on the displacement variable u, which is given by

$$\begin{aligned} f_u &= \frac{\mathrm{d}}{\mathrm{d}u}\left[\int_{S_{FT}} (u\bar{p}_x + v\bar{p}_y)\,\mathrm{d}s\right] \\ &= \int_{S_{FT}} \bar{p}_x\,\mathrm{d}s \end{aligned} \tag{2.22}$$

On S_{FU}, the displacements are constrained to suit the prescribed conditions. Similarly, along S_C, the displacements are consistent loads on appropriate stress variables over S_{CU}. Let $f_{\tau_{xy}}$ be the consistent load on the stress variable τ_{xy}, which becomes

$$\begin{aligned} f_{\tau_{xy}} &= \frac{\mathrm{d}}{\mathrm{d}\tau_{xy}}\left[\int_{S_{CU}} \bar{u}(l\tau_{xx} + m\tau_{xy}) + \bar{v}(l\tau_{xy} + m\tau_{yy})\,\mathrm{d}s\right] \\ &= \int_{S_{CU}} (u\bar{m} + \bar{v}l)\,\mathrm{d}s \end{aligned} \tag{2.23}$$

On S_{CT}, the stresses are constrained to suit the friction condition. If the element is oriented in a local normal–tangential coordinate system, then the constraint condition is simply setting the tangential shear stress to a fraction of the compressive normal stress. Further details may be found in [11, 14].

2.2.2.1 *Bonded contact for known contact surface S_C*

In the bonded or adhering type of contact where the tangential shear stress is less than the allowable fraction of the compressive normal stress, relative slip does not occur between the surfaces of contact and continuum is established across the interface. In the case where the body is against a rigid surface, the normal and tangential displacements within the contact region are both zero. In the finite element model, this results in zero consistent loads on the stress variables on the contact boundary. The appropriate form of the functional is simply obtained by dropping the S_{CT} boundary integral in equation (2.21).

2.2.2.2 Sliding contact for known contact surface S_C

In the sliding type of contact where relative slip does occur, the tangential shear stress is constrained to the compressive normal stress multiplied by the Coulomb coefficient of friction μ. When μ is zero, the contact is unbonded. The appropriate form of the functional is the full form of equation (2.21). It is convenient to work in local normal–tangential coordinates so that both displacements and stresses are defined with respect to the boundary normal. Continuum is maintained across the region of contact in the normal direction. The tangential displacements of the nodes slipping relative to each other enter the matrix equation as consistent loads on the tangential shear stress variables. The need for an iterative procedure is obvious as the tangential displacements in the equation formulation should agree with those in the solution.

2.2.2.3. Stick-slip contact for known contact surface S_C

In the stick-slip type of contact, an iterative scheme is employed to determine the unknown adhering and sliding regions. The designated contact nodes in the finite element representation are assumed to be adhering nodes in the first solution of the matrix equation. The tangential stresses at these nodes are then compared with the normal stresses to check against the assumption of adhesion. If at any of these nodes the tangential shear stress exceeds μ times the compressive normal stress, the friction condition is applied at that node constraining the stress variables. These are nodes with the 'sliding' conditions. The functional is changed for the different boundaries.

The reformulated equations are solved iteratively for the displacements in the sliding contact region. When convergence is within tolerance, the conditions at the contact nodes are examined for comparison with the assumptions made in the previous solution step. At those nodes assumed to be adhering, the stresses are compared with the Coulomb friction condition as described above. At the nodes assumed to be sliding, the tangential relative slip should be in the opposite direction to the assumed tangential shear stress. If not, the node is released from the friction constraint and redefined to be an adhering node. A revision in the contact condition of any contact node prompts the next step in the iterative scheme. A solution to the problem is obtained when the contact conditions in the solution coincide with those in the previous assumption.

2.2.2.4. Contact for unknown contact surface S_C

In the finite element method, the portions of the boundaries expected to come into contact are designated as the contact boundary on which lie designated contact nodes. In the case of two bodies coming into contact, these nodes on opposite sides of the contact interface constitute contact node pairs. Continuous progressive contact becomes discretized successive closing or opening of the gaps between the nodes of contact node pairs, changing the type of contact at nodes already in contact. The problem reduces to determination of the load increments required to bring the two bodies from one contact stage to the next.

At the end of the previous load vector increment ΔR_m, the contact conditions at the ith contact node pair are C^i_m, denoting open or closed, and A^i_m, denoting adhering or sliding conditions at the mth stage, and the cumulative load vector is R_m and the cumulative solution vector is r_m. The problem, then, is to establish an incremental load ΔR_{m+1} that would cause a change in the contact condition of one and only one node pair.

Assuming a test load increment ΔQ, the system with contact conditions C^i_m, A^i_m is solved to obtain a test increment solution vector Δq. Suppose k^i_{m+1} is the multiplicative factor for ΔQ to invoke a change in the contact condition at the ith node pair, then k_{m+1} of all node pairs. Then

$$\Delta R_{m+1} = k_{m+1} \times \Delta Q \tag{2.24}$$

The possible k^i_{m+1} values are summarized in [11, 14].

Having determined the multiplicative factor k_{m+1}, the contact conditions over the contact region are updated. At the ith node pair, where the open or closed condition is to change, the appropriate boundary conditions are applied. Node pairs coming into adhering conditions from sliding conditions, and vice versa, will also have their boundary conditions accordingly constrained.

To account for the geometrically non-linear nature of the problem, the coordinates of the nodes are updated at each contact stage in addition to the rectification of boundary conditions. The nodes in a node pair that are in sliding contact are forced to occupy the same point in space by averaging their coordinates at the end of each increment. A solution to the progressive contact problem is obtained when the sum of the load increments reaches the given load level.

2.3 NUMERICAL EXAMPLE

The foregoing formulation and ideas have been checked out on several numerical examples, using a triangular element with linear interpolations for both stresses and displacements. The nodal variables were then two displacement and three stress components, yielding a total element size of 15. The assembly of the elemental equations for a gridwork of elements simply follows the standard finite element procedure, and this automaticaly enforces continuity of both stresses and displacements between adjacent elements.

2.3.1 Plane stress square plate with parabolically varying end loads

A square plate with parabolically varying end loads is depicted in Figure 2.1. This problem has an exact series solution and hence the energy convergence may be studied. Using the symmetry about the x and y axes, only one-quarter of the plate ABCD is analysed with uniform grids, as shown. A consistent load vector is used for the load on edge CD.

Numerical results for some typical stresses and displacements at points A,B,C, and D, and the strain energy, are presented in Table 2.3, along with

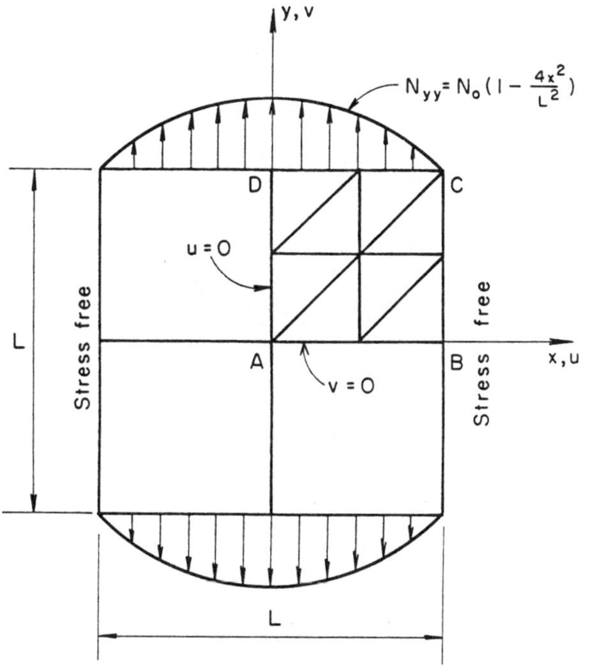

Figure 2.1 Parabolically loaded plane stress problem ($N-4$)

Table 2.3 Numerical results for square plate with parabolic edge loads

Finite Element grid	$\frac{-10Etu_B}{(1-\nu^2)N_0L}$	$\frac{10^2Etu_C}{(1-\nu^2)N_0L}$	$\frac{10Etv_C}{(1-\nu^2)N_0L}$	$\frac{10Etv_D}{(1-\nu^2)N_0L}$	$\frac{-10N_{xxA}}{N_0}$	$\frac{10N_{yyA}}{N_0}$	$\frac{10N_{yyB}}{N_0}$	$\frac{10^2N_{xxB}}{N_0}$	$\frac{10N_{xxC}}{N_0}$	$\frac{10t^2U}{(1-\nu^2)L^2N_0^2}$
2×2	1.4440	−0.7670	1.4534	4.8498	2.0123	9.1696	3.5427	0.9348	0.7266	2.74633
4×4	1.4982	1.3373	1.2886	5.0167	1.5842	8.9227	4.0288	3.0654	−0.3742	2.78996
6×6	1.5123	1.6488	1.2751	5.0472	1.4987	8.7416	4.1372	1.5472	−0.2530	2.79275
Exact	1.5199	1.7837	1.2773	5.0735	1.4095	8.5905	4.1067	0.0000	0.0000	2.79357

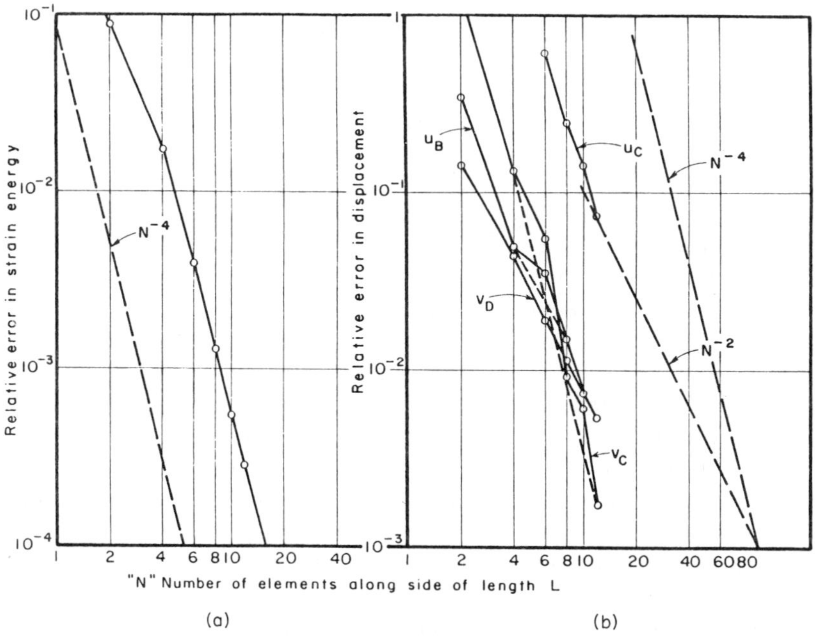

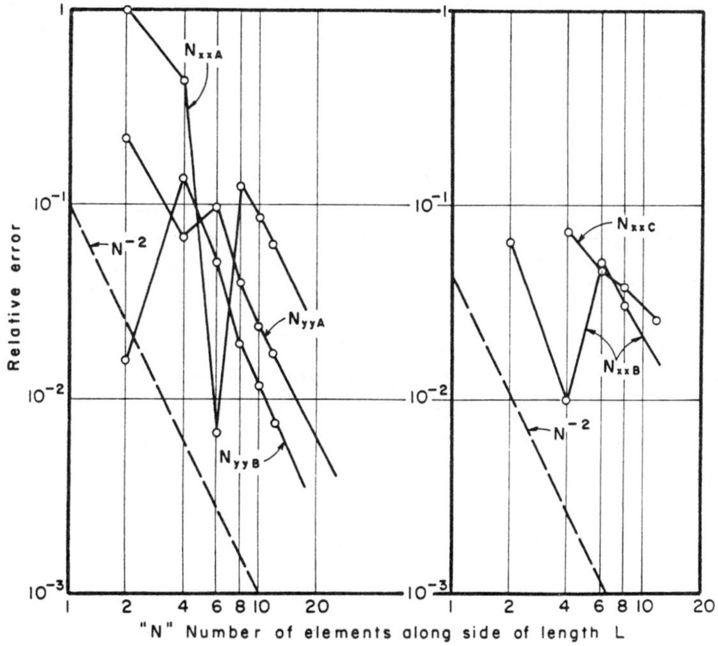

Figure 2.2 (a) Strain energy convergence for parabolically loaded square plate. (b) Displacement convergence for parabolically loaded square plate. (c) Stress convergence for parabolically loaded square plate

the exact values. Note that, in the present formulation, the stress boundary conditions are left as natural conditions which are recovered only in the limit of large N. Although all the quantities are converging to the exact values, some of them overshoot and oscillate about them. However, we note that the strain energy is definitely converging from below. Convergence plots for these quantities are shown in Figure 2.2.

The energy convergence rate shown in Figure 2.2(a) appears to be very close to N^{-4}, as predicted. Further, the convergence rate for the stresses shown in Figure 2.2(c) appears to be close to N^{-2}, again as predicted. However, peculiar kinks are observed in these plots, but these can be associated with the fact that certain stresses are fortuitously close to their exact values for some grids, and that the relative error changes sign (absolute is plotted). The same phenomenon was reported in [15], using QST displacement elements. Figure 2.2(b) shows the convergence of displacements indicating faster convergence for u_C and v_C (close to N^{-4}) than u_B and v_D (close to N^{-2}), but, of course, there is no analogous prediction for these rates. The error in u_C looks rather large at first, but u_C itself is one order smaller than the other displacements. Hence, if its error were non-dimensionalized with respect to the largest displacement v_D, it would appear very small. Again, some kinks are observed in the convergence plots which are analogous to those for stresses. In conclusion, we note that the theoretical convergence rates were completely verified by the numerical results from this example and, further, that good engineering accuracy was obtained for both stresses and displacements, even for relative coarse element gridworks.

2.3.2 Stress concentration around a circular hole

The next example is that of a square plate with a circular hole in the middle (Figure 2.3), loaded by a uniform uniaxial stress τ_0. The diameter of the hole is one-eighth of the plate width and the plate is of unit thickness. The plane strain state is analysed for both isotropic and orthotropic cases. The grid and the symmetry boundary conditions used for one-quarter of the plate are shown in Figure 2.4. This is essentially the same grid as used in [16] for CST displacement elements.

The stress predictions from both the mixed elements and the CST displacement elements are plotted in Figure 2.5, along with the analytical solutions for the infinite plate case (Timoshenko and Goodier [17] for isotropic and Savin [18] for orthotropic); the stress concentration results on the hole are tabulated in Table 2.4. Note that the CST stresses are plotted at the element centroids since they are discontinuous between elements, whereas the mixed ones are nodal ones which, of course, are continuous. Both stress predictions are quite good away from the hole, but the mixed ones are far superior right at the hole where the CST ones have to be

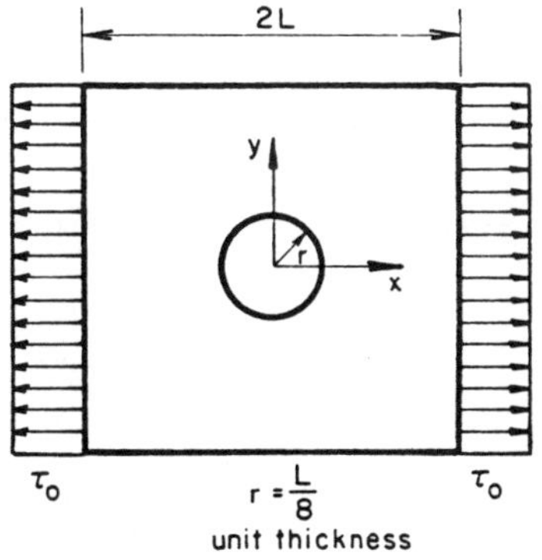

Isotropic: $E = 1.0,\ \nu = 0.1,\ G = \dfrac{E}{2(1+\nu)}$

Orthotropic: $E_1 = 1.0,\ E_2 = 3.0,\ \nu_1 = 0.1,\ \nu_2 = 0,\ G_{12} = 0.42$

Figure 2.3 Square plate with a circular hole at centre, isotropic and orthotropic

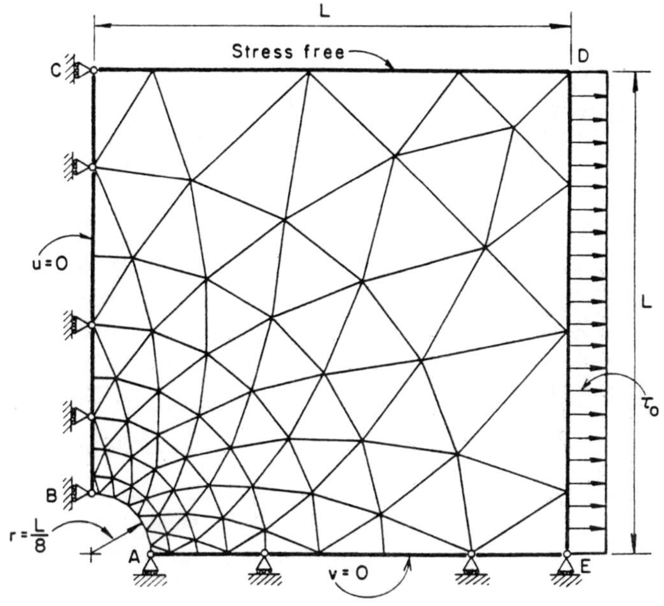

Figure 2.4 Finite element grid for the square plate with a circular hole in the middle

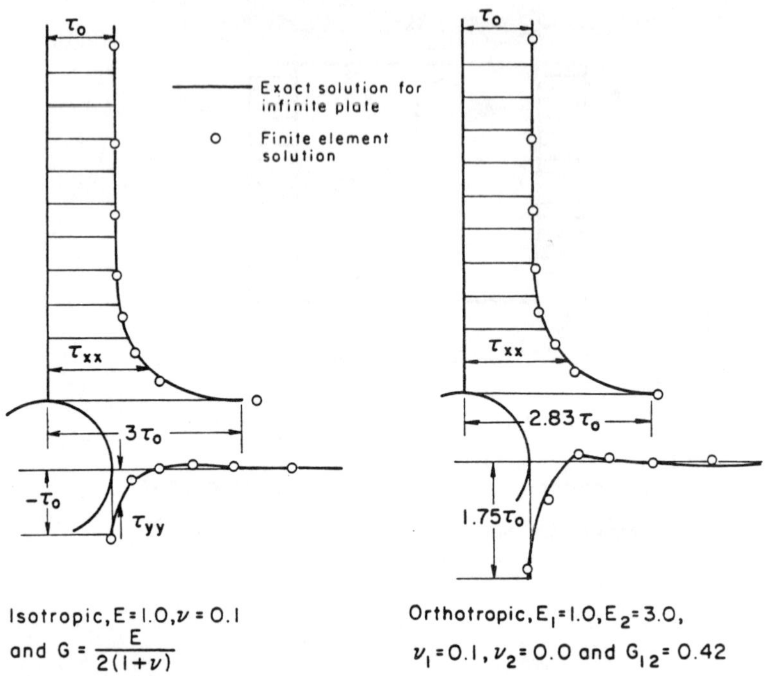

Figure 2.5 Comparison of theoretical and finite element results for infinite plate with a circular hole in the middle

extrapolated from the centroids. In particular, the CST badly underpredicts the orthotropic τ_{yy} stress concentration at the hole. On the other hand, both elements overpredict the τ_{xx} stress concentration (using an extrapolation of the CST one), but this is to be expected for the finite plate compared to the infinite one.

The foregoing comparison of CST and mixed elements is somewhat unfair since, for the same grid, the mixed elements have many more degrees of freedom. However, it does highlight the advantages of having a continuous stress field with unique nodal values. Furthermore, the mixed elements do not require the extremely fine grids that are needed by CST elements in regions of stress concentration or steep gradients.

Table 2.4 Stresses at edges of hole in square plate under unaxial tension

	Isotropic		Orthotropic	
Method	τ_{xx}/τ_0	τ_{yy}/τ_0	τ_{xx}/τ_0	τ_{yy}/τ_0
CST	2.910	−0.878	2.920	−0.888
Mixed	3.263	−1.093	3.085	−1.646
Infinite plate solution	3.000	−1.000	2.83	−1.75

2.3.3 Plane stress square plate with symmetric edge cracks

The problem of a square plate with symmetric edge cracks (mode type I) is considered next. Figure 2.6(a) shows the problem description and Figure 2.6(b) illustrates the finite element idealization of a quarter of the plate considered using the symmetry. The problem is solved using mixed finite

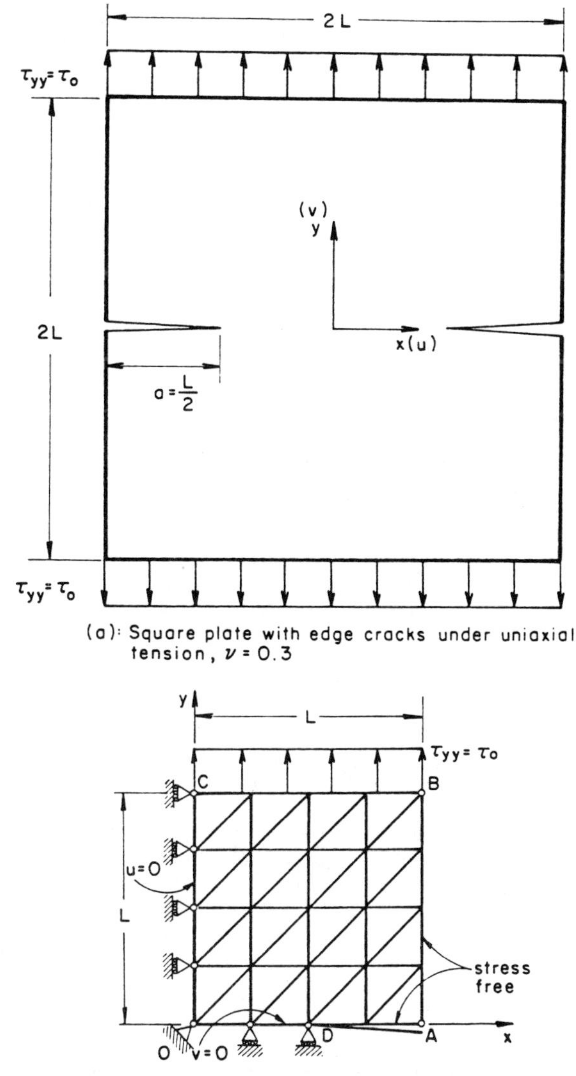

(a): Square plate with edge cracks under uniaxial tension, $\nu = 0.3$

(b): Finite element idealization of the quarter plate along with boundary conditions (N = 4).

Figure 2.6 Plane stress square plate with symmetric edge cracks

Table 2.5 Numerical results for plane stress square plate with symmetric edge cracks.

(a) τ_{yy} continuous at crack tip

N	$\frac{Ev_A}{\tau_0 L}$	$\frac{Ev_C}{\tau_0 L}$	$\frac{\tau_{xxD}}{\tau_0}$	$\frac{\tau_{yyD}}{\tau_0}$	$\frac{EU^*}{\tau_0^2 L^2 t}$
4	1.7771	1.1737	1.4203	2.3751	3.122624
8	1.8312	1.1849	2.1230	3.4100	3.196404
12	1.8272	1.1869	2.6225	4.1357	3.212178

(b) τ_{yy} discontinuous at crack tip

N	$\frac{Ev_A}{\tau_0 L}$	$\frac{Ev_C}{\tau_0 L}$	$\frac{\tau_{xxD}}{\tau_0}$	$\frac{\tau_{yyD}}{\tau_0}$	$\frac{EU^*}{\tau_0^2 L^2 t}$
4	1.7749	1.1719	1.4257	3.0437	3.113852
8	1.8273	1.1837	2.1493	4.3664	3.191367
12	1.8241	1.1862	2.6562	5.2798	3.208874

* Exact value: $\frac{UE}{\tau_0^2 L^2 t} = 3.228$ (Tong and Pian [19]).

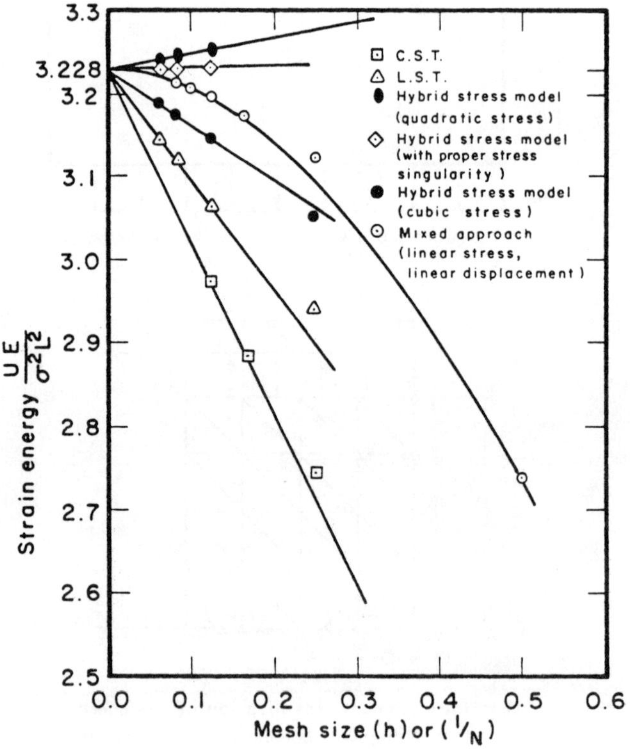

Figure 2.7 Plots of strain energy versus mesh size for plane stress problem with symmetric edge cracks

$$E = 21000 \text{ ksi} \quad , \quad h = \ell = 1.0$$
$$\nu = 0.3 \quad , \quad k = 200 \text{ ksi/in}$$

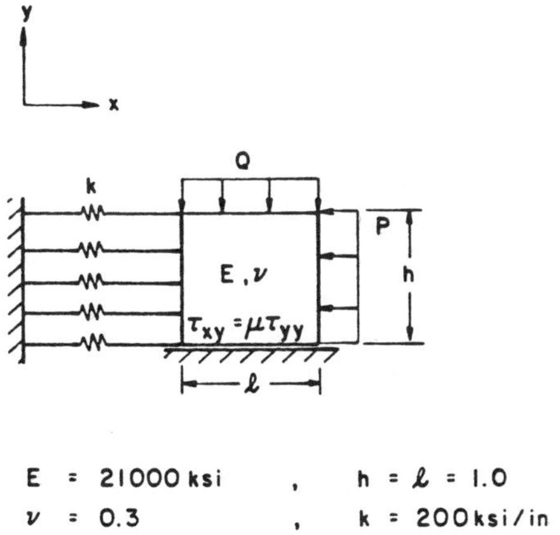

Figure 2.8 Sliding contact problem

elements for various grid sizes for two cases. The stress τ_{yy} is kept continuous across point D (the crack tip) in the first case and, in the second case, an extra node is introduced along the x axis next to the original one at D; only u, v, τ_{xx}, and τ_{xy} degrees of freedom are equated at the two nodes, thus allowing τ_{yy} to be discontinuous across point D, the crack tip.

The numerical results for both cases are presented in Table 2.5. The strain energy is converging from below in both cases, while the peak stress τ_{yyD} at the crack tip is about 20 per cent. higher when the normal stress is discontinuous across the point D than for the case when it is continuous. The plots of strain energy versus the mesh size appear in Figure 2.7. The shapes of the curves in both cases are very similar and exhibit faster convergence than just linear, as might have been expected. Figure 2.7 also shows a comparison with the solutions obtained using various other elements. The present mixed element definitely shows a faster strain energy convergence than the constant stress triangles, the linear stress triangles, and the hybrid stress rectangles with cubic stress distribution within the element and quadratic displacements along the boundaries. The convergence rate is indicated by the plot of the relative error in strain energy versus N, the number of elements along the edge of OA (Figure 2.8). It can be observed that the convergence rate approaches N^{-2} as N gets larger, for both cases. It is clearly faster than N^{-1}, indicating some sort of cancellation of error in the energy product due to singular stress terms. Further, a slightly larger error is observed in the case of discontinuous normal stress at the crack tip.

2.3.4 Sliding contact problem

The contact formulation of Section 2.2.2 is applied to the general sliding contact problem shown in Figure 2.8. The figure represents an elastic block of unit thickness lying on a rigid horizontal base adjacent to an elastic

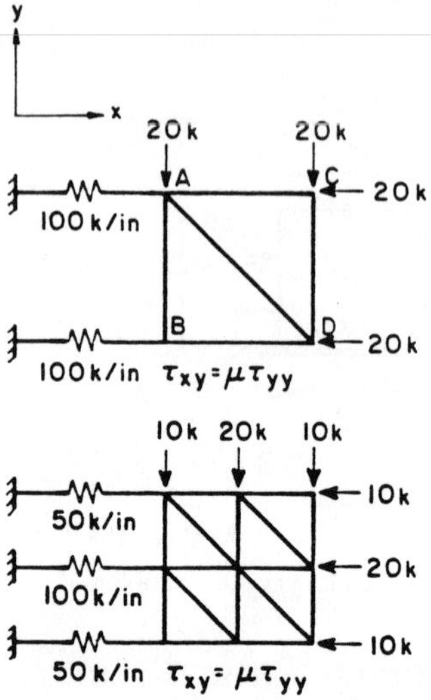

Figure 2.9 Typical finite element grids for sliding contact problem

foundation on the left, and is subject to uniform vertical and horizontal loads Q and P. In the present analysis P and Q were each 40 k.s.i. (kips per square inch) and μ was 0.2. This problem has no simple solution and therefore it is solved with a number of finite element grids of progressive refinement, as shown in Figure 2.9.

Table 2.6 Convergence of strain energy with iterations for sliding contact problem

	Strain energy (kin)			
Iteration number	1×1	2×2	4×4	8×8
0	0.0524672	0.0124315	0.0122534	0.0119737
1	0.0524830	0.0162960	0.0167065	0.0166729
2	0.0524830	0.0163187	0.0167490	0.0167151
3	0.0524830	0.0163188	0.0167493	0.0167151

Table 2.7 Convergence of displacements with grid refinement for sliding contract problem

Grid	$v_B(10^{-4}$ in)	$v_D(10^{-4}$ in)	$v_A(10^{-4}$ in)	$v_C(10^{-4}$ in)
1×1	−4.03884	−5.19759	−19.6863	−8.38099
2×2	−2.46810	−1.14094	−17.7952	−9.23910
4×4	−2.22208	−0.97027	−17.7549	−9.15021
8×8	−1.96500	−0.50836	−17.9635	−9.19802
Exact	0.0	0.0	?	?

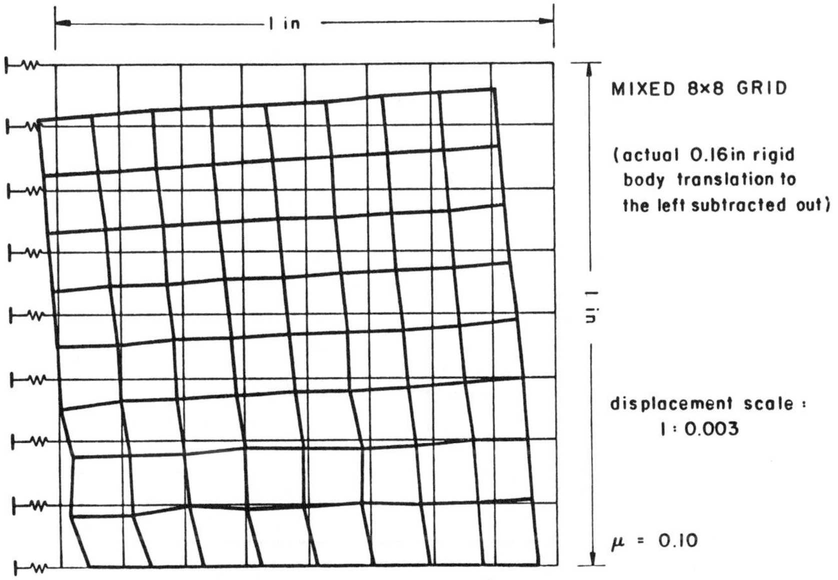

Figure 2.10 The deformed block for sliding contact problem

For each finite element grid, the solution is obtained iteratively. The convergence with iterations is illustrated in Table 2.6 where the numerical results obtained for the strain energy in the block from each finite element grid are tabulated. A rapid rate of convergence with iteration is observable.

Although the exact continuum solution to the problem is unknown, the boundary conditions that must be satisfied by such a solution are definite. For instance, along the bottom edge BD, the vertical displacement v should be zero. However, in the present method, this condition is a natural boundary condition and therefore is only achieved in the limit of zero element size. The actual numerical convergence of these displacements at B and D is shown in Table 2.7, along with those at A and C for comparison. It is seen that v_B and v_D from the 8×8 grid are still about 8 per cent. of v_A and v_C, but apparently are converging towards zero. The deformed shape of the block as obtained from the 8×8 grid is shown in Figure 2.10. Note that the average displacement of the elastic foundation has been subtracted out.

The convergence of strain energy with finite element grid refinement is illustrated in Figure 2.11 where the values of strain energy from the fifth iteration are plotted against the number of elements per unit height. The convergence is obviously very rapid but not monotonic in that there is apparent overshoot. The normal and shear stresses on the bottom face of

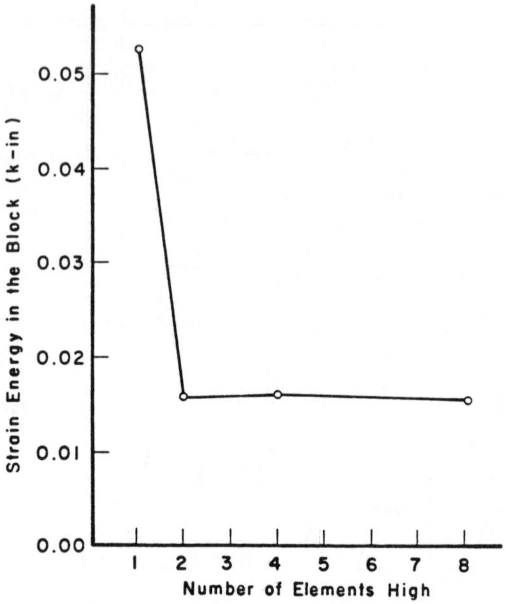

Figure 2.11 Strain energy versus grid size for sliding contact problem

the block are shown in Figure 2.12. Again the convergence is obvious. The rigid block solution shown for a rough comparison is obtained from basic mechanics.

2.3.5 Stick-slip problem

A unit thickness block again rests on a rigid base and is subject to a uniform vertical load Q as shown in Figure 2.13. The block expands in the x direction as a result of Poisson's effect. The frictional stresses which develop between the block and the base may be sufficient to prevent slippage in some parts of the contact region (adhering portion), while at other parts slippage will occur (sliding portion).

For comparison, the problem is also solved with a potential energy formulated finite element, namely the quadratic strain triangular element (QST). The method for the QST is the one used in previous works for potential energy formulated finite element models. Using the symmetry, half the block is modelled with $1 \times 1, 2 \times 2, 3 \times 3, 4 \times 4$, and 6×6 grids of both the mixed and QST elements. The 1×1 grids are shown in Figure 2.14.

The calculations began with all contact nodes assumed to be adhering. How these were subsequently modified by the iterations is illustrated in

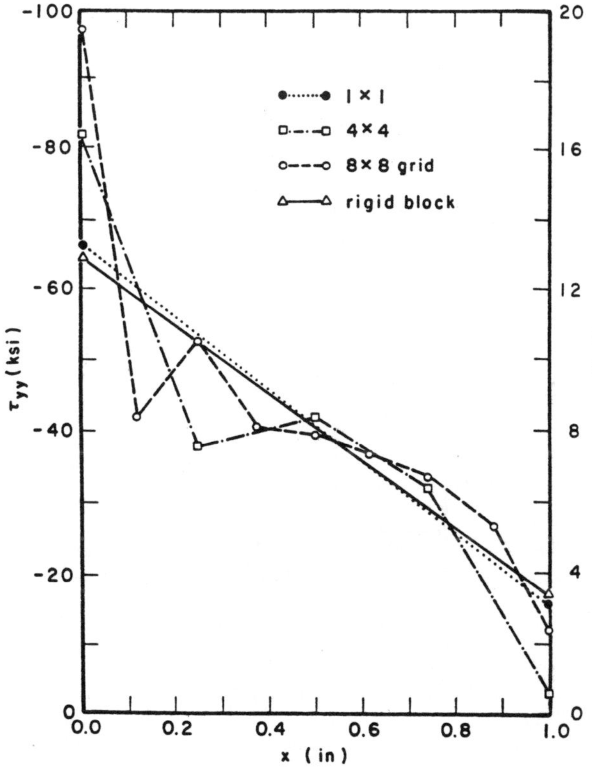

Figure 2.12 Stress distributions on the contact face

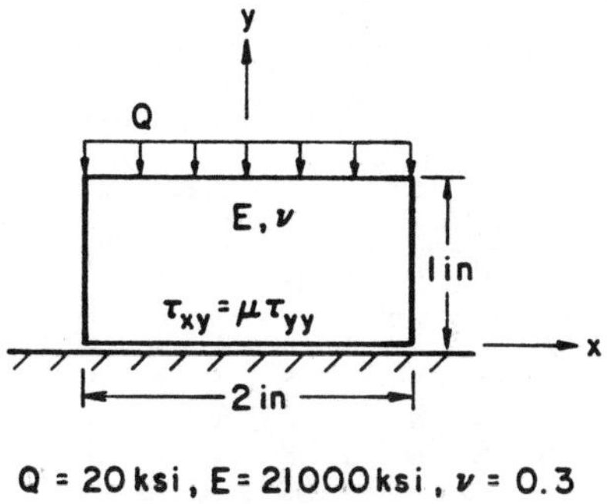

Figure 2.13 Stick-slip contact problem

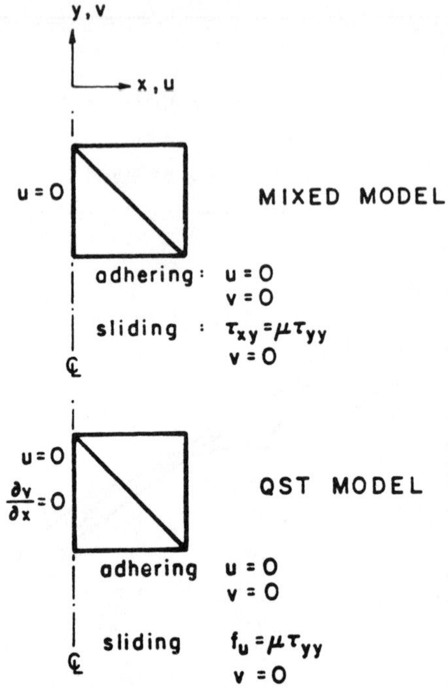

Figure 2.14 The QST and mixed finite element grids for the stick-slip problem

Table 2.8 from the 6×6 grid results for $\mu = 0.10$. The points a to g refer to the contact nodes from the centreline to the outer edge respectively. Clearly the convergence with iterations was very rapid. In general, this boundary condition convergence was always slightly faster for the mixed elements than for the QST ones.

The change from adhering to sliding conditions which apparently occurred

Table 2.8 Iteration convergence of boundary conditions for stick-slip contact, $\mu = 0.10$

	QST							Mixed						
Iteration	Nodes							Nodes						
number	a	b	c	d	e	f	g	a	b	c	d	e	f	g
0	A	A	A	A	A	A	A	A	A	A	A	A	A	A
1	A	A	A	A	S	S	S	A	A	A	S	S	S	S
2	A	A	A	S	S	S	S	A	A	S	S	S	S	S
3	A	A	S	S	S	S	S	A	A	S	S	S	S	S
4	A	A	S	S	S	S	S							

A = adherind node, S = sliding node.

Table 2.9 Effect of friction coefficient μ on boundary conditions for stick-slip contact

	QST											Mixed											
μ/x	0	0.1	0.2	0.3	0.4	0.5	0.6	0.7	0.8	0.9	1.0	0	0.1	0.2	0.3	0.4	0.5	0.6	0.7	0.8	0.9	1.0	Grid
0.05	A			S		S		S			S	A			S		S		S			S	
0.10	A			S		S		S			S	A			A		S		S			S	4
0.15	A			A		S		S			S	A			A		A		S			S	×
0.20	A			A		A		A			S	A			A		A		A			S	4
0.05	A		S	S		S		S	S		S	A		S	S		S	S		S		S	
0.10	A		A	A		A		S	S		S	A		A	A		A	S		S		S	6
0.15	A		A	A		A		S	S		S	A		A	A		A	S		S		S	×
0.20	A		A	A		A		A	A		S	A		A	A		A	A		S		A	6

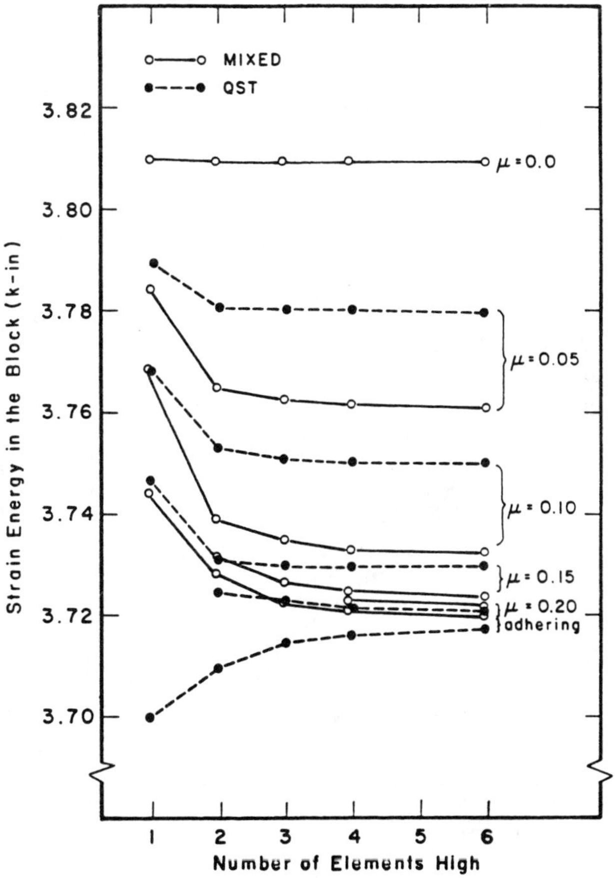

Figure 2.15 Strain energy versus grid size for the stick-slip problem

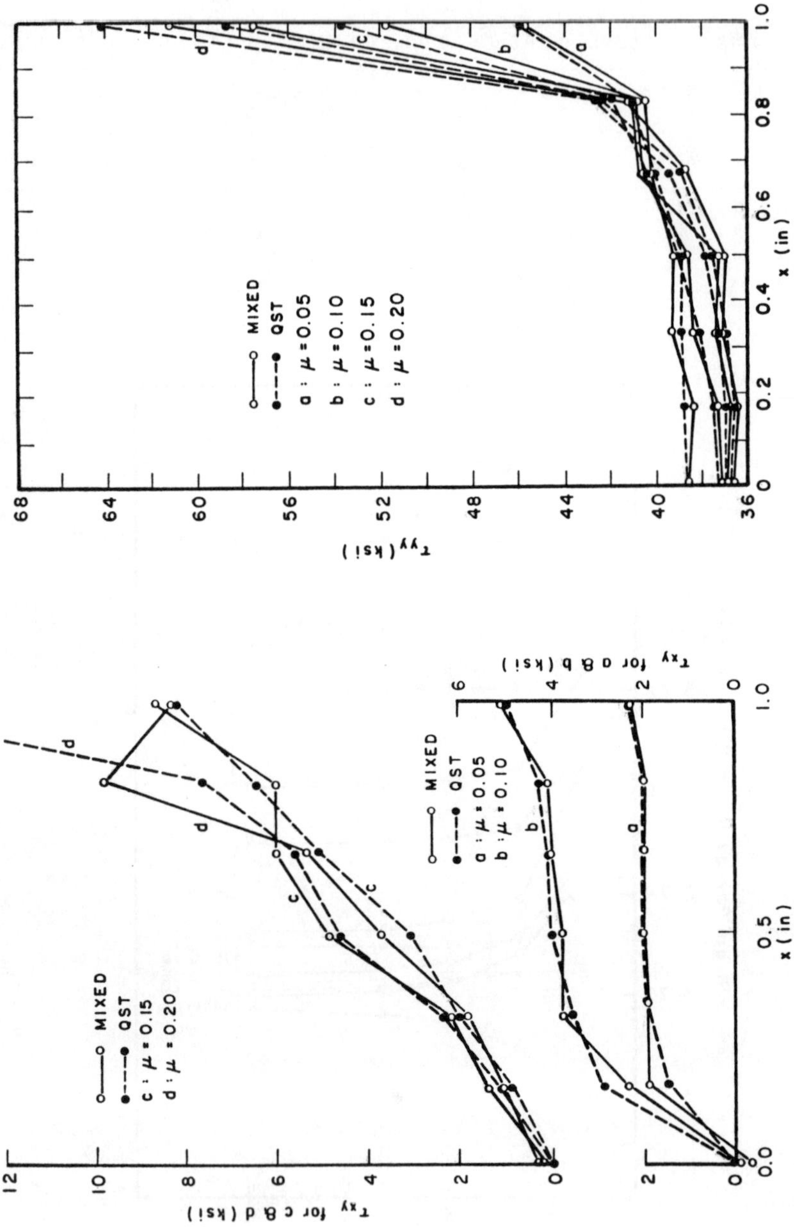

Figure 2.16 (a) Shear stress distributions on contact face for the stick-slop problem. (b) Normal stress distributions on contact face for the stick-slip problem

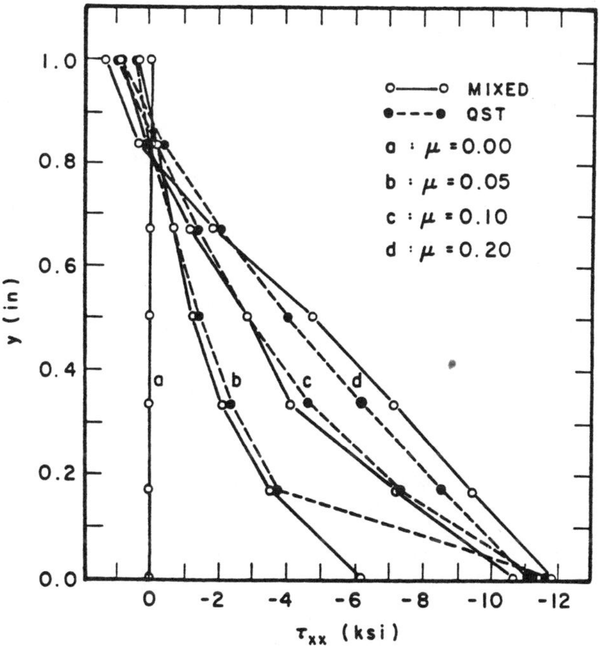

Figure 2.17 Normal stress distribution along vertical centreline of stick-slip problem

between nodes b and c for $\mu = 0.10$ (Table 2.8) is actually a function of μ. This is illustrated in Table 2.9 where the 4×4 and 6×6 grid results are tabulated for $\mu = 0.05$, 0.10, 0.15, and 0.20. The dashed line just indicates the locus of the changeover point. In general, the mixed and QST results agreed very well in this respect.

Strain energy convergence plots are shown in Figure 2.15 for various values of μ as obtained from both the mixed and QST elements. Note that the QST *adhering* curve exhibits monotonic convergence from below as predicted by the potential energy theorem. However, when sliding occurs, the QST convergence is apparently from above. Further, it appears that the mixed method always converges from above. Again, the agreement between the two methods is quite good, thereby effectively verifying each other.

The stress distributions on the contact face at different values of μ obtained from the 6×6 grid analyses with both methods are plotted in Figure 2.16(a) and (b). Also shown in Figure 2.17 is a plot of the τ_{xx} distribution on the centreline of the block for different values of μ. Again, they indicate good agreement between the two methods.

Table 2.10 Typical displacement results for stick-slip contact, $\mu = 0.10$

Grid	$v_B(10^{-4}\text{ in})$	$v_D(10^{-4}\text{ in})$	$v_A(10^{-4}\text{ in})$	$v_C(10^{-4}\text{ in})$
2×2	−0.0268	−1.0109	−18.2195	−19.6268
3×3	−0.0398	−0.8373	−18.2916	−19.6408
4×4	0.1148	−0.0765	−18.2556	−19.6417
6×6	0.0490	−0.6726	−19.2799	−19.6060
Exact	0.0	0.0	?	?

Table 2.11 Typical stress results for stick-slip contact, $\mu = 0.10$

Grid	τ_{xxC}(k.s.i.)	τ_{xyC}(k.s.i.)	τ_{xyA}(k.s.i.)	τ_{xyB}(k.s.i.)	τ_{xxB}(k.s.i.)
2×2	−1.1712	−0.1878	0.7993	−1.1472	−7.9515
3×3	−0.0021	−0.6525	−0.4497	−0.8181	−9.7737
4×4	−0.5938	−0.2282	0.1347	−1.0028	−10.5017
6×6	−0.1526	−0.2238	−0.2334	−0.0827	−10.6588
Exact	0.0	0.0	0.0	0.0	?

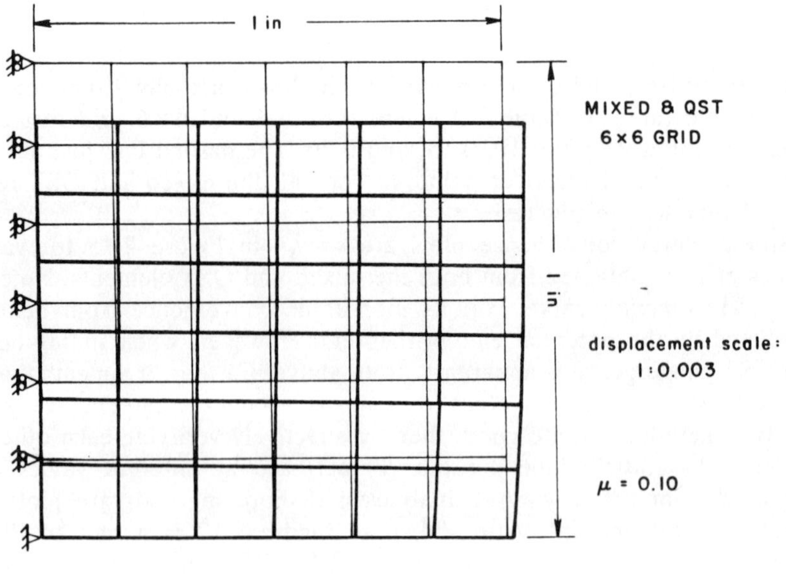

Figure 2.18 The deformed block in the stick-slip problem

Again, as in the sliding problem, the exact continuum solution to the problem is unknown, but the boundary conditions that must be satisfied by such a solution are known. The numerical convergence of the displacements associated with natural boundary conditions at B and D are shown in Table 2.10 for $\mu = 0.10$. It is seen that v_B and v_D are respectively about 3 and 4 per cent. of v_A and v_C. The convergence of the stresses τ_{xx} and τ_{xy} to the stress-free condition at C and that of τ_{xy} to zero at A and B by the symmetry are shown in Table 2.11 also for $\mu = 0.10$. They are about 2 per cent. of τ_{xxB}, which is also tabulated for comparison. The deformed shape of the block as obtained from the 6×6 grid is shown in Figure 2.18.

The system of equations of the finite element model is linear with respect to the loading and non-linear with respect to the Coulomb coefficient of friction μ. Hence the position of the adhering–sliding changeover point on the contact face is independent of the applied loading but dependent of μ.

Although not shown here, the Hertz problem of contact between two elastic cylinders was also analysed with the present mixed formulation. Quite good results were also obtained for this problem. See [11, 14] for details.

2.4 CONCLUDING REMARKS

Our experiences with applying the mixed finite element method to plane elasticity problems have been very satisfactory. We found that the use of displacement and stress approximations which satisfied the following completeness criterion:

(a) the displacement approximations should include all rigid body and constant strain modes,
(b) the stress approximations should include all constant stress modes,
(c) the strains from the stress approximations should possess at least all the strain modes that are present in the strains derived from the displacement approximations,

did indeed provide convergence in the energy sense. Further, the error in strain energy was found to be of the same order as the mean square error in the approximate stress solution. Hence, the strain energy convergence rate is generally higher with the mixed element than with the displacement element incorporating the same displacement interpolation.

The plane stress triangular element, using linear interpolations for both displacements and stresses, yielded a predicted mean square error of $O(N^{-2})$ in stresses and, hence, an error in strain energy of $O(N^{-4})$. In the numerical application of this element, the energy convergence rate was indeed found to be $O(N^{-4})$ for the plane stress square plate with parabolic end loads. In comparison, the corresponding displacement element yields an energy convergence rate of only $O(N^{-2})$.

The results from the application to the square plate with symmetric edge cracks indicated a strain energy convergence rate close to $O(N^{-2})$ whereas only $O(N^{-1})$ was expected. Indeed, higher order displacement or hybrid finite elements yield a convergence rate of only $O(N^{-1})$.

In the application to contact problems, we found that the different boundary conditions required on the contact surface could be incorporated directly into the Reissner principle. In some cases, the stresses became forced boundary conditions while the displacements became natural. Then an iterative procedure was required to find the final solution. However, only a few such iterations were required.

In stick-slip problems where the conditions on the contact boundary are sought, iterations are again required. It was observed that only a few iterations were needed to attain the final boundary conditions. It was also observed that it took more iterations for the displacement or potential energy method in which the friction condition could not be applied directly as a constraint condition. Nevertheless, the two formulations yielded comparable results.

REFERENCES

1. L. R. Herrmann, 'A bending analysis of plates', *Proc. Conf. Matrix Methods in Structural Mechanics*, AFFDL-TR-66-80, 577 (1966).
2. K. Hellan, 'Analysis of elastic plates in flexure by a simplified finite element method', *Acta Polytechnica Scandinavia*, Civil Engineering Series No. 46, Trondheim (1967).
3. R. S. Dunham and K. S. Pister, 'A finite element application of the Hellinger–Reissner variational theorem', *Proc. Conf. Matrix Methods in Structural Mechanics*, AFFDL-TR-68-150, 471 (1968).
4. J. T. Oden, 'Some contributions to the mathematical theory of mixed finite element approximations', in *Theory and Practice in Finite Element Structural Analysis* (Eds. Y. Yamada and R. H. Gallagher), University of Tokyo Press, 1973.
5. J. N. Reddy and J. T. Oden, 'Mathematical theory of mixed finite element approximations', *Q. Appl. Math.*, **33,** 255 (1975).
6. J. T. Oden and J. N. Reddy, 'On mixed finite element approximations', *SIAM J. Num. Anal.*, **13,** 392 (1976).
7. I. Babuska, J. T. Oden, and J. K. Lee, 'Mixed-hybrid finite element approximations of second-order elliptic boundary-value problems', *Comp. Meth. Eng. Appl. Mech.*, **14,** 1 (1978).
8. P. A. Raviart and J. M. Thomas, 'A mixed finite element for 2nd order elliptic problems' (to appear).
9. J. T. Oden and J. K. Lee, 'Theory of mixed and hybrid finite element approximations in linear elasticity', *Proc. IUTAM/IUM Symp. Applications of Methods of Functional Analysis to Problems of Mechanics*, Marseilles, France, September 1975.
10. F. A. Mirza, *Convergence of Mixed Methods in Continuum Mechanics and Finite*

Element Analysis, Ph.D. thesis, University of British Columbia, Vancouver, 1977.

11. J. Tseng, *Application of the Mixed Finite Element Method to the Elastic Contact Problem*, M.A.Sc. thesis, University of British Columbia, 1980.
12. F. A. Mirza and M. D. Olson, 'Energy convergence and evaluation of stress intensity factor K_1 for stress singular problems by mixed finite element method', *Int. J. Fracture*, **14**, 555–573 (1978).
13. F. A. Mirza and M. D. Olson, 'The mixed finite element method in plane elasticity', *Int. J. Num. Meth. in Eng.*, **15**, 273–289 (1980).
14. J. Tseng and M. D. Olson, 'The mixed finite element method applied to two-dimensional elastic contact problems', *Int. J. Num. Meth. in Eng.* (in press).
15. G. R. Cowper, G. M. Lindberg, and M. D. Olson, 'A shallow shell finite element of triangular shape', *Int. J. Solids and Structures*, **4**, 1133 (1970).
16. O. C. Zienkiewicz, Y. K. Cheung, and K. G. Stagg, 'Stresses in anisotropic media with particular reference to problems of rock mechanics', *J. Strain Anal.*, **1**, 172 (1966).
17. S. P. Timoshenko and J. N. Goodier, *Theory of Elasticity*, 3rd ed., McGraw-Hill, 1970.
18. G. N. Savin, *Stress Concentrations around Holes*, Pergamon, 1961.
19. P. Tong and T. H. H. Pian, 'On the convergence of the finite element method for problems with singularity', *Int. J. Solids and Structures*, **9**, 313–321 (1973).

Hybrid and Mixed Finite Element Methods
Edited by S. N. Atluri, R. H. Gallagher, and O. C. Zienkiewicz

Chapter 3

Recent Studies in Hybrid and Mixed Finite Element Methods in Mechanics

S. N. Atluri, P. Tong, and H. Murakawa

3.1 INTRODUCTION

In this chapter we present a summary of some of the recent basic developments in hybrid and mixed finite element methods in linear and non-linear solid mechanics, and to a limited extent in fluid mechanics.

Following a summary of notation used in this chapter in Section 3.2, we present in Section 3.3 an account of the rationale and origin of hybrid and mixed-hybrid finite element methods in linear solid mechanics. It is shown that the Hu–Washizu mixed variational principle in solid mechanics can be modified, in its application to a finite element assemblage, in two alternate ways to introduce the constraints of interelement displacement compatibility and traction reciprocity as *a posteriori* constraints through Lagrange multipliers. It is shown that a wide variety of hybrid and mixed hybrid finite elements can be constructed through the above two general modified principles.

In Section 3.4 we illustrate the use and versatility of hybrid finite element methods in solving problems with constraint. The specific example chosen here is the analysis of the steady, slow flow of a viscous, incompressible fluid.

In Section 3.5, which is written in a spirit similar to that in Section 3.3, we consider finite deformation (large rotations as well as stretches) problems of inelastic solids. Variational formulations governing the rates of displacement, strain, and stress are considered. Various stress and conjugate strain measures, and their rates, in a finitely deformed solid are introduced. The most general mixed hybrid finite element variational principles involving alternate stress rates and conjugate strain rates are presented. As examples, modified complementary energy rate principles governing hybrid stress finite elements are discussed in detail.

With the limited aim of presenting the basic underlying ideas in this chapter, no attempt is made herein at a comprehensive survey of pertinent literature. The other chapters in this book, in fact, serve as supplementary sources of references to further literature on the subjects dealt with in this chapter.

3.2 NOTATION

A fixed rectangular Cartesian system is used.

$\mathbf{u}$	bold type denotes a vector
V	italic sans-serif type denotes a second-order tensor
$\mathbf{a}=\mathsf{B}\cdot\mathbf{C}$	implies $a_i = B_{ik}C_k$
$\mathsf{B}:\mathsf{C}$	implies $B_{ij}C_{ij}$
$\mathbf{x}$	position vector of a particle in a solid in its undeformed configuration, C_0
∇_0	gradient operator in C_0, $\nabla_0 = \mathbf{e}_i\, \partial/\partial x_i$
$\mathbf{y}$	position vector of a particle in a solid in its deformed configuration, say C_N
∇_N	$\mathbf{e}_i\, \partial/\partial y_i$
$\boldsymbol{\tau}$	true (Cauchy) stress
$\boldsymbol{\varepsilon}$	infinitesimal strain in a linear theory
g	Green–Lagrange finite strain
$\mathbf{u}$	displacement vector
t	I Piola–Kirchhoff stress tensor
s	II Piola–Kirchhoff stress tensor
r	symmetrized Biot–Lure stress tensor
$\boldsymbol{\sigma}$	Kirchhoff stress tensor
$(\dot{\boldsymbol{\tau}}, \dot{\boldsymbol{\sigma}})$	material derivatives of $(\boldsymbol{\tau}, \boldsymbol{\sigma})$
$\dot{\boldsymbol{\sigma}}^*$	co-rotational rate of $\boldsymbol{\sigma}$
$\dot{\boldsymbol{\varepsilon}}$	Eulerean strain rate

3.3 HYBRID AND MIXED MODELS IN LINEAR SOLID MECHANICS

In linear solid mechanics, it is well known that the principle of minimum potential energy involves a compatible displacement field alone as a variable and has as its Euler–Lagrange (EL) equations the conditions of balance of momenta and the traction boundary conditions. Likewise the principle of minimum complementary energy involves a stress field, which is equilibrated and satisfies the traction boundary conditions, alone as a variable and has as its EL equations the kinematic compatibility conditions and displacement boundary conditions. These principles are often referred to as the 'primal' and the 'dual primal' principles of linear solid mechanics. If, in the principle of potential energy, the conditions of kinematic compatibility and displacement boundary conditions are introduced as conditions of constraint through

Lagrange multipliers (which turn out to be stresses and surface tractions respectively), one then obtains the so-called Hu–Washizu principle [1, 2]. Likewise, if the condition of equilibrium of stresses is introduced as a constraint condition through a Lagrange multiplier field (which turns out to be displacement) in the complementary energy principle, one is led to the so-called Hellinger–Reissner principle [3, 4]. Thus, the Hu–Washizu and Hellinger–Reissner principles which involve more than one field in the continuum as variables, some of which play the roles of Lagrange multipliers to enforce certain constraint conditions, are often referred to as mixed variational principles.

Suppose now that one considers the solution of a linear boundary value problem, for the solid continuum, by the finite element method. In the straightforward application of the potential energy principle to such a finite element assemblage, it is evident that the admissible displacement field in each element should be such that it is not only continuous within the element but also compatible at the interelement boundaries (i.e.b), even though the strains may be discontinuous at the interelement boundaries. The variational principle then leads, as its EL equations, to (a) not only equilibrium within the element but also (b) interelement traction reciprocity (action and reaction being equal at the element interfaces). Likewise, in the straightforward application of the complementary energy principle to the finite element assemblage, it is seen that the admissible stress field should not only be equilibrated within each element but also satisfy the interelement traction reciprocity condition, even though the stresses may be discontinuous at the interelement boundaries. The finite element complementary energy principle then leads, as its EL equations, to (a) not only kinematic compatibility within each element but also (b) interelement displacement compatibility. The above requirements of interelement displacement continuity and traction reciprocity on the admissible displacement fields and stress fields respectively may not provide sufficient flexibility in the finite element solution of several problems in linear solid mechanics, such as plate bending, shells, multilayered composites, problems with singularities as in fracture mechanics, etc. To gain this flexibility in application, one may relax the conditions of displacement continuity and/or traction reciprocity at interelement boundaries on the admissible displacement or stress fields, but introduce these as *a posteriori* constraints into the respective finite element variational principles through Lagrange multiplier fields introduced at the interelement boundaries. This is the underlying concept in the so-called hybrid finite element methods, which owe their development to the pioneering work of Pian, whose 1964 paper [5] was the first step in the direction of what are now called hybrid methods.

We now give a general framework for hybrid and mixed hybrid finite element methods in linear solid mechanics. From this general theory, several special types of hybrid finite element methods can be formulated. Consider a solid continuum Ω with boundary S. Let S_u and S_t be the segments of S where displacements and tractions respectively are prescribed, for instance. Let $\boldsymbol{\nabla}_0$ be the gradient operator in the undeformed body which, under the assumption of infinitesimal deformation, is indistinguishable from the deformed body. The boundary value problem in linear solid mechanics can be stated as:

$$\boldsymbol{\nabla}_0 \cdot \boldsymbol{\tau} + \bar{\mathbf{F}} = 0 \text{ (LMB)}; \qquad \boldsymbol{\tau} = \boldsymbol{\tau}^{\mathrm{T}} \text{ (AMB)} \tag{3.1}$$

$$\boldsymbol{\varepsilon} = \tfrac{1}{2}[(\boldsymbol{\nabla}\mathbf{u}) + (\boldsymbol{\nabla}\mathbf{u})^T] \text{ (compatibility)} \tag{3.2}$$

$$\boldsymbol{\tau} = \partial A/\partial \boldsymbol{\varepsilon} \text{ (constitutive law)} \tag{3.3}$$

$$n \cdot \boldsymbol{\tau} = \bar{\mathbf{t}} \text{ (traction boundary condition) at } S_t \tag{3.4}$$

$$\mathbf{u} = \bar{\mathbf{u}} \text{ (displacement boundary condition) at } S_u \tag{3.5}$$

wherein LMB and AMB stand respectively for linear and angular momentum balance, $\mathbf{n}$ is a unit vector outward normal to S_t, and A is the strain energy density per unit volume. For a linear elastic solid A is quadratic in $\boldsymbol{\varepsilon}$. We assume that (3.3) can be inverted to express $\boldsymbol{\varepsilon}$ in terms of $\boldsymbol{\tau}$ and thereby establish a Legendre contact transformation:

$$\boldsymbol{\tau} : \boldsymbol{\varepsilon} - A(\boldsymbol{\varepsilon}) = B(\boldsymbol{\tau}) \tag{3.6}$$

wherein B is the complementary energy density. It is now well known [1, 2] that equations (3.1) to (3.5) are the Euler–Lagrange equations corresponding to the stationarity of the mixed variational functional:

$$\begin{aligned} \mathrm{HW}[\mathbf{u}, \boldsymbol{\varepsilon}, \boldsymbol{\sigma}] = & \int_\Omega \{A(\boldsymbol{\varepsilon}) + \boldsymbol{\tau} : [\tfrac{1}{2}\langle(\boldsymbol{\nabla}\mathbf{u}) + (\boldsymbol{\nabla}\mathbf{u})^T\rangle - \boldsymbol{\varepsilon}] - \bar{\mathbf{F}} \cdot \mathbf{u}\}\, \mathrm{d}\Omega \\ & - \int_{S_t} \bar{\mathbf{t}} \cdot \mathbf{u}\, \mathrm{d}s + \int_{S_u} \mathbf{n} \cdot \boldsymbol{\tau} \cdot (\mathbf{u} - \bar{\mathbf{u}})\, \mathrm{d}s \end{aligned} \tag{3.7}$$

or, equivalently, upon using the divergence theorem,

$$\begin{aligned} \mathrm{HW}[\mathbf{u}, \boldsymbol{\varepsilon}, \boldsymbol{\sigma}] = & \int_\Omega \{A(\boldsymbol{\varepsilon}) - (\boldsymbol{\nabla}_0 \cdot \boldsymbol{\tau} + \bar{\mathbf{F}}) \cdot \mathbf{u} - \boldsymbol{\tau} : \boldsymbol{\varepsilon}\}\, \mathrm{d}\Omega \\ & - \int_{S_t} (\bar{\mathbf{t}} - \mathbf{n} \cdot \boldsymbol{\tau}) \cdot ds + \int_{S_u} \mathbf{n} \cdot \boldsymbol{\tau} \cdot \bar{\mathbf{u}}\, \mathrm{d}s \end{aligned} \tag{3.8}$$

Consider the solid to be discretized into N finite elements Ω_m, $m = 1, \ldots, N$ $(\Omega = \sum_1^N \Omega_m)$. Let the boundary of Ω_m be $\partial\Omega_m$. It is seen that, in general,

$$\partial\Omega_m = \rho_m + S_{tm} + S_{um} \tag{3.9}$$

where ρ_m is the interelement boundary, S_{tm} and S_{um} are segments of $\partial\Omega_m$

where tractions and displacements respectively are prescribed. It is seen that for elements which do not adjoin the boundary S of the solid, $S_{tm} = S_{um} = 0$. With this notation, it is clear that the fields $\mathbf{u}$, $\boldsymbol{\tau}$, and $\boldsymbol{\varepsilon}$ in each Ω_m must satisfy the following necessary conditions in order that the solution generated for the finite element assemblage would approach that for an otherwise continuous solid:

$$\boldsymbol{\nabla}_0 \cdot \boldsymbol{\tau} + \bar{\mathbf{F}} = 0; \qquad \boldsymbol{\tau} = \boldsymbol{\tau}^T \text{ in } \Omega_m \tag{3.10}$$

$$\boldsymbol{\tau} = \partial A / \partial \boldsymbol{\varepsilon} \text{ in } \Omega_m \tag{3.11}$$

$$\boldsymbol{\varepsilon} = \tfrac{1}{2}[(\nabla_0 \mathbf{u}) + (\boldsymbol{\nabla}_0 \mathbf{u})^T] \text{ in } \Omega_m \tag{3.12}$$

$$\mathbf{n} \cdot \boldsymbol{\tau} \equiv \mathbf{t} = \bar{\mathbf{t}} \text{ at } S_{tm} \tag{3.13}$$

$$\mathbf{u} = \bar{\mathbf{u}} \text{ at } S_{um} \tag{3.14}$$

$$\mathbf{u}^+ = \mathbf{u}^- \text{ at } \rho_m \tag{3.15}$$

$$(\mathbf{n} \cdot \boldsymbol{\tau})^+ + (\mathbf{n} \cdot \boldsymbol{\tau})^- = 0 \text{ at } \rho_m \tag{3.16}$$

Equations (3.10) to (3.14) are analogous to their continuum counterparts, (3.1) to (3.5) respectively. Noting that the superscripts (+) and (−) denote, arbitrarily, the two sides of ρ_m in the limit as ρ_m is approached, (3.15) can be seen to be the statement of interelement displacement compatibility. Likewise, noting that $\mathbf{n}^+ = -\mathbf{n}^-$ at ρ_m, it is seen that (3.16) is the statement of interelement traction reciprocity (Newton's third law). In connection with (3.14) we make the following observation. In two or three dimensional elasticity problems, there are two or three independent displacement components respectively. In such cases it is necessary that these displacements be continuous at ρ_m. However, in problems such as thin plates and shells, these displacement components are usually subject to certain plausible approximations such as those in Kirchhoff–Love hypotheses. In the plate problem, for instance, if x_α $(\alpha = 1, 2)$ are the inplane cartesian coordinates and x_3 is normal to the plate mid-surface, the Kirchhoff hypotheses impose a kinematic constraint on u_α such that $u_\alpha = -x_3\, \partial u_3 / \partial x_\alpha$. Thus in order for all the three components u_i $(i = 1, 2, 3)$ to be continuous at ρ_m, it is necessary that not only u_3 but also $\partial u_3 / \partial x_\alpha$ be continuous at ρ_m. This problem has plagued the developers of finite elements ever since the method came into vogue.

We now attempt to 'construct' a finite element variational principle whose Euler–Lagrange equations are equations (3.10) to (3.16) respectively. If one were to replace the domain and boundary integrals in (3.7) and (3.8) simply by the sum of the respective integrals for each finite element, i.e. write

$$\begin{aligned} \mathrm{FHW}(\mathbf{u}, \boldsymbol{\varepsilon}, \boldsymbol{\sigma}) = \sum_m \Bigg(\int_{\Omega_m} \{A(\boldsymbol{\varepsilon}) + \boldsymbol{\tau} : [\tfrac{1}{2}\langle(\boldsymbol{\nabla}\mathbf{u}) + (\boldsymbol{\nabla}\mathbf{u})^T\rangle - \boldsymbol{\varepsilon}] - \bar{\mathbf{F}} \cdot \mathbf{u}\}\, \mathrm{d}\Omega \\ - \int_{S_{tm}} \bar{\mathbf{t}} \cdot \mathbf{u}\, \mathrm{d}s + \int_{S_{um}} \mathbf{n} \cdot \boldsymbol{\tau} \cdot (\mathbf{u} - \bar{\mathbf{u}})\, \mathrm{d}s \Bigg) \end{aligned} \tag{3.17}$$

then one would find that δHW = 0 leads to, as its EL equations, (3.10) to (3.14) but not (3.15). Also, the recovery of (3.16) would depend on the fact that $\mathbf{u}^+ = \mathbf{u}^-$ at ρ_m *a priori*.

If $\mathbf{u}^+ = \mathbf{u}^-$ at ρ_m is *not* satisfied *a priori*, then it is necessary to introduce this as a constraint, through a Lagrange multiplier, into (3.17). This, as will be shown in the following, can be achieved in one of two ways. In the first way, the constraint $\mathbf{u}^+ = \mathbf{u}^-$ at ρ_m is enforced directly through a Lagrange multiplier field $\tilde{\mathbf{T}}_\rho$ at ρ_m. It can be shown easily that the field $\tilde{\mathbf{T}}_\rho$ has the meaning of interelement traction field, and as such should obey, *a priori*, the requirement that $(\tilde{\mathbf{T}}_\rho)^+ + (\tilde{\mathbf{T}}_\rho)^- = 0$ at ρ_m. This can be done easily by choosing $\tilde{\mathbf{T}}_\rho$ such that it is equal and opposite for two elements adjoining at a common boundary. The thus modified functional can be written as

$$(\mathrm{MFHW})_1(\mathbf{u}, \boldsymbol{\varepsilon}, \boldsymbol{\tau}, \tilde{\mathbf{T}}_0) = \mathrm{FHW}(\mathbf{u}, \boldsymbol{\varepsilon}, \boldsymbol{\tau}) - \sum_m \int_{\Omega_m} \tilde{\mathbf{T}}_\rho \cdot \mathbf{u}\, ds \tag{3.18}$$

where FHW is defined in (3.17). It is seen that the condition of stationarity of the functional in (3.18) leads to

$$\begin{aligned}\delta(\mathrm{MFHW})_1 = 0 = \sum_m \Bigg(\int_{\Omega_m} & \{(\partial A/\partial \boldsymbol{\varepsilon} - \boldsymbol{\tau}) : \delta\boldsymbol{\varepsilon} + \delta\boldsymbol{\tau} : \langle \tfrac{1}{2}[(\boldsymbol{\nabla}_0 \mathbf{u}) + (\boldsymbol{\nabla}_0 \mathbf{u})^T] - \boldsymbol{\varepsilon}\rangle \\ & - (\boldsymbol{\nabla}_0 \cdot \boldsymbol{\tau} + \bar{\mathbf{F}}) \cdot \delta\mathbf{u}\}\, d\Omega \\ & + \int_{S_{tm}} (\mathbf{t} - \bar{\mathbf{t}}) \cdot \delta\mathbf{u}\, ds + \int_{S_{um}} \delta\mathbf{t} \cdot (\bar{\mathbf{u}} - \mathbf{u})\, ds - \int_{\rho_m} \delta\tilde{\mathbf{T}}_\rho \cdot \mathbf{u}\, ds \\ & + \int_{\rho_m} (\mathbf{t} - \tilde{\mathbf{T}}_\rho) \cdot \delta\mathbf{u}\, ds \Bigg)\end{aligned} \tag{3.19}$$

The first five terms on the right-hand side of (3.19) can be seen to lead to equations (3.10) to (3.14) respectively. The last two terms deserve some explanation. Consider an example of two planar finite elements adjoining at the line AB, and consider the sixth term on the right-hand side of (3.19). It is then see that

$$\sum_m \int_{\rho_m} \delta\tilde{\mathbf{T}}_\rho \cdot \mathbf{u}\, ds = \int_{AB} [(\delta\tilde{\mathbf{T}}_\rho \cdot \mathbf{u})^+ + (\delta\tilde{\mathbf{T}}_\rho \cdot \mathbf{u})^-]\, ds \tag{3.20}$$

However, since $\tilde{\mathbf{T}}_\rho^+ + \tilde{\mathbf{T}}_\rho^- = 0$ at ρ_m *a priori*, it is seen that the vanishing of the term in equation (3.20) implies $(\mathbf{u}^+ - \mathbf{u}^-) = 0$ *a posteriori*, or the required equation (3.15). Likewise, in the above example, the seventh term on the right-hand side of (3.19) can be written as

$$\sum_m \int_{\rho_m} (\mathbf{t} - \tilde{\mathbf{T}}_\rho) \cdot \delta\mathbf{u}\, ds = \int_{AB} [(\mathbf{t} - \tilde{\mathbf{T}}_\rho)^+ \cdot \delta\mathbf{u}^+ + (\mathbf{t} - \tilde{\mathbf{T}}_\rho)^- \cdot \delta\mathbf{u}^-]\, ds \tag{3.21}$$

Since $\mathbf{u}^+ \neq \mathbf{u}^-$ *a priori*, it is seen that the vanishing of the right-hand side of equation (3.20) implies $\mathbf{t}^+ - \tilde{\mathbf{T}}_\rho^+ = 0$ at AB *and* $\mathbf{t}^- - \tilde{\mathbf{T}}_\rho^- = 0$ at AB. Thus, this implies $\mathbf{t}^+ + \mathbf{t}^- = 0$ at ρ_m, the required equation (3.16).

An alternative way of enforcing the constraint $\mathbf{u}^+ = \mathbf{u}^-$ *a posteriori* is to first introduce a unique element boundary displacement field $\tilde{\mathbf{u}}_\rho$ at ρ_m and then enforce the constraints $\mathbf{u}^+ = \tilde{\mathbf{u}}_\rho$ and $\mathbf{u}^- = \tilde{\mathbf{u}}_\rho$ separately. Thus, unlike the first way shown above, one can introduce two separate and arbitrary Lagrange multiplier fields, $\mathbf{T}_\rho^+$ and $\mathbf{T}_\rho^-$ respectively. Once again, these Lagrange multiplier fields have the meaning of element boundary traction fields. The thus-modified finite element variational functional can be written as

$$(\text{MFHW})_2(\mathbf{u}, \boldsymbol{\varepsilon}, \boldsymbol{\tau}, \tilde{\mathbf{u}}_\rho, \mathbf{T}_\rho) = \text{FHW}(\mathbf{u}, \boldsymbol{\varepsilon}, \boldsymbol{\tau}) - \sum_m \int_{\rho_m} \mathbf{T}_\rho \cdot (\mathbf{u} - \tilde{\mathbf{u}}_\rho)\, ds \quad (3.22)$$

where FHW is defined in equation (3.17).

It is seen that the condition of stationarity of the functional in (3.22) leads to

$$\delta(\text{MFHW})_2 = 0 = [\text{the first five terms on the r.h.s. of (3.19)}]$$
$$-\sum_m \int_{\rho_m} \delta\mathbf{T}_\rho \cdot (\mathbf{u} - \tilde{\mathbf{u}}_\rho)\, ds - \sum_m \int_{\rho_m} \delta\mathbf{u} \cdot (\mathbf{T}_\rho - \mathbf{t})\, ds$$
$$+\sum_m \int_{\rho_m} \mathbf{T}_\rho \cdot \delta\tilde{\mathbf{u}}_\rho\, ds \quad (3.23)$$

Again, considering the two-element example as above, it is seen that the vanishing of the second term on the right-hand side of (3.23) leads to $\mathbf{u}^+ = \tilde{\mathbf{u}}_\rho$ and $\mathbf{u}^- = \tilde{\mathbf{u}}_\rho$ at ρ_m, since $\delta\mathbf{T}_\rho^+$ and $\delta\mathbf{T}_\rho^-$ can be arbitrary at ρ_m. Likewise, since $\delta\mathbf{u}^+$ and $\delta\mathbf{u}^-$ can be arbitrary at ρ_m, the vanishing of the third term on the right-hand side of equation (3.19) leads to the conditions: $\mathbf{T}_\rho^+ = \mathbf{t}^+$ and $\mathbf{T}_\rho^- = \mathbf{t}^-$. Finally, since $\delta\tilde{\mathbf{u}}_\rho^+ = \delta\tilde{\mathbf{u}}_\rho^-$ at ρ_m, the vanishing of the fourth term on the right-hand side of (3.19) leads to the condition $\mathbf{T}_\rho^+ + \mathbf{T}_\rho^- = 0$. As a result, the required constraints that $\mathbf{u}^+ = \mathbf{u} = \tilde{\mathbf{u}}_\rho$ and $\mathbf{t}^+ + \mathbf{t}^- = 0$ at π_m are satisfied.

The modified finite element Hu–Washizu functionals of (3.18) and (3.22) and the attendant variational principles of (3.19) and (3.23) respectively may now be referred to as the most general mixed hybrid finite element variational principles. Both of these general principles have as their Euler–Lagrange equations the complete set of the necessary finite element field equations, viz. (3.10) to (3.16). By satisfying some of the equations (3.10) to (3.16) *a priori* and demanding the rest to be the *a posteriori* equations arising out of the stationarity of a reduced functional, one can develop a variety of special hybrid and/or mixed-hybrid finite element methods. Such special cases have been summarized in [6, 7, 8]. In the following we consider only one example due to its special merits and due to historical reasons.

3.3.1 Hybrid stress method

Suppose that in the functional of (3.22) one eliminates: (a) the strain $\boldsymbol{\varepsilon}$ though the contact transformation, $A(\boldsymbol{\varepsilon})-\boldsymbol{\tau}:\boldsymbol{\varepsilon}=-B(\boldsymbol{\tau})$; (b) $\mathbf{u}$ as a variable in Ω by satisfying, *a priori*, the momentum balance conditions, $\boldsymbol{\nabla}_0\cdot\boldsymbol{\tau}+\bar{\mathbf{F}}=0$ and $\boldsymbol{\tau}=\boldsymbol{\tau}^T$; (c) $\mathbf{T}_\rho^+$ and $\mathbf{T}_\rho^-$ as variables at ρ_m by setting $(\mathbf{n}\cdot\boldsymbol{\tau})^+=\mathbf{T}_\rho^+$ and $(\mathbf{n}\cdot\boldsymbol{\tau})^-=\mathbf{T}_\rho^-$ at ρ_m. One obtains the reduced functional

$$\mathrm{HS}(\boldsymbol{\tau},\tilde{\mathbf{u}}_\rho,\mathbf{u}_s)=\sum_m\left[\int_{\Omega_m}-B(\boldsymbol{\tau})\,\mathrm{d}\Omega-\int_{S_{\sigma m}}(\bar{\mathbf{t}}_i-\mathbf{n}\cdot\boldsymbol{\tau})\cdot\mathbf{u}_s\,\mathrm{d}s\right.$$
$$\left.+\int_{S_{um}}(\mathbf{n}\cdot\boldsymbol{\tau})\cdot\bar{\mathbf{u}}\,\mathrm{d}s+\int_{\rho_m}\mathbf{n}\cdot\boldsymbol{\tau}\cdot\tilde{\mathbf{u}}_\rho\,\mathrm{d}s\right]\tag{3.24}$$

The Euler–Lagrange equations of the above functional are (3.12) to (3.16). It is seen that the variable $\mathbf{u}_s$ serves as a Lagrange multiplier to enforce the traction boundary condition, (3.13). This is sometimes convenient in a numerical development to satisfy the traction boundary conditions accurately [9].

On the other hand, one can further simplify (3.24) by choosing a displacement field $\tilde{\mathbf{u}}$ all along the element boundary $\partial\Omega_m$, such that $\tilde{\mathbf{u}}=\bar{\mathbf{u}}$ at S_{um}, $\tilde{\mathbf{u}}=\tilde{\mathbf{u}}_\rho$ at ρ_m, and $\tilde{\mathbf{u}}=\mathbf{u}_S$ at $S_{\sigma m}$. Thus one obtains the functional

$$\mathrm{HSP}(\boldsymbol{\tau},\tilde{\mathbf{u}})=\sum_m\left[\int_{\Omega_m}-B(\boldsymbol{\tau})\,\mathrm{d}\Omega+\int_{\partial\Omega_m}\mathbf{n}\cdot\boldsymbol{\tau}\cdot\tilde{\mathbf{u}}\,\mathrm{d}s-\int_{S_{\sigma m}}\bar{\mathbf{t}}\cdot\tilde{\mathbf{u}}\,\mathrm{d}s\right]\tag{3.25}$$

If, in fact, it is easier to assume $\mathbf{u}$ over each Ω_m such that $\mathbf{u}^+=\tilde{\mathbf{u}}=\mathbf{u}^-$ at ρ_m *a priori*, since $\boldsymbol{\tau}$ is equilibrated, upon use of the divergence theorem one may write (3.25) equivalently as

$$\mathrm{HSP}(\boldsymbol{\tau},\mathbf{u})=\sum_m\left\{\int_{\Omega_m}[-B(\boldsymbol{\tau})+\boldsymbol{\tau}:\tfrac{1}{2}\langle(\boldsymbol{\nabla}_0\mathbf{u})^T+(\boldsymbol{\nabla}_0\mathbf{u})\rangle-\bar{\mathbf{F}}\cdot\mathbf{u}]\,\mathrm{d}\Omega-\int_{S_{\sigma m}}\bar{\mathbf{t}}\cdot\mathbf{u}\,\mathrm{d}s\right\}\tag{3.26}$$

Equation (3.26) can also be viewed as an application of the unmodified Hellinger–Reissner [3, 4] functional to a finite element assemblage, which demands that $\mathbf{u}^+=\mathbf{u}^-=\tilde{\mathbf{u}}_\rho$ *a priori* at ρ_m in order for (3.10) to (3.16) to be satisfied.

Equation (3.26), which is equivalent to (3.25) under the above-stated conditions, was first used by Pian [5]. Equation (3.25), which can be seen to have the meaning of the application of a complementary energy principle to a finite element assembly wherein interelement traction reciprocity is enforced *a posteriori* through a Lagrange multiplier, and which has a certain special appeal in situations where in element displacement fields that satisfy interelement compatibility *a priori* are difficult to select, was later used in [7, 8]. The interpretation of Fraeijs de Veubeke's 'equilibrium' stress model based on (3.25) is given in [7].

In the development of a 'finite element' based on (3.25), a great deal of physical insight is necessary in selecting the appropriate field $\boldsymbol{T}$ for a given field $\tilde{\mathbf{u}}$ at the boundary, to avoid the so-called kinematic deformation modes, i.e. those deformation modes which in addition to the rigid body modes perform zero work on the equilibrated field $\boldsymbol{T}$ [7, 10]. A more mathematical statement on the stability and convergence of this method is contained in [11, 12, 13]. Also, in the formulation of an element stiffness matrix based on (3.25), it becomes necessary to invert a matrix, say $\boldsymbol{H}$, which corresponds to the stress energy in the element, $\frac{1}{2}\boldsymbol{\beta}^T\boldsymbol{H}\boldsymbol{\beta}$, where $\boldsymbol{\beta}$ are the chosen stress parameters. Some ideas in reducing the computational work involved in forming $\boldsymbol{H}^{-1}$ were recently put forward [14, 15].

For a more comprehensive account of the various other possible hybrid and mixed hybrid methods, the reader is referred to [6, 7, 8, 16, 17].

3.4 HYBRID METHODS IN PROBLEMS WITH CONSTRAINT

A great deal has appeared in recent literature concerning problems with constraint such as incompressibility. Some such problems arise in incompressible elastic (linear or non-linear) solids, slow viscous flow of incompressible fluids, and in elastic–plastic solids at large strain since plastic flow is primarily distortional in nature.

Of several solution methods proposed, the currently most popular ones appear to be those based on the so-called selective reduced–integration–penalty (SRIP) methods [18, 19]. The basic idea of SRIP methods is to incorporate the incompressibility condition into the constitutive equations via a penalty parameter. Thus, the finite element system is obtained solely from momentum balance equations; thus they involve only nodal velocities as solution variables. Recent investigations of these methods [9] indicate that some of the RIP methods may be unstable. Attempts have been made at 'averaging' or 'filtering' the pressure solutions to stabilize these methods, but the methods still appear to be sensitive to singularities and mesh distortions.

In the following we present a hybrid finite element scheme based on a modified complementary energy principle, with the deviatoric fluid stress and element boundary velocity fields as the assumed field variables. The hydrostatic pressure and deviatoric fluid stresses are subject to the constraint of momentum balance. Thus the hydrostatic pressure field, which is arbitrary in each element, is determined from the assumed deviatoric stress field to within an arbitrary constant. The assumed element boundary velocity field serves as a Lagrange multiplier to enforce the traction reciprocity constraint at interelement boundaries. It is shown [20, 21] that a finite element system of equations would have as unknowns: (a) the finite element nodal velocities, and (b) the 'constant term' in the arbitrarily varying pressure field over each element.

We will consider the steady, slow flow of a viscous, incompressible fluid. The statement of the boundary value problem is as follows: let ∇_0 be the gradient operator in spatial coordinates, $\boldsymbol{\tau}$, $\boldsymbol{\tau}'$, p respectively the total, deviatoric, and hydrostatic components of fluid stresses, $\mathbf{u}$ the velocity, $\boldsymbol{V}$ the velocity strain tensor, and μ the coefficient of viscosity. Then the field equations are

$$\nabla_0 \cdot \mathbf{u} = 0 \tag{3.27a}$$

$$\nabla_0 \cdot \boldsymbol{\tau} + \rho \bar{\mathbf{F}} = 0 \tag{3.27b}$$

$$\boldsymbol{\tau} = \boldsymbol{\tau}^T \text{ in } \Omega \tag{3.27c}$$

$$\boldsymbol{V} = \tfrac{1}{2}[(\nabla_0 \mathbf{u}) + (\nabla_0 \mathbf{u})^T] \tag{3.28}$$

$$\boldsymbol{\tau} = \partial A / \partial \boldsymbol{V} \tag{3.29a}$$

$$A = -p\boldsymbol{V} : \boldsymbol{I} + \mu \boldsymbol{V} : \boldsymbol{V} \tag{3.29b}$$

$$\mathbf{n} \cdot \boldsymbol{\tau} = \bar{\mathbf{t}} \text{ at } S_t \tag{3.30a}$$

$$\mathbf{u} = \bar{\mathbf{u}} \text{ at } S_u \tag{3.30b}$$

From (3.29a, b) it is seen that

$$\boldsymbol{\tau} = -p\boldsymbol{I} + 2\mu \boldsymbol{V} \tag{3.31}$$

where $\boldsymbol{I}$ is the identity tensor. We now write

$$\boldsymbol{\tau} = -p\boldsymbol{I} + \boldsymbol{\tau}' \tag{3.32}$$

such that

$$\boldsymbol{\tau}' = 2\mu \boldsymbol{V} \tag{3.33}$$

We consider the contact transformation to express the stress working density of the fluid in terms of stresses. Thus, let

$$-B(\boldsymbol{\tau}) = A(p, \boldsymbol{V}) - \boldsymbol{\tau} : \boldsymbol{V} \tag{3.34}$$

Using (3.29) and (3.31) in (3.34) we obtain

$$B(\boldsymbol{\tau}) = \frac{1}{4\mu} \boldsymbol{\tau}' : \boldsymbol{\tau}' \tag{3.35}$$

We rewrite (3.27b, c) in the form

$$\nabla_0 \cdot \boldsymbol{\tau}' - \nabla_0 p + \rho \bar{\mathbf{F}} = 0, \qquad \boldsymbol{\tau}' = (\boldsymbol{\tau}')^T \tag{3.36}$$

We designate the fields $\boldsymbol{\tau}'$ and p that satisfy (3.36) and (3.30a) as being admissible in the present formulation. Then it can be shown [20] that the stationarity of the functional

$$C = -\int_{\Omega} \frac{1}{4\mu} \boldsymbol{\tau}' : \boldsymbol{\tau}' \, d\Omega + \int_{S_u} \mathbf{n} \cdot [-p\boldsymbol{I} + \boldsymbol{\tau}'] \cdot \bar{\mathbf{u}} \, ds \tag{3.37}$$

for all admissible stress fields $\boldsymbol{\tau}'$ and p leads to the Euler–Lagrange equations (3.27a), (3.28), and (3.30b) respectively. By incorporating the interelement traction reciprocity condition

$$\{\mathbf{n} \cdot [-p\boldsymbol{I} + \boldsymbol{\tau}']\}^{+} + \{\mathbf{n} \cdot [-p\boldsymbol{I} + \boldsymbol{\tau}']\}^{-} = 0 \text{ at } \rho_m \tag{3.38}$$

as a constraint condition in (3.37) through Lagrange multipliers $\tilde{\mathbf{u}}_{\rho}$ at ρ_m, one can develop a finite element hybrid functional,

$$\mathrm{MC}(\boldsymbol{\tau}', p, \tilde{\mathbf{u}}_{\rho}) = \sum_m \left\{ -\int_{\Omega_m} \frac{1}{4\mu} \boldsymbol{\tau}' : \boldsymbol{\tau}' \, d\Omega + \int_{\partial\Omega_m} \mathbf{n} \cdot [-p\boldsymbol{I} + \boldsymbol{\tau}'] \cdot \tilde{\mathbf{u}} \, ds \right.$$
$$\left. - \int_{S_{tm}} \bar{\mathbf{t}} \cdot \tilde{\mathbf{u}} \, ds \right\} \tag{3.39}$$

wherein, as before, $\tilde{\mathbf{u}} \equiv \bar{\mathbf{u}}$ at S_{um}, and the traction boundary condition, $\mathbf{n} \cdot \boldsymbol{\tau} = \bar{\mathbf{t}}$, is also introduced as an *a posteriori* constraint for added flexibility.

It is observed that in (3.39) the hydrostatic pressure enters only in the element boundary integrals. In the element property development, one starts by assuming an arbitrary polynomial approximation for each of the components of $\boldsymbol{\tau}'$. Because of the form of the stress energy density $(\boldsymbol{\tau}' : \boldsymbol{\tau}')/4\mu$, it is positive definite for any $\boldsymbol{\tau}'$. Thus, if $\boldsymbol{\beta}$ are the coefficients in the chosen $\boldsymbol{\tau}'$, one can write:

$$\int_{\Omega_m} \frac{1}{4\mu} \boldsymbol{\tau}' : \boldsymbol{\tau}' \, d\Omega = \tfrac{1}{2} \boldsymbol{\beta}^{T} \boldsymbol{H} \boldsymbol{\beta} \tag{3.40}$$

It is seen that in addition to being positive definite, because each component of $\boldsymbol{\tau}'$ can be assumed arbitrarily, $\boldsymbol{H}$ is not fully populated. Moreover, $\boldsymbol{H}$ can be seen to have a special [diagonal submatrix] structure, which makes it easier to find $\boldsymbol{H}^{-1}$. From the chosen $\boldsymbol{\tau}'$, one can determine p from the equation:

$$\boldsymbol{\nabla}_0 p = \boldsymbol{\nabla}_0 \cdot \boldsymbol{\tau}' + \rho \bar{\mathbf{F}} \tag{3.41}$$

Since p is determined from (3.41) to within a constant, say α_m ($m = 1, 1, \ldots, N$ elements), we may write:

$$p = \alpha_m + \boldsymbol{D}\boldsymbol{\beta} \text{ in } \Omega_m \tag{3.42}$$

Finally, the velocity field at $\partial\Omega_m$ is assumed as

$$\tilde{\mathbf{u}} = \boldsymbol{L}\boldsymbol{q} \text{ at } \partial\Omega_m \tag{3.43}$$

Using (3.40) to (3.43) in (3.39), one obtains

$$\mathrm{MC}=\sum_m \left(-\tfrac{1}{2}\boldsymbol{\beta}^T\boldsymbol{H}\boldsymbol{\beta}+\mathbf{q}^T\boldsymbol{G}\boldsymbol{\beta}-\mathbf{q}^T\mathbf{S}\alpha_m-\mathbf{Q}_1^T\mathbf{q}\right) \tag{3.44}$$

We observe that $\boldsymbol{\beta}$ as well as α_m are arbitrary and independent for each element. Thus, the stationarity of MC with respect to $\boldsymbol{\beta}$ leads to

$$\boldsymbol{\beta}=\boldsymbol{H}^{-1}\boldsymbol{G}^T\mathbf{q} \tag{3.45}$$

Thus, (3.44) can be written as

$$\mathrm{MC}=\sum \left(\tfrac{1}{2}\mathbf{q}^T\boldsymbol{K}\mathbf{q}-\mathbf{q}^T\mathbf{S}\alpha_m-\mathbf{Q}^T\mathbf{q}\right) \tag{3.46}$$

where

$$\boldsymbol{K}=\boldsymbol{G}\boldsymbol{H}^{-1}\boldsymbol{G}^T$$

Denoting by $\boldsymbol{\alpha}$ the vector of coefficients α_m ($m=1,2,\ldots,N$ elements) it is seen that the finite element system of equations for the flow domain have the form:

$$\boldsymbol{K}\mathbf{q}^*-\boldsymbol{S}^*\boldsymbol{\alpha}=\boldsymbol{Q} \tag{3.47a}$$

$$\boldsymbol{S}^{*T}\mathbf{q}^*=0 \tag{3.47b}$$

Thus, as seen from equations (3.47), even though p can be assumed in each Ω_m to be an arbitrary polynomial, it is only the constant term (α_m) of this polynomial in each Ω_m that is the solution variable in the finite element system of equations. A variety of finite elements based on the above formulation, for two-dimensional flow problems, have been developed and reported in [20, 21]. In these developments, all numerical integrations were performed using 'full' quadrature rules. In a variety of problems involving singularities and mesh distortions, it was shown [20, 21] that the present methods lead to highly accurate solutions for both velocities and hydrostatic pressure. A detailed mathematical study of stability and convergence of the above method was presented in [13].

The advantage of assumed stress finite elements based on a modified (hybrid) complementary energy principle were reported: (a) for incompressible finite elasticity problems [22], (b) in finite strain elastoplasticity and creep (where inelastic strain rates are incompressible) [23], and (c) for nearly incompressible linear elastic materials [24, 25]. We view this ability of handling problems with constraint as a highly salutory feature of the hybrid methods, and one that deserves further mathematical analysis.

3.5 HYBRID AND MIXED FINITE ELEMENT METHODS IN NON-LINEAR SOLID MECHANICS

For simplicity let us consider a fixed rectangular Cartesian coordinate system. A particle in the undeformed body C_0 has the position vector

$\mathbf{x} = x_\alpha \mathbf{e}_\alpha$ where $\mathbf{e}_\alpha$ are unit Cartesian bases. The gradient operator ∇_0 in C_0 is $\nabla_0 = \mathbf{e}_\alpha\, \partial/\partial x_\alpha$. The position vector of the same particle in a deformed configuration, C_N, is $\mathbf{y} = y_i \mathbf{e}_i$. The gradient operator in C_N is $\nabla_N = \mathbf{e}_i\, \partial/\partial y_i$. The deformation gradient tensor is $\boldsymbol{F} = (\nabla_0 \mathbf{y})^T$ such that $F_{i\alpha} = y_{i,\alpha} = \partial y_i/\partial x_\alpha$. The non-singular $\boldsymbol{F}$ has the polar decomposition $\boldsymbol{F} = \boldsymbol{\alpha} \cdot \boldsymbol{U} \equiv \boldsymbol{\alpha} \cdot (\boldsymbol{I} + \boldsymbol{h})$, where $\boldsymbol{\alpha}$ is an orthogonal rotation, $\boldsymbol{I}$ is the identity tensor, and $\boldsymbol{h}$ is a symmetric, positive definite, 'stretch' tensor. The Green–Lagrange strain is $\boldsymbol{g} = \frac{1}{2}(\boldsymbol{F}^T \cdot \boldsymbol{F} - \boldsymbol{I}) = \frac{1}{2}(\boldsymbol{e} + \boldsymbol{e}^T + \boldsymbol{e}^T \cdot \boldsymbol{e})$, where $\boldsymbol{e} = (\nabla_0 \mathbf{u})^T$; $\mathbf{u} = \mathbf{y} - \mathbf{x}$.

We now introduce the stress measures: (a) the true or Cauchy stress $\boldsymbol{\tau}$; (b) the Kirchhoff stress $\boldsymbol{\sigma} = J\boldsymbol{\tau}$, where J is the absolute determinant of $\boldsymbol{F}$; (c) the first Piola–Kirchhoff stress tensor $\boldsymbol{t}$; (d) the second Piola–Kirchhoff stress tensor $\boldsymbol{s}$; and (e) the symmetrized Biot–Lure stress tensor (also sometimes referred to as the Jaumann stress tensor) $\boldsymbol{r}$. As shown from first principles in [26], the above are related as:

$$\boldsymbol{\tau} = \frac{1}{J}\boldsymbol{F} \cdot \boldsymbol{t} = \frac{1}{J}\boldsymbol{F} \cdot \boldsymbol{s} \cdot \boldsymbol{F}^T = \frac{1}{J}\boldsymbol{\sigma} \tag{3.48}$$

$$\boldsymbol{t} = \boldsymbol{s} \cdot \boldsymbol{F}^T = J(\boldsymbol{F}^{-1} \cdot \boldsymbol{\tau}); \qquad \boldsymbol{s} = J(\boldsymbol{F}^{-1} \cdot \boldsymbol{\tau} \cdot \boldsymbol{F}^{-T}) \tag{3.49}$$

$$\boldsymbol{r} = \tfrac{1}{2}(\boldsymbol{t} \cdot \boldsymbol{\alpha} + \boldsymbol{\alpha}^T \cdot \boldsymbol{t}^T) = \tfrac{1}{2}[\boldsymbol{s} \cdot (\boldsymbol{I} + \boldsymbol{h}) + (\boldsymbol{I} + \boldsymbol{h}) \cdot \boldsymbol{s}] \tag{3.50}$$

The tensors $\boldsymbol{\tau}$, $\boldsymbol{\sigma}$, $\boldsymbol{s}$, and $\boldsymbol{r}$ are symmetric, while $\boldsymbol{t}$ is unsymmetric.

3.6 RATE FORMULATIONS

We now consider the rate analysis of finite deformation (finite stretches as well as rotations) of an inelastic (or elastic) solid with a general rate-type constitutive law. In doing so, one can choose an arbitrary reference frame. In practice, however, two choices, one the so-called total Lagrangean (TL) and the other the so-called updated Lagrangean (UL) reference frames, are appealing. Since the choice of a reference frame does not, per se, affect the theoretical or computational developments, we discuss here the details of an UL formulation, since the rate constitutive relations of an inelastic solid depend, naturally, on the current value of true (Cauchy) stress, etc.

In the UL formulation, the solution variables in the generic state C_{N+1} are referred to the configuration of the body in the immediately preceding state, C_N, which is presumed known. Let y_i be the current spatial coordinates of a material particle in C_N and let $\nabla_N = \mathbf{e}_i\, \partial/\partial y_i$ be the gradient operator in C_N. Let $\dot{\mathbf{u}}$ be the rate of deformation (velocity) from C_N and let $\boldsymbol{\tau}$ be the Cauchy stress in C_N. We define the velocity gradient $\dot{\boldsymbol{e}} = (\nabla_N \dot{\mathbf{u}})^T$ and write $\dot{\boldsymbol{e}} = \dot{\boldsymbol{\varepsilon}} + \dot{\boldsymbol{\omega}}$, where $\dot{\boldsymbol{\varepsilon}} = \frac{1}{2}[(\nabla_N \dot{\mathbf{u}})^T + (\nabla_N \dot{\mathbf{u}})]$ and $\dot{\boldsymbol{\omega}} = \frac{1}{2}[(\nabla_N \dot{\mathbf{u}})^T - (\nabla_N \dot{\mathbf{u}})]$ are the strain and spin rates respectively. Let $\dot{\boldsymbol{\tau}}$, and $\dot{\boldsymbol{\sigma}}(= J[\dot{\boldsymbol{\tau}} + (\dot{\boldsymbol{\varepsilon}} : \boldsymbol{I})\boldsymbol{\tau}])$ (where $J = \rho_0/\rho_N$, ρ_0 and ρ_N being mass densities in C_0 and C_N respectively) be the material rates of the Cauchy and Kirchhoff stresses respectively. As is well known, these stress rates are not objective. Let $\dot{\mathbf{t}}$, $\dot{\mathbf{r}}$, and $\dot{\mathbf{s}}$ be the appropriate stress rates referred to C_N, i.e., for instance, $\dot{\mathbf{s}}\,\Delta t = \mathbf{s}_N^{N+1} - \boldsymbol{\tau}$ where $\mathbf{s}_N^{N+1}$ is the second Piola–Kirchhoff stress in C_{N+1} as referred to (and measured per unit area in) C_N. It can be shown [27] that

$$\dot{\mathbf{s}} = \frac{\dot{\boldsymbol{\sigma}} - \dot{\mathbf{e}} \cdot \boldsymbol{\sigma} - \boldsymbol{\sigma} \cdot \dot{\mathbf{e}}^T}{J} \qquad (\boldsymbol{\sigma} = J\boldsymbol{\tau}) \tag{3.51}$$

$$\dot{\mathbf{t}} = \frac{(\dot{\boldsymbol{\sigma}} - \dot{\mathbf{e}} \cdot \boldsymbol{\sigma})}{J} \tag{3.52}$$

$$\dot{\mathbf{r}} = \tfrac{1}{2}[\dot{\mathbf{t}} + \dot{\mathbf{t}}^T + \boldsymbol{\tau} \cdot \dot{\boldsymbol{\omega}} + \dot{\boldsymbol{\omega}}^T \cdot \boldsymbol{\tau}] \equiv \dot{\mathbf{s}} + \tfrac{1}{2}(\boldsymbol{\tau} \cdot \dot{\boldsymbol{\varepsilon}} + \dot{\boldsymbol{\varepsilon}} \cdot \boldsymbol{\tau}) \qquad (\dot{\boldsymbol{\omega}} = -\dot{\boldsymbol{\omega}}^T) \tag{3.53}$$

As is now well recognized, the principle of objectivity is met, in writing the rate constitutive law of the material, by postulating the constitutive relation between the objective strain rate $\dot{\boldsymbol{\varepsilon}}$ and the objective co-rotational (or at times referred to as the Zaremba, or the rigid body, or the Jaumann) rate of Kirchhoff stress, denoted here by $\dot{\boldsymbol{\sigma}}^*$. It is well known that

$$\dot{\boldsymbol{\sigma}}^* = \dot{\boldsymbol{\sigma}} - \dot{\boldsymbol{\omega}} \cdot \boldsymbol{\sigma} + \boldsymbol{\sigma} \cdot \dot{\boldsymbol{\omega}} \tag{3.54}$$

It is to be noted that the above stress rate remains objective, in the usual sense of calculus, in the limit $\Delta t \rightarrow 0$. However, in a computational procedure, since Δt is finite, simple algorithms can be derived to maintain objectivity of incremental constitutive relations over finite time steps Δt [28]. In view of (3.51) to (3.54), it is seen that

$$\dot{\mathbf{s}} = \frac{\dot{\boldsymbol{\sigma}}^* - \dot{\boldsymbol{\varepsilon}} \cdot \boldsymbol{\sigma} - \boldsymbol{\sigma} \cdot \dot{\boldsymbol{\varepsilon}}}{J} \tag{3.55}$$

$$\dot{\mathbf{t}} = \frac{\dot{\boldsymbol{\sigma}}^* - \dot{\boldsymbol{\varepsilon}} \cdot \boldsymbol{\sigma} - \boldsymbol{\sigma} \cdot \dot{\boldsymbol{\omega}}}{J} \tag{3.56}$$

$$\dot{\mathbf{r}} = \frac{\dot{\boldsymbol{\sigma}}^* - \tfrac{1}{2}(\dot{\boldsymbol{\varepsilon}} \cdot \boldsymbol{\sigma} + \boldsymbol{\sigma} \cdot \dot{\boldsymbol{\varepsilon}})}{J} \tag{3.57}$$

Suppose now that V is the postulated potential for $\dot{\boldsymbol{\sigma}}^*$ in terms of $\dot{\boldsymbol{\varepsilon}}$ such that

$$\dot{\boldsymbol{\sigma}}^* = \frac{\partial V}{\partial \dot{\boldsymbol{\varepsilon}}} \tag{3.58}$$

Examples of constitutive relations of the type (3.58), for instance, in classical elastoplasticity, rate-sensitive viscoplasticity, and creep can be found in [27, 29]. In view of (3.55) to (3.57) one can define:

$$\dot{\mathbf{s}} = \partial W/\partial \dot{\boldsymbol{\varepsilon}}; \qquad W = V - \boldsymbol{\sigma} : (\dot{\boldsymbol{\varepsilon}} \cdot \dot{\boldsymbol{\varepsilon}}) \tag{3.59}$$

$$\dot{\mathbf{t}} = \partial U/\partial \dot{\mathbf{e}}^T; \qquad U = W + \tfrac{1}{2}\boldsymbol{\sigma} : (\dot{\mathbf{e}}^T \cdot \dot{\mathbf{e}}) \tag{3.60}$$

$$\dot{\mathbf{r}} = \partial Q/\partial \dot{\boldsymbol{\varepsilon}}; \qquad Q = V - \tfrac{1}{2}\boldsymbol{\sigma} : (\dot{\boldsymbol{\varepsilon}} \cdot \dot{\boldsymbol{\varepsilon}}) \tag{3.61}$$

Further, one can consider Legendre contact transformations of the type:

$$\dot{\mathbf{s}} : \dot{\boldsymbol{\varepsilon}} - W(\dot{\boldsymbol{\varepsilon}}) = S^*(\mathbf{s}) \tag{3.62}$$

$$\dot{\mathbf{t}}^T : \mathbf{e} - U(\mathbf{e}) = E^*(\dot{\mathbf{t}}) \tag{3.63}$$

and

$$\dot{\boldsymbol{r}} : \dot{\boldsymbol{\varepsilon}} - Q(\dot{\boldsymbol{\varepsilon}}) = R^*(\dot{\boldsymbol{r}}) \tag{3.64}$$

such that

$$\partial S^*/\partial \dot{\mathbf{s}} = \dot{\boldsymbol{\varepsilon}} \tag{3.65a}$$

$$\partial E^*/\partial \dot{\mathbf{t}} = \dot{\mathbf{e}} \tag{3.65b}$$

$$\partial R^*/\partial \dot{\boldsymbol{r}} = \dot{\boldsymbol{\varepsilon}} \tag{3.65c}$$

Now we consider the solid in C_N to be composed of a finite number of finite elements. Let Ω be the domain in C_N, Ω_m be the finite element, $\partial\Omega_m$ its boundary, and ρ_m the interelement boundary. The rate form of field equations applicable to the finite element assemblage in C_N can be stated thus:

LMB:

$$\boldsymbol{\nabla}_N \cdot [\dot{\mathbf{s}} + \boldsymbol{\tau} \cdot (\boldsymbol{\nabla}_N \dot{\mathbf{u}})] + \rho_N \dot{\mathbf{B}} = 0 \tag{3.66a}$$

or

$$\boldsymbol{\nabla}_N \cdot \dot{\mathbf{t}} + \rho_N \dot{\mathbf{B}} = 0 \tag{3.66b}$$

AMB:

$$\dot{\mathbf{s}} = \dot{\mathbf{s}}^T \tag{3.67a}$$

or

$$(\boldsymbol{\nabla}_N \dot{\mathbf{u}})^T \cdot \boldsymbol{\tau} + \dot{\mathbf{t}} = \dot{\mathbf{t}}^T + \boldsymbol{\tau} \cdot (\boldsymbol{\nabla}_N \dot{\mathbf{u}}) \tag{3.67b}$$

or, equivalently

$$\dot{\boldsymbol{\omega}} \cdot \boldsymbol{\tau} + \dot{\boldsymbol{\varepsilon}} \cdot \boldsymbol{\tau} + \dot{\mathbf{t}} = \dot{\mathbf{t}}^T + \boldsymbol{\tau} \cdot \dot{\boldsymbol{\varepsilon}} + \boldsymbol{\tau} \cdot \dot{\boldsymbol{\omega}}^T \tag{3.67c}$$

Compatibility:

$$\dot{\mathbf{e}} = \dot{\boldsymbol{\varepsilon}} + \dot{\boldsymbol{\omega}} = (\boldsymbol{\nabla}_N \dot{\mathbf{u}})^T \tag{3.68a}$$

or

$$\dot{\boldsymbol{\varepsilon}} = \tfrac{1}{2}[(\boldsymbol{\nabla}_N \dot{\mathbf{u}})^T + (\boldsymbol{\nabla}_N \dot{\mathbf{u}})] \tag{3.68b}$$

TBC

$$\dot{\mathbf{t}} \equiv \mathbf{n}^* \cdot [\dot{\mathbf{s}} + \boldsymbol{\tau} \cdot (\boldsymbol{\nabla}_N \dot{\mathbf{u}})] \equiv \mathbf{n}^* \cdot \dot{\mathbf{t}} = \bar{\mathbf{t}} \text{ at } S_{\sigma m} \tag{3.69}$$

DBC:

$$\dot{\mathbf{u}} = \dot{\bar{\mathbf{u}}} \text{ at } S_{um} \tag{3.70}$$

$$\dot{\mathbf{u}}^+ = \dot{\mathbf{u}}^- \text{ at } \rho_m \tag{3.71}$$

and

$$(\mathbf{n}^* \cdot \dot{\mathbf{t}})^+ + (\mathbf{n}^* \cdot \dot{\mathbf{t}})^- = 0 \text{ at } \rho_m \tag{3.72}$$

In the above, $\mathbf{n}^*$ is the unit outward normal to $\partial\Omega_m$ in C_N.

Assuming that equation (3.71) is satisfied *a priori* in each case, the most general (Hu–Washizu type) mixed rate variational principles in terms of the above alternate conjugate stress rates and strain rates can be stated [30, 31,

27] as the stationarity conditions of the functionals:

$$\text{FHW}(\dot{\mathbf{u}}, \dot{\boldsymbol{\varepsilon}}, \dot{\mathbf{s}}) = \sum_m \Big(\int_{\Omega_m} \{W(\dot{\boldsymbol{\varepsilon}}) - \rho_N \dot{\mathbf{B}} \cdot \dot{\mathbf{u}} + \tfrac{1}{2}\boldsymbol{\tau} : \langle(\nabla_N \dot{\mathbf{u}}) \cdot (\nabla_N \dot{\mathbf{u}})^T\rangle$$
$$- \dot{\mathbf{s}} : \langle \dot{\boldsymbol{\varepsilon}} - \tfrac{1}{2}[(\nabla_N \dot{\mathbf{u}})^T + (\nabla_N \dot{\mathbf{u}})]\rangle \, d\Omega - \int_{S_{\sigma m}} \dot{\bar{\mathbf{t}}} \cdot \dot{\mathbf{u}} \, ds$$
$$- \int_{S_{un}} \dot{\mathbf{t}} \cdot (\dot{\mathbf{u}} - \dot{\bar{\mathbf{u}}}) \, ds \Big) \tag{3.73}$$

$$\text{FHW}(\dot{\mathbf{u}}, \dot{\mathbf{e}}, \dot{\mathbf{t}}) = \sum_m \Big\{ \int_{\Omega_m} [U(\dot{\mathbf{e}}) - \rho_N \dot{\mathbf{B}} \cdot \dot{\mathbf{u}} + \dot{\mathbf{t}}^T : \langle(\nabla_N \dot{\mathbf{u}})^T - \dot{\mathbf{e}}\rangle] \, d\Omega$$
$$- \int_{1S_{\sigma m}} \dot{\bar{\mathbf{t}}} \cdot \dot{\mathbf{u}} \, ds - \int_{S_{um}} \dot{\mathbf{t}} \cdot (\dot{\mathbf{u}} - \dot{\bar{\mathbf{u}}}) \, ds \Big\} \tag{3.74}$$

$$\text{FHW}(\dot{\mathbf{u}}, \dot{\boldsymbol{\varepsilon}}, \dot{\boldsymbol{\omega}}, \dot{\mathbf{t}}) = \sum_m \Big\{ \int_{\Omega_m} [Q(\dot{\boldsymbol{\varepsilon}}) + \tfrac{1}{2}\boldsymbol{\tau} : (\dot{\boldsymbol{\omega}}^T \cdot \dot{\boldsymbol{\omega}}) + \boldsymbol{\tau} : (\dot{\boldsymbol{\omega}}^T \cdot \dot{\boldsymbol{\varepsilon}})$$
$$- \rho_N \dot{\mathbf{B}} \cdot \dot{\mathbf{u}} + \dot{\mathbf{t}}^T : \langle(\nabla_N \dot{\mathbf{u}})^T - \dot{\boldsymbol{\varepsilon}} - \dot{\boldsymbol{\omega}}\rangle] \, d\Omega - \int_{S_{\sigma m}} \dot{\bar{\mathbf{t}}} \cdot \dot{\mathbf{u}} \, ds$$
$$- \int_{S_{um}} \dot{\mathbf{t}} \cdot (\dot{\mathbf{u}} - \dot{\bar{\mathbf{u}}}) \, ds \Big\} \tag{3.75}$$

In (3.75) $\dot{\boldsymbol{\omega}}$ is a skew-symmetric tensor field, while $\dot{\boldsymbol{\varepsilon}}$ is symmetric. Observing the *a priori* constraint that $\delta\dot{\mathbf{u}}^+ = \delta\dot{\mathbf{u}}^-$ at ρ_m *a priori*, it can be shown [30] that the stationarity of each of these functionals leads to the Euler equations (3.66) to (3.70) and (3.72).

As in the linear theory, one may relax the constraint $\dot{\mathbf{u}}^+ = \dot{\mathbf{u}}^-$ at ρ_m *a priori*, and introduce it as an *a posteriori* constraint through the method of Lagrange multipliers in one of the two ways: (a) introduce the constraint $\dot{\mathbf{u}}^+ = \dot{\mathbf{u}}^-$ at ρ_m directly through unique Lagrange multipliers $\dot{\tilde{\mathbf{t}}}_\rho$ at ρ_m such that $\dot{\tilde{\mathbf{t}}} + \dot{\tilde{\mathbf{t}}}^- = 0$ at ρ_m *a priori*, or (b) introduce the separate constraints $\dot{\mathbf{u}}^+ = \tilde{\mathbf{u}}_\rho$ and $\dot{\mathbf{u}}^- = \tilde{\mathbf{u}}_\rho$ (where $\tilde{\mathbf{u}}_\rho$ is a unique field at ρ_m) through independent Lagrange multipliers $\dot{\mathbf{t}}_\rho^+$ and $\dot{\mathbf{t}}_\rho^-$ respectively wherein $\dot{\mathbf{t}}_\rho^+ + \dot{\mathbf{t}}_\rho^- \neq 0$ at ρ_m *a priori*. Modifying each of the general finite element principles in (3.73) to (3.75) respectively, in each of the above two ways, we obtain the following most general finite element mixed hybrid variational functionals:

$$(\text{MFHW})_1(\dot{\mathbf{u}}, \dot{\boldsymbol{\varepsilon}}, \dot{\mathbf{s}}, \tilde{\mathbf{t}}_\rho) = \text{FHW}(\dot{\mathbf{u}}, \dot{\boldsymbol{\varepsilon}}, \dot{\mathbf{s}}) - \sum_m \int_{\rho_m} \dot{\tilde{\mathbf{t}}}_\rho \cdot \dot{\mathbf{u}} \, ds \tag{3.76}$$

$$(\text{MFHW})_2(\dot{\tilde{\mathbf{u}}}, \dot{\boldsymbol{\varepsilon}}, \dot{\mathbf{s}}, \tilde{\mathbf{u}}_\rho, \dot{\mathbf{t}}_\rho) = \text{FHW}(\dot{\mathbf{u}}, \dot{\boldsymbol{\varepsilon}}, \dot{\mathbf{s}}) - \sum_m \int_{\rho_m} \dot{\mathbf{t}}_\rho (\dot{\mathbf{u}} - \dot{\tilde{\mathbf{u}}}_\rho) \, ds \tag{3.77}$$

$$(\text{MFHW})_1(\dot{\mathbf{u}}, \dot{\boldsymbol{\varepsilon}}, \dot{\mathbf{t}}, \dot{\tilde{\mathbf{t}}}\rho) = \text{FHW}(\dot{\mathbf{u}}, \dot{\boldsymbol{\varepsilon}}, \dot{\mathbf{t}}) - \sum_m \int_{\rho_m} \dot{\tilde{\mathbf{t}}}_\rho \cdot \dot{\mathbf{u}} \, ds \tag{3.78}$$

$$(\text{MFHW})_2(\dot{\mathbf{u}}, \dot{\boldsymbol{e}}, \dot{\boldsymbol{t}}, \tilde{\mathbf{u}}_\rho, \mathbf{t}_\rho) = \text{FHW}(\dot{\mathbf{u}}, \dot{\boldsymbol{e}}, \dot{\boldsymbol{t}}) - \sum_m \int_{\rho_m} \dot{\mathbf{t}}_\rho(\dot{\mathbf{u}} - \dot{\tilde{\mathbf{u}}}_\rho)\, ds \tag{3.79}$$

$$(\text{MFHW})_1(\dot{\mathbf{u}}, \dot{\boldsymbol{e}}, \dot{\boldsymbol{\omega}}, \dot{\tilde{\mathbf{t}}}_\rho) = \text{FHW}(\dot{\mathbf{u}}, \dot{\boldsymbol{\varepsilon}}, \dot{\boldsymbol{\omega}}, \dot{\boldsymbol{t}}) - \sum_m \int_{\rho_m} \dot{\tilde{\mathbf{t}}}_\rho \cdot \dot{\mathbf{u}}\, ds \tag{3.80}$$

$$(\text{MFHW})_2(\dot{\tilde{\mathbf{u}}}, \dot{\boldsymbol{\varepsilon}}, \dot{\boldsymbol{\omega}}, \dot{\boldsymbol{t}}, \tilde{\mathbf{u}}_\rho, \tilde{\mathbf{t}}_\rho) = \text{FHW}(\dot{\mathbf{u}}, \dot{\boldsymbol{\varepsilon}}, \dot{\boldsymbol{\omega}}, \dot{\boldsymbol{t}}) - \sum_m \int_{\rho_m} \dot{\mathbf{t}}_\rho(\dot{\mathbf{u}} - \dot{\tilde{\mathbf{u}}}_\rho)\, ds \tag{3.81}$$

The terms $\text{FHW}(\dot{\mathbf{u}}, \dot{\boldsymbol{\varepsilon}}, \dot{\mathbf{s}})$, $\text{FHW}(\dot{\mathbf{u}}, \dot{\boldsymbol{e}}, \dot{\boldsymbol{t}})$, and $\text{FHW}(\dot{\mathbf{u}}, \dot{\boldsymbol{\varepsilon}}, \dot{\boldsymbol{\omega}}, \dot{\boldsymbol{t}})$, occurring above, are defined in (3.73) to (3.75) respectively.

A wide variety of hybrid and mixed hybrid finite element methods can be developed from (3.76) to (3.81) and their specializations. These specializations are such that some of the equations (3.66) to (3.72) are satisfied *a priori*, while the rest are *a posteriori* conditions arising out of the stationarity of the specialized functionals. Such general and special hybrid and mixed hybrid finite element methods are elaborately discussed in [27, 30].

We now discuss only a few examples of such specialized hybrid finite element methods. Suppose that, in (3.77): (a) one satisfies, *a priori*, the constraints as given by (3.66a), (3.67a), and (3.69), (b) one eliminates $\dot{\boldsymbol{\varepsilon}}$ as a variable from (3.77) through the contact transformation (3.62), and (c) one chooses $\dot{\mathbf{t}}_\rho$ at ρ_m such that $\dot{\mathbf{t}}_\rho^+ = \langle \mathbf{n}^* \cdot [\dot{\mathbf{s}} + \boldsymbol{\tau} \cdot (\boldsymbol{\nabla}_N \dot{\mathbf{u}})] \rangle^+$; $\dot{\mathbf{t}}_\rho^- = \langle \mathbf{n}^* \cdot [\dot{\mathbf{s}} + \boldsymbol{\tau} \cdot (\boldsymbol{\nabla}_N \dot{\mathbf{u}}] \rangle^-$ at ρ_m, then one obtains a hybrid finite complementary energy functional:

$$\text{HS}(\dot{\mathbf{u}}, \dot{\mathbf{s}}, \dot{\tilde{\mathbf{u}}}_\rho) = \sum_m \left(\int_{\Omega_m} \{-S^*(\dot{\mathbf{s}}) - \tfrac{1}{2}\boldsymbol{\tau}_N : [(\boldsymbol{\nabla}_N \dot{\mathbf{u}}) \cdot (\boldsymbol{\nabla}_N \dot{\mathbf{u}})^T]\}\, d\Omega \right.$$
$$\left. + \int_{S_{um}} \dot{\mathbf{t}} \cdot \dot{\tilde{\mathbf{u}}}\, ds + \int_{\rho_m} \mathbf{n}^* \cdot [\dot{\mathbf{s}} + \boldsymbol{\tau} \cdot (\boldsymbol{\nabla}_N \dot{\mathbf{u}})] \cdot \dot{\tilde{\mathbf{u}}}_\rho\, ds \right) \tag{3.82}$$

As seen from above, the complementary energy density rate, for the present finite deformation problem, involves not only $\dot{\mathbf{s}}$ but also $\dot{\mathbf{u}}$ in each Ω_m, unlike in the linear theory (see 3.24). The stationarity of the functional in (3.82) leads to (3.68b), (3.70), (3.72), and (3.71) in the form $\dot{\mathbf{u}}^+ = \tilde{\mathbf{u}}_\rho = \dot{\mathbf{u}}^-$ at ρ_m.

Some observations regarding the above finite element method are made: (a) Even though both $\dot{\mathbf{s}}$ and $\dot{\mathbf{u}}$ appear in the integral over Ω_m, $\dot{\mathbf{u}}$ in Ω_m does not play the role of a Lagrange multiplier. Further $\dot{\mathbf{u}}$ chosen in Ω_m need not be interelement compatible, *a priori*. (b) The *a priori* constraint condition of momentum balance, (3.66a), involves both $\dot{\mathbf{s}}$ and $\dot{\mathbf{u}}$. Thus the selection of symmetric $\dot{\mathbf{s}}$ such that (3.66a) is satisfied *a priori* is not an altogether easy proposition. Several interesting ways of satisfying (3.66a) were discussed in [27, 32]. Equation (3.82) was used in a somewhat less general form [33] in formulating a stress hybrid finite element method and was applied to solve the problem of buckling of a shallow arch. The shortcomings of this approach [33] were later discussed in [30]. Applications of (3.82) wherein $\dot{\mathbf{s}}$

was subject to a simpified constraint, viz. $\nabla_N \cdot \dot{\mathbf{s}} + \rho_N \mathbf{B} = 0$ were presented in [34].

Suppose now that in (3.79): (a) one satisfies, *a priori*, (3.66b), thereby eliminating $\dot{\mathbf{u}}$ as a variable in Ω_m; (b) one also satisfies (3.69) *a priori* and eliminates $\dot{\mathbf{e}}$ as a variable through the contact transformation, (3.63), and (c) one chooses $\tilde{\mathbf{t}}_\rho^{\pm} = [\mathbf{n}^* \cdot \dot{\mathbf{t}}]^{\pm}$ a priori at ρ_m and then obtains the reduced functional

$$\mathrm{HS}(\dot{\boldsymbol{t}}, \dot{\tilde{\mathbf{u}}}_\rho) = \sum_m \left\{ \int_{\Omega_m} [-E^*(\dot{\mathbf{t}})\, d\Omega + \int_{S_{um}} \mathbf{n}^* \cdot \dot{\boldsymbol{t}} \cdot \dot{\bar{\mathbf{u}}}\, ds + \int_{\rho_m} \mathbf{n}^* \cdot \dot{\boldsymbol{t}} \cdot \dot{\tilde{\mathbf{u}}}_\rho\, ds \right\} \quad (3.83)$$

The stationary condition of the above functional, while leading to (3.68a), (3.70), and (3.72) as the Euler–Lagrange equations, does not, unfortunately, lead to the angular momentum balance, (3.67c). Note that the AMB was not satisfied *a priori* either. For this reason and also due to the problems with establishing $E^*(\dot{\mathbf{t}})$ for a general material behaviour, as elaborated upon in [22, 27], the development of a hybrid stress method based on equation (3.83) appears in general to be invalid.

Now we present what appears to be the most rational and consistent hybrid stress finite element complementary energy principle. Suppose that in (3.81): (a) one satisfies (3.66b) and (3.69) *a priori*, (b) one eliminates $\dot{\boldsymbol{\varepsilon}}$ as a variable through the contact transformation, (3.65), and (c) one chooses $\dot{\tilde{\mathbf{t}}}_\rho^{\pm} = [\mathbf{n}^* \cdot \dot{\mathbf{t}}]^{\pm}$ *a priori* at ρ_m and then obtains a reduced functional

$$\mathrm{HS}(\dot{\boldsymbol{t}}, \dot{\tilde{\mathbf{u}}}_\rho, \dot{\boldsymbol{\omega}}) = \sum_m \left\{ \int_{\Omega_m} [-R^*(\dot{\boldsymbol{r}}) + \tfrac{1}{2}\boldsymbol{\tau} : (\dot{\boldsymbol{\omega}}^T \cdot \dot{\boldsymbol{\omega}}) - \dot{\boldsymbol{t}}^T : \dot{\boldsymbol{\omega}}]\, dV \right.$$

$$\left. + \int_{S_{um}} (n^* \cdot \dot{\mathbf{t}}) \cdot \dot{\bar{\mathbf{u}}}\, ds + \int_{\rho_m} (\mathbf{n}^* \cdot \dot{\mathbf{t}}) \cdot \dot{\tilde{\mathbf{u}}}_\rho\, ds \right\} \quad (3.84)$$

In the above, by definition,

$$\dot{\boldsymbol{r}} = \tfrac{1}{2}(\dot{\boldsymbol{t}} + \dot{\boldsymbol{t}}^T + \boldsymbol{\tau} \cdot \dot{\boldsymbol{\omega}} + \dot{\boldsymbol{\omega}}^T \cdot \boldsymbol{\tau}) \quad (3.85)$$

The stationary conditions of the functional in (3.84) leads, as Euler equations, to: (a) the angular momentum balance, (3.67c); (b) (3.68a); (c) (3.70); (d) (3.72); and (e) (3.71) through the relation, $\dot{\mathbf{u}}^+ = \dot{\mathbf{u}}^- = \dot{\tilde{\mathbf{u}}}_\rho$ at ρ_m.

The hybrid finite element complementary energy rate principle in the form of (3.84), involving as it does (a) the rate of the first Piola–Kirchhoff stress, (b) the rate of spin, and (c) the interelement boundary velocities, is of special appeal in analysing non-linear structural mechanics problems. This is because finite deformation of thin beams, plates, and shells usually involves large rotations, but only small or moderate stretches. In such problems, treatment of rigid rotations of material elements as direct variables is of considerable advantage. Considerable work has been published recently concerning the use of (3.84), or its total Lagrangean reference frame counterpart, in analysing finite deformation (large rotations as well as stretches) of (a) compressible as well as incompressible materials, (b) rate-sensitive as well as rate-insensitive materials characterized by viscoplasticity, creep, or classical elastoplasticity, and (c) beams, plates, and shells undergoing instability as well [22, 23, 29, 30, 35–40].

A new shell theory, valid for large mid-plane rotations as well as stretches, motivated in part by the search for a variational principle of the type in (3.84), was recently presented [26]. In this theory, the mid-plane deformation gradient is decomposed, in two alternate ways, into pure stretch and rigid rotation. The bending strains, which may be arbitrary in magnitude, depend, in this theory, only on the rotation tensor [26]. A systematic application of this new shell theory, through hybrid stress finite elements, is currently under way.

ACKNOWLEDGEMENTS

The support of USAFOSR under grant 81-0057B is gratefully acknowledged. The authors thank Dr. A. Amos for his encouragement. Ms. Margarete Eiteman deserves a special note of thanks for her careful assistance in the preparation of this manuscript.

REFERENCES

1. H. C. Hu, 'On some variational principles in the theory of elasticity and the theory of plasticity', *Scientia Sinica* (*Peking*), **4,** 33–54 (1955).
2. K. Washizu, 'On the variational principles of elasticity and plasticity', Report 25–18, Cont. Nsori-07833, M.I.T., March 1955.
3. E. Hellinger, 'Die allgemeine Ausatze der Mechanik der Kontinuua', in Vol. 4[4] of *Encyklopedie der Mathematischen Wissenschaften* (eds. F. Klein and C. Muller), Tebner, Leipzig, 1914.
4. E. Reissner, 'On a variational theorem in elasticity', *J. Maths. Physics*, **29,** 207–210 (1950).
5. T. H. H. Pian, 'Derivation of element stiffness matrices by assumed stress distributions', *AIAA J.*, **2,** 1333–1336 (1964).
6. S. N. Atluri, 'On "hybrid" finite element models in solid mechanics', in *Advances in Computer Methods for Partial Differential Equations* (Ed. R. Vichnevetsky), pp. 346–356, AICA, Rutgers University (USA)/University of Ghent (Belgium), 1975.
7. T. H. H. Pian and P. Tong, 'Basis of finite element methods for solid continua', *Int. J. Num. Meth. in Eng.*, **1,** 3–28 (1969).
8. P. Tong and T. H. H. Pian, 'A variational principle and convergence of a finite element method based on assumed stress distributions', *Int. J. Solids and Structures*, **5,** 463–472 (1969).
9. S. N. Atluri and H. C. Rhee, 'On traction boundary conditions in the hybrid stress finite element method', *AIAA J.*, **16,** 529–533 (1978).
10. B. Fraeijs de Veubeke, *Upper and Lower Bounds in Matrix Structural Analysis*, AGARDograph 72, pp. 165–201, Pergamon, Oxford, 1964.

11. F. Brezzi, 'On the existence, uniqueness, and approximation of saddle-point problems arising from Lagrange multipliers', *Rev. Francaise Automat. Informat. Recherche Operationnelle, Ser. Rouge Anal. Num.*, **R-2,** 129–151 (1974).
12. I. Babuska, J. T. Oden, and J. K. Lee, 'Part I, Mixed-hybrid finite element approximations of second-order elliptic boundary value problems', *Computer Meth. in Appl. Mech. and Eng.*, **11,** 175–206 (1977), and 'Part II, Weak-hybrid methods', **14,** 1–22 (1978).
13. L.–a. Ying and S. N. Atluri, 'A hybrid finite-element method for Stokes flow: Part II—Stability and convergence studies', *Computer Meth. in Appl. Mech. and Eng.* (in press).
14. P. Tong, 'On the construction of three dimensional hybrid stress elements', *Int. J. Num. Meth. in Eng.*, (to appear).
15. T. H. H. Pian and D.-P. Chen, 'Alternate ways for formulation of hybrid stress elements', *Int. J. Num. Meth. in Eng.* (to appear).
16. G. Horrigmoe, 'Nonlinear finite element models in solid mechanics', Report 76-2, Div. of Structural Mech., The Norwegian Inst. of Technology, Norway, August 1976.
17. J. P. Wolf, *Generalized Stress Models for Finite-Element Analysis,* Doctoral dissertation, ETH, Zurich, 1974.
18. T. J. R. Hughes, W. K. Lui, and A. Brooks, 'Finite element analysis of incompressible fluid flow by the penalty function formulation', *J. of Computational Physics,* **30,** 1–60 (1979).
19. J. T. Oden, N. Kikuchi, and S. W. Song, 'Penalty-finite element methods for the analysis of Stokesian flows', *Computer Methods in Applied Mech. and Eng.* (in press).
20. C. Bratianu and S. N. Atluri, 'A hybrid finite element method for Stokes flow: Part I—Formulation and numerical studies', *Computer Methods in Appl. Mech. and Eng.* (in press).
21. C. Bratianu, *Hybrid and Mixed Finite Elements Models for Viscous, Incompressible Fluid Flow,* Ph.D. thesis, Georgia Institute of Technology, 1980.
22. H. Murakawa and S. N. Atluri, 'Finite elasticity solutions using hybrid finite elements based on a complementary energy principle', *J. App. Mech. ASME,* **45,** 539–547 (1978).
23. H. Murakawa and S. N. Atluri, 'Finite elasticity solutions using hybrid finite elements based on a complementary energy principle, Part II—Incompressible materials', *J. Appl. Mech., ASME,* **46,** 71–78 (1979).
24. P. Tong, 'An assumed stress hybrid finite element method for incompressible and near-incompressible materials', *Int. J. Solids and Structures,* **5,** 455–461 (1969).
25. T. H. H. Pian and S. W. Lee, 'Notes on finite elements for nearly incompressible materials', *AIAA J.*, **14,** 824–826 (1976).
26. S. N. Atluri, 'Alternate stress and conjugate strain measures, and mixed variational formulations involving rigid rotations, for computational analyses of finitely deformed solids, with application to plates and shells, Part I: Theory', *Computers and Structures* (in press).
27. S. N. Atluri, 'On some new general and complementary energy theorems for the rate problems of classical, finite strain elasto-plasticity', *J. Structural Mechanics,* **8,** 36–66 (1980).
28. R. Rubinstein and S. N. Atluri, 'Objectivity of incremental constitutive relations over finite time steps in computational finite deformation analyses', *Computer Meth. in Appl. Mech. and Eng.* (in press).

29. S. N. Atluri and H. Murakawa, 'New general and complementary energy theorems, finite strain rate–sensitive inelasticity, and finite elements: Some computational studies', in *Nonlinear Finite Element Analysis in Structural Mechanics* (Eds. W. Wunderlich, E. Stein, and K.-J. Bathe), pp. 28–48, Springer-Verlag, 1981.
30. S. N. Atluri and H. Murakawa, 'Hybrid finite element models in nonlinear solid mechanics', in *Finite Elements in Nonlinear Mechanics*, (Eds. P. G. Bergan *et al.*), Vol. 1, pp. 3–41, TAPIR Press, Norway, 1977.
31. H. Murakawa, *Incremental Hybrid Finite Element Methods for Finite Deformation Problems* (with special emphasis on the complementary energy principle), Ph.D. thesis, Georgia Institute of Technology, 1978.
32. S. N. Atluri, 'Rate complementary energy principles, finite strain plasticity problems; and finite elements', in *Variational Methods in the Mechanics of Solids* (Ed. S. Nemat-Nasser), pp. 363–368, Pergamon Press, 1980.
33. S. N. Atluri, 'On the hybrid stress finite element model in incremental analysis of large deflection problems', *Int. J. Solids and Structures*, **9,** 1177–1191 (1973).
34. P. Boland and T. H. H. Pian, 'Large deflection analysis of thin elastic structures by the assumed stress finite element models', *Computers and Structures*, **7,** 1–12 (1977).
35. S. N. Atluri, 'On rate principles for finite strain analysis of elastic and inelastic nonlinear solids', In *Recent Research on Mechanical Behavior of Solids* (Prof. H. Miyamoto's 60th Anniversity Volume), pp. 79–107, University of Tokyo Press, Tokyo, 1979.
36. H. Murakawa and S. N. Atluri, 'Finite element solutions of finite strain elastic–plastic problems, based on a new complementary energy rate principle', in *Advances in Computer Methods for Partial Differential Engineering* (Eds. R. Vichnevetsky and B. Stepleman), pp. 53–61, IMAS, Rutgers University, 1979.
37. N. Fukuchi and S. N. Atluri, 'Finite deformation analysis of shells: A complementary energy–hybrid method', in *Nonlinear Finite Element Analysis of Shells*, (Eds. T. J. R. Hughes *et al.*), Vol. 48, pp. 233–249, ASME AMD, 1981.
38. K. W. Reed and S. N. Atluri, 'Viscoplasticity and creep: A finite deformation analysis using stress-based finite elements', in *Advances in Aerospace Structures and Materials* (Eds. S. S. Wang, *et al.*), pp. 211–221, ASME, AD-01, 1981.
39. H. Murakawa, K. W. Reed, S. N. Atluri, and R. Rubinstein, 'Stability analysis of structures via a new complementary energy method', *Computers and Structures*, **13,** 11–18 (1981).
40. H. Murakawa and S. N. Atluri, 'Finite deformations, finite rotations, and stability of plates: A complementary energy–finite element analysis', *Proc. of 22nd AIAA/ASME/AHS Structures, Structural Dynamics and Materials Conf.* Atlanta, Ga. April 1981, pp. 7–15 (1981).

Hybrid and Mixed Finite Element Methods
Edited by S. N. Atluri, R. H. Gallagher, and O. C. Zienkiewicz

Chapter 4

Hybrid and Hellinger–Reissner Plate and Shell Finite Elements

F. N. Rigby, J. J. Webster, and R. D. Henshell

4.1 NOTATION

σ_{ij}	stress tensor
U_i	displacement vector (volume)
$\tilde{U}_i$	displacement vector (boundary)
F_i	body force vector
T_i	boundary tractions
dV	elemental volume
ds	elemental surface
da	elemental area
$\boldsymbol{C}_{ijkl}$	compliance tensor
$\boldsymbol{\sigma}$	list of stresses or stress resultants
Q	matrix of stress functions
$\boldsymbol{\beta}$	arbitrary stress coefficients
N	shape functions
$\mathbf{U}$	displacements
$\tilde{\mathbf{U}}$	boundary displacements
$\boldsymbol{\sigma}'$	derivatives of stresses
$\mathbf{F}$	body loading vector
ϕ_1	interpolation functions
ϕ_2	interpolation functions
$\mathbf{T}$	boundary tractions
G	matrix relating stress derivatives to stress coefficients
R	matrix relating boundary tractions to stress coefficients
C	compliance matrix
E	Young's modulus of elasticity
ν	Poisson's ratio
h	plate or shell thickness
P	matrix of polynomial terms
$\mathbf{a}-\mathbf{g}$	lists of arbitary constants

i, j, k	unit vector in x, y, z directions
e_b, e_t	shell edge normal and tangential unit vectors
e_n	shell normal unit vector

Other symbols used are defined in the text.
A suffix n denotes nodal values.

4.2 BASIC THEORY

The Hellinger–Reissner principle can be written in Cartesian tensor notation, for an assemblage of n finite elements as (see [1])

$$\pi_H = \sum_n \left\{ \int_V [-\tfrac{1}{2}\boldsymbol{C}_{ijkl}\sigma_{ij}\sigma_{kl} + \tfrac{1}{2}\sigma_{ij}(U_{i,j} + U_{j,i}) - F_i U_i]\,\mathrm{d}V - \int_{S\sigma} \bar{T}_i \tilde{U}_i\,\mathrm{d}s \right\} \quad (4.1)$$

After integration by parts this equation becomes

$$\pi_H = \sum_n \left\{ \int_V (-\tfrac{1}{2}\boldsymbol{C}_{ijkl}\sigma_{ij}\sigma_{kl} - \sigma_{ij,j}U_i - F_i U_i)\,\mathrm{d}V + \int_s T_i \tilde{U}_i\,\mathrm{d}s - \int_{S\sigma} T_i \tilde{U}_i\,\mathrm{d}s \right] \quad (4.2)$$

Equations (4.1) and (4.2) depend on three independent fields:

σ_{ij}	the stress field
U_i	the volume displacement field
$\tilde{U}_i$	the boundary surface displacement field

For the finite element presentation we define the matrix quantities:

$$\begin{aligned} \boldsymbol{\sigma} &= Q\boldsymbol{\beta} \\ \mathbf{U} &= N\mathbf{U}_n \\ \tilde{\mathbf{U}} &= \tilde{N}\mathbf{U}_n \\ \boldsymbol{\sigma}' &= G\boldsymbol{\beta} \\ \bar{F} &= \phi_1 \bar{F}_n \\ \bar{\mathbf{T}} &= \phi_2 \bar{T}_n \\ \bar{\mathbf{T}} &= R\boldsymbol{\beta} \end{aligned} \quad (4.3)$$

Using equations (4.3) and dropping the element summation symbol for clarity, equation (4.2) becomes

$$\pi_H = \tfrac{1}{2}\boldsymbol{\beta}^T H\boldsymbol{\beta} - \boldsymbol{\beta}^T E\mathbf{U}_n + \boldsymbol{\beta}^T T\mathbf{U}_n - F_n^T \mathbf{U}_n \quad (4.4)$$

where

$$
\begin{aligned}
H &= \int_V Q^T C Q \, \mathrm{d}V \\
E &= \int_V G N \, \mathrm{d}V \\
T &= \int_S R \tilde{N} \, \mathrm{d}s \\
F_n &= \int_V \bar{F}^T \phi_1^T N \, \mathrm{d}V + \int_{S\bar{\sigma}} \bar{T}^T \phi_2^T \tilde{N} \, \mathrm{d}s
\end{aligned}
\tag{4.5}
$$

The Hellinger–Reissner principle of equation (4.4) depends now on two fields ($\boldsymbol{\beta}$ and $\mathbf{U}_n$) which can vary independently. Finding the extremum of π_H with respect to $\boldsymbol{\beta}$ and $\mathbf{U}_n$ we obtain the standard stiffness expression:

$$\mathbf{S}\mathbf{U}_n = F_n \tag{4.6}$$

where

$$S = (T - E)^T H^{-1} (T - E) \tag{4.7}$$

For the hybrid element the matrix G of equation (4.3) is null and equation (4.7) becomes:

$$S = T^T H^{-1} T \tag{4.8}$$

With the hybrid formulation, the stress functions of equation (4.3) (Q) are assume to satisfy the homogeneous form of the stress equilibrating equations. An alternative approach [2] is to augment the stress function expression with terms which satisfy the non-homogeneous equations as

$$\boldsymbol{\sigma} = Q\boldsymbol{\beta} + Q_F \boldsymbol{\beta}_F \tag{4.9}$$

A consistent body loading vector can now be obtained by using equation (4.9) in equation (4.2) to obtain

$$F_n = T^T H^{-1} H_F \boldsymbol{\beta}_F - T_F^T \boldsymbol{\beta}_F + \int_{S\bar{\sigma}} \bar{T}^T \phi_2^T \tilde{N} \, \mathrm{d}s \tag{4.10}$$

where

$$H_F = \int_V Q^T C Q_F \, \mathrm{d}V$$

and

$$T_F = \int_s R \tilde{N} \, \mathrm{d}s$$

4.2.1 A thin plate hybrid finite element

A flat six-noded triangular thin plate element is shown in Figure 4.1, the mid-side nodes allowing the sides of the element to be curved. Each node has three degrees of freedom, i.e. the normal translation (U_z) and two rotations about the coordinate axes (ϕ_x, ϕ_y). The stress resultants satisfy the homogeneous equations:

$$
\begin{aligned}
\frac{\partial M_x}{\partial x} - \frac{\partial M_{xy}}{\partial y} - Qx = 0 \\
\frac{\partial M_y}{\partial y} - \frac{\partial M_{yx}}{\partial x} - Qy = 0 \\
\frac{\partial Q_x}{\partial x} + \frac{\partial Q_y}{\partial y} = 0
\end{aligned}
\tag{4.11}
$$

with $M_{xy} = M_{yx}$. The stress resultant assumption used for the element is

$$
\begin{Bmatrix} M_x \\ M_y \\ M_{xy} \\ Q_x \\ Q_y \end{Bmatrix} =
\left[\begin{array}{ccccccccccccccccc}
1 & x & y & x^2 & y^2 & & & & & & & & & & \tfrac{1}{2}x^2 & xy & \\
 & & & y^2 & & 1 & x & y & x^2 & & & & & & -\tfrac{1}{2}y^2 & & xy \\
 & & & 2xy & & & & & & 1 & x & y & x^2 & y^2 & & & \\
 & 1 & & & & & & & & & & -1 & & -2y & x & & \\
 & & & & & & & 1 & & & -1 & & 2x & & -y & & x
\end{array}\right] \{\beta\}
$$

$$
\boldsymbol{\sigma} = Q\boldsymbol{\beta} \tag{4.12}
$$

In equation (4.12) the number of stress coefficients is seventeen, which is sufficient to prevent spurious rigid body modes [3].

The compliance matrix relating bending curvatures to bending moments is [4]

$$
C = \frac{1}{D(1-\nu^2)} \begin{bmatrix} 1 & -\nu & \\ -\nu & 1 & \\ & & 2/(1-\nu) \end{bmatrix} \tag{4.13}
$$

where

$$
D = \frac{Eh^3}{12(1-\nu^2)}
$$

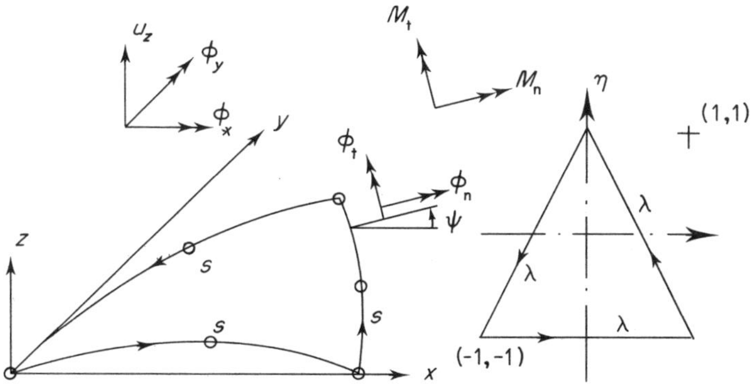

Figure 4.1 The triangular thin plate hybrid finite element

The complementary strain energy matrix H is formed using equations (4.12) and (4.13). Integration is performed on the area of the plate as stress resultants are used instead of stress in the basic formulation. A standard mapping from the x, y plane to an auxiliary plane (ξ, n) is used to facilitate integration.

The boundary displacements shown in Figure 4.1 cannot be represented by simple polynomial terms if rigid body motion of the side is to be included. The approach used here is to represent the boundary displacements in terms of some arbitrary parameters, the first three of which explicitly represent rigid body motion of the side. Thus we have

$$\begin{Bmatrix} U_z \\ \phi_t \\ \phi_n \end{Bmatrix} = \begin{bmatrix} 1 & y & -x \\ 0 & -\sin\psi & \cos\psi \\ 0 & \cos\psi & \sin\psi \end{bmatrix} \{a\}_{1-3} \tag{4.14}$$

where ψ is the angle between the x axis and the local normal to the side and $\{a\}$ are arbitrary parameters. With nine degrees of freedom along the side, six more terms are used. Since we have

$$\phi_n = \frac{\partial U_z}{\partial s}$$

we can write

$$\begin{Bmatrix} U_z \\ \phi_t \\ \phi_n \end{Bmatrix} = \begin{bmatrix} \lambda^2 & \lambda^3 & \lambda^4 & \lambda^5 & & \\ & & & & \lambda & \lambda^2 \\ 2c\lambda & 3c\lambda^2 & 4c\lambda^3 & 5c\lambda^4 & & \end{bmatrix} \{a\}_{4-9} \tag{4.15}$$

where λ is a side length parameter in the ξ, n space and

$$c = \frac{d\lambda}{ds} = \frac{\partial\lambda}{\partial x}\frac{dx}{ds}$$

$$\frac{\partial\lambda}{\partial x} = \frac{d\lambda}{d\xi}\frac{d\xi}{\partial x} + \frac{d\lambda}{d\eta}\frac{\partial n}{\partial x}$$

Equations (4.14) and (4.15) can be merged to give

$$\mathbf{U} = P\mathbf{a} \tag{4.16}$$

Equation (4.16) is evaluated at each node on the side to obtain an expression for the arbitrary parameters **a**. When the boundary displacements are related to the coordinate directions (x, y, z) we obtain

$$\mathbf{U} = \tilde{N}\mathbf{U}_n$$

Boundary tractions are obtained by considering equilibrium with the internal stress resultants and after boundary integration the matrix T of equation (4.5) can be formed. The resulting stiffness matrix (equation 4.8) follows in a standard fashion.

A consistent body loading vector for the case of constant pressure can be obtained in the manner indicated in equations (4.9) and (4.10) using

$$Q_F\boldsymbol{\beta} = \frac{p}{3}\begin{bmatrix} x(1-x/4) \\ y(1-y/4) \\ xy \\ (1-3x/2) \\ (1-3y/2) \end{bmatrix}\{1\} \tag{4.17}$$

where p is the pressure acting normal to the element surface.

4.2.2 Some hybrid plate element results

Figure 4.2 shows the performance of the hybrid plate element on a simply supported plate under a concentrated central load (P). The present element is indicated by E18. Also shown are the results of various other elements. 'HT' is a three-noded hybrid by Edwards [5], 'Q4' is a curvilinear displacement element of Sullivan [6], 'ACM' is an equilibrium model of Fraejis de Veubeker [7], and 'Q' a high-precision displacement element of Cowper *et al.* [8].

Also shown is an element indicated by E15 which is the present formula-

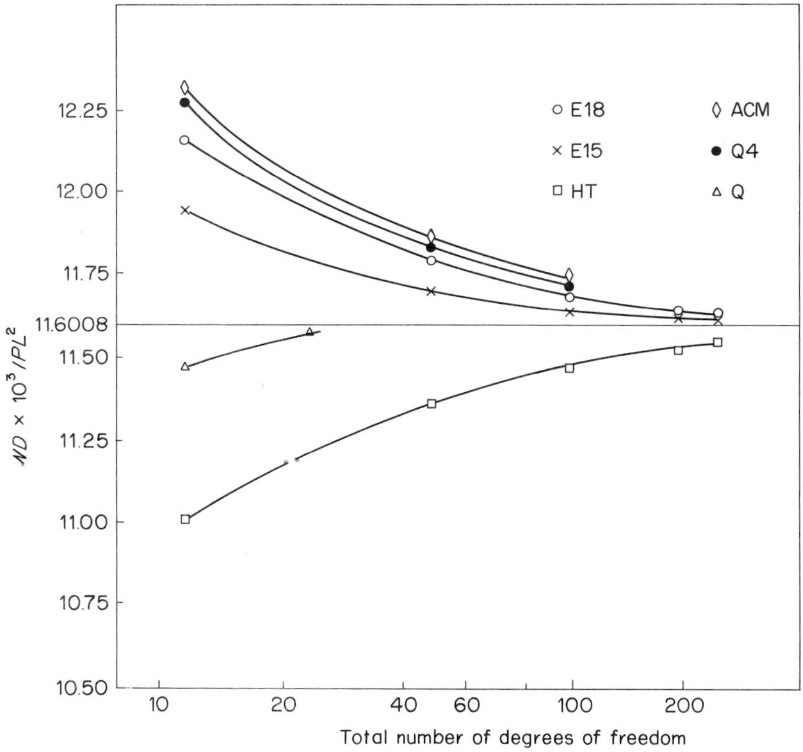

Figure 4.2 Convergence of elements on a simply supported plate under a concentrated load

tion with a reduced number of parameters in the boundary displacement assumption (equation 4.15). In this version the (λ^5) variation is not used and hence the ϕ_n rotation at a mid-side node does not appear as a degree of freedom. The element then only has fifteen degrees of freedom. Figure 4.3 shows for the same problem the maximum bending moment on a plate centreline for a mesh with only two elements along the centreline (a quarter of the plate being analysed). The performances of E18 and E15 are shown.

The use of consistent loading as given by equations (4.10) and (4.17) produced good results on a series of test problems, typically on a 2×2 mesh of triangles on one-quarter of a simply supported square plate with a length to thickness ratio of 20 : 1, an error of 0.2 per cent. in the central deflection for a constant pressure load. An alternative method shown in equation (4.5) of finding a loading vector which gives the same work done on the structure as the pressure load gave an error of 1.1 per cent. This pattern of improvement with a consistent loading vector was typical of many tests.

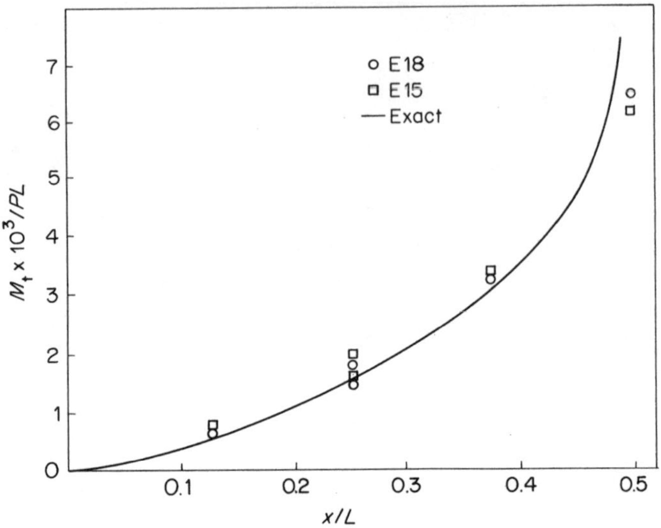

Figure 4.3 Maximum bending moment on centreline of simply supported plate

With the addition of an area displacement assumption for the evaluation of loading vectors using the 'work done' method the inclusion of a mass matrix for the evaluation of natural frequencies is quite straightforward. The displacement assumption used is that given by Sullivan [6] for his six-noded displacement element which consists of an eighteen-term polynomial giving some quintic variations.

Table 4.1 Vibration of a simply supported plate

Mode	1×1 mesh error (%)	2×2 mesh error (%)	3×3 mesh error (%)
1, 1	21.48	0.588	0.147
1, 2	2.88	−0.088	−0.02
2, 2	30.32	−0.64	−0.110
1, 3	1.54	−0.941	−0.058
2, 3	29.1	−0.266	−1.17
2, 4	49.42	−2.45	−0.34
5, 5	32.21	−6.57	−2.7

Table 4.1 shows some results for a simply supported square plate (length to thickness ratio of 20:1) with one-quarter of the plate being modelled with various mesh densities.

4.3 HELLINGER–REISSNER THICK PLATE ELEMENTS

The Hellinger–Reissner principle of equation (4.1) can be used to form a thick plate finite element if we use a simple compatible displacement field. The chosen stress function expression is the simple non-equilibrating form:

$$\begin{Bmatrix} M_{xx} \\ M_{yy} \\ M_{xy} \\ Q_x \\ Q_y \end{Bmatrix} = \begin{bmatrix} P & & & & \\ & P & & & \\ & & P & & \\ & & & P & \\ & & & & P \end{bmatrix} \{\beta\}$$

$$\boldsymbol{\sigma} = Q\boldsymbol{\beta} \tag{4.18}$$

where $P = [1, \xi, \eta, \xi\eta, \ldots]$, and its length is equal to the number of nodes in the element. Thus the complementary strain energy matrix can be written

$$H = \int_A \begin{bmatrix} P^T P C_{11} & P^T P C_{12} \cdots \\ P^T P C_{21} & \cdot \\ \cdot & \cdot \end{bmatrix} \mathrm{d}a \tag{4.20}$$

where C_{ij} is an element of the compliance matrix and

$$H^{-1} = \begin{bmatrix} Gd_{11} & Gd_{12} & \cdots \\ Gd_{21} & \cdot & \cdot \\ \cdot & \cdot & \end{bmatrix} \tag{4.21}$$

where $G = [\int P^T P d(\text{area})]^{-1}$ and d_{ij} is an element of $[C]^{-1}$. It can be seen that it is only necessary to invert a comparatively small matrix in the formation of H^{-1}.

The second term in equation (4.1) is the product of a stress and strain; stresses are expressed in equation (4.18). If the displacements are expressed indpendently in the form:

$$\begin{aligned} U_z &= P\mathbf{a} \\ \phi_x &= P\mathbf{b} \\ \phi_y &= P\mathbf{c} \end{aligned} \tag{4.22}$$

and if in addition we have

$$\begin{aligned} x &= P\mathbf{d} \\ y &= P\mathbf{e} \\ z &= P\mathbf{f} \end{aligned} \tag{4.23}$$

where **a** to **f** are arbitrary constants (an isoparametric form giving rise to a compatible displacement field), then the generalized strains for the thick plate element [9] are

$$\begin{Bmatrix} K_{xx} \\ K_{yy} \\ K_{xy} \\ \bar{\gamma}_{xy} \\ \bar{\gamma}_{yz} \end{Bmatrix} = \begin{bmatrix} & & -P,x \\ & P,y & \\ & -P,x & P,y \\ P,x & & -P \\ -P,y & P & \end{bmatrix} \begin{Bmatrix} \mathbf{a} \\ \mathbf{b} \\ \mathbf{c} \end{Bmatrix} \tag{4.24}$$

$$\boldsymbol{\varepsilon} = B\mathbf{g}$$

where a comma denotes partial differentiation with respect to the individual suffix. Hence the second term in equation (4.1) can be expressed as

$$\boldsymbol{\beta}^T \int_A Q^T B d(\text{area})\mathbf{g} \tag{4.25}$$

The list of arbitrary parameters in equation (4.25), **g**, can be evaluated in the standard fashion by using equations (4.22) at the element nodes.

Equation (4.25) results in a matrix similar to the matrix T in equation (4.5) except that integration is not a surface integral, and the standard form of equation (4.6) can be used with H^{-1} of equation (4.21) to form the stiffness matrix:

$$S = T^T H^{-1} T \tag{4.26}$$

The form of the constituent matrices of the stiffness matrix equation (4.26) leads to a method whereby numerical integration is not required.

Firstly, the matrix G of equation (4.21):

$$G = \left[\left[\int P^T P d(\text{area})\right]\right]^{-1}$$

comprises terms of the form:

$$I = \int_A \xi^m \eta^n \, d\xi \, d\eta \tag{4.27}$$

when standard isoparametric mappings are used. For a quadrilateral shape and when the standard two-unit square is used for the ξ, η space:

$$\begin{aligned} I &= 0 \text{ for } n \text{ or } m \text{ odd} \\ &= 4/(m+1)(n+1) \text{ for } n \text{ and } m \text{ even} \end{aligned} \tag{4.28}$$

A similar form can be found [10] for a triangular form. Hence the identification of the various terms in the formation of matrix G allows the use of equation (4.28).

The formation of the matrix T of equation (4.26) developed from equation (4.25) leads to submatrices of the form:

$$G_x = \left[\int P^T P_{,x} d(\text{area})\right]$$

$$G_y = \left[\int P^T P_{,y} (\text{area})\right] \tag{4.29}$$

The derivatives of P with respect to the coordinate directions (x, y) can be expressed as, for example,

$$P_{,x} = P_{,\xi}\frac{\partial \xi}{\partial x} + P_{,\eta}\frac{\partial \eta}{\partial x} \tag{4.30}$$

By algebraically inverting the Jacobian matrix relating (x, y) space to (ξ, η) space equation (4.30) becomes

$$P_{,x} = \frac{P_{,\xi}\, \partial y/\partial \eta - P_{,\eta}\, \partial y/\partial \eta}{|J|}$$

where $|J|$ is the determinant of the Jacobian matrix and hence G_x of equation (4.29) becomes

$$G_x = \int P^T P_{,x}\, |J|\, d\xi\, d\eta$$

$$= \int P^T\left(P_{,\xi}\frac{\partial y}{\partial \eta} - P_{,\eta}\frac{\partial y}{\partial \eta}\right) d\xi\, d\eta \tag{4.31}$$

All the terms in equation (4.31) can be written in the form of equation (4.27) and thus can be explicitly integrated. The matrix G_y of equation (4.29) follows in a similar manner and all the constituents of the stiffness expression of equation (4.26) can be evaluated.

A mass matrix can be developed in the standard manner from the kinetic energy expression:

$$\text{KE} = \tfrac{1}{2}\rho h^3 \int_A \dot{U}_z d(\text{area}) + \tfrac{1}{24}\rho h^2 \int_A (\dot{\phi}_x + \dot{\phi}_y) d(\text{area}) \tag{4.32}$$

where ρ is the density and h is the plate thickness. The second term in equation (4.32) gives the contribution to the kinetic energy of the effects of rotary inertia.

4.4 SOME HELLINGER–REISSNER THICK PLATE VIBRATION RESULTS

Tables 4.2 and 4.3 show the percentage errors in natural frequencies of vibration for the symmetric–symmetric modes of a simply supported square

Table 4.2 Percentage errors in natural frequencies for the symmetrical modes of a simply supported thick plate, length to thickness ratio of 20:1

Frequency number	Element type				Thin plate
	3 noded	4 noded	6 noded	8 noded	
1	58.3	19.1	1.01	0.25	0.93
2	38.0	35.0	3.9	3.7	4.4
3	53	28.3	12.5	4.8	7.7
4	52	50.0	11.4	12.54	10.95
	49 nodes	49 nodes	139 nodes	108 nodes	

5×5 mesh

Table 4.3 Percentage errors in natural frequencies for the symmetrical modes of a simply supported thick plate, length to thickness ratio of 10:1

Frequency number	Element type				Thin plate
	3 noded	4 noded	6 noded	8 noded	
1	18.8	5.4	0.34	0.108	3.5
2	14.48	12.73	1.77	1.81	16.2
3	23.98	9.3	5.8	2.1	27.08
4	23.70	22.04	6.1	7.08	36.81
	49 nodes	49 nodes	139 nodes	108 nodes	

5×5 mesh

plate with length to thickness ratios of 20:1 and 10:1 respectively (one-quarter of the plate being modelled by a 5×5 mesh). Results are given for various element types in the isoparametric family. The results given by Mindlin [11] are used as the basis of comparison. Also shown are the variations when using a thin plate analysis.

It is clear that good thick plate results are exhibited and some acceptable results for plates with a length to thickness ratio of 20:1 are obtained. However, because of the relatively low-order polynomial representation of curvatures (equation 4.24) only the eight-noded element is reasonable.

The Hellinger–Reissner formulation has allowed an efficient thick plate element to be produced without using numerical integration. This immediately precludes the use of reduced integration [12] which may extend the range of usefulness of the element. Furthermore, it can be shown [12] that the present isoparametric formulation is identical to the corresponding displacement approach when the elements are rectangular.

4.4.1 A thin curved shell based on the Hellinger–Reissner principle

The development of a general shell element from the previous simpler form leads one naturally to the use of the Hellinger–Reissner principle. The fact

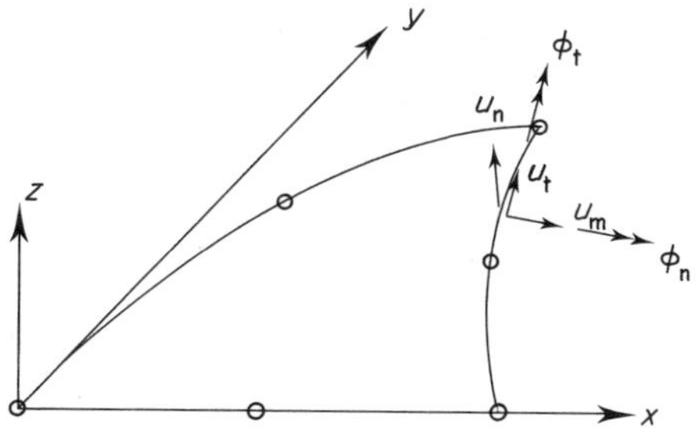

Figure 4.4 Six-noded general shell element showing boundary displacements on edge of shell

that a hybrid element requires an equilibrating stress assumption for its formulation results in intractable difficulties. The Hellinger–Reissner principle, on the other hand, allows a non-equilibrating stress assumption, as shown in equation (4.18). However, since the problem of a compatible displacement assumption is present, the form of the Hellinger–Reissner principle which is most useful is equation (4.2) requiring, in addition to the stress assumption, an area displacement assumption and a boundary displacement assumption.

The thin shell element which has been developed is shown in Figure 4.4 and is a six-node triangular shell element. The nodes at the vertices and the single mid-side node on each side allow a surface to be generated, the height of which (z coordinate) is a quadratic function of the base plane coordinate (x, y). As this surface is amenable to analytic geometry [13], the lines of maximum and minimum curvature (principal coordinates) can be used to present the shell equations in their relatively simple form. However, higher order elements (e.g. the eight-noded quadrilateral) cannot be similarly formed.

The shell theory on which the element is based is that due to Novozhilov [14]. Following Novozhilov, the two principal twisting moments are set equal and if one omits quantities of the order of h/R where h is the shell thickness and R is a typical shell radius, then one can set the two inplane shear stress resultants equal to each other. The stress resultants in this form are recommended by Love [15] as a first approximation. The complete list of stress resultants in the principal coordinate system is

$$\boldsymbol{\sigma}^T = [N_{11}N_{22}N_{12}Q_1Q_2M_{11}M_{22}M_{12}] \tag{4.33}$$

where the N's refer to inplane stress resultants, the Q's refer to transverse shear forces, and the M's refer to bending and twisting moments.

With the approximations in the form of the stress resultants chosen, it is not possible to satisfy identically the six stress equilibrium equations, but

Novozhilov states that the errors implied by non-satisfaction do not produce inaccuracies exceeding those due to the initial assumptions on which the shell theory is based. Furthermorc, the use of the Hellinger–Reissner principle will not satisfy the equilibrium conditions with a non-equilibriating stress assumption except in some averaged sense (due to the second term in equation 4.2).

The form of the stress assumption is the same as shown in equation (4.18) where the stress resultants are expressed independently. The polynomial form chosen was

$$P=[1,\ \xi,\ \eta,\ \xi\eta,\ \xi^2,\ \eta^2]$$

The formation of the complementary strain energy matrix H of equation (4.5) is in a similar manner to equations (4.20) and (4.21) when the corresponding compliance matrix relating inplane strains, curvatures, and twists to the stress resultants are used [14]. One point to make is that since the stress resultants are assumed in an independent form, it is necessary to include some complementary strain energy term relating to the transverse shear stress. Otherwise the matrix H and H^{-1} will not be compatible to the size of the boundary work matrix T (equations 4.5) and hence the formation of the stiffness matrix of equation (4.7) would not be possible. To overcome this difficulty the transverse shear stress/strain expression for the thick plate element was used. Tests showed that convergence to thin plate results was not affected and ill-conditioned matrices were not produced. With this stress assumption the equilibrium matrix E of equation (4.5) can be found using the stress equilibrium equations [14] for the thin shell theory.

The boundary displacements which are shown in Figure 4.4 are considered in a similar manner to the hybrid plate element presented earlier, namely the first six terms of the displacement assumption were used to represent rigid body motion about the origin of the element coordinates as

$$\begin{Bmatrix} u_m \\ u_t \\ u_n \\ \phi_n \\ \phi_t \end{Bmatrix} = \begin{bmatrix} R_1 & R_3 \\ & \\ 0 & R_2 \end{bmatrix} \{a\} \tag{4.34}$$

where
$$R_1 = \begin{bmatrix} i\cdot eb & j\cdot eb & k\cdot eb \\ i\cdot et & j\cdot et & k\cdot et \\ i\cdot en & j\cdot en & k\cdot en \end{bmatrix}$$

$$R_2 = \begin{bmatrix} i\cdot eb & j\cdot eb & k\cdot eb \\ i\cdot et & j\cdot et & k\cdot et \end{bmatrix}$$

where the dot represents vector inner products

$$R_3 = \begin{bmatrix} eb & \cdot & R_4 \\ et & \cdot & R_4 \\ en & \cdot & R_4 \end{bmatrix}$$

and the matrix R_4 is

$$R_4 = \begin{bmatrix} 0 & Z & -Y \\ -Z & 0 & X \\ Y & -X & 0 \end{bmatrix}$$

Since there are three nodes along a side of each node, each node having five degrees of freedom, we have fifteen parameters to construct the boundary displacement assumption, six having been used to represent rigid body motion.

It can be shown [10] that the vector rotation normal to the side of the element can be expressed as

$$\phi_n = \frac{\partial u_n}{\partial s} - \frac{u_n}{R_t} - \frac{u_t}{R_n} \tag{4.35}$$

where $1/R_t$ represents the twist of the line segment and $1/R_n$ represents the normal curvature of the line. The extra parameter used to construct the displacement assumption can be written in terms of the length parameter λ in the ξ, η space:

$$\begin{Bmatrix} u_m \\ u_t \\ u_n \\ \phi_m \\ \phi_t \end{Bmatrix} = \begin{bmatrix} \lambda^2 & & & & & & & & \\ & \lambda & \lambda^2 & & & & & & \\ & & & \lambda^2 & \lambda^3 & \lambda^4 & \lambda^5 & & \\ n\lambda^2 & m\lambda & m\lambda^2 & 2p\lambda & 3p\lambda^2 & 4p\lambda^3 & 5p\lambda^4 & & \\ & & & & & & & \lambda & \lambda^2 \end{bmatrix} \{a\} \tag{4.36}$$

where $n = -1/R_t$, $m = -1/R_n$, $p = \mathrm{d}\lambda/\mathrm{d}s$.

Expressions (4.34) and (4.36) can then be evaluated at each node on the side to obtain expressions for the arbitrary constants **a** and hence the boundary shape functions. Each side is considered in turn.

The construction of the area displacements function follows a similar pattern, with the first six parameters representing rigid body motion of an interior point in the shell and the allowable extra variations using higher order terms in the ξ, η space. This results in quadratic variation of principal inplane deflections and quintical variation of the normal shell displacement expressed in terms of the ξ, η space [10]. The normal rotations about the principal directions are then derived using the shell equations:

$$\phi_1 = \frac{1}{A_2}\frac{\partial u_n}{\partial \alpha_1} - \frac{u_2}{R_2}$$
$$\phi_2 = \frac{1}{A_1}\frac{\partial u_m}{\partial \alpha_2} + \frac{u_1}{R_1} \tag{4.37}$$

where A_1, A_2 are the Lamé parameters which describe the principal coordinate system α_1, α_2, u_1, u_2 are the principal inplane coordinates, and R_1, R_2 are the principal radii of curvature.

The integration of the constituent matrices of the stiffness matrix equation (4.5) is performed numerically using Gaussian quadrature in conjunction with ξ, η space transformations.

4.4.2 Some curved thin shell results

The pinched cylindrical shell problem has been used extensively for tests on shell elements. Ashwell *et al.* [16, 17] has shown that elements even with the inclusion of rigid body motion perform badly on 'thin shells'—in particular those problems exhibiting inextensional deformation.

Table 4.4 shows some results for the deflection under the load of a pinched cylindrical shell with free ends. Using symmetry, only one-eighth of the shell was modelled, using triangular elements of regular mesh densities. The two problems designated 'thick' and 'thin' have radius to thickness ratios of 53 and 320 respectively. It can be seen that the present formulation performs very badly on these tests compared to the accepted values shown, in a like manner to the element of Cantin and Clough [18, 19]. This is due to the simple polynomial forms of the displacement assumption which cannot represent the necessary relations between displacements which are required for inextensional deformation. This has been shown by Morley [20] for cylindrical shells.

Figure 4.5 shows a pinched spherical shell with a radius to thickness ratio of 50 and a mesh used by Cowper *et al.* [21] to analyse the problem. Figures 4.5 and 4.6 show various stress resultant results for a mesh density parameter $n = 5$. The results are compared to an analytical solution by Flugge [22]

Table 4.4 Deflection under load on pinched circular cylinder shell (free ends)

Cylindrical mesh ($m \times n$)	Thick	Thin
4×1	0.0921	0.00513
4×2	0.0972	0.00525
3×3	0.0575	0.00380
Accepted value	0.1139	0.02439

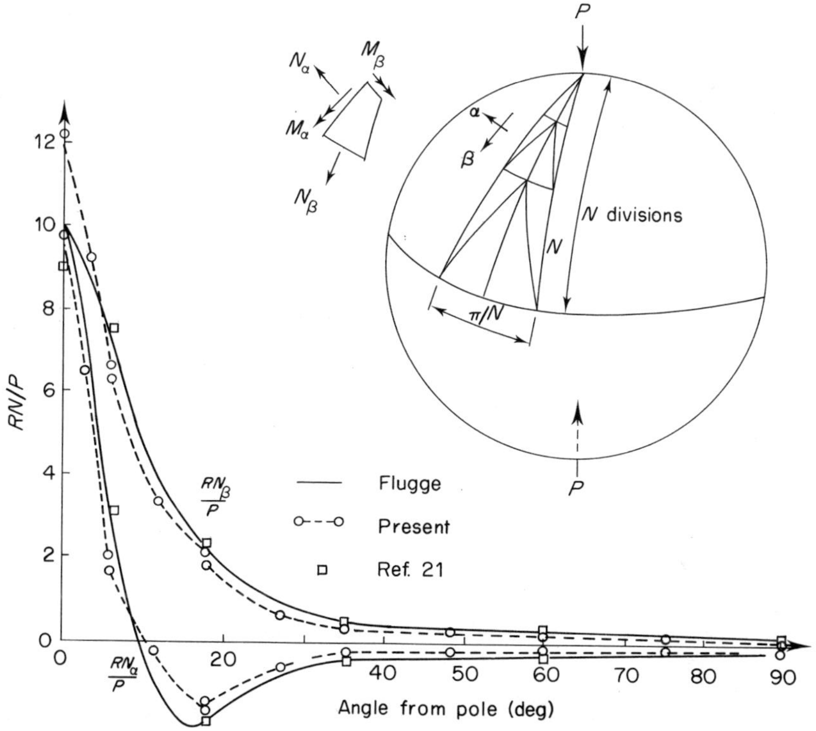

Figure 4.5 Variation of membrane stress resultants in pinched spherical shell ($N = 5$)

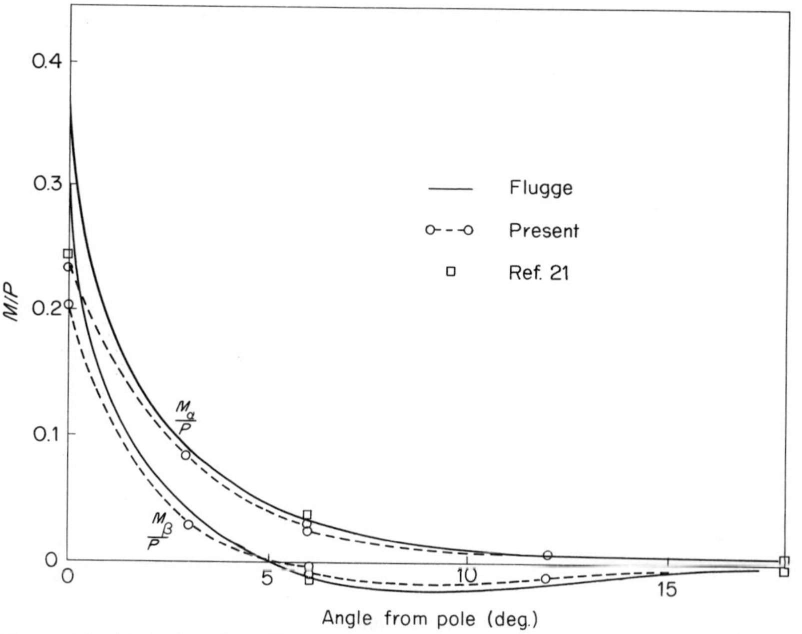

Figure 4.6 Variation of bending moments in vicinity of pole in pinched spherical shell ($N = 5$)

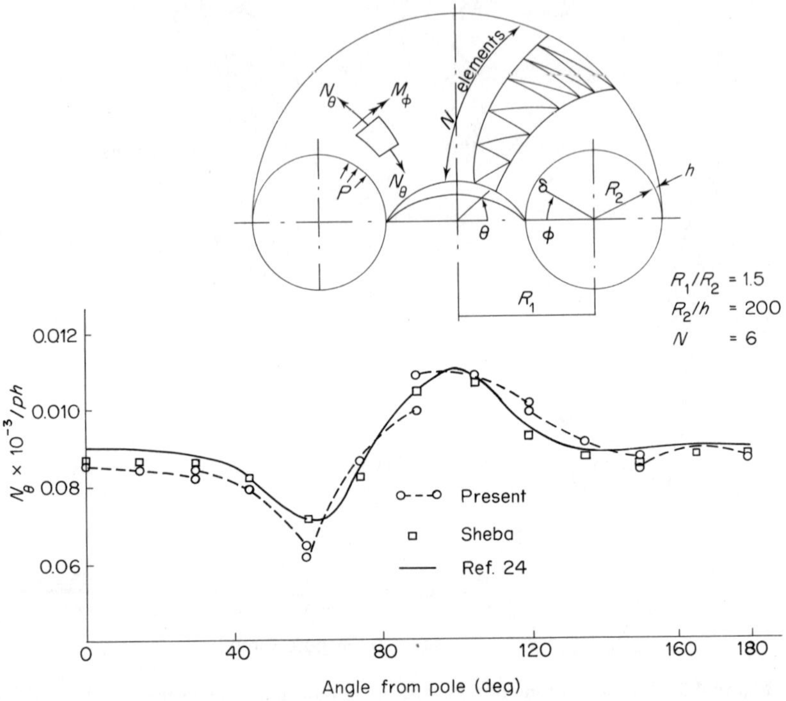

Figure 4.7 Variation of membrane stress in a torus under an internal pressure

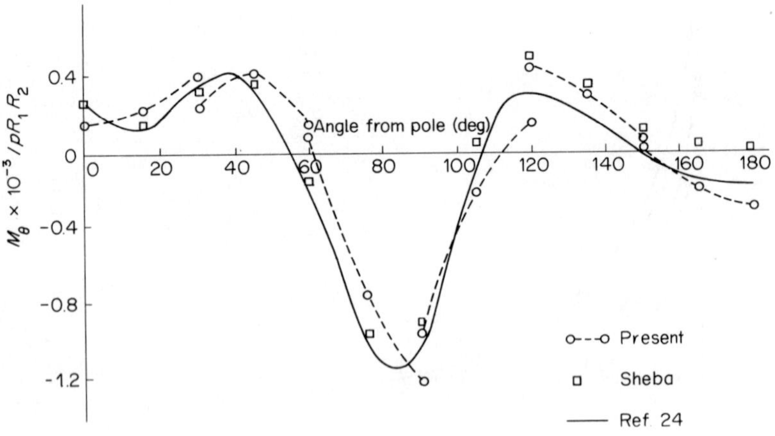

Figure 4.8 Variation of bending moment in a torus under an internal pressure

and the 'CURSHL' element of Cowper *et al.* [21] based upon Koiter–Sanders shell theory.

Figure 4.7 shows a torus under internal pressure and a mesh used for its investigation. This problem, although axisymmetric, is a useful test as the shell has regions of both positive and negative Gaussian curvature. Figures 4.7 and 4.8 show variations of stress resultants over half of the toriod. The results are compared to an analytical investigation by Kalnins [23] and finite element analysis with the SHEBA element by Argyris and Lockner [24].

The results presented here generally show acceptable behaviour, but it is clear that no advantage over other particular formulations ensues from the rather complicated approach using the Hellinger–Reissner formulation for the present thin shell formulation.

4.4.3 Conclusions

The Hellinger–Reissner variation principle has proved to be a useful basis for the development of finite elements. The flexibility in choosing stress function expansions which do not satisfy equilibrium conditions has been shown to allow economical evaluation of many terms which previously have been costly to evaluate in computer time. However, other aspects relating to boundary conditions and integration are complicated and expensive for a general case. The plate elements presented which are based on Hellinger–Reissner and complementary energy principles have been shown to be acceptable and sufficiently economical to justify their general use. The development of these plate elements to the general curved shell has not, on the present formulation, produced a significantly better element than others currently available, particularly since inextensional deformations are not represented and there are difficulties in extending the formulation to the quadrilateral shape.

REFERENCES

1. T. H. H. Pian, 'Finite element methods by variational principles with relaxed continuity requirements', *Proc. Int. Conf. Variational Methods in Engineering*, Southampton, Vol. 1, 1972.
2. P. Tong and T. H. H. Pian, 'A variational principle and convergence of a finite element method basd on assumed stress distributions', *Int. J. Solids and Structures*, **5**, 463 (1969).
3. R. D. Henshell, 'On hybrid finite elements', in *The Mathematics of Finite Elements and Applications*, Academic Press, 1973.
4. S. Timoshenko and S. Woinowsky-Krieger, *Theory of Plates and Shells*, McGraw-Hill, 1959.
5. G. Edwards, *Cylindrical Shell Hybrid Finite Element*, Ph.D. thesis, Nottingham University, 1974.

6. C. J. Sullivan, *Finite Element Analysis of Box Structures*, Ph.D. thesis, Nottingham University, 1975.
7. B. Fracjis de Veubeke, 'An equilibrium model for plate bending', *Int. J. Solids and Structures*, **4,** 447–468 (1968).
8. G. R. Cowper, E. Kosko, G. M. Lindberg, and M. D. Olsen, 'Static and dynamic applications of a high precision triangular plate bending element, *AIAA J.*, **7,** No. 10 (1969).
9. J. O. Makoju, *Finite Elements for Thick Plates*, M.Phil. thesis Nottingham University, 1976.
10. F. N. Rigby, *The Development of General Shell Finite Elements*, Ph.D. thesis, Nottingham University, 1978.
11. R. D. Mindlin, 'Influence of rotary inertia and shear on flexural motions of isotropic elastic plates', *J. Appl. Mech.*, **18,** 31–38 (1951).
12. O. C. Zienkiewicz, R. L. Taylor, and J. Too, 'Reduced integration techniques in general analysis of plates and shells', *Int. J. Num. Meth. in Eng.*, **3,** 275–290 (1971).
13. L. P. Eisenhart, *Differential Geometry*, Ginn and Company, 1956.
14. V. Novozhilov, *The Theory of Thin Shells*, P. N. Nordhoff Gronmger, Netherlands, 1964.
15. A. E. Love, *Mathematical Theory of Elasticity*, 4th ed., Cambridge University Press, 1927.
16. D. G. Ashwell and A. B. Sabir, 'Limitations of certain curved finite elements when applied to arches', *Int. J. Mech. Sci.*, **13,** 133–139 (1971).
17. D. G. Ashwell, A. B. Sabir, and T. M. Roberts, 'Further studies in the application of curved finite elements to circular arches, *Int. J. Mech. Sci.*, **13,** 507–517 (1971).
18. G. Cantin and R. W. Clough, 'A curved cylindrical shell finite element', *AIAA J.*, **6,** 1057–1062 (1968).
19. G. A. Fonder and R. W. Clough, 'A curved cylindrical shell finite element', *AIAA J.*, **6,** 1057–1062 (1968).
20. L. S. D. Morley, *Analysis of Developable Shells with Special Reference to the Finite Element Method and Circular Cylinders*, RAE Technical Report 74180.
21. G. R. Cowper, G. M. Lindberg, and M. D. Olsen, 'Comparison of Two High Precision Triangular Bending Elements for the Analysis of Arbitrary Deep Shells,' *Proc. of Third Air Force Congress on Matrix Methods in Structural Mechanics*, AFFDL-TR-71-160, Ohio, 1968.
22. W. Flugge, *Stresses in Shells*, Springer-Verlag, Berlin, 1962.
23. A. Kalnins, 'Analysis of shells of revolution subjected to symmetrical and non-symmetrical loads, *J. Appl. Mech.*, **31,** 317–347 (1972).
24. J. H. Argyris and N. Lockner, 'On the application of the SHEBA shell element', *Computer Meth. of Appl. Mech.*, **1,** 317–347 (1972).

Hybrid and Mixed Finite Element Methods
Edited by S. N. Atluri, R. H. Gallagher, and O. C. Zienkiewicz

Chapter 5

New Hybrid Stress Models in the Limit Analysis of Solids and Structures

Tadahiko Kawai

5.1 INTRODUCTION

In order to overcome difficulties encountered in the finite element non-linear analysis, five years ago the present author proposed a family of new discrete elements which are called the rigid bodies–spring models (RBSM). Basic studies of these elements have proved them to be very powerful in the limit analysis of structures or solids in general, but it was also found that convergency of the elastic solutions may not be guaranteed except in the case of the beam and some other elements [1]. Kondou pointed out that the RBSM incremental solutions obtained by Yamada's method can always be the best upper bound solution for the assumed mesh pattern [2].

It is, however, not sufficient to guarantee the accuracy of the RBSM solutions unless the corresponding lower bound solution is obtained. For this purpose Takeuchi and Kawai have developed a general and practical method for obtaining the approximate lower bound solution from the RBSM incremental solution [3]. Although this method may be acceptable from the practical point of view, the rigorous mathematical proof cannot be made because an approximate determination of the stress fields in individual elements is inevitable in calculating the lower bound solutions.

It is well known that the essential feature of the plastic deformation is the 'slip' movement, which is characterized by the displacement discontinuity across the slip lines or surfaces. At this point the stress model is naturally superior to the displacement model in an analysis of inelastic problems, although use of the stress model is quite limited in the present day practice. It should also be mentioned that the lower bound collapse loads may be determined by using the stress models. With this in mind, Kondou has developed a series of new, simplified elements which are especially suitable for limit analyses of beam and plate structures by using the hybrid

Hellinger–Reissner variational principle. Based on the hybrid complementary energy principles Watanabe derived another type of new element in which thc effect of shear deformation could be taken into account in a bending analysis of beam and plate problems.

5.2 A FAMILY OF SIMPLIFIED PLATE-BENDING ELEMENTS TO BE DERIVED FROM THE HYBRID HELLINGER–REISSNER VARIATIONAL PRINCIPLE

Kondou developed a series of simplified elements which are useful in an analysis of the plate-bending problems in the elastic as well as elastoplastic range by using the Hellinger–Reissner variational principle [4].

Following Washizu, a functional of the hybrid Hellinger–Reissner variational principle for an analysis of the plate-bending problems is given as follows [5]:

$$\Pi_{RH} = \sum_e \iint_{S_\sigma} \left[-B(M_x, M_y, M_{xy}) - M_x \frac{\partial^2 w}{\partial x^2} - M_y \frac{\partial^2 w}{\partial y^2} - 2M_{xy} \frac{\partial^2 w}{\partial x\, \partial y} - \bar{p} \right] dx\, dy$$

$$+ \sum H_{ab} + \int_{C_\sigma} \left(-\bar{V}_z w + \bar{M}_n \frac{\partial w}{\partial n} + \bar{M}_{ns} \frac{\partial w}{\partial s} \right) ds$$

$$+ \int_{Cu} \left[-V_n (w - \bar{w}) + M_n \left(\frac{\partial w}{\partial n} - \frac{\partial \bar{w}}{\partial n} \right) + M_{ns} \left(\frac{\partial w}{\partial s} - \frac{\partial \bar{w}}{\partial s} \right) \right] ds \tag{5.1}$$

where

$$H_{ab} = \int_{C_{ab}} M_n^{(a)} \left(\frac{\partial w^{(a)}}{\partial n^{(a)}} + \theta \right) ds + \int_{C_{ab}} M_n^{(b)} \left(\frac{\partial w^{(b)}}{\partial n^{(b)}} - \theta \right) ds \tag{5.2}$$

Here the following standard notations in the plate-bending problems are employed:

Moment–curvature relationship:

$$\begin{aligned} M_x &= -D\left(\frac{\partial^2 w}{\partial x^2} + \nu \frac{\partial^2 w}{\partial y^2} \right) \\ M_y &= -D\left(\nu \frac{\partial^2 w}{\partial x^2} + \frac{\partial^2 w}{\partial y^2} \right) \\ M_{xy} &= -D(1-\nu) \frac{\partial^2 w}{\partial x\, \partial y} \end{aligned} \tag{5.3a}$$

where $D = \dfrac{Eh^3}{12(1-\nu^2)}$ = the bending rigidity of a given plate

Complementary energy density:

$$B(M_x, M_y, M_{xy}) = \frac{1}{2}\frac{12}{Eh^3}[(M_x + M_y)^2 + 2(1+\nu)(M_{xy}^2 - M_x M_y)] \tag{5.3b}$$

Boundary forces:

$$\begin{aligned} M_n &= M_x l^2 + 2M_{xy} lm + M_y m^2 \\ M_{ns} &= -(M_x - M_y)lm + M_{xy}(l^2 - m^2) \\ V_l &= \left(\frac{\partial M_x}{\partial x} - \frac{\partial M_{xy}}{\partial y}\right)l + \left(\frac{\partial M_{xy}}{\partial x} + \frac{\partial M_y}{\partial y}\right)m \end{aligned} \tag{5.3c}$$

$\bar{p}$ is intensity of a given distributed load, $\bar{M}_n$, $\bar{M}_{ns}$, $\bar{V}_l$ are the prescribed boundary forces as shown in Figure 5.1, and C_σ and C_u are boundary curves where the mechanical and geometrical boundary conditions are prescribed respectively. $\sum_e$ implies summation with respect to all individual elements, while $\sum$ implies summation of all element boundary curves, and θ is the angular displacement on the interelement boundary.

The associated condition to equation (5.1) is the continuity of the plate deflection w on the interelement boundary and variation should be taken with respect to w and (M_x, M_y, M_{xy}). For this purpose the following linear displacement field $w(x, y)$ is considered (see Figure 5.2):

$$w^{(a)}(x, y) = \lfloor 1, x, y \rfloor \mathbf{C}^{-1}\mathbf{w} \tag{5.4a}$$

$$\text{where} \quad \mathbf{C} = \begin{bmatrix} 1 & x_1 & y_1 \\ 1 & x_2 & y_2 \\ 1 & x_3 & y_3 \end{bmatrix}, \quad \mathbf{w} = \begin{Bmatrix} w_1 \\ w_2 \\ w_3 \end{Bmatrix} \tag{5.4b}$$

A constant moment field $(M_x^{(a)}, M_y^{(a)}, M_{xy}^{(a)})$ is also assumed together with constant angular displacements $(\theta_i, \theta_j, \theta_k)$ along the element boundaries.

Substituting equations (5.4a,b) into equations (5.1) and (5.2), the following stationary conditions can be derived with respect to θ_i and the stress resultants $(M_x^{(a)}, M_y^{(a)}, M_{xy}^{(a)})$ respectively:

$$\frac{\partial \Pi_{RH}}{\partial \theta_i} = 0: \qquad M_n^{(a)} = M_n^{(b)} \tag{5.5}$$

$$\frac{\partial \Pi}{\partial M_x^{(a)}} = \frac{\partial \Pi}{\partial M_y^{(a)}} = \frac{\partial \Pi}{\partial M_{xy}^{(a)}} = 0:$$

$$\begin{Bmatrix} \dfrac{\partial^2 w^{(a)}}{\partial x^2} \\ \dfrac{\partial^2 w^{(a)}}{\partial y^2} \\ \dfrac{\partial^2 w^{(a)}}{\partial x\, \partial y} \end{Bmatrix} = \begin{bmatrix} l_i^2 & l_j^2 & l_k^2 \\ m_i^2 & m_j^2 & m_k^2 \\ 2l_i m_i & 2l_j m_j & 2l_k m_k \end{bmatrix} \begin{Bmatrix} \dfrac{2\phi_i}{h_i} \\ \dfrac{2\phi_j}{h_j} \\ \dfrac{2\phi_k}{h_k} \end{Bmatrix} \tag{5.6}$$

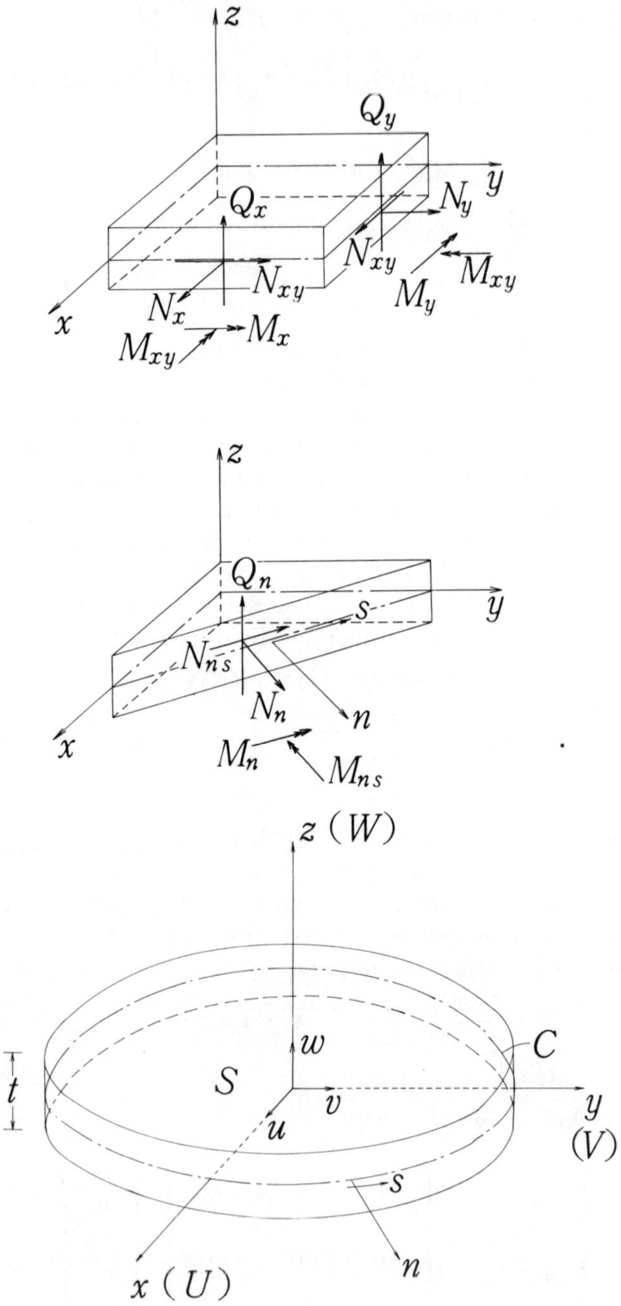

Figure 5.1 Coordinate system of a plate and definition of stress resultants

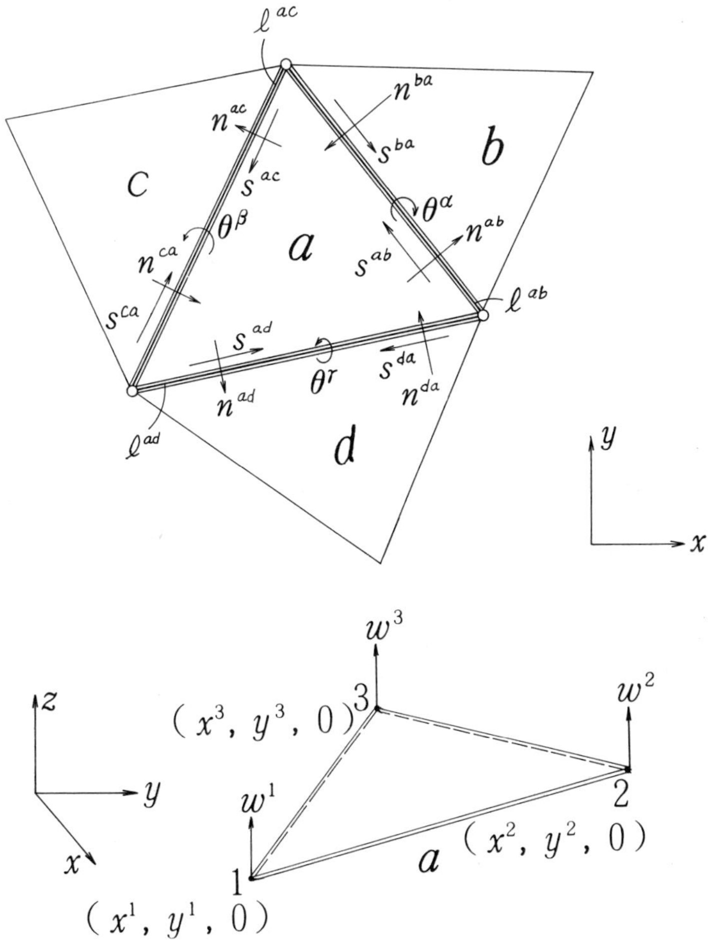

Figure 5.2 A new plate-bending element

where $\quad l_i = \cos(n_i, x)$

$$m_i = \cos(n_i, y) \tag{5.7}$$

and $\quad \phi_i = \theta_i + \dfrac{\partial w^{(a)}}{\partial n_i^{(a)}}, \qquad \text{etc.}$

The direction of θ_i may be defined arbitrarily as long as θ_i is common for two adjacent elements on the same boundary. It should be mentioned here that equation (5.3a) is used in the derivation of equation (5.6).

If the moment field satisfying equation (5.5) is used, θ_i can be eliminated from the functional Π_{RH} and equation (5.2) can be given by

$$H_{ab} = \int_{C_{ab}} M_n^{(a)} \left(\frac{\partial w^{(a)}}{\partial n^{(a)}} + \frac{\partial w^{(b)}}{\partial n^{(b)}} \right) ds \tag{5.8}$$

From the functional corresponding to this case the Hermann plate element can be derived [6].

On the other hand, if the stationary condition equation (5.6) is substituted together with equation (5.3a) into equation (5.1), the moments $(M_x^{(a)}, M_y^{(a)}, M_{xy}^{(a)})$ will be eliminated from Π_{RH} and therefore Π_{RH} turns out to be functional of w only. Thus Π_{RH} will be transformed into the functional for the minimum potential energy principle from which the conventional plate-bending element of the displacement type can be obtained.

It is interesting to note that the Hermann plate element and this plate element are essentially the same, the former being derived by using the stationary condition equation (5.5), while the latter by using equation (5.6). Since the Herrmann plate element is a mixed model, it is not convenient to apply to the non-linear analysis, but this difficulty can be completely eliminated by using the latter model.

Substituting equations (5.6) and (5.3a) into Π_{RH} and performing the energy integration the following equation can be derived:

$$\Pi_{RH}^{(e)} = \tfrac{1}{2}\mathbf{M}_n^{(a)} \begin{bmatrix} \frac{2\Delta^{(a)}}{h_i} & & \\ & \frac{2\Delta^{(a)}}{h_j} & \\ & & \frac{2\Delta^{(a)}}{h_k} \end{bmatrix} \boldsymbol{\phi}^{(a)} \tag{5.9}$$

where $\Pi_{RH}^{(e)}$ is the functional Π_{RH} of the individual element and $\Delta^{(a)}$ is the element area, $\mathbf{M}_n^T = \lfloor M_{ni}, M_{nj}, M_{nk} \rfloor$. It can be concluded that $\Pi_{RH}^{(e)}$ represents the strain energy of a specific plate element and is equal to $\frac{1}{2}$ (boundary moment $\mathbf{M}_n$)$\times$(the corresponding relative angular displacement). Thus the strain energy of a given plate is automatically lumped on the element boundaries.

This feature will play a very important role in the collapse load analysis of a bent plate. In what follows the RBSM plate element will be derived by applying the further simplification to the solution obtained so far.

Combining the moment–curvature and curvature–displacement relations with respect to the local coordinates $(n_i^{(a)}, S_i^{(a)})$ as shown in Figure 5.3, the following moment–rotation matrix can be derived after some calculations:

$$\mathbf{M}_n^{(a)} = \begin{Bmatrix} M_{n_i}^{(a)} \\ M_{n_j}^{(a)} \\ M_{n_k}^{(a)} \end{Bmatrix} = \begin{bmatrix} \bar{\mathbf{D}}^{(a_i)} & \mathbf{T}^{(a_i)} \\ \bar{\mathbf{D}}^{(a_j)} & \mathbf{T}^{(a_j)} \\ \bar{\mathbf{D}}^{(a_k)} & \mathbf{T}^{(a_k)} \end{bmatrix} \begin{Bmatrix} \dfrac{2\phi_i}{h_i} \\ \dfrac{2\phi_i}{h_i} \\ \dfrac{2\phi_i}{h_i} \end{Bmatrix} \tag{5.10}$$

where

$$\mathbf{D}^{(a_i)} = \begin{bmatrix} D_{11}^{(a_i)} & D_{12}^{(a_i)} & D_{13}^{(a_i)} \\ D_{21}^{(a_i)} & D_{22}^{(a_i)} & D_{23}^{(a_i)} \\ D_{31}^{(a_i)} & D_{32}^{(a_i)} & D_{33}^{(a_i)} \end{bmatrix}, \qquad \text{etc.} \tag{5.11}$$

and

$$\mathbf{T}^{(a_i)} = \left[\begin{array}{c|c|c} 1 & \bar{l}_{ij}^2 & \bar{l}_{ik}^2 \\ \hline 0 & \bar{m}_{ij}^2 & \bar{m}_{ik}^2 \\ \hline 0 & 2\bar{l}_{ij}\bar{m}_{ij} & 2\bar{l}_{ik}\bar{m}_{ik} \end{array}\right], \qquad \text{etc.} \tag{5.12}$$

$$\bar{l}_{ij} = \cos\,(n_i^{(a)}, n_j^{(a)}), \qquad \bar{m}_{ij} = \cos\,(n_i^{(a)}, s_j^{(a)})$$

Neglecting the off-diagonal elements in the above matrix, equation (5.10) will be simplified as follows:

$$\mathbf{M}_n^{(a)} = \begin{bmatrix} D_{11}^{(a_i)} & & \\ & D_{11}^{(a_j)} & \\ & & D_{11}^{(a_k)} \end{bmatrix} \begin{Bmatrix} \dfrac{2\phi_i}{h_i} \\ \dfrac{2\phi_j}{h_j} \\ \dfrac{2\phi_k}{h_k} \end{Bmatrix} = \begin{bmatrix} \dfrac{2D_{11}^{(a_i)}}{h_i} & & \\ & \dfrac{2D_{11}^{(a_j)}}{h_j} & \\ & & \dfrac{2D_{11}^{(a_k)}}{h_k} \end{bmatrix} \boldsymbol{\phi}^{(a)} \tag{5.13}$$

Substituting equation (5.13) into equation (5.9) the following functional $\Pi_{\mathrm{RH}}^{(\mathrm{e})}$ will be obtained:

$$\Pi_{\mathrm{RH}}^{(\mathrm{e})} = \tfrac{1}{2}\boldsymbol{\phi}^{(a)\mathrm{T}} \begin{bmatrix} \dfrac{2}{h_i} D_{11}^{(a_i)} \dfrac{2\Delta^{(a)}}{h_i} & & \\ & \dfrac{2}{h_j} D_{11}^{(a_j)} \dfrac{2\Delta^{(a)}}{h_j} & \\ & & \dfrac{2}{h_k} D_{11}^{(a_k)} \dfrac{2\Delta^{(a)}}{h_k} \end{bmatrix} \boldsymbol{\phi}^{(a)} \tag{5.14}$$

Since $2\Delta^{(a)}/h_i$, etc., are the lengths of each boundary side, it can be

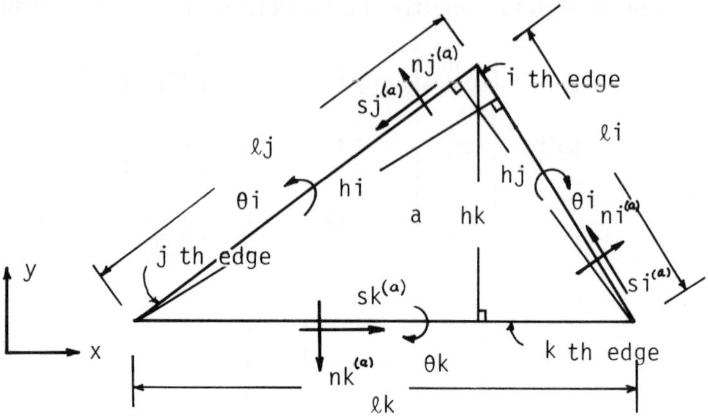

Figure 5.3 A new plate-bending model

concluded that equation (5.13) will be interpreted as a moment-rotation matrix for the rotational springs whose intensity is given by $2D_{11}^{(a_i)}/h_i$, etc.

It is also seen that there is no coupling among ϕ_i, ϕ_j, and ϕ_k and equation (5.13) can be obtained by simply adding the individual relations. Therefore θ_i, θ_j, θ_k can be easily eliminated by the matrix condensation. For example, consider the ith edge (see Figure 5.4). The function $\Pi_{RH}^{(e_i)}$ can be given as follows:

$$\Pi_{RH}^{(e_i)} = \frac{1}{2}\left(\theta_i + \frac{\partial w^{(a)}}{\partial n_i^{(a)}}\right)^2 \frac{2}{h_i^{(a)}} D_{11}^{(a_i)} \frac{2\Delta^{(a)}}{h_i^{(a)}} + \frac{1}{2}\left(-\theta_i + \frac{\partial w^{(b)}}{\partial n_i^{(b)}}\right)^2 \frac{2}{h_i^{(b)}} D_{11}^{(b_i)} \frac{2\Delta^{(b)}}{h_i^{(b)}} \quad (5.15)$$

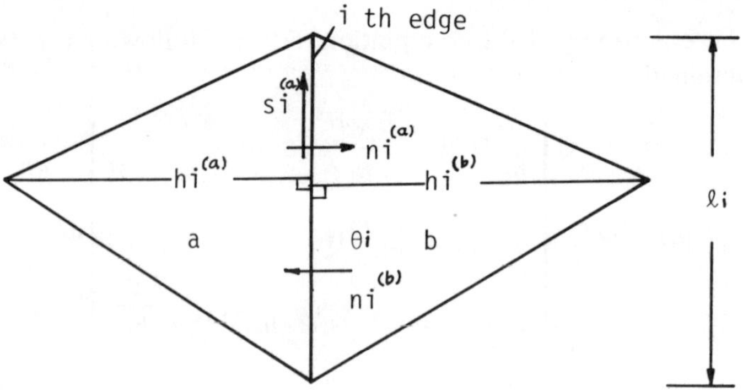

Figure 5.4 RBSM plate-bending model

from which the following stationary condition of $\Pi_{\text{RH}}^{(e_i)}$ with respect to θ_i is obtained:

$$\theta_i = \frac{-\dfrac{1}{h_i^{(a)}} D_{11}^{(a_i)} \dfrac{\partial w^{(a)}}{\partial n_i^{(a)}} + \dfrac{1}{h_i^{(b)}} D_{11}^{(b_i)} \dfrac{\partial w^{(b)}}{\partial n_i^{(b)}}}{\dfrac{1}{h_i^{(a)}} D_{11}^{(a_i)} + \dfrac{1}{h_i^{(b)}} D_{11}^{(b_i)}} \tag{5.16}$$

Substituting equation (5.16) into equation (5.15), $\Pi_{\text{RH}}^{(e_i)}$ is finally obtained in the following form:

$$\Pi_{\text{RH}}^{(e_i)} = \frac{1}{2} k^{(ab)} \left(\frac{\partial w^{(a)}}{\partial n_i^{(a)}} + \frac{\partial w^{(b)}}{\partial n_i^{(b)}} \right)^2 \tag{5.17}$$

where

$$k^{(ab)} = \frac{2 \dfrac{1}{h_i^{(a)}} D_{11}^{(a_i)} \dfrac{1}{h_i^{(b)}} D_{11}^{(b_i)}}{\dfrac{1}{h_i^{(a)}} D_{11}^{(a_i)} + \dfrac{1}{h_i^{(b)}} D_{11}^{(b_i)}} L_i \tag{5.18}$$

and

$$L_i = \frac{2\Delta^{(a)}}{h_i^{(a)}} = \frac{2\Delta^{(b)}}{h_i^{(b)}} = \text{the length of the } i\text{th edge}$$

It can be seen from equation (5.17) that the function $\Pi_{\text{RH}}^{(e)}$ will eventually represent the strain energy expression of the rotational spring system of the intensity $k^{(ab)}$ which is distributed along the interelement boundaries' ith, jth, and kth edges.

If the plate is made of isotropic elastic material,

$$k^{(ab)} = \frac{2D}{h_i^{(a)} + h_i^{(b)}} L_i \qquad \text{where} \quad D_{11}^{(a_i)} = D_{11}^{(b_i)} = D \tag{5.19}$$

Thus it can be concluded that the RBSM plate-bending element with the spring constant given by equation (5.19) can be derived by neglecting the off-diagonal elements of the 'moment–curvature' matrix given by equation (5.10) in the previous formulation. Due to such elimination of the off-diagonal elements, however, reliability of the RBSM elastic solutions becomes questionable. As the stiffness is lumped on the boundary edges in the RBSM plate element this model can be used to define the collapse load of a bent plate.

Furthermore, the displacement parameter is only one (w) for each node and therefore analysis can be made easily from the elastic range to the collapse stage with a very short computing time. Thus it can be concluded that the RBSM elements can be very useful in practice if the application is carefully made.

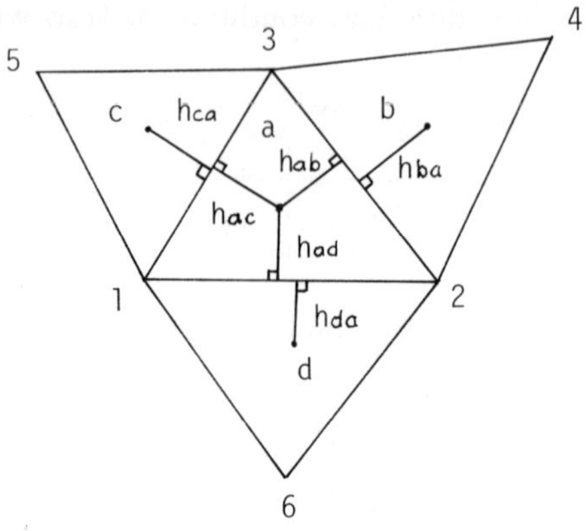

Figure 5.5 Simplified hybrid displacement model for the plate-bending problem

Finally, it is worth while to note that a simplified hybrid displacement model can be derived from the previous formulation by taking the weighted mean operation for θ_i as follows. Referring the Figure 5.5, the following weighted mean operation is introduced for θ:

$$\theta_i = \frac{1}{h_i^{(a)} + h_i^{(b)}} \left(\frac{\partial w^{(a)}}{\partial n_i^{(a)}} h_i^{(b)} - \frac{\partial w^{(b)}}{\partial n_i^{(b)}} h_i^{(a)} \right), \qquad \text{etc.} \tag{5.20}$$

Substitution of equation (5.20) into equations (5.6) and (5.3a) in the previous formulation will yield a new simplified plate-bending element with only three nodal parameters, i.e. the lateral displacement w at each vertex, and naturally the non-linear terms of this element become extremely simple because the first derivative of the displacement is constant throughout the element domain. The slope continuity across the interelement boundaries is secured in the average sense due to equation (5.20). Therefore it can be concluded that this simplified plate element is especially useful in an analysis of the large deflection problems of elastic plates.

It is also worth while to mention Watanabe and Kawai's work on the analysis of the two-dimensional creeping flow in a square cavity [7]. Considering that the two-dimensional creeping flow is governed by the homogeneous biharmonic equation of the stream function, he carried out the viscous flow analysis successfully up to the Reynolds number $R = 9{,}000$ by using Kondou's simplified element. Since verification studies of this simplified element were described in a recent paper written by Kondou, Shiina, and Kawai [8], only one numerical example will be given in the following figures. A study was made on the elastic buckling analysis of a simply supported rectangular plate subjected to a constant edge shear.

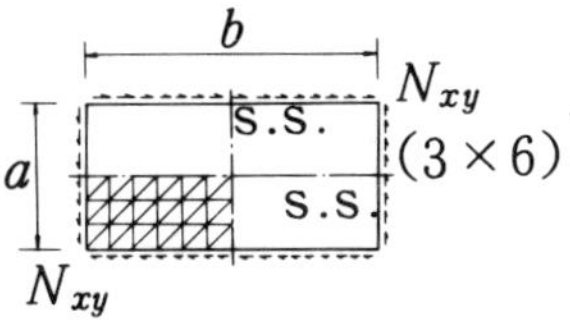

(i) mesh pattern used

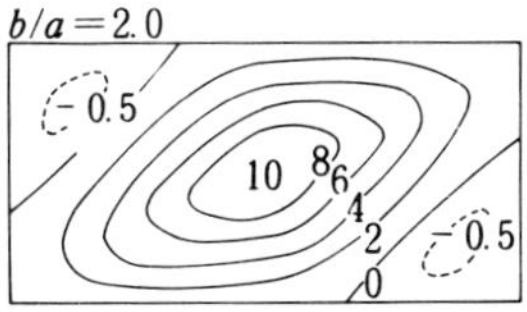

(ii) calculated buckling mode

Figure 5.6 Buckling analysis of a simply supported rectangular plate under shear

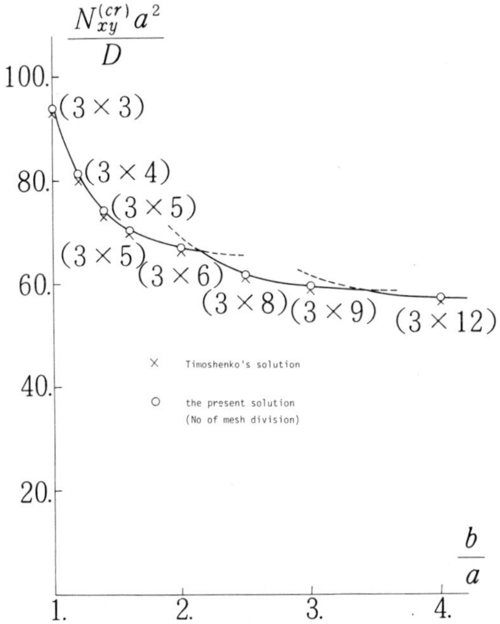

Figure 5.7 Buckling analysis of a simply supported rectangular plate under shear (critical stress versus aspect ratio)

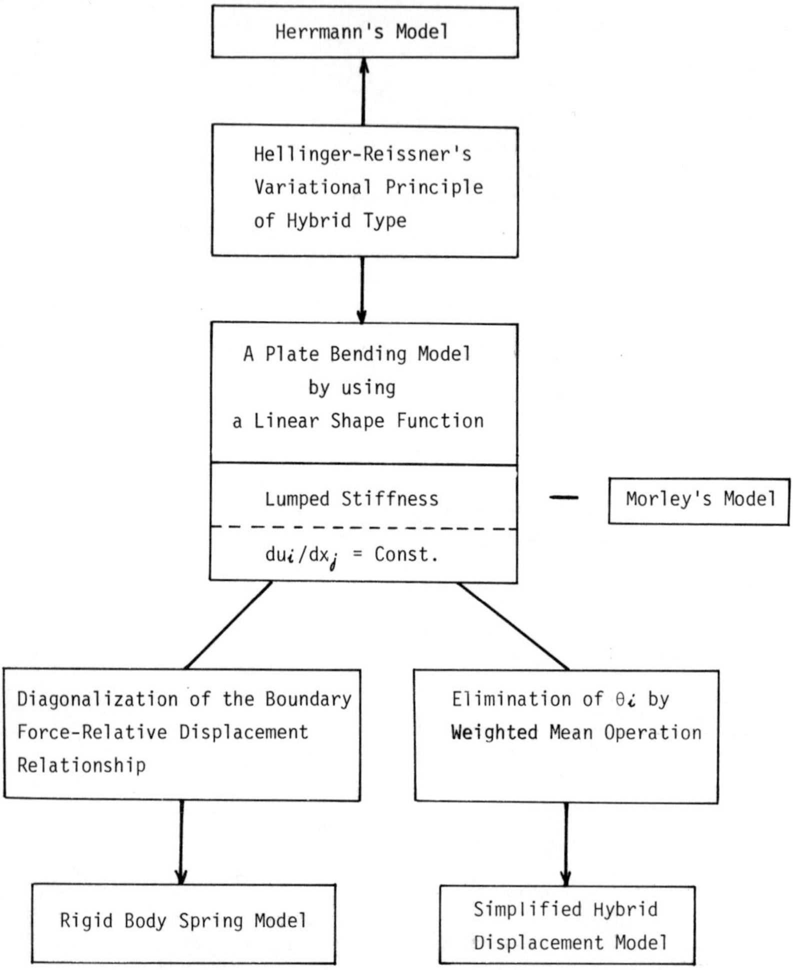

Figure 5.8 A series of plate-bending models

In Figure 5.6 the buckling mode of a rectangular plate whose aspect ratio $b/a = 2$ is shown, together with the assumed mesh pattern, and Figure 5.7 shows the accuracy of the solutions of rectangular plates with various aspect ratios versus the number of mesh divisions. It can be seen from this figure that if an appropriate mesh pattern is selected, solutions of sufficient accuracy from the engineering application point of view can be obtained for the intractable problem as described here. It should also be mentioned that recently Mukudai, Matsuo, and Kondou [9] conducted a successful numerical analysis of the inelastic buckling test of a plate girder under simple shear by using Kondou's element.

In conclusion it can be stated that a series of simple plate-bending elements can be derived based on the hybrid Hellinger–Reissner principle together with the linear displacement field for w and constant moment field in which a constant angular displacement is introduced along each edge independently. The process of derivation of various elements can be illustrated as shown in Figure 5.8.

5.3 HYBRID STRESS PLATE MODELS AS DERIVED FROM THE PRINCIPLE OF THE HYBRID COMPLEMENTARY ENERGY

Using the principle of the hybrid complementary energy, Watanabe and Kawai derived a family of hybrid stress models for the bending as well as inplane problems of plates [10]. The functional $\Pi_{CH}^{(e)}$ for the principle of the hybrid complementary energy can be given by referring to Figure 5.9 as follows:

$$\Pi_{CH}^{(e)} = \iint_{\Omega_e} B(\sigma_{ii})\, dx\, dy - \int_{\partial\Omega_e} T_i u_i\, ds \tag{5.21}$$

where $B(\sigma_{ij})$ is the complementary energy defined with respect to an element domain Ω_e from the stress field which satisfies the equation of equilibrium and the surface traction T_i can be expressed by $T_i = \sigma_{ij} n_j$ where n_i is the unit normal drawn outward to each boundary edge; u_i is essentially the Lagrange

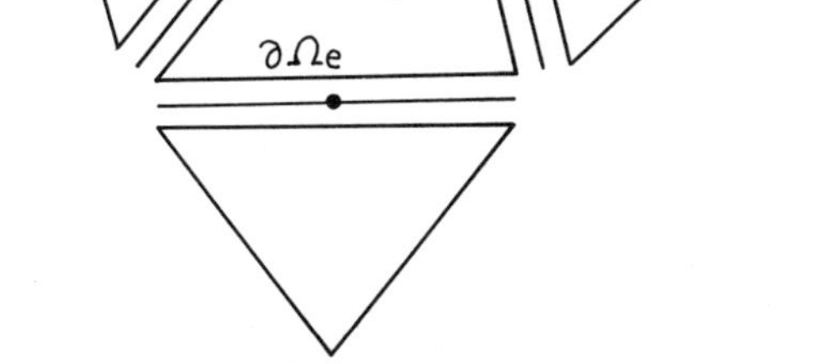

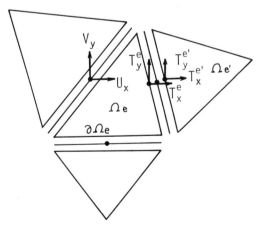

Figure 5.9 One element domain Ω_e bounded by $\partial\Omega_e$ and three adjacent elements

multiplier to be introduced so as to make T_i continuous along the boundary edge, but it is obvious that u_i physically represents the displacement.

The following linear stress field which satisfies the equilibrium condition is assumed:

$$\begin{Bmatrix} \sigma_x \\ \sigma_y \\ \tau_{xy} \end{Bmatrix} = \left[\begin{array}{c:c:c:c:c:c:c} 1 & x & y & & & & \\ \hdashline & & & 1 & x & y & \\ \hdashline & -y & & & & -x & 1 \end{array}\right] \begin{Bmatrix} a \\ b \\ c \\ d \\ e \\ f \\ g \end{Bmatrix} \tag{5.22}$$

or

$$\boldsymbol{\sigma} = \mathbf{B}\mathbf{a} \tag{5.23}$$

Assuming the plane strain condition the strain vector $\boldsymbol{\varepsilon}^T = [\varepsilon_x, \varepsilon_y, \gamma_{xy}]$ is given by

$$\boldsymbol{\varepsilon} = \mathbf{C}\boldsymbol{\sigma} \tag{5.24}$$

where

$$\mathbf{C} = \frac{1+\nu}{E}\begin{bmatrix} 1-\nu & \nu & 0 \\ \nu & 1-\nu & 0 \\ 0 & 0 & 2 \end{bmatrix} \tag{5.25}$$

The complementary energy of an individual element can be given from equations (5.23) and (5.24) as follows:

$$\iint_{\Omega_e} B(\sigma_{ij})\, dx\, dy = \tfrac{1}{2}\mathbf{a}^T\mathbf{G}\mathbf{a} \tag{5.26}$$

where **G** is given by

$$\mathbf{G} = \iint_{\Omega_e} \mathbf{B}^T\mathbf{C}\mathbf{B}\, dx\, dy \tag{5.27}$$

Using the assumed displacement on the element boundary edge the second term of equation (5.21) can be expressed by the following equation:

$$\int_{\partial\Omega_e} T_i u_i\, ds = \mathbf{a}^T\mathbf{L}\mathbf{u} \tag{5.28}$$

Expressions for **L** and **u** will be given later.

Substituting equations (5.26) and (5.28) into equation (5.21) $\Pi_{CH}^{(e)}$ can be expressed as follows:

$$\Pi_{CH}^{(e)} = \tfrac{1}{2}\mathbf{a}^T\mathbf{G}\mathbf{a} - \mathbf{a}^T\mathbf{L}\mathbf{u} \tag{5.29}$$

Since the stress field given by equation (5.22) is assumed in each element independently, **a** can be eliminated from equation (5.29) by taking the stationary conditions of equation (5.29):

$$-\Pi_{\mathrm{CH}}^{(e)} = \tfrac{1}{2}\mathbf{u}^T\mathbf{L}^T\mathbf{G}^{-1}\mathbf{L}\mathbf{u} \tag{5.30}$$

Thus $\Pi_{\mathrm{CH}}^{(e)}$ can be expressed only in terms of the displacement assumed along the element boundary edges and therefore it is now reduced to the strain energy expressed by the boundary displacements as in the displacement method. Consequently the element stiffness matrix $\mathbf{K}^{(e)}$ can be given by

$$\mathbf{K}^{(e)} = \mathbf{L}^T\mathbf{G}^{-1}\mathbf{L} \tag{5.31}$$

Now it is assumed that the element boundary edge is assumed to exhibit the rigid body displacement in this hybrid stress element.

Denoting the rigid body displacements at the mid-span by u_x, v_y, θ, the displacements (U_x, V_y) at an arbitrary point can be given by

$$\begin{aligned} U_x &= U_x - (y - y_1)\theta \\ V_y &= V_y + (x - x_1)\theta \end{aligned} \tag{5.32}$$

where x_1, y_1 are the coordinates of the mid-span.

Then equation (5.28) will be calculated as follows:

$$\begin{aligned} \int_{\partial\Omega_e} T_i u_i \, \mathrm{d}s &= \int_{\partial\Omega_e} [(\sigma_x l + \tau_{xy} m)U_x + (\tau_{xy} l + \sigma_y m)V_y] \, \mathrm{d}s \\ &= \int_{\partial\Omega_e} \{(\sigma_x l^2 + 2\tau_{xy} lm + \sigma_y m^2)U_n \\ &\quad + [(-\sigma_x + \sigma_y)lm + \tau_{xy}(l^2 - m^2)]V_s\} \, \mathrm{d}s \end{aligned} \tag{5.33}$$

where $U_n = (lu_x + mv_y) - (s - s_1)\theta$, $V_s = -mu_x + lv_y$, and n, s are the local coordinates of the mid-span (see Figure 5.10).

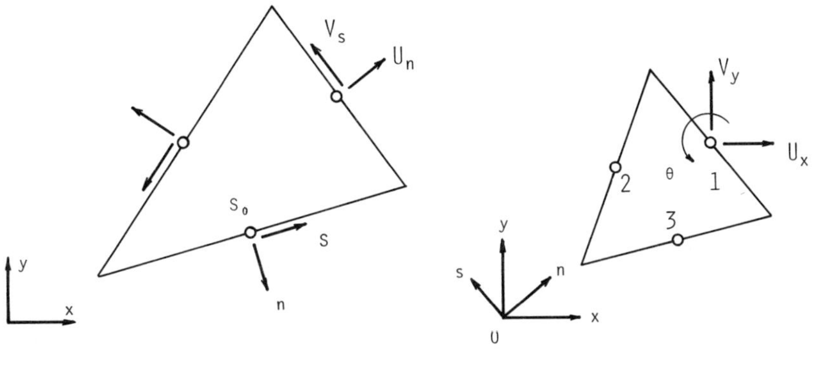

Figure 5.10 Two-dimensional hybrid stress model

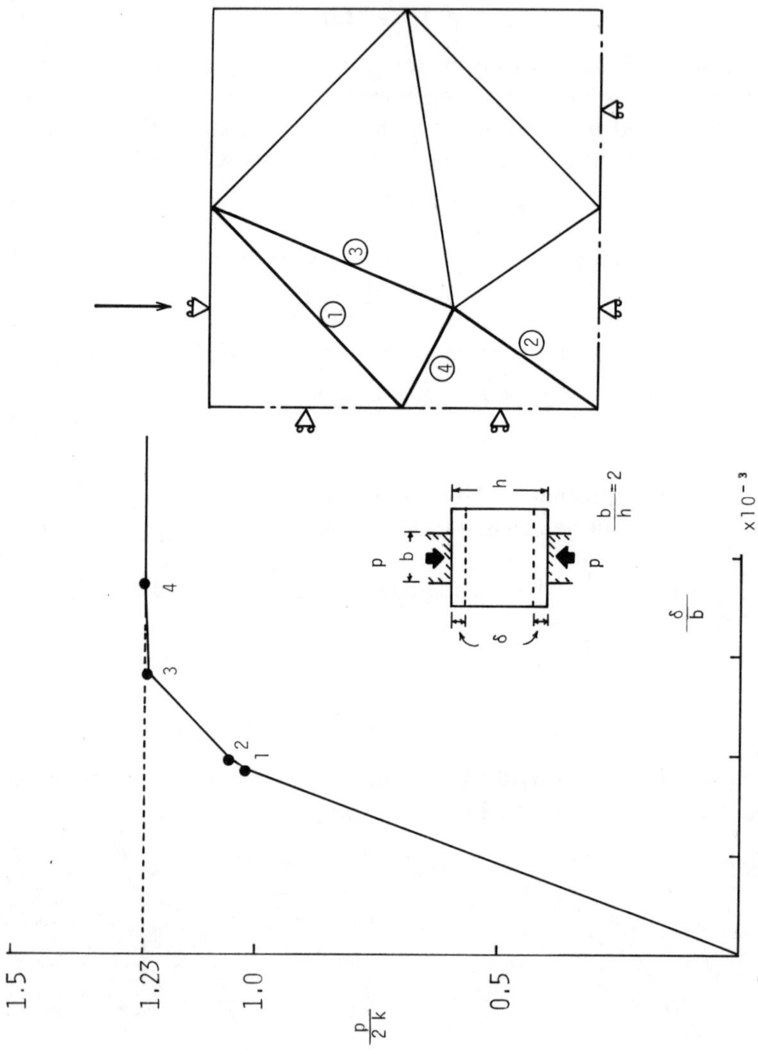

Figure 5.11 Slip line analysis of the punch problem using the hybrid stress model

It is further transformed into the work of the equivalent boundary stress vector (σ_n, τ_{ns}, t) defined with respect to the local coordinates in which t represents a torque:

$$\int_{\partial\Omega_e} T_i u_i \, \mathrm{d}s = \int_{\partial\Omega_e} (\sigma_n u_n + \tau_{ns} u_s + t\theta) \, \mathrm{d}s = \mathbf{a}^{\mathrm{T}} \mathbf{L} \mathbf{u} \tag{5.34}$$

where $\mathbf{u} = \lfloor u_n, v_s, \theta \rfloor$. Now $\mathbf{L}$ can be obtained from equation (5.34) as follows:

$$\mathbf{L} = \int_{\partial\Omega_e} \left[\begin{array}{c:c:c} l^2 & -lm & -l^2(s-s_1) \\ sl^2 - 2ylm & -xlm - y(l^2 - m^2) & -(xl^2 - 2ylm)(s-s_1) \\ yl^2 & -xlm & -yl^2(s-s_1) \\ m^2 & lm & -m^2(s-s_1) \\ xm^2 & xlm & -xm^2(s-s_1) \\ -2xlm + ym^2 & ylm - x(l^2 - m^2) & -(-2xlm + ym^2)(s-s_1) \\ 2lm & l^2 - m^2 & 2lm(s-s_1) \end{array} \right] \tag{5.35}$$

Watanabe carried out a numerical analysis of the punch problem, as shown in Figure 5.11, which was studied by Kawai and Toi using the RBSM model [11]. The analysis was carried out using the same mesh pattern and it was found that the sequence of the slip lines growth versus the corresponding external load was different from Kawai and Toi's result, but he was obtained the same collapse load. Figure 5.12 is the enlarged displacement

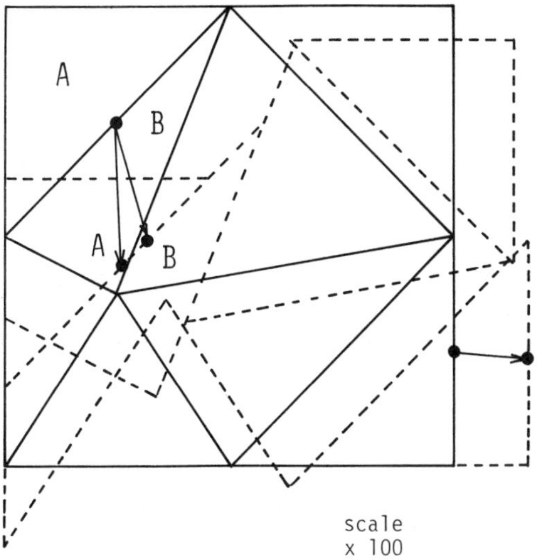

Figure 5.12 Deformation at collapse mode (the effect of displacement θ is not included)

pattern from which the relative displacement can be seen along the sliding edge.

It should also be mentioned that the complicated overlapping of the common vertices of the triangular elements is due to incomplete satisfaction of the displacement continuity, and the angular displacement θ is not shown in Figure 5.12. The external work done due to the unit enforced displacement W can be given by

$$W = R_m L_0$$

where L_0 is the length of the loading edge and R_n is the reaction force.

Denoting the relative slip along each slip line by Δv, the internal dissipative energy can be given by

$$U = \sum K\,|\Delta v_i|\,L_i$$

where $\sum$ implies the summation with respect to all slip lines and L_i is the length of the individual edges. It is interesting to note that the same collapse load of the RBSM solutions can be obtained by equating U and W.

This is a good example showing that the collapse load can be calculated immediately if the final collapse mode is known *a priori.* Similarly, Watanabe derived a simple plate-bending element including the effect of the shear deformation by assuming a simple stress field under equilibrium in the plate element and applying the principle of complementary energy.

For the inplane stress field (σ_x, σ_y, τ_{xy}) a linear distribution is assumed and shearing stresses τ_{xz}, τ_{yz} are assumed in order to satisfy the equilibrium condition with these stresses. Neglecting the body forces the following stress fields are assumed:

$$\begin{aligned}
\sigma_x &= z(a + bx + cy) \\
\sigma_y &= z(d + ex + fy) \\
\tau_{xy} &= z(g + hx + iy) \\
\tau_{xz} &= \frac{b+i}{2}\left(\frac{h^2}{4} - z^2\right) \\
\tau_{yz} &= \frac{h+f}{2}\left(\frac{h^2}{4} - z^2\right)
\end{aligned} \tag{5.36}$$

where $a \sim i$ imply the generalized stressed from which $\mathbf{G}$ defined by equation (5.27) can be obtained (see Figure 5.13).

The line integral along the element boundary edges defined by equation (5.28) can be given by the following equation:

$$\iint_{\partial\Omega_e} T_i u_i \,ds\,dz = \iint_{\partial\Omega_e} \{(\sigma_x l^2 + 2\tau_{xy} lm + \sigma_y m^2) U_n + [(-\sigma_x + \sigma_y) lm + \tau_{xy}(l^2 - m^2)] V_s + (\tau_{xz} l + \tau_{yz} m) w\}\,ds\,dz \tag{5.37}$$

The displacement (U_x, V_y, W) along the element boundary edge can be expressed by the rigid body displacement (w, θ_x, θ_y) of a given element as follows:

$$
\begin{aligned}
U_x &= z\theta_y \\
V_y &= -z\theta_x \\
W &= w + (y - y_1)\theta_x - (x - x_1)\theta_y
\end{aligned}
\tag{5.38}
$$

where the neutral plane is defined by $z = 0$ and x_1, y_1 are the coordinates of the mid-span point.

Tangential as well as normal components on the element boundary will be given by

$$U_n = z\theta_s, \qquad V_s = -z\theta_n \tag{5.39}$$

where θ_n, θ_s are the rotation angle around n axis and s axis respectively and are defined as follows:

$$
\begin{aligned}
\theta_n &= l\theta_x + m\theta_y \\
\theta_s &= -m\theta_x + l\theta_y
\end{aligned}
\tag{5.40}
$$

Solving the above equation with respect to θ_x and θ_y and substituting them into the third equation of (5.38), the following equation is obtained:

$$W = w + (s - s_1)\theta_n \tag{5.41}$$

It should be noted here that the element boundary edge can be specified by $n = n_1$. Equations (5.38) and (5.39) are the final displacement fields assumed on the element boundaries expressed by w, θ_n, θ_s which are defined with the z, n, s local coordinates. Substituting these equations with equation (5.36) into equation (5.37), **L** defined by equation (5.28) can be calculated. The hinge lines can be expressed by the method described previously.

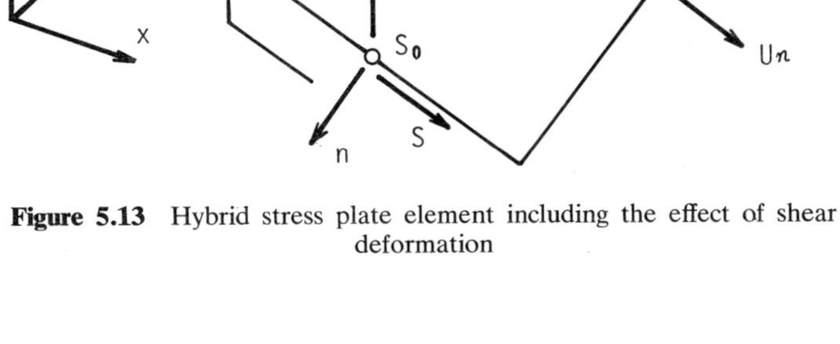

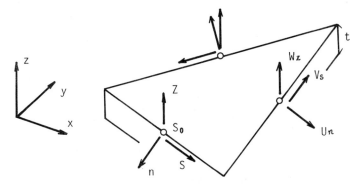

Figure 5.13 Hybrid stress plate element including the effect of shear deformation

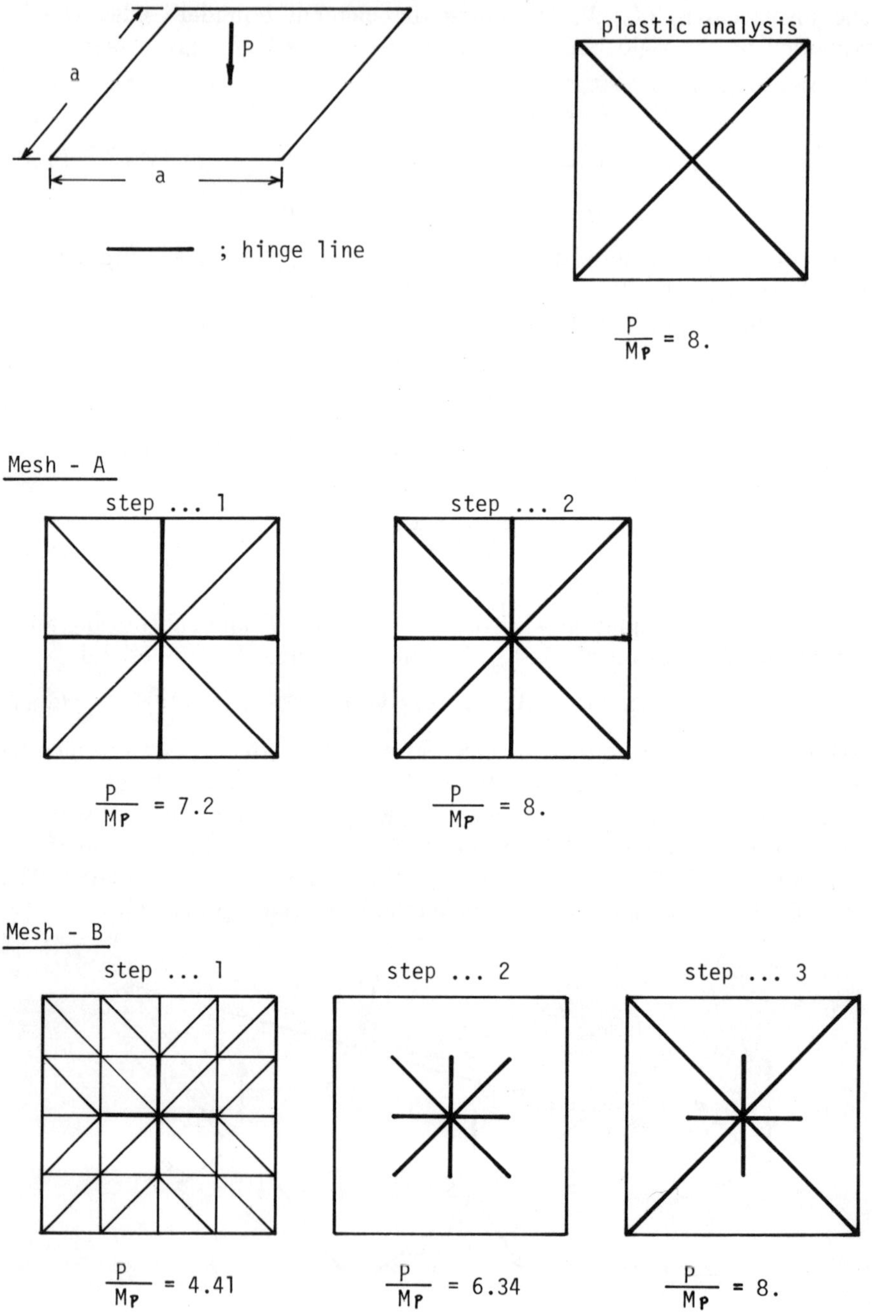

Figure 5.14 Hinge line analysis of a simply supported square plate under a concentrated load

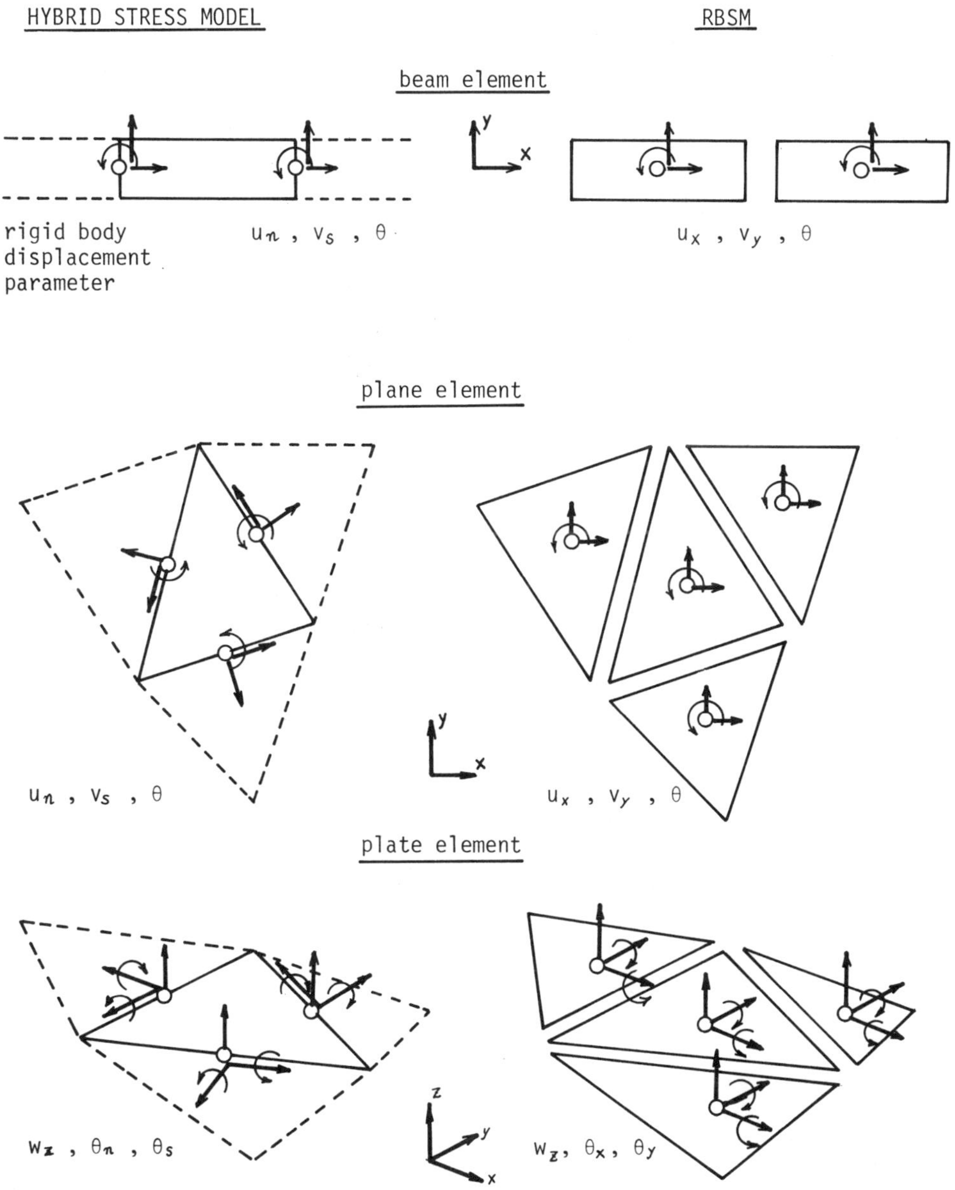

Figure 5.15 Comparisons of the rigid displacement parameters

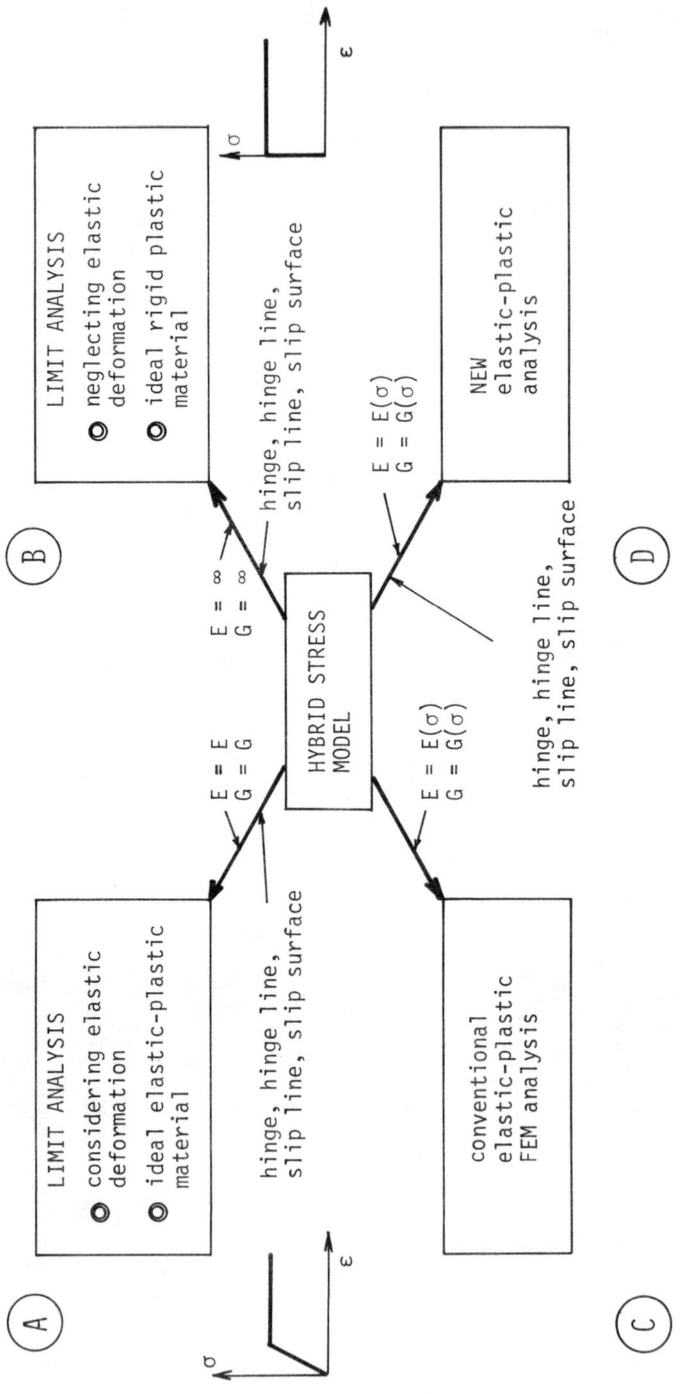

Figure 5.16 Four different elastic–plastic analyses

For a verification study of this element, a result is shown in Figure 5.14 on the limit load analysis of a simply supported square plate subjected to a concentrated load at the centre. It should be mentioned here that the concentrated load was equally divided to all the edges connected to the loading point by substituting the coordinates into equation (5.41) in this calculation, since the loading point was not located at any node of the discretized plate model.

In conclusion, the plate elements derived by Watanabe are incompatible displacement models which satisfy the displacement continuity only at the mid-span points. They correspond to the RBSM plate elements as shown in Fig. 5.15, which will yield only the upper bound solutions of true collapse loads. It should, however, be mentioned that an almost unlimited number of the hybrid stress models can be obtained because the element stress field as well as the displacement field along the boundary edge are by no means restricted to those assumed in the above formulations. The more elaborate elements may be derived by using the hybrid stress methods.

Finally, Watanabe discussed four possible elastoplastic analyses from the standpoint of the hybrid stress method as shown in Figure 5.16. The method proposed by him belongs to method A where the element deformation as well as the displacement discontinuity along interelement boundaries can be considered, and it is intended to obtain the upper bound of the collapse loads by assuming that the material is perfectly elastoplastic.

Method B may also exist where deformation of the elements is not considered and therefore the material is assumed to be perfectly rigid plastic (Young's modulus is considered infinitely large). The RBSM belongs to this method where transmission of forces are treated approximately.

In method C discontinuity of the displacements such as slip is neglected, but yielding in the element domain is considered. The conventional finite element method is naturally included in method C. A better elastoplastic analysis may be expected by using the present hybrid stress models than the conventional FEM, because the hybrid stress models whose nodes are located at the mid-span points may satisfy the force equilibrium among elements more rigorously than the conventional FEM where element nodes are usually taken at vertices of the elements.

Finally, method D is a completely new method of analysis where consideration of the pertinent yield criterion in the element domain and introduction of the displacement discontinuity are made at the same time. This kind of method may be very effective in analysis of polycrystalline structures.

5.4 CONCLUSION

It is pointed out that the stress models are essentially superior to the displacement models in the inelastic analysis of solids where consideration of the slip lines or surfaces is inevitable in order to obtain the limit loads.

For this purpose a series of new stress models have been developed based on the hybrid variational principles in the author's laboratory. On focusing attention to the platc problems some typical elements proposed by Kondou and Watanabe are described together with the results of their verification studies.

ACKNOWLEDGEMENTS

The author would like to express his appreciation to Dr. Yutaka Toi of the University of Tokyo, Dr. Kazuo Kondou of Hiroshima University, Dr. Masaaki Watanabe of the Mitsubishi Research Institute, and Mr. Norio Takeuchi of the Institute of Industrial Science, University of Tokyo, for their cooperation in carrying out the present studies. He also wishes to express his thanks to Miss Sueko Suzuki for typing the manuscript and drawing the figures.

REFERENCES

1. T. Kawai, 'Some considerations on the finite element method', *Int. J. Num. Meth. in Eng.*, **16,** 81–120 (1980).
2. K. Kondou, *Basic Studies on the Discrete Analysis of Beam and Plate Elements by Using the Lower Order Shape Functions,* Doctorate thesis submitted to the University ot Tokyo (in Japanese), 1977.
3. N. Takeuchi and T. Kawai, 'A method for the error evaluation of the new discrete limit analysis solutions', *J. Seisan Kenkyu, Institute of Industrial Sci., Univ. of Tokyo* (in Japanese), **33,** No. 2 (1981).
4. K. Kondou, 'A new discrete method on collapse load analysis of bending plates', *Bull. Faculty of Engineering, Hiroshima University,* **29,** No. 1 (1980).
5. K. Washizu, *Introduction to the Variational Principles in Elasticity, a Series of Textbooks on the Computational Structural Engineering* (in Japanese), II-3-A, Baifukan Press, Tokyo, 1972.
6. L. R. Herrmann, 'Finite element bending analysis of plates', *Proc. ASME,* **93,** No. EM5 (1967).
7. M. Watanabe and T. Kawai, 'A discrete method of analysis of fluid dynamics problem (2nd Report)—Finite element analysis of incompressible viscous flows by using stream function with triangular linear element', *J. Soc. Naval Architects of Japan,* **143** (1978).
8. K. Kondou, S. Shiina, and T. Kawai, 'A new discrete plate bending model by using a lower order shape function', *J. Soc. Naval Architects of Japan,* **143,** 259–265 (1978).
9. Y. Mukudai, A. Matsuo, and K. Kondou, 'Local buckling analysis of a plate girder by using a new discrete plate bending model', *Proc. of the Research Reports of the Chugoku Branch, the Architectural Institute of Japan,* **7,** No. 2 (1980).
10. M. Watanabe and T. Kawai, 'Treatment of slip lines, hinges and hinge lines using new hybrid stress models', *J. Soc. of Naval Architects of Japan,* **147,** 309–317 (1980).
11. T. Kawai and Y. Toi, 'A new element in discrete analysis of plane strain problems', *J. Seisan Kenkyu, Institute of Industrial Sci., Univ. of Tokyo,* **29,** No. 4 (1977).

Hybrid and Mixed Finite Element Methods
Edited by S. N. Atluri, R. H. Gallagher, and O. C. Zienkiewicz

Chapter 6

Bilinear Mindlin Plate Elements

Robert L. Spilker and Ted Belytschko

6.1 INTRODUCTION

The development of plate-bending elements based on independent transverse displacement and cross-section rotations (corresponding to Mindlin plate theory [1]), in conjunction with the assumed displacement model, has received considerable attention because the displacement interpolation must satisfy only C^0 continuity. Thus, standard Serendipity and Lagrange interpolation families can be used. In order to avoid 'locking' (severe element stiffening) in the thin plate limit, reduced or selective reduced integration concepts have been applied. However, spurious zero energy modes and/or continued locking are found in most of these assumed displacement elements (see, for example, [2 to 5]).

To circumvent the difficulties of spurious modes and/or continued locking, it would appear that non-conventional schemes must be employed. For example, Hughes and coworkers [6, 7] make use of different orders of interpolations for transverse displacement and cross-section rotations. Here, we will examine two schemes which produce non-locking elements without potentially harmful zero energy modes: (a) use of the hybrid stress model and (b) addition of a stabilization matrix to the assumed displacement model. Emphasis for both schemes will be on the four-node plate element.

Since its introduction by Pian in 1964 [8], the hybrid stress model has been used successfully for a variety of developments including linear and non-linear behaviour, static and dynamic loading. Briefly, the hybrid stress model is based on a modified complementary energy principle and requires the independent interpolation of intraelement equilibrating stresses and compatible element boundary (or intraelement, if convenient) displacements. Stress parameters are eliminated at the element level and a stiffness matrix results.

An area where hybrid stress elements are particularly advantageous is the analysis of plates. For thin plates, Kirchhoff-type triangular and quadrilateral elements are possible since C' continuity interpolations are more easily constructed on the element boundaries. Examples of element developments and/or numerical comparisons are found in [9 to 12]. For moderately thick or thick plates transverse shear effects can be incorporated by assuming independent transverse displacement and cross-section rotations and by including all components of stress. Developments of plate elements in this category are found in [13 to 16]; some of these elements have also been extended to the analysis of moderately thick laminated composite plates [13, 17 to 19].

Results obtained in these earlier studies of hybrid stress Mindlin-type plate elements suggested that some of these hybrid stress elements were 'stiff' when applied to thin plate problems (i.e. typical $h/L=0.01$) while others were not. Furthermore, some of these elements exhibited rank deficiency and hence spurious zero energy modes. However, no particular emphasis was given to behaviour in the thin plate limit, and no explanation of the sporadic element performance was given. Recently, Spilker and Munir [20 to 22] have reexamined hybrid stress elements of this type to demonstrate the constraints imposed as $h \rightarrow 0$. A rationale has also been suggested for selection of the element stress field so that the resulting element is non-locking and the stiffness is of the correct rank. Here, we summarize those developments and illustrate the behaviour obtained for the bilinear plate element.

The suppression of kinematic modes by means of stabilization matrices is also examined. Although various methods of construction of stabilization matrices have been reported, we will here confine our attention to those obtained by taking a linear combination of a locking element and a rank deficient element. In particular, for the four-node bilinear displacement element, it has been verified by numerical experiments that if the scaling is properly chosen, an element which does not lock and exhibits no spurious behaviour may easily be developed.

6.2 HYBRID STRESS PLATE ELEMENTS

6.2.1 Definition of element matrices

The formulation and matrix definitions for Mindlin-type hybrid stress plate elements can be found elsewhere [20 to 22]. However, for the sake of completeness, the initial assumptions and final matrix definitions will be repeated. The hybrid stress functional, π_{mc}, may be stated in the form (for no body forces)

$$\pi_{\mathrm{mc}}=\sum_n \left(\frac{1}{2}\int_{V_n} \boldsymbol{\sigma}^T \mathbf{S} \boldsymbol{\sigma}\, \mathrm{d}V - \int_{V_n} \boldsymbol{\sigma}^T \hat{\boldsymbol{\varepsilon}}\, \mathrm{d}V + \int_{S_{\sigma_n}} \mathbf{u}^T \bar{\mathbf{T}}\, \mathrm{d}s\right) \tag{6.1}$$

where $\boldsymbol{\sigma}$ is the vector of stress components, $\hat{\boldsymbol{\varepsilon}}$ is the vector of strain components obtained, via the strain–displacement relations, from displacements $\mathbf{u}$, $\mathbf{S}$ is the material property matrix ($\boldsymbol{\varepsilon}=\mathbf{S}\boldsymbol{\sigma}$ where $\boldsymbol{\varepsilon}$ are the actual

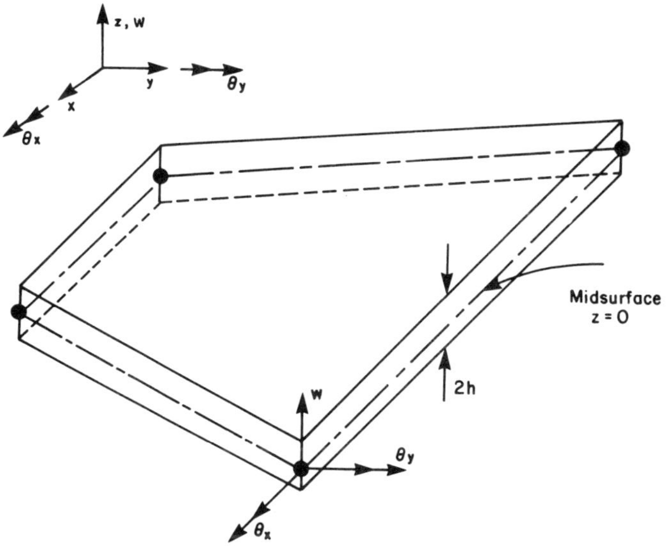

Figure 6.1 Geometry and degrees of freedom for a four-node isoparametric plate-bending element

strains), V_n is the volume of the nth element, and S_{σ_n} is the portion of the element boundary on which tractions $\bar{\mathbf{T}}$ are prescribed. Note also that the second term in equation (6.1) replaces the usual boundary surface integral. This is appropriate since C^0 continuity intraelement displacement interpolations can be easily constructed for the present elements.

The plate element is assumed to lie in the xy plane, with $z = 0$ corresponding to the geometric mid-surface (Figure 6.1). The through-thickness displacement behaviour is expressed in terms of the transverse displacement w and cross-section rotations θ_x and θ_y, about the x and y axes respectively, in the form:

$$\begin{aligned} u(x, y, z) &= z\theta_y(x, y) \\ v(x, y, z) &= -z\theta_x(x, y) \\ w(x, y, z) &= w(x, y) \end{aligned} \tag{6.2}$$

The strains $\hat{\boldsymbol{\varepsilon}}$ can be related to displacements $\mathbf{u}$ via the strain–displacement relations. Alternatively, the effects of z can be extracted from $\hat{\boldsymbol{\varepsilon}}$ yielding strains $\hat{\hat{\boldsymbol{\varepsilon}}}$ which are related to w, θ_x, and θ_y. From equations (6.2), $\hat{\boldsymbol{\varepsilon}}$ will be related to $\hat{\hat{\boldsymbol{\varepsilon}}}$ by

$$\begin{aligned} \lceil \hat{\boldsymbol{\varepsilon}}_x \hat{\boldsymbol{\varepsilon}}_y \hat{\gamma}_{xy} \rfloor &= z \lceil \hat{\hat{\varepsilon}}_x \hat{\hat{\varepsilon}}_y \hat{\hat{\gamma}}_{xy} \rfloor \\ \lceil \hat{\gamma}_{xz} \hat{\gamma}_{yz} \rfloor &= \lceil \bar{\gamma}_{xz} \bar{\gamma}_{yz} \rfloor \\ \hat{\varepsilon}_z &= 0 \end{aligned} \tag{6.3}$$

In view of the first of equations (6.3), the inplane stresses should be assumed linear in z. Since stresses must satisfy the three-dimensional homogeneous equilibrium equations in hybrid stress elements, transverse shear stresses must be of the order z^2 and transverse normal stresses of the order z^3. In addition, transverse shear stresses are zero at the upper/lower plate surface and transverse normal stress is zero at the lower plate surface, corresponding to a transversely loaded plate. Corresponding to these assumptions, the stresses can be expressed in the form (for a plate of thickness $2h$):

$$\begin{aligned}
\sigma_x(x, y, z) &= z\bar{\sigma}_x(x, y) \\
\sigma_y(x, y, z) &= z\bar{\sigma}_y(x, y) \\
\sigma_{xy}(x, y, z) &= z\bar{\sigma}_{xy}(x, y) \\
\sigma_{xz}(x, y, z) &= \tfrac{1}{2}(h^2 - z^2)\left(\frac{\partial\bar{\sigma}_x}{\partial x} + \frac{\partial\bar{\sigma}_{xy}}{\partial y}\right) = \tfrac{1}{2}(h^2 - z^2)\bar{\sigma}_{xz}(x, y) \\
\sigma_{yz}(x, y, z) &= \tfrac{1}{2}(h^2 - z^2)\left(\frac{\partial\bar{\sigma}_{xy}}{\partial x} + \frac{\partial\bar{\sigma}_y}{\partial y}\right) = \tfrac{1}{2}(h^2 - z^2)\bar{\sigma}_{yz}(x, y) \\
\sigma_z(x, y, z) &= \tfrac{1}{6}(z^3 - 3h^2 z - 2h^3)\left(\frac{\partial\bar{\sigma}_{xz}}{\partial x} + \frac{\partial\bar{\sigma}_{yz}}{\partial y}\right) = \tfrac{1}{6}(z^3 - 3h^2 z - 2h^3)\bar{\sigma}_z(x, y)
\end{aligned} \tag{6.4}$$

Equations (6.2), (6.3), and (6.4) are substituted into π_{mc} and all through-thickness integrations are carried out analytically. The resulting expression for π_{mc} is

$$\pi_{\text{mc}} = \sum_n \left[\frac{2h^3}{3}\left(\frac{1}{2}\int_{A_n} \bar{\boldsymbol{\sigma}}^T \bar{\mathbf{S}} \bar{\boldsymbol{\sigma}}\, \mathrm{d}A - \int_{A_n} \bar{\boldsymbol{\sigma}}^T \hat{\boldsymbol{\varepsilon}}\, \mathrm{d}A\right) + \int_{A_n} wp(x, y)\, \mathrm{d}A\right] \tag{6.5}$$

where A_n is the mid-surface area of the element, $p(x, y)$ is the prescribed transverse load distribution, and $\bar{\mathbf{S}}$ is given by (assuming the ordering $\bar{\boldsymbol{\sigma}}^T = \lfloor \bar{\sigma}_x \bar{\sigma}_y \bar{\sigma}_z \bar{\sigma}_{xy} \bar{\sigma}_{xz} \bar{\sigma}_{yz} \rfloor$,

$$\bar{\mathbf{S}} = \frac{1}{E}\begin{bmatrix}
1 & & & & & \\
-\nu & 1 & & & \text{SYM} & \\
\nu\left(\dfrac{2h^2}{5}\right) & \nu\left(\dfrac{2h^2}{5}\right) & \dfrac{52}{105}h^4 & & & \\
0 & 0 & 0 & 2(1+\nu) & & \\
0 & 0 & 0 & 0 & \dfrac{4h^2}{5}(1+\nu) & \\
0 & 0 & 0 & 0 & 0 & \dfrac{4h^2}{5}(1+\nu)
\end{bmatrix} \tag{6.6}$$

The transverse displacement and rotations are interpolated within the element in terms of nodal parameters, $\mathbf{q}$,

$$\begin{Bmatrix} w \\ \theta_x \\ \theta_y \end{Bmatrix} = \mathbf{N}\mathbf{q} \tag{6.7}$$

where $\mathbf{N}(\xi, \eta)$ is expressed in the parent (ξ, η) coordinates using any of the standard C^0 interpolation functions. The usual isoparametric mapping is applied so that

$$\begin{aligned} x &= \sum_i N_i(\xi, \eta)x_i \\ y &= \sum_i N_i(\xi, \eta)y_i \end{aligned} \tag{6.8}$$

where x_i, y_i are the global coordinates of node i. The strains $\hat{\boldsymbol{\varepsilon}}$ can then be related to $\mathbf{q}$ by

$$\hat{\boldsymbol{\varepsilon}} = \mathbf{D}\begin{Bmatrix} w \\ \theta_x \\ \theta_y \end{Bmatrix} = \mathbf{DNq} = \frac{1}{|J|}\mathbf{B}^*\mathbf{q} \tag{6.9}$$

where $|J|$ is the Jacobian of the coordinate transformation (equation 6.8) and $\mathbf{D}$ is the differential operator matrix defined in detail in [20].

The stresses $\bar{\boldsymbol{\sigma}}$ are interpolated in terms of stress parameters, $\boldsymbol{\beta}$, in the form

$$\bar{\boldsymbol{\sigma}} = \mathbf{P}\boldsymbol{\beta} \tag{6.10}$$

In the construction of $\mathbf{P}(x, y)$, $\bar{\sigma}_x$, $\bar{\sigma}_y$, and $\bar{\sigma}_{xy}$ are first interpolated in terms of $\boldsymbol{\beta}$. The remaining stresses, $\bar{\sigma}_{xz}$, $\bar{\sigma}_{yz}$, and $\bar{\sigma}_z$, are then related to $\boldsymbol{\beta}$ using the last three of equations (6.4), which guarantees that equilibrium is satisfied. The result is collected in matrix form to yield $\mathbf{P}$. Also, $\mathbf{P}(\xi, \eta)$ can be obtained by using equations (6.8).

Equations (6.7), (6.9), and (6.10) are substituted into equations (6.5) and the following matrices are defined (using the mapping of equation 6.8):

$$\mathbf{H} = \frac{2h^3}{3}\iint_{-1}^{1} \mathbf{P}^T\bar{\mathbf{S}}\mathbf{P}\,|J|\,\mathrm{d}\xi\,\mathrm{d}\eta \tag{6.11a}$$

$$\mathbf{G} = \frac{2h^3}{3}\iint_{-1}^{1} \mathbf{P}^T\mathbf{B}^*\,\mathrm{d}\xi\,\mathrm{d}\eta \tag{6.11b}$$

$$\mathbf{Q} = \iint_{-1}^{1} \mathbf{N}^T p(x, y)\,|J|\,\mathrm{d}\xi\,\mathrm{d}\eta \tag{6.11c}$$

where $\mathbf{Q}$ is the element nodal load vector.

With these definitions, π_{mc} now appears in the standard form, so that $\boldsymbol{\beta}$ may be eliminated at the element level by

$$\mathbf{B} = \mathbf{H}^{-1}\mathbf{G}\mathbf{q} \tag{6.12}$$

and the elements stiffness matrix is given by

$$\mathbf{k} = \mathbf{G}^T\mathbf{H}^{-1}\mathbf{G} \tag{6.13}$$

6.2.2 Constraints in the thin plate limit

The contributions of plate thickness, h, have been identified explicitly in the formulation of the last subsection. It is first observed that no numerical difficulties are encountered in the formation of $\mathbf{k}$ as $h \to 0$, independent of the interpolations used for $\bar{\boldsymbol{\sigma}}$ and displacement/rotations. The contributions of $\bar{\sigma}_z$, $\bar{\sigma}_{xz}$, and $\bar{\sigma}_{yz}$ in $\mathbf{H}$ will become negligible (in view of the form of $\mathbf{S}$ given by equation 6.6) as $h \to 0$. However, since these stresses are related to the same $\boldsymbol{\beta}$ as the inplane stresses, the symmetric $\mathbf{H}$ matrix will remain positive definite. All terms in $\mathbf{G}$ receive the same contribution from h, and therefore the rank of $\mathbf{k}$, which is controlled by $\mathbf{G}$, is unaffected as $h \to 0$.

Constraints are, however, imposed on the element as $h \to 0$. These constraints have been identified in [20] by examining the Euler equations of π_{mc} as $h \to 0$ and correspond to

$$\int_{A_n} \lfloor \delta\bar{\sigma}_{xz} \; \delta\bar{\sigma}_{yz} \rfloor \begin{Bmatrix} \hat{\bar{\gamma}}_{xz} \\ \hat{\bar{\gamma}}_{yz} \end{Bmatrix} \mathrm{d}A = 0 \tag{6.14}$$

for $\delta\bar{\sigma}_{xz} \neq 0$ and $\delta\bar{\sigma}_{yz} \neq 0$. Since $\bar{\sigma}_{xz}$ and $\bar{\sigma}_{yz}$ are related to $\boldsymbol{\beta}$ and $\hat{\bar{\gamma}}_{xz}$ and $\hat{\bar{\gamma}}_{yz}$ are related to $\mathbf{q}$, equation (6.14) could be expressed as

$$\delta\boldsymbol{\beta}^T \int_{A_n} \mathbf{P}_s^T \mathbf{B}_s^* \, \mathrm{d}A\mathbf{q} = \delta\boldsymbol{\beta}^T \mathbf{G}_s \mathbf{q} = 0 \tag{6.15}$$

where $\mathbf{P}_s$ and $\mathbf{B}_s^*$ are the portions of $\mathbf{P}$ and $\mathbf{B}^*$ respectively corresponding to transverse shear stresses/strains. Equation (6.15) must be satisfied for $\delta\boldsymbol{\beta}^T \neq 0$, and therefore the constraints imposed as $h \to 0$ are given by

$$\mathbf{G}_s \mathbf{q} = \mathbf{0} \tag{6.16}$$

The number of constraints imposed as $h \to 0$ is equal to the number of independent equations in equation (6.16), which will always be less than the total number of β's. In view of equations (6.14) and (6.16), the number and type of constraints imposed as $h \to 0$ is controlled by the form of $\bar{\sigma}_{xy}(x, y)$ and $\bar{\sigma}_{yz}(x, y)$, which motivates the rationale of [20] for selection of stress interpolations so that no *artificial* constraints are imposed as $h \to 0$.

Constraint counting has proven useful in some constrained media problems (e.g. the 'constraint index' of Malkus and Hughes [23]). For the present elements, a measure termed the 'rotational constraint index' (RCI) [21] is believed useful. The RCI is computed by considering a hypothetical square plate, with two adjacent edges ideally clamped, modelled by an $N \times N$ mesh of elements. The RCI is then given by

$$\mathrm{RCI} = \frac{n_{\theta\,\mathrm{dof}} - n_{\theta\mathrm{b}}}{N^2} - n_{\theta\mathrm{e}} \tag{6.17}$$

where $n_{\theta\,\mathrm{dof}}$ is the number of θ degrees of freedom in the $N \times N$ mesh, $n_{\theta\mathrm{b}}$ is the number of θ degrees of freedom constrained on the adjacent clamped edges, and $n_{\theta\mathrm{e}}$ is the number of θ-only constraints per element as $h \to 0$ (equal to the total number of independent constraints in equation (6.16) minus the number of w contributions). An $\mathrm{RCI} \leqslant 0$ suggests that the element will lock.

6.2.3 Displacement and stress interpolations for four-node elements

Four-node isoparametric elements (Figure 6.1) are considered so that the displacement interpolations are given by

$$\begin{aligned} w &= \sum_{i=1}^{4} N_i(\xi, \eta) w_i \\ \theta_x &= \sum_{i=1}^{4} N_i(\xi, \eta) \theta_{x_i} \\ \theta_y &= \sum_{i=1}^{4} N_i(\xi, \eta) \theta_{y_i} \end{aligned} \tag{6.18}$$

where $N_i(\xi, \eta)$, $i = 1, 2, 3, 4$, represent the bilinear shape functions and w_i, θ_{x_i}, and θ_{y_i} are the degrees of freedom at node i.

In order to assess the constraints as $h \to 0$ using equation (6.16), it is convenient to consider an illustrative square element of side length 2 (so that $\xi = x$, $\eta = y$, and $|J| = 1$) and rewrite the displacement interpolation in the form

$$\begin{aligned} \theta_y &= \alpha_1 + \alpha_2 x + \alpha_3 y + \alpha_4 xy \\ \theta_x &= \alpha_5 + \alpha_6 x + \alpha_7 y + \alpha_8 xy \\ w &= \alpha_9 + \alpha_{10} x + \alpha_{11} y + \alpha_{12} xy \end{aligned} \tag{6.19}$$

In this form, the α_i are generalized displacement parameters which can be uniquely related to the nodal degrees of freedom. Using equations (6.19), the transverse shear strains (from displacements) are then (for the illustrative square element)

$$\begin{aligned} \hat{\hat{\gamma}}_{xz} &= \left(\frac{\partial w}{\partial x} + \theta_y\right) = (\alpha_1 + \alpha_{10}) + \alpha_2 x + (\alpha_3 + \alpha_{12}) y + \alpha_4 xy \\ \hat{\hat{\gamma}}_{yz} &= \left(\frac{\partial w}{\partial y} - \theta_x\right) = (\alpha_{11} - \alpha_5) + (\alpha_{12} - \alpha_6) x - \alpha_7 y - \alpha_8 xy \end{aligned} \tag{6.20}$$

Strictly speaking, in order that the element stiffness be invariant (i.e. independent of the location and orientation of the global xy system chosen) the stress field should be based on complete polynomials [24]. The minimum complete polynomials for $\bar{\sigma}_x$, $\bar{\sigma}_y$, and $\bar{\sigma}_{xy}$ which will preserve invariance and maintain correct rank is quadratic, so that

$$\begin{aligned}
\bar{\sigma}_x &= \beta_1 + \beta_2 x + \beta_3 y + \beta_4 xy + \beta_5 x^2 + \beta_6 y^2 \\
\bar{\sigma}_y &= \beta_7 + \beta_8 x + \beta_9 y + \beta_{10} xy + \beta_{11} x^2 + \beta_{12} y^2 \\
\bar{\sigma}_{xy} &= \beta_{13} + \beta_{14} x + \beta_{15} y + \beta_{16} xy + \beta_{17} x^2 + \beta_{18} y^2 \\
\bar{\sigma}_{xz} &= (\beta_2 + \beta_{15}) + (2\beta_5 + \beta_{16})x + (\beta_4 + 2\beta_{18})y \\
\bar{\sigma}_{yz} &= (\beta_9 + \beta_{14}) + (\beta_{10} + 2\beta_{17})x + (2\beta_{12} + \beta_{16})y \\
\bar{\sigma}_z &= 2(\beta_5 + \beta_{12} + \beta_{16})
\end{aligned} \tag{6.21}$$

The constraint conditions as $h \to 0$ for an illustrative square element can be evaluated using equations (6.20), $\bar{\sigma}_{xz}$ and $\bar{\sigma}_{yz}$ from equations (6.21), and equation (6.16), and are

$$\begin{aligned}
(\beta_2 + \beta_{15})&: \quad \alpha_1 + \alpha_{10} = 0 \\
(2\beta_5 + \beta_{16})&: \quad \alpha_2 = 0 \\
(\beta_4 + 2\beta_{18})&: \quad \alpha_3 + \alpha_{12} = 0 \\
(\beta_9 + \beta_{14})&: \quad \alpha_{11} - \alpha_5 = 0 \\
(\beta_{10} + 2\beta_{17})&: \quad \alpha_{12} - \alpha_6 = 0 \\
(2\beta_{12} + \beta_{16})&: \quad \alpha_7 = 0
\end{aligned} \tag{6.22}$$

where, for convenience, the β_i associated with each constraint condition have been identified.

The set of constraints imposed via equations (6.22) as $h \to 0$ are excessive. This can be seen by computing the RCI. For the four-node elements $n_{\theta\,\mathrm{dof}} = 2(N+1)^2$ and $n_{\theta\mathrm{b}} = 2(2N+1)$ so that the first term in equation (6.17) is 2. Equations (6.22) represent six constraints in terms of three w contributions (that is α_{10}, α_{11}, and α_{12}) so that $n_{\theta\mathrm{e}} = 3$ and RCI $= -1$; locking should be expected. It is also of interest to consider the form of the constraints. The first, third, fourth, and fifth conditions in equations (6.22) represent realistic Kirchhoff constraints between w and θ degrees of freedom which should be imposed as $h \to 0$ (balanced terms). However, the second and sixth of equations (6.22) are unrealistic (non-Kirchhoff) constraints which may tend to overconstrain element behaviour as $h \to 0$. This is the motivation for the rationale suggested in [20]; stress distributions should be selected so that only realistic Kirchhoff constraints are imposed. This suggests that the xy terms in $\bar{\sigma}_{xz}$ and $\bar{\sigma}_{yz}$ should contain a constant and y term and $\bar{\sigma}_{yz}$ should contain a constant and x term.

In view of the excessive constraints implied by the use of equations (6.21), several reduced stress fields will be considered here. The reduced stress fields and corresponding four-node elements are summarized as follows.

6.2.3.1 *Element LH3*

This 9β element is obtained by setting β_4 to β_6, β_{10} to β_{12}, and β_{16} to β_{18} to zero. The element is invariant and has a RCI = 3. However, the element possesses two spurious zero energy modes: a θ mode which is constrained in an assemblage of elements and a w mode which remains in an assemblage of elements can can act up in selected examples [14].

6.2.3.2 *Element LH4*

This 11β element is obtained by setting β_5, β_6, β_{11}, β_{12}, β_{16}, β_{17}, and β_{18} to zero. The element stiffness is of correct rank and RCI = 1. However, the element is not naturally invariant. Note that the stress field of this element follows exactly the rationale described earlier.

6.2.3.3 *Element LH5*

This 12β element is obtained by setting β_5, β_6, β_{11}, β_{12}, β_{17}, and β_{18} to zero. The element stiffness is of correct rank and RCI = 0; locking may be expected in some problems. The element is not naturally invariant.

6.2.3.4 *Element LH11*

This 12β element is obtained by setting β_6, β_{11}, β_{16}, β_{17}, and β_{18} to zero and requiring that $\beta_{12} = -\beta_5$. The element stiffness is of correct rank and RCI = 0. Locking may be expected for some problems, but it should be noted that the *form* of the constraints for this element differ from those of element LH5; elements LH5 and LH11 have been included here to illustrate the effects of constraints on element behaviour as $h \rightarrow 0$.

It should be noted that an invariant four-node element with correct rank can be obtained by eliminating $\bar{\sigma}_z$ in equations (6.21) according to $\beta_{16} = -(\beta_5 + \beta_{12})$. The resulting 17β element has a value of RCI = 0 with constraint conditions identical to element LH11; this potential element has therefore not been included in this study.

6.2.3.5 *Element DISP*

This element is the assumed displacement four-node element of Hughes, Taylor, and Kanoknulchui [3] which employs a 2×2 Gauss rule for flexural contributions and a one-point rule for transverse shear contributions (i.e. selective reduced integration). The element possesses two spurious zero energy modes as in LH3. It would appear that this element has a value of RCI = 3.

6.2.4 Example problems and numerical results

To illustrate element performances, limited results are presented for a square plate of side length $2L$ and total thickness $2h$, for which a uniform $N_{el} \times N_{el}$ mesh is used in one quadrant (Figure 6.2). Boundary conditions correspond to simply supported (SS) or clamped (C); either a uniform transverse load (UL) or centre concentrated load (CL) is applied. For all cases, $L = 5$ in, material properties are $E = 10^7$ p.s.i., $\nu = 0.3$, and h is varied to achieve needed L/h values. Predicted values are normalized by thin plate theory solutions [25].

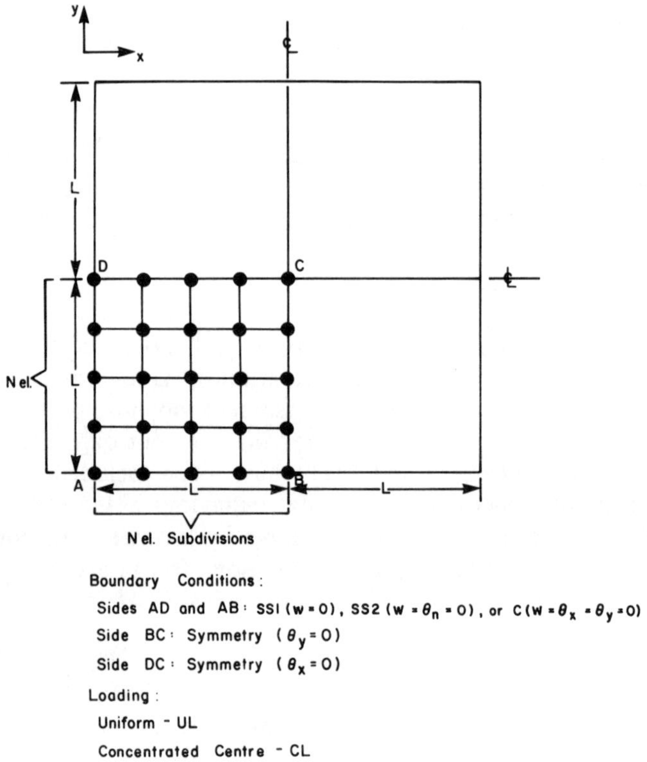

Figure 6.2 Mesh, boundary conditions, and loading for the square plate problems

The effects of plate thickness ratio on predicted centre transverse displacement, using a 4×4 element mesh, for the four cases SS-UL, SS-CL, C-UL, and C-CL are shown in Figures 6.3(a) to (d). Elements LH3 and LH4 approach the thin-plate theory (TPT) solution for $(L/h) = 10^2$ and are unaffected as L/h is increased to arbitrarily large values. Elements LH5 and LH11 produce excessively stiff solutions for $(L/h) > 10^2$ in the SS and/or C cases. The error is, however, independent of L/h which suggests that increased mesh refinement can lead to a converged solution. Recall that LH5 and LH11 impose a single (different) artificial constraint, and each has a value of RCI = 0. It may also be concluded that the optimal value of RCI is 1 for four-node elements; values of RCI > 1 (LH3) are accompanied by spurious zero energy modes, and values of RCI < 1 produce stiff solutions (LH5 and LH11).

Convergence of the centre transverse displacement for the four cases and elements LH3, LH4, and DISP are shown in Figures 6.4(a) to (d). Here a thin plate ($L/h = 10^2$) is considered. All elements yield rapidly converging solutions and, in general, LH4 is superior to LH3. Note that in all cases the hybrid stress elements converge from above whereas DISP converges from below.

Further results obtained using these four-node elements are found in [20]. Note that it should be expected that these four-node elements will produce optimal stress–strain results at the element centroid [24]; stress results in [20] also suggest good intraelement distributions for the dominant stresses. Of the hybrid stress elements, LH4 would appear to be the best choice because of correct rank and non-locking performance. The lack of invariance of LH4 could be important in some applications, and should be studied.

Finally, it should be noted that the apparent difficulties in obtaining a four-node hybrid stress element which is non-locking, of correct rank, *and* naturally invariant do not occur in eight-node elements. A study of eight-node hybrid stress elements possessing all of these favourable characteristics is found in [26].

6.3 STABILIZATION OF RANK DEFICIENT ELEMENTS

As can be seen from the numerical examples in Section 6.2, rank deficient elements often offer the most accuracy. Therefore, they are quite appealing from a practical viewpoint. However, for certain geometries and support conditions, the rank deficiency of these elements leads to either singular assembled stiffness matrices or to assembled matrices which are so nearly singular that they exhibit solutions with large spatial oscillations.

One remedy for this difficulty is to add a stabilization matrix to rank deficient stiffnesses so that correct rank is obtained yet locking is avoided.

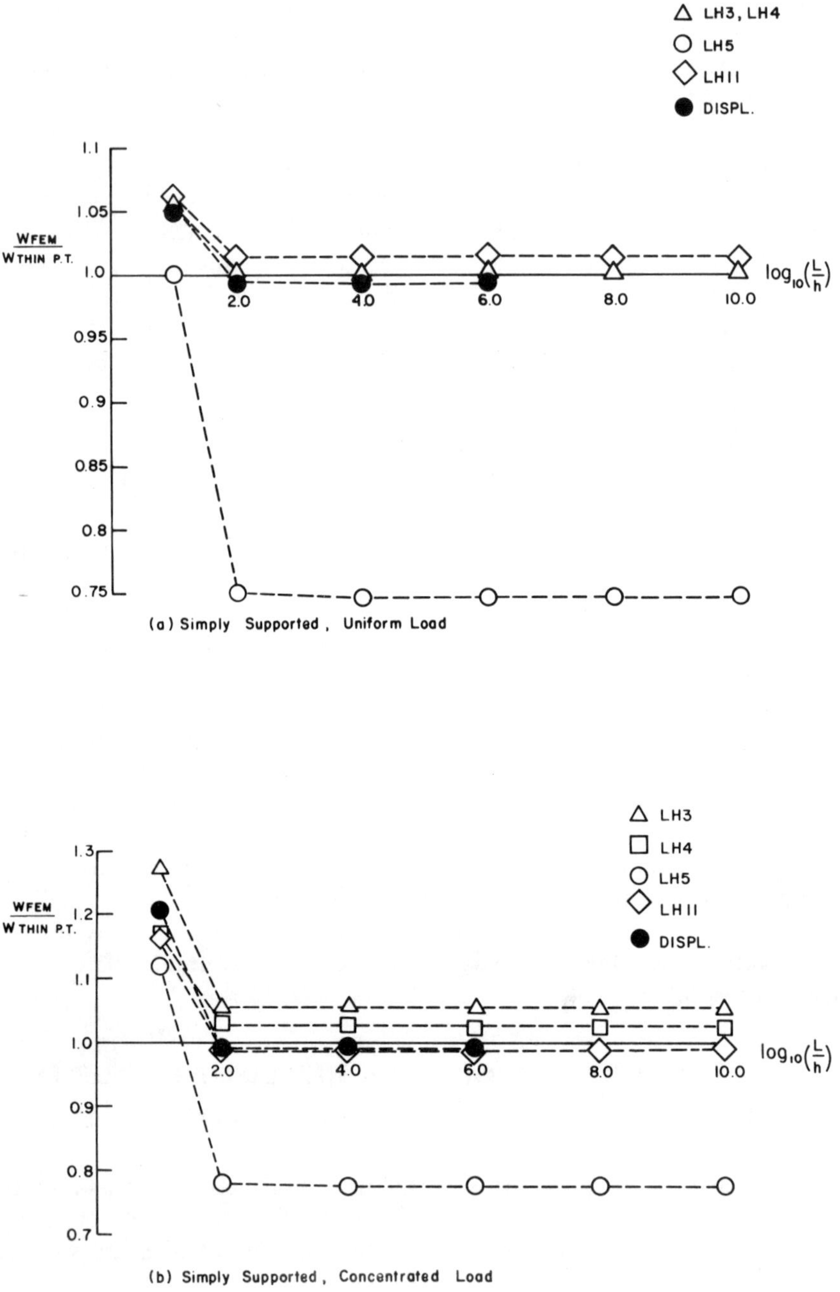

Figure 6.3 Predicted plate centre transverse displacement, normalized by

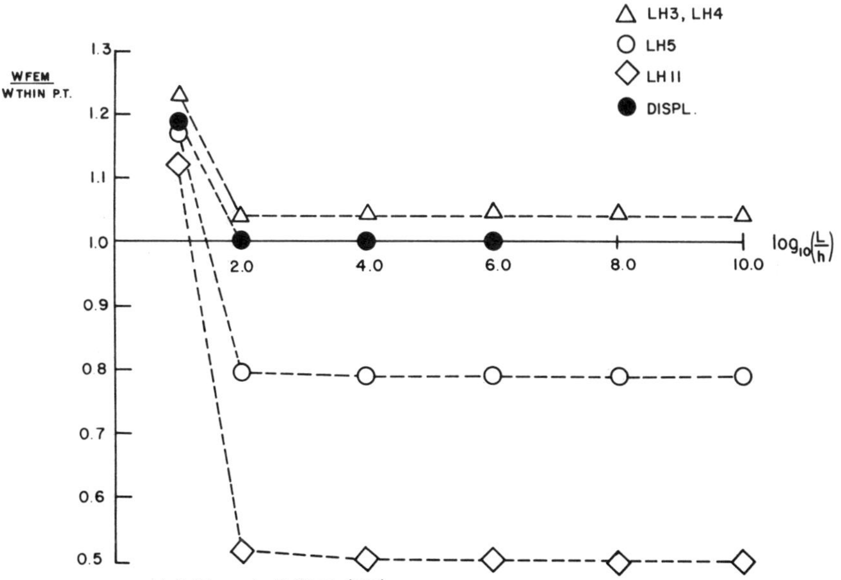

(c) Clamped, Uniform Load

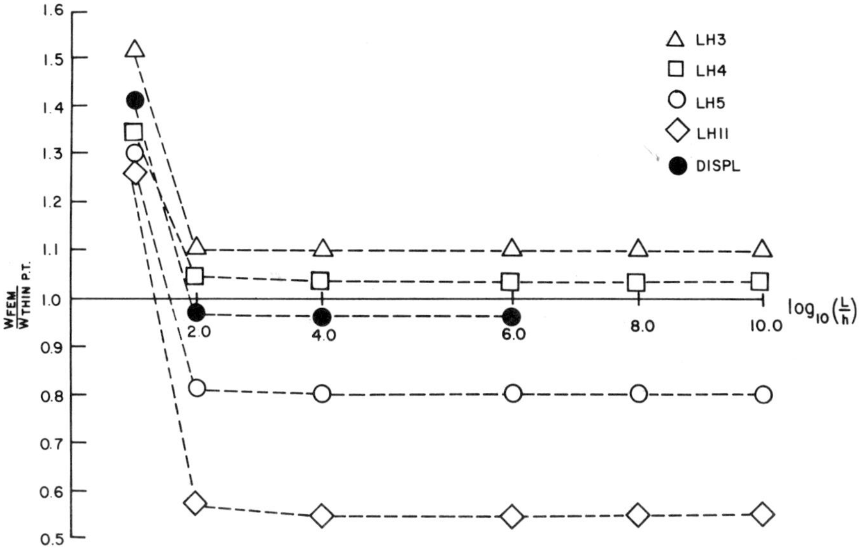

(d) Clamped, Concentrated Load

the thin plate theory solution, as $h \to 0$. A 4×4 mesh of four-node elements

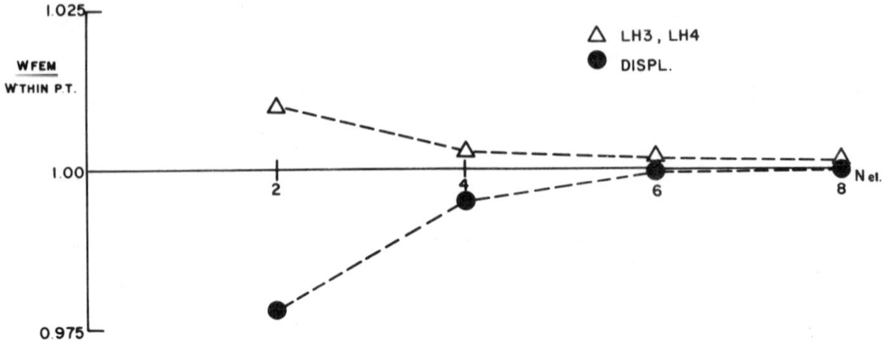

(a) Simply Supported, Uniform Load

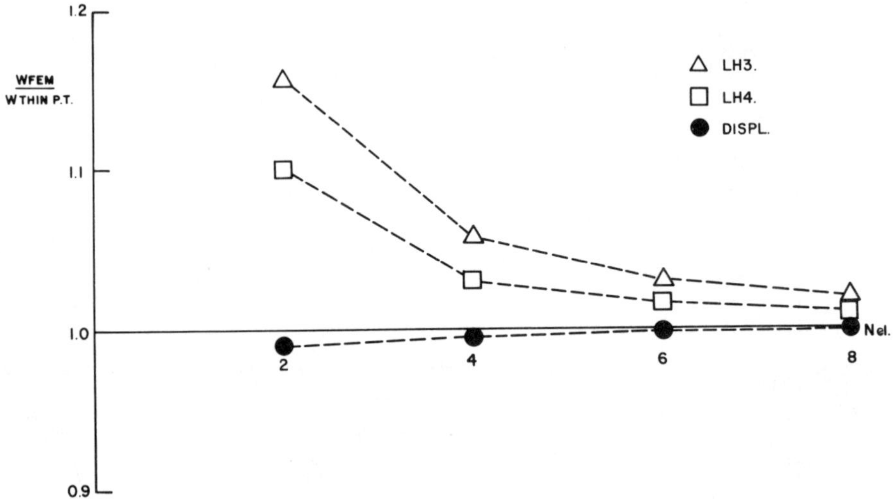

(b) Simply Supported, Concentrated Load

Figure 6.4 Convergence of the plate centre transverse displacement normalized

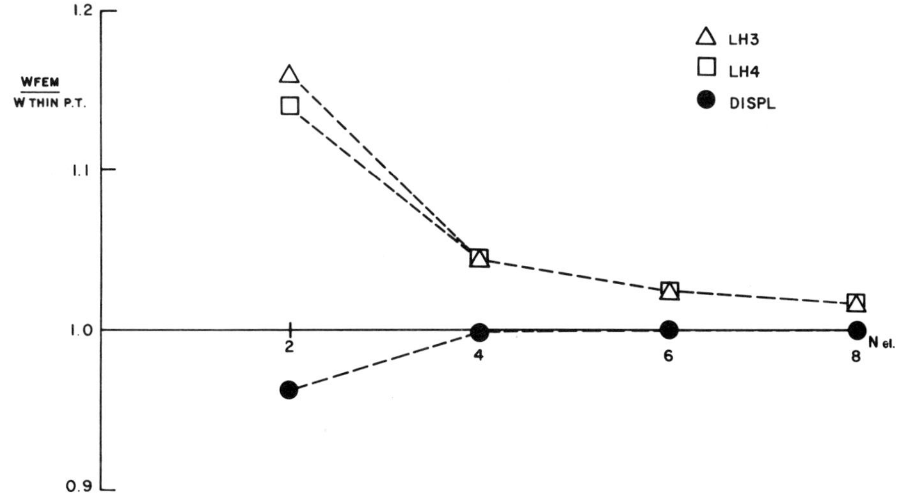

(c) Clamped, Uniform Load

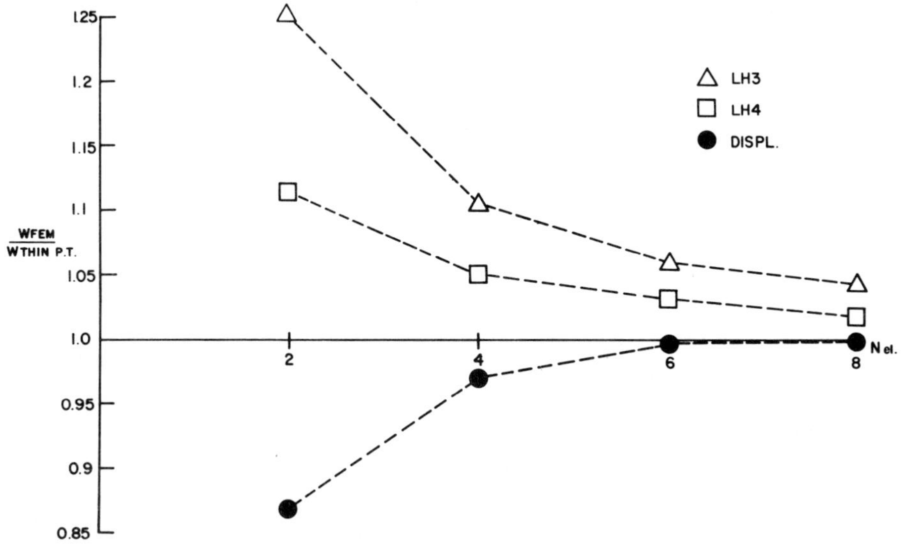

(d) Clamped, Concentrated Load

by the thin plate theory solution, for the square plate problems; $L/h = 10^2$

For example, consider the S1 element from [3] which was called DISP in the previous section. The stiffness matrix of this element is obtained by using full integration (2×2) on the bending terms $\mathbf{K}_B$ and reduced integration (1×1) on the shear terms $\mathbf{K}_s$. We can write this as

$$\mathbf{K}=\mathbf{K}_B^{(2\times 2)}+\mathbf{K}_s^{(1\times 1)} \tag{6.23}$$

This stiffness matrix has a rank deficiency of 2.

When the element stiffness is obtained by fully integrating both the bending and shear terms, the stiffness matrix is not rank deficient but it locks. Correct rank can therefore be obtained by taking a linear combination of these stiffnesses of the form

$$\mathbf{K}=\mathbf{K}_B^{(2\times 2)}+(1-\varepsilon)\mathbf{K}_s^{(1\times 1)}+\varepsilon\mathbf{K}_s^{(2\times 2)} \tag{6.24}$$

so the stabilization matrix $\mathbf{K}_H$ becomes

$$\mathbf{K}_H=\varepsilon(\mathbf{K}_s^{(2\times 2)}-\mathbf{K}_s^{(1\times 1)}) \tag{6.25}$$

The procedure of equation (6.24) was first proposed by Kavanagh and Key [27], although no results were reported, and subsequently Cook [28] concluded that the procedure leads to poor results. Perhaps the factor ε which was used in [28] was too large and caused locking.

Recently, however, it has been shown by Belytschko, Tsay, and Liu [29] that rank deficiency and locking can be avoided for all element geometries which were tested if the stabilization parameter is chosen by

$$\varepsilon=r\frac{h^2}{A} \tag{6.26}$$

where h is the thickness, A the area of the plate, and r is a parameter of order 10^{-2} to 10^{-1}.

This is illustrated in Figure 6.5, which shows the deflection of two adjacent nodes for a square plate supported only at the four corners and loaded by a concentrated load. The results for $r=0$ are not shown since the assembled stiffness matrix is then singular. When r is of order of magnitude 10^{-3}, the solution exhibits spatial oscillations, as can be seen from the large difference in the displacement of the adjacent nodes and the fact that the mid-point deflects less than the adjacent node. For $0.01<r<0.2$, the solution is almost independent of r and oscillations are eliminated. Finally, for values of r of order 1 or larger, the mesh locks.

Similar results are also obtained for the circular plate problem, in which the shortcomings of the SI element were first noted in [3]. These results are shown in Figure 6.6. For $r=0$, the results are identical to those of the element S1, which exhibits large oscillations in the normal deflection w. For the other three values of r shown in Figure 6.6, which are all of order of

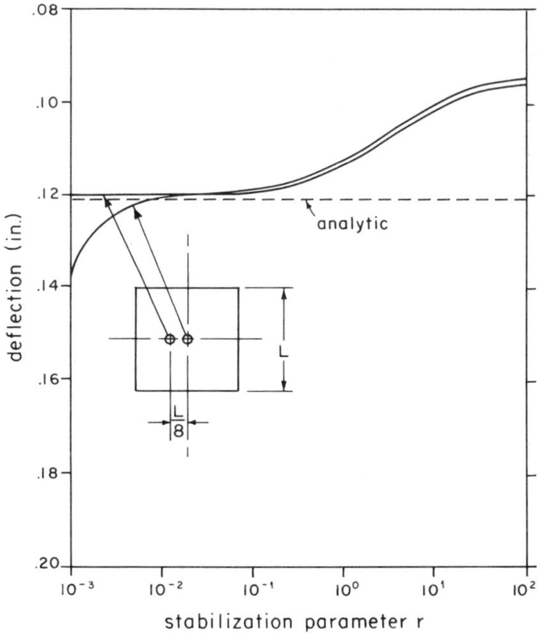

Figure 6.5 Deflections of nodes 1 and 2 for corner-supported plate subjected to a concentrated load

magnitude 10^{-2} to 10^{-1}, the oscillations are eliminated and the effects on the solution are minimal.

Similar procedures could be used for the elements presented in Section 6.2. For example, element LH5 is of correct rank but locks, while element LH3 converges well but is rank deficient. If we take the stiffness of LH3 and add a stabilization matrix $\mathbf{K}_{\mathrm{H}}$ of the form

$$\mathbf{K}_{\mathrm{H}} = \varepsilon(\mathbf{K}_{\mathrm{LH7}} - \mathbf{K}_{\mathrm{LH3}}) \tag{6.27}$$

then both rank deficiency and locking should be avoided.

Stabilization matrices can also be constructed by other methods. The major requirement is that the matrix gives zero forces in rigid body motion and the correct forces in a patch test. Procedures such as those exemplified by equations (6.25) and (6.27) satisfy this requirement because the linear combination of two stiffness matrices which meet the patch test will also satisfy it.

For simple elements, it is often possible to construct the stabilization matrix directly from the kinematic modes. This has been accomplished for the four-node element [30] with full reduced integration (one-point on

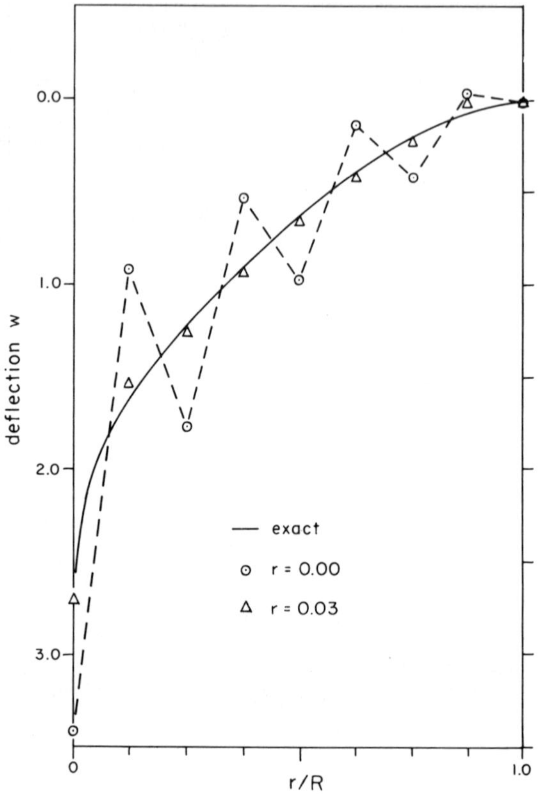

Figure 6.6 Deflections for circular plate along radius from centre to outside edge for concentrated load *P* at centre; *D* is flexural rigidity, *R* the radius of the plate

bending and shear) and the four-node quadrilateral and eight-node hexahedral continuum elements [31]. These procedures are important in nonlinear and explicitly integrated dynamic problems because the evaluation of the constitutive equations at the quadrature points consumes a large fraction of the computer time. If the number of quadrature points can be reduced from four to one without the introduction of troublesome kinematic modes, an advantage is definitely gained.

REFERENCES

1. R. D. Mindlin, 'Influence of rotary inertia and shear on flexural motion of isotropic elastic plates', *J. Appl. Mech.*, **18,** 31–38 (1951).
2. O. C. Zienkiewicz, R. L. Taylor, and J. M. Too, 'Reduced integration techniques in general analysis of plates and shells', *Int. J. Num. Meth. in Eng.*, **3,** 275–290 (1971).

3. T. J. R. Hughes, R. L. Taylor, and H. Kanoknulchui, 'A simple and efficient element for plate bending', *Int. J. Num. Meth. in Eng.*, **11,** 1529–1543 (1977).
4. T. J. R. Hughes, M. Cohen, and M. Haroun, 'Reduced and selective integration techniques in the finite element analysis of plates', *Nuclear Engineering and Design*, **46,** 203–222 (1978).
5. D. E. L. Pugh, E. Hinton, and O. C. Zienkiewicz, 'A study of quadrilateral plate-bending elements with "reduced" integration', *Int. J. Num. Meth. in Eng.*, **12,** 1059–1079 (1978).
6. T. J. R. Hughes and M. Cohen, 'The "heterosis" finite element for plate bending', *Computers and Structures*, **9,** 445–450 (1978).
7. T. J. R. Hughes and T. E. Tezduyar, 'Finite elements based upon Mindlin plate theory with particular reference to the four-node bilinear isoparametric element', *J. Appl. Mech.*, **48,** 587–596 (1981).
8. T. T. H. Pian, 'Derivation of element stiffness matrices by assumed stress distribution', *AIAA J.*, **2,** 1333–1336 (1964).
9. T. H. H. Pian, 'Element stiffness matrices for boundary compatibility and for prescribed boundary stresses', *Proc. Conf. Matrix Meth. in Structural Mech.*, AFFDL-TR-66-80, 457–477 (1966).
10. T. H. H. Pian and P. Tong, 'Rationalization in deriving element stiffness matrix assumed-stress approach', *Proc. Second Conf. Matrix Meth. in Structural Mech.*, AFFDL-R-68-150, 441–449 (1969).
11. T. H. H. Pian, 'Hybrid elements', in *Numerical and Computer Methods in Structural Mechanics* (Eds. S. J. Fenves, N. Perrone, A. R. Robinson, and W. C. Schnobrich), pp. 59–78, Academic Press, New York, 1973.
12. J. L. Batoz, K. J. Bathe, and L. W. Ho, 'A study of three-node triangular plate bending elements', *Int. J. Num. Meth. in Eng.*, **15,** 1771–1812 (1980).
13. S. T. Mau and E. A. Witmer, 'Static, vibration, and thermal stress analysis of laminated plates and shells by the hybrid-stress finite element method with transverse shear deformation affects included', MIT-ASRL-TR-169-2 (1972).
14. T. H. H. Pian and S. T. Mau, 'Some recent studies in assumed-stress hybrid models', in *Advances in Computational Methods in Structural Mechanics and Design* (Eds. J. T. Oden, R. W. Clough, and Y. Yamamoto), University of Alabama Press, Alabama, 1972.
15. R. D. Cook, 'Some elements for analysis of plate bending', *J. Eng. Mech. Div., ASCE*, **98,** 1453–1470 (1972).
16. R. D. Cook and S. G. Ladkany, 'Observations regarding assumed-stress hybrid plate elements', *Int. J. Num. Meth. in Eng.*, **8,** 513–519 (1974).
17. S. T. Mau, P. Tong, and T. H. H. Pian, 'Finite element solutions for laminated thick plates', *J. Comp. Mat.*, **6,** 304–311 (1972).
18. A. J. Barnard, 'A sandwich plate finite element', in *The Mathematics of Finite Elements and Applications* (Ed. J. R. Whiteman), Academic Press, London, 1973.
19. R. L. Spilker, S. C. Chou, and O. Orringer, 'Alternate hybrid-stress elements for analysis of multilayer composite plates', *J. Comp. Mat.*, **11,** 51–70 (1977).
20. R. L. Spilker and N. I. Munir, 'The hybrid stress model for thin plates', *Int. J. Num. Meth. in Eng.*, **15,** 1239–1260 (1980).
21. R. L. Spilker and N. I. Munir, 'A hybrid stress quadratic displacement Mindlin plate bending element', *Computers and Structures*, **12,** 11–21 (1980).
22. R. L. Spilker and N. I. Munir, 'A serendipity cubic displacement hybrid stress element for thin and moderately thick plates', *Int. J. Num. Meth. in Eng.*, **15,** 1261–1278 (1980).

23. D. S. Malkus and T. J. R. Hughes, 'Mixed finite element methods—reduced and selective integration techniques: A unification of concepts', *Computer Methods in Applied Mechanics and Engineering*, **15,** 63–81 (1978).
24. R. L. Spilker, S. M. Maskeri, and E. Kania, 'Plane isoparametric hybrid-stress elements: Invariance and optimal sampling', *Int. J. Num. Meth. in Eng.*, **17,** 1469–1496 (1981).
25. S. Timoshenko and S. Woinowsky-Krieger, *Theory of Plates and Shells*, 2nd ed., McGraw-Hill, New York, 1959.
26. R. L. Spilker, 'Invariant 8-node hybrid-stress elements for thin and moderately-thick plates', *Int. J. Num. Meth. in Eng.* (to appear).
27. K. T. Kavanagh and S. W. Key, 'A note on selective and reduced integration techniques in the finite element method', *Int. J. Num. Meth. in Eng.*, **4,** 148–150 (1972).
28. R. D. Cook, 'More on reduced integraton and isoparametric elements', *Int. J. Num. Meth. in Eng.*, **5,** 141–148 (1972).
29. T. Belytschko, C. S. Tsay, and W. K. Liu, 'A stabilization matrix for the bilinear Mindlin plate element', *Computer Methods in Applied Mechanics and Engineering* (in press).
30. T. Belytschko and C. S. Tsay, 'Stabilization procedures for the 4-node shell element' (submitted for publication).
31. D. P. Flanagan and T. Belytschko, 'A uniform strain hexahedron and quadrilateral with orthogonal hourglass control', *Int. J. Num. Meth. in Eng.*, **17,** 679–706 (1981).

Hybrid and Mixed Finite Element Methods
Edited by S. N. Atluri, R. H. Gallagher, and O. C. Zienkiewicz

Chapter 7

On a Mixed Method Related to the Discrete Kirchhoff Assumption

Fumio Kikuchi

7.1 INTRODUCTION

The discrete Kirchoff assumption is a promising idea available for finite element analysis of plate-bending problems. In this approach, we start from a plate-bending theory with the transverse shear deformation taken into account. Then the lateral deflection and the rotations of the plate can be considered as independent displacement components. However, the strain energy due to transverse shear is not counted in the calculation of the total potential energy to derive stiffness matrices. To compensate for errors induced by such an approximation, the Kirchhoff assumption is imposed in a discrete way: the rotations are forced to coincide with the first-order derivatives of the lateral deflection at some selected points in the finite elements. This type of method was first proposed by Wempner, Oden and Kross [1], and has been effectively employed for various problems of plates and shells. In fact, the 9 d.o.f. triangular plate-bending element based on this idea has won very high estimation in the recent work of Batoz, Bathe, and Ho [2]; this method is generally easy in implementation and gives fairly good results in most cases. theoretical analysis has also been performed for this approach applied to some linear and non-linear problems [3 to 5].

However, it is not necessarily easy to improve convergence orders of methods in this category. In this note, we will propose a mixed variational principle related to the discrete Kirchhoff assumption, and present a finite element model of high convergence order. We will make error analysis of the model and give some numerical results. It is to be noted that we can compute shearing forces of plates as the Lagrange multipliers for the constraint conditions of the Kirchhoff assumption. Such stress resultants are quite important in stress analysis of plates made of composite materials. Furthermore, the multipliers may be eliminated before hand at element level, and the method can be implemented as the usual displacement method.

7.2 PRELIMINARIES AND A MIXED VARIATIONAL FORMULATION

Let Ω be a convex polygonal domain of R^2 with boundary $\partial\Omega$. The independent variable of R^2 is denoted by $x=\{x_1, x_2\}$. As for function spaces on Ω, we will use the usual Sobolev spaces $H^m(\Omega)$ and $H_0^m(\Omega)$ with the common Sobolev norm $\|\cdot\|_m$ $(m=0,1,2,\ldots)$. More especially, the norm and the inner product of $L_2(\Omega)=H^0(\Omega)$ are designated by $\|\cdot\|$ and $(\cdot,\cdot)$, respectively. Furthermore, we will also use the following function spaces and norms:

$$U=H_0^1(\Omega)\times L_2(\Omega)\times L_2(\Omega)$$
$$\|u\|_U=\|w\|_1+\|\theta_1\|+\|\theta_2\| \qquad \text{for } u=\{w,\theta_1,\theta_2\}\in U$$
$$V=\{H_0^1(\Omega)\}^3\subset U$$
$$\|u\|_V=\|w\|_1+\|\theta_1\|_1+\|\theta_2\|_1 \qquad \text{for } u=\{w,\theta_1,\theta_2\}\in V$$
$$W=\{L_2(\Omega)\}^2$$
$$\|\lambda\|_W=\|p_1\|+\|p_2\| \qquad \text{for } \lambda=\{p_1,p_2\}\in W$$

As the notations of partial derivatives, we will use ∂_i, ∂_{ij}, and ∂_{ijk} in place of $\partial/\partial x_i$, $\partial^2/\partial x_i\partial x_j$, and $\partial^3/\partial x_i\partial x_j\partial x_k$ for $1\leq i, j, k\leq 2$. Define a subset Z of V by

$$Z=\{u=\{w,\theta_1,\theta_2\}\in V \mid \theta_i=\partial_i w \text{ for } i=1,2\}$$

We also need some bilinear forms:

$$\langle\cdot,\cdot\rangle: H^2(\Omega)\times H^2(\Omega)\to R^1$$
$$\langle w,\bar{w}\rangle=D\{\partial_{11}w,\partial_{11}\bar{w})+(\partial_{22}w,\partial_{22}\bar{w})+\nu(\partial_{11}w,\partial_{22}\bar{w})+\nu(\partial_{22}w,\partial_{11}\bar{w})$$
$$+2(1-\nu)(\partial_{12}w,\partial_{12}\bar{w})\} \qquad \text{for } w,\bar{w}\in H^2(\Omega)$$
$$a(\cdot,\cdot): V\times V\to R^1$$
$$a(u,\bar{u})=D\{(\partial_1\theta_1,\partial_1\bar{\theta}_1)+(\partial_2\theta_2,\partial_2\bar{\theta}_2)+\nu(\partial_1\theta_1,\partial_2\bar{\theta}_2)+\nu(\partial_2\theta_2,\partial_1\bar{\theta}_1)$$
$$+\frac{1-\nu}{2}(\partial_2\theta_1+\partial_1\theta_2,\partial_2\bar{\theta}_1+\partial_1\bar{\theta}_2)\}$$
$$\text{for } u=\{w,\theta_1,\theta_2\},\ \bar{u}=\{\bar{w},\bar{\theta}_1,\bar{\theta}_2\}\in V$$
$$b(\cdot,\cdot): U\times W\to R^1$$
$$b(u,\lambda)=\sum_{i=1}^{2}(\theta_i-\partial_i w, p_i) \qquad \text{for } u=\{w,\theta_1,\theta_2\}\in U,\ \lambda=\{p_1,p_2\}\in W$$

In the above, D and ν are constants such that $D>0$, $\frac{1}{2}>\nu>0$. Usually, D and ν are called the bending stiffness and Poisson's ratio respectively. Notice that the above bilinear forms are all continuous over their domains of definition.

The classical problem of bending of clamped plates is stated as follows.

Problem 7.1 *Given $f \in L_2(\Omega)$, find $w \in H_0^2(\Omega)$ such that*

$$\langle w, \bar{w} \rangle = (f, \bar{w}) \qquad \text{for all } \bar{w} \in H_0^2(\Omega) \tag{7.1}$$

Physically, w and f are the lateral deflection of the plate and the distributed lateral force respectively (cf. Chap. 8 of Washizu [6]).

We can show that the bilinear form $\langle \cdot, \cdot \rangle$ is $H_0^2(\Omega)$-elliptic, i.e. we can find a positive constant $C = C(\Omega, D, \nu)$ such that

$$\langle w, w \rangle \geqslant C \|w\|_2^2 \qquad \text{for all } w \in H_0^2(\Omega) \tag{7.2}$$

Consequently, the solution w of Problem 7.1 exists uniquely in $H_0^2(\Omega)$ for each given $f \in L_2(\Omega)$ (cf. Chap. 1 of Ciarlet [7]). Furthermore, w belongs to $H^3(\Omega) \cap H_0^2(\Omega)$ with the estimation

$$\|w\|_3 \leqslant C^* \|f\| \tag{7.3}$$

where C^* is a positive constant similar to C. (Hereafter, we will use C, C^*, C_1, etc., as generic positive constants.) Here it is essential that Ω is a convex polygonal domain (see Kondrat'ev [8]). If Ω is of a more special shape (for example, of non-obtuse polygonal shape), then w belongs to $H^4(\Omega) \cap H_0^2(\Omega)$ with

$$\|w\|_4 \leqslant C^{**} \|f\| \tag{7.4}$$

as was proved by Mitzutani [9]. Hereafter, we will usually assume that Ω is chosen so that the above holds.

To obtain a different kind of variational formulation to the present problem, we first introduce two functions θ_i $(i = 1, 2)$ by

$$\theta_i = \partial_i w \qquad \text{for } w \in H_0^2(\Omega) \tag{7.5}$$

Then we can easily check that $u = \{w, \theta_1, \theta_2\}$ with the above relation belonging to Z. Conversely, if $u = \{w, \theta_1, \theta_2\} \in Z$, then we have $w \in H_0^2(\Omega)$. Therefore, Problem 7.1 is equivalent to:

Problem 7.2 *Given $f \in L_2(\Omega)$, find $u = \{w, \theta_1, \theta_2\} \in Z$ such that*

$$a(u, \bar{u}) = (f, \bar{w}) \qquad \textit{for all } \bar{u} = \{\bar{w}, \bar{\theta}_1, \bar{\theta}_2\} \in Z \tag{7.6}$$

Remark 7.1 Physically, θ_i $(i = 1, 2)$ are rotations of the normal to the middle surface of the plate. In a strict sense, the rotations are not necessarily equal to the first-order derivatives of the lateral deflection w, and hence the relation (7.5) is an assumption made to deduce the present plate-bending theory from the three-dimensional theory of elasticity. This assumption is usually called the *Kirchhoff assumption.* In some plate-bending theories, the rotations can be independent of w (cf. Chap. 8 of Washizu [6]).

Let us introduce the Lagrange multiplier for the linear constraint $u \in Z$ in Problem 7.2. Then we have the following mixed formulation [10].

Problem 7.3 *Given $f \in L_2(\Omega)$, find $\{u, \lambda\} \in V \times W$ such that*

$$a(u, \bar{u}) + b(\bar{u}, \lambda) = (f, \bar{w}) \qquad \text{for all } \bar{u} = \{\bar{w}, \bar{\theta}_1, \bar{\theta}_2\} \in V \tag{7.7}$$

$$b(u, \bar{\lambda}) = 0 \qquad \text{for all } \bar{\lambda} \in W \tag{7.8}$$

The function $\lambda \in W$ is the Lagrange multiplier in this case. In the sequel, we will see that Problem 7.3 is equivalent to Problem 7.2 (or Problem 7.1).

Let us first prove that the solution $\{u, \lambda\} \in V \times W$ of Problem 7.3, if it exists, must satisfy Problem 7.2. From (7.7), we have $a(u, \bar{u}) = (f, \bar{w})$ for all $\bar{u} \in Z$, since $b(\bar{u}, \lambda) = 0$ for $\bar{u} \in Z$. On the other hand, (7.8) implies that $u \in Z$, and hence we can conclude that u of Problem 7.3 is also a solution of Problem 7.2.

As was already noted, the first component w of the solution $u = \{w, \theta_1, \theta_2\}$ of Problem 7.2 belongs to $H^3(\Omega)$. Therefore, we can apply Green's formula to the left-hand side of (7.7) to show that $\lambda = \{p_1, p_2\}$ must satisfy

$$p_1 = D\left[\partial_{11}\theta_1 + \nu\, \partial_{12}\theta_2 + \frac{1-\nu}{2}(\partial_{22}\theta_1 + \partial_{12}\theta_2)\right] = D(\partial_{111}w + \partial_{122}w) \in L_2(\Omega) \tag{7.9}$$

$$p_2 = D\left[\partial_{22}\theta_2 + \nu\, \partial_{12}\theta_1 + \frac{1-\nu}{2}(\partial_{12}\theta_1 + \partial_{11}\theta_2)\right] = D(\partial_{222}w + \partial_{112}w) \in L_2(\Omega) \tag{7.10}$$

Conversely, let us show that the solution $u \in Z$ of Problem 7.2, together with $\lambda = \{p_1, p_2\} \in W$ defined by (7.9) and (7.10), satisfies Problem 7.3. The relation (7.8) is obvious, while (7.7) is checked by using Green's formula again. Consequently, we can conclude the equivalence between Problems 7.2 and 7.3. The uniqueness and the existence of the solution $\{u, \lambda\}$ now follow from Problem 7.1.

Remark 7.2 Physically, the functions p_1 and p_2 defined by (7.9) and (7.10) are shearing forces of plates (cf. Chap. 8 of Washizu [6]). For plates made of composite materials, it is often important to evaluate these quantities.

Remark 7.3 In some mixed formulations, certain regularity conditions such as (7.3) are required to show the existence and the uniqueness of the solution. We can observe that the solution $\{u, \lambda\}$ of Problem 7.3 is the stationary point of the variational functional

$$I[u, \lambda] = \tfrac{1}{2}a(u, u) + b(u, \lambda) - (f, w) \qquad \text{for } \{u, \lambda\} = \{\{w, \theta_1, \theta_2\}, \{p_1, p_2\}\} \in V \times W$$

7.3 THE DISCRETE PROBLEM AND ABSTRACT ERROR ESTIMATES

In order to solve Problem 7.3 numerically, we should prepare appropriate finite dimensional subspaces V^h of V and W^h of W. Then the approximate problem to solve is:

Problem 7.4 *Given $f \in L_2(\Omega)$, find $\{u_h, \lambda_h\} \in V^h \times W^h$ such that*

$$a(u_h, \bar{u}_h) + b(\bar{u}_h, \lambda_h) = (f, \bar{w}_h) \qquad \text{for all } \bar{u}_h = \{\bar{w}_h, \bar{\theta}_{h1}, \bar{\theta}_{h2}\} \in V^h \tag{7.11}$$

$$b(u_h, \bar{\lambda}_h) = 0 \qquad \text{for all } \bar{\lambda}_h \in W^h \tag{7.12}$$

We will discuss the solvability and the error estimation of the above mixed scheme. to this end, define:

$$Z^h = \{u_h \in V^h \mid b(u_h, \lambda_h) = 0; \forall \lambda_h \in W^h\} \tag{7.13}$$

$$S(V^h) = \sup_{\substack{u_h \in V^h \\ u_h \neq 0}} \frac{\|u_h\|_V}{\|u_h\|_U} (<\infty) \tag{7.14}$$

We will employ the following two hypotheses to apply the general theory of Babuska and Aziz [11] and Brezzi [12], or rather its variant due to Bercovier and Pironneau [13].

Hypothesis 7.1 *There exists a positive constant C_{h1} such that*

$$\sup_{\substack{u_h \in V^h \\ u_h \neq 0}} \frac{b(u_h, \lambda_h)}{\|u_h\|_U} \geqslant C_{h1} \|\lambda_h\|_W \qquad \text{for all } \lambda_h \in W^h \tag{7.15}$$

Hypothesis 7.2 *The bilinear form $a(\cdot, \cdot)$ is Z^h-elliptic: there exists a positive constant C_{h2} such that*

$$a(u_h, u_h) \geqslant C_{h2} \|u_h\|_V^2 \qquad \text{for all } u_h \in Z^h \tag{7.16}$$

In the above, the positive constants C_{h1} and C_{h2} may depend on V^h and W^h, but are independent of u_h and λ_h themselves.

Under these preparations, we have the following results.

Proposition 7.1 *For each $f \in L_2(\Omega)$, Problem 7.4 has a unique solution $\{u_h, \lambda_h\} \in V^h \times W^h$ with the estimation*

$$\|u_h\|_V \leqslant C_{h2}^{-1} \|f\|, \qquad \|\lambda_h\|_W \leqslant C_{h1}^{-1}(1 + CC_{h2}^{-1} S(V^h)) \|f\| \tag{7.17}$$

where $C > 0$ is independent of f, V^h, and W^h.

Proof The above is a direct consequence of the Babuska–Brezzi theory (see, for example, Proposition 2.1 of Brezzi [12]).

Proposition 7.2 *For each $f \in L_2(\Omega)$, consider the solutions $\{u, \lambda\} \in V \times W$ of Problem 7.3 and $\{u_h, \lambda_h\} \in V^h \times W^h$ of Problem 7.4. Then*

$$\|u_h - u\|_V \leq (1 + CC_{h2}^{-1}) \inf_{u_h^* \in V^h} \{\|u_h^* - u\|_V + CC_{h1}^{-1}C_{h2}^{-1}S(V^h)\|u_h^* - u\|_U\}$$
$$+ C_{h2}^{-1} \inf_{\lambda_h^* \in W^h} \left\{ \sup_{\substack{v_h \in Z^h \\ v_h \neq 0}} \frac{b(v_h, \lambda_h^* - \lambda)}{\|v_h\|_V} \right\} \tag{7.18}$$

$$\|\lambda_h - \lambda\|_W \leq (1 + CC_{h1}^{-1}) \inf_{\lambda^* \in W^h} \|\lambda_h^* - \lambda\|_W + CC_{h1}^{-1}S(V^h)\|u_h - u\|_V \tag{7.19}$$

where $C > 0$ is independent of f, V^h, and W^h.

Proof We can prove the above in a manner similar to that of Theorem 1 of Bercovier and Pironneau [13]. We will only sketch the proof.

First we have

$$a(u_h - u_h^\dagger, u_h - u_h^\dagger) = a(u - u_h^\dagger, u_h - u_h^\dagger) + b(u_h - u_h^\dagger, \lambda - \lambda_h) \qquad \text{for all } u_h^\dagger \in V^h$$

Choosing $u_h^\dagger$ from Z^h in the above, we find

$$a(u_h - u_h^\dagger, u_h - u_h^\dagger) = a(u - u_h^\dagger, u_h - u_h^\dagger) + b(u_h - u_h^\dagger, \lambda - \lambda_h^*)$$
$$\text{for all } u_h^\dagger \in Z^h, \qquad \text{for all } \lambda_h^* \in W^h$$

from which we obtain

$$\|u_h - u_h^\dagger\|_V \leq C_{h2}^{-1}\left\{C_1\|u_h^\dagger - u\|_V + \sup_{\substack{v_h \in Z^h \\ v_h \neq 0}} \frac{b(v_h, \lambda_h^* - \lambda)}{\|v_h\|_V}\right\}$$
$$\text{for all } u_h^\dagger \in Z^h, \text{ for all } \lambda_h^* \in W^h \tag{a}$$

Given $u_h^* \in V^h$, consider $\{u_h^\dagger, \lambda_h^\dagger\} \in V^h \times W^h$ of

$$a(u_h^\dagger, \bar{u}_h) + b(\bar{u}_h, \lambda_h^\dagger) = a(u_h^*, \bar{u}_h)$$
$$b(u_h^\dagger, \bar{\lambda}_h) = 0 \qquad \text{for all } \bar{u}_h \in V^h, \qquad \text{for all } \bar{\lambda}_h \in W^h$$

Then $u_h^\dagger \in Z^h$, and we have

$$a(u_h^\dagger - u_h^*, u_h^\dagger - u_h^*) = b(u_h^* - u_h^\dagger, \lambda_h^\dagger) = b(u_h^* - u, \lambda_h^\dagger)$$

from which we obtain the following estimations:

$$\|u_h^\dagger - u_h^*\|_V^2 \leq C_2C_{h2}^{-1}\|u_h^* - u\|_U\|\lambda_h^\dagger\|_W, \qquad \|\lambda_h^\dagger\|_W \leq C_3C_{h1}^{-1}S(V^h)\|u_h^\dagger - u_h^*\|_V$$

Combining the above gives

$$\|u_h^\dagger - u_h^*\|_V \leq C_4C_{h1}^{-1}C_{h2}^{-1}S(V^h)\|u_h^* - u\|_U \tag{b}$$

The estimation (7.18) now follows from (a) and (b).

To derive (7.19), notice the relation

$$b(\bar{u}_h, \lambda_h - \lambda_h^*) = a(u - u_h, \bar{u}_h) + b(\bar{u}_h, \lambda - \lambda_h^*) \qquad \text{for all } \bar{u}_h \in V^h, \quad \text{for all } \lambda_h^* \in W^h$$

Evaluating the both sides of the above, we have

$$\|\lambda_h - \lambda_h^*\|_W \leq C_5 C_{h1}^{-1}\{S(V^h)\,\|u_h - u\|_V + \|\lambda_h^* - \lambda\|_W\}$$

Now we can easily deduce (7.19), and the proof is completed.

7.4 A FINITE ELEMENT MODEL

We will give an example of finite element model for the discrete problem, Problem 7.4. Let $\{T^h\}_{h>0}$ be a family of triangulations of Ω, i.e. each T^h is a set of (closed) triangles contained by $\bar{\Omega}$ = closure of Ω. A representative triangle in T^h will be designated as T, and the index $h>0$ is the maximum side length of all triangles in T^h. We assume that $\{T^h\}$ is regular in the sense:

(a) Any side of $T \in T^h$ is either a portion of $\partial\Omega$ or coincides with a side of another triangle in T^h.
(b) $\bigcup_{T \in T^h} T = \bar{\Omega}$.
(c) Any $T \in T^h$ contains a circle of radius Kh for a certain $K>0$ independent of h.

We will construct $V^h = V^h(T^h)$ and $W^h = W^h(T^h)$. We denote by $\{L_1, L_2, L_3\}$ the area coordinate functions for $T \in T^h$. Let us define some function sets related to $T \in T^h$:

$P^k(T)$ = totality of polynomials of degree $\leq k$ over T $(k = 0, 1, 2, \ldots)$
$Q^2(T)$ = totality of functions over T of the form

$$q = \sum_{\substack{i+j+k=2\\ i,j,k\geq 0}} \alpha_{ijk} L_1^i L_2^j L_3^k + \alpha_{111} L_1 L_2 L_3$$

$Q^3(T)$ = totality of functions over T of the form

$$q = \sum_{\substack{i+j+k=3\\ i,j,k\geq 0\\ i\neq j, i\neq k}} \beta_{ijk} L_1^i L_2^j L_3^k + (\beta_{211} L_1 + \beta_{121} L_2 + \beta_{112} L_3) L_1 L_2 L_3$$

The space W^h is defined simply by

$$W^h = \{\lambda_h = \{p_{h1}, p_{h2}\} \in W \,|\, p_{hi}|_T \in P^0(T) \text{ for } i = 1, 2; \text{ for all } T \in T^h\}$$

To define V^h, we utilize the notations in Chap. 2 of Ciarlet [7]. We determine the first component w_h of $u_h = \{w_h, \theta_{h1}, \theta_{h2}\} \in V^h$. The local space P_T and the local degrees of freedom (or nodal parameters) Σ_T of w_h in

$T \in T^h$ are respectively given by

$$P_T = Q^3(T)$$
$$\Sigma_T = \{q(a_i), \partial_1 q(a_i), \partial_2 q(a_i), \partial_n q(b_i) \text{ for } i = 1, 2, 3; q \in P_T\}$$

where a_i and b_i for $i = 1, 2, 3$ are three vertices and three mid-points of sides of T respectively, and ∂_n denotes the derivative in the outward normal direction of $\partial T =$ boundary of T. For vertices and mid-points on $\partial\Omega$, the corresponding nodal parameters should be all equated to zero to approximate the zero Dirichlet conditions imposed on functions of $H_0^2(\Omega)$. We can show the unisolvence of the above choice. The function w_h is now constructed as follows:

(a) $w_h|_T \in P_T$ for any $T \in T^h$.
(b) At vertices and mid-points of T^h, the values of the corresponding nodal parameters are common to all traingles joining there. Notice that the sign of $\partial_n w_h$ along the interface of two triangles must be adjusted appropriately.

We can show that $w_h \in H_0^1(\Omega)$ but $w_h \notin H_0^2(\Omega)$ in general, and hence w_h is non-conforming in the usual sense.

Next, we employ the following P_T and Σ_T to define both θ_{h1} and θ_{h2}:

$$P_T = Q^2(T)$$
$$\Sigma_T = \left\{q(a_i), q(b_i) \text{ for } i = 1, 2, 3, \int_T q(x)\,dx \Big/ \int_T dx; q \in P_T\right\}$$

We can show the unisolvence of the above. The functions θ_{h1} and θ_{h2} are constructed as follows:

(a) $\theta_{hi}|_T \in P_T$ for $i = 1, 2$; $\forall T \in T^h$.
(b) The nodal values of θ_{hi} $(i = 1, 2)$ at vertices and mid-points are common to all triangles joining there.

Now we introduce the *discrete Kirchhoff assumption* to w_h, θ_{h1}, and θ_{h2}:

$$\theta_{hi} = \partial_i w_h \ (i = 1, 2) \text{ at all vertices and mid-points of } T^h \qquad (7.20)$$

Then $\theta_{hi} \in H_0^1(\Omega)$ for $i = 1, 2$. Notice that the degrees of freedom $\int_T \theta_{hi}(x)\,dx / \int_T dx$ $(i = 1, 2; T \in T^h)$ are still independent of w_h. Thus we have determined the form of $u_h = \{w_h, \theta_{h1}, \theta_{h2}\} \in V^h \subset V$ completely.

Finally, we express the condition $u_h \in Z^h$ explicitly. Since $\lambda_h \in W^h$ is constant in each $T \in T^h$, (7.13) amounts to

$$\int_T \theta_{hi}(x)\,dx = \int_T \partial_i w_h(x)\,dx \ (i = 1, 2) \qquad \text{for all } T \in T^h \qquad (7.21)$$

which may be also regarded as additional conditions of the discrete Kirchhoff assumption. If the above are satisfied, θ_{h1} and θ_{h2} are completely subject to w_h.

Remark 7.4 As may be easily seen, the conditions (7.20) and (7.21) can be imposed elementwise. Therefore, in the computer implementation of Problem 7.4, we need not obtain u_h and λ_h at the same time. Instead, we should first find $u_h \in Z^h$ such that

$$a(u_h, \bar{u}_h) = (f, \bar{w}_h) \qquad \text{for all } \bar{u}_h = \{\bar{w}_h, \bar{\theta}_{h1}, \bar{\theta}_{h2}\} \in Z^h$$

which is nothing but the discrete analogue of Problem 7.2. To solve the above, we can rely upon the standard displacement method and the direct stiffness method.

To calculate $\lambda_h \in W^h$ from the above $u_h \in Z^h$, we should equate $\bar{w}_h$ to zero in (7.11). Then we have

$$\sum_{T \in T^h} \left\{ \sum_{i=1}^{2} \int_T p_{hi} \bar{\theta}_{hi} \, dx \right\} + a(u_h, \bar{u}_h) = 0 \quad \text{for all } \bar{u}_h = \{0, \bar{\theta}_{h1}, \bar{\theta}_{h2}\} \in V^h \quad \text{(a)}$$

In the above, $\bar{\theta}_{hi}$ $(i = 1, 2)$ must be of the form

$$\theta_{hi}|_T = \alpha_i(T) L_1 L_2 L_3 \qquad \text{for all } T \in T^h$$

where $\alpha_i(T)$ $(i = 1, 2;\ T \in T^h)$ are coefficients that can take arbitrary values. Since λ_h is constant in each $T \in T^h$ and $\int_T L_1 L_2 L_3 \, dx \neq 0$ for any $T \in T^h$, we can compute element values of λ_h by the use of (a).

7.5 ERROR ESTIMATES

This section is devoted to error analysis of the finite element model presented in the preceding section. Since the complete proofs are lengthy, we will see only the outline of the process.

For each $T \in T^h$, we denote by $\|\cdot\|_T$ and $(\cdot, \cdot)_T$ the norm and the inner product of $L_2(T)$ respectively (strictly, we should use the notation $L_2(T^{\text{int}})$ with $T^{\text{int}} =$ interior of T). Similar notations such as $\langle \cdot, \cdot \rangle_T$ and $a_T(\cdot, \cdot)$ will be frequently employed in this section. furthermore, we use

$$\langle w_h, \bar{w}_h \rangle_h = \sum_{T \in T^h} \langle w_h, w_h \rangle_T$$

where w_h and $\bar{w}_h$ are the first components of $u_h \in V^h$ and $\bar{u}_h \in V^h$ respectively, and the notations w_h and $\bar{w}_h$ in the right-hand side should be interpreted as $w_h|_T$ and $\bar{w}_h|_T$, which belong to $H^2(T)$.

First, let us prove Hypotheses 7.1 and 7.2.

Proposition 7.3 *For the finite element model introduced in the preceding section, Hypothesis 7.1 holds with $C_{h1} > 0$ independent of h.*

Proof For each non-zero $\lambda_h = \{p_{h1}, p_{h2}\} \in W^h$, consider $u_h \in V^h$ of the form $u_h = \{0, \theta_{h1}, \theta_{h2}\}$ (the first component is zero!) with

$$(\theta_{hi} - p_{hi}, \bar{p}_{hi}) = 0 \ (i = 1, 2) \qquad \text{for all } \bar{\lambda}_h = \{\bar{p}_{h1}, \bar{p}_{h2}\} \in W^h$$

which in this case is equivalent to

$$(\theta_{hi} - p_{hi}, \bar{p}_{hi})_T = 0 \ (i = 1, 2) \qquad \text{for all } T \in T^h$$

Taking notice of Remark 7.4, we have

$$\theta_{hi}|_T = \frac{\left(\int_T p_{hi}\, dx\right) L_1 L_2 L_3}{\int_T L_1 L_2 L_3\, dx}$$

Now $b(u_h, \lambda_h)$ for the above u_h and λ_h is evaluated as

$$b(u_h, \lambda_h) = \sum_{i=1}^{2} (\theta_{hi}, p_{hi}) = \sum_{i=1}^{2} \|p_{hi}\|^2 \geqslant \tfrac{1}{2} \|\lambda_h\|_W^2$$

On the other hand,

$$\|u_{hi}\|_T = \frac{\left(\int_T L_1^2 L_2^2 L_3^2\, dx\right)^{1/2}}{\int_T L_1 L_2 L_3\, dx} \left|\int_T p_{hi}\, dx\right|$$

$$\leqslant \frac{\left(\int_T L_1^2 L_2^2 L_3^2\, dx\right)^{1/2} \left(\int_T dx\right)^{1/2}}{\int_T L_1 L_2 L_3\, dx} \|p_{hi}\|_T = C\, \|p_{hi}\|_T$$

for $C > 0$ independent of T and h. Thus we have

$$\|u_h\|_U = \|\theta_{h1}\| + \|\theta_{h2}\| \leqslant C^*\, \|\lambda_h\|_W$$

and Hypothesis 7.1 is shown with $C_{h1} > 0$ independent of h.

Proposition 7.4 *For finite element model in the preceding section, Hypothesis 7.2 holds with $C_{h2} > 0$ independent of h.*

Proof The complete proof may be performed as in the proof of Lemma 3.5 of Kikuchi [3]. We will only indicate essential points.

We should first show that there exists a positive constant C such that

$$a_T(u_h, u_h) \geqslant C \langle w_h, w_h \rangle_T \qquad \text{for all } u_h = \{w_h, \theta_{h1}, \theta_{h2}\} \in Z^h, \text{ for all } T \in T^h$$

To prove the above, it is essential to show that $\langle w_h, w_h \rangle_T = 0$ when $a_T(u_h, u_h) = 0$ for $u_h \in Z^h$. From $a_T(u_h, u_h) = 0$, we have

$$\theta_{h1}|_T = \alpha_1 + \alpha_3 x_2, \qquad \theta_{h2}|_T = \alpha_2 - \alpha_3 x_1$$

where α_i $(i=1,2,3)$ are appropriate constants. Define θ_{hs} and $\partial_s w_h$ on ∂T by

$$\theta_{hs} = -n_2\theta_{h1} + n_1\theta_{h2}, \qquad \partial_s w_h = -n_2\partial_1 w_h + n_1\partial_2 w_h$$

where $\{n_1, n_2\}$ is the unit outward normal on ∂T. In the present finite element model, θ_{hs} coincides with $\partial_s w_h$ on ∂T. Then, applying Green's formula, we have

$$0 = \oint_{\partial T} \partial_s w_h \, \mathrm{d}s = \oint_{\partial T} \theta_{hs} \, \mathrm{d}s = \int_T (-\partial_2\theta_{h1} + \partial_1\theta_{h2}) \, \mathrm{d}x = \int_T (-2\alpha_3) \, \mathrm{d}x$$

where $\mathrm{d}s$ is the line element on ∂T. Therefore $\alpha_3 = 0$ and $\theta_{hi}|_T = \alpha_i$ $(i=1,2)$. Taking account of the relation $\partial_s w_h = \theta_{hs}$ on ∂T with (7.20) and (7.21), we can find that $w_h|_T = \alpha_0 + \alpha_1 x_1 + \alpha_2 x_2$ for an appropriate constant α_0. For such w_h holds clearly $\langle w_h, w_h\rangle_T = 0$. Thus we can conclude that $a(u_h, u_h) \geqslant C\langle w_h, w_h\rangle_h$ for any $u_h \in Z^h$. We can also derive the Poincare–Friedrichs type of inequality for non-conforming functions (cf. Stummel [14]):

$$\langle w_h, w_h\rangle_h \geqslant C^* \|w_h\|_1^2 \qquad \text{for all } u_h = \{w_h, \theta_{h1}, \theta_{h2}\} \in V^h$$

Applying now the Korn inequality $a(u, u) \geqslant C^{**}(\|\theta_1\|_1^2 + \|\theta_2\|_1^2)$ for any $u \in V$, we can deduce Hypothesis 7.2 with $C_{h2} > 0$ independent of h.

As the second step, we should prepare some results to evaluate the approximation capability of the finite element space $V^h \times W^h$. To this end, we can apply the standard interpolation theory for Sobolev spaces (cf. Chap. 2 of Ciarlet [7]).

Let $\{u, \lambda\} = \{\{w, \theta_1, \theta_2\}, \{p_1, p_2\}\} \in V \times W$ be the solution of Problem 7.3 for $f \in L_2(\Omega)$. Then $w \in H_0^2(\Omega) \cap H^3(\Omega)$, $\theta_i \in H_0^1(\Omega) \cap H^2(\Omega)$, and $p_i \in L_2(\Omega)$ for $i=1,2$. Thus, for the finite element space $V^h \times W^h$ introduced in the preceding section, we can define the *interpolate* $\{u_h^*, \lambda_h^*\} = \{\{w_h^*, \theta_{h1}^*, \theta_{h2}^*\}, \{p_{h1}^*, p_{h2}^*\}\} \in V^h \times W^h$ of $\{u, \lambda\}$ as follows:

(a) w_h^*: $\left.\begin{aligned} &w_h^*(a_i) = w(a_i), \partial_1 w_h^*(a_i) = \partial_1 w(a_i) \\ &\partial_2 w_h^*(a_i) = \partial_2 w(a_i), \partial_n w_h^*(b_i) = \partial_n w(b_i) \end{aligned}\right\}$ $(i=1,2,3)$; for all $T \in T^h$.

(b) θ_{hi}^* $(i=1,2)$: these are determined from the above w_h^* so that (7.20) and (7.21) hold.

(c) p_{hi}^* $(i=1,2)$: $\displaystyle\int_T p_{hi}^* \, \mathrm{d}x = \int_T p_i \, \mathrm{d}x$ $(i=1,2)$; for all $T \in T^h$.

Notice that the above are all well defined for each $\{u, \lambda\}$ and that u_h^* thus determined actually belongs to Z^h. Then we can show:

Proposition 7.5 *For each $f \in L_2(\Omega)$, consider the solution $\{u, \lambda\} = \{\{w, \theta_1, \theta_2\}, \{p_1, p_2\}\}$ of Problem 7.3 and the above-defined interpolate $\{u_h^*, \lambda_h^*\} = \{\{w_h^*, \theta_{h1}^*, \theta_{h2}^*\}, \{p_{h1}^*, p_{h2}^*\}\} \in V^h \times W^h$. Then we can find $C > 0$,* *independent of f and h, such that*

$$\|w_h^* - w\|_\alpha \leq Ch^{\beta-\alpha} \|f\| \qquad (\alpha = 0, 1) \tag{7.22}$$

$$\|\theta_{hi}^* - \theta_i\|_\alpha \leq Ch^{\beta-\alpha-1} \|f\| \qquad (i = 1, 2; \alpha = 0, 1) \tag{7.23}$$

$$\|p_{hi}^* - p_i\| \leq Ch^{\beta-3} \|f\| \qquad (i = 1, 2) \tag{7.24}$$

where $\beta = 3$ in general and $\beta = 4$ if (7.4) holds.

Proposition 7.6 *For any $u_h = \{w_h, \theta_{h1}, \theta_{h2}\} \in Z^h$, we have*

$$\|\theta_{hi} - \partial_i w_h\| \leq Ch \|u_h\|_V \qquad (i = 1, 2) \tag{7.25}$$

for a certain $C > 0$ independent of u_h and h.

Proof For $u_h \in Z^h$, θ_{hi} may be regarded as a kind of interpolate of $\partial_i w_h$, since both (7.20) and (7.21) hold. Thus the use of interpolation theory gives

$$\|\theta_{hi} - \partial_i w_h\| \leq C^* h \langle w_h, w_h \rangle_h$$

Notice here the fact that $\langle w_h, w_h \rangle_T$ is equivalent to $\sum_{i,j=1}^2 (\partial_{ij} w_h, \partial_{ij} w_h)_T$ for any $T \in T^h$. Now the desired estimation follows from

$$\langle w_h, w_h \rangle_h \leq C^{**} \|u_h\|_V$$

which may be derived in the same manner as in the proof of Proposition 7.4.

We can now obtain the fundamental error estimates of the finite element solution with the aid of Proposition 7.2.

Theorem 7.1 *Let $V^h \times W^h$ be the finite element space introduced in the preceding section. For each $f \in L_2(\Omega)$, consider the solutions $\{u, \lambda\} = \{\{w, \theta_1, \theta_2\}, \{p_1, p_2\}\} \in V \times W$ of Problem 7.3 and $\{u_h, \lambda_h\} = \{\{w_h, \theta_{h1}, \theta_{h2}\}, \{p_{h1}, p_{h2}\}\} \in V^h \times W^h$ of Problem 7.4. Then we can find a positive constant C, independent of f and h, such that*

$$\|u_h - u\|_V = \|w_h - w\|_1 + \sum_{i=1}^{2} \|\theta_{hi} - \theta_i\|_1 \leq Ch^\gamma \|f\| \tag{7.26}$$

$$\|\lambda_h - \lambda\|_W = \sum_{i=1}^{2} \|p_{hi} - p_i\| \leq Ch^{\gamma-1} \|f\| \tag{7.27}$$

where $\gamma = 1$ in general and $\gamma = 2$ when (7.4) holds.

Proof To prove the above, we can use the interpolate $\{u_h^*, \lambda_h^*\}$ of $\{u, \lambda\}$ in the right-hand sides of (7.18) and (7.19). Then we have from Proposition 7.5

$$\|u_h^* - u\|_V \leq C^* h^\gamma \|f\|, \qquad \|u_h^* - u\|_U \leq C^* h^{\gamma+1} \|f\|, \qquad \|\lambda_h^* - \lambda\|_W \leq C^* h^{\gamma-1} \|f\|$$

Thus the only thing left for us to do is to evaluate $S(V^h)$ and

$$A_h = \sup_{\substack{v_h \in Z^h \\ v_h \neq 0}} \frac{b(v_h, \lambda_h^* - \lambda)}{\|v_h\|_V}$$

From the regularity of $\{T^h\}$, we can show that $S(V^h) \leq C^{**}h^{-1}$, while A_h may be estimated by taking notice of the relation

$$b(v_h, \lambda_h^* - \lambda) \leq \sum_{i=1}^{2} \|\theta'_{hi} - \partial_i w'_h\| \cdot \|p_{hi}^* - p_i\| \qquad \text{for } v_h = \{w'_h, \theta'_{h1}, \theta'_{h2}\} \in Z^h$$

Applying (7.25) to the above, we have

$$b(v_h, \lambda_h^* - \lambda) \leq C^{***}h \|v_h\|_V \|\lambda_h^* - \lambda\|_W \qquad \text{for all } v_h \in Z^h$$

or $A_h \leq C^{***}h \|\lambda_h^* - \lambda\|_W$, and the proof is completed.

As is well known, we may expect to get higher order estimates of errors in some other norms if we utilize the so-called Aubin–Nitsche trick:

Theorem 7.2 *With the same notation as in Theorem 7.1, we have*

$$\|w_h - w\| \leq Ch^{2\gamma} \|f\| \tag{7.28}$$

$$\|u_h - u\|_U \leq Ch^{3\gamma/2} \|f\| \tag{7.29}$$

Proof Again, we will only sketch the proof. To derive (7.28), define $f' \in L_2(\Omega)$ by $f' = w_h - w$, and let $\{u', \lambda'\} \in V \times W$ be the solution of Problem 7.3 for this f'. Then

$$\begin{aligned}\|f'\|^2 &= (f', w_h - w) = a(u', u_h - u) + b(u_h - u, \lambda') \\ &= a(u' - \bar{u}_h, u_h - u) + b(u' - \bar{u}_h, \lambda_h - \lambda) + b(u_h - u, \lambda' - \bar{\lambda}_h)\end{aligned}$$

for any $\{\bar{u}_h, \bar{\lambda}_h\} \in V^h \times W^h$. Choosing $\{\bar{u}_h, \bar{\lambda}_h\} \in V^h \times W^h$, the interpolate of $\{u', \lambda'\}$ in the above, we get

$$\begin{aligned}a(u' - \bar{u}_h, u_h - u) &\leq C^* \|\bar{u}_h - u'\|_V \|u_h - u\|_V \leq C^{**}h^\gamma \|f'\| h^\gamma \|f\| \\ b(u' - \bar{u}_h, \lambda_h - \lambda) &\leq C^* \|\bar{u}_h - u'\|_U \|\lambda_h - \lambda\|_W \leq C^{**}h^{\gamma+1} \|f'\| h^{\gamma-1} \|f\| \\ b(u_h - u, \lambda' - \bar{\lambda}_h) &= b(u_h - u_h^*, \lambda' - \bar{\lambda}_h) + b(u_h^* - u, \lambda' - \bar{\lambda}_h) \\ &\leq C^*\{h \|u_h - u_h^*\|_V + \|u_h^* - u\|_U\} \|\bar{\lambda}_h - \lambda'\|_W \\ &\leq C^{**}h^{\gamma+1} \|f\| h^{\gamma-1} \|f'\|\end{aligned}$$

where we have used the interpolate $u_h^* \in Z^h$ of u as well and utilized (7.25). The estimation (7.28) is a direct consequence of the above.

To derive (7.29), define $B_{hi} = \|\partial_i(w_h - w)\|$ for $i = 1, 2$. Then

$$\begin{aligned}B_{hi}^2 &= (\partial_i(w_h - w), \partial_i(w_h - w)) \\ &= (\partial_i(w_h - w), \partial_i w_h - \partial_i w_h^* - \theta_{hi} + \theta_{hi}^* + \partial_i(w_h^* - w) + \theta_{hi} - \theta_{hi}^*) \\ &= (\partial_i(w_h - w), \partial_i w_h - \partial_i w_h^* - \theta_{hi} + \theta_{hi}^*) + (\partial_i(w_h - w), \partial_i(w_h^* - w)) \\ &\quad + (w_h - w, \partial_i(\theta_{hi}^* - \theta_{hi}))\end{aligned}$$

where Green's formula has been employed to obtain the last equality. Evaluating the three terms in the right-hand side of the above, we find

$$(\partial_i(w_h - w), \partial_i w_h - \partial_i w_h^* - \theta_{hi} + \theta_{hi}^*) \leq C^* B_{hi} h \, \|u_h^* - u_h\|_V \leq C^{**} h^{\gamma+1} B_{hi} \, \|f\|$$

$$(\partial_i(w_h - w), \partial_i(w_h^* - w)) \leq C^* B_{hi} h^{\gamma+1} \, \|f\|$$

$$(w_h - w, \partial_i(\theta_{hi}^* - \theta_{hi})) \leq C^* h^{2\gamma} \, \|f\| \, h^{\gamma} \, \|f\|$$

Thus we have

$$B_{hi}^2 \leq C_1 \, \|f\| \, (h^{\gamma+1} B_{hi} + h^{3\gamma} \, \|f\|)$$

which implies $B_{hi} \leq C_2 h^{3\gamma/2} \, \|f\|$ since $3\gamma \leq 2\gamma + 2$. Similarly, we can obtain

$$\begin{aligned} \|\theta_{hi} - \theta_i\| &\leq \|\theta_{hi} - \theta_{hi}^* - \partial_i w_h + \partial_i w_h^*\| + \|\theta_{hi}^* - \theta_i\| + \|\partial_i(w_h - w_h^*)\| \\ &\leq C_1(h \, \|u_h^* - u_h\|_V + h^{\gamma+1} \, \|f\| + h^{3\gamma/2} \, \|f\|) \leq C_2 h^{3\gamma/2} \, \|f\| \end{aligned}$$

and (7.29) is also deduced.

Remark 7.5 The error estimates obtained above are of the same orders in h as those of the HCT conforming element with the identical degrees of freedom (cf. Chap. 6 of Ciarlet [7]). Therefore, we need not necessarily use this type of conforming plate-bending element with very complicated interpolation functions. When $\gamma = 1$, we cannot conclude convergence of λ_h to λ from the estimation (7.27). However, we can still assure the weak convergence of λ_h to λ in W as $h \to 0$ when w of Problem 7.1 belongs only to $H^3(\Omega) \cap H_0^2(\Omega)$, since we can show the existence of $\{\bar{\lambda}_h \in W^h\}_{h>0}$ for each $\bar{\lambda} \in W$ such that $\|\bar{\lambda}_h - \bar{\lambda}\|_W \to 0$ as $h \to 0$.

7.6 NUMERICAL RESULTS

Numerical solutions are obtained in the case where $D = 1.0$, $\nu = 0.3$, and Ω is a unit square defined by $\Omega = \{\{x_1, x_2\} \in R^2 \mid |x_i| < \frac{1}{2} \ (i = 1, 2)\}$. As for boundary conditions, we consider the clamped edge condition as well as the simply supported one. The load term f is taken as the uniform load $f(x) \equiv 1$ or the central concentrated one $f(x) \equiv \delta(x)$ (= Dirac's delta function). In the latter case of loading condition, $f \notin L_2(\Omega)$ and (7.3) does not hold.

Remark 7.6 Since Ω is of non-obtuse polygonal shape, (7.4) holds in this case. For the simply supported edge condition, the solution w of Problem 7.1 must be found in $H^2(\Omega) \cap H_0^1(\Omega)$, and V should be replaced by $H_0^1(\Omega) \times H^1(\Omega) \times H^1(\Omega)$. Since Ω is a square domain, we can use the Fourier double series to show that (7.4) is valid even for this edge condition.

Figure 7.1 shows examples of mesh patterns employed in our numerical computations. Only a quadrant portion of Ω is analysed with the symmetric properties of the problems taken into account. We first decompose the

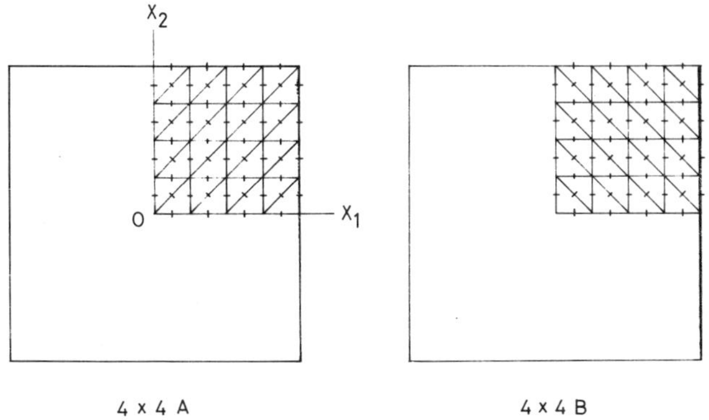

Figure 7.1 Examples of mesh patterns employed in the analysis ($N=4$)

Table 7.1 Numerical results for w_h at the centre of Ω

(a) Uniform distributed load $f(x)\equiv 1$

Boundary condition	Clamped		Simply supported	
Mesh type	A	B	A	B
$N=$ 1	0.7236441E-3	0.4189519-E	4.166667E-3	3.703768E-3
2	1.212608E-3	1.162063E-3	4.073719E-3	4.019489E-3
3	1.252880E-3	1.239306E-3	4.065748E-3	4.050648E-3
4	1.260992E-3	1.256052E-3	4.063812E-3	4.057824E-3
5	1.263456E-3	1.261235E-3	4.063101E-3	4.060220E-3
10	1.265198E-3	1.265013E-3	4.062434E-3	4.062164E-3
Exact	1.26532E-3		4.062353E-3	

(b) Central concentrated load $f(x)=\delta(x)$

Boundary condition	Clamped		Simply supported	
Mesh type	A	B	A	B
$N=$ 1	2.285192E-3	2.285192E-3	1.041667E-2	0.9446841E-2
2	5.080605E-3	5.041774E-3	1.130776E-2	1.114594E-2
3	5.424860E-3	5.393838E-3	1.147082E-2	1.141037E-2
4	5.517820E-3	5.498233E-3	1.152737E-2	1.149667E-2
5	5.555408E-3	5.542307E-3	1.155357E-2	1.153525E-2
10	5.599294E-3	5.595821E-3	1.158884E-2	1.158492E-2
Exact	5.605E-3		1.160084E-2	

(The calculations were performed in double precision arithmetic on FACOM M-200.)

quadrant portion into $N \times N$ uniform square meshes where N is the number of squares along each side of the quadrant portion, and then divide each small square into two triangles by the use of a diagonal with a common direction. Corresponding to the possible directions of diagonals, we consider two sequences of meshes denoted by A and B. In the present type of triangulations, we have $h = \sqrt{2}/(2N)$.

There exist four possible combinations of boundary and loading conditions, each of which is tested in our numerical calculations. Table 7.1 shows the numerical results for the lateral deflection at the centre of Ω. The exact values are due to Timoshenko and Woinowsky-Krieger [15]. We can see that the present method gives generally good results in all the cases treated here. In order to observe the details of convergence behaviours, we plot the errors $|w_h(x) - w(x)|$ at the centre of Ω versus various values of N. The results are illustrated in Figures 7.2 and 7.3 in the case of the uniform load $f(x) \equiv 1$. The results for the concentrated load are not included here, because our theoretical analysis does not apply to this case due to lack of regularity of solutions. The slopes of the plots in logarithmic scales appear to lie between -3 and -4, and this result does not contradict the estimation (7.28) which implicates that the error of w_h in $L_2(\Omega)$-norm is bounded by $Ch^4 = C^*N^{-4}$.

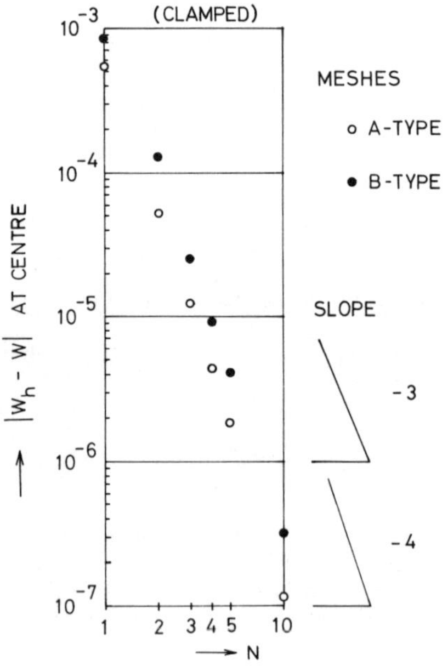

Figure 7.2 Convergence behaviour of w_h at centre ($f(x) \equiv 1$, clamped)

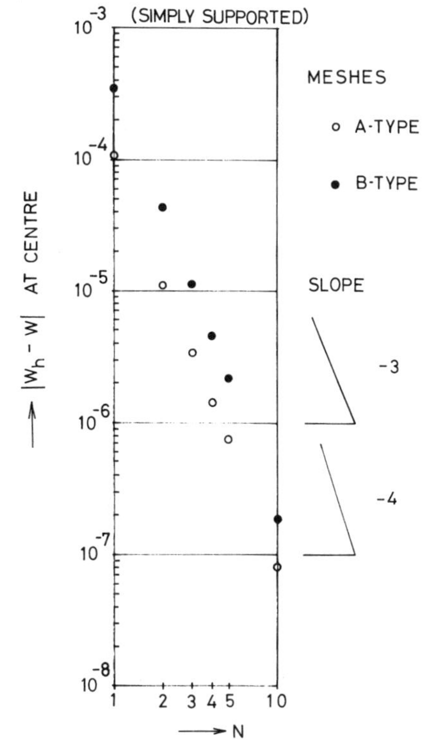

Figure 7.3 Convergence behaviour of w_h at centre ($f(x) \equiv 1$, simply supported)

7.7 CONCLUDING REMARKS

We have performed error analysis of a mixed finite element model related to the discrete Kirchhoff assumption. The accuracy of the numerical solutions is evaluated to be at the same level as the 12 d.o.f. HCT conforming element with high accuracy. In the present model, we can also compute the shearing forces of the plates as the Lagrange multipliers for the linear constraint of the Kirchhoff assumption. Numerical results are also obtained to see the accuracy of the approximate solutions.

The potentiality of the discrete Kirchhoff approach has not been fully recognized until quite recently. However, as is reported by Batoz, Bathe, and Ho [2], the simplest triangular element based on this approach is now available in the computer program ADINA. We have shown that the improvement of the accuracy of the simplest element may be achieved by the proposed mixed formulation.

REFERENCES

1. G. A. Wempner, J. T. Oden, and D. Kross, 'Finite-element analysis of thin plates', *J. Eng. Mech. Div., Proc. ASCE*, **1968,** No. EM6, 1273–1294 (1968).
2. J.-L. Batoz, K.-J. Bathe, and L.-W. Ho, 'A study of three-node triangular plate bending elements', *Int. J. Num. Meth. in Eng.*, **15,** No. 2, 1771–1812 (1980).
3. F. Kikuchi, 'On a finite element scheme based on the discrete Kirchhoff assumption', *Num. Math.*, **24,** No. 3, 211–231 (1975).
4. F. Kikuchi, 'The discrete Kirchhoff hypothesis applied to nonlinear problems', *Theoretical and Applied Mechanics*, **25,** 555–565 (1977).
5. F. Kikuchi, 'The discrete Kirchhoff assumption for large deflection analysis of plates', *ISAS Report, University of Tokyo*, **44,** No. 2, 11–22 (1979).
6. K. Washizu, *Variational Methods in Elasticity and Plasticity*, Pergamon Press, Oxford, 1974.
7. P. G. Ciarlet, *The Finite Element Method for Elliptic Problems*, North-Holland, Amsterdam, 1978.
8. V. A. Kondrat'ev, 'Boundary problems for elliptic equations in domains with conical or angular points', *Trans. Moscow Math. Soc.*, **16,** 227–313 (1967).
9. A. Mizutani, 'On the regularity of solutions of biharmonic equations in domains with angular points: Notes on a paper of Kondrat'ev (in Japanese)', *Kokyu-roku Research Inst. Math. Sci., Univ. Kyoto*, **329,** 2–9 (1978).
10. F. Kikuchi, 'On a finite element model based on the discrete Kirchhoff hypothesis (in Japanese)', *Computation and Analysis*, **6,** No. 1, 60–67 (1974).
11. I. Babuska and A. K. Aziz, 'Survey lectures on the mathematical foundations of the finite element method', in *The Mathematical Foundations of the Finite Element Method with Applications to Partial Differential Equations* (Ed. A. K. Aziz), pp. 3–359, Academic Press, New York, 1972.
12. F. Brezzi, 'On the existence, uniqueness and approximation of saddle-point problems arising from Lagrangian multipliers', *RAIRO, Série Analyse Numérique*, **8,** No. 2, 129–151 (1974).
13. M. Bercovier and O. Pironneau, 'Error estimates for finite element method solution of the Stokes problem in the primitive variables', *Num. Math.*, **33,** No. 2, 211–224 (1979).
14. F. Stummel, 'Basic compactness properties of nonconforming and hybrid finite element spaces', *RAIRO, Série Analyse Numérique*, **14,** No. 1, 81–115 (1980).
15. S. Timoshenko and S. Woinowsky-Krieger, *Theory of Plates and Shells*, McGraw-Hill, New York, 1959.

Hybrid and Mixed Finite Element Methods
Edited by S. N. Atluri, R. H. Gallagher, and O. C. Zienkiewicz

Chapter 8

A Variational Finite Strip Method with Mixed Variables

Bo Fransson and Alf Samuelsson

8.1 INTRODUCTION

In the same way as the ordinary finite element method can be classified as a Ritz–Galerkin method, the ordinary finite strip method as described by Cheung in 1968 [1] can be classified as a Kantorovich method, from 1933 (see [2] and [3]). According to this method the solution to an elliptic partial differential equation defined on a rectangular domain in the xy plane is written as a finite sum of products of assumed functions $\varphi_k(y)$ and unknown functions $f_k(x)$. Taking the minimum of the corresponding functional gives a set of ordinary differential equations in the unknown functions $f_k(x)$.

In the finite strip method already devised by Kantorovich the rectangular domain is divided into rectangular strips represented by lines (Figure 8.1a). The functions $f_k(x)$ are localized to these lines and the functions $\varphi_k(y)$ are interpolarion functions.

A somewhat different strip method has been used by Kärrholm, in 1956 [4], who establishes an ordinary differential equation for each strip by approximating the y derivatives of all orders in the partial differential equation by finite differences.

In this paper a mixed finite strip method is presented as developed by Fransson in his doctoral thesis [5]. A similar approach is developed by Wunderlich [6]. One advantage with this mixed version is that all types of boundary conditions can be handled in a straightforward manner. As the ordinary differential equations for the strips are of the first order a solution by simple successive integrations can be obtained directly. Also rectangular openings and internal supports can be handled.

The mixed finite strip method has in [5] not only been successfully applied to planar structures but also to non-planar ones, such as folded plates and multistorey buildings. Here, only inplane and laterally loaded elastic plates will be treated.

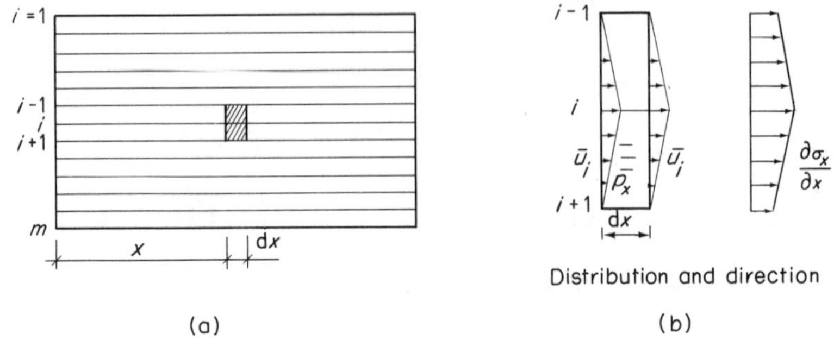

Figure 8.1 Finite strips, virtual displacements $\bar{u}_i$

8.2 INPLANE LOADED PLATES

8.2.1 Introduction

The influence of the choice of principal unknowns on the system of ordinary differential equations for inplane rectangular plates, which have been subdivided into strips by longitudinal lines, will be studied here. A comparison will be made between pure displacement, pure force, and mixed formulations.

It is assumed that the material is homogeneous, isotropic and linearly elastic, that the thickness is small and constant all over the plate, and that the deformations are small.

8.2.2 Mixed finite strip method

Let for a rectangular inplane loaded plate both stresses $\sigma_x(x, y)$, $\sigma_y(x, y)$, and $\tau_{xy}(x, y)$ and displacements $u(x, y)$ and $v(x, y)$ be treated as principal unknowns. Divide the plate in the x direction into strips defined by m longitudinal middle lines (Figure 8.1.a). Let any principal unknown, denoted $f(x, y)$, be expanded in a finite series

$$f(x, y) = \sum_{i=1}^{m} f_i(x)\varphi_i(y) \tag{8.1}$$

where the functions $f_i(x)$ are unknown and the function $\varphi_i(y)$ are prescribed. The linearly independent functions $\varphi_i(y)$ can be regarded as generalized coordinates. The unknown functions $f_i(x)$ are generalized stresses or displacements.

Assume now the same local and dimensionless function $\varphi_i(y)$ for all the principal unknowns and let this function be linear, equal to one at line i and zero at the adjacent lines. Equilibrium and compatibility equations are set up by use of the principles of virtual work. Alternatively, the method with weighted residuals could be used.

8.2.3 Equilibrium equations

In accordance with the finite element method the equilibrium equations for line i are established by application of the principle of virtual work on a small element dx. The generalized coordinate $\varphi_i(y)$ is used as virtual displacement. As an example the equilibrium equation in the longitudinal direction is derived. The virtual displacement is then $\bar{u}_i = \varphi_i(y)$. Equating external and internal work yields (see Figure 8.1b)

$$\int \frac{\partial \sigma_x}{\partial x} \bar{u}_i t \, \mathrm{d}y + \int p_x \bar{u}_i \, \mathrm{d}y = \int \tau_{xy} \frac{\mathrm{d}\bar{u}_i}{\mathrm{d}y} t \, \mathrm{d}y \tag{8.2}$$

The term to the right is the virtual internal work due to shear stress.

8.2.4 Compatibility equations

The compatibility equations of line i are set up by the principle of complementary virtual work. Virtual stresses instead of virtual displacement are then exerted upon the small element dx (see Figure 8.2). The equations are obtained by equating the external and the internal work. Only one virtual generalized stress is applied at a time. The elements dx and any part thereof must be in equilibrium in the virtual stress state, which sometimes calls for virtual external loads.

The shear compatibility equation will be derived here as an example. The virtual stress is then $\bar{\tau}_i = \varphi_i(y)$. The virtual load $\bar{p}_{xi} = -(\mathrm{d}\bar{\tau}_i/\mathrm{d}y)t$ is required to maintain equilibrium. The work equation is

$$\int \frac{\partial v}{\partial x} \bar{\tau}_i t \, \mathrm{d}y + \int u \bar{p}_{xi} \, \mathrm{d}y = \int \gamma_{xy} \bar{\tau}_i t \, \mathrm{d}y \tag{8.3}$$

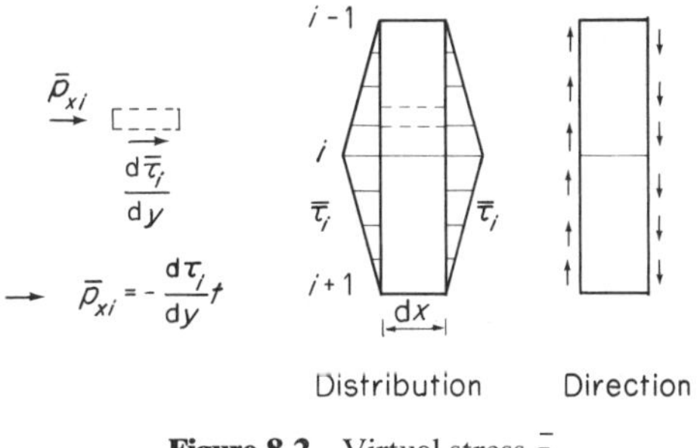

Figure 8.2 Virtual stress $\bar{\tau}_i$

8.2.5 Set of ordinary differential equations

The basic system of equations for inplane loaded plates using the mixed method is obtained by insertion of Hooke's law and the series expansions of the unknowns (8.1) into the equilibrium and compatibility equations of all lines i followed by integration. The basic system contains ordinary differential equations of the first order. It can be written in matrix form as

$$
\left[\begin{array}{cc|ccc}
 & & [A] & & \\
 & & & & [A] \\
\hline
[A] & & & & \\
 & & & & \\
 & [A] & & &
\end{array}\right]
\begin{Bmatrix} \{u\} \\ \{v\} \\ \{\sigma_x\} \\ \{\sigma_y\} \\ \{\tau\} \end{Bmatrix}'
-
\left[\begin{array}{cc|ccc}
 & & & & -[B_u]^T \\
 & & & -[B_v]^T & \\
\hline
 & & [F_e] & [F_\nu] & \\
 & [B_v] & [F_\nu] & [F_e] & \\
[B_u] & & & & [F_g]
\end{array}\right]
\begin{Bmatrix} \{u\} \\ \{v\} \\ \{\sigma_x\} \\ \{\sigma_y\} \\ \{\tau\} \end{Bmatrix}
= -
\begin{Bmatrix} \{P_x\} \\ \{P_y\} \\ \{0\} \\ \{0\} \\ \{0\} \end{Bmatrix}
\tag{8.4}
$$

where all subvectors are of type

$$\{f(x)\}^T = \{f_1(x), f_2(x), \ldots, f_i(x), \ldots, f_m(x)\}$$

The large coefficient matrices in (8.4) are thus square and of order $5m$, where m is the number of longitudinal lines. They have been divided into four parts by dashed lines. The first large coefficient matrix is symmetrical. The two parts along the main diagonal of the second of the large matrices are symmetrical. The lower of its off-diagonal parts is equal to the negative transpose of the upper one.

The submatrices $[A]$, $[B]$, and $[F]$ are thus square and of order m. They are three-diagonal and contain only constant elements. The matrix $[A]$ is symmetrical and positive definite.

Since the unknown stress vector $\{\sigma_y(x)\}$ is not differentiated in (8.4) it can be solved directly from the fourth vector equation

$$\{\sigma_y\} = -[F_e]^{-1}([B_v]\{v\} + [F_\nu]\{\sigma_x\}) \tag{8.5}$$

After substitution of (8.5) into (8.4) and change of order of the equations, the following set of equations is obtained:

$$\left[\begin{array}{cc:cc} [A] & & & \\ & [A] & & \\ \hdashline & & [A] & \\ & & & [A] \end{array}\right] \begin{Bmatrix} \{\sigma_x\} \\ \{v\} \\ \{u\} \\ \{\tau\} \end{Bmatrix}' - \left[\begin{array}{cc:cc} & & & -[B_u]^T \\ & & [B_u] & [F_g] \\ \hdashline [F_{e\nu}] & [B_\nu] & & \\ -[B_\nu]^T & [F_{ev}] & & \end{array}\right] \begin{Bmatrix} \{\sigma_x\} \\ \{v\} \\ \{u\} \\ \{\tau\} \end{Bmatrix} = - \begin{Bmatrix} P_x \\ 0 \\ 0 \\ P_y \end{Bmatrix} \tag{8.6}$$

where $[F_{e\nu}] = (1-\nu^2)[F_e]$

$[F_{ev}] = [B_v]^T[F_e]^{-1}[B_v]$

$[B_\nu] = \nu[B_v]$

The convenient form of the large coefficient matrices in (8.6) is obtained by separating the unknown functions into two groups. We observe that (σ_x, u) and (τ, v) are pairs of associated variables.

Two boundary conditions must be satisfied at each end of a line i. They can be directly formulated in the undifferentiated unknown functions of the mixed method. Some homogeneous boundary conditions are shown in

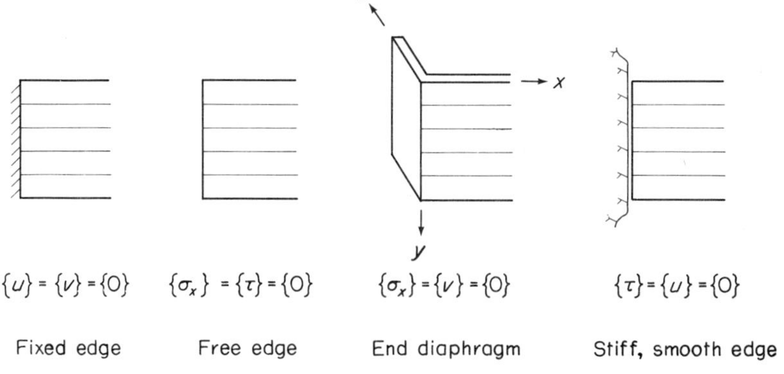

Figure 8.3 Homogeneous boundary conditions

Figure 8.3. Different boundary conditions can be prescribed at different locations of the same transversal edge.

8.2.6 Longitudinal boundary lines

The boundary conditions at the longitudinal edges can be satisfied within the system of equations of the mixed method. The derivation of boundary relations will be exemplified on the equations of longitudinal equilibrium and shear compatibility.

Suppose that line 1 is supported by continuous shear springs with the uniform stiffness S_x (see Figure 8.4). Apply the virtual displacement $\bar{u}_1 = \varphi_1(y)$. The principle of virtual work gives

$$\int \frac{\partial \sigma_x}{\partial x} \bar{u}_1 t \, dy + \int p_x \bar{u}_1 \, dy = \int \tau_{xy} \frac{d\bar{u}_1}{dy} t \, dy + S_x \bar{u}_1 \bar{u}_1(0) t \tag{8.7}$$

Consider then the virtual stress $\bar{\tau}_1 = \varphi_1(y)$ in Figure 8.5. The principle of complementary virtual work yields

$$\int \frac{\partial v}{\partial x} \bar{\tau}_1 t \, dy - \int u \frac{\partial \bar{\tau}_1}{\partial y} t \, dy = \int \frac{\tau}{G} \bar{\tau}_1 t \, dy + \frac{\tau_1}{S_x} \bar{\tau}_1(0) t \tag{8.8}$$

Homogeneous boundary conditions of the first line can be obtained from (8.7) and (8.8) as special cases. The spring stiffness is equal to zero for a free edge. The last term of (8.7) then vanishes and τ_1/S_x of the last term of (8.8) should be replaced by u_1. The boundary conditions for a fixed longitudinal line is obtained by replacing $S_x u_1$ of the last term of (8.7) by τ_1 and by cancelling the last term of (8.8), as the springs are infinitely stiff in that case. The first element of the main diagonal of the matrix $[B_u]$ in (8.4) can be determined from (8.8). Thus,

$$\begin{aligned} B_u(1, 1) &= 0,5t \qquad \text{(free edge)} \\ B_u(1, 1) &= -0,5t \qquad \text{(fixed (edge)} \end{aligned} \tag{8.9}$$

The end elements of the main diagonal of the matrix $[B_u]$ are thus

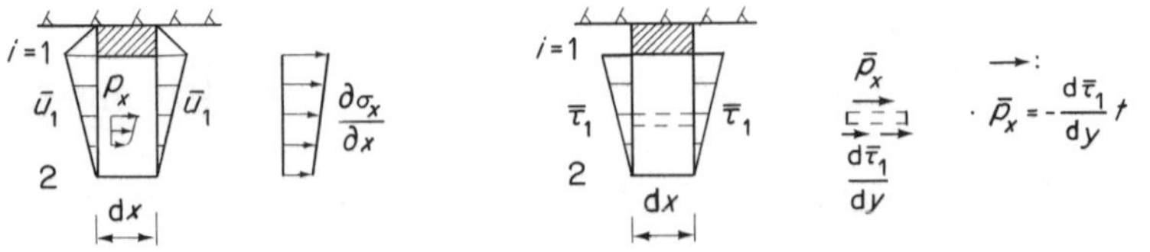

Figure 8.4 Shear spring fixed to rigid support. Virtual displacement $\bar{u}_1$ and virtual stress $\bar{\tau}_1$

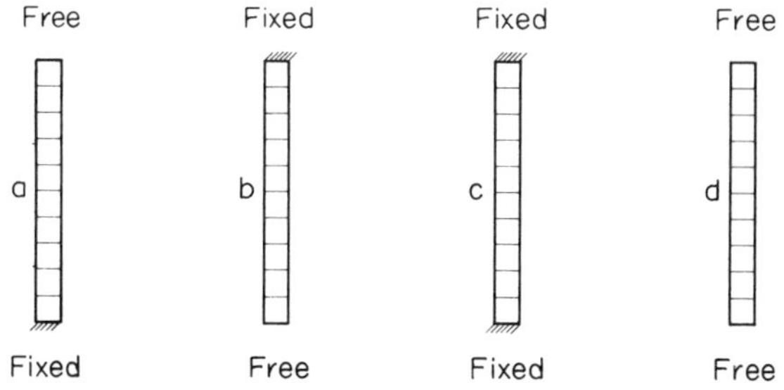

Figure 8.5 Homogeneous boundary conditions on longitudinal edges

influenced by the boundary conditions at the longitudinal edges. Similar derivations can easily be made also for the other longitudinal boundary line and for the matrix $[B_v]$.

The ordinary differential equations have been established above for a small transversal element dx. Four different combinations of homogeneous boundary conditions exist for the element dx (see Figure 8.5). It is supported in a stable and statically determinate way in the cases a and b in the figure. The matrices $[B]$ are then non-singular. The element dx is, however, supported in a statically and kinematically indeterminate way in the cases c and d respectively, which corresponds to singular matrices $[B]$. By insertion of fictitious springs as in the equations (8.7) and (8.8) a singular matrix $[B]$ can always be changed so that it becomes non-singular while the true boundary conditions are preserved by modifications of constitutive matrices $[F]$ with the fictitious spring stiffness. This procedure will increase the flexibility and the applicability of the mixed method and it is often valuable for composite structures as folded plates.

8.2.7 Displacement and force formulations

It is possible to eliminate any of the vector functions $\{u\}$, $\{v\}$, $\{\sigma_v\}$, $\{\sigma_y\}$, $\{\tau\}$ from the equations of the mixed method. A resulting system of ordinary differential equations with any desired combination of stresses and displacements as principal unknowns can thus be obtained.

8.2.8 Comparisons between finite strip formulations

The formulations with both displacements and stresses, with only displacements, and with only stresses as principal unknowns will be compared. The

resulting differential equations may be summarized as:
(a) Mixed Formulation:

$$\begin{bmatrix}[A] & & & \\ & [A] & & \\ & & [A] & \\ & & & [A]\end{bmatrix}\begin{Bmatrix}\{\sigma_x\}\\ \{v\}\\ \{u\}\\ \{\tau\}\end{Bmatrix}' - \begin{bmatrix} & & & -[B_u]^T\\ & & [B_u] & [F_g]\\ [F_{ev}] & [B_v] & & \\ -[B_v]^T & [F_{ev}] & & \end{bmatrix}\begin{Bmatrix}\{\sigma_x\}\\ \{v\}\\ \{u\}\\ \{\tau\}\end{Bmatrix} = -\begin{Bmatrix}\{P_x\}\\ \{0\}\\ \{0\}\\ \{P_y\}\end{Bmatrix} \tag{8.10}$$

(b) Displacement formulation:

$$\begin{bmatrix}[A_u] & \\ & [A_v]\end{bmatrix}\begin{Bmatrix}\{u\}\\ \{v\}\end{Bmatrix}'' + \begin{bmatrix} & [C]\\ -[C]^T & \end{bmatrix}\begin{Bmatrix}\{u\}\\ \{v\}\end{Bmatrix}' + \begin{bmatrix}[D_u] & \\ & [D_v]\end{bmatrix}\begin{Bmatrix}\{u\}\\ \{v\}\end{Bmatrix} = -\begin{Bmatrix}\{P_x\}\\ \{P_y\}\end{Bmatrix} \tag{8.11}$$

(c) Force formulation:

$$[A_F]\{F\}'''' + [C_F]\{F\}'' + [D_F]\{F\} = \{P_F\} \tag{8.12}$$

In the equations F is the line trace of the stress function. All the subvectors and submatrices in the equations above are of order m. The content of the submatrices is given in [5].

Consider first the number of *unknown vector functions.* The equations above must be modified to suit the solution methods and it is the number of unknowns of the modified systems that is of primary interest. To suit successive integrations, the systems of the displacement method and of the force method must be transformed into a system of the same form as the mixed formulation above and with the same number of principal unknown vectors. To suit Fourier analysis the mixed method and the displacement method can be modified so that only one vector function remains as the principal unknown. The resulting system will then have the same form as the force formulation above. The number of unknown functions of all the formulations is thus the same with respect to these solution methods.

All displacements and stresses are unknown in the mixed method. It is consequently easy to satisfy any *boundary conditions.* The derivations of equations for the longitudinal edges have been made so that the boundary conditions there are included in the coefficient matrices. In the displacement method it is more difficult to satisfy boundary conditions for stresses at the transversal edges, because they become coupled and contain terms with the first derivatives of the displacements. The boundary conditions of the longitudinal edges can, however, also be satisfied within the system of equations in this case. In the force method, it is very difficult to fulfil the boundary conditions for plates that are supported in a statically indeterminate way. If one of the longitudinal edges is fixed, redundant stress functions must be applied there and be determined from complicated geometrical conditions. Thus as far as the boundary conditions are concerned the mixed method is to be prefered.

The *stresses* are often of primary interest. In the displacement method they are obtained from expressions including one differentiation of displacement components. In the force method the stresses are obtained after double differentiation of the stress function.

All the stresses are unknown in the mixed formulation. If the solution is obtained with successive integrations they are obtained directly, undifferentiated. Since any differentiation means a loss of accuracy the mixed method gives the most accurate values of the stresses.

The mixed method thus seems to be preferable both to the fulfilment of the boundary conditions and to the accuracy of the stresses. Further, it gives the simplest and most flexible system of equations.

8.3 LATERALLY LOADED PLATES

8.3.1 Introduction

It will be shown that the set of ordinary differential equations for a laterally loaded rectangular plate is of the same principal form as for inplane loaded plates.

The principal unknowns in the mixed method are generalized forces and displacement. The forces consist of the bending moments $M_x(x, y)$, $M_y(x, y)$, the twisting moment $M_{xy}(x, y)$, and the shear forces $T_x(x, y)$ and $T_y(x, y)$. The displacements are the lateral deflection $w(x, y)$ and the rotations of sections parallel with the x and y directions. The rotations are denoted by $r_x(x, y)$ and $r_y(x, y)$. The unknowns are developed in finite series as in (8.1). The shear deformations $\gamma_x(x, y)$ and $\gamma_y(x, y)$ caused by the shear forces $T_x(x, y)$ and $T_y(x, y)$ are considered.

8.3.2 Set of ordinary differential equations

The basic set of equations for laterally loaded plates can, according to [5], be written in matrix form as

$$
\left[\begin{array}{ccc|ccccc}
 & & & [A] & & & & \\
 & & & & & [A] & & \\
 & & & & & & & [A] \\
\hline
[A] & & & & & & & \\
 & [A] & & & & & & \\
 & & [A] & & & & & \\
 & & & & & & & \\
 & & & & & & &
\end{array}\right]
\left\{\begin{array}{c}
\{w\} \\ \{r_x\} \\ \{r_y\} \\ \{T_x\} \\ \{T_y\} \\ \{M_x\} \\ \{M_y\} \\ \{M_{xy}\}
\end{array}\right\}'
$$

$$
-\left[\begin{array}{ccc|ccccc}
 & & & & -[B_w]^T & & & \\
 & & & [A] & & & & -[B_x]^T \\
 & & & & [A] & & -[B_y]^T & \\
\hline
 & -[A] & & [F_g] & & & & \\
[B_w] & & -[A] & & [F_g] & & & \\
 & & & & & [F_b] & [F_\nu] & \\
 & & [B_y] & & & [F_\nu] & [F_b] & \\
 & [B_x] & & & & & & [F_t]
\end{array}\right]
\left\{\begin{array}{c}
\{w\} \\ \{r_x\} \\ \{r_y\} \\ \{T_x\} \\ \{T_y\} \\ \{M_x\} \\ \{M_y\} \\ \{M_{xy}\}
\end{array}\right\}
= -
\left\{\begin{array}{c}
\{P_z\} \\ \{Q_x\} \\ \{Q_y\} \\ \{0\} \\ \{0\} \\ \{0\} \\ \{0\} \\ \{0\}
\end{array}\right\}
\tag{8.13}
$$

The set of equations for laterally loaded plates (8.13) is of the same principal form as for inplane loaded plates (8.4). The same solution methods and computer programs can therefore be used in the two cases.

In the classical Kirchhoff plate theory the shear deformations $\gamma_x(x, y)$ and $\gamma_y(x, y)$ are neglected. Corresponding eliminations in (8.13) will give a resulting set of differential equations with $\{M_x(x)\}$, $\{w(x)\}$, $\{M_x(x)\}'$, and $\{w(x)\}'$ as unknowns.

8.4 SOLUTION METHODS

8.4.1 Introduction

The set of ordinary differential equations established in a finite strip procedure is commonly solved by use of orthogonal series with terms satisfying the boundary conditions, thus separating the set of equations (see [7, 8]). A method with combined integration and use of Fourier series has in [5] been applied to the mixed formulations of plates with ordinary differential equations of the second order. The integrations increase the rate of convergence of the series substantially.

A more versatile method for solution of a set of differential equations of the first order as in the mixed formulation (8.6) is the method with successive integrations described below. The unknown boundary values at one end must then be determined by the boundary conditions at the other end. This method is referred to in literature by many different names, such as the initial parameter method, transfer matrix method, matrix progression method, and line solution method. It will here be called the method of successive integrations to emphasize its large range of applicability. Arbitrary boundary conditions, internal supports, rectangular openings, and varying properties along the strips can thus easily be handled.

8.4.2 Single-step successive integration

The set of ordinary differential equations for inplane loaded plates (8.6) can be reformulated as

$$\{f\}' = [G]\{f\} + \{g\} \tag{8.14}$$

where $\{f(x)\}^T = \{\sigma_x(x), \tau(x), u(x), v(x)\}$ is unknown and $\{g(x)\}$ contains the load. The system consists of $4m$ equations, where m is the number of longitudinal lines.

An equivalent integral equation to (8.14) is

$$\{f(x)\} = \{c\} + \int_0^x ([G]\{f(t)\} + \{g(t)\})\,dt \tag{8.15}$$

where the vector $\{c\}$ contains the initial values of the unknowns, i.e.

$$\{c\} = \begin{bmatrix} \{c_\sigma\} \\ \{c_\tau\} \\ \{c_u\} \\ \{c_v\} \end{bmatrix} \begin{bmatrix} \{\sigma_x\} \\ \{\tau\} \\ \{u\} \\ \{v\} \end{bmatrix} \text{ at } x = 0 \tag{8.16}$$

The equation (8.15) can be solved by successive integrations using the recurrence formula

$$\{f(x)\}_n = \{c\} + \int_0^x ([G]\{f(t)\}_{n-1} + \{g(t)\}]\,dt \tag{8.17}$$

The approximate solution $f(x)$ obtained by n_i successive integrations can, if the load vector $\{g(x)\}$ is constant, be written formally as

$$\{f\} = [H]\{c\} + \{h\} \tag{8.18}$$

where
$$[H(x)] = \sum_{n=0}^{n_i} [G]^n \frac{x^n}{n!}$$

$$[G]^0 = [I]$$

$$\{h(x)\} = \left(\sum_{n=1}^{n_i} [G]^{n-1} \frac{x^n}{n!} \right) \{g\}$$

The values of $2m$ of the elements of $\{c\}$ in (8.16) are known from the boundary conditions at the initial edge, $x=0$. The other $2m$ values should be determined so that the boundary conditions at the far edge, $x=l_x$, are fulfilled. An algebraic set of equations, from which the unknown part of $\{c\}$ can be calculated, is established by the use of (8.18).

The integration above is performed in one step. The integration can, however, be performed in several steps by dividing the structure into several parts by transversal lines. Varying properties along the strips can then be allowed.

In the multistep method the matrix $[H]$ of (8.18) is determined for each transversal part. The calculation starts with the first element. The initial values of the second element is obtained by matrix multiplication with the matrix $[H]_1$ on the initial vector $\{c\}_1$ with due regard to the external load within the first transversal element. By successively adding the transversal elements the far end of the structure is finally reached. The unknown initial values are then determined so that the boundary conditions at the far end are satisfied. By repeating the calculation from the initial end using the determined initial values all the displacements and all section forces can be determined all over the structure.

For a structure, which is geometrically uniform in the longitudinal direction, all the elements are made identical by a subdivision into equal transversal parts. Then only one matrix $[H]$ needs to be established from (8.18). In addition, the number of successive integrations to reach the far end can be reduced by using the following property of the matrix $[H]$:

$$[H(x_1+x_2)]=[H(x_1)][H(x_2)] \tag{8.19}$$

8.4.3 Numerical stability

Numerical investigation of the successive integration procedure has shown that the method is stable only if the longitudinal length l_x of a structure is less than a certain critical value l_c. It was found that the critical length depends on the type of structure, on the strip subdivision, and on the number of integrations n_i in (8.18).

Consider a plate the length of which is longer than the critical length l_c. Divide the plate by transversal lines into intervals, whose lengths are less than l_c. The multistep successive integration can thus be used for each interval provided that the inknown initial values are of a local character. A numerically stable method can then be obtained if the originally unknown initial values are replaced by new ones located at the initial end of each large interval, as the integration continues along the strips. The corresponding solution procedure is described in [5].

It is important to determine the critical length correctly at the start to avoid reanalysis due to numerical instability of the solution. The critical length for a certain type of structure can be determined by a numerical investigation. Obtained values for inplane and laterally loaded plates are

$$l_c \geq 5 b_{\min} \qquad \text{(inplane loaded)}$$
$$l_c \geq 10 b_{\min} \qquad \text{(laterally loaded)} \tag{8.20}$$

where $b_{\min}$ is the smallest strip width. The necessary number of terms in the summation (8.18) depends on the length of the small transversal element. If its length is equal to the width of the smallest strip ten terms are sufficient.

8.5 NUMERICAL EXAMPLES

8.5.1 Continuous deep beam

The continuous deep beam in Figure 8.6 has been calculated by the method of successive integrations using two different strip subdivisions. A finite element calculation using a mesh of 10×21 for the left half of the structure was also performed. The displacements were the principal unknowns and they varied linearly within the rectangular elements. The results are compared in Figure 8.7. The agreement between the finite element method and

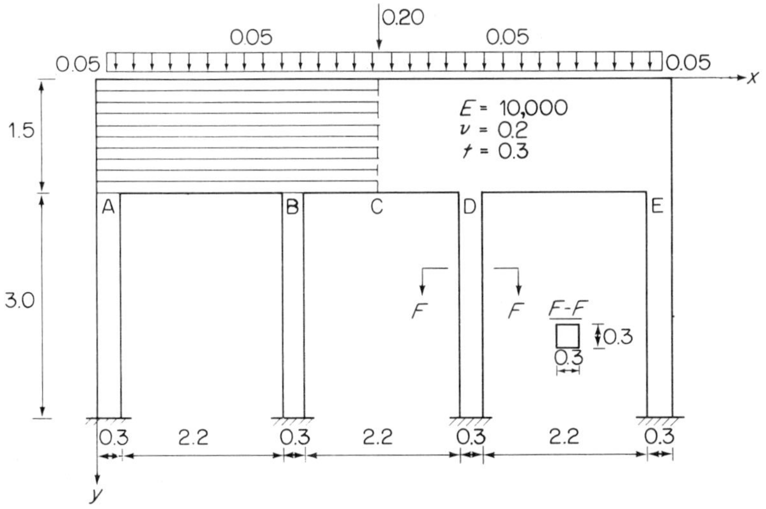

Figure 8.6 Continuous deep beam subjected to uniform and concentrated transversal load at the top. Strip subdivision

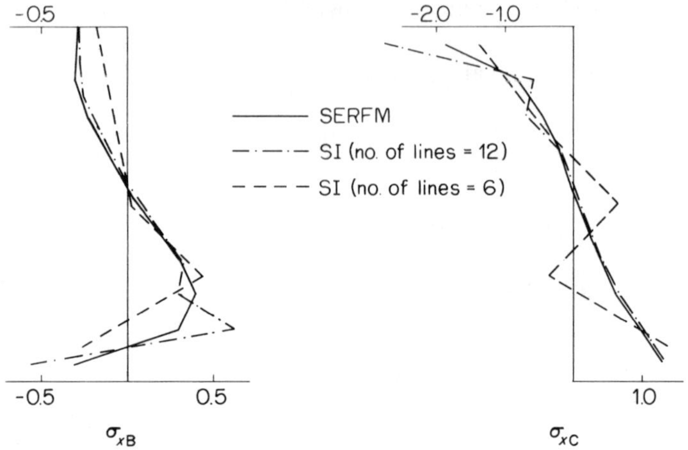

Figure 8.7 Variation of longitudinal stress above the internal support and at mid-span of the deep beam in Figure 8.6

the method of successive integrations (SI) with twelve lines is also good close to the stress concentrations if the mean values at the middle of the strips are used there for the finite strip method.

8.5.2 Rectangular plate with uniform transversal load

The efficiency of the finite strip method will be compared with that of the finite element method for the laterally loaded plate in Figure 8.8. The plate and the load are symmetrical about a transversal line through the centre. Both half the plate (h), taking advantage of the symmetry, and the whole plate (w) have been calculated to study the influence of the length to width

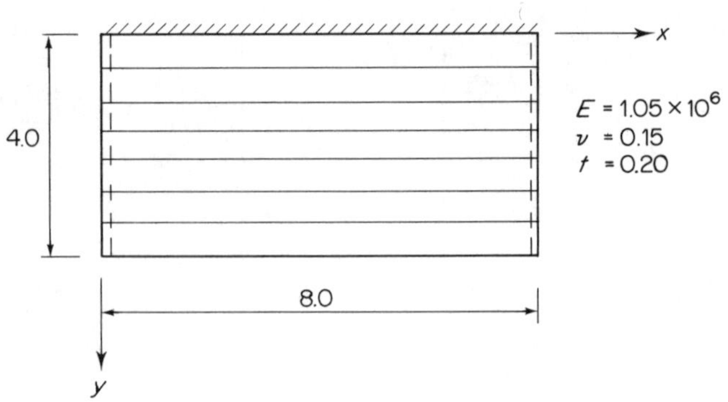

Figure 8.8 Rectangular plate subjected to uniform lateral load $p_z(x, y) = 1.0$. Strip subdivision

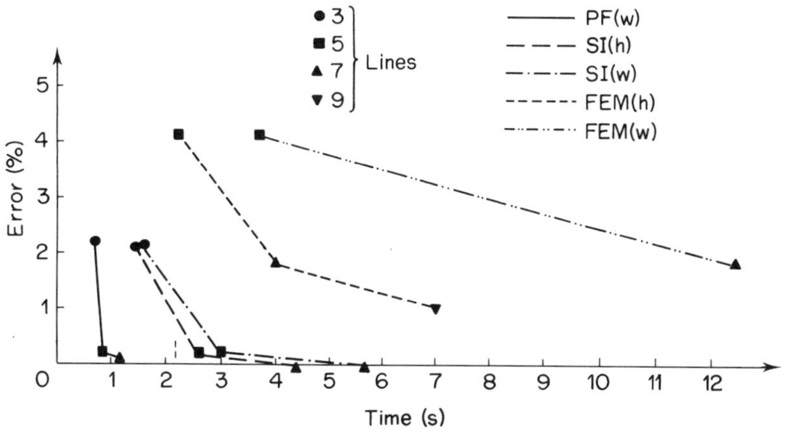

Figure 8.9 Comparison between the finite strip method and the finite element method in calculation of the displacement $w(4.0, 4.0)$

ratio of a structure. Strip subdivisions with 3, 5, 7, and 9 longitudinal lines were used.

The plate has been analysed by pure Fourier series (PF) and by successive integrations (SI). A mixed finite element programme was used. Finite element meshes were obtained by dividing the strips into square elements by transversal lines. In the FEM programme the twisting moment was constant and the bending moments varied linearly within the element, while the lateral displacement varied as a bilinear polynomial.

The results obtained for the displacement w and for the bending moment M_x at (4.0, 4.0) are shown in Figures 8.9 and 8.10 respectively. The

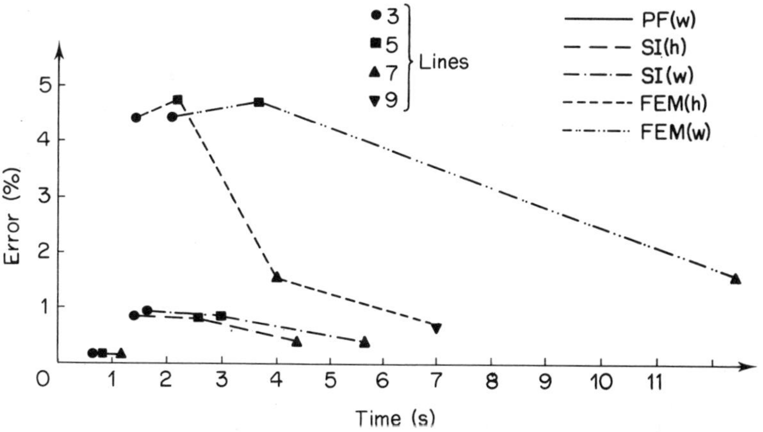

Figure 8.10 Comparison between the finite strip method and the finite element method in calculation of the bending moment M_x (4.0, 4.0)

computer time (central processing time) in second and the percentage deviation from the values of the methods of successive integration with twelve lines for the different methods and subdivisions are marked in the figures. It is obvious from the figures that for the investigated cases:

(a) The finite strip method gives more accurate results than the finite element method for each subdivision used.
(b) The Fourier method is by far the most efficient of the solution methods.
(c) The method of successive integration gives slightly longer computer time than the finite element method for a square plate with uniform properties along the strips for each subdivision used.
(d) The method of successive integration competes, nevertheless, favourably with the finite element method for the uniform square plate, as a subdivision with a smaller number of lines can be used for the finite strip method due to its better accruacy.
(e) The method of successive integration becomes even more advantageous when the length to width ratio of the plate increases.

8.6 CONCLUSIONS

For laterally loaded plates, mixed finite element methods have been found useful. The main reasons are that the degree of the polynomial shape functions can be kept low and that displacements and stresses can be obtained with the same accuracy. The same arguments are valid for mixed finite strip methods.

Two further arguments can, however, be added in favour of mixed strip methods, namely that arbitrary boundary conditions can be easily treated and that the partial differential equations can be reduced to a set of first-order ordinary differential equations making solution methods applicable which can handle, for example, internal supports and openings efficiently. In addition, for inplane loaded plates, the mixed finite strip method is found to be useful and indeed superior to non-mixed methods.

REFERENCES

1. Y. K. Cheung, 'Finite strip method in the analysis of elastic plates with two opposite ends', *Proc. Inst. Civ. Engrs.*, **40,** 1–7 (1968).
2. L. V. Kantorovich, *A Direct Method for Calculation of the Minimum of a Double Integral* (in Russian), No. 5, Isvestija Academy of Science, 1933.
3. L. V. Kantorovich and V. I. Krylov, *Approximate Methods of Higher Analysis*, Interscience Publishers, New York, 1958.
4. G. Kärrholm, *Parallellogram Plates Analysed by Strip Method*, Chalmers University of Technology, Diss, Göteborg, 1956.

5. B. Fransson, *A Generalized Finite Strip Method for Plate and Wall Structures*, Chalmers University of Technology, Department of Structural Mechanics Publication 77:1 (thesis), Göteborg, 1977.
6. W. Wunderlich, 'Ein verallgemeinertes Variationsverfahren zur vollen oder teilweisen Diskretisierung mehrdimensionaler Elastizitätsprobleme', *Ingenieurarchiv*, **39,** 230–247, 1970.
7. Y. K. Cheung, *Finite Strip Method in Structural Analysis*, Pergamon, Oxford, 1976.
8. Y. C. Loo and A. R. Cusens, *The Finite Strip Method in Bridge Engineering*, Viewpoint, London, 1978.

Hybrid and Mixed Finite Element Methods
Edited by S. N. Atluri, R. H. Gallagher, and O. C. Zienkiewicz

Chapter 9

String Net Function Approximation and Quasi-Conforming Technique

Tang Limin, Chen Wanji, and Liu Yingxi

9.1 INTRODUCTION

The approximate solutions of the finite element method are obtained from certain kinds of interpolating functions and the success of this method is due to the use of piecewise polynomials with variational constraint. These are very well done in one-dimensional cases and also in two- and three-dimensional cases for C^0 problems only. However, in the two-dimensional cases and higher, when the fundamentals involve second-order derivatives, the interelement compatibility problem will arise. Many researches have been worked on the modification of variational principles while very few have dealt with the approximating functions themselves. In fact, it is very difficult, or rather impossible, to describe a series of surfaces of successive smoothness with prescribed curvatures by the piecewise polynomials. After all, the piecewise polynomials are nowhere the proper functions for surface fitting, and any function space which has been used has to be enlarged.

A simple and more natural way to describe a surface, proposed here, is to use the 'string net function approximation'. With this function, the so-called 'quasi-conforming element technique', a unified approach by using the element boundary integral to obtain the strain discretization in the element has been developed. Then the element stiffness matrices can be obtained as usual without altering the conventional computing frame. The quasi-conforming element technique is a general method which treats the conforming, non-conforming, and hybrid elements in a simple unified way. This method was initially inspired by Pian's pioneering work [1], but the theory developed here emphasizes changing the idea of interpolating functions, which seems more fundamental. However, this method is related to the so-called 'generalized hybrid model' which can also be derived from the Hu–Washizu principle.

9.2 STRING NET FUNCTION APPROXIMATION

Let the domain on which the surface (representing the displacement) is defined be divided into many subdomains (elements) as usual (Figure 9.1).

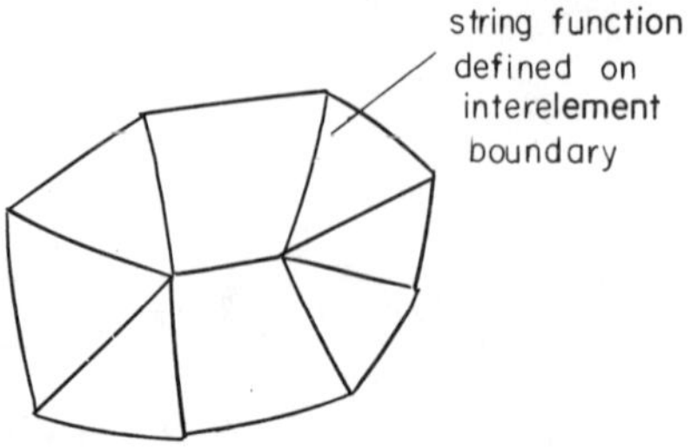

Figure 9.1 Finite element domain

However, now the functions describing the surface are only defined on the interelement boundaries and leave the functions inside the elements undefined explicitly. For surface fitting, this method is much easier than the use of piecewise polynomials which are defined on each element and leave the interelement boundaries incompatible. The string net approximation is actually what people have already been doing in many engineering practices for surface fitting. So what we need is only a set/net of one-dimensional string functions and some nodal values on it. This is a natural extension of a one-dimensional spline to a two-dimensional one. Thus there will be no more problems of interelement conformity. What kinds of nodal values of the surface should be used will be determined by the boundary integral of the derivatives (strains) of the surface (displacement).

We shall discuss the two-dimensional cases only; the three-dimensional cases are similar and straightforward.

9.3 DISCRETIZATION OF DERIVATIVES (STRAINS) IN AN ELEMENT AND QUASI-CONFORMING ELEMENT TECHNIQUE

Instead of assuming the shape functions of displacements we start on discretizing the derivatives (strains) in an element. For the two-dimensional cases, the derivatives $\boldsymbol{\varepsilon}$ of the surface (displacement w) are

$$\boldsymbol{\varepsilon} = \left(\frac{\partial w}{\partial x}, \frac{\partial w}{\partial y}\right)^T, \quad \left(\frac{\partial^2 w}{\partial x^2}, \frac{\partial^2 w}{\partial y^2}, 2\frac{\partial^2 w}{\partial x\, \partial y}\right)^T, \tag{9.1}$$

depending on what kinds of problems are prescribed. The derivative is

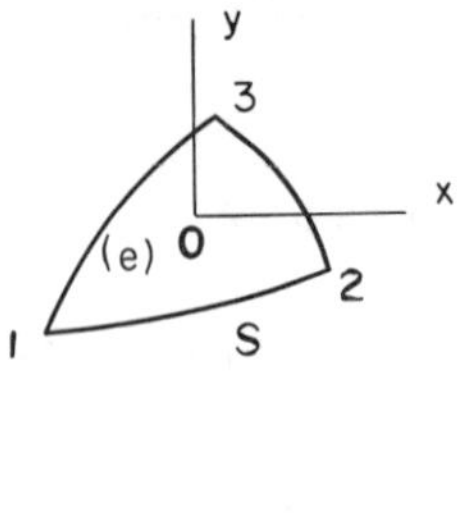

Figure 9.2 A single element (e) with boundary line S where O is the centroid of the element

supposed to be a set of piecewise functions defined on each element separately. These functions are continuous in elements but discontinuous across the interelement boundaries. They can be expressed by Taylor's expansion, for example, in element (e) (Figure 9.2):

$$\frac{\partial w}{\partial x}=\left(\frac{\partial w}{\partial x}\right)_0+\left(\frac{\partial^2 w}{\partial x^2}\right)_0 x+\left(\frac{\partial^2 w}{\partial x\,\partial y}\right)_0 y+\left(\frac{\partial^3 w}{\partial x^3}\right)_0 \frac{x^2}{2}+\left(\frac{\partial^3 w}{\partial y\,\partial x^2}\right)_0 xy$$
$$+\left(\frac{\partial^3 w}{\partial y^2\,\partial x}\right)_0 \frac{y^2}{2}+\cdots \tag{9.2}$$

Equation (9.2) can also be expressed approximately by a polynomial with n truncated terms:

$$\frac{\partial w}{\partial x}=\alpha_0+\alpha_1 x+\alpha_2 y+\cdots=\sum_i^{n-1}\alpha_i\varphi_i \tag{9.3}$$

where φ_i are the basis trial functions. Multiply both sides by the basis test functions

$$\psi_i=1,\,x,\,y,\,x^2,\ldots \tag{9.4}$$

and integrate over the element domain (e); we then have equations to determine the coefficients α_i as follows:

$$\sum_i^{n-1}\alpha_i\iint_{(e)}\varphi_i\psi_j\,\mathrm{d}x\,\mathrm{d}y=\iint_{(e)}\frac{\partial w}{\partial x}\psi_j\,\mathrm{d}x\,\mathrm{d}y \tag{9.5}$$

The double integral on the right-hand side of the equation can always be converted to the line integral by Green's theorem:

$$\iint_{(e)}\frac{\partial w}{\partial x}\psi_j\,\mathrm{d}x\,\mathrm{d}y=\oint_S w\psi_j n_x\,\mathrm{d}S-\iint_{(e)}w\frac{\partial\psi_j}{\partial x}\,\mathrm{d}x\,\mathrm{d}y \tag{9.6}$$

where S is the boundary line of the element, $n_x=\cos(n,x)$ with n directing to the outward normal of the boundary lines. Similarly, for second-order derivatives, they are

$$\iint_{(e)}\frac{\partial^2 w}{\partial x_1\,\partial x_2}\psi_j\,\mathrm{d}x_1\,\mathrm{d}x_2=\frac{1}{2}\left\{\oint_S\left(\frac{\partial w}{\partial x_2}n_{x_1}+\frac{\partial w}{\partial x_1}n_{x_2}\right)\psi_j\,\mathrm{d}S\right.$$
$$\left.-\oint_S\left(\frac{\partial\psi_j}{\partial x_1}n_{x_2}+\frac{\partial\psi_j}{\partial x_2}n_{x_1}\right)w\,\mathrm{d}S+2\iint_{(e)}w\frac{\partial^2\psi_j}{\partial x_1\,\partial x_2}\mathrm{d}x_1\,\mathrm{d}x_2\right\} \tag{9.7}$$

The line integrals on the right-hand sides of (9.6) and (9.7) can be evaluated approximately by the string net function $\tilde{W}$ on the boundaries just mentioned, and the double integrals can be also evaluated approximately by an internal function $\hat{W}$ without concern for the interelement compatibility.

Both of them must be evaluated with sufficient accuracy and are expressed in terms of nodal values

$$\boldsymbol{\delta} = \left(W_k, \left(\frac{\partial w}{\partial x}\right)_K, \left(\frac{\partial w}{\partial y}\right)_K, \ldots \right)^T$$

where K is the index of nodal number. Then we shall obtain

$$\boldsymbol{\varepsilon} = \mathbf{B}\boldsymbol{\delta} \tag{9.8}$$

where $\boldsymbol{\beta}$ is a discretization matrix. This completes the discretization in an element which are expressed by the nodal displacement parameters. Then the stiffness matrix can be obtained as usual:

$$\mathbf{K}^e = \iint_{(e)} \mathbf{B}^T\mathbf{D}\mathbf{B}\, dx\, dy \tag{9.9}$$

where $\mathbf{D}$ is the elastic constant matrix.

We have named these techniques as the 'quasi-conforming element technique' [2, 3, 4] because the uses of the string net function for fulfilling the requirement of element boundary conformity are carried out actually under the integral sign. This method can also be derived using the Hu–Washizu principle and the formulation may be called the 'generalized hybrid model' with four variables σ_{ij}, ε_{ij}, $\tilde{w}$, and $\hat{w}$, in which the first two correspond to ψ_i and ϕ_i respectively [5, 6, 7]. The quasi-conforming element technique is simple, straightforward, and more fundamental. It is entirely a mathematical method to describe a surface by the string net function with prescribed curvatures so that it means more than a pure mechanical formulation of elasticity. These methods are best illustrated by the following examples.

9.4 CONFORMING, NON-CONFORMING, AND DIVERGENT ELEMENT

Example 9.1. Suppose the strains of a triangular element (Figure 9.3) are required to be constants only. They are

$$\boldsymbol{\varepsilon} = \begin{Bmatrix} \dfrac{\partial w}{\partial x} \\ \dfrac{\partial w}{\partial y} \end{Bmatrix} = \begin{Bmatrix} \alpha_0 \\ \beta_0 \end{Bmatrix} \tag{9.10}$$

Multiply both sides of (9.10) by 1 and follow with (9.5) and (9.6); then we obtain

$$\begin{aligned} \alpha_0 A = \alpha_0 \iint_{(e)} 1\, dx\, dy &= \iint_{(e)} \frac{\partial w}{\partial x} 1\, dx\, dy = \oint_S w n_x\, dS \\ &= (n_x)_{12} \int_1^2 w\, ds + (n_x)_{23} \int_2^3 w\, ds + (n_x)_{31} \int_3^1 w\, dS \end{aligned} \tag{9.11}$$

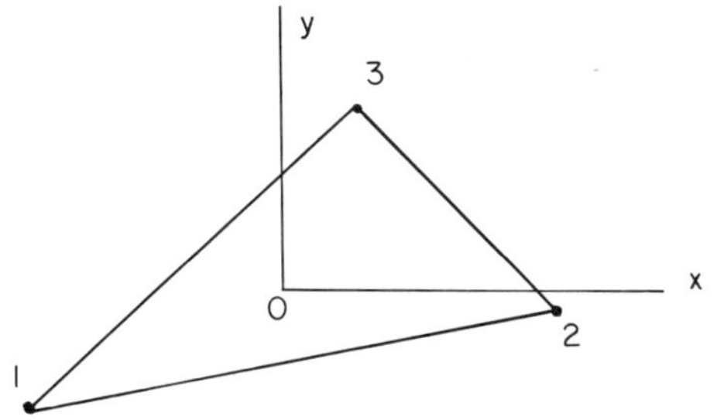

Figure 9.3 Triangular element with three nodes at the vertices

where $A = \iint_{(e)} dx\, dy$ is the area of the triangle. The line integral can be evaluated approximately by numerical integration. For example, by the trapezoidal rule, we have

$$\int_1^2 w\, dS = \frac{h_{12}}{2}(w_1 + w_2) + O(h_{12}^3) \tag{9.12}$$

where h_{12} is the length of boundary line 1–2 and $O(h_{12}^3)$ represents the order of error. If we notice that

$$(n_x)_{12} = \frac{y_{21}}{h_{12}}, \qquad (n_y)_{12} = \frac{x_{12}}{h_{12}} \tag{9.13}$$

where $y_{ij} = y_i - y_j$, $x_{ij} = x_i - x_j$, then we obtain the equation (9.8) $\boldsymbol{\varepsilon} = \mathbf{B}\boldsymbol{\delta}$ as

$$\begin{Bmatrix} \alpha_0 \\ \beta_0 \end{Bmatrix} = \frac{1}{2A}\begin{Bmatrix} y_{23} & y_{31} & y_{12} \\ x_{32} & x_{13} & x_{21} \end{Bmatrix}\begin{Bmatrix} w_1 \\ w_2 \\ w_3 \end{Bmatrix} \tag{9.14}$$

What we have done in (9.12) is that at the same time we have assumed there is the string function on the line 1–2 and a linear function with nodal values w_1 and w_2.

If (9.14) is checked by Taylor's expansion we shall find that

$$\alpha_0 = \left(\frac{\partial w}{\partial x}\right)_0 + O(h), \qquad \beta_0 = \left(\frac{\partial w}{\partial y}\right)_0 + O(h) \tag{9.15}$$

as expected, where $O(h)$ represents the order of the truncated terms with h being the longest length of the triangle. So far, no displacement function in the element has been concerned. But by integrating the strains directly the displacement function is given by

$$w = \int_0^{(x,y)} \frac{\partial w}{\partial x} dx + \frac{\partial w}{\partial y} dy = w_0 + \alpha_0 x + \beta_0 y \tag{9.16}$$

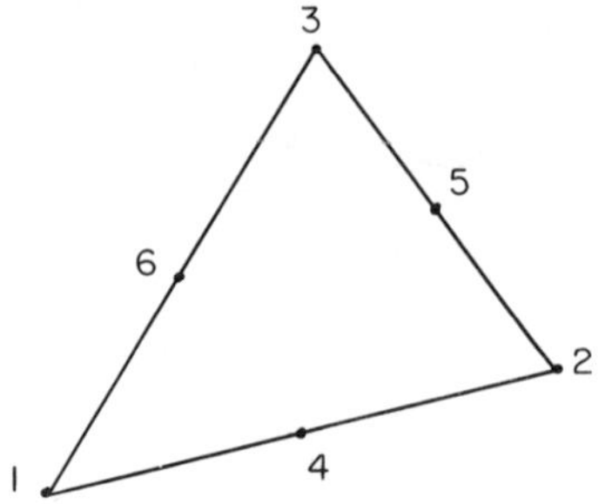

Figure 9.4 Triangular element with nodes at the middle of the edges

This is a linear piecewise polynomial and it fits the string function on the boundaries exactly. Therefore it is a conforming element in the conventional sense. In this case, the string net function coincides with the conforming piecewise polynomial.

Obviously the quasi-conforming technique can be applied to the elements with curved boundaries as well simply by changing the length of a straight line to that of a curved one; it does not matter what kind of displacement function there should be in the element.

Example 9.2 If the integration (9.12) is evaluated by the rectangular rule (Figure 9.4) as

$$\int_1^2 w \, dS = w_4 h_{12} + O(h_{12}^3) \tag{9.17}$$

then we obtain a non-conforming element with only one mode at the middle of each edge of the triangle. This kind of element passes the patch test, but their assembly may become kinematically unstable.

Example 9.3 If the nodes in the last example are not at the middle of the edges (Figure 9.5), then the order of accuracy of the integral (9.17) becomes lower as

$$\int_1^2 w \, dS = w_4 h_{12} + O(h_{12}^2) \tag{9.18}$$

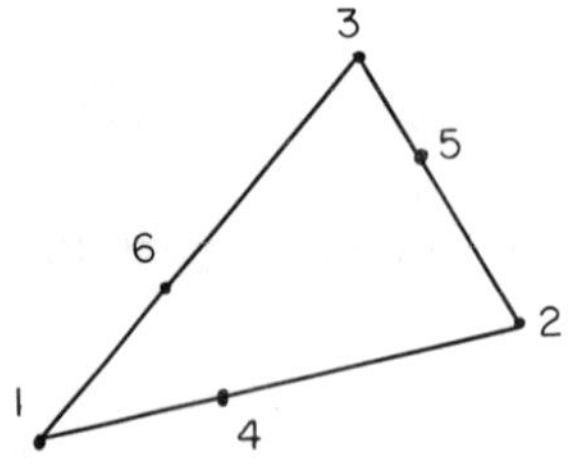

Figure 9.5 Triangular element with nodes not at the middle of the edges and not at the vertices either

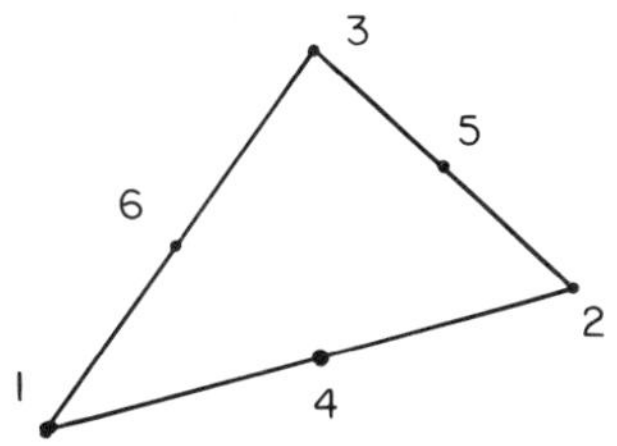

Figure 9.6 Triangular element with six nodes

The strain will become such as

$$\alpha_0 = \frac{1}{2A}(y_{21}w_4 + y_{32}w_5 + y_{13}w_6) + O(1) \tag{9.19}$$

This kind of element will become divergent.

Example 9.4 If more terms in the expression of derivatives are involved such as

$$\frac{\partial w}{\partial x} = \alpha_0 + \alpha_1 x + \alpha_2 y, \qquad \frac{\partial w}{\partial y} = \beta_0 + \beta_1 x + \beta_2 y \tag{9.20}$$

then the number of nodes on the boundary must be increased (Figure 9.6) to keep sufficient accuracy of the boundary integrals (in this case Simpson rule will be used) and the double integral

$$\iint_{(e)} w \, dx \, dy \tag{9.21}$$

should also be evaluated with sufficient accuracy.

9.5 SIX-PARAMETER PLATE-BENDING ELEMENT

Example 9.5 The thin plate bending strain is given by

$$\boldsymbol{\varepsilon} = \left(\frac{\partial^2 w}{\partial x^2}, \frac{\partial^2 w}{\partial y^2}, 2\frac{\partial^2 w}{\partial x \, \partial y}\right) \tag{9.22}$$

If only constant strain is required, put

$$\frac{\partial^2 w}{\partial x^2} = \alpha_0, \qquad \frac{\partial^2 w}{\partial y^2} = \beta_0, \qquad 2\frac{\partial^2 w}{\partial x \, \partial y} = \gamma_0 \tag{9.23}$$

then (9.7) yields

$$\begin{aligned} \alpha_0 A &= \iint_{(e)} \frac{\partial^2 w}{\partial x^2} 1 \, dx \, dy = \oint_S (n_x)^2 \frac{\partial w}{\partial n} dS - \oint_S n_x n_y \, dw \\ &= (n_x)_{12}^2 \int_1^2 \frac{\partial w}{\partial n} dS - (n_x)_{12}(n_y)_{12}(w_2 - w_1) + \cdots \text{ cyclic terms} \end{aligned} \tag{9.24}$$

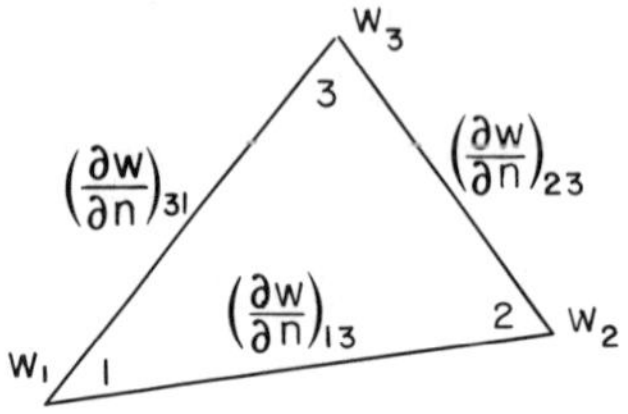

Figure 9.7 Six-parameter plate-bending element

The approximate integration here can be evaluated as

$$\int_1^2 \frac{\partial w}{\partial n} \mathrm{d}S = \left(\frac{\partial w}{\partial n}\right)_{12} h_{12} + O(h_{12}^3) \tag{9.25}$$

Then (9.24) may be written as

$$\alpha_0 = \frac{1}{A}\left[(n_x)_{12}(n_y)_{12}(w_1 - w_2) + (n_x)_{12}^2 h_{12}\left(\frac{\partial w}{\partial n}\right)_{12} + \cdots \text{cyclic terms}\right] + O(h) \tag{9.26}$$

Similarly, for β_0 and γ_0, the highly non-conforming Morley element (Figure 9.7) can be easily derived by the straight quasi-conforming technique.

9.6 NINE-PARAMETER PLATE-BENDING ELEMENT AND RANK OF ELEMENT MATRIX

Example 9.6 If the integration (9.25) in the above example is replaced by using the trapezoidal rule, we obtain a nine-parameter element (Figure 9.8). However, this is a rank deficient element.

The rule of keeping the element matrix from not being rank deficient is

$$M - K \geqslant N - G \tag{9.27}$$

where M is the number of constants of strain, K is the number of dependent

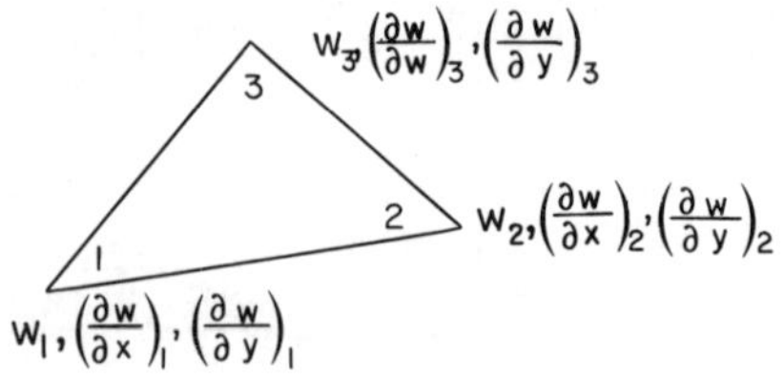

Figure 9.8 Nine-parameter plate-bending element

constants of strain due to the differential compatibility relations which are

$$2\frac{\partial}{\partial y}\left(\frac{\partial^2 w}{\partial x^2}\right)=\frac{\partial}{\partial x}\left(2\frac{\partial^2 w}{\partial x\,\partial y}\right), \qquad 2\frac{\partial}{\partial x}\left(\frac{\partial^2 w}{\partial y^2}\right)=\frac{\partial}{\partial y}\left(2\frac{\partial^2 w}{\partial x\,\partial y}\right) \tag{9.28}$$

N is the number of parameters, i.e. the degrees of freedom of displacements, and G is the number of rigid body displacements.

In Example 9.6, with $M=3$, $K=0$, $N=9$, $G=3$, it is seen that the rule (9.27) is not satisfied.

Example 9.7 In the case to obtain a nine-parameter element, we have to put the strains in the form of

$$\frac{\partial^2 w}{\partial x^2}=\alpha_0+\alpha_1 x+\alpha_2 y, \qquad \frac{\partial^2 w}{\partial y^2}=\beta_0+\beta_1 x+\beta_2 y, \qquad 2\frac{\partial^2 w}{\partial x\,\partial y}=\gamma_0+\gamma_1 x+\gamma_2 y \tag{9.29}$$

There are $M=9$ constants in total. Since the relations (9.28) yield $K=2$ dependent constants

$$2\alpha_2=\gamma_1, \qquad 2\beta_1=\gamma_2$$

the rule (9.27) is being satisfied as

$$9-2>9-3$$

The discretization of α_1 will be obtained as follows. Multiply x on both sides of the first equation of (9.29). This becomes

$$\begin{aligned}\alpha_1\iint_{(e)} x^2\,dx\,dy &= \iint_{(e)}\frac{\partial^2 w}{\partial x^2}x\,dx\,dy\\ &=\oint_S n_x^2 x\frac{\partial w}{\partial n}\,dS-\oint_S n_x n_y x\frac{\partial w}{\partial S}\,dS-\oint_S n_x w\,dS\\ &=(n_x)_{12}\int_1^2\frac{\partial w}{\partial n}x\,ds-(n_x)_{12}(n_y)_{12}(x_2w_2-x_1w_1)\\ &\quad -[(n_x)_{12}(n_x)_{12}^2+(n_x)_{12}]\int_1^2 w\,ds+\cdots\text{cyclic terms}\end{aligned} \tag{9.30}$$

The following formulae can be used for the approximate integration:

$$\int_1^2 w\,dS=\frac{h_{12}}{2}(w_1+w_2)+\frac{h_{12}^2}{12}\left[\left(\frac{\partial w}{\partial S}\right)_1-\left(\frac{\partial w}{\partial S}\right)_2\right]+O(h_{12}^5) \tag{9.31}$$

$$\int_1^2\frac{\partial w}{\partial n}x\,dS=\frac{h_{12}}{3}\left[\frac{1}{2}\left(\frac{\partial w}{\partial n}\right)_1+\left(\frac{\partial w}{\partial n}\right)_2\right]+O(h_{12}^4) \tag{9.32}$$

Since (9.30) should be divided by the quantity $\iint_{(e)} x^2\,dx\,dy$ which has the order of $O(h_{12}^4)$ so that α_1 will have the order of error of $O(1)$, this means that the α_1 will not converge to $(\partial^2 w/\partial x^3)_0$ as they should. However, in the first equation of (9.29), the term $\alpha_1 x$ is of order $O(h)$; thus it does not effect the discretization of α_0 which converges to $(\partial^2 w/\partial x^2)_0$. Therefore, for an element of arbitrary shape, the terms $\alpha_1 x, \alpha_2 y, \ldots, \gamma_2 y$ are used to fill the rank of matrix. They may be omitted when evaluating the strains or moments after the total stiffness matrix has been solved. However, in a symmetrical network with regular shaped elements (see Figure 9.10) they may make a contribution to the accuracy due to the fact that lower order error terms of elements may cancel each other. It can be seen that increasing the number of terms of strains in (9.29) further will not do any good for the solution.

This kind of nine-parameter plate-bending element in finite element computation gives exactly the same numerical results as that of Pian's hybrid stress model. It seems that this is the best one among all kinds of nine-parameter plate-bending elements [8].

9.7 TWELVE-PARAMETER PLATE-BENDING ELEMENT

Example 9.8 The twelve-parameter element (Figure 9.9) should use the complete second-degree polynomial as the expression of its strains to keep it rank sufficient. This kind of element is accurate to linear strain. They pass the higher order patch test (third-degree polynomial) and are actually an Hermite-form finite difference pattern for a regular shape network (Figure 9.10), i.e. every final equation of the finite element method at the interior point is a discretization of the governing fourth-order partial differential equation of plate bending with an error of order $O(h^2)$.

By increasing the number of terms of strains further, we can obtain fifteen, eighteen, and at last the twenty-one parameter full compatible plate-bending element, the so-called high precision one [9]. Therefore, a series of plate-bending elements, both conforming and non-conforming, are obtained by a unified treatment of the quasi-conforming technique.

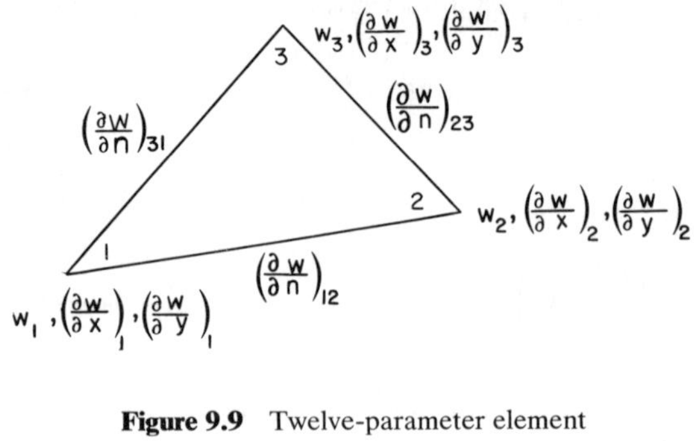

Figure 9.9 Twelve-parameter element

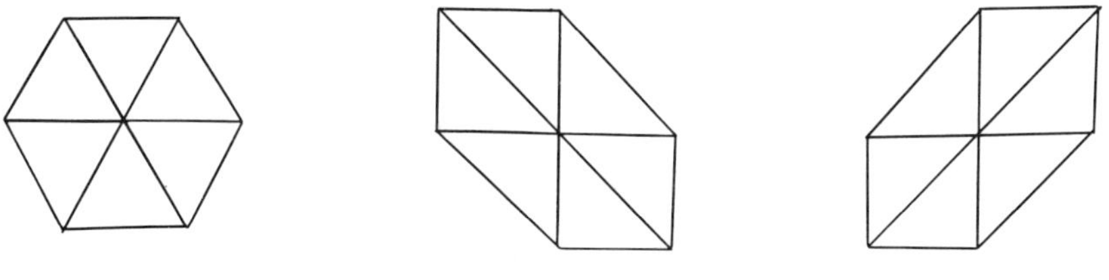

Figure 9.10 Regular shape network

In practice, it is convenient for computation to use the area coordinates L_i $(i = 1, 2, 3)$ in the expression of strains in place of the Cartesian ones. For example, the strains may be written as

$$\frac{\partial^2 w}{\partial x^2} = \left(\frac{\partial^2 w}{\partial x^2}\right)_1 L_1 + \left(\frac{\partial^2 w}{\partial x^2}\right)_2 L_2 + \left(\frac{\partial^2 w}{\partial x^2}\right)_3 L_3 = a_1 L_1 + a_2 L_2 + a_3 L_3 \tag{9.33}$$

$$\frac{\partial^2 w}{\partial x^2} = a_1 L_1^2 + a_2 L_2^2 + a_3 L_3^2 + a_4 L_1 L_2 + a_5 L_2 L_3 + a_6 L_3 L_1 \tag{9.34}$$

Then the formulation and computation of stiffness matrices become very convenient and no inversion of matrix is needed [10].

9.8 RECTANGULAR PLATE-BENDING ELEMENT

Example 9.9 The Adini twelve-parameter plate-bending element (Figure 9.11) can be easily derived by the quasi-conforming technique. The error of discretization of the strains of the element should be deduced from Taylor's expansion because of the regular shape of the element. In such a case the line integral errors of different sides may be somewhat cancelled by each other. The Adini element is not the only one of its sort. There are still many varieties of twelve-parameter rectangular elements that can be obtained by quasi-conforming techniques which vary the terms of strains and the types of approximate integral [11].

It is worth noting that when a nine-parameter triangular element is connected to a rectangular one (Figure 9.12), the compatibility is very easy to achieve as they have the same string function on their mutual boundary.

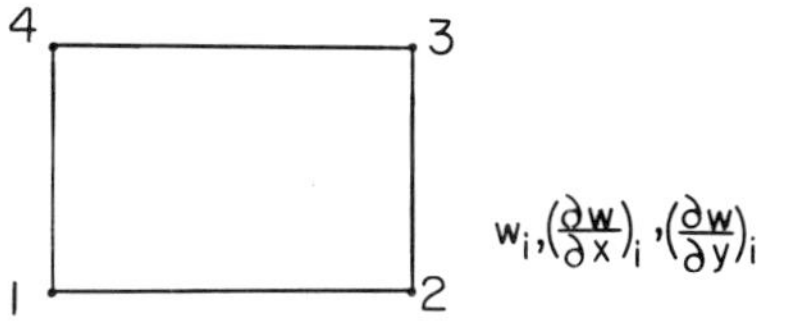

Figure 9.11 Twelve-parameter rectangular element

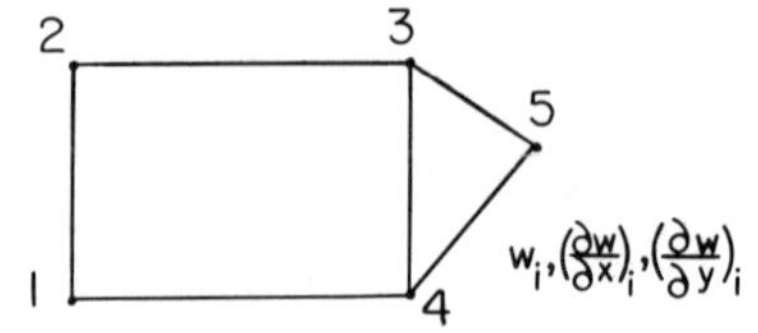

Figure 9.12 A triangular element neighbouring a rectangular one

9.9 SHALLOW CIRCULAR ARCH ELEMENT

Example 9.10 The strains of a shallow circular arch (Figure 9.13) are

$$\varepsilon = \frac{\mathrm{d}U}{\mathrm{d}x} - \frac{w}{R}, \qquad \kappa = \frac{\mathrm{d}^2 w}{\mathrm{d}x^2} \tag{9.35}$$

where u is the tangential displacement, w the normal displacement, ε the tangential strain, κ the curvature strain, R the radius of the circular arch. The trouble of this kind of problem is that both u and w are coupled in the expression of ε. If we put, for example,

$$u = a_0 + a_1 x, \qquad w = b_0 + b_1 x + b_2 x^2 + b_3 x^3 \tag{9.36}$$

as the conventional method, we shall obtain an absurd expression of strain such as

$$\varepsilon = a_1 - \frac{1}{R}(b_0 + b_1 x + b_2 x^2 + b_3 x^3) \tag{9.37}$$

which has more degrees of freedom than it should have. It will not be able to represent the rigid body modes [12].

By the use of the quasi-conforming technique [13], it is natural to start interpolating the strains first. Put

$$\varepsilon = \alpha_0, \qquad \kappa = \beta_0 + \beta_1 x \tag{9.38}$$

After multiplying 1 by ε and 1, x by κ and integrating, we finally obtain

$$\begin{aligned}
\alpha_0 &= \frac{1}{2L}(u_2 - u_1) - \frac{1}{2R}(w_1 + w_2) - \frac{L}{6R}\left[\left(\frac{\mathrm{d}w}{\mathrm{d}x}\right)_1 - \left(\frac{\mathrm{d}w}{\mathrm{d}x}\right)_2\right] \\
\beta_0 &= \frac{1}{2L}\left[\left(\frac{\mathrm{d}w}{\mathrm{d}x}\right)_1 - \left(\frac{\mathrm{d}w}{\mathrm{d}x^2}\right)\right] \\
\beta_1 &= \frac{3}{2}\left\{\frac{1}{L^3}(w_1 - w_2) + \frac{1}{L^2}\left[\left(\frac{\mathrm{d}w}{\mathrm{d}x}\right)_1 - \left(\frac{\mathrm{d}w}{\mathrm{d}x}\right)_2\right]\right\}
\end{aligned} \tag{9.39}$$

so that in each node i we need u_i, w_i, $(\mathrm{d}w/\mathrm{d}x)_i$ as its nodal values.

For a numerical test, putting $L = 5$, $R = 50$, $E = 10^5$, and the thickness of

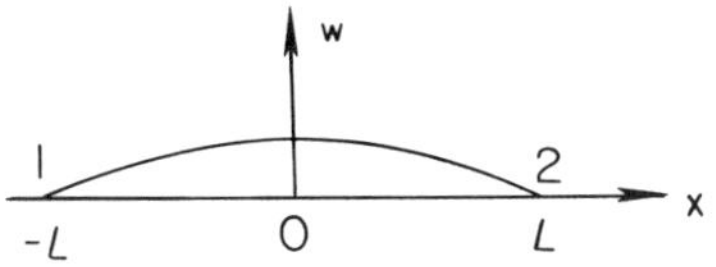

Figure 9.13 A shallow circular arch element

the arch $h = 1$, we obtain the *characteristic* values of the element stiffness matrix as follows:

5200.000000	1618.283286	20803.938936
0.000000	0.000000	0.000000

It is seen that the representations of rigid body modes are satisfied exactly.

9.10 SHELL ELEMENTS

Starting by interpolating the strains using similar techniques to those in the last example of arch, some thin shell elements have been worked out. These are:

(a) A triangular element for double curvature shallow shells [14] with nodal values u_i, v_i, w_i, $(\partial w/\partial x)_i$, $(\partial w/\partial y)_i$ (Figure 9.14), where u_i, v_i, w_i are components of the displacement of node i.
(b) A rectangular element for cylindrical shells [15] with nodal values u_i, v_i, w_i, $(\partial w/\partial \alpha)_i$, $(\partial w/\partial \beta)_i$ $(i = 1, 2, 3, 4)$ (Figure 9.15).
(c) A curved element for shells of revolution with nodal values u_i, w_i, $\beta_i = (dw/dS)_i$ (Figure 9.16).

In this element, since all the strains are assumed as polynomials at first, it is natural, when R approaches zero, that no singularity will occur at the vertices of the shell [16].

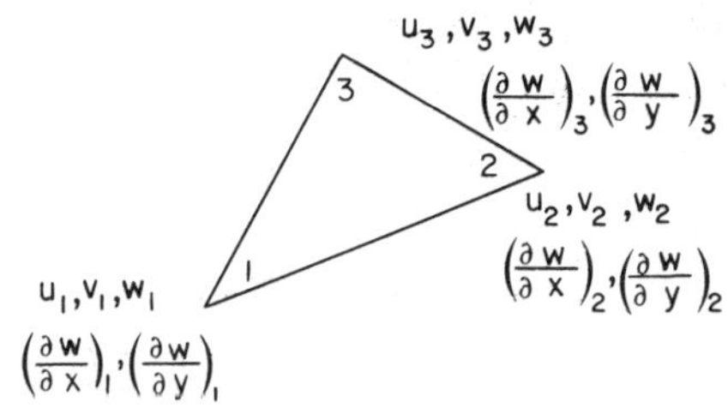

Figure 9.14

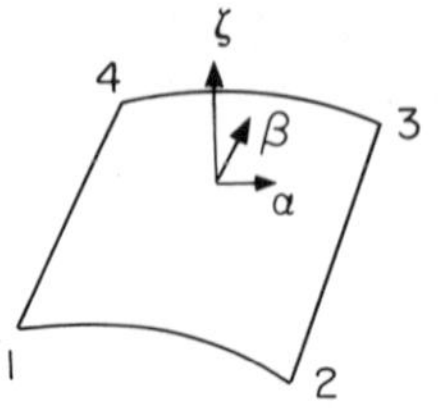

Figure 9.15

All the above elements describe the rigid body modes very well, and the numerical results of some examples using them are satisfactory.

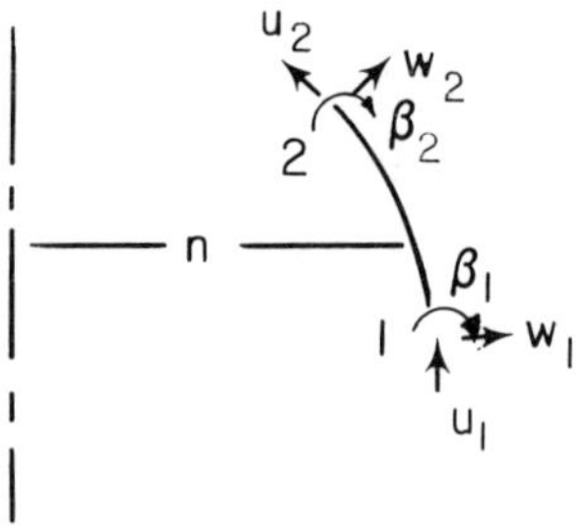

Figure 9.16

9.11 ISOPARAMETRIC ELEMENTS

The strains of the isoparametric element are in the $\xi\eta$ plane. For example, the expression of strain is given by

$$\boldsymbol{\varepsilon} = \left[\frac{\partial u}{\partial \xi}, \frac{\partial u}{\partial \eta}, \frac{\partial v}{\partial \xi}, \frac{\partial v}{\partial \eta}\right]^T \tag{9.40}$$

Put

$$\frac{\partial u}{\partial \xi} = \alpha_1 + \alpha_2 \xi + \alpha_3 \eta, \ldots \tag{9.41}$$

Multiplying both sides by 1, ξ, η and integrating over the element on the $\xi\eta$ plane, we have

$$\begin{aligned}
\int_{-1}^{1}\!\!\int \frac{\partial u}{\partial \xi} 1 \, \mathrm{d}\xi \, \mathrm{d}\eta &= \oint u n_\xi \, \mathrm{d}S \\
\int_{-1}^{1}\!\!\int \frac{\partial u}{\partial \xi} \xi \, \mathrm{d}\xi \, \mathrm{d}\eta &= \oint u \xi n_\xi \, \mathrm{d}S - \int_{-1}^{1}\!\!\int u \, \mathrm{d}\xi \, \mathrm{d}\eta \\
\int_{-1}^{1}\!\!\int \frac{\partial u}{\partial \xi} \eta \, \mathrm{d}\xi \, \mathrm{d}\eta &= \oint u \eta n_\xi \, \mathrm{d}S
\end{aligned} \tag{9.42}$$

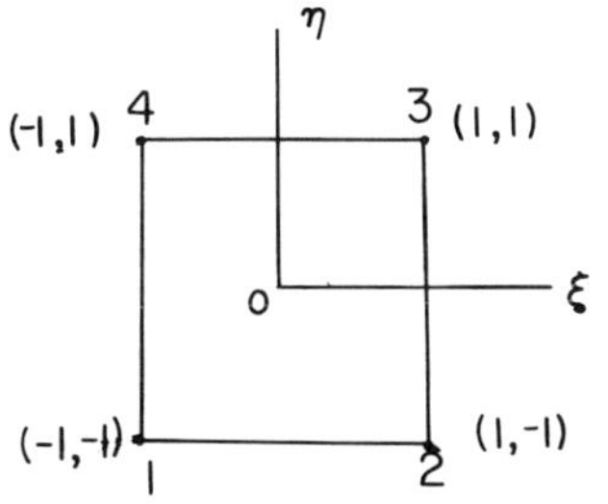

Figure 9.17

According to the quasi-conforming element technique, for the four-node element (Figure 9.17) we obtain the usual Q_4 element. By adding two internal degrees of freedom,

$$\begin{aligned} u_c(\xi, \eta) &= \beta_1 f_1(\xi) f_2(\eta) \\ v_c(\xi, \eta) &= \beta_2 f_1(\xi) f_2(\eta) \end{aligned} \tag{9.43}$$

where $f_1(\xi)$ and $f_2(\eta)$ are even functions of ξ and η respectively, and substituting them into (9.42), we can obtain a new element Q_{c5}. We increase further by two more degrees of freedom and use the condition:

$$\iint_{(e)} \boldsymbol{\varepsilon}_0 \, dx \, dy = 0 \tag{9.44}$$

where $\boldsymbol{\varepsilon}_0$ is the non-conforming part of the strains. This condition is equivalent to

$$\iint \frac{\partial u}{\partial x} dx \, dy = \oint u n_x \, dS = 0, \qquad \iint \frac{\partial u}{\partial y} dx \, dy = \oint u n_y \, dS = 0 \tag{9.45}$$

It can easily be seen that the Q_6 element [17] does not satisfy the above relations and does not pass the patch test.

Suppose now the adding functions are

$$u_S(\xi, \eta) = \begin{cases} a_1\xi^2 & \text{on } 1\text{–}2 \\ a_2\eta^2 & \text{on } 2\text{–}3 \\ a_3\xi^2 & \text{on } 2\text{–}3 \\ a_4\eta^2 & \text{on } 4\text{–}1 \end{cases}, \qquad v_S(\xi, \eta) = \begin{cases} a_1\xi^2 & \text{on } 1\text{–}2 \\ a_2\eta^2 & \text{on } 2\text{–}3 \\ a_3\xi^2 & \text{on } 3\text{–}4 \\ a_4\eta^2 & \text{on } 4\text{–}1 \end{cases} \tag{9.46}$$

Substituting these into (9.45) and using the isoparametric transformation, we obtain

$$a_1 = (y_{23}x_{34} - x_{23}y_{34}), \qquad a_2 = (y_{21}x_{34} - x_{21}y_{34}),$$
$$a_3 = (y_{23}x_{21} - x_{23}y_{21}), \qquad a_4 = 0 \tag{9.47}$$

This result is the same as that in [18]. Substituting (9.47) into (9.42), we can obtain a new Q_{c6} element. This kind of element can represent the pure bending mode when the elements are rectangular in shape, and they pass the patch test when they are in arbitrary shapes. The numerical test of this kind of element is satisfactory [19].

The extension to three-dimensional cases is similar [20].

REFERENCES

1. T. H. H. Pian, 'Derivation of element stiffness matrices by assumed stress distributions', *AIAA J.*, **2,** No. 7, 1333–1336 (1964).
2. Tang Limin, 'Some basic problems of the finite element method', *J.DIT,** **2** (1979).
3. Tang Limin, Chen Wanji, and Liu Yingxi, 'Quasi-conforming elements for finite element analysis', *J.DIT,* **2** (1980).
4. Tang Limin, Lu Hexiang, Chen Wanji, and Liu Yingxi, 'Quasi-conforming element technique for the finite element method', *Proc. Second Int. Congr. on Numerical Methods for Engineering DUNOD,* pp. 569–572, 1980.
5. Chen Wanji and Liu Yingxi, 'The quasi-conforming element models and the generalized variational principle', *J.DIT,* **3** (1980).
6. Chen Wanji, *Generalized Hybrid Model* (unpublished).
7. Tang Limin, Chen Wanji, and Liu Yingxi, 'Quasi-conforming element and generalized variational illegalities', paper presented at *CMES, CSTAM, CMS Symp. on Finite Element Method,* Hefei, China, 19–23 May 1981.
8. Tao Zhengguo, 'Behavior of quasi-conforming 9 parameter triangular plate bending element', Postgraduate project at DIT, 1980.
9. Jiang Heyang, 'Derivation of higher order plate bending element by quasi-conforming element method', Postgraduate project at DIT, 1980.
10. Chen Wanji, Liu Yingxi, and Tang Limin, 'The formulation of quasi-conforming elements', *J.DIT,* **2** (1980).
11. Shi Guangyu, 'Rectangular quasi-conforming element of 12 nodes for plate bending', Postgraduate project at DIT, 1980.
12. S. W. Lee and T. H. H. Pian, 'Improvement of plate and shell finite element by mixed formulations', *AIAA J.*, **16,** No. 1, 29–34 (1978).
13. Lu Hexiang, 'Quasi-conforming element applied to arch element', Technical paper of DIT, 1980.
14. Lu Hexiang and Liu Yingxi, 'Quasi-conforming element technique applied to double curvature shallow shell', *J.DIT,* **1** (1981).
15. Liu Yingxi, Lu Hexiang, and Tang Limin, 'Quasi-conforming cylindrical shell element', *J.DIT* (to be published) (1981).
16. Jin Wugen, 'Quasi-conforming curved element for shells of revolution', Postgraduate project at DIT, 1980.
17. R. L. Taylor, P. L. Beresford, and E. L. Wilson, 'A nonconforming element for stress analysis', *Int. J. Num. Meth. in Eng.*, **10,** No. 6 (1976).
118. E. L. Wachspress, 'Incompatible quadrilateral basis functions', *Int. J. Num. Meth. in Eng.*, **12,** No. 4 (1978).
19. Chen Wanji and Tang Limin, 'Isoparametric quasi-conforming element', *J.DIT,* **1** (1981).
20. Chen Wanji, *On a High Precision 8 node Brick Element* (unpublished).

* *Journal of the Dalian Institute of Technology* (in China).

Hybrid and Mixed Finite Element Methods
Edited by S. N. Atluri, R. H. Gallagher, and O. C. Zienkiewicz

Chapter 10

A Mixed Method for Non-Linear Plate and Shallow Shell Problems

E. Giencke

10.1 INTRODUCTION

Plates and shells are usually analysed by the finite element method, especially by the deformation method [1 to 3]. In this case the most interesting normal forces and bending moments are obtained as differences of the unknown deformations, which may result in a considerable loss in accuracy and in convergence if inelastic material behaviour is involved. To avoid such a loss in accuracy, we introduce the interior 'forces' and 'strains', the stretching forces n_x, n_y, n_{xy}, and the corresponding strains ε_x, ε_y, ε_{xy}, as well as the moments m_x, m_y, m_{xy} and the corresponding curvatures κ_x, κ_y, κ_{xy} immediately as unknowns [4 to 6]. These forces and strains as well as the geometrical curvatures a_x, a_y, a_{xy} of the shell and the loads p_x, p_y, p_z will be approximated linearly from point to point by introducing the values at the nodal points into the analysis. The structure may be divided into triangular or rectangular elements. In this paper the guideline of the analysis for triangular elements is represented with some examples.

Since the interior forces and strains do not satisfy either the equilibrium or the compatibility conditions, both conditions have to be formulated as in a mixed method by means of the principles of virtual displacements and forces. This will be done in a very simple manner by using, for example, rigid body motions of the elements as virtual displacements. In this case the interior forces do work only along the gridlines; therefore it is quite simple to establish the equations because only line integrals and not surface integrals must be determined. The formulation of the equations becomes so easy that matrix formulation is not necessary.

In this way we get the difference equations belonging to the differential equations of the shell. Thus it is possible to establish the difference equations in the same mechanical manner as the finite element equations, namely by the principles of virtual work, and furthermore to find difference equations also for quite irregular grids. The equations are established elementwise, as in the finite element method.

In the case of the deformation method the stiffness matrices for stretching and bending have unfortunately a quite different structure, and it is difficult to develop simple stiffness matrices for double-curved shell elements with unrestrained rigid body motions. Furthermore, the stretching and bending strains have a different order of approximation and are discontinuous perpendicular to the boundaries of the elements. In this method the stretching and bending strains are approximated in the same order of accuracy and have no discontinuities at the boundaries of the elements. By using the duality between bending and stretching the effort for establishing the computer codes is reduced. Both problems are treated with the same programme which must be extended slightly for application to shallow shells and for involving large deflection and non-linear material behaviour. To demonstrate the duality, the differential equations for stretching will be arranged on the left-hand side and those for bending on the right-hand side. We start with the equations for constant stiffnesses and linear elastic behaviour and extend them subsequently to plates with large deflections and variable stiffnesses for preparing the solution for plates with non-linear material behaviour. The differential equations are first summarized to point out later on the close connection between the differential and the finite equations [5].

10.2 LINEAR BASIC EQUATIONS (CONSTANT STIFFNESSES)

(a) STRETCHING (b) BENDING

The interior 'forces' (Figure 10.1)

$$n_x,\ n_y,\ n_{xy} \qquad m_x,\ m_y,\ m_{xy} \tag{10.1a, b}$$

have to satisfy the conditions of equilibrium (buckling forces s_x, s_y, s_{xy})

$$\begin{aligned} &n'_x + n^{\cdot}_{xy} + p_x = 0 \qquad && m''_x + 2m'^{\cdot}_{xy} + m^{\cdot\cdot} + s_x\kappa_x + 2s_{xy}\kappa_{xy} + s_y\kappa_y \\ &n'_{xy} + n^{\cdot}_y + p_y = 0 && \quad + a_x n_x + 2a_{xy}n_{xy} + a_y n_y + p_z = 0 \end{aligned} \tag{10.2a, b}$$

and the 'strains'

$$\begin{aligned} \varepsilon_x &= u' - wa_x \qquad & \kappa_x &= -w'' \\ \varepsilon_y &= v^{\cdot} - wa_y & \kappa_y &= -w^{\cdot\cdot} \\ 2\varepsilon_{xy} &= u^{\cdot} + v' - 2wa_{xy} & \kappa_{xy} &= -w'^{\cdot} \end{aligned} \tag{10.3a, b}$$

that of compatibility

$$\varepsilon''_y - 2\varepsilon'^{\cdot}_{xy} + \varepsilon^{\cdot\cdot}_x - a_x\kappa_y - 2a_{xy}\kappa_{xy} - a_y\kappa_x = 0 \qquad \begin{aligned} \kappa'_y - \kappa^{\cdot}_{yx} &= 0 \\ \kappa'_{xy} - k_x &= 0 \end{aligned} \tag{10.4a, b}$$

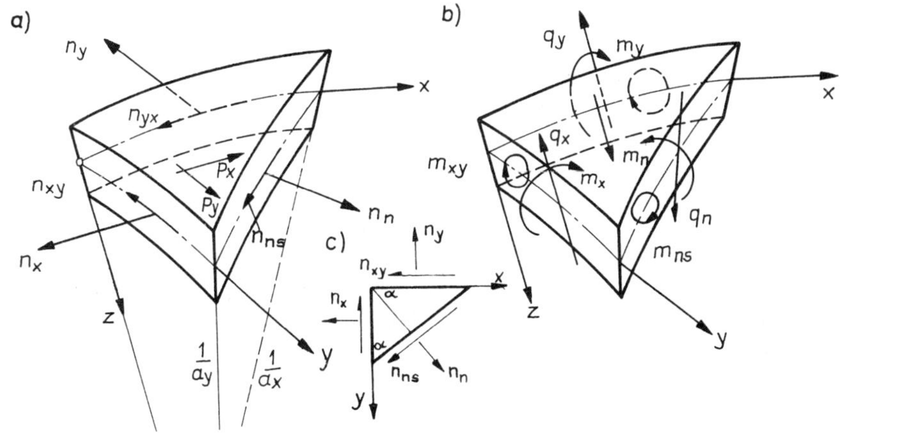

Figure 10.1 Interior forces and moments, external loads. (a) Stretching forces, (b) bending moments, (c) transformation

Only the stretching compatibility equation (10.4a) and the bending equilibrium equation (10.2b) are coupled for shallow shells and for plates with large defections. The duality between the stretching forces n and the curvatures κ as well as between the stretching strains ε and the moments m becomes evident by comparing the equations (10.2) and (10.4). To involve the advantage of this duality the stretching strains ε and the moments m shall be substituted by the 'unknowns' n and κ, using the stress–strain relationships

$$\begin{bmatrix} \varepsilon_x - \alpha\bar{T} \\ \varepsilon_y - \alpha\bar{T} \\ \varepsilon_{xy} \end{bmatrix} = \begin{bmatrix} f_x & f_{xy} & \cdot \\ f_{xy} & f_y & \cdot \\ \cdot & \cdot & f_t \end{bmatrix} \begin{bmatrix} n_x \\ n_y \\ n_{xy} \end{bmatrix} \quad \begin{bmatrix} m_x \\ m_y \\ m_{xy} \end{bmatrix} = \begin{bmatrix} b_x & b_{xy} & \cdot \\ b_{xy} & b_y & \cdot \\ \cdot & \cdot & b_t \end{bmatrix} \begin{bmatrix} \kappa_x - \alpha\tilde{T} \\ \kappa_y - \alpha\tilde{T} \\ \kappa_{xy} \end{bmatrix} \qquad (10.5\text{a, b})$$

where $\bar{T}(x, y)$ is the average value and $\tilde{T}(x, y)$ the gradient of the temperature. Thus only six unknowns of n, κ remain. Further, it is possible to eliminate the shearing forces n_{xy} and the twisting curvatures κ_{xy} in the stretching compatibility equation (10.4a) and the bending equilibrium equation (10.2b) by means of the equations (10.2a) and (10.4b):

$$2\varepsilon_{xy} = 2f_t n_{xy}^{\prime\cdot} = -f_t(n_x'' + n_y^{\cdot\cdot} + p_x' + \dot{p}_y) \qquad 2m_{xy}^{\prime\cdot} = 2b_t\kappa_{xy}^{\prime\cdot} = b_t(\kappa_y'' + \kappa^{\cdot\cdot}) \qquad (10.6)$$

The basic terms (for a plane plate) in the equations (10.4a) and (10.2b) then become

$$
\begin{aligned}
&f_y n_y + \bar{f}_t n_x'' + \bar{f}_t n_y^{\cdot\cdot} + f_x n^{\cdot\cdot} \cdots && b_x \kappa_x'' + \bar{b}_t \kappa_y'' + \bar{b}_t \kappa_x^{\cdot\cdot} + b_y \kappa_y^{\cdot\cdot} \cdots \\
&\quad = \alpha(\bar{T}'' + \bar{T}^{\cdot\cdot}) - f_t(p_x' + p_y^{\cdot}) && \quad = \alpha(b_x \tilde{T}'' + b_y \tilde{T}^{\cdot\cdot}) - p
\end{aligned}
\tag{10.7}
$$

where

$$\bar{f}_t = f_t + f_{xy} \qquad \bar{b}_t = b_t + b_{xy} \tag{10.8}$$

are the effective shearing flexibility or the twisting stiffness respecively. The formulation (10.7) yields in the isotropic case the well-known Laplace equation for the sum of the normal forces and of the bending curvatures:

$$f\Delta(n_x + n_y - \alpha\bar{T}/f) = -f_t(p_x' + p_y^{\cdot}) \qquad b\Delta(\kappa_x + \kappa_y - \alpha\tilde{T}) = -p_z \tag{10.9}$$

The corresponding finite formulation is of great advantage in the numerical analysis because of their significant numerical stability. Furthermore, the equations (10.2a) and (10.4b) may be transformed into second-order equations by eliminating the shearing and twisting terms in one of them:

$$n_x'' - n_y^{\cdot\cdot} + p_x' - p_x^{\cdot} = 0 \qquad \kappa_y'' - \kappa_x^{\cdot\cdot} = 0 \tag{10.10}$$

$$\Delta n_{xy} + (n_x + n_y)'^{\cdot} + p_x^{\cdot} + p_y^{\cdot} = 0 \qquad \Delta\kappa_{xy} - (\kappa_x + \kappa_y)'^{\cdot} = 0 \tag{10.11}$$

Thus it is often possible first to calculate the normal forces and bending curvatures by equations (10.7) and (10.10) and in a second step the shearing forces and the twisting curvatures by equation (10.11).

10.3 FINITE EQUATIONS (CONSTANT STIFFNESSES)

The mechanical interpretation of the finite equations becomes obvious, if at first both the interior forces and the strains remain in the analysis and later on the stretching strains ε and the bending moments m are eliminated by the stress–strain relationship (10.5). As before we have to establish one stretching compatibility equation and one bending equilibrium equation and two stretching equilibrium equations and two bending compatibiity equations. In the finite range we do it by the principles of virtual displacements and forces which are equivalent to the equilibrium and compatibility equations.

The virtual states are chosen as simple as possible in a similar manner, as it is well known in the bending of beams. For deriving the equilibrium condition between the bending moments M and the distributed load p in a beam, the virtual displacement δw in Figure 10.2(a) is enforced upon the two neighbouring elements. Because the elements rotate as rigid bodies by $\delta\beta = 1/h$ about the outer points, only external work arises from the mo-

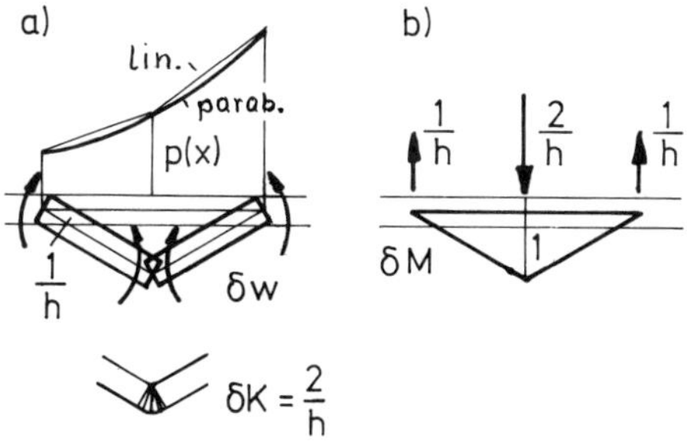

Figure 10.2 Virtual states for bending of beams (a) Virtual displacement δw, (b) virtual forces δP

ments M in the nodal points and from the load $\int p\, \delta w\, \mathrm{d}x$:

$\delta W_e = 0$:

$$\boxed{1 \quad -2 \quad 1}\ \underset{(w)}{M/h} +$$

$(\delta W_e^* = \delta W_i^*)$:

$$\left.\begin{array}{ll} \boxed{1 \quad 4 \quad 1}\ \underset{(\kappa)}{ph/6} & \text{(linearly approximated)} \\ \boxed{1 \quad 10 \quad 1}\ \underset{(\kappa)}{ph/12} & \text{(parabolically approximated)} \end{array}\right\} = 0 \qquad (10.12)$$

This equation is the multipoint difference equation belonging to the differential equation:

$$\text{Equilibrium: } M'' + p = 0 \qquad (\text{Compatibiity: } w'' + \kappa = 0) \qquad (10.13)$$

Usually in difference equations only the centre-point load enters into the equation. The load coefficients are different if the load is approximated linearly or parabolically. The compatibility equation has the same structure as the equilibrium condition, as well as in the infinitesimal (equation 10.13) or in the finite range values within the brackets in equation 10.12), if we introduce as virtual forces the self-equilibrating force group δP in Figure 10.2(b) to which the interior virtual moment δM belongs. The external work δW_e^* of the virtual loads at the true deflection w is equal to the interior work δW_i^* of the virtual moments δM at the true curvature κ. In the following this formulation will be extended first to a plane plate and later on to shallow shells.

10.3.1 Equations for the plane plate

We start with the *equilibrium* equation for *bending* by enforcing the virtual displacement δw in Figure 10.3(a) at the considered nodal point 1. Each

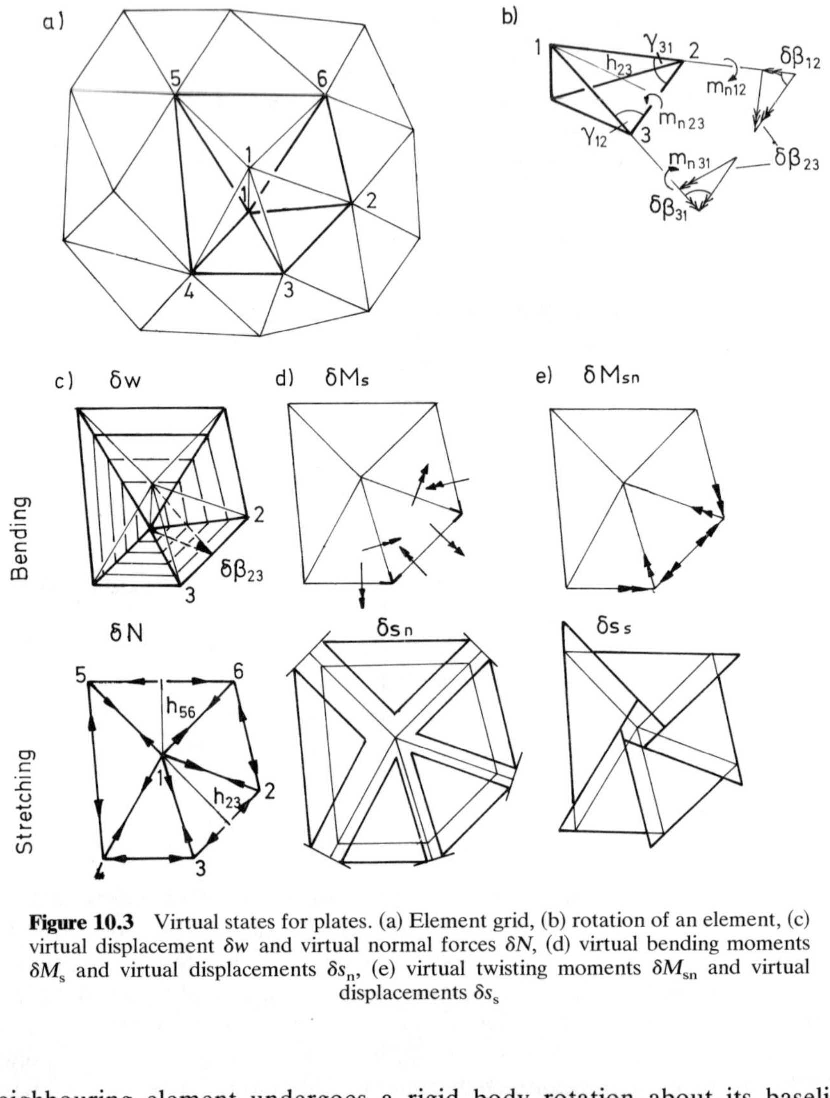

Figure 10.3 Virtual states for plates. (a) Element grid, (b) rotation of an element, (c) virtual displacement δw and virtual normal forces δN, (d) virtual bending moments δM_s and virtual displacements δs_n, (e) virtual twisting moments δM_{sn} and virtual displacements δs_s

neighbouring element undergoes a rigid body rotation about its baseline opposite to point 1. Each neighbouring element undergoes a rigid body rotation about its baseline opposite to point 1. At each element boundary a constant virtual kink $\delta\beta$ arises at which the real bending moments m_n normal to the edges do work as external edge moments. Since the virtual displacements at a gridline between two neighbouring elements agree with each other, the shearing forces q_n and the twisting moments m_{ns} at the gridlines do not participate in the virtual work. Furthermore, the vertical load p_z works at the virtual displacement δw; thus the whole virtual work is

$$\delta W_e = \int m_n(s)\, \delta\beta\, ds + \iint p_z\, \delta w\, dx\, dy \tag{10.14}$$

An interior virtual work does not exist because of the rigid body motion of the elements.

Due to the virtual displacement $\delta w = 1$ in point 1 the triangular element 123 rotates about its baseline 23 by the angle

$$\delta\beta_{23} = 1/h_{23}(= -\delta N_{23}) \tag{10.15}$$

where h_{23} is the height of the triangle (Figure 10.3b). The kinks on the other lines are equal to those at the baseline multiplied by the cosine of the angle between these lines and the baseline (Figure 10.3b):

$$\delta\beta_{31} = -\cos\gamma_{12}/h_{23}(= -\delta N_{31}),\ \delta\beta_{12} = -\cos\gamma_{31}/h_{23}(= -\delta N_{12}) \tag{10.16}$$

Due to equation (10.14) the kinks $\delta\beta$ are positive in the sense of the bending moments m_n. In Figure 10.3(b) the moments and kinks at line 31 and 12 are opposite to each other; therefore $\delta\beta_{31}$ and $\delta\beta_{12}$ become negative. Since the virtual work depends linearly on the amplitude of the virtual displacement, the amplitude may be chosen equal to 1 for simplifying the analysis. In reality it is infinitesimally small; thus no elongations arise in a plane plate by the virtual displacement δw. Since the kinks $\delta\beta$ are constant at each line the virtual work (10.14) of the bending moments is equal to the work of their average values $\bar{m}_{n,ik}$ on the gridlines ik:

$$\delta W_{e123} = \left(\frac{l_{23}}{h_{23}}\,\bar{m}_{n23} - \frac{l_{31}}{h_{23}}\,\bar{m}_{n31}\cos\gamma_{12} - \frac{l_{12}}{h_{23}}\,\bar{m}_{n12}\cos\gamma_{31}\right) + \iint p\,\delta w\, dx\, dy = 0 \tag{10.17}$$

The moments m_n are marked both by the index of the point and that of the gridline because the moments on the different gridlines through a point differ from each other. We obtain the total virtual work by summarizing the work in the elements surrounding point 1. Equation (10.17) is still exact but it becomes approximately, if the distribution of the bending moments m_n and of the load p_z is approached linearly,

$$\bar{m}_{nik} = \frac{(m_{ni} + m_{nk})_{ik}}{2}, \qquad \iint p\,\delta w\, dx\, dy = \frac{2p_1 + p_2 + p_3}{12F} \tag{10.18}$$

It is also possible to formulate the virtual work of the bending moment m_n as an interior work by interpreting the virtual kinks $\delta\beta$ as concentrated virtual bending curvatures $\delta K_n = -\delta\beta_n$ which have the same values as the kinks but the opposite sign (Figure 10.2a):

$$\delta W_{\mathrm{i}} = \delta W_{\mathrm{e}}: \qquad \sum (\bar{m}_n l\, \delta K)_{ik} = \iint p\, \delta w\, \mathrm{d}x\, \mathrm{d}y \tag{10.19}$$

The *stretching compatibility* equation results from the principal of virtual forces. The most simple virtual equilibrium force system, restricted to the neighbouring elements of nodal point 1, are the normal forces δN in Figure 10.3(c). These virtual normal forces δN do work on the real extensions ε_s along the gridlines

$$\delta W_{\mathrm{i}}^{*} = \int \delta N \varepsilon_s(s)\, \mathrm{d}s = \sum (\delta N \bar{\varepsilon}_s l)_{ik} \tag{10.20}$$

This is the only work done in a plane plate; an external work does not exist. As in the case of the plate equilibrium equation only one kind of force does work, the virtual normal forces δN. These virtual forces have the same values as the virtual kinks $\delta\beta$ in equations (10.15) and (10.16) due to the virtual displacement δw. About the nodal point 1 in Figure 10.3(c) there exists equilibrium in the moments if to each normal force at the baselines belongs a moment equal to 1. Therefore all normal forces at the baselines have to be reciprocal to the height h of the triangular element as the kink $\delta\beta_{23}$. If these forces are compression forces the forces at the gridlines to the centre point 1 must be tension forces with the same values as the kinks $\delta\beta$ due to δw (10.16). Thus we get the same equation as the equilibrium equation (10.19) for plate bending if the bending moments m_n are substituted by the extensions ε_s and the work of the external load is cancelled. In this way the duality between stretching and bending is also confirmed in the finite equations.

Next the two *bending compatibility* equations according to the differential equations (10.4b) will be formulated also by the principle of virtual forces. Therefore, two self-equilibrating virtual moment groups, namely a group of bending moments δM_s (Figure 10.2d) and a group of torsional moments δM_{sn} (Figure 10.2e), are enforced to the plate. These moment systems are linear independently to each other, they act only on the gridlines; and have the same values as the virtual normal forces δN in the stretching problem, as discussed before, since the torsional moment vectors δM_{sn} agree with the virtual force vectors δN and the bending moments δM_s are only rotated about 90°. Virtual work arises only along the gridlines; it is the virtual work of the virtual bending moments δM_s on the real curvature κ_s and in the twisting case the work of the virtual torsional moments δM_{sn} on the real twisting curvatures κ_{sn}:

$$\delta W_{\mathrm{i}}^{*} = \int \delta M_s \kappa_s\, \mathrm{d}s = 0 \qquad \text{and} \qquad \delta W_{\mathrm{i}}^{*} = \int \delta M_{sn} \kappa_{sn}\, \mathrm{d}s = 0 \tag{10.21}$$

External work does not exist. The equations are of the same form as the bending equilibrium equation (10.14), m_n is substituted by κ_s or κ_{sn}, and the virtual work of the load is cancelled.

Finally, we have to formulate the two *equilibrium* conditions for the *stretching* forces by the principle of virtual displacements. Due to virtual inplane displacements in Figure 10.2(d) and (e), the elements are shifted as rigid bodies such that in the first case only discontinuities δs_n or concentrated strains $\delta E_n = -\delta s_n$ normal to the gridlines and in the second case only discontinuities δs_s or concentrated shearing strains $\delta E_{sn} = -\delta s_s$ parallel to the gridlines result. These virtual displacements are obtained if in the first case the elements are shifted only normal to the baselines of the elements without gaps parallel to the gridlines and in the second case, vice versa. Since the virtual concentrated strains δE have the same values as the virtual moments δM because of the duality between stretching and bending, the stretching equilibrium equations are of the same form as the bending compatibility equations (10.21), only load terms due to the work of the tangential loads p_x and p_y on the virtual displacements δu and δv must be added.

One might object that the virtual displacements are not geometrically compatible, because there are gaps δs, $\delta\beta$ in the displacements and slopes. But the principle of virtual displacements only demands that the virtual displacements must be compatible within the elements. If there are discontinuities at the boundaries then the real interior forces are acting there as external loads. In each virtual state there exist only one kind of virtual forces or concentrated strains working at one kind of real strains or forces which have always the same form since the geometrical and the statical equations are of the same type as in a beam (equation 10.13). The coefficients (equation 10.17) in the six equations derived before become a cyclical setup [6]:

$$\frac{l_{23}}{h_{23}} = \frac{x_{23}^2 + y_{23}^2}{2F_{123}}, \qquad -\frac{l_{31}\cos\gamma_{12}}{h_{23}} = \frac{x_{31}x_{23} + y_{31}y_{23}}{2F_{123}} \tag{10.22}$$

if they are expressed by the differences of the point coordinates

$$x_{ik} = x_i - x_k, \qquad y_{ik} = y_i - y_k \tag{10.23}$$

In the virtual work all elements adjoining the investigated nodal point are to involve. The unknown forces and strains normal or parallel to the edges have to be transformed into the global xy system:

$$\begin{bmatrix} m_n \\ m_s \\ m_{ns} \end{bmatrix} \begin{bmatrix} n_n \\ n_s \\ n_{ns} \end{bmatrix} \begin{bmatrix} \kappa_s \\ \kappa_n \\ -\kappa_{sn} \end{bmatrix} \begin{bmatrix} \varepsilon_s \\ \varepsilon_n \\ -\varepsilon_{sn} \end{bmatrix} = \begin{bmatrix} c^2 & s^2 & 2cs \\ s^2 & c^2 & -2cs \\ -cs & cs & c^2-s^2 \end{bmatrix} \begin{bmatrix} m_x \\ m_y \\ m_{xy} \end{bmatrix} \begin{bmatrix} n_x \\ n_y \\ n_{xy} \end{bmatrix} \begin{bmatrix} \kappa_y \\ \kappa_x \\ -\kappa_{xy} \end{bmatrix} \begin{bmatrix} \varepsilon_y \\ \varepsilon_x \\ -\varepsilon_{xy} \end{bmatrix} \tag{10.24}$$

where the cosine and sine functions may be expressed by the coordinate differences (10.23)

$$c = \cos \alpha_{ik} = y_{ki}/l_{ik} = -y_{ik}/l_{ik}, \qquad s = \sin \alpha_{ik} = x_{ik}/l_{ik} \tag{10.25}$$

Finally, the stress–strain relationships (10.5) are used to substitute the bending moments m by the elastic curvatures κ and the stretching strains ε by the stretching forces n. Thus only the six unknowns

$$n_x, n_y, n_{xy}, \kappa_y, \kappa_x, -\kappa_{xy}$$

remain.

10.3.2 Combination of the basic equations

If each principle of virtual work is true, both principles may be used simultaneously by introducing in the case of plate bending the virtual displacement δw (Figure 10.2c) and the virtual moments δM_s (Figure 10.2d) multiplied by $(1-\nu)b$. The entire virtual work is then the interior work δW_i of the real bending moments m_n on the concentrated curvarutes δK_n and that of the virtual bending moments $(1-\nu)b\,\delta M_s$ on the real curvatures κ_s and finally the external work δW_e of the vertical load p_z at the virtual displacement δw:

$$\delta W_i + \delta W_i^* = \delta W_e: \qquad \int m_n\, \delta K_n\, ds + \int (1-\nu)b\, \delta M_s \kappa_s\, ds = \iint p\, \delta w\, dx\, dy \tag{10.26}$$

Since δM_s agrees with δK_n and the combination

$$m_n + (1-\nu)b\kappa_s = b(\kappa_n + \kappa_s) = b(\kappa_x + \kappa_y)$$

is invariant in isotropic plates, then the virtual work can be condensed to

$$\int \underbrace{[m_n + (1-\nu)b\kappa_s]}_{b(\kappa_x + \kappa_y)}\, \delta K_n = \iint p\, \delta w\, dx\, dy \tag{10.27}$$

which is the finite equation to the differential equation (10.9).

For plate stretching the virtual states may be combined in a similar manner [6]. For orthotropic plates or shells a similar procedure (Section 10.3.4) is also favourable, because the accuracy of the equations is thereby improved.

10.3.3 Additional shell terms

For shallow shells the stretching equilibrium equations and the bending compatibility equations derived before are also valid. Only some additional terms must be added to the bending equilibrium equation (10.14) and to the stretching compatibility equation (10.20), if the virtual displacement δw is enforced upon the shallow shell normal to its middle surface. Besides the virtual kinks $\delta\beta$, from which we get the bending terms as in equation (10.14), virtual stretching strains $\delta\varepsilon$ arise, which are proportional to δw and to the geometrical curvature a:

$$\delta\varepsilon_x = -a_x\,\delta w, \qquad \delta\varepsilon_y = -a_y\,\delta w, \qquad \delta\varepsilon_{xy} = -a_{xy}\,\delta w \tag{10.28}$$

At these virtual strains $\delta\varepsilon$ work will be done by the real stretching forces n:

$$\begin{aligned}\delta W_i &= \iint (n_x\,\delta\varepsilon_x + 2n_{xy}\,\delta\varepsilon_{xy} + n_y\,\delta\varepsilon_y)\,\mathrm{d}x\,\mathrm{d}y\\ &= -\iint (a_x n_x + 2a_{xy}n_{xy} + a_y n_y)\,\delta w\,\mathrm{d}x\,\mathrm{d}y\end{aligned} \tag{10.29}$$

This interior work must be added to (10.14). If the real stretching forces n are also approximated linearly and the geometrical curvatures a_k are constant within each element, the virtual work expressed by equation (10.29) is of the same form as the virtual work of load p_z in equation (10.18):

$$\delta W_{i123} = \frac{\sum (2a_{k1}n_{k1} + a_{k2}n_{k2} + a_{k3}n_{k3})}{12F_{123}} \qquad (k = x,\, y,\, xy) \tag{10.30}$$

We get the same results if the combination an is approximated linearly from point to point and not a and n separately. The buckling terms are established in the same way, as is seen from the differential equation (10.2b).

Also in the stretching compatibility equation we have to add additional terms since in a shell the gridlines are curved and the virtual normal forces δN in Figure 10.3(c) themselves do not satisfy the equilibrium condition. In addition, virtual moments are necessary:

$$\delta m_y = a_x\,\delta\Phi, \qquad \delta m_x = a_y\,\delta\Phi, \qquad \delta m_{xy} = a_{xy}\,\delta\Phi \tag{10.31}$$

where $\delta\phi$ is the Airy stress function belonging to the force system δN. Because of the duality between stretching and bending $\delta\Phi$ is equal to δw. The virtual moments (equation 10.31) do work on the real elastic curvatures κ:

$$\Delta\,\delta W_i^* = \iint (\delta m_x \kappa_x + 2\delta m_{xy}\kappa_{xy} + \delta m_y \kappa_y)\,\mathrm{d}x\,\mathrm{d}y \tag{10.32}$$

This interior work must be added to equation (10.20). The finite representation is analogous to equation (10.30).

10.3.4 Formal derivation of the equations

Till now the equations have been derived in a vivid manner. In this section it is done by introducing virtual standard states (Figure 10.4), which are the x

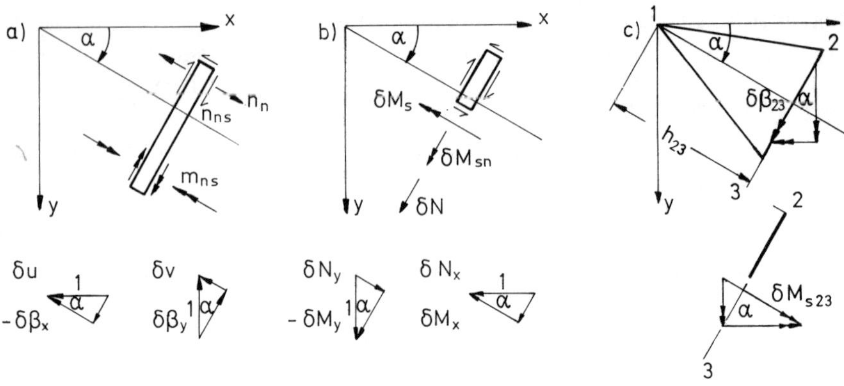

Figure 10.4 Virtual standard states. (a) Virtual displacement, (b) virtual forces, (c) combination of the standard states

and y components of the virtual states used before (Figure 10.3). To the virtual standard displacements (Figure 10.4a) belong the concentrated virtual strains

$$\left.\begin{matrix}\delta u;\,\delta v\\ \delta\beta_y;\,\delta\beta_x\end{matrix}\right\}=1:\qquad \left.\begin{matrix}\delta E_n\\ \delta K_n\end{matrix}\right\}=c;\,s\qquad \left.\begin{matrix}\delta E_{ns}\\ \delta K_{ns}\end{matrix}\right\}=-s;\,c$$

(for c, s, see equations 10.25) (10.33)

At these virtual strains the real interior forces do work:

$$\delta W_{\mathrm{i}}=\int(n_n\,\delta E_n+n_{ns}\,\delta E_{ns}+m_n\,\delta K_n+m_{ns}\,\delta K_{ns})\,\mathrm{d}s \tag{10.34}$$

Since the virtual strains are constant at each gridline the interior work at a gridline may be represented by these virtual strains and the average value of the interior forces $\bar{n}$, $\bar{m}$ as in equation (10.19). The interior work $2\,\delta W_{\mathrm{i}}$ at a gridline ik then becomes, by applying the transformation (10.24):

$$\left.\begin{matrix}\delta u\\ \delta\beta_y\end{matrix}\right\}=1:\qquad l\left(\begin{matrix}\bar{n}_n\\ \bar{m}_n\end{matrix}\,c-\begin{matrix}\bar{n}_{ns}\\ \bar{m}_{ns}\end{matrix}\,s\right)=l\left(\begin{matrix}\bar{n}_x\\ \bar{m}_x\end{matrix}\,c+\begin{matrix}\bar{n}_{xy}\\ \bar{m}_{xy}\end{matrix}\,s\right)=-\begin{matrix}\bar{n}_x\\ \bar{m}_x\end{matrix}\,y_{ik}+\begin{matrix}\bar{n}_{xy}\\ \bar{m}_{xy}\end{matrix}\,x_{ik}$$

$$\left.\begin{matrix}\delta v\\ \delta\beta_x\end{matrix}\right\}=1:\qquad l\left(\begin{matrix}\bar{n}_n\\ \bar{m}_n\end{matrix}\,s+\begin{matrix}\bar{n}_{ns}\\ \bar{m}_{ns}\end{matrix}\,c\right)=l\left(\begin{matrix}\bar{n}_y\\ \bar{m}_y\end{matrix}\,s+\begin{matrix}\bar{n}_{xy}\\ \bar{m}_{xy}\end{matrix}\,c\right)=\begin{matrix}\bar{n}_y\\ \bar{m}_y\end{matrix}\,x_{ik}-\begin{matrix}\bar{n}_{xy}\\ \bar{m}_{xy}\end{matrix}\,y_{ik} \tag{10.35}$$

If the distribution of the interior forces is approximated linearly from point to point the average values $\bar{n}$, $\bar{m}$ will be expressed by the nodal point values (10.18) and we obtain finally the interior work of the triangle 1, 2, 3 by summarizing the work (10.35) at the gridlines and changing *ijk* cyclically:

$$\left.\begin{matrix}\delta u\\ \delta\beta_y\\ \delta v\\ \delta\beta_x\end{matrix}\right\}=1: \qquad 4\,\delta W_i = y_{ij}\begin{bmatrix} n_x\\ m_x\\ n_{xy}\\ m_{xy}\end{bmatrix}_k - x_{ij}\begin{bmatrix} n_{xy}\\ m_{xy}\\ n_y\\ m_y\end{bmatrix}_k \tag{10.36}$$

To formulate compatibility conditions standard forces (Figure 10.4b) are enforced on the triangle. They produce the interior forces

$$\left.\begin{matrix}\delta N_y\,;\,\delta N_x\\ \delta M_x\,;\,\delta M_y\end{matrix}\right\}=1: \qquad \left.\begin{matrix}\delta N\\ \delta M_s\end{matrix}\right\}=c\,;\,s \qquad \left.\begin{matrix}\delta N_{sn}\\ \delta M_{sn}\end{matrix}\right\}=s\,;\,-c \tag{10.37}$$

concentrated at the gridlines. The interior work (principle of virtual forces)

$$\delta W_{\mathrm{i}}^{*} = \int(\delta N\varepsilon_s + \delta N_{sn}\varepsilon_{sn} + \delta M_s\kappa_s + \delta M_{sn}\kappa_{sn})\,\mathrm{d}s \tag{10.38}$$

at the gridline *ik* becomes, in analogy to (10.35),

$$\left.\begin{matrix}\delta N_y\\ \delta M_x\end{matrix}\right\}=1: \qquad l\left(\frac{\bar{\varepsilon}_s}{\bar{\kappa}_s}c+\frac{\bar{\varepsilon}_{sn}}{\bar{\kappa}_{sn}}s\right)=l\left(\frac{\bar{\varepsilon}_y}{\bar{\kappa}_y}c-\frac{\bar{\varepsilon}_{xy}}{\bar{\kappa}_{xy}}s\right)=-\frac{\bar{\varepsilon}_y}{\bar{\kappa}_y}y_{ik}-\frac{\bar{\varepsilon}_{xy}}{\bar{\kappa}_{xy}}x_{ik}$$

$$\left.\begin{matrix}\delta N_x\\ \delta M_y\end{matrix}\right\}=1: \qquad l\left(\frac{\bar{\varepsilon}_s}{\bar{\kappa}_s}s-\frac{\bar{\varepsilon}_{sn}}{\bar{\kappa}_{sn}}c\right)=l\left(\frac{\bar{\varepsilon}_x}{\bar{\kappa}_x}s-\frac{\bar{\varepsilon}_{xy}}{\bar{\kappa}_{xy}}c\right)=\frac{\bar{\varepsilon}_x}{\bar{\kappa}_x}x_{ik}+\frac{\bar{\varepsilon}_{xy}}{\bar{\kappa}_{xy}}y_{ik} \tag{10.39}$$

and finally for the triangular element 123:

$$\left.\begin{matrix}\delta N_y\\ \delta M_x\\ \delta N_x\\ \delta M_y\end{matrix}\right\}=1: \qquad 4\,\delta W_{\mathrm{i}}^{*} = y_{ij}\begin{bmatrix} \varepsilon_y\\ \kappa_y\\ -\varepsilon_{xy}\\ -\kappa_{xy}\end{bmatrix}_k - x_{ij}\begin{bmatrix} -\varepsilon_{xy}\\ -\kappa_{xy}\\ \varepsilon_x\\ \kappa_x\end{bmatrix}_k \tag{10.40}$$

The equations (10.36) and (10.40) are favourable for virtual work since they are formulated in the global *xy* system and not in the individual *sn* system of the edges.

Due to the virtual displacement δw in Figure 10.3(a), each element undergoes the rotation $\delta\beta_{23}=1/h_{23}$ (10.15) about the baseline 2, 3, and we get the standard rotations $\delta\beta_y$ and $\delta\beta_x$ (Figure 10.4c):

$$\delta\beta_y = -\delta\beta_{23}\cos\alpha = \delta\beta_{23}\frac{y_{23}}{l_{23}}=\frac{y_{23}}{2F_{123}}, \qquad \delta\beta_x = -\delta\beta_{23}\sin\alpha = -\frac{x_{23}}{2F_{123}}$$

$$(\delta M_x = -\delta M_s) \qquad\qquad (\delta M_y = -\delta M_s) \tag{10.41}$$

where the relationship (10.25) for $\sin\alpha$ and $\cos\alpha$ as well as that for the element area

$$2F_{123} = (hl)_{23} \tag{10.42}$$

is involved. After combining the virtual standard work (10.35) with the coefficients (10.41), the interior work in the element 1, 2, 3 becomes due to the virtual displacement δw:

$$\begin{aligned} 4\delta W_i F_{123} &= y_{23}(y_{ij}\, m_{xk} - x_{ij}\, m_{xyk}) - x_{23}(y_{ij}\, m_{xyk} - x_{ij}\, m_{yk}) \\ (\delta W_i^*) &\qquad\quad (\kappa_y) \qquad (-\kappa_{xy}) \qquad\qquad (-\kappa_{xy}) \quad (\kappa_x) \end{aligned} \tag{10.43}$$

In a similar way the virtual moment $\delta M_s = -1/h_{23}$ at the baselines 2, 3 due to the moment system in Figure 10.2(d) may be represented by the standard 'forces' δM_x and δM_y (values within the brackets in equation 10.41) to which belongs the interior work δW_i^* in equation (10.43).

The twisting moments $m_{xy} = b_t \kappa_{xy}$ and the twisting curvatures κ_{xy} cancel each other in equation (10.43) if the second is multiplied by the torsional stiffness b_t and the two equations are summarized:

$$4(\delta W_i + b_t\, \delta W_i^*) F_{123} = y_{23} y_{ij} \underbrace{(m_x + b_t \kappa_y)_k}_{b_x \kappa_x + \bar{b}_t \kappa_y} k + x_{23} x_{ij} \underbrace{(m_y + b_t \kappa_x)_k}_{b_y \kappa_y + b_t \kappa_x} \tag{10.44}$$

Thus the twisting curvature is eliminated out of the plate-bending equation also for orthotropic plates, as we have done before (equation 10.27) for an isotropic plate by using both principles of virtual work simultaneously. Equation (10.44) is the finite one belonging to the differential equation (10.7).

The finite equation belonging to the differential equations (10.10) and (10.11) are derived by combining the standard force systems (10.40) in the following manner:

$$(y_{23}\, \delta M_x + x_{23}\, \delta M_y)/F_{123}$$

$$\rightarrow 4\, \delta W_i^* F_{123} = y_{23}(y_{ij}\kappa_{yk} + x_{ij}\kappa_{xyk}) - x_{23}(y_{ij}\kappa_{xyk} + x_{ij}\kappa_{xk}) \tag{10.45a}$$

$$(x_{23}\, \delta M_x - y_{23}\, \delta M_y)/F_{123}$$

$$\rightarrow 4\, \delta W_i^* F_{123} = (y_{23} y_{ij} + x_{23} x_{ij})\kappa_{xyk} + (x_{23} y_{ij}\kappa_{yk} + y_{23} x_{ij}\kappa_{xk}) \tag{10.45b}$$

Comparing the finite equations (10.44) and (10.45) with the corresponding differential equations (10.7) to (10.11) the relationship between the differential quotients and the corresponding difference quotients

$$\begin{aligned} &4F_{123}(\ \ )'' = y_{23} y_{ij}(\ \ )_k, \qquad 4F_{123}(\ \ )^{\cdot\cdot} = x_{23} x_{ij}(\ \ )_k \\ &8F_{123}(\ \ )'^{\cdot} = -(y_{23} x_{ij} + x_{23} y_{ij})(\ \ )_k \text{ changing } ijk \text{ cyclically} \end{aligned} \tag{10.46}$$

become evident. In equation (10.45a) the twisting curvatures κ_{xy} at the interior point also cancel each other from element to element since there are no discontinuities in the twisting curvatures at the gridlines. Only in the equations at a boundary point do the twisting values at the edge points on the boundary remain. The stretching equations are derived in a similar manner, only the two principles of virtual work must be exchanged, since an equilibrium bending equation is dual to a compatibility stretching equation, and vice versa.

10.4 Linear stress concentration problems

The first stress concentration problem is a rectangular built-in plate with an interrupted support on the middle line as it is found in double-walled ship structures if the frame webs are not welded continuously to the ship walls (Figure 10.5a). At the point where the support ceases a severe stress concentration arises. Therefore a refined element grid is chosen there. The results are compared with those from the usual finite element analysis [7] in which the plate is divided three times more but the results from our coarse grid (solid line in Figure 10.5a) are better, especially near the stress concentration point. In the finite element analysis the stress peak is rounded out (dotted line in Figure 10.5a).

In the second example (Figure 10.5b), a plate loaded by a concentrated moment *M*, there are peaks and jumps in the stress distribution. The plate is simply supported on three edges and elastically supported by a girder on the edge where the concentrated moment is introduced. In Figure 10.5(b) the numerical and analytical results are compared. The bending stiffness *B* of the edge girder is chosen one half of the whole plate stiffness *ba* in a plate section. The numerical solution approximates quite well the peaks and jumps in the moment distribution. The third example exists in a steam separator with hexagonally arranged pipes which are connected by small upright sheets (Figure 10.5c). The normal force *P* in the sheet must be transferred to the opposite sheet by circumferential pipe-wall bending. Because of the symmetry we have to consider only a 90° part of the shell. In Figure 10.5(c) the element grid and the distribution of the circumferential moments m_x in the most strained section AA is demonstrated and compared with FEM results which also round out the expected moment peak at the connection.

10.5 GEOMETRIC NON-LINEAR PROBLEMS

Panels of plates and shells in thin-walled structures are often loaded as well as in the lateral and the longitudinal directions since they act in a double sense. For example, the upper and lower walls of a taileron (Figure 10.6) have to transfer the distributed lifting loads by bending to the ribs and spars. Furthermore, these walls are strained by the normal and shearing forces due to the bending of the whole box beam. If the liting load *p* increases more and more, the buckling load of the wall panels will be exceeded and the panels will therefore be severely deflected. This, combined with a stretching

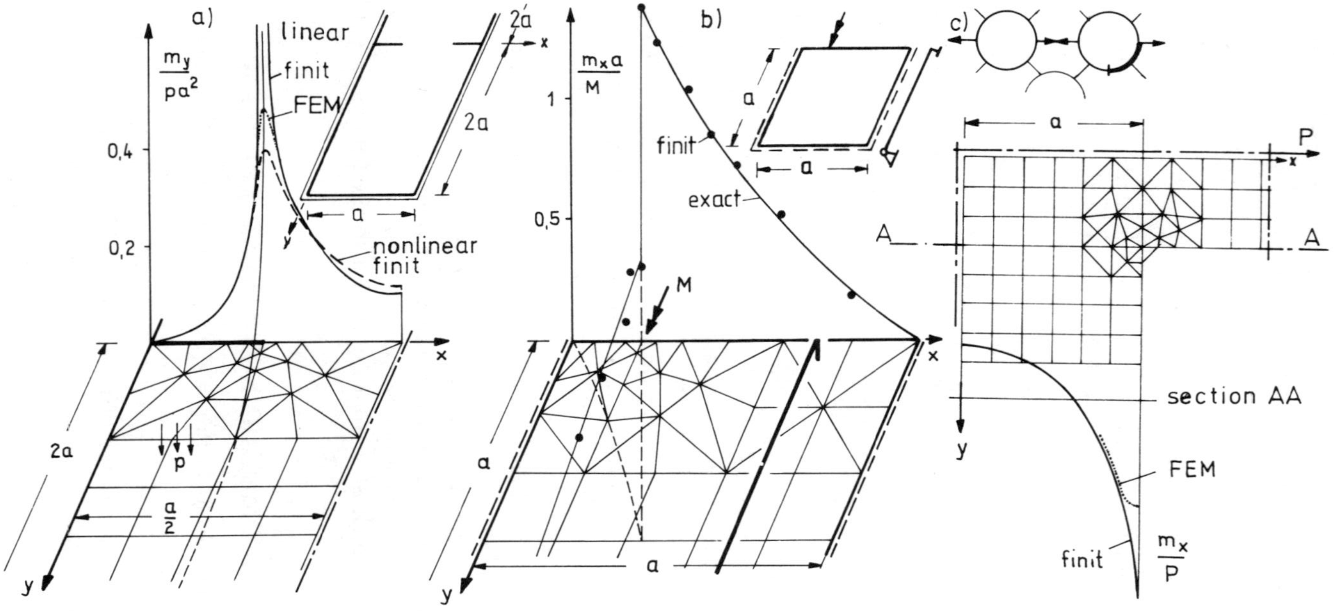

Figure 10.5 Stress concentrations problems. (a) Built in plate with an interrupted interior support, (b) plate with an elastically supported edge ($B/ba = 0.5$) loaded by a concentrated moment, (c) cylindrical shell loaded on an attached sheet

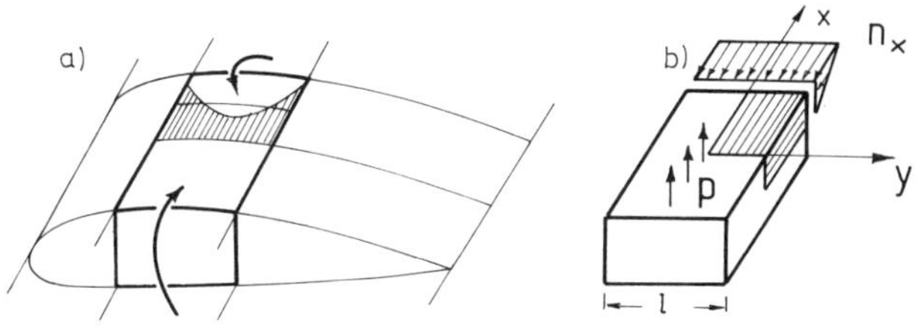

Figure 10.6 Box beam of a taileron. (a) System, (b) idealized box beam cell

of the middle surface, will cause the compression forces in the interior of the panel to be reduced and those near the supporting edges to be increased (Figure 10.6a). This well-known problem was solved long ago [8], but if carbon fibre restrained resins (CFC) are used it must be solved once more since the bending stresses due to the postbuckling phenomena are too involved because of the brittle material behaviour.

A similar problem exists in moulded resin plates which are used as bottoms in table drawers (see Figure 10.8a). They are stiffened by close-meshed stringers in the longitudinal and the cross directions to obtain a sufficient bending rigidity. Because of the low Youngs module of the resin large deflections arise even at smaller loads. The cover sheet has to transfer the loads by plate bending to the stringers and is strained in addition by normal compression forces due to the bending of the stringers. Thus we have also a combined plate and stretching problem as in the taileron demonstrated earlier.

For involving the influence of the large deflection only the equilibrium equation for plate (10.2b) and the compatibility equation for plate stretching (10.4a) have to be extended slightly, if the deflection w is large in comparison to the inplane deformations u, v. Therefore the geometrical curvature in the shell terms of the plate-bending equation is extended by the elastic curvatures: $a \rightarrow a + \kappa$. Furthermore, the additional stretching of the middle face due to the large deflections $w''w^{\cdot\cdot} - w'^2 = \kappa_x\kappa_y - \kappa_{xy}^2$ is involved in the stretching compatibility equation by adding the half value of the elastic curvature to the geometrical curvature: $a \rightarrow a + \kappa/2$. Thus the equations (10.7) become, for an isotropic plate [9],

$$b\Delta(\kappa_x + \kappa_y) + \boxed{a_x + \kappa_x}\, n_x + \boxed{a_y + \kappa_y}\, n_y + 2\boxed{a_{xy} + \kappa_{xy}}\, n_{xy} = -p_z$$

$$f\Delta(n_x + n_y) - \boxed{a_x + \kappa_x/2}\, \kappa_y - \boxed{a_y + \kappa_y/2}\, \kappa_x - 2\boxed{a_{xy} + \kappa_{xy}/2}\, \kappa_{xy} = 0$$

These equations are solved step by step where the original stretching and bending terms are fixed and only the shell terms (framed in equations 10.47) are changed by the additional elastic curvatures κ. The iteration converges rapidly if the expected curvature is estimated from the results analysed in the steps before.

Let us first look to the walls of the CFC taileron in Figure 10.6. The substantial load-carrying element is the box beam [10]. The essentials about the postbuckling behaviour of the wall panels become evident if the problem is solved for only one cell of the box beam. To involve the interaction with the other cells we have to demand that the cross-sections at the ribs remain plane. The box beam moment M will be transferred to the neighbouring cell by the normal wall forces n_x in the se tion at the rib, to which belongs the resulting normal force N:

$$M = \int n_x z \, \mathrm{d}s, \qquad N = \int n_x \, \mathrm{d}s = 0 \tag{10.48}$$

The upper and lower walls of the taileron are loaded by the lifting forces p. The problem becomes symmetric to the x and y axes if a rectangular cross-section, a constant bending moment M, and a constant lifting load p are assumed. Furthermore, the stress distribution is postulated at first antisymmetrically to the middle of the spar webs, i.e. only one-eighth of the box is considered (Figure 10.6b). The non-dimensional bending moment $\bar{m}$ and the stretching forces $\bar{n}$

$$\bar{m} = \frac{m}{pl^2}, \qquad \bar{n} = \frac{ni}{pl^2} \tag{10.49}$$

are introduced to obtain more common results, where

$$i = \sqrt{(bf)} \tag{10.50}$$

is the inertia radius which represents the ratio between the bending and stretching stiffnesses. Furthermore, two additional parameters are introduced. The first, the non-dimensional box beam moment

$$\xi = \frac{M}{M_0} \tag{10.51}$$

represents the exceeding of the upper panel buckling load where M_0 is the critical moment when the upper panel is buckling. The second parameter

$$\lambda = \frac{pl^4}{b\pi^4 i} = \frac{w_0}{i} \tag{10.52}$$

represents the ratio between the panel deflection w_0 and the plate thickness, or better the inertia radius i. This parameter is a measure of the non-linearity of the problem. If the moment M results from the lifting load the parameter λ also grows linearly with the buckling load parameter ξ.

In Figure 10.7(a) and (b) the distribution of the normal forces $\bar{n}_x$ and the

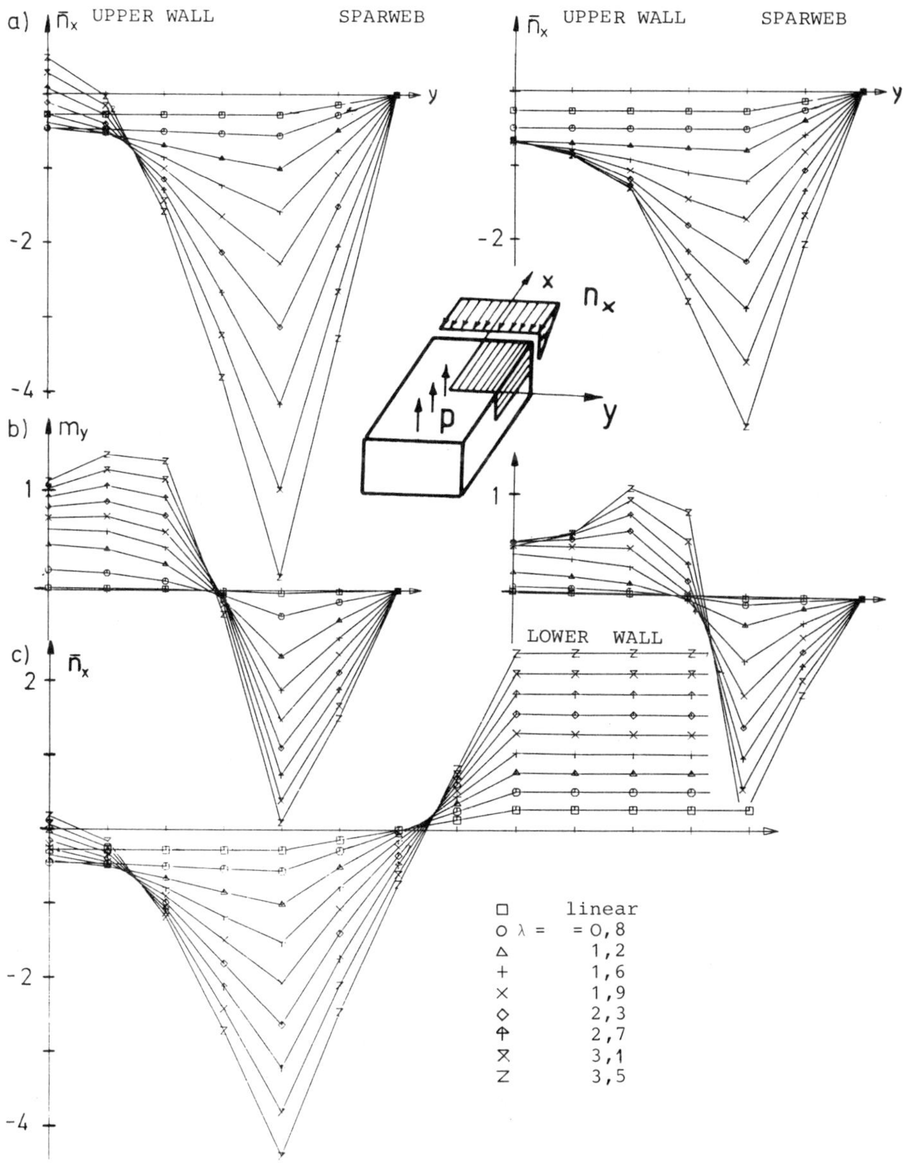

Figure 10.7 Distribution of the interior forces in the walls of a box beam due to a constant load p (parameter $\lambda = \xi$). (a) Normal forces $\bar{n}_x$, (b) bending moments $\bar{m}_y$, (c) changing of the normal forces $\bar{n}_x(y)$ due to a different non-linear behaviour of the upper and the lower wall

bending moment $\bar{m}_y$ in the middle section of the cell ($x=0$) is demonstrated, where ξ and λ are equal to each other and are changed between 0.8, and 3.5. The y coordinate is wound off and starts at the centreline of the upper panel followed by the spar web. If the moment M or the value ξ is small ($\xi<1$) the normal forces are constant in the upper wall and linear in the spar web (Figure 10.7a). With increasing moment M or parameter ξ these forces decrease at the centreline and increase at the stiff edge between the upper wall and the spar web. The stress distribution in the spar web remains linearly up to large values ξ. In Figure 10.7 the results for two limit situations are given on the left-hand side if the upper wall is simply supported at the rib, which means that the rib-bending stiffness is much smaller than that of the wall, and on the right-hand side if the wall is clamped at the rib, which is nearly performed by a sandwich rib. In the second case the reloading of the normal forces $\bar{n}_x$ is smaller since the buckling load of the wall panel increases due to the clamping condition at the rib. The bending moments $\bar{m}_y$ at the edge between the wall and the spar web increases rapidly with increasing parameter ξ, but the peak values are not reduced by the clamping condition at the rib; in both cases they are nearly the same. The lateral load p is small, since to a parameter $\lambda=1$ belongs a deflection w_0 only a quarter of the inertia radius ($i=t/\sqrt{12}$). The large deflections are induced, however, by the postbuckling behaviour.

The assumption of any asymmetric stress distribution in the spar web is not realistic since in the upper wall there are compression forces and in the lower one there are tension forces due to box beam bending. The results in Figure 10.7(c) involve the different behaviour of these two walls by considering a quarter of the box beam cell. With increasing box beam moment M the interior of the upper wall panel is reloaded, but the normal force distribution in the lower one remains nearly constant. Thus the neutral axis in the spar web is shifted nearer to the lower wall. From the comparison between the results in Figure 10.7(a) (right-hand side) and (c) it becomes evident that the changing of the stress distribution in the upper wall due to the postbuckling behaviour may be analysed only for the simpler system in Figure 10.7(a).

The results for the resin plate (Figure 10.8) which is stiffened on one side by close-meshed stringers are similar since the cover sheet is strained by the normal forces due to the bending of the stringers as well as by the moments due to the bending of the cover sheet. However, the normal forces in the cover sheet are smaller since the neutral axis of the stringers is near to the sheet [10]. The reloading of the normal forces due to the postbuckling behaviour therefore occurs due to larger parameters λ (betwen 20 and 100) as in the box beams discussed earlier to which the deflections w_0/t between 7 and 42 belong. In Figure 10.8(d) the distribution of the normal forces $\bar{n}_x$ and the bending moments $\bar{m}_y$ in the cover sheet (section I) of a plate which

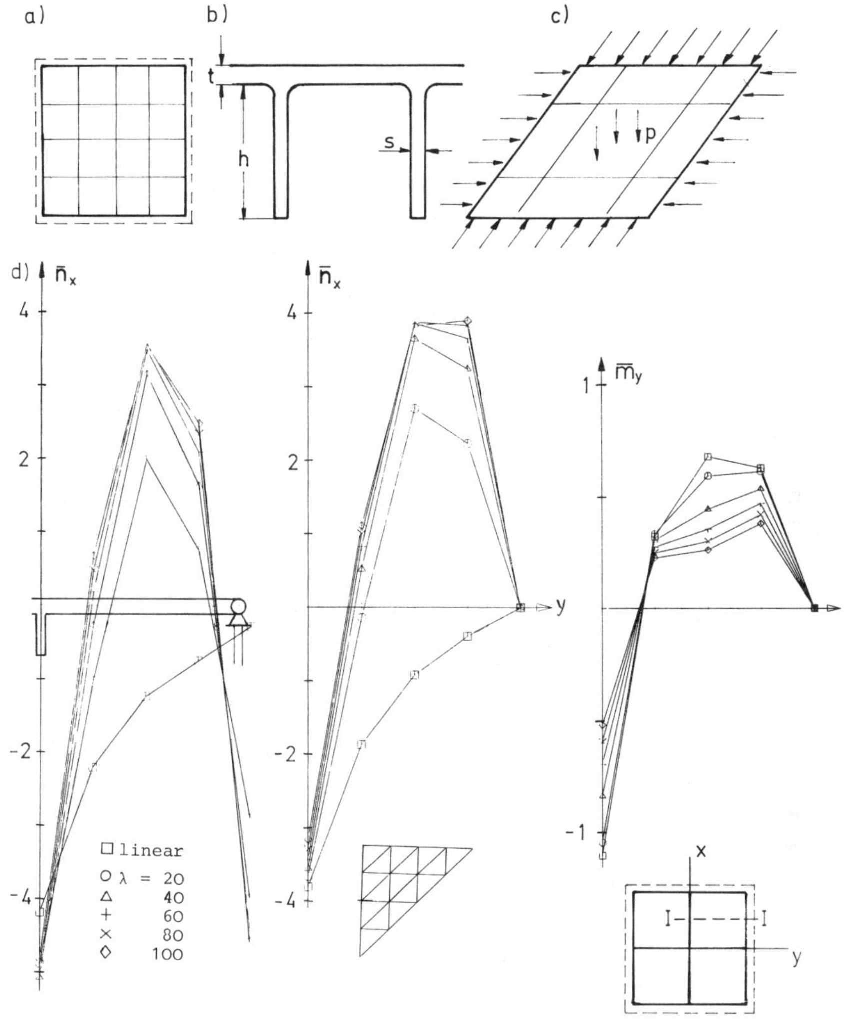

Figure 10.8 Moulded resin plate eccentric stiffened. (a) Stiffened plate, (b) stringer, (c) loading of the cover sheet, (d) distribution of the interior forces in section I of the cover plate

is stiffened by one stringer in both directions is represented. At the boundary the plate is simply supported in respect to the bending deformations since the side walls in a box are mostly stiff enough in their plane but are weak against bending perpendicular to their plane. In respect to the inplane deformations of the plate two conditions are discussed. First there is asumed a stringer which simulates the apparent stretching resistance of the side wall. As a second condition this stringer is considered as infinite rigid, which means that the side wall prohibits deformations parallel to the boundary edge ($\varepsilon_s \equiv 0$); thus the normal forces parallel and perpendicular to the edge vanish there. Due to the symmetry it is sufficient to consider only one-eighth of the whole plate (Figure 10.8d).

In the distribution of the normal forces $\bar{n}_x$ (parallel to the stringer) there is a peak at the stringer (Figure 10.8d) which vanishes in the linear case monotonously with increasing distance from the central stringer. Due to increasing loading, appreciable stretching of the middle surface in the cover panels arises and the normal forces there change to tension as in a membrane. By this membrane effect the bending of the panel is reduced (Figure 10.8d) and the apparent flange of the cover plate belonging to the central stringer becomes smaller. For large loadings there are two peaks in the normal force distribution, one at the interior stringer and the other at the boundary stringer (Figure 10.8d). The peak at the boundary vanishes if the stringer there is assumed as rigid. Finally, we obtain a limit distribution for the normal forces $\bar{n}_x$ and the bending moments $\bar{m}_y$ since they are standardized in equation (10.49) by the load p.

10.6 PLATES WITH VARIABLE STIFFNESS

In plates with variable thickness or with a physical non-linear behaviour of the Hencky type the plate stiffness b and the flexibility f varies within each element. The basic equations (10.1) to (10.5) remain valid. The essentials become evident if we restrict ourselves to plate bending of an isotropic plate. The plate-stretching problem is dual to the bending problem. Only the differential equation (10.9) for the bending curvatures which are derived from the equilibrium condition (10.2b) by introducing the stress–strain relationship (10.5b) changes for an isotropic plate to

$$\Delta b(\kappa_x + \kappa_y) - (1-\nu)[b''\kappa_y - 2b'^{\cdot}\kappa_{xy} + b^{\cdot\cdot}\kappa_x] = -p \tag{10.53}$$

The twisting curvature κ_{xy} is nearly eliminated completely by using the plate compatibility equations (10.3b):

$$2m_{xy}'^{\cdot} = [2b(1-\nu)\kappa_{xy}]'^{\cdot} = (1-\nu)[(b\kappa_y)'' + (b\kappa_y)^{\cdot\cdot} - b''\kappa_y - b^{\cdot\cdot}\kappa_x] \tag{10.54}$$

Equation (10.53) is similar to equation (10.9); only the terms in the brackets are added. The advantage of this formulation is that the first significant term is invariant in simply supported plates, since the sum of the curvatures $\kappa_x + \kappa_y$ vanishes at the boundaries and is nearly invariant in plates with other boundary conditions.

In Figure 10.9 the influence of the additional terms is demonstrated for a simply supported plate whose stiffness changes by an exponential function only in the y direction. In this special situation an analytical solution of the

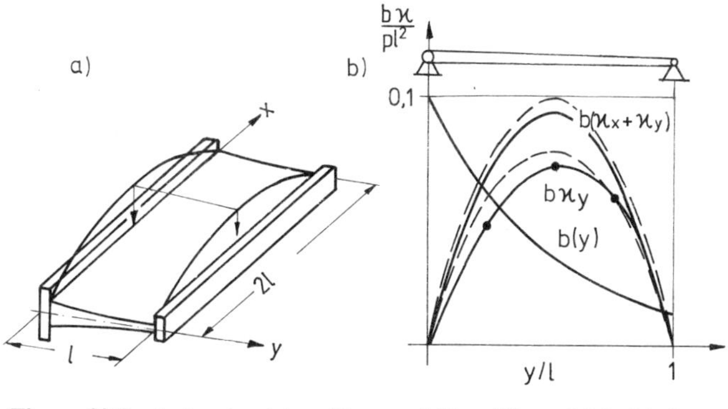

Figure 10.9 Isotropic plate with a variable stiffness $b(y)$. (a) System and load, (b) results.

— exact, ● numerical } results, -- results of the equation $\Delta(b\kappa) = -p$

problem exists. The difference between the complete solution (solid line) and the solution of the approximate equation $\Delta b(\kappa_x + \kappa_y) = -p$ is small (dashed line), although the plate thickness is doubled from one edge to the other or the stiffness becomes eight times larger. This formulation is also of great advantage in physical non-linear problems since due to the non-linearity only the bending equilibrium and the stretching compatibility condition are changed and these equations are not sensitive to variations in the stiffnesses. In Figure 10.9(b) some numerical results are also given when the span l of the plate is subdivided into four elements. These equations must now be transferred to the finite ones. The bending equilibrium equation (10.19) is also valid in a plate with varying stiffness. However, if we simultaneously use the principle of virtual forces for eliminating the twisting moments, the equation (10.26) has to be extended. Because the stiffness b_t now varies from point to point, we have to enforce a virtual self-equilibrium moment sytem δM_s which is equal to $b_t(s)\,\delta K_n$ with variable virtual moments along the gridlines. Since the virtual moments have to fulfil the equilibrium conditions, virtual moments δm within the elements are also necessary:

$$\delta m_y = -\frac{b'_t y_{23}}{2F_{123}}, \qquad \delta m_{yx} = -\frac{b'_t x_{23}}{2F_{123}}; \quad \delta m_x = \frac{b^{\cdot}_t x_{23}}{2F_{123}}, \qquad \delta m_{xy} = \frac{b^{\cdot}_t y_{23}}{2F_{123}} \qquad (10.55)$$

They depend on the gradients b'_t and $b^{\cdot}_t$ of the plate-twisting stiffness $b_t = \bar{b} = (1-\nu)b$. If b_t is approximated linearly within each element the derivatives are approached by

$$b'_t = \frac{y_{ij}\bar{b}_k}{2F_{123}} \qquad \text{and} \qquad b^{\cdot}_t = -\frac{x_{ij}\bar{b}_k}{2F_{123}} \qquad (10.56)$$

with cyclically changing ijk. The additional virtual work within the elements becomes

$$12\Delta\ \delta W_{\mathrm{i}}^{*} F_{123} = -[y_{ij}(y_{23}\Sigma\kappa_{yl} + x_{23}\Sigma\kappa_{xyl}) + x_{ij}(x_{23}\Sigma\kappa_{xl} + y_{23}\Sigma\kappa_{xyl})]\bar{b}_k \qquad (10.57)$$

This work is similar to that induced by the load p_z and the buckling forces s. By comparing equation (10.57) with the corresponding differential equation (10.53) the close connection betwen these equations becomes evident.

10.7 NON-LINEAR MATERIAL BEHAVIOUR

If the material is non-linear, of the Hencky type, and there is no reloading within the structure, the equations derived earlier may be used. In the non-linear case an iteration process is necessary until the plate stiffness b coincides with the special stress situation in each point. In the more common problems of the Prandtl–Reuss type we have to use a stress–strain relationship in the strain and stress increments which are assumed to be divisible into elastic and plastic parts ($\mathrm{d}\varepsilon = \mathrm{d}\varepsilon_{\mathrm{e}} + \mathrm{d}\varepsilon_{\mathrm{p}}$). If the material is incompressible due to plastic deformations the non-linear component of the strains depends only on the deviator stress component $\tilde{\sigma}_{ik}$:

$$2G\ \mathrm{d}\tilde{\varepsilon}_{ik} = 2G(\mathrm{d}\tilde{\varepsilon}_{ik\mathrm{e}} + \mathrm{d}\tilde{\varepsilon}_{ik\mathrm{p}}) = \mathrm{d}\tilde{\sigma}_{ik} + 2\tilde{\sigma}_{ik}G\ \mathrm{d}\lambda \qquad (10.58)$$

and the rate $\mathrm{d}\lambda$ of the non-linear module T_{p} changes with the second-order stress invariant σ_v:

$$\mathrm{d}\lambda = \frac{3\ \mathrm{d}\sigma_v}{2\sigma_v T_{\mathrm{p}}(\sigma_v)} \qquad (10.59)$$

After expressing the deviator components by the complete stresses and strains, the stress–strain relationship due to the plastic deformation is that for an anisotropic plate:

$$\begin{bmatrix} \mathrm{d}\varepsilon_x \\ \mathrm{d}\varepsilon_y \\ \mathrm{d}\gamma \end{bmatrix}_{\mathrm{p}} = \begin{bmatrix} f_{11} & f_{12} & f_{13} \\ f_{12} & f_{22} & f_{23} \\ f_{13} & f_{23} & 2f_{33} \end{bmatrix}_{\mathrm{p}} \begin{bmatrix} \mathrm{d}n_x \\ \mathrm{d}n_y \\ \mathrm{d}n_{xy} \end{bmatrix} \qquad \left(f_{ik\mathrm{p}} = \frac{9}{4} \frac{\tilde{n}_i \tilde{n}_k}{n_v^2 T_{\mathrm{p}}(n_v)} \right) \qquad (10.60)$$

with coupling terms between the normal strains and the shearing stresses, and vice versa. For non-linear materials we have to introduce the strain increments of their linear and their non-linear parts. Substituting the shearing forces n_{xy} for the normal forces n_x, n_y, by equation (10.6) the compatibility equation becomes

$$\begin{aligned} &[f_{11}\,\mathrm{d}n_x + (f_{12}+f_{33})\,\mathrm{d}n_y]^{\cdot\cdot} + [(f_{12}+f_{33})\,\mathrm{d}n_x + f_{22}\,\mathrm{d}n_y]'' \\ &\quad -[(f_{33}'' - f_{13}'^{\cdot})\,\mathrm{d}n_x + (f_{33}^{\cdot\cdot} - f_{23}'^{\cdot})\,\mathrm{d}n_y + (2f_{33}'^{\cdot} - f_{13}^{\cdot\cdot} - f_{23}'')\,\mathrm{d}n_{xy}] \\ &\quad -2[f_{13}\,\mathrm{d}n_x + f_{23}\,\mathrm{d}n_y]'^{\cdot} - \dot{f}_{13}\,\mathrm{d}n_x' - f_{23}'\,\mathrm{d}n_y^{\cdot} + f_{13}'\,\mathrm{d}n_x^{\cdot} + f_{23}^{\cdot}\,\mathrm{d}n_y' = 0 \end{aligned} \qquad (10.61)$$

In the flexibilities f_{ik} both the elastic and plastic values are involved. The first row in equation (10.61) coincides with that in equation (10.7), the second row belongs to the additional terms in equation (10.53), and in the last row there are the coupling terms between the normal strains and the shearing forces, and vice versa. The corresponding finite equation may be derived as in the previous section. The plate-bending differential equation is similar to equation (10.61). In Figure 10.5(a) some results for a stress concentration problem with non-linear material are given. In stress concentration problems there is only a small 'non-linear' region near to the stress peak, but in structures with large non-linear parts good results may be expected due to the nearly invariant behaviour of the strain sum $\varepsilon_x + \varepsilon_y$ and the moment sum $m_x + m_y$, which are the essential terms in the stretching compatibility or the bending equilibrium equations respectively.

REFERENCES

1. O. C. Zienkiewicz, *The Finite Element Method in Engineering Science*, McGraw-Hill, 1977.
2. B. M. Fraeijs de Veubeke, *Matrix Methods of Structural Analysis*, AGARDograph 72, Pergamon Press, Oxford, 1964.
3. J. T. Oden, *Finite Elements of Non-linear Continua*, McGraw-Hill, 1972.
4. E. Giencke, 'Ein einfaches und genaues finites Verfahren zur Berechnung von orthotropen Scheiben und Platten', *Stahlbau*, **36,** 260–268 and 303–315 (1967).
5. E. Giencke, 'The mechanical interpretation of high accuracy multipoint difference methods for plates and shells', *IUTAM Symp. High Speed Computing of Elastic Structures*, Liége 1971. 'Les Congrés et colloques de Université de Liége', Vol. 61.
6. E. Giencke, 'A simple mixed method for plate and shell problems', *Nuclear Eng. and Design*, **29,** 141–155 (1974).
7. K.-H. Hapel, J. Sperlich, and M. Kohler, 'Untersuchungen an einem Kirchhoffschen Plattenfeld mit einspringender Berandung unter konstantem Druck', *Schiff und Hafen*, **30,** 702–708 (1978).
8. K. Marguerre, 'Die mittragende Breite der gedrückten Platte', *Luftfahrtforschung*, **XIV,** 121 (1937).
9. K. Marguerre, 'Zur Theorie der gekrümmten Platte großer Formänderungen', *Luftfahrtforschung*, **1939,** 413 (1939).
10. S. Worm, *Platten mit großen Deformationen*, Diplomarbeit TU, Berlin, 1980.

Hybrid and Mixed Finite Element Methods
Edited by S. N. Atluri, R. H. Gallagher, and O. C. Zienkiewicz

Chapter 11

Mixed Models for Plates and Shells: Principles—Elements—Examples

W. Wunderlich

11.1 INTRODUCTION

For the approximate solution of problems in structural mechanics and related fields the finite element method has proven to be a powerful tool of analysis. For reasons of historical development, discretization was most often performed by inserting trial functions for the displacements into appropriate global forms of the basic equations (e.g. into the principle of virtual work). These trial functions have to satisfy certain subsidiary conditions, which follow from the principle used. In the case of plates and shells this leads to the requirement of high-order interpolation functions and results in sophisticated element models with a large number of unknowns. The side conditions can be relaxed by including them in the principle itself. This leads to the various generalized principles which are the basis of a number of mixed or hybrid finite element models.

While in the hybrid approach [1] the boundary and interelement conditions are relaxed, in the mixed approach equilibrium and kinematic equations within the body and on its boundary are approximated independently [2]. Thus, both static and kinematic variables are employed as unknowns simultaneously, and as a result the corresponding trial functions can be chosen independently as well. Due to the fact that only low-order derivatives appear in the generalized principles the requirement of C_0 continuity is normally sufficient, and simple interpolation functions may be used. Also, because numerical evaluation of the integrals can be avoided, an efficient calculation of the element matrices is possible.

Unlike in the hybrid approach—where the element matrices are condensed to a stiffness matrix on the element level—in the mixed approach the element matrix is generally assembled to the global coefficient matrix by the usual superposition process. The solution of the resulting algebraic equations immediately also yields the static variables which are of basic interest in the analysis and design of a structure.

In the application of mixed models to plates and shells a number of advantages may be noted. Firstly, the underlying principles are transparent in that their mechanical content is easy to interpret. This aspect is especially important in the non-linear case. Moreover, linear interpolation functions are often sufficient to give satisfactory results for practical applications. From a numerical point of view the choice of the global principle and the variables to be used as well as the proper modelling of the boundaries and the consistent inclusion of the loads need special consideration.

In this chapter emphasis is first given to the formulation of the principles underlying a consistent theory of thin shells formulated entirely in terms of associated (mixed) tensor components. These offer a number of advantages, in that the components of the stress resultants and displacements acting normal to an element face are identical with their respective physical components, and the constitutive equations of the shell can be expressed in closed form. This theory has been derived earlier by the author [3] utilizing the general three-dimensional Hellinger–Reissner principle. Its reduced, two-dimensional form will be employed here as the basis of a certain class of mixed shell elements. Next, since in the mixed approach the element properties are strongly dependent on the choice of the unknowns and on the order of the trial functions, attention will also be focused on the numerical behaviour of different models. A comparison will be made of cylindrical shell calculations for which various principles and interpolation functions have been used.

A different kind of mixed approach is based on the direct numerical integration of the fundamental relationships written in the form of a system of ordinary first-order differential equations. It can be utilized with great advantage in all cases in which the local equations can be reduced to a set of ordinary differential equations. Prominent examples are shells of revolution and beams, and cases which are amenable to a solution by the finite strip method. Applied to an element of such a structure, this procedure results in an integral matrix with prescribed *a priori* error bounds which by an appropriate choice of variables is identical to the transfer matrix of the element. After converting it to the corresponding stiffness matrix on the element level the system analysis can be performed in the same way as in the usual stiffness approach. This alternative procedure is outlined, and its application to linear and geometrically non-linear problems is demonstrated for the case of a spherical shell. It is shown that it yields highly accurate and reliable solutions. Nevertheless, it should be noted that its application is limited to the mentioned class of problems.

At the present time a final statement on the efficiency, accuracy, and convergence of mixed models is not yet possible. Advantages and shortcomings, however, are already clearly apparent and trends from the viewpoint of numerical applications can be stated; these merit a thorough discussion despite the fact that strict mathematical convergence criteria are still lacking.

Nevertheless, these seem promising points compared to the widely used displacement method, especially in the linear and non-linear applications to plates and shells.

11.2 BASIC EQUATIONS—GENERALIZED VARIATIONAL PRINCIPLES

11.2.1 Thin shells and plates

As the starting point for the development of mixed-type finite element models generalized variational principles are usually employed. Depending on the choice of the unknown variables, various forms can be constructed. They may be characterized by their equivalent local forms and their subsidiary conditions. Or vice versa, they may be constructed by casting the local equations in a global form and applying Gauss' divergence theorem. For the case of thin shells and plates the corresponding global forms of the basic equations are summarized in Table 11.1 in which a first approximation of the linear shell theory [3] is incorporated, written in associated (mixed-type) tensor components. The variables are inplane forces n^{α}_{β}, the moments m^{α}_{β}, the shearing forces q_{α}, and the conjugate inplane strains α^{β}_{α}, the change of curvature β^{β}_{α}, the shear strains γ^{α}, as well as the displacements v^{α}, v^{3} and the rotations φ^{α} of a point on the middle surface S. In the boundary

Table 11.1 Modules for construction of variational principles for thin shells

		Global [local] forms
A	Equilibrium	$\int_S \{\delta u^{\beta}[n^{\alpha}_{\beta}\|_{\alpha} - (b^{\alpha}_{\lambda} - 2H\delta^{\alpha}_{\lambda})m^{\lambda}_{\beta} - b^{\alpha}_{\beta}q_{\alpha} + \bar{p}_{\beta}] + \delta v^{3}(q^{\alpha}\|_{\alpha} + b^{\lambda}_{\alpha}n^{\alpha}_{\lambda} - Km^{\lambda}_{\lambda} + \bar{p}_3)$ $+ \delta\varphi^{\beta}(m^{\alpha}_{\beta}\|_{\alpha} - q_{\beta})\}\,dS$
B	Kinematics	$\int_S \{\delta n^{\alpha}_{\beta}(v^{\beta}\|_{\alpha} - b^{\beta}_{\alpha}v_3 - \alpha^{\beta}_{\alpha}) + \delta q_{\alpha}(\varphi^{\alpha} + v_3\|^{\alpha} + b^{\alpha}_{\lambda}v^{\lambda} - \gamma^{\alpha})$ $+ \delta m^{\alpha}_{\beta}[\varphi^{\beta}\|_{\alpha} + (b^{\lambda}_{\alpha} - 2H\delta^{\lambda}_{\alpha})v^{\beta}\|_{\lambda} + K\delta^{\beta}_{\alpha}v_3 - \beta^{\beta}_{\alpha}]\}\,dS$
M	Constitutive laws	$\int_S \left[\delta\alpha^{\beta}_{\alpha}\left(n^{\alpha}_{\beta} - \frac{\partial W}{\partial\alpha^{\beta}_{\alpha}}\right) + \delta\gamma^{\alpha}\left(q_{\alpha} - \frac{\partial W}{\partial\gamma^{\alpha}}\right) + \delta\beta^{\beta}_{\alpha}\left(m^{\alpha}_{\beta} - \frac{\partial W}{\partial\beta^{\beta}_{\alpha}}\right)\right] dS$
M*		$\int_S \left[\delta n^{\alpha}_{\beta}\left(\alpha^{\beta}_{\alpha} - \frac{\partial W^*}{\partial n^{\alpha}_{\beta}}\right) + \delta q_{\alpha}\left(\gamma^{\alpha} - \frac{\partial W^*}{\partial q_{\alpha}}\right) + \delta m^{\alpha}_{\beta}\left(\beta^{\beta}_{\alpha} - \frac{\partial W^*}{\partial m^{\alpha}_{\beta}}\right)\right] dS$
A_r	Static boundary conditions	$\int_{s_p} [\delta v^{t}(n_t - \bar{n}_t) + \delta v^{3}(q_n - \bar{q}_n) + \delta\varphi^{t}(m_t - \bar{m}_t) + \delta\varphi^{n}(m_n - \bar{m}_n)]\,ds$
B_r	Kinematic boundary conditions	$\int_{s_p} [\delta n_t(v^{t} - \bar{v}^{t}) + \delta q_n(v^{3} - \bar{v}^{3}) + \delta m_t(\varphi^{t} - \bar{\varphi}^{t}) + \delta m_n(\varphi^{n} - \bar{\varphi}^{n})]\,ds$

conditions the values of the variables in the directions tangential (index t) and normal (index n) to the edges are used. The equations are valid for the linear theory of thin shells with averaged shear deformations included. Under the Kirchhoff–Love hypothesis the shear forces and rotations are no longer independent, leading to second-order derivatives in the local equations. For plates the curvatures of the surface b^{α}_{β} vanish, and bending and stretching terms are uncoupled. The modules for the construction of generalized variational principles for plates under the Kirchhoff hypothesis are given in Table 11.2. In the boundary conditions modified shear forces s_n and a discrete corner term have to be introduced to account for the constraint of the Kirchhoff hypothesis on the boundaries.

The equivalence of the global statement of equilibrium (A) and static boundary conditions (A_r) to the principle of virtual work can be shown by performing a transformation using Gauss' divergence theorem; this leads, for the thin shells, to

$$
\begin{aligned}
\text{(C)}\quad & \int_S (\delta\alpha^{\beta}_{\alpha} n^{\alpha}_{\beta} + \delta\beta^{\beta}_{\alpha} m^{\alpha}_{\beta} + \delta\gamma^{\alpha} q_{\alpha})\,\mathrm{d}S - \int_S (\delta v^{\beta}\bar{p}_{\beta} + \delta v^{3}\bar{p}_{3})\,\mathrm{d}S \\
& - \int_{s_p} (\delta v^{\mathrm{t}}\bar{n}_{\mathrm{t}} + \delta v^{3}\bar{q}_{\mathrm{n}} + \delta\varphi^{\mathrm{t}}\bar{m}_{\mathrm{t}} + \delta\varphi^{\mathrm{n}}\bar{m}_{\mathrm{n}})\,\mathrm{d}s \\
& - \int_{s_u} \underline{(\delta v^{\mathrm{t}} n_{\mathrm{t}} + \delta v^{\mathrm{n}} q_{\mathrm{n}} + \delta\varphi^{\mathrm{t}} m_{\mathrm{t}} + \delta\varphi^{\mathrm{n}} m_{\mathrm{n}})\,\mathrm{d}s = 0}
\end{aligned}
$$

Table 11.2 Modules for construction of variational principles: plate under Kirchhoff—hypothesis

		Global [local] forms
A	Equilibrium	$\int_S [\delta v^{\beta}(n^{\alpha}_{\beta}\vert_{\alpha} + \bar{p}_{\beta}) + \delta v^{3}(m^{\alpha}_{\beta}\vert^{\beta}_{\alpha} + \bar{p})]\,\mathrm{d}S$
B	Kinematics	$\int_S [\delta n^{\alpha}_{\beta}(v^{\beta}\vert_{\alpha} - \alpha^{\beta}_{\alpha}) + \delta m^{\alpha}_{\beta}(v^{3}\vert^{\beta}_{\alpha} - \beta^{\beta}_{\alpha})]\,\mathrm{d}S$
M	Constitutive laws	$\int_S \left[\delta\alpha^{\beta}_{\alpha}\left(n^{\alpha}_{\beta} - \frac{\partial W}{\partial\alpha^{\beta}_{\alpha}}\right) + \delta\beta^{\beta}_{\alpha}\left(m^{\alpha}_{\beta} - \frac{\partial W}{\partial\beta^{\beta}_{\alpha}}\right)\right]\mathrm{d}S$
M*		$\int_S \left[\delta n^{\alpha}_{\beta}\left(\alpha^{\beta}_{\alpha} - \frac{\partial W^*}{\partial n^{\alpha}_{\beta}}\right) + \delta m^{\alpha}_{\beta}\left(\beta^{\beta}_{\alpha} - \frac{\partial W^*}{\partial m^{\alpha}_{\beta}}\right)\right]\mathrm{d}S$
A_r	Static boundary conditions	$\int_{s_p} [\delta v^{\mathrm{t}}(n_{\mathrm{t}} - \bar{n}_{\mathrm{t}}) + \delta v^{3}(s_{\mathrm{n}} - \bar{s}_{\mathrm{n}}) + \delta\varphi^{\mathrm{t}}(m_{\mathrm{t}} - \bar{m}_{\mathrm{t}})] + (\delta v^{3} m_{\mathrm{n}})$
B_r	Static boundary conditions	$\int_{s_u} [\delta n_{\mathrm{t}}(v^{\mathrm{t}} - \bar{v}^{\mathrm{t}}) + \delta s_{\mathrm{n}}(v^{3} - \bar{v}^{3}) + \delta m_{\mathrm{t}}(\varphi^{\mathrm{t}} - \bar{\varphi}^{\mathrm{t}})] + (\delta m_{\mathrm{n}} v^{3})$

The underlined terms are usually omitted. This is due to the subsidiary conditions which, for the principle of virtual work (C), state that the kinematic boundary conditions (B_r) must be satisfied. Also, the strains and stress resultants are replaced by the displacements and rotations, since (B) and (M) must be satisfied identically.

In the same manner the equivalence of the global forms of the kinematical equations (B) and boundary conditions (B_r) to the principle of complementary virtual work can be stated:

$$(\mathrm{D}_1) \quad \int_S (\delta n^\alpha_\beta \alpha^\beta_\alpha + \delta m^\alpha_\beta \beta^\beta_\alpha + \delta\varphi_\alpha \gamma^\alpha)\,\mathrm{d}S + \int_{s_u} (\delta n_t \bar{v}^t + \delta q_n \bar{v}^3 + \delta m_t \bar{\varphi}^t + \delta m_n \bar{\varphi}^n)\,\mathrm{d}s$$

$$(\mathrm{D}_2) \quad + \underline{\int_{s_p} (\delta n_t v^t + \delta q_n v^3 + \delta m_t \varphi^t + \delta m_n \varphi^n)\,\mathrm{d}s}$$

$$+ \underline{\int_S \{[\delta n^\alpha_\beta|_\alpha - (b^\alpha_\lambda - 2H\delta^\alpha_\lambda)\delta m^\lambda_\beta|_\alpha - b^\alpha_\beta \delta q_\alpha] v^\beta}$$

$$\underline{+ [\delta q^\alpha|_\alpha + b^\lambda_\alpha \delta n^\alpha_\lambda - K\delta m^\lambda_\lambda] v^3 + [\delta m^\alpha_\beta|_\alpha - \delta q_\beta]\varphi^\beta\}\,\mathrm{d}S} = 0.$$

Again the underlined terms are usually omitted under the subsidiary condition of statically admissible stresses [satisfying (A) and (A_r)]. Here, the stresses are also replaced by the stress resultants through the constitutive relations (M*).

Various generalized principles may now be constructed depending on the equations which are to be included in the principle and on the unknowns to be employed. With displacements, stresses, and strains as independent variables the combination of (C), (M), and (B) yields the so-called Hu–Washizu principle [4], which, written for shells, reads

$$(\mathrm{CBM}) \quad \delta J = \delta\Big\{\int W(\varepsilon) + (v^\beta|_\alpha - b^\beta_\alpha v^3 - \alpha^\beta_\alpha) n^\alpha_\beta$$

$$+ [\varphi^\beta|_\alpha - (b^\beta_\lambda - 2H\delta^\beta_\lambda) v^\lambda|_\alpha + K\delta^\beta_\alpha v^3 - \beta^\beta_\alpha] m^\alpha_\beta$$

$$+ (\varphi^\alpha + v^3|^\alpha + b^\alpha_\lambda v^\lambda - \gamma^\alpha) q_\alpha - v^\beta \bar{p}_\beta - v^3 \bar{p}_3\Big\}\,\mathrm{d}S$$

$$- \int_{s_p} (\delta v^t \bar{n}_t + \delta v^3 \bar{q}_n + \delta\varphi^t \bar{m}_t + \delta\varphi^n \bar{m}_n)\,\mathrm{d}s$$

$$- \int_{s_u} \delta[n_t(v^t - \bar{v}^t) + q_n(v^3 - \bar{v}^3) + m_t(\varphi^t - \bar{\varphi}^t) + m_n(\varphi^n - \bar{\varphi}^n)]\,\mathrm{d}s$$

$$= 0$$

Writing the surface integral in a matrix operator form reveals the structure of the principle more clearly, and also brings out the structure of the corresponding element matrices after discretization:

$$\begin{matrix}\text{(C)}\\ \text{(B)}\\ \text{(M)}\end{matrix} \quad \int_S \delta\mathbf{z}^T \begin{bmatrix} \mathbf{0} & {}_u\mathbf{D}^T & \mathbf{0} \\ \boldsymbol{D}_u & \mathbf{0} & -\mathbf{I} \\ \mathbf{0} & -\mathbf{I} & \mathbf{E} \end{bmatrix} \begin{bmatrix} \mathbf{u} \\ \mathbf{s} \\ \boldsymbol{\varepsilon} \end{bmatrix} \mathrm{d}S \qquad \begin{matrix} \mathbf{u} = \{v^\alpha & v^3 & \varphi^\alpha\} \\ \mathbf{s} = \{n^\alpha_\beta & q_\beta & m^\alpha_\beta\} \\ \boldsymbol{\varepsilon} = \{\alpha^\beta_\alpha & \gamma_\alpha & \beta^\beta_\alpha\} \end{matrix} \tag{11.1}$$

The coefficient matrix is symmetric due to the conjugate operators **D** appearing in both the kinematic equations (B) and the virtual work expression (C). The index of **D** points to the variable on which the derivation is to be performed, for example $\delta(\mathbf{D}_u^T\mathbf{u}) = \delta\mathbf{u}^T{}_u\mathbf{D}$. For elastic material properties the matrix **E** contains terms which relate the respective strain components to the membrane forces ($\mathbf{E}_M$), to the bending moments ($\mathbf{E}_B$), and to the shear forces ($\mathbf{E}_Q$). **I** stands for the unit matrix. This principle is rarely used as a starting point of finite element approximations because, due to the large number of variables, it results in large element matrices which are sparsely populated. Its merit lies in the fact that it contains all basic equations of the problem and is a good basis for a consistent formulation.

Imposing the constitutive equations (M) as subsidiary conditions reduces the independent variables to the displacements **u** and stresses **s**. As equilibrium (A) and virtual work (C), on the one hand, and kinematics (B) and complementary virtual work (D), on the other, are equivalent statements, four combinations are possible [5]: A+B=AB, C+D=CD, C+B=CB, and A+D=AD. This is illustrated for the case of a plate under Kirchhoff hypothesis in Table 11.3. As the corresponding operator matrices of the surface integral show, only the combination CB and AD are symmetric:

$$\begin{matrix}\text{(C)}\\ \text{(B)}\end{matrix} \quad \int_S \delta\mathbf{z}^T \begin{bmatrix} \mathbf{0} & {}_u\mathbf{D}^T \\ \mathbf{D}_u & -\mathbf{E}^{-1} \end{bmatrix} \begin{bmatrix} \mathbf{u} \\ \mathbf{s} \end{bmatrix} \mathrm{d}S \tag{11.2}$$

$$\begin{matrix}\text{(A)}\\ \text{(D)}\end{matrix} \quad \int_S \delta\mathbf{z}^T \begin{bmatrix} \mathbf{0} & \mathbf{D}_s^T \\ {}_s\mathbf{D} & -\mathbf{E}^{-1} \end{bmatrix} \begin{bmatrix} \mathbf{u} \\ \mathbf{s} \end{bmatrix} \mathrm{d}S \tag{11.3}$$

The form (CB) is usually called the Hellinger–Reissner [6, 7] principle (but see also Prange [8]), and for the linear theory of thin shells reads

$$\begin{aligned}
\text{(CB)} \quad \delta J_{\mathrm{CB}} = \delta\Big\{ \int_S & \Big\{ -W^*(s) + (v^\beta|_\alpha - b^\beta_\alpha v^3) n^\alpha_\beta \\
& + [\varphi^\beta|_\alpha - (b^\beta_\lambda - 2H\delta^\beta_\lambda) v^\lambda|_\alpha + K\delta^\beta_\alpha v^3] m^\alpha_\beta \\
& + (v^3|^\alpha - b^\alpha_\beta v^\beta + \varphi^\alpha) q_\alpha - v^\beta \bar{p}_\beta - v^3 \bar{p} \Big\} \mathrm{d}S \\
& - \int_{S_p} (\delta v^{\mathrm{t}} n_{\mathrm{t}} + \delta v^3 \bar{q}_{\mathrm{n}} + \delta\varphi^{\mathrm{t}} \bar{m}_{\mathrm{t}} + \delta\varphi^{\mathrm{n}} \bar{m}_{\mathrm{n}}) \, \mathrm{d}s \\
& - \int_{S_u} \delta[n_{\mathrm{t}}(v^{\mathrm{t}} - \bar{v}^{\mathrm{t}}) + q_{\mathrm{n}}(v^3 - \bar{v}^3) + m_{\mathrm{t}}(\varphi^{\mathrm{t}} - \bar{\varphi}^{\mathrm{t}}) + m_{\mathrm{n}}(\varphi^{\mathrm{n}} - \bar{\varphi}^{\mathrm{n}})] \, \mathrm{d}s = 0
\end{aligned}$$

Table 11.3 Variational forms for plates under Kirchhoff hypothesis

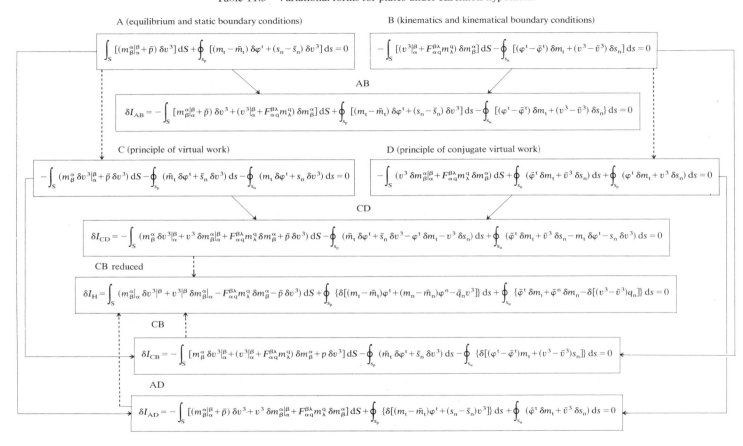

A (equilibrium and static boundary conditions)

$$\int_S [(m^{\alpha}_{\beta}|^{\beta}_{\alpha} + \bar{p})\,\delta v^3]\,dS + \oint_{s_p} [(m_t - \bar{m}_t)\,\delta\varphi^t + (s_n - \bar{s}_n)\,\delta v^3]\,ds = 0$$

B (kinematics and kinematical boundary conditions)

$$-\int_S [(v^3|^{\beta}_{\alpha} + F^{\beta\lambda}_{\alpha q} m^q_{\lambda})\,\delta m^{\alpha}_{\beta}]\,dS - \oint_{s_u} [(\varphi^t - \bar{\varphi}^t)\,\delta m_t + (v^3 - \bar{v}^3)\,\delta s_n]\,ds = 0$$

AB

$$\delta I_{AB} = -\int_S [m^{\alpha}_{\beta}|^{\beta}_{\alpha} + \bar{p})\,\delta v^3 + (v^3|^{\beta}_{\alpha} + F^{\beta\lambda}_{\alpha q} m^q_{\lambda})\,\delta m^{\alpha}_{\beta}]\,dS + \oint_{s_p} [(m_t - \bar{m}_t)\,\delta\varphi^t + (s_n - \bar{s}_n)\,\delta v^3]\,ds - \oint_{s_u} [(\varphi^t - \bar{\varphi}^t)\,\delta m_t + (v^3 - \bar{v}^3)\,\delta s_n\}\,ds = 0$$

C (principle of virtual work)

$$-\int_S (m^{\alpha}_{\beta}\,\delta v^3|^{\beta}_{\alpha} + \bar{p}\,\delta v^3)\,dS - \oint_{s_p} (\bar{m}_t\,\delta\varphi^t + \bar{s}_n\,\delta v^3)\,ds - \oint_{s_u} (m_t\,\delta\varphi^t + s_n\,\delta v^3)\,ds = 0$$

D (principle of conjugate virtual work)

$$-\int_S (v^3\,\delta m^{\alpha}_{\beta}|^{\beta}_{\alpha} + F^{\beta\lambda}_{\alpha q} m^q_{\lambda}\,\delta m^{\alpha}_{\beta})\,dS + \oint_{s_u} (\bar{\varphi}^t\,\delta m_t + \bar{v}^3\,\delta s_n)\,ds + \oint_{s_p} (\varphi^t\,\delta m_t + v^3\,\delta s_n)\,ds = 0$$

CD

$$\delta I_{CD} = -\int_S (m^{\alpha}_{\beta}\,\delta v^3|^{\beta}_{\alpha} + v^3\,\delta m^{\alpha}_{\beta}|^{\beta}_{\alpha} + F^{\beta\lambda}_{\alpha q} m^q_{\lambda}\,\delta m^{\alpha}_{\beta} + \bar{p}\,\delta v^3)\,dS - \oint_{s_p} (\bar{m}_t\,\delta\varphi^t + \bar{s}_n\,\delta v^3 - \varphi^t\,\delta m_t - v^3\,\delta s_n)\,ds + \oint_{s_u} (\bar{\varphi}^t\,\delta m_t + \bar{v}^3\,\delta s_n - m_t\,\delta\varphi^t - s_n\,\delta v^3)\,ds = 0$$

CB reduced

$$\delta I_H = \int_S (m^{\alpha}_{\beta}|_{\alpha}\,\delta v^3|^{\beta} + v^3|^{\beta}\,\delta m^{\alpha}_{\beta}|_{\alpha} - F^{\beta\lambda}_{\alpha q} m^q_{\lambda}\,\delta m^{\alpha}_{\beta} - \bar{p}\,\delta v^3)\,dS + \oint_{s_p} \{\delta[(m_t - \bar{m}_t)\varphi^t + (m_n - \bar{m}_n)\varphi^n - \bar{q}_n v^3]\}\,ds + \oint_{s_u} \{\bar{\varphi}^t\,\delta m_t + \bar{\varphi}^n\,\delta m_n - \delta[(v^3 - \bar{v}^3) q_n]\}\,ds = 0$$

CB

$$\delta I_{CB} = -\int_S [m^{\alpha}_{\beta}\,\delta v^3|^{\beta}_{\alpha} + (v^3|^{\beta}_{\alpha} + F^{\beta\lambda}_{\alpha q} m^q_{\lambda})\,\delta m^{\alpha}_{\beta} + p\,\delta v^3]\,dS - \oint_{s_p} (\bar{m}_t\,\delta\varphi^t + \bar{s}_n\,\delta v^3)\,ds - \oint_{s_u} \{\delta[(\varphi^t - \bar{\varphi}^t) m_t + (v^3 - \bar{v}^3) s_n]\}\,ds = 0$$

AD

$$\delta I_{AD} = -\int_S [(m^{\alpha}_{\beta}|^{\beta}_{\alpha} + \bar{p})\,\delta v^3 + v^3\,\delta m^{\alpha}_{\beta}|^{\beta}_{\alpha} + F^{\beta\lambda}_{\alpha q} m^q_{\lambda}\,\delta m^{\alpha}_{\beta}]\,dS + \oint_{s_p} \{\delta[(m_t - \bar{m}_t)\varphi^t + (s_n - \bar{s}_n) v^3]\}\,ds + \oint_{s_u} (\bar{\varphi}^t\,\delta m_t + \bar{v}^3\,\delta s_n)\,ds = 0$$

Table 11.4 Operator matrices for principle (CB) for thin shells

	v^1	v^2	v^3	φ^1	φ^2	
δn^1_1	D_1		$-b^1_1$			$\mathbf{M}^T$
δn^1_2		D_1	$-b^2_1$			
δn^2_1	D_2		$-b^1_2$			
δn^2_2		D_2	b^2_2			
δm^1_1	$-b^2_2D_1 + b^2_1D_2$		K	D_1		$\mathbf{B}^T$
δm^1_2		$-b^2_2D_1 + b^2_1D_2$			D_1	
δm^2_1	$-b^1_1D_2 + b^1_2D_1$			D_2		
δm^2_2		$-b^1_1D_2 + b^1_2D_1$	K		D_2	
δq_1	b^1_1	b^1_2	d^1	1		
δq_2	b^2_1	b^2_2	d^2		1	

$\mathbf{D}_u$ $\mathbf{u}$ $\quad d_\alpha(v^3) = v^3_{,\alpha}$

$\boldsymbol{\varepsilon}$	=	n^1_1	n^1_2	n^2_1	n^2_2	m^1_1	m^1_2	m^2_1	m^2_2
$D\alpha^1_1$		1			$-v$				
$\bar{D}\alpha^1_2$			$1+\nu$						
$D\alpha^2_1$				$1+\nu$			$\mathbf{0}$		
$D\alpha^2_2$	=	$-\nu$			1				
$B\beta^1_1$						1			$-\nu$
$B\beta^1_2$							$1+\nu$		
$B\beta^2_1$			$\mathbf{0}$					$1+\nu$	
$B\beta^2_2$						$-\nu$			1

$\boldsymbol{\varepsilon} = \mathbf{E}^{-1}\,\mathbf{s}$

$D = Et$

$B = Et^3/12$

and $\frac{5}{6}Gt\gamma_\alpha = q_\alpha$

Table 11.4 gives the operator matrices of this equation. With the inclusion of shear deformations the principle only contains first-order derivatives. This is one prominent advantage of the mixed model as the discretization need only be of low order. With the adaption of the Kirchhoff hypothesis the influence of shear terms is neglected:

$$\gamma^\beta = v^3|^\beta + b^\beta_\lambda v^\lambda + \varphi^\beta = 0$$

thus leading to second-order derivatives in the bending expressions:

$$\int_S m^{\alpha}_{\beta}\varphi^{\beta}|_{\alpha}\, dS = -\int_S m^{\alpha}_{\beta}(v^3|^{\beta} + b^{\beta}_{\lambda}v^{\lambda})|_{\alpha}\, dS \qquad (11.4)$$

However, the advantage of first-order derivatives can be retained by applying Gauss' theorem once, for example, in

$$\int_S m^{\alpha}_{\beta}\varphi^{\beta}\,|_{\alpha}\, dS = -\int_S m^{\alpha}_{\beta}\,|_{\alpha}\,\varphi^{\beta}\, dS + \int_s (m_t\varphi^t + m_n\varphi^n)\, ds$$

It should be noted that this transformation leads to an additional boundary integral. For the plate under Kirchhoff hypothesis the reduced form of the principle is shown in Table 11.3. Transforming (CB) and (AD) leads to the same expressions. The corresponding bending terms of the principles for thin shells can be handled in a similar fashion. The unknowns of this reduced principle are

$$\mathbf{z}_R = \{v^{\alpha} \quad v^3 \quad n^{\alpha}_{\beta} \quad m^{\alpha}_{\beta}\} \qquad \text{for } \alpha, \beta = 1, 2$$

while the operator matrix $\mathbf{B}_u$ for the bending part of the subdivided matrix

$$\int_S \delta\mathbf{z}_R^T \begin{bmatrix} \mathbf{0} & {}_u\mathbf{M}^T & {}_u\mathbf{B}^T \\ \mathbf{M}_u & -{}^M\mathbf{E}^{-1} & \mathbf{0} \\ \mathbf{B}_u & \mathbf{0} & {}^B\mathbf{E}^{-1} \end{bmatrix} \begin{bmatrix} \mathbf{v} \\ \mathbf{n} \\ \mathbf{m} \end{bmatrix} dS \qquad (11.5)$$

is given in Table 11.5.

This formulation is well suited for discretization procedures. Its main feature is that it allows independent trial functions for displacements, membrane forces, and moments of the same or of different order. Shear

Table 11.5 Operator matrix $\mathbf{B}_u$ of principle (11.5) for thin shells

	v^1	v^2	v^3
δm^1_1	$-b^2_2 D_1 + b^2_1 D_2 + {}_1Db^1_1$	${}_1Db^1_2$	$K + {}_1Dd^1$
δm^1_2	${}_1Db^2_1$	$-b^2_2 D_1 + b^2_1 D_2 + {}_1Db^2_2$	${}_1Dd^2$
δm^2_1	$-b^1_1 D_2 + b^1_2 D_1 + {}_2Db^1_1$	${}_2Db^1_2$	${}_2Dd^1$
δm^2_2	${}_2Db^2_1$	$-b^1_1 D_2 + b^1_2 D_1 + {}_2Db^2_2$	$K + {}_2Dd^2$

$\mathbf{B}_u\mathbf{u}$ $\qquad D_{\alpha}(v^{\beta}) = v^{\beta}|_{\alpha},\ d_{\alpha}(v^3) = v^3_{,\alpha}$

forces and rotations are now dependent variables and are determined after the solution of the main problem.

Further reduction of the formulations leads to additional possibilities. Imposing admissibility of the kinematic membrane variables in (11.5) leads to the operator matrix

$$\int_S \delta[\mathbf{v} \quad \mathbf{m}] \begin{bmatrix} {}_u\mathbf{M}^M\mathbf{E}\mathbf{M}_u & {}_u\mathbf{B}^T \\ \mathbf{B}_u & -{}^B\mathbf{E}^{-1} \end{bmatrix} \begin{bmatrix} \mathbf{v} \\ \mathbf{m} \end{bmatrix} \mathrm{d}S \tag{11.6}$$

which, after discretization, is basically identical to that of a displacement model as far as the membrane effects are concerned, and to a mixed model for the bending part. Further elimination of the moments leads to the usual displacement formulation

$$\int_S \delta\mathbf{v}^T \underbrace{{}_u\mathbf{D}\mathbf{E}\mathbf{D}_u}_{\mathbf{k}^D} \mathbf{v}\, \mathrm{d}S \tag{11.7}$$

in which the structure of the operator $\mathbf{k}^D$-already reflects that of the stiffness matrix $\mathbf{k}$ generated by inserting trial functions for the unknowns.

11.2.2 Shells of revolution

As shells of revolution are often used in practical applications it is useful to give the specialized equations for this type separately. Incidentally, here it is also possible to use Fourier harmonics as trial functions in the circumferential direction. Table 11.6 gives the operator matrix $\mathbf{B}_u$ for the bending terms. There, the usual contravariant components are used which allow advantage to be taken of the symmetry of the twisting moment. In this case the elasticity matrix $\mathbf{E}$ contains the metric coefficients instead of the Kronecker delta symbols.

The finite element approximation obtained by inserting trial functions into the principle leads here to mixed two-dimensional elements or, if a Fourier decomposition in the circumferential direction is used, to ring elements for each harmonic.

Another mixed-type approach to shells of revolution should be mentioned [9] which leads to ring elements of high accuracy. Instead of trial functions on which the quality of the solution of a finite element approximation depends, it employs a direct procedure for the numerical integration of

Table 11.6 Operator matrix $\mathbf{B}_u$ for shells of revolution

	v^1	v^2	v^3
δm^{11}	${}_1Db_1^1 - b_2^2D_1$	0	$K + {}_1Dd_1$
δm^{12}	${}_2Db_1^1 - b_1^1D_2$	${}_1Db_2^2 - b_2^2D_1$	${}_1Dd_2 + {}_2Dd_1$
δm^{22}	0	${}_2Db_2^2 - b_1^1D_2$	$K + {}_2Dd_2$

$\mathbf{B}_u\mathbf{u}$

ordinary differential equations. Here, these are the local basic equations associated with the global principles, provided that they can be reduced to systems of first-order differential equations of the form

$$\mathbf{z}' = \mathbf{A}\mathbf{z} - \mathbf{p}, \qquad \mathbf{z} = \{\mathbf{u} \quad \mathbf{s}\} \tag{11.8}$$

In the case of shells of revolution this form may be obtained by a Fourier decomposition of the variables in the circumferential direction (index 1), where the distribution is assumed to be symmetric for $\mathbf{z}^{(s)}$ and antisymmetric for $\mathbf{z}^{(a)}$:

$$\mathbf{z}^{(s)} = \{v^2, v^3, \varphi^2, n_1^1, n_2^2, m_1^1, m_2^2, q_2\}$$
$$\mathbf{z}^{(a)} = \{v^1, \varphi^1, n_2^1, m_2^1, q_1\}$$

The respective local equations are equivalent to the principle (CB) with inclusion of shear deformations and only contain first derivatives. They can be brought into the form (11.8), when those variables are used as unknowns for which boundary conditions can be satisfied directly:

$$\mathbf{z} = \{v^1 \quad v^2 \quad v^3 \quad \varphi^2 \,\vdots\, s_1^2 \quad n_2^2 \quad s_2 \quad m_2^2\}$$

Here s_2 stands for the transverse shear force and s_1^2 for membrane shear, both modified so as to satisfy the Kirchhoff hypothesis. For details see [9]. After solving (11.8) the additional variables φ^1, n_1^1, m_1^2, q_1 can be obtained from a set of algebraic equations. The coefficient matrix **A** for the specific linear shell theory considered here is given in Table 11.7.

11.2.3 *Inclusion of non-linear effects*

The application of mixed models in the non-linear regime requires the formulation of the basic principles in incremental form. For three-dimensional continua several such forms have been given in [10]. Compared to the linear theory, additional terms appear which result from the influence of the stresses $\overset{0}{\mathbf{s}}$ and displacement $\overset{0}{\mathbf{u}}$ of the current or fundamental state to which the incremental change is referred ('initial stresses and displacements'). For the volume integral of the generalized principle (CB) (Hellinger–Reissner type) this leads to the following expression in operator form:

$$\int_{V_0} [\delta\mathbf{u}^T\,\delta\mathbf{s}^T]\left\{\begin{bmatrix}\overset{0}{\mathbf{N}} & {}_u\overset{0}{\mathbf{R}}{}^T\\ \overset{0}{\mathbf{R}}_u & \mathbf{0}\end{bmatrix} + \begin{bmatrix}\mathbf{0} & {}_u\mathbf{D}^T\\ \mathbf{D}_u & -\mathbf{E}^{-1}\end{bmatrix}\right\}\begin{bmatrix}\mathbf{u}\\ \mathbf{s}\end{bmatrix}\mathrm{d}V_0$$

$$= \int_{V_0} \delta\mathbf{z}^T \quad \{ \quad \mathbf{c}_G^D \quad + \quad \mathbf{c}_L^D \quad \} \ \mathbf{z}\,\mathrm{d}V_0 \tag{11.9}$$

Table 11.7 Differential system

$v^{(1)}$	Γ	$-d_1$	$-\frac{B}{D}\Gamma(3B_2^2-B_1^1)d_1$
$v^{(2)}$	$-\nu d_1$	$-\nu\Gamma$	$B_2^2+\nu B_1^1$
$v^{(3)}$		$-B_2^2$	
$\varphi^{(2)}$	$\nu B_2^2 d_1$	$-\nu\Gamma(B_1^1-B_2^2)$	$\nu B_1^1(B_1^1-B_2^2)+\nu d_1^2$
$r\tilde{n}_{(1)}^{(2)}$	$-Drd_1^2$	$-Dr\Gamma d_1$	$Drd_1 \times\left[B_1^1+\frac{B}{D}B_2^2(B_1^1B_1^1-d_1^2)\right]$
$r\tilde{n}_{(2)}^{(2)}$	$Dr\Gamma d_1$	$Dr\times\left[\Gamma\Gamma\left(1+\frac{B}{D}B_1^1B_1^1\right)-\frac{B}{D}\frac{2}{1+\nu}B_1^1B_1^1d_1^2\right]$	$-Dr\Gamma\times\left[B_1^1\left(1+\frac{B}{D}B_1^1B_1^1\right)+\frac{B}{D}d_1^2\left(B_1^1\frac{3+\nu}{1+\nu}-B_2^2\right)\right]$
$rs^{(2)}$	$-Drd_1\times\left[B_1^1+\frac{B}{D}B_2^2(B_1^1B_1^1-d_1^2)\right]$	$-Dr\Gamma\times\left[B_1^1\left(1+\frac{B}{D}B_1^1B_1^1\right)+\frac{B}{D}d_1^2\left(B_1^1\frac{3+\nu}{1+\nu}-B_2^2\right)\right]$	$Dr\left[B_1^1B_1^1\left(1+\frac{B}{D}B_1^1B_1^1\right)+\frac{B}{D}d_1^2\left(d_1^2-\frac{2}{1+\nu}\Gamma\Gamma+2B_1^1(B_1^1-B_2^2)\right)\right]$
$r\tilde{m}_{(2)}^{(2)}$	$-Br\Gamma B_2^2 d_1$	$Br\left[\Gamma\Gamma(B_1^1-B_2^2)-\frac{2}{1+\nu}B_1^1d_1^2\right]$	$-Br\Gamma\times\left[B_1^1(B_1^1-B_2^2)+\frac{3+\nu}{1-\nu}d_1^2\right]$

$D=Eh \qquad B=\frac{Eh^3}{12} \qquad d_1^n=\frac{\partial^n}{\partial s_1^n}$

In (11.9) $\overset{0}{\mathbf{N}}$ contains the initial stresses $\overset{0}{\mathbf{s}}$ and $\overset{0}{\mathbf{R}}_u$ contains the initial displacements $\overset{0}{\mathbf{u}}$. The formulation of the equations governing an increment in the step-by-step solution of the complete non-linear problem requires that the incremental variables be referred to the same configuration. Normally this is either the undeformed ('total Lagrangian' approach) or the current, fundamental state ('updated Lagrangian' formulation). As in the for shells of revolution

$$
\left[\begin{array}{c|c|c|c|c}
-\dfrac{B}{D}(3B_2^2-B_1^1)d_1 & 2\dfrac{1+\nu}{Dr} & & & \\
\hline
-\nu\dfrac{B}{D}\Gamma(B_1^1-B_2^2) & & \dfrac{1-\nu^2}{Dr} & & \dfrac{1-\nu^2}{Dr}\times(B_1^1-B_2^2) \\
\hline
-1 & & & & \\
\hline
-\nu\Gamma & & \dfrac{1-\nu^2}{Dr}\times(B_1^1-B_2^2) & & \dfrac{1-\nu^2}{Br} \\
\hline
Br\Gamma B_2^2 d_1 & -\Gamma & -\nu d_1 & & \nu B_2^2 d_1 \\
\hline
Br\times\left[\Gamma\Gamma(B_1^1-B_2^2)-\dfrac{2}{1+\nu}B_1^1 d_1^2\right] & -d_1 & \nu\Gamma & B_2^2 & \nu\Gamma\times(B_1^1-B_2^2) \\
\hline
-Br\Gamma\times\left[B_1^1(B_1^1-B_2^2)+\dfrac{3+\nu}{1+\nu}d_1^2\right] & -\dfrac{B}{D}\Gamma d_1\times(3B_2^2-B_1^1) & -B_2^2-\nu B_1^1 & & -\nu B_1^1\times(B_1^1-B_2^2)-\nu d_1^2 \\
\hline
Br\left(\Gamma\Gamma-\dfrac{2}{1+\nu}d_1^2\right) & -\dfrac{B}{D}(3B_2^2-B_1^1)d_1 & \nu\dfrac{B}{D}\Gamma\times(B_1^1-B_2^2) & 1 & \nu\Gamma
\end{array}\right]
\left[\begin{array}{c}
v^{(1)} \\ \hline v^{(2)} \\ \hline v^{(3)} \\ \hline \varphi^{(2)} \\ \hline r\tilde{n}^{(2)}_{(1)} \\ \hline r\tilde{n}^{(2)}_{(2)} \\ \hline rs^{(2)} \\ \hline r\tilde{m}^{(2)}_{(2)}
\end{array}\right]
\left[\begin{array}{c}
0 \\ \hline 0 \\ \hline 0 \\ \hline 0 \\ \hline rp_{(1)} \\ \hline rp_{(2)} \\ \hline rp_{(3)} \\ \hline 0
\end{array}\right]
$$

$$d_2=\frac{\partial}{\partial s_2} \qquad \Gamma=\frac{\Gamma_{12}^1}{\sqrt{a_{22}}}$$

latter the geometry must be updated at each step, the operator matrix $\overset{0}{\mathbf{R}}$ containing $\overset{0}{\mathbf{u}}$ is not present if this reference frame is used. Consideration of the full geometrically non-linear expressions for plates and shells shows that they are rather involved. In most applications only the important terms are included, leading to a variety of approximations. For shells the membrane forces $\overset{0}{n}{}^{\alpha}_{\beta}$ of the fundamental state are predominant, and the influence of the

Table 11.8 Operator matrix $\overset{0}{\mathbf{N}}$ of (11.10) for cylindrical shell

u	v	w
$\left[\frac{1}{R^2} + {}_1\partial\partial_1 \middle\vert {}_2\partial\partial_2 \middle\vert {}_2\partial\partial_1 + {}_1\partial\partial_2\right]\overset{0}{\mathbf{n}}$	0	$\left[-{}_1\partial\frac{1}{R} + \frac{1}{R}\partial_1 \;\; 0 \;\; -{}_2\partial\frac{1}{R} + \frac{1}{R}\partial_2\right]\overset{0}{\mathbf{n}}$
0	$[{}_1\partial\partial_2 \vert {}_1\partial\partial_2 \vert {}_2\partial\partial_1 + {}_1\partial\partial_2]\overset{0}{\mathbf{n}}$	0
$\left[{}_1\partial\frac{1}{R} - \frac{1}{R}\partial_1 \;\; 0 \;\; {}_2\partial\frac{1}{R} - \frac{1}{R}\partial_2\right]\overset{0}{\mathbf{n}}$	0	$\left[\frac{1}{R^2} + {}_1\partial\partial_1 \middle\vert {}_2\partial\partial_2 \middle\vert {}_2\partial\partial_1 + {}_1\partial\partial_2\right]\overset{0}{\mathbf{n}}$

$\overset{0}{\mathbf{n}}{}^T = [n^{11} \;\; n^{22} \;\; n^{12}]$ membrane forces of fundamental state

boundary moments $\overset{0}{m}{}^{\alpha}_{\beta}$ and initial displacements $\overset{0}{v}{}^i$ is often neglected. For the variational principle (CB) this leads to the addition of the following term:

$$\delta \int_S \overset{0}{n}{}^{\alpha}_{\beta}\,{}^{\mathrm{NL}}\alpha^{\beta}_{\alpha}\, \mathrm{d}S = \delta \int_S \mathbf{u}^T \overset{0}{\mathbf{N}} \mathbf{u}\, \mathrm{d}S \tag{11.10}$$

For the case of a cylindrical shell the matrix $\overset{0}{\mathbf{N}}$ of (11.10) is given on Table 11.8. The usual approximation for small strains and moderate rotations in the tangent plane, whereby

$${}^{\mathrm{NL}}\alpha^{\beta}_{\alpha} = \tfrac{1}{2}\varphi_{\alpha}\varphi^{\beta} \tag{11.11}$$

is represented by the coefficient (3, 3) of Table 11.8.

For the case of shells of revolution, the local equations for this type of non-linear shell theory may again be written as a system of first-order differential equations. Compared to (11.8) the coefficient matrix is supplemented by a matrix $\overset{0}{\mathbf{A}}_G$ containing the non-linear terms. For axisymmetric $\overset{0}{n}{}^{\alpha\beta}$ it is given in Table 11.9. For non-axisymmetric fundamental states special considerations regarding the coupling of the Fourier harmonics in the non-linear range are necessary.

The influence of non-linear material behaviour may be included in the mixed approach as well. In the framework of the continuum the incremental principles valid for elastoplastic material were formulated in [10]. In addition to the usual forms, the plastic multipliers appearing in the expression for the plastic strains below may be chosen as additional unknowns. For the extended principle (CB) the corresponding operator matrix of the homogenous part reads, see [10]:

$$\int_{V_0} \begin{matrix} \delta\mathbf{u}^T \\ \delta\mathbf{s}^T \\ \delta\lambda \end{matrix} \begin{bmatrix} \overset{0}{\mathbf{N}} & {}_u\mathbf{D}^T + {}_u\mathbf{R}^T & \mathbf{0} \\ \mathbf{D}_u + \overset{0}{\mathbf{R}}_u & -\mathbf{E}^{-1} & -\dfrac{\partial f}{\partial \overset{0}{\mathbf{s}}{}^T} \\ \mathbf{0} & -\dfrac{\partial f}{\partial \overset{0}{\mathbf{s}}} & h \end{bmatrix} \begin{bmatrix} \mathbf{u} \\ \mathbf{s} \\ \lambda \end{bmatrix} \mathrm{d}V_0 \tag{11.12}$$

Table 11.9 Submatrices for geometrically non-linear part $\overset{0}{\mathbf{A}}_G$ of differential system (11.8) for shells of revolution

$$\overset{0}{\mathbf{A}}_1=\begin{bmatrix} b^1_1\overset{0}{\varphi}_{(2)} & 0 & \dfrac{-n}{r\overset{0}{\varphi}_2} & 0 \\ 0 & 0 & 0 & -\overset{0}{\varphi}_2 \\ 0 & 0 & 0 & 0 \\ 0 & 0 & 0 & -b^1_1\overset{0}{\varphi}_{(2)} \end{bmatrix}, \qquad \overset{0}{\mathbf{A}}_G=\begin{bmatrix} \overset{0}{\mathbf{A}}_1 & \mathbf{0} \\ \overset{0}{\mathbf{A}}_3 & -\overset{0}{\mathbf{A}}{}^T_1 \end{bmatrix}$$

$$\overset{0}{\mathbf{A}}_3=\begin{bmatrix} r b^1_1 b^1_1 \overset{0}{n}{}^{(1)}_{(1)} & 0 & -n b^1_1 \overset{0}{n}{}^{(1)}_{(1)} & n\dfrac{2B}{1+\nu} b^1_1 b^1_1 \overset{0}{\varphi}_{(2)} \\ 0 & 0 & 0 & 0 \\ -n b^1_1 \overset{0}{n}{}^{(1)}_{(1)} & 0 & \dfrac{n^2}{r\overset{0}{n}{}^{(1)}_{(1)}} & \dfrac{-n^2}{r}\dfrac{2B}{1+\nu} b^1_1\overset{0}{\varphi}_{(2)} \\ n\dfrac{2B}{1+\nu} b^1_1 b^1_1 \overset{0}{\varphi}_{(2)} & 0 & \dfrac{-n^2}{r\dfrac{2B}{1+\nu} b^1_1\overset{0}{\varphi}_{(2)}} & r\overset{0}{n}{}^{(2)}_{(2)} \end{bmatrix}$$

n = number of Fourier harmonics

with

$$\boldsymbol{\varepsilon}^P=\frac{\partial f}{\partial \overset{0}{\mathbf{s}}{}^T}\lambda \qquad \text{(normality rule)}$$

and

$$h=\frac{\partial f}{\partial \overset{0}{\mathbf{s}}{}^T}+\frac{\partial f}{\partial \overset{0}{\boldsymbol{\varepsilon}}{}^P} \qquad \text{(hardening parameter)}$$

In the application to shells and plates the propagation of the plastic zones through the thickness has to be traced, for example, by subdivision into layers or integration points. The direct use of the static variables as unknowns in the mixed models is especially advantageous for the non-linear material description as these variables do not have to be calculated by an additional differentiation of the displacement field, which normally results in a loss of accuracy and influences the whole non-linear solution process.

11.2.4 Inclusion of dynamic effects

For the application of the generalized principles to dynamic problems the influence of the kinetic energy has to be incorporated. This leads to generalized Hamilton's or Toupin's principles or the corresponding Lagrangian function. The displacement field is assumed to take on prescribed values at times t_1 and t_2. After elimination of time and by making the usual assumptions for harmonic vibrations, the operator matrix of the extended principle reads

$$\int_V \left\{ \begin{bmatrix} \mathbf{0} & {}_u\mathbf{D} \\ \mathbf{D}_u & -\mathbf{E}^{-1} \end{bmatrix} - \omega^2 \begin{bmatrix} \mathbf{M} & \mathbf{0} \\ \mathbf{0} & \mathbf{0} \end{bmatrix} \right\} \begin{bmatrix} \mathbf{u} \\ \mathbf{s} \end{bmatrix} \mathrm{d}V \tag{11.13}$$

where $\mathbf{M}$ is the mass matrix. Like the other operators it contains only first-order derivatives, so that the kinetic and potential energies can be approximated in the same manner.

Another difference to the displacement approach lies in the possibility to reduce the size of the problem in a formal condensation technique which, in contrast to the usual Ritz technique, does not require series expansions and truncations. Equation (11.13) shows that the stress variables $\mathbf{s}$ can be eliminated in such a way that in shell problems the number of unknowns is reduced to one-third and the resulting eigenvalue problem is not influenced by simplifications and assumptions which are necessary in the displacement method. When making comparisons it should be kept in mind that the number of parameters of a mixed plate or shell model is not larger than that of a comparable displacement model. The reduction is physically meaningful as the mass is not connected with the stresses and they in turn do not contribute to the kinetic energy. The results of the application of this reduction to the examples of a plate and a cylindrical shell are discussed in [11].

11.3 DISCRETIZATION

In discretizing the generalized principles, interpolation functions over an element region can be chosen for the variables independently; e.g. for displacements and stresses

$$\mathbf{u} = \mathbf{\Phi}\hat{\mathbf{u}} \qquad \text{and} \qquad \boldsymbol{\varepsilon} = \mathbf{s}_p + \mathbf{\Psi}\hat{\mathbf{s}} \tag{11.14}$$

with $\hat{\mathbf{u}}$, $\hat{\mathbf{s}}$ as unknown parameters. The element matrices are obtained by applying the derivatives in the operator matrix to the trial functions $\mathbf{\Phi}$, $\mathbf{\Psi}$, and by performing the integration over each element:

$$\mathbf{c} = \begin{bmatrix} \mathbf{0} & \mathbf{g}^T \\ \mathbf{g} & -\mathbf{f} \end{bmatrix} \tag{11.15}$$

It is noted that the operator matrices already reflect the structure of the element matrix after discretization. Like in the stiffness method the assembly of the element matrices to the system matrix is achieved by the summation process implied by the addition of work terms in the principle:

$$\sum_m (\delta \mathbf{z}^T \mathbf{c}\mathbf{z})$$

However, due to the zero submatrix in **c** the system matrix is, in general, semidefinite and requires a modification of those solution procedures which are used for the positive definite systems of the displacement method. In this respect it is feasible to order the equations in such a way that those containing the material law (with the positive definite submatrix **f**) are treated first and, if necessary, to provide for local interchange of rows and columns, e.g. in a modified Cholesky procedure [11]].

The development of mixed models in the context of the finite element approach and their application to plates and shells began in the 1960s. It was preceded by the development of the transfer matrix method for beams ten years earlier [12]. This is essentially a mixed method, too, and substitutes a boundary value problem by several initial value processes. Although this direct transfer matrix solution of a whole system is limited and can be numerically quite sensitive, the use of the transfer matrix of an element, evaluated by direct numerical integration, is an effective alternative for the derivation of a very accurate element stiffness matrix which may then be used for the assembly of the system matrix. This procedure solves the system (11.8) and results in an integral matrix which gives the change of the variables from one edge of an element to the other. This is, however, also the definition of the transfer matrix of an element. It may be evaluated by any integration scheme applicable to systems of ordinary differential equations. For $\mathbf{A}=$constant the formal solution of (11.8) is $\mathbf{U}=e^{\mathbf{A}h}$, where h is the length of the element and can be calculated directly from the corresponding monotonically convergent series expansion with any desired relative error bound. The transfer matrix of an element may be easily transformed into the stiffness matrix **k** using the four submatrices of **U**:

$$\mathbf{k}=\left[\begin{array}{c|c} -\mathbf{U}_2^{-1}\mathbf{U}_1 & \mathbf{U}_2^{-1} \\ \hline \mathbf{U}_3-\mathbf{U}_4\mathbf{U}_2^{-1}\mathbf{U}_1 & \mathbf{U}_4\mathbf{U}_2^{-1} \end{array}\right] \tag{11.16}$$

This permits the numerically stable solution of the whole structure to be calculated by any finite element programme.

This treatment avoids any trial functions, and is used successfully in arbitrary beam or special shell problems [9]. An extension of it suitable also for variable coefficients of matrix **A** may be regarded as an approach for the automatic selection of trial functions.

11.4 ELEMENT MODELS FOR PLATES

Mixed finite element models for plates with the Kirchhoff assumptions have been proposed mostly on the basis of the reduced variational form (Table 11.3) with the lateral deflection and the moments as unknowns. This form contains only first derivatives and allows C_0 continuity in contrast to the C_1

Table 11.10 Order of polynomials in mixed plate models

Author	Shape	w	$\mathbf{m}$	
Herrmann [13]	T	1	0	
Visser [14]	T	2	1	
Bron/Dhatt [15]	Q	1, 2 1, 2	1, 2 1, 2	
Poceski [16]	T Q	1 1	1 1	Four parameters
Akay [17]	Q T	1, 2 1, 2	1, 2 1, 2	Isoparametric

T = triangle, Q = quadratical; 0, 1, 2, . . . order of polynomials

continuity necessary for displacement models. Table 11.10 shows the degree of the polynomials used by some authors.

Most often linear interpolation is employed allowing exceedingly simple evaluation of the element matrices. Mixed discretization of form (CB) with the method of strips was performed by the author in [5]. The results also brought out the boundary layer effects of a theory which accounts for shear deformations.

Comparisons made in different investigations show that linear interpolation with mixed models is sufficient for quadrilateral or nearly rectangular shapes, and performs as well as the compatible displacement rectangle due to Schäfer [18], Bogner, Fox, and Schmit [19]. It is noted that in general the evaluation of the element matrices is faster for mixed than for displacement elements. Considering the accuracy of the models it was observed that triangular mixed models on the basis of nodal parameters are not so effective as quadrilaterals. In the case of linear interpolations they may even not converge to the real solution due to the missing twist term in the trial functions for the displacement.

11.5 GENERAL SHELLS

Various curved shell elements of mixed type have been investigated. Most of them use the reduced variational form for the bending part just as in the case of plates. For the membrane part, either both displacements and forces or the membrane displacements only are chosen as unknowns.

In Table 11.11 some of the elements proposed in the literature are listed, from which some [11, 25, 26] are applied also for geometrically non-linear problems. As only derivatives of first order appear in the basic principles, linear interpolation can be used for the displacement and stress variables of

Table 11.11 Order of polynomials in mixed shell elements

Author	Shape	v^3	$\mathbf{m}$	v^α	$\mathbf{n}$	
Prato [20]	T	1	1	1	—	Shallow shell theory
Connor and Will [21]	T	1,2	1	1,2	—	Shallow shell theory
Visser [22]	T	2	1	2	—	Shallow shell theory
Altmann and Iguti [23]	Q	1	1	1	1	Cylinder
Noor and Andersen [24]	T, Q	2	2	2	2	With shear
Tahiani and Lachance [25]	T	1, 2	1, 2	1, 2	1, 2	Triangular coordinates
Harbord [26]	Q	1, 2	1, 2	1, 2	1, 2	
Talaslidis and Wunderlich [11]	Q	1, 2	1, 2	1, 2	1, 2	

the shell theory employed. The evaluation of integrals remains very simple for elements with nearly rectangular shape, and numerical integration is not necessary. The same holds for higher order interpolation where symbolic manipulation and symmetry considerations may be used to reduce the effort significantly [24]. It is also observed that the performance of mixed-type triangular elements is not as good as that of quadrilateral ones. In practical applications the latter should be preferred, and triangles should be used only in small irregular regions if necessary.

Some comparisons of different mixed element types are shown for the example of a cylindrical shell. The cylindrical shell is subject to sinusoidally distributed loads along the generator and has certain boundary conditions along its section curves which frequently occur in practice. This allows a comparison with the solutions of the special procedure for shells of revolution (equation 11.8) for which a Fourier decomposition of the variable is necessary and which here is called model R. For the discretization of the example with two-dimensional mixed elements different trial functions and principles are used:

(a) Model LL: principle (11.5) with the unknowns $\mathbf{z}^T = [v^\alpha v^3 n^{\alpha\beta} m^{\alpha\beta}]$, linear interpolation between the nodes, 36 parameters
(b) Model QQ: like model LL, but quadratic interpolation, 72 parameters
(c) Model KL: like model LL, but only constant interpolation for the membrane forces $n^{\alpha\beta}$, 27 (24) parameters

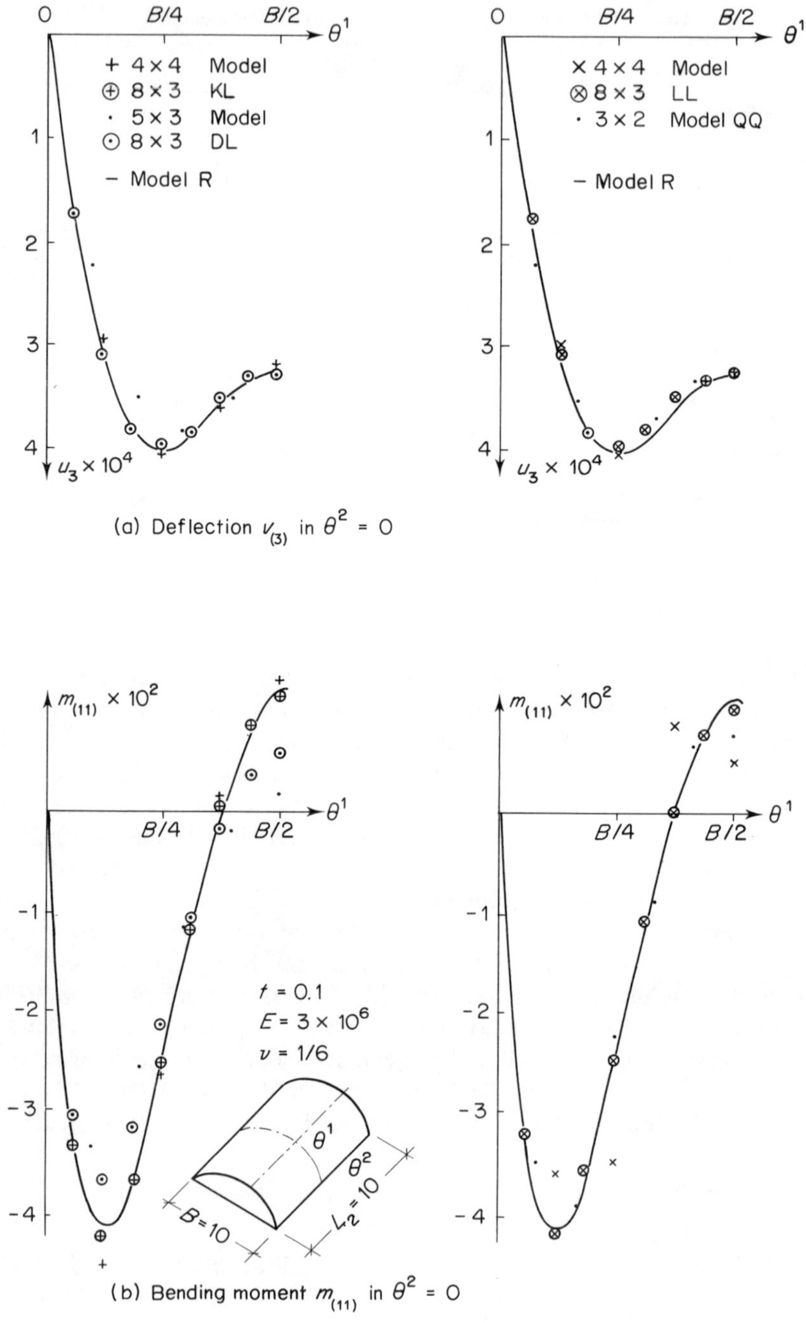

Figure 11.1 Cylindrical shell under lateral load $\bar{p} = 1.0 \sin \pi/L\theta^2$

(d) Model DL: principle (11.6) with unknowns $\mathbf{z}^T = [v^\alpha v^3 m^{\alpha\beta}]$ and linear interpolation, displacement model only for the membrane part, 18 parameters

In addition, the analysis was performed with different numbers of elements. From the various results, Figure 11.1(a) shows the behaviour of the displacement around the middle section and Figure 11.1(b) that of the bending moment in the direction of the generator in the same section. The solid line gives the result of model R and can be regarded as exact because the element matrix was calculated with a given *a priori* relative error of 10^{-6}. The approximation of the deflection is rather good for all models except for DL (5×3). This difference is still more pronounced in the tangential deflections (not shown) and the other variables. For the bending moment, which varies considerably, the deviations of the models are even more pronounced, but apart from model DL they still give satisfactory results, especially for grids which are not too coarse in the θ^1 direction. In this and other examples it became apparent that the model DL with linear interpolation is, in general, not satisfactory, because the displacement model for the membrane part needs higher interpolation to include the rigid body effects properly. On the other hand, the direct approximation of the membrane forces with merely constant interpolation (model KL) yields considerably better results. They are also comparable to those of model LL, which are even more sensitive and show weak oscillations around the exact solution. This behaviour is typical of many mixed models, especially as far as the nodal parameters of the membrane part are concerned, and seems to be the result of a specific type of propagation of numerical errors. Comparing quadratic (QQ) to linear (LL) interpolations, it is seen that the increase of the parameters in QQ is about as effective as the corresponding condensation of the mesh in the case LL.

In summary, the quality of approximation of the simple mixed elements investigated is good for the bending and the membrane parts of the shell, with the exception of model DL. The number of unknowns is not higher than in the stiffness approach, and the evaluation of the element matrices is very simple. It should also be noted that model R gives reliable results of high quality. From our experiences its use can be recommended within its range of applicability.

11.6 SHELLS OF REVOLUTION

Special mixed-type elements for shells of revolution employing Fourier decomposition along the meridian were proposed in different forms:

(a) Continuum elements with trial functions either across the whole thickness (resulting in 'discrete' shell theories) or across individual layers (giving 'finite surface methods') [2]. Kinematic assumptions can be incorporated, for example, in the form of a discrete Kirchhoff hypothesis. These then correspond to 'degenerate' displacement elements.

(b) Thin shell elements with trial functions (linear to cubic) along the meridian, e.g. for the displacements and moments, on the basis of the variational form of the type (11.5) and combined with appropriate interpolation of the geometry [27 to 29]. In this context functionals which include shear deformations, like (CB), were also used [30].

(c) Thin shell elements of high accuracy [9] on the basis of the differential system (11.8) as described earlier. Similar procedures were also used in connection with other methods, e.g. in [31, 32, 33]. It should be emphasized here that the Fourier decomposition is not limited to linear problems. It is equally applicable to the case of non-linear and incremental formulations where the different harmonics are coupled, which leads to large coefficient matrices [29]. This can be avoided by treating all coupled terms as additional, pseudoload quantities, and by solving the non-linear equations iteratively [34]. Thus, the homogenenous part of the equations remains uncoupled and is the same as in the linear case. This 'initial stress approach' is limited to the prebuckling range, however.

As an example of the application in the non-linear range the results of the bifurcation analysis of a spherical cap are shown. The same problem was analysed in [35] on the basis of a shallow shell theory. Figure 11.2 shows the

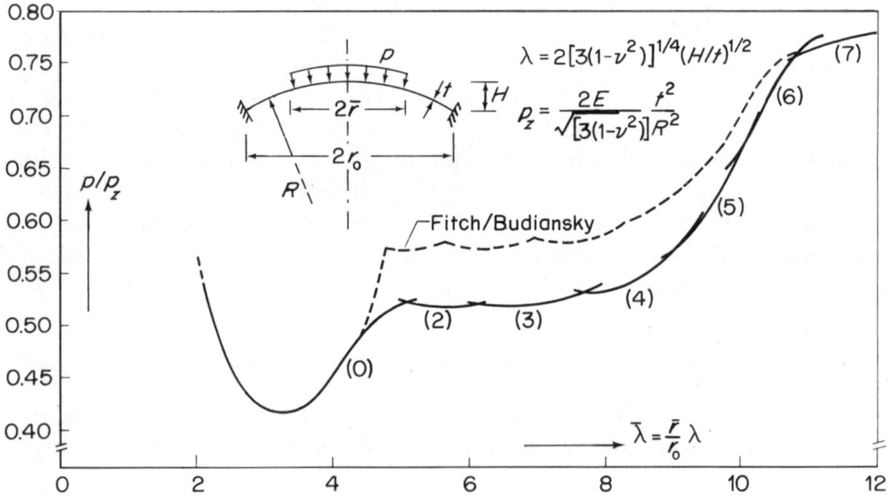

Figure 11.2 Bifurcation load of spherical cap

critical loads for different loading regions and the corresponding harmonics of the eigenfunction which, for example, for the fully loaded cap is $n = 7$. In comparison with the results of [35] a remarkable difference in parts with smaller critical harmonics is observed. It may be explained by the differences in the shell theories employed, mainly by those due to the assumptions for shallow shells. Additional calculations included the effects of geometric imperfections in incremental non-linear analysis. Reductions of the critical load of more than 45 per cent were found to be caused by imperfections in the shape of the eigenfunction and amplitudes of about 25 per cent of the shell thickness.

11.7 PROPERTIES AND ADVANTAGES OF MIXED MODELS

11.7.1 Element properties

The assessment of mixed models based on generalized principle can be supplemented by considering the solution of the eigenvalue problem of an element. In this way additional information about its static and kinematic behaviour is obtained. The eigenvalues and eigenforms of any given element

$$\left\{\begin{bmatrix} \mathbf{0} & \mathbf{g}^T \\ \mathbf{g} & -\mathbf{f} \end{bmatrix} - \lambda \begin{bmatrix} \mathbf{I} & \mathbf{0} \\ \mathbf{0} & \mathbf{I} \end{bmatrix}\right\} \begin{bmatrix} \mathbf{u} \\ \mathbf{s} \end{bmatrix} = \mathbf{0} \tag{11.17}$$

may be classified as follows:

(a) $\lambda_i = 0$: then, the eigenfunctions are solutions of $\mathbf{g}\mathbf{u}_i = \mathbf{0}$, with $\mathbf{s}_i = 0$, and should describe the rigid body modes of the problem.

(b) $\lambda_i < 0$: the eigenfunctions are solutions of the homogenous equations of equilibrium $\mathbf{g}^T\mathbf{s}_i = \mathbf{0}$, with $\mathbf{u}_i = 0$. This case gives information on how well the trial functions for the stress resultants approximate equilibrium. The respective eigenvectors are also eigenvectors of $(\mathbf{f} - \boldsymbol{\lambda}\mathbf{I})\mathbf{z} = \mathbf{0}$.

(c) $\lambda_i < 0$, $\mathbf{u}_i \neq 0$, $\mathbf{s}_i \neq 0$: the eigenfunctions correspond to the solution of the force method and show how well the trial functions satisfy the non-homogenous equilibrium equations *a priori*.

(d) $\lambda_i > 0$, $\mathbf{u}_i \neq 0$, $\mathbf{s}_i \neq 0$: the eigenfunctions correspond to the solutions of the displacement method and show how well the trial functions satisfy the kinematic equations *a priori*.

11.7.2 System requirements convergence

On the system level the whole coefficient matrix has a structure equivalent to that of an element, and reflects certain relationships between static and kinematic equations. For a statically indeterminate problem, like the bending of a shell, the matrix $\mathbf{G}$ of the global equilibrium equations

$$\mathbf{G}^T\mathbf{S} = \mathbf{P}$$

should have a positive defect d in order to allow for non-trivial solutions. This implies that, in the discretized system, the number of stress parameters n_s should be larger than the number of displacement parameters n_u:

$$n_s - n_u > 0 \tag{11.18}$$

This requirement has to be observed in the case that polynomials are chosen for the displacements which are of higher order than those for the stresses. Test runs for such cases have shown that for certain boundary conditions and mesh size the requirement (11.18) may be violated, and consequently unstable solutions were obtained. Nevertheless, as for the reduced functionals (11.5) the number of the stress resultants exceeds that of the displacements considerably, and the order of the polynomials for the forces might very well be smaller. It was observed in model KL of the shell example that this was advantageous for the performance of this model.

The question of convergence of mixed models has attracted many engineers and mathematicians (see, for example, [36, 37, 38]). Results are mainly given for simpler thin shell problems. Although the basic principles have stationary rather than extremum character, careful numerical investigations show convergence to the true solution. It was also observed that the mixed model is not stiffer than a corresponding displacement model and not weaker than an equilibrium model. The stationarity of these saddle-point functionals does not allow bounds to be given for the energy as is possible for linear displacement models, but in both cases the problem of local error bounds for the variables is a rather unresolved question, especially for shells.

11.7.3 Some advantages of mixed models

In summarizing the present understanding and results available in the field of mixed finite element analysis of shells and plates several conclusions regarding the properties of the mixed models can be drawn.

The generalized principles contain the basic equations in their natural canonical form. The formulation is clear and straightforward, a fact which is important in more involved problem areas (like shell theory) or in the non-linear range.

First-order derivatives appear in the principles equally for membrane, bending, and dynamic parts, so that a uniform approximation is achieved. Simple trial functions can be chosen due to the C_0 requirement and linear interpolation is sufficient in many cases.

The integration of these functions is easy to perform and numerical integration can be avoided in general. This results in an effective evaluation of the element matrices.

The boundary and interelement conditions can be represented properly and no difficulties caused by higher derivatives arise.

Stress variables as the main design parameters are direct results of the solution of the algebraic system and need not be obtained by differentiation of the displacements, which usually results in loss of accuracy. This is also of advantage in physically non-linear analysis, in which yield conditions, etc., are expressed in terms of stresses.

The implementation of mixed models in computer programs is very easy to accomplish.

As drawbacks, a larger number of unknown parameters compared to displacement models is sometimes mentioned. It must be emphasized that this generally does not hold for plates and shells where in displacement models as well derivatives of variables are included as parameters.

Nevertheless, there is still an attractive variety of open questions left in connection with mixed models for plates and shells. Only a few examples can be mentioned here, such as the treatment of the combination with other structural members, evaluation of the possible advantages in non-linear applications which may result from the special structure of mixed models, investigation of the stability and convergence of the solution, and local error bounds or practical estimates for the variables of plate and shell models.

REFERENCES

1. T. H. H. Pian, 'Hybrid models', *Numerical and Computer Methods in Structural Mechanics* (Eds. S. J. Fenves *et al.*), pp. 59–78, Academic Press, New York, 1971.
2. W. Wunderlich, 'Ein verallgemeinertes Variationsverfahren zur vollen oder teilweisen Diskretisierung mehrdimensionaler Elastizitätsprobleme', *Ing.-Arch.*, **39,** 230–247, 1970.
3. W. Wunderlich, 'On a consistent shell theory in mixed tensor formulation', *Proc. Third IUTAM Symposium on Shell Theory, Tbilisi, 1978: Theory of Shells*, pp. 607–633, North Holland, 1980.
4. K. Washizu, *Variational Methods in Elasticity and Plasticity*, Pergamon Press, 1968.
5. W. Wunderlich, 'Discretisation of structural problems by a generalized variational approach', *Proc. IASS Symposium, Hawaii, 1971: Hydromechanically Loaded Shells*, Part 1, pp. 779–793, University Press of Hawaii, 1971.
6. E. Hellinger, 'Der allgemeine Ansatz der Mechanik der Kontinua', *Enzyklopädie der mathematischen Wissenschaften*, **4,** Part 4, 602–694 (1914).
7. E. Reissner, 'On a variational theorem in elasticity', *J. Math. Phys.*, **29,** 90–95 (1950).
8. G. Prange, *Das Extremum der Formänderungsarbeit*, Habilitation thesis, Hannover, 1916.
9. W. Wunderlich, 'Differentialsystem und Übertragungsmatrizen der Biegetheorie allgemeiner Rotationsschalen', *Schriftenreihe des Lehrstuhls für Stahlbau*, Vol. 4, TH Hannover, 1966.
10. W. Wunderlich, 'Incremental formulation of the generalized variational approach in structural mechanics', *Variational Methods in Engineering* (Eds. C. A. Brebbia and H. Tottenham), Vol. 2, pp. 109–123, Southampton, 1973.

11. D. Talaslidis and W. Wunderlich, 'Static and dynamic analysis of Kirchhoff shells based on a mixed finite element formulation', *Comp. Struct.*, **10,** 239–249 (1979).
12. S. Falk, 'Die Berechnung des beliebig gestützten Durchlaufträgers nach dem Reduktionsverfahren', *Ing.-Arch.*, **24,** 216–232 (1956).
13. L. R. Herrmann, 'Finite-element bending analysis for plates', *Proc. ASCE, Eng. Mech. Div.*, **93,** No. EM5, 13–26 (1967).
14. W. Visser, 'A refined mixed-type plate bending element', *AIAA J.*, **7,** 1801–1803 (1969).
15. J. Bron and G. Dhatt, 'Mixed quadrilateral elements for bending', *AIAA J.*, **10,** 1359–1361 (1972).
16. A. Poceski, 'A mixed finite element method for bending of plates', *Int. J. Num. Meth. in Eng.*, **9,** 3–15 (1975).
17. H. U. Akay, 'An investigation of first- and second-order mixed plate-bending elements', *Int. J. Meth. Eng.*, **15,** 351–360 (1980).
18. H. Schäfer: 'Eine einfache Konstruktion von Koordinatenfunktionen für die numerische Lösung zweidimensionaler Randwertprobleme nach Rayleigh/Ritz', *Ing.-Arch.*, **35,** 73–81 (1966).
19. F. K. Bogner, R. C. Fox, and L. A. Schmit: 'The generation of interelement compatible stiffness and mass matrices by the use of interpolation formulae', *Proc. Conf. Matrix Methods in Struct. Mech.*, Wright-Patterson AFB, Ohio, 1965.
20. C. A. Prato, 'Shell finite element method via Reissner's principle', *Int. J. Solids Struct.*, **5,** 1119–1133 (1969).
21. J. Connor and D. Will, 'A mixed finite element shallow shell formulation', *Recent Advances in Matrix Methods of Structural Analysis and Design* (Eds. R. H. Gallagher, Y. Yamada, and J. T. Oden), pp. 105–137, University of *Alabama* Press, 1971.
22. W. Visser, 'The application of a curver mixed-type shell element', *IUTAM Symposium on High Speed Computing of Elastic Structures*, Liege, 1970, Vol. 1, pp. 321–356.
23. W. Altmann and F. Iguti, 'A thin cylindrical shell finite element based on a mixed formulation', *Comp. Struct.*, **6,** 149–155 (1976).
24. A. K. Noor and C. M. Andersen, 'Mixed isoparametric finite element models of laminated composite shells', *Comp. Meth. Appl. Mech. Eng.*, **11,** 255–280 (1977).
25. C. Tahiani and L. Lachance, 'Linear and non-linear analysis of thin shallow shells by mixed finite elements', *Comp. Struct.*, **5,** 167–177 (1975).
26. R. Harbord, *Berechnung von Schalen mit endlichen Verschiebungen—Gemischte Finite Elemente*, Doctoral thesis, TU Braunschweig, 1973.
27. P. L. Gould and S. K. Sen, 'Refined mixed method finite elements for shells of revolution', *Proc. Third Conf. on Matrix Methods in Structural Mechanics*, 1971, AFFDL TR-71-160, pp. 397–422.
28. S. Y. Barony and H. Tottenham, 'The analysis of rotational shells using a curved ring element and the mixed variational formulation', *Int. J. Num. Meth. in Eng.*, **10,** 861–872 (1976).
29. A. S. L. Chan and V. M. Trbojevic, 'Thin shell finite element by the mixed method formulation—Part 1', *Comp. Meth. Appl. Mech.*, **9,** 337–367 (1976).
30. E. Hofbauer, 'Zur Berechnung von Rotationsschalen mit gemischten Variationsprinzipien und Ringelementen für eine beliebige statische Belastung', *Ing.-Arch.*, **47,** 129–137 (1978).
31. J. E. Goldberg, 'Computer analysis of shells', *Proc. Symp. on the Theory of Shells*, University of Houston, Texas, 1967, pp. 5–22.

32. A. Kalnins, 'Analysis of curved thin-walled shells of revolution', *AIAA J.*, **6,** 584–588 (1968).
33. D. Bushnell, 'Analysis of buckling and vibration of ring-stiffened, segmented shells of revolution', *Int. J. Solids Struct.*, **6,** 157–181 (1970).
34. H. J. Rensch and W. Wunderlich, 'A semi-analytical finite element process for nonlinear elastoplastic analysis of arbitrarily loaded shells of revolution', *Proc. SMIRT 6*, Paris, 1981 (to appear).
35. J. R. Fitch and B. Budiansky, 'Buckling and postbuckling behavior of spherical caps under axisymmetric load', *AIAA J.* **8,** 686–693 (1970).
36. D. Talaslidis, 'On the convergence of a mixed finite element approximation for cylindrical shells', *ZAMM,* **59,** 431–436 (1979).
37. J. N. Reddy and J. T. Oden, 'Mixed finite element approximations of linear boundary value problems', *Quart. Appl. Math.*, **33,** 255–280 (1975).
38. J. Babuska, J. T. Oden, and J. K. Lee, 'Mixed-hybrid finite element approximations of second-order elliptic boundary—value problems', *Comp. Meth. Appl. Mech. Eng.*, **11,** 175–206 (1977).

Hybrid and Mixed Finite Element Methods
Edited by S. N. Atluri, R. H. Gallagher, and O. C. Zienkiewicz

Chapter 12

Estimate of Dynamic Properties of Composites by Mixed Finite Elements Methods

S. Nemat-Nasser

12.1 INTRODUCTION

Mixed variational principles provide powerful means for estimating overall properties of elastic composites. The field equations (linearized) for problems of this kind have a common feature: the differential equations involve coefficients with large variations or with discontinuities (i.e. partial differential equations with rough coefficients). Mixed variational methods are singularly suited to provide solutions for problems of this kind, because they allow for a wider class of functions on which the stationary value of the corresponding functional is sought. In the finite element approximation, departure from a minimum potential energy and, therefore, assumed displacement fields was pioneered by Pian [1] who obtained element stiffness matrices on the basis of an assumed stress distribution. It was again Pian [2] who brought into focus the significance of relaxed continuity requirements in the application of variational principles to finite element estimation (see also [3, 4, and 5]. Other pioneering works relating to the application of variational principles with several independent fields to finite element estimations are by Herrmann [6, 7], Prager [8, 9], Dunham and Pister [10], Tong [11, 12], Atluri [13, 14], Oden and Reddy [15], Mang and Gallagher [16], and others; see other contributions to this volume and references cited therein, as well as the Nemat-Nasser [17] and Zienkiewicz [18].

In my association with the late Professor Prager, who had a keen interest in effective computational methods, I became interested in mixed variational methods during the late 1960's. Then I had the good fortune of having Professor Reissner as my immediate colleague, and therefore it was natural that I would carefully study Reissner's variational principles [19, 20], Hellinger's work [21], as well as Hu's [22] and Washizu's [23, 24] contributions. Then, in the summer of 1970, Professor Lee brought to my attention his interesting work with Professors Kohn and Krumhansl [25] on the application of the minimum energy principle of elasticity to estimating the eigenfrequencies of composites with periodic structures. It was during my discussions with Professor Lee that I came to realize the significance of Reissner's principle in connection with estimating dynamic properties of elastic composites. The following year I spent a month in La Jolla where Professor Lee presented the results of his work at an ASME meeting, and during a discussion he encouraged me to work on my idea of a more general variational method [26] for that class of problems. The results of this effort were indeed rewarding and, in fact, astonishing, as a mixed variational principle with both displacement and stress fields varied, provided estimates for the first two eigenvalues with less than 1 per cent error, using only a three-term Rayleigh–Ritz approximation, whereas the corresponding results obtained by the minimum potential energy (i.e. the Rayleigh quotient) method has a substantial error even for the first eigenvalue.* Moreover, the actual computational effort was less for the mixed variational approach. The numerical calculations were done by Dr. H.-T. Tang, then at the University of California, San Diego. I repeated the calculations myself. Also, Mr. K.-N. Lee, a graduate student at Northwestern University, performed an independent calculation; arriving at exactly the same results. Since then, in collaboration with my associates, we have exploited the method which we call the 'new quotient', and have applied it to estimate dynamic properties of composite beams, plates, and laminated and fibre-reinforced composites [27 to 32].

Later on, Babuška and Osborne [33] produced an elegant proof which showed that under rather relaxed conditions, the method of the new quotient for one-dimensional composites also has a convergence rate which is about one order of magnitude better than the corresponding one for the Rayleigh quotient. Unfortunately, I am not aware of a similar proof for two- and three-dimensional composites. Moreover, in these cases, exact solutions to sample problems do not exist. Therefore, whether or not the new quotient retains its effectiveness for these higher dimensional problems remains an open question; in the case of plate bending, however, a sample problem shows that the new quotient is still equally effective (see [34]).

In this review article, I will briefly summarize the method of the new quotient in connection with three-dimensional elastic composites, outline both finite element as well as the usual Rayleigh–Ritz approximation schemes, and give an illustrative example.

* For example, for harmonic waves propagating normal to a layered medium consisting of two alternating materials of equal thickness with a mass density ratio of 3 and an elastic modulus ratio of 100, the exact, the new quotient estimate, and the Rayleigh quotient estimate for the first eigenvalue respectively are 0.098, 0.098, and 0.237, for the dimensionless wave number $Q=0.5$, where three exponential approximating functions (see 12.12) are used (see [27]).

12.2 HARMONIC WAVES IN ELASTIC COMPOSITES

For harmonic waves in elastic composites with periodic structures one considers a representative unit cell with suitable boundary conditions. For a fibre-reinforced composite with fibre of a rectangular cross-section, this is illustrated in Figure 12.1. In general, it is assumed that the unbounded composite consists of a collection of completely bonded unit cells that are repeated in all directions, Then the mass density $\rho(\mathbf{x})$ and the elasticity tensor $\mathbf{C}(\mathbf{x})$ satisfy

$$\rho(\mathbf{x}) = \rho(\mathbf{x} + m'\mathbf{l}^{\beta}), \qquad \mathbf{C}(\mathbf{x}) = \mathbf{C}(\mathbf{x} + m'\mathbf{l}^{\beta}) \tag{12.1}$$

where m' is an integer and $\mathbf{l}^{\beta}$ $(\beta = 1, 2, 3)$ defines the three edges of the unit cell. Here and in the sequel it is assumed that the unit cell is in the form of a parallelepiped, occupying region $\mathcal{R}$ having boundary $\partial\mathcal{R}$. Further, the collection of all interior surfaces which separate different material constituents within the cell is denoted by Σ, where Σ may admit corners that are formed by the intersection of two or three smooth surfaces.

For harmonic waves, the stress and displacement fields are in the form of $\boldsymbol{\sigma}(\mathbf{x}) \exp(\pm i\omega t)$, $\mathbf{u}(\mathbf{x}) \exp(\pm i\omega t)$, where ω is the wave frequency. Hence the basic field equations become

$$\boldsymbol{\nabla} \cdot \boldsymbol{\sigma} + \lambda\rho\mathbf{u} = \mathbf{0}, \qquad \lambda = \omega^2 \tag{12.2}$$

$$\boldsymbol{\sigma} = \mathbf{C} : \boldsymbol{\nabla}\mathbf{u} \tag{12.3}$$

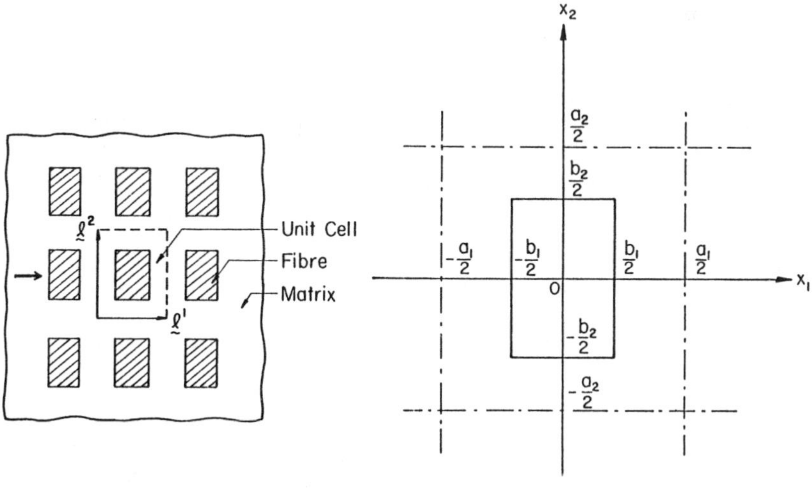

Figure 12.1

with the quasi-periodicity conditions

$$\begin{aligned}\mathbf{u}(\mathbf{x}+\mathbf{l}^\beta) &= \mathbf{u}(\mathbf{x}) \exp(i\mathbf{q}\cdot\mathbf{l}^\beta)\\ \boldsymbol{\sigma}(\mathbf{x}+\mathbf{l}^\beta) &= \boldsymbol{\sigma}(\mathbf{x}) \exp(i\mathbf{q}\cdot\mathbf{l}^\beta)\end{aligned} \qquad i=\sqrt{-1} \tag{12.4}$$

where $\mathbf{q}$ is the wave vector and direct notation is used.*

Equations (12.2) and (12.3) may be regarded as Euler–Lagrange equations of the 'new quotient':

$$\lambda_N = \frac{\langle \boldsymbol{\sigma}, \nabla\mathbf{u}\rangle + \langle \nabla\mathbf{u}, \boldsymbol{\sigma}\rangle - \langle \mathbf{D}:\boldsymbol{\sigma}, \boldsymbol{\sigma}\rangle}{\langle \rho\mathbf{u}, \mathbf{u}\rangle} \tag{12.5}$$

where

$$\langle \boldsymbol{\sigma}, \nabla\mathbf{u}\rangle = \int_{\mathcal{R}} \boldsymbol{\sigma} : \nabla\mathbf{u}^* \, dV, \qquad \mathbf{D} = \mathbf{C}^{-1} \tag{12.6}$$

The asterisk denotes the complex conjugate and the integration is carried out over a typical unit cell.

To see why the new quotient (12.5) should yield better estimates for the eigenfrequencies ($\omega^2 = \lambda$), compare with the Rayleigh quotient,

$$\lambda_R = \frac{\langle \mathbf{C}:\nabla\mathbf{u}, \nabla\mathbf{u}\rangle}{\langle \rho\mathbf{u}, \mathbf{u}\rangle} \tag{12.7}$$

and observe that the latter involves products of displacement gradients, whereas in (12.5) $\nabla\mathbf{u}$ occurs *linearly*. Hence the stationary values of (12.5) may be sought on a larger function class. Indeed, some components of $\nabla\mathbf{u}$ are discontinuous across Σ, in such a manner that the tractions remain continuous:

$$[\boldsymbol{\sigma}(\mathbf{x}^+) - \boldsymbol{\sigma}(\mathbf{x}^-)]\cdot\mathbf{n} = \mathbf{0} \qquad \text{for } \mathbf{x} \text{ on } \Sigma \tag{12.8}$$

where $\mathbf{n}$ is a unit vector on Σ. This fact presents some difficulties when continuous and continuously differentiable approximating functions are used in connection with the Rayleigh quotient, whereas $\boldsymbol{\sigma}$ and $\mathbf{u}$ may be estimated by even analytic functions in (12.5); continuity conditions (12.8) are then automatically satisfied. Observe that (12.8) and quasi-periodicity conditions (the second equation in 12.4) are *not essential* boundary conditions and may be suppressed in a Rayleigh–Ritz minimization procedure. When the approximation is performed on a complete function space, the result would converge to the exact eigenvalue (see [35]). However, for composites with the discontinuous elasticity tensor $\mathbf{C}(\mathbf{x})$ such a procedure is inefficient.

* With respect to a fixed rectangular Cartesian coordinate system with the coordinate unit base vector $\mathbf{e}_j$ and coordinate axes x_j ($j = 1, 2, 3$), we have $\nabla \equiv \mathbf{e}_j\, \partial/\partial x_j$ so that (12.2) and (12.3) become $\partial\sigma_{jk}/\partial x_k + \lambda\rho u_j = 0$, $\sigma_{jk} = C_{jkmn}\, \partial u_m/\partial x_n$, where repeated indices are summed.

For approximate solution consider

$$\bar{\mathbf{u}} = \sum_{\alpha,\beta,\gamma=1}^{M'} \mathbf{U}^{(\alpha\beta\gamma)} f^{(\alpha\beta\gamma)}(\mathbf{x}) \tag{12.9}$$

$$\bar{\boldsymbol{\sigma}} = \sum_{\alpha,\beta,\gamma=1}^{M'} \mathbf{S}^{(\alpha\beta\gamma)} f^{(\alpha\beta\gamma)}(\mathbf{x}) \tag{12.10}$$

and upon rendering λ_N stationary with respect to the unknown coefficients $\mathbf{U}^{(\alpha\beta\gamma)}$ and $\mathbf{S}^{(\alpha\beta\gamma)}$ we obtain

$$\langle(\nabla \cdot \bar{\boldsymbol{\sigma}} + \lambda_N \rho \bar{\mathbf{u}}), f^{(\alpha\beta\gamma)}\rangle = \mathbf{0} \qquad \alpha, \beta, \gamma = 1, 2, \ldots, M' \tag{12.11}$$

$$\langle \mathbf{D} : \bar{\boldsymbol{\sigma}} - \nabla \bar{\mathbf{u}}), f^{(\alpha\beta\gamma)}\rangle = \mathbf{0} \tag{12.12}$$

There are $3M'^3$ linear equations in (12.11) and $6M'^3$ ones in (12.12) for a given function space characterized by the coordinate functions $\{f^{(\alpha\beta\gamma)}(\mathbf{x})\}$, $\alpha, \beta, \gamma = 1, 2, \ldots, M'$. For example, one may consider

$$f^{(\alpha\beta\gamma)}(\mathbf{x}) = \exp\{i[\mathbf{q} \cdot \mathbf{x} + 2\pi(\alpha x_1 + \beta x_2 + \gamma x_3)]\} \tag{12.13}$$

which satisfies the quasi-periodicity conditions and is a complete set with $\alpha, \beta, \gamma = 1, 2, \ldots, \infty$. In general, for a fixed M', the $6M'^3$ equations (12.12) may be solved for the coefficients $\mathbf{S}^{(\alpha\beta\gamma)}$ in terms of $\mathbf{U}^{(\alpha\beta\gamma)}$. Upon substitution into (12.11), a set of $3M'^3$ equations for the unknowns $\mathbf{U}^{(\alpha\beta\gamma)}$ results. The vanishing of the determinant of the corresponding coefficients then yields estimates for the eigenfrequencies.

Let $\{\lambda_N^{(p)}\}$, $p = 1, 2, \ldots, 3M'^3$, be the set of eigenvalues obtained in the above-stated manner, and denote the corresponding displacements by $\{\bar{\mathbf{u}}^{(p)}(\mathbf{x})\}$ and the stresses calculated using (12.3) by $\{\bar{\boldsymbol{\sigma}}^{(p)}(\mathbf{x})\}$. Then it can be shown that (see [38])

$$\langle \rho \bar{\mathbf{u}}^{(p)}, \bar{\mathbf{u}}^{(q)} \rangle = \delta_{pq} \tag{12.14}$$

$$\langle \mathbf{D} : \bar{\boldsymbol{\sigma}}^{(p)}, \bar{\boldsymbol{\sigma}}^{(q)} \rangle = \{\lambda_N^{(p)} \lambda_N^{(q)}\}^{1/2}\, \delta_{pq} \tag{12.15}$$

where the displacement field is normalized so that

$$\langle \rho \bar{\mathbf{u}}^{(p)}, \bar{\mathbf{u}}^{(p)} \rangle = 1 \tag{12.16}$$

Note that (12.14) and (12.15) hold under rather general conditions.

12.3 FINITE ELEMENT APPROXIMATION

Finite element calculations of waves in composites have been presented by Yang and Lee [37], Golub, Jenning, and Yang [38], Orris and Petyt [39], and Minagawa, Nemat-Nasser, and Yamada [32]. Some of the results of [32] are summarized here.

Consider two-dimensional (plane strain) problems, i.e. fibre-reinforced composites. Then the unit cell may be divided into a finite number of triangular subregions. Designate each node by a pair of numbers (α, β). The finite element equations are obtained systematically by the introduction of a 'unit field' [40]. For piecewise linear fields, a unit field $f^{(\alpha\beta\gamma)}(\mathbf{x})$ is a function which has unit value at node (α, β), zero value at all other nodes, and varies linearly over all triangles which meet at node (α, β). The displacement and stress fields now become

$$\bar{\mathbf{u}} = \sum_{\alpha=1}^{M} \sum_{\beta=1}^{N} \mathbf{U}^{(\alpha\beta)} f^{(\alpha\beta)}(\mathbf{x}) \tag{12.17}$$

$$\bar{\boldsymbol{\sigma}} = \sum_{\alpha=1}^{M} \sum_{\beta=1}^{N} \mathbf{S}^{(\alpha\beta)} f^{(\alpha\beta)}(\mathbf{x}) \tag{12.18}$$

Note that while (12.17) and (12.18) resemble (12.9) and (12.10), they stand for completely different approximating functions.

As a specific example consider a unit cell of dimensions a_1 and a_2, and let $\mathbf{U}_1$, $\mathbf{U}_2$, $\mathbf{S}_{11}$, $\mathbf{S}_{12}$, and $\mathbf{S}_{22}$ be vectors with $M \times N$ components U_1^I, U_2^I, S_{11}^I, S_{12}^I, and S_{22}^I, where $I = \alpha + (\beta - 1)M$ $(\alpha = 1, 2, \ldots, M; \beta = 1, 2, \ldots, N)$. Construct $2M \times N$ and $3M \times N$ dimensional vectors:

$$\mathbf{U} = [\mathbf{U}_1, \mathbf{U}_2]^T, \qquad \mathbf{S} = [\mathbf{S}_{11}, \mathbf{S}_{12}, \mathbf{S}_{22}]^T \tag{12.19}$$

where T denotes transpose. The quasi-periodicity conditions (12.4) become

$$\begin{aligned} \mathbf{U}^{MP} &= \mathbf{U}^{1P} \exp(iq_1 a_1), & \mathbf{U}^{QN} &= \mathbf{U}^{Q1} \exp(iq_2 a_2) \\ \mathbf{S}^{MP} &= \mathbf{S}^{1P} \exp(iq_1, a_1), & \mathbf{S}^{QN} &= \mathbf{S}^{Q1} \exp(iq_2 a_2) \end{aligned} \tag{12.20}$$

where $P = 1, 2, \ldots, N$ and $Q = 1, 2, \ldots, M$. These can easily be expressed in matrix form as follows [32]. Set

$$\mathbf{U} = \mathbf{E}\bar{\mathbf{U}}, \qquad \mathbf{S} = \mathbf{F}\bar{\mathbf{S}} \tag{12.21}$$

where

$$\mathbf{E} = \left[\begin{array}{c|c} \hat{\mathbf{E}} & \\ \hline & \hat{\mathbf{E}} \end{array}\right], \qquad \mathbf{F} = \left[\begin{array}{c|c|c} \hat{\mathbf{E}} & & \\ \hline & \hat{\mathbf{E}} & \\ \hline & & \hat{\mathbf{E}} \end{array}\right] \tag{12.22}$$

$$\hat{\mathbf{E}} = \left[\begin{array}{ccc} \mathbf{T} & & \\ & \ddots & \\ & & \mathbf{T} \\ \hline c_2\mathbf{T} & & \end{array}\right], \qquad \mathbf{T} = \left[\begin{array}{c} \mathbf{I} \\ \hline \mathbf{c}_1 \end{array}\right] \tag{12.23}$$

Here $c_1 = \exp(iq_1 a_1)$, $c_2 = \exp(iq_2 a_2)$, $\mathbf{I}$ is the $(M-1) \times (M-1)$ unit matrix, $\mathbf{c}_1$ is the $(M-1)$-dimensional vector $\{c_1, 0, \ldots, 0\}$, and blanks are zeros.

By direct substitution, the new quotient (12.5) is expressed in matrix form. The detailed calculations are given by [32] and will not be repeated here. These authors also consider square meshes with interior nodes. They then compare numerical results for symmetric and asymmetric triangular, as

Table 12.1 Phase velocities of harmonic waves in fibre-reinforced composites with fibres of square cross-sections ($\theta = 3$, $\gamma = 100$, $\bar{n}_2 = 0.36$)

Q		Asymmetric triangular mesh ($M = N = 6$)	Symmetric triangular mesh ($M = N = 6$)	Method of exponential functions ($M' = 1$)
1.0	Trans.	0.9725	0.9710	0.9724
	Long.	2.0829	2.0835	2.0860
2.0	Trans.	0.9422	0.9407	0.9419
	Long.	1.9416	1.9690	1.9520

well as square meshes. A set of typical results is given in Table 12.1, where the dimensionless quantities are

$$\theta = \frac{\rho_2}{\rho_1}, \qquad \gamma = \frac{\mu_2}{\mu_1} \tag{12.24}$$

where ρ_1 and μ_1 are the mass density and shear modulus of the fibre and ρ_2 and μ_2 are those of the matrix;

$$Q = a_2\sqrt{(q_1^2 + q_2^2)} \tag{12.25}$$

is the dimensionless wave number; and

$$\bar{n}_2 = \frac{(N_2 - N_1)(M_2 - M_1)}{(N-1)(M-1)} \tag{12.26}$$

is a parameter which represents the surface fraction of the fibre within the cell. Note that the mesh pattern is chosen in such a manner that N_1 to N_2 rows of nodes and M_1 to M_2 columns of nodes are placed on the interface between the fibre and the matrix.

ACKNOWLEDGEMENT

This work has been supported by the U.S. Army Research Office under Grant DAA29-79-0168 to Northwestern University.

REFERENCES

1. T. H. H. Pian, 'Derivation of element stiffness matrices by assumed stress distribution', *AIAA J.*, **2,** 1333–1336 (1964).
2. T. H. H. Pian, 'Finite element methods by variational principles with relaxed continuity requirements', in *Variational Methods in Engineering* (Eds. C. A. Brebbia and H. Tottenham) pp. 3/1–3/24, Southampton University Press, 1973.
3. T. H. H. Pian and P. Tong, 'Basis of finite elements methods for solid continua', *Int. J. Num. Meth. in Eng.*, **1,** 3–28 (1969).

4. T. H. H. Pian and P. Tong, 'Finite element methods in continuum mechanics', in *Advances in Applied Mechanics* (Ed. C. S. Yih), Vol. 12, pp. 1–58, Academic Press, 1972.
5. T. H. H. Pian, 'Formulation of finite element methods for solid continua', in *Recent Advances in Matrix Methods of Structural Analysis and Design* (Eds. R. H. Gallagher, Y. Yamada, and J. T. Oden), pp. 49–83, University of Alabama in Huntsville Press, 1971.
6. L. R. Herrmann, 'Bending analysis for plates', *First Conf. on Matrix Methods in Structural Mechanics*, Wright–Patterson AFB, pp. 577–602, 1965.
7. L. R. Herrmann, 'Finate element bending analysis for plates', *J. Eng. Div. ASCE*, No. EM5, 13–26 (1967).
8. W. Prager, 'Variational principles of linear elastostatics for discontinuous displacements, strains, and stresses', in *Recent Progress in Applied Mechanics*, (Eds. B. Broberg, J. Hult, and F. Niordson), The Folke–Odqvist Volume, pp. 463–474, Almqvist and Wiksell, Stockholm, 1967.
9. W. Prager, 'Numerical methods of stress analysis', in *Advances in Computers* (Eds. F. L. Alt and M. Rubinoff), Vol. 10, pp. 253–273, 1970.
10. R. S. Dunham and K. S. Pister, 'A finite element application of the Hellinger–Reissner variational theorem', *Second Conf. on Matrix Methods in Structural Mechanics*, Wright–Patterson AFB, pp. 471–487, 1968.
11. P. Tong, 'An assumed stress hybrid finite element method for incompressible and near-incompressible material', *Int. J. Num. Meth. in Eng.*, **1,** 3–28 (1969).
12. P. Tong, 'New displacement hybrid finite-element models for solid continua', *Int. J. Num. Meth. in Eng.*, **2,** 73–83 (1970).
13. S. N. Atluri, 'On hybrid stress finite element model for incremental analysis of large deflection problems', *Int. J. Solids Struct.*, **9,** 1177–1191 (1973).
14. S. N. Atluri, 'Rate complementary energy principles; finite strain plasticity problems; and finite elements', in *Variational Methods in the Mechanics of Solids* (Ed. S. Nemat-Nasser), pp. 363–367, Pergamon Press, Oxford, 1980.
15. J. T. Oden and J. N. Reddy, 'Mathematical theory of mixed finite element approximations', *Quart. Appl. Math.*, **33,** 255–280 (1975).
16. H. A. Mang and R. H. Gallagher, 'A critical assessment of the simplified hybrid displacement method', *Int. J. Num. Meth. in Eng.*, **11,** 145–167 (1977).
17. S. Nemat-Nasser, 'General variational principles in nonlinear and linear elasticity with applications', in *Mechanics Today*, Vol. 1, pp. 214–261, Pergamon Press, 1974.
18. O. C. Zienkiewicz, 'The finite element method: From intuition to generality', *Appl. Mech. Rev.*, **23,** 249–256 (1970).
19. E. Reissner, 'On a variational theorem in elasticity', *J. Math. Phys.*, **29,** 90–95 (1950).
20. E. Reissner, 'On a variational theorem for finite elastic deformations', *J. Math. Phys.*, **32,** 129–135 (1953).
21. E. Hellinger, 'Die allgemeinen Ansätze der Mechanik der Kontinua', *Enz. math. Wis.*, **4,** 602–694 (1914).
22. H.-C. Hu, 'On some variational principles in the theory of elasticity and the theory of plasticity', *Scientia Sinica*, **4,** 33–54 (1955).
23. K. Washizu, 'On the variational principles of elasticity and plasticity', Techn. Report 25-18, M.I.T., 1955.
24. K. Washizu, *Variational Methods in Elasticity and Plasticity*, Pergamon Press, New York, 1968.
25. W. Kohn, J. A. Krumhansl, and E. H. Lee, 'Variational methods for dispersion

relations and elastic properties of composite materials', *J. Appl. Mech.*, **39,** 327–336 (1972).
26. S. Nemat-Nasser, 'On variational methods in finite incremental elastic deformation problems with discontinuous fields', *Q. Appl. Math.*, **30,** 143–156 (1972).
27. S. Nemat-Nasser, 'General variational methods for waves in elastic composites', *J. Elasticity*, **2,** 73–90 (1972).
28. S. Nemat-Nasser and F. C. L. Fu, 'Harmonic waves in layered composites: Bounds on frequencies', *J. Appl. Mech.*, **41,** and *Trans. Am. Soc. Mech. Engrs.*, **96,** Ser. E, 288–290 (1974).
29. S. Nemat-Nasser and K. N. Lee, 'Finite-element formulations for elastic plates by general variational statements with discontinuous fields', in *Developments in Mechanics*, Vol. 7, 13*th Midwestern Mech. Conf. Proc.*, pp. 979–995, 1973.
30. S. Nemat-Nasser and K.-W. Lang, 'Eigenvalue problems for heat conduction in composite materials', *Iranian J. Science and Technol.*, **7,** 243–260 (1979).
31. S. Nemat-Nasser and S. Minagawa, 'Harmonic waves in layered composites: Comparison among several schemes', *J. Appl. Mech.*, **42,** and *Trans. Am. Mech. Engrs.*, **97,** Ser. E, 699–704 (1975).
32. S. Minagawa, S. Nemat-Nasser, and M. Yamada, 'Finite-element analysis of harmonic waves in layered and fiber-reinforced composites', *Int. J. Num. Meth. in Eng.*, **17,** 1335–1353 (1981).
33. I. Babuška and J. E. Osborn, 'Numerical treatment of eigenvalue problems for differential equations with discontinuous coefficients', *Math. of Comput.*, **32,** 991–1023 (1978).
34. K.-W. Lang and S. Nemat-Nasser, 'Vibration and stability of rectangular strip-plates, *J. Sound and Vibration*, **61,** 9–24 (1978).
35. H. F. Weinberger, 'Variational methods for eigenvalue approximation', *Regional Conf. Series in Applied Mathematics*, SIAM, Philadelphia, 1974.
36. S. Nemat-Nasser, F. C. L. Fu and S. Minagawa, 'Harmonic waves in one-, two- and three dimensional composites: Bounds for eigenfrequencies', *Int. J. Solid Structures*, **11,** 617–642 (1975).
37. W. H. Yang and E. H. Lee, 'Modal analysis of floquet waves in composite materials', *J. Appl. Mech.*, **41,** 429–433 (1974).
38. G. H. Golub, L. Jenning, and W. H. Yang, 'Waves in periodically structured media', *J. Comput. Physics*, **17,** 349–357 (1975).
39. R. M. Orris and M. Petyt, 'Finite element study of harmonic wave propagation in periodic structures', *J. Sound and Vibration*, **33,** 223–236 (1974).
40. M. Tanrikulu and W. Prager, 'Consistent finite difference equations for thin elastic disks of variable thickness', *Int. J. Solids Struct.*, **3,** 197–205 (1967).

Hybrid and Mixed Finite Element Methods
Edited by S. N. Atluri, R. H. Gallagher, and O. C. Zienkiewicz

Chapter 13

Finite Element Properties, Based Upon Elastic Potential Interpolation

J. F. Besseling

13.1 INTRODUCTION

In common finite element programmes the evaluation of the matrices, required in a finite element analysis, is for all but the simplest elements carried out by numerical integration. It has frequently been noted that it is erroneous to think that the errors thus introduced in the evaluation of the functionals should be reduced as much as possible. A reduced order of integration may actually improve the accuracy and convergence of solutions, though this fact of experience does not seem to be much of a topic in mathematical convergence and error analysis up to this day. Also, it may not be such a fruitful starting point to consider the problem in terms of functional analysis without taking due note of the underlying mechanical concepts.

One of the main features of the finite element method is the possibility of mesh refinement in regions with pronounced deformation gradients in order to improve the accuracy of solutions without unnecessarily large systems of unknowns. This mesh refinement can be carried out even with the boundary conditions taken from an analysis with a coarser mesh, provided the boundaries are sufficiently far from the region to be reanalysed. This is due to the fact that in a proper finite element representation, in which all infinitesimal rigid body modes of the individual finite elements are represented in the displacement interpolations, the loads that an element exerts on the system form a self-equilibrating system if the element does not carry external loads. Errors in these loads are therefore self-equilibrating systems and these generally damp out quickly, as is known in the mechanics of deformable bodies since the work of De St. Venant. The essential role played by the rigid modes is best expressed by the principle of virtual work in the version which in [1] is attributed to Piola. It requires the virtual work of the external forces to be zero for all infinitesimal rigid body motions. Absence of virtual deformations due to virtual displacements can be expressed as subsidiary conditions. Introduced into the virtual work condition with the aid of multipliers the resulting contribution is to be interpreted as the virtual work of deformation, where the multipliers are components of a stress field.

We may conclude that for the derivation of a discrete model it is essential to define for each finite element the necessary and sufficient conditions for rigid body motions in terms of the nodal virtual displacements; or equivalently to define the deformation modes by properly chosen generalized strain measures. The dual stress multipliers must be related to these strains by constitutive equations in order to arrive at a system of equations by which the physical deformation problem can be solved. Here independent of the representation of the rigid body motions the problem of the best approximation of the physical properties of a finite element arises. In this paper we restrict ourselves to purely elastic materials for which we employ the notion of the elastic potential.

The elastic potential per unit undeformed volume is a function of the components of a proper strain tensor. In the case of non-linear elasticity, such as rubber elasticity, this functional dependence is quite complicated. For a finite element, rather than resort to numerical integration in each step of the deformation process, we may determine first an approximation of the elastic potential of the finite element as a whole. From an explicit dependence of the elastic potential of a finite element on the generalized strains the subsidiary conditions of Piola's virtual work principle are obtained as a first variation of the elastic potential of the finite elements, which replaces the scalar product of the virtual deformations and the dual stress multipliers. In other words, the derivatives of the elastic potential of the finite element with respect to the generalized strains provide the constitutive expressions for the generalized stresses and no more numerical integration is called for. Also, the incremental equations are obtained directly by differentiation.

In order to bring out the fact that finite element theory is the discrete analogue of the continuum theory we bring in the next section a summary of the latter.

3.2 PRINCIPLE OF VIRTUAL WORK AND ELASTIC POTENTIAL

Let us consider a body B of mass density ρ, subjected to surface tractions $\mathbf{t}$, a volume force density $\rho\mathbf{f}$, and an acceleration field in an inertial system $\ddot{\mathbf{u}}$. Written with rectangular Cartesian components of the vector quantities involved, the principle of virtual work states as a necessary condition for any motion of the body B

$$\int_{\partial B} t_i\,\delta u_i\,\mathrm{d}A+\int_B \rho f_i\,\delta u_i\,\mathrm{d}V=\int_B \rho\ddot{u}_i\,\delta u_i\,\mathrm{d}V \tag{13.1}$$

for all rigid virtual displacement fields δu_i. A rigid virtual displacement field is a field of infinitesimal, kinematically admissible (i.e. continuous and piecewise differentiable) displacements satisfying the condition

$$\delta(\mathrm{d}l^2) = \left(\frac{\partial \delta u_i}{\partial x_j} + \frac{\partial \delta u_j}{\partial x_i}\right) \mathrm{d}x_i \, \mathrm{d}x_j = 0 \tag{13.2}$$

where $\mathrm{d}l$ is the length of an infinitesimal line element in the body B. Obviously,

$$\delta\varepsilon_{ij} = \frac{1}{2}\left(\frac{\partial \delta u_i}{\partial x_j} + \frac{\partial \delta u_j}{\partial x_i}\right) = 0 \qquad \text{for all} \quad O(p_0, \varepsilon) \in B \tag{13.3}$$

represents necessary and sufficient conditions for displacement fields δu_i to be rigid. We observe that $\delta\varepsilon_{ij}$ is a symmetric tensor and hence the conditions (13.3) can be introduced into the variational statement of the principle of virtual work by means of a symmetric field of multipliers t_{ij}. Thus (13.1) is transformed into

$$\int_{\partial B} t_i \, \delta u_i \, \mathrm{d}A + \int_B \rho f_i \, \delta u_i = \int_B (\rho \ddot{u}_i \, \delta u_i + t_{ij} \, \delta\varepsilon_{ij}) \, \mathrm{d}V \tag{13.4}$$

which now must hold for all kinematically admissible virtual displacements δu_i.

By applying Green's transformation we derive from (13.4)

$$t_i = t_{ij} n_j \quad \text{on} \quad \partial B \tag{13.5}$$

and

$$\frac{\partial t_{ij}}{\partial x_j} + \rho f_i = \rho \ddot{u}_i \qquad \text{in} \quad B \tag{13.6}$$

which identifies t_{ij} as the stress tensor of Cauchy. Hence $t_{ij}\delta\varepsilon_{ij}$ in the right-hand side of (13.4) may rightly be called the virtual work of deformation per unit volume.

The tensor $\delta\varepsilon_{ij}$ cannot be derived as a first variation from a measure of strain in the case of finite displacement gradients. Therefore if all work of deformation is reversibly stored as an elastic potential (an appropriate definition of pure elasticity), the expression $t_{ij}\delta\varepsilon_{ij}$ for the virtual work of deformation per unit volume cannot be considered to be identical to the first variation of this potential. However, such an identity is needed for the derivation of the constitutive equations. We must look for proper measures of strain which can be used as state variables.

In the case of elastic material behaviour, deformations with respect to an initial geometry of the body are such state variables. Any function of these variables specifies as an elastic potential a physically admissible, particular kind of material behaviour. It is then appropriate to express the volume integrals in (13.4) in terms of the initial geometry B_0 of the body. We define a_i as the Cartesian coordinates of the material points in this initial geometry. They identify as material coordinates the material points during the deformation process.

The difference of the square of a line element in the deformed and in the undeformed geometry is defined by

$$dl^2 - dl_0^2 = 2E_{\alpha\beta}\, da_\alpha\, da_\beta \tag{13.7}$$

where $E_{\alpha\beta}$ is the Lagrangian strain tensor:

$$E_{\alpha\beta} = \frac{1}{2}\left(\frac{\partial x_k}{\partial a_\alpha}\frac{\partial x_k}{\partial a_\beta} - \delta_{\alpha\beta}\right) = \frac{1}{2}\left(\frac{\partial u_\alpha}{\partial a_\beta} + \frac{\partial u_\beta}{\partial a_\alpha} + \frac{\partial u_k}{\partial a_\alpha}\frac{\partial u_k}{\partial a_\beta}\right) \tag{13.8}$$

A rigid virtual displacement field can now be defined by the condition that the first variation of the strain tensor (13.8) vanishes in the neighbourhoods $O(p_0, \varepsilon)$ of all points of the body:

$$\delta E_{\alpha\beta} = \frac{1}{2}\left(\frac{\partial \delta u_\alpha}{\partial a_\beta} + \frac{\partial \delta u_\beta}{\partial a_\alpha} + \frac{\partial u_k}{\partial a_\alpha}\frac{\partial \delta u_k}{\partial a_\beta} + \frac{\partial \delta u_k}{\partial a_\alpha}\frac{\partial u_k}{\partial a_\beta}\right) = 0 \qquad \text{for all } O(p_0, \varepsilon) \in B_0 \tag{13.9}$$

Using the equality $\rho\, dV = \rho_0\, dV_0$, and again with the aid of a symmetric tensor field of multipliers $\sigma_{\alpha\beta}$, we express the principle of virtual work by

$$\int_{\partial B} t_i \delta u_i\, dA + \int_{B_0} \rho_0 t_i \delta u_i\, dV_0 = \int_{B_0} (\rho_0 \ddot{u}_i \delta u_i + \sigma_{\alpha\beta}\delta E_{\alpha\beta})\, dV_0 \tag{13.10}$$

which must hold for all kinematically admissible virtual displacement fields.

The tensor $\sigma_{\alpha\beta}$ is the so-called pseudo stress tensor of Kirchhoff, related to the stress tensor of Cauchy by

$$\sigma_{\alpha\beta} = \frac{\rho_0}{\rho}\frac{\partial a_\alpha}{\partial x_i} t_{ij} \frac{\partial a_\beta}{\partial x_j} \tag{13.11}$$

This relation follows from the identities

$$\begin{aligned} \int_B t_{ij}\delta\varepsilon_{ij}\, dV &= \int_{B_0} \frac{\rho_0}{\rho}\frac{\partial a_\alpha}{\partial x_i} t_{ij} \frac{\partial a_\beta}{\partial x_j}\frac{\partial x_k}{\partial a_\alpha}\frac{\partial \delta u_k}{\partial a_\beta}\, dV_0 \\ &= \int_{B_0} \frac{\rho_0}{\rho}\frac{\partial a_\alpha}{\partial x_i} t_{ij} \frac{\partial a_\beta}{\partial x_j}\delta E_{\alpha\beta}\, dV_0 \end{aligned} \tag{13.12}$$

If we introduce the constitutive assumption of pure elasticity, namely that all work of deformation is stored as an elastic potential, we have

$$\int_{B_0} \sigma_{\alpha\beta}\dot{E}_{\alpha\beta}\, dV_0 = \int_{B_0} \rho_0 \frac{\partial e}{\partial E_{\alpha\beta}} \dot{E}_{\alpha\beta}\, dV_0 \qquad \text{for all } \dot{E}_{\alpha\beta} \tag{13.13}$$

from which we deduce

$$\sigma_{\alpha\beta} = \rho_0 \frac{\partial e}{\partial E_{\alpha\beta}} \tag{13.14}$$

In case e is a quadratic form in $E_{\alpha\beta}$ we speak of physical linearity, in contrast with the geometrical non-linearity of the expressions for $E_{\alpha\beta}$ (13.8). The general case of physical and geometrical non-linearity in the deformation of an elastic body is practically only encountered in rubber elasticity. An expression for ρe, which approximates the behaviour of many rubbers, is given by

$$\rho_0 e = \tfrac{1}{2}C\varepsilon_v^2 + \tfrac{1}{2}G[g^{1/3}C_{\alpha\alpha}^{-1} - 3 + \mu(g^{-1/3}C_{\alpha\alpha} - 3)^m] \tag{13.15}$$

where $\varepsilon_v = g^{1/2} - 1$. Here g represents the determinant of the tensor components $C_{\alpha\beta} = 2E_{\alpha\beta} + \delta_{\alpha\beta}$ and C and G are the well-known compressibility and shear moduli from small strain elasticity (for rubber $C \gg G$). The two additional material constants μ and m for this isotropic material must give a description of the large strain behaviour of rubber. Typical values for μ are between 0.01 and 0.05 and for m between 2 and 3. For the Mooney–Rivlin material holds $m = 1$, but then the inflection point, which usually is found in the stress–strain curve of the tensile test, cannot be described.

It is clear that the essential non-linearities in (13.8), as well as those resulting from (13.15), preclude any exact analytical approach to the solution of boundary value problems in all but the simplest cases of homogeneous fields. Finite element theory, however, provides a discrete analogue of the continuum theory and the resulting algebraic equations can be solved.

13.3 THE FINITE DIMENSIONAL DEFORMATION SPACE OF FINITE ELEMENTS

In the finite element method at interelement boundary points, lines, and surfaces the adjoining elements must have common displacements in order to ensure continuity of the structure. If these displacements are to be approximated by a finite set of functions, the displacements at the interelement boundaries and at the natural boundary of the structure can be taken as interpolations between nodal displacements at the interface between any two elements or between an element and the surroundings of the structure. Continuity of the structure is then ensured for all values of the nodal displacements and the displacement field is represented by a finite dimensional vector space $\mathcal{U}^e$ with elements in $\mathcal{R}^n$.

We recall that in the formulation of the principle of virtual work rigid virtual displacements had to be characterized. The interpolations between nodal displacements must be such that all possible rigid motions of the finite elements are represented. Even for elements with curved boundaries this requirement is met by the isoparametric concept [2]. For the transformation to natural coordinates ζ_α of the finite element and for the interpolation between nodal displacements, the same interpolation functions are used. In the case of material finite elements the coordinates of the material points of a finite element are then given by

$$x_i^e = \psi_N(\zeta_\alpha)(a_{N_i}^e + u_{N_i}^e), \qquad \sum_{N=1}^{n} \psi_N(\zeta_a) = 1 \tag{13.16}$$

Infinitesimal rigid body motions of a finite element are now defined by a subspace of $\mathcal{U}^e$, which will be called the vector space of degrees of freedom $\mathcal{V}^e$ [3]. It is appropriate to call the complementary subspace $\mathcal{E}^e$ the deformation space of the finite element:

$$\mathcal{U}^e = \mathcal{V}^e + \mathcal{E}^e \tag{13.17}$$

The conditions for a rigid body motion are expressed by requiring $\delta E_{\alpha\beta} = 0$ in all points of the finite element. However, this condition needs to be applied only in a number of points and for a number of components such that for the finite element under consideration a number of linearly independent conditions is obtained equal to the dimension of its deformation space $\mathcal{E}^e$. In other words, for the finite element model we may derive the contribution of the subsidiary conditions in the variational statement (13.10),

$$\int_{B_0^e} \sigma_{\alpha\beta} \delta E_{\alpha\beta} \, \mathrm{d}V_0 \tag{13.18}$$

from a strain distribution with parameters obtained by local application of (13.8). The number of these strain parameters is determined by the number of nodal displacements $u_{N_i}^e$ minus the number of degrees of freedom of the finite element as a rigid body. The strain parameters are by application of (13.8) expressed as non-linear functions of the nodal displacements with the aid of the interpolations (13.16):

$$\varepsilon_i = D_i(u_k) \tag{13.19}$$

The strain parameters thus defined are elements of a vector of generalized strains of the finite element, $\boldsymbol{\varepsilon}^e$, for which $\delta\boldsymbol{\varepsilon}^e \in \mathcal{E}^c$. Instead of (13.18) we now may add $\sigma_i \delta\varepsilon_i$ to the virtual work condition in order to characterize rigid body motions. It is clear that, just as in the continuum approach, constitutive equations must be given for the stress multipliers σ_i before any deformation problem can be solved. But first we shall discuss as an example the choice of the strain parameters ε_i for the well-known TRIM-6 element (Figure 13.1).

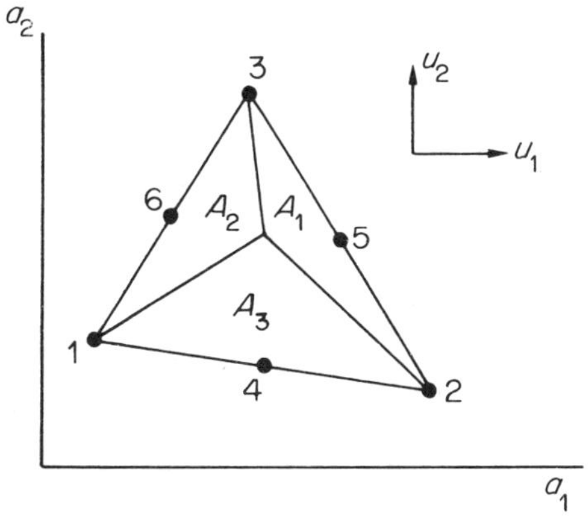

Figure 13.1 TRIM-6 element

In terms of the triangular coordinates:

$$\zeta_1 = \frac{A_1}{A}, \qquad \zeta_2 = \frac{A_2}{A}, \qquad \zeta_3 = \frac{A_3}{A} \tag{13.20}$$

we can easily formulate the interpolation functions in (13.16):

$$[\psi_N(\zeta_\alpha)] = [(2\zeta_1 - 1)\zeta_1 (2\zeta_2 - 1)\zeta_2 (2\zeta_3 - 1)\zeta_3 4\zeta_1\zeta_2 4\zeta_2\zeta_3 4\zeta_3\zeta_1] \tag{13.21}$$

With the aid of the coordinate differences

$$\begin{aligned} \alpha_1 &= a_{31} - a_{21}, & \beta_1 &= a_{22} - a_{32} \\ \alpha_2 &= a_{11} - a_{31}, & \beta_2 &= a_{32} - a_{12} \\ \alpha_3 &= a_{21} - a_{11}, & \beta_3 &= \alpha_{12} - a_{22} \end{aligned} \tag{13.22}$$

we get for the derivatives with respect to a_1 and a_2 the following expressions:

$$\frac{\partial u_i}{\partial a_1} = \frac{1}{2A} [(4\zeta_1 - 1)\beta_1 \, (4\zeta_2 - 1)\beta_2 \, (4\zeta_3 - 1)\beta_3 \, 4(\zeta_2\beta_1 + \zeta_1\beta_2)$$
$$\times 4(\zeta_3\beta_2 + \zeta_2\beta_3) 4(\zeta_1\beta_3 + \zeta_3\beta_1)] \mathbf{u}_i \tag{13.23}$$

$$\frac{\partial u_i}{\partial a_2} = \frac{1}{2A} [(4\zeta_1 - 1)\alpha_1 \, (4\zeta_2 - 1)\alpha_2 \, (4\zeta_3 - 1)\alpha_3 \, 4(\zeta_2\alpha_1 + \zeta_1\alpha_2)$$
$$\times 4(\zeta_3\alpha_2 + \zeta_2\alpha_3) 4(\zeta_1\alpha_3 + \zeta_3\alpha_1)] \mathbf{u}_i$$

For motions in its plane the TRIM-6 element has a $2 \times 6 - 3 = 9$ dimensional deformation space. As suitable strain parameters suggest themselves the strain components E_{11}, E_{12}, E_{22} appear in each of the three corner points. Also the same strain components, but now in the points

$(\frac{2}{3}, \frac{1}{6}, \frac{1}{6})$, $(\frac{1}{6}, \frac{2}{3}, \frac{1}{6})$, $(\frac{1}{6}, \frac{1}{6}, \frac{2}{3})$, can be chosen as a basis for the deformation space of the finite element. In both cases the local strain values can with the aid of six geometric vectors, $\mathbf{d}_i$ $(i = 1, \ldots, 6)$, be expressed in terms of the nodal displacements $\mathbf{u}^e_{Ni}$ by substitution of the local coordinates into (13.23), followed by application of (13.8):

$$\mathbf{E}_{11} = \left|\begin{matrix} \mathbf{d}_1^T \\ \mathbf{d}_2^T \\ \mathbf{d}_3^T \end{matrix}\right| \mathbf{u}_1 + \begin{bmatrix} \frac{1}{2}\mathbf{u}_1^T\mathbf{d}_1\mathbf{d}_1^T\mathbf{u}_1 + \frac{1}{2}\mathbf{u}_2^T\mathbf{d}_1\mathbf{d}_1^T\mathbf{u}_2 \\ \frac{1}{2}\mathbf{u}_1^T\mathbf{d}_2\mathbf{d}_2^T\mathbf{u}_1 + \frac{1}{2}\mathbf{u}_2^T\mathbf{d}_2\mathbf{d}_2^T\mathbf{u}_2 \\ \frac{1}{2}\mathbf{u}_1^T\mathbf{d}_3\mathbf{d}_3^T\mathbf{u}_1 + \frac{1}{2}\mathbf{u}_2^T\mathbf{d}_3\mathbf{d}_3^T\mathbf{u}_2 \end{bmatrix}$$

$$2\mathbf{E}_{12} = \left|\begin{matrix} \mathbf{d}_1^T \\ \mathbf{d}_2^T \\ \mathbf{d}_3^T \end{matrix}\right| \mathbf{u}_2 + \left|\begin{matrix} \mathbf{d}_4^T \\ \mathbf{d}_5^T \\ \mathbf{d}_6^T \end{matrix}\right| \mathbf{u}_1 + \begin{bmatrix} \frac{1}{2}\mathbf{u}_1^T(\mathbf{d}_1\mathbf{d}_4^T + \mathbf{d}_4\mathbf{d}_1^T)\mathbf{u}_1 + \frac{1}{2}\mathbf{u}_2^T(\mathbf{d}_1\mathbf{d}_4^T + \mathbf{d}_4\mathbf{d}_1^T)\mathbf{u}_2 \\ \frac{1}{2}\mathbf{u}_1^T(\mathbf{d}_2\mathbf{d}_5^T + \mathbf{d}_5\mathbf{d}_2^T)\mathbf{u}_1 + \frac{1}{2}\mathbf{u}_2^T(\mathbf{d}_2\mathbf{d}_5^T + \mathbf{d}_5\mathbf{d}_2^T)\mathbf{u}_2 \\ \frac{1}{2}\mathbf{u}_1^T(\mathbf{d}_3\mathbf{d}_6^T + \mathbf{d}_6\mathbf{d}_3^T)\mathbf{u}_1 + \frac{1}{2}\mathbf{u}_2^T(\mathbf{d}_3\mathbf{d}_6^T + \mathbf{d}_6\mathbf{d}_3^T)\mathbf{u}_2 \end{bmatrix} \tag{13.24}$$

$$\mathbf{E}_{22} = \left|\begin{matrix} \mathbf{d}_4^T \\ \mathbf{d}_5^T \\ \mathbf{d}_6^T \end{matrix}\right| \mathbf{u}_2 + \begin{bmatrix} \frac{1}{2}\mathbf{u}_1^T\mathbf{d}_4\mathbf{d}_4^T\mathbf{u}_1 + \frac{1}{2}\mathbf{u}_2^T\mathbf{d}_4\mathbf{d}_4^T\mathbf{u}_2 \\ \frac{1}{2}\mathbf{u}_1^T\mathbf{d}_5\mathbf{d}_5^T\mathbf{u}_1 + \frac{1}{2}\mathbf{u}_2^T\mathbf{d}_5\mathbf{d}_5^T\mathbf{u}_2 \\ \frac{1}{2}\mathbf{u}_1^T\mathbf{d}_6\mathbf{d}_6^T\mathbf{u}_1 + \frac{1}{2}\mathbf{u}_2^T\mathbf{d}_6\mathbf{d}_6^T\mathbf{u}_2 \end{bmatrix}$$

When the strain components in the corner points are chosen as a basis for the deformation space then we have, for $\mathbf{d}_i$,

$$\begin{aligned} 2A\mathbf{d}_1^T &= |3\beta_1 \quad -\beta_2 \quad -\beta_3 \quad 4\beta_2 \quad 0 \quad 4\beta_3| \\ 2A\mathbf{d}_2^T &= |-\beta_1 \quad 3\beta_2 \quad -\beta_3 \quad 4\beta_1 \quad 4\beta_3 \quad 0| \\ 2A\mathbf{d}_3^T &= |-\beta_1 \quad -\beta_2 \quad 3\beta_3 \quad 0 \quad 4\beta_2 \quad 4\beta_1| \\ 2A\mathbf{d}_4^T &= |3\alpha_1 \quad -\alpha_1 \quad -\alpha_1 \quad 4\alpha_2 \quad 0 \quad 4\alpha_3| \\ 2A\mathbf{d}_5^T &= |-\alpha_1 \quad 3\alpha_1 \quad -\alpha_1 \quad 4\alpha_1 \quad 4\alpha_3 \quad 0| \\ 2A\mathbf{d}_6^T &= |-\alpha_1 \quad -\alpha_1 \quad 3\alpha_1 \quad 0 \quad 4\alpha_2 \quad 4\alpha_1| \end{aligned} \tag{13.25}$$

The expressions for $\mathbf{d}_i$ corresponding to the $(\frac{2}{3}, \frac{1}{6}, \frac{1}{6})$, $(\frac{1}{6}, \frac{2}{3}, \frac{1}{6})$, $(\frac{1}{6}, \frac{1}{6}, \frac{2}{3})$ points are given by

$$\begin{aligned} 6A\mathbf{d}_1^T &= |5\beta_1 \quad -\beta_2 \quad -\beta_3 \quad (2\beta_1+8\beta_2) \quad (2\beta_2+2\beta_3) \quad (8\beta_3+2\beta_1)| \\ 6A\mathbf{d}_2^T &= |-\beta_1 \quad 5\beta_2 \quad -\beta_3 \quad (8\beta_1+2\beta_2) \quad (2\beta_2+8\beta_2) \quad (2\beta_3+2\beta_1)| \\ 6A\mathbf{d}_3^T &= |-\beta_1 \quad -\beta_2 \quad 5\beta_3 \quad (2\beta_1+2\beta_2) \quad (8\beta_2+2\beta_3) \quad (2\beta_3+8\beta_1)| \\ 6A\mathbf{d}_4^T &= |5\alpha_1 \quad -\alpha_2 \quad -\alpha_3 \quad (2\alpha_1+8\alpha_2) \quad (2\alpha_2+2\alpha_3) \quad (8\alpha_3+2\alpha_1)| \\ 6A\mathbf{d}_5^T &= |-\alpha_1 \quad 5\alpha_2 \quad -\alpha_3 \quad (8\alpha_1+2\alpha_2) \quad (2\alpha_2+8\alpha_3) \quad (2\alpha_3+2\alpha_1)| \\ 6A\mathbf{d}_6^T &= |-\alpha_1 \quad -\alpha_2 \quad 5\alpha_3 \quad (2\alpha_1+2\alpha_2) \quad (8\alpha_2+2\alpha_3) \quad (2\alpha_3+8\alpha_1)| \end{aligned} \tag{13.26}$$

with

$$\boldsymbol{\varepsilon}^T = |\mathbf{E}_{11}^T \quad \mathbf{E}_{22}^T \quad 2\mathbf{E}_{12}^T| \tag{13.27}$$

The strain parameters ε_i are defined as non-linear functions of the nodal displacements. We can also write

$$\varepsilon_i = D^0_{i,k} u_k + \tfrac{1}{2} D^0_{u,kl} u_k u_l \tag{13.28}$$

If the TRIM-6 element is used as an element in three-dimensional space the deformation space defined by (13.27) is deficient since deformations perpendicular to the plane of the element are not represented. The nodal displacements $\mathbf{u}_3$ perpendicular to the plane give rise to extra quadratic terms of the same structure as those already presented in (13.24), and they lead to a contribution in $\sigma_i \delta\varepsilon_i$ of the form

$$\sigma_i D_{i,kl} u_{k3} \delta u_{l3} \tag{13.29}$$

Thus the membrane stresses σ_i appear in the equations for the equilibrim normal to the plane of the triangle as soon as $D_{i,kl} u_{k3} u_{l3} \neq 0$. For certain loads there can be equilibrium of the TRIM-6 elements in the deflected state under pure membrane action.

For a complete triangular plate and shell element we refer to [4].

13.4 DETERMINATION OF THE ELASTIC POTENTIAL AND FINITE ELEMENT

In the most common type of finite element analysis the interpolation functions (13.16) are substituted directly into the variational condition (13.10). Then we obtain with a summation over all finite elements

$$\sum_e \left\{ \int_{\partial B'} t_k \psi_N \, dA + \int_{B_0} \rho_0 f_k \psi_N \, dV_0 - \ddot{u}_{Mk} \int_{B_0} \rho_0 \psi_M \psi_N \, dV_0 \right.$$
$$\left. + \int_{B_0} \left[\rho_0 \frac{\partial e}{\partial E_{\alpha\beta}} \psi_{N_i} \frac{\partial \zeta_i}{\partial a_\beta} \left(\delta_{\alpha k} + \frac{\partial \psi_M}{\partial \zeta_j} \frac{\partial \zeta_j}{\partial a_\alpha} u_{Mk} \right) \right] dV_0 \right\} \delta u_{Nk} = 0$$
$$\text{for all } \delta\mathbf{u} \tag{13.30}$$

In the case of geometrical and physical non-linearity the resulting equations are highly non-linear and for numerical solutions we shall also need the incremental form

$$\sum_e \left\{ \Delta \int_{\partial B'} t_k \psi_N \, dA + \int_{B_0} \rho_0 \Delta f_k \psi_N \, dV_0 - \Delta\ddot{u}_{Mk} \int_{B_0} \rho_0 \psi_M \psi_N \, dV_0 \right.$$
$$+ \Delta u_{Pl} \int_{B_0} \left[\psi_{Pm} \frac{\partial \zeta_m}{\partial a_\gamma} \left(\delta_{\delta l} + \frac{\partial \psi_Q}{\partial \zeta_n} \frac{\partial \zeta_n}{\partial a_\delta} u_{Ql} \right) \rho_0 \frac{\partial^2 e}{\partial E_{\alpha\beta} \partial E_{\gamma\delta}} \right.$$
$$\left. \left. \times \psi_{N_i} \frac{\partial \zeta_i}{\partial a_\beta} \left(\delta_{\alpha k} + \frac{\partial \psi_M}{\partial \zeta_j} \frac{\partial \zeta_j}{\partial a_\alpha} u_{Mk} \right) \right] dV_0 \right\} = 0 \tag{13.31}$$

We shall not enter here into the difficulties which are encountered in rubber elasticity because of $C \gg G$. It suffices to say that for rubber membranes for

$\varepsilon_v = 0$ the elastic potential can be expressed as a function of E_{11}, E_{22}, E_{12}, while in the case of three-dimensional states of stress the ill-conditioning of the equations due to $C \gg G$ can be circumvented by elimination of ε_v by the introduction of its dual, the pressure p, with the aid of the multiplier method.

Clearly the equations (13.30) and (13.31) require extensive numerical integration, and in the case of physical non-linearity the numerical integrations have to be carried out in each step of the loading process insofar as the terms originating from the elastic potential are concerned. Exceptions are only the constant strain elements and this is of course what makes them so popular in non-linear analysis. They also avoid the principal difficulty that for finite displacement gradients the contribution (13.18)

$$\sum_e \int_{B_0^e} \sigma_{\alpha\beta}\, \delta E_{\alpha\beta}\, \mathrm{d}V_0$$

in the variational condition no longer expresses the subsidiary requirement of zero virtual deformation for each finite element separately, unless a reduced numerical integration is carried out that takes properly into account the dimension of the deformation space. We prefer the formulation of the subsidiary conditions in the application of the principle of virtual work with the aid of the concept of generalized strains and generalized stresses. Instead of (13.18) $\sigma_i \delta\varepsilon_i$ is added to the virtual work condition in order to characterize the rigid body motions. For the derivation of the constitutive equations for σ_i we introduced in [5] and subsequent papers the approach based upon strain interpolation. A set of functions of the natural coordinates is defined such that the deformed state of each finite element over its whole domain is represented by

$$E^e_{\alpha\beta} = \varepsilon_i \phi_{i\alpha\beta}(\zeta_k) \tag{13.32}$$

For the quadratic elastic potential of linear elasticity the elastic potential of the finite element as a whole is now easily calculated from

$$E = \int_{B_0} \tfrac{1}{2}\varepsilon_i \phi_{i\alpha\beta} S_{\alpha\beta\gamma\delta} \phi_{j\gamma\delta} \varepsilon_j \,\mathrm{d}V_0 = \tfrac{1}{2}\boldsymbol{\varepsilon}^{\mathrm{T}} \mathbf{S} \boldsymbol{\varepsilon} \tag{13.33}$$

The identity of pure elasticity, analogous to (13.13),

$$\sigma_i \dot{\varepsilon}_i = \frac{\partial E}{\partial \varepsilon_i} \dot{\varepsilon}_i \qquad \text{for all } \dot{\varepsilon}_i \tag{13.34}$$

provides the constitutive equations of the finite element in the form

$$\sigma_i = \frac{\partial E}{\partial \varepsilon_i} = S_{ij} \varepsilon_j \tag{13.35}$$

In the case of non-linear elasticity an explicit expression for the elastic potential of a finite element with its generalized strains ε_i as parameters is not easily obtained. With functions $\rho_0 e$ like (13.15) for rubber elasticity, numerical integration in each step of the loading process is called for, not only in

$$\sigma_i \delta\varepsilon_i = \delta\varepsilon_i \int_{B_0} \rho_0 \frac{\partial e}{\partial \varepsilon_i} \, \mathrm{d}V_0 \tag{13.36}$$

but, if incremental equations are needed, also in

$$\Delta\sigma_i \delta\varepsilon_i = \Delta\varepsilon_j \, \delta\varepsilon_i \int_{B_0} \rho_0 \frac{\partial^2 e}{\partial \varepsilon_i \, \partial \varepsilon_j} \, \mathrm{d}V_0 \tag{13.37}$$

In any case the elastic potential distribution of a finite element can only be described approximately on the basis of a finite dimensional deformation space. However, as long as the local expression for $\rho_0 e$ can be represented by uniform mesh refinement in the limit, convergence to the exact distribution of the elastic potential is ensured. Therefore simple interpolation for the elastic potential itself also provides a basis for the determination of the elastic potential of the finite element as a whole.

If we express local values of $\rho_0 e$ in terms of the generalized strains ε_i, if necessary with the aid of the interpolations (13.32), the elastic potential distribution in a finite element can be approximated by interpolation by means of functions of the natural coordinates:

$$\rho_0 e = \rho_0 e_i \chi_i(\zeta_\alpha), \qquad e_i = e_i(\varepsilon_k) \tag{13.38}$$

An explicit expression for the elastic potential of the finite element as a whole is then easily calculated from

$$E = e_i \int_{B_0} \rho_0 \chi_i(\zeta_\alpha) \, \mathrm{d}V_0 \tag{13.39}$$

Particularly simple is, for instance, the expression for the elastic potential of a triangular element, obtained by linear interpolation with three local values of e. For a plate thickness t and a surface area A it reads

$$E = \tfrac{1}{3} A t (e_1 + e_2 + e_3) \tag{13.40}$$

Instead of the numerical integrations in (13.36) and (13.37) we now have simply

$$\sigma_i \delta\varepsilon_i = \frac{\partial E}{\partial \varepsilon_i} \delta\varepsilon_i, \qquad \Delta\sigma_i \delta\varepsilon_i = \Delta\varepsilon_j \frac{\partial^2 E}{\partial \varepsilon_i \, \partial \varepsilon_j} \delta\varepsilon_i \tag{13.41}$$

13.5 A SIMPLE ILLUSTRATION

Some insight into the merits of the elastic potential interpolation method as compared to the traditional finite element method can be obtained from the shear linear elastic analysis of a square plate subjected to bending by a shear force F. This is shown in Figure 13.2 where

$$\tau = \frac{3}{2}\frac{F}{at}\left[1-\left(\frac{2y}{a}\right)^2\right]$$

$$\sigma = 12\frac{F}{at}\frac{xy}{a^2}$$

$$\left.\begin{array}{l} \dfrac{x}{a}=1 \quad u=0 \\ \\ \dfrac{x}{a}=0 \quad v=0 \end{array}\right\}$$

$$\left.\begin{array}{l} \dfrac{x}{a}=1 \\ \\ \dfrac{y}{a}=\pm\dfrac{1}{2} \end{array}\right\} u=0$$

$$\frac{u}{a}=\frac{F}{Eat}\left[6\frac{x^2y}{a^3}-(4+2\nu)\frac{y^3}{a^3}-\frac{10-\nu}{2}\frac{y}{a}\right]$$

$$\frac{v}{a}=\frac{F}{Eat}\left[-6\nu\frac{xy^2}{a^3}-2\frac{x^3}{a^3}+\frac{1}{2}(16+5\nu)\frac{x}{a}-\frac{1}{2}(12+5\nu)\right]$$

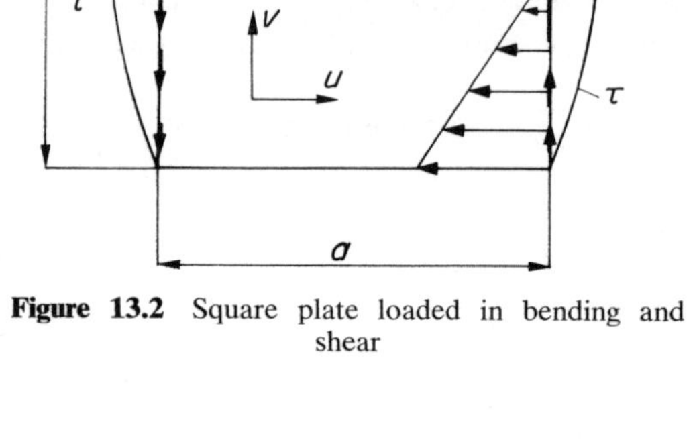

Figure 13.2 Square plate loaded in bending and shear

From an analysis by means of the TRIM-6 element with a 12×6 triangular mesh in half the plate we can compare the results of the traditional method with results according to interpolation of the elastic potential. In both cases the elastic potential of the TRIM-6 element is a quadratic expression in the strain parameters ε_i:

$$E = \tfrac{1}{2}\boldsymbol{\varepsilon}^T\mathbf{S}\boldsymbol{\varepsilon} \tag{13.42}$$

In the case of geometrical linearity and linear elasticity there is no difference between the method of strain interpolation and the traditional direct integration of (13.18), though in the latter approach the matrix **S** is usually not calculated. Instead the contribution of a finite element to the system of equations for the nodal displacements,

$$\mathbf{D}^{e^T}\mathbf{S}^e\mathbf{D}^e = \mathbf{K}^e, \qquad \boldsymbol{\varepsilon}^e = \mathbf{D}^e\mathbf{u}^e \tag{13.43}$$

is evaluated directly. We prefer to determine the matrix $\mathbf{S}^e$ separately. In this case it is given by

$$\mathbf{S}^e = \frac{2GAt}{1-\nu}\begin{bmatrix} \mathbf{H} & \nu\mathbf{H} & \mathbf{0} \\ \nu\mathbf{H} & \mathbf{H} & \mathbf{0} \\ \mathbf{0} & \mathbf{0} & \frac{1-\nu}{2}\mathbf{H} \end{bmatrix}, \qquad \mathbf{H} = \frac{1}{12}\begin{bmatrix} 2 & 1 & 1 \\ 1 & 2 & 1 \\ 1 & 1 & 2 \end{bmatrix} \tag{13.44}$$

where the strain components E_{11}, E_{22}, $2E_{12}$ in the three corner points form the basis of the deformation space.

The stiffness matrix resulting from a linear interpolation of the elastic potential is simply

$$\mathbf{S}^e = \frac{2GAt}{3(1-\nu)}\begin{bmatrix} \mathbf{I} & \nu\mathbf{I} & \mathbf{0} \\ \nu\mathbf{I} & \mathbf{I} & \mathbf{0} \\ \mathbf{0} & \mathbf{0} & \frac{1-\nu}{2}\mathbf{I} \end{bmatrix} \tag{13.45}$$

if the e_1, e_2, e_3 in (13.40) are taken in the points where the strain components E_{11}, E_{22}, $2E_{12}$ are evaluated to form a basis of the deformation space.

If the e_1, e_2, e_3 in the three corner points are chosen, it must be expected that by linear interpolation of the quadratic elastic potential distribution of the TRIM-6 element unnecessarily large errors are made. Indeed, analysis of the problem shown above with a 12×6 mesh leads to errors in the nodal displacements of up to 5 per cent, as compared to the 0.1 per cent error obtained with the strain interpolation and the traditional method. If we take, however, the e_1, e_2, e_3 in the $(\frac{2}{3}, \frac{1}{6}, \frac{1}{6})$, $(\frac{1}{6}, \frac{2}{3}, \frac{1}{6})$, $(\frac{1}{6}, \frac{1}{6}, \frac{2}{3})$ points then the elastic potential interpolation method gives results that have the same accuracy as the results obtained with the strain interpolation and the traditional method: 0.1 per cent error in the maximum displacement, 1.5 per cent error in σ_{max}, and 2 per cent error in τ_{max}.

The example also brings out the correspondence between the elastic potential interpolation method and the method of reduced order of integration with a judicious choice of integration points [2]. The elastic potential interpolation method stresses more clearly, however, the role of the finite dimensional deformation space of the individual finite elements.

REFERENCES

1. C. Truesdell and R. Toupin, 'The classical field theories', in *Encyclopedia of Physics*, Vol. III/1, p. 596, Springer-Verlag, Berlin, Göttingen, Heidelberg, 1960.
2. O. C. Zienkiewicz, *The Finite Element Method*, 3rd ed., McGraw-Hill, London, 1977.
3. J. F. Besseling, 'Finite element methods: Trends in solid mechanics', *Proc. Symp. Dedicated to 65th Birthday of W. T. Koiter*, Delft University Press, Sijthoff and Noordhoff Int. Publ., 1979.
4. J. F. Besseling, 'Another look at the application of the principle of virtual work with particular reference to finite plate and shell elements', *Proc. European–USA Workshop on Finite Elements*, Bochum, July 1980, (to be published).
5. J. F. Besseling, 'Post-buckling and nonlinear analysis by the finite element as a supplement to a linear analysis', *ZAMM*, **55,** T3–T15 (1975).

Hybrid and Mixed Finite Element Methods
Edited by S. N. Atluri, R. H. Gallagher, and O. C. Zienkiewicz

Chapter 14

Hybrid Approximations of Non-Linear Plate-Bending Problems

F. Brezzi

14.1 INTRODUCTION

The assumed stresses hybrid methods for elasticity problems and in particular for plate-bending problems were first introduced by Pian [1] and proved to be very successful in a number of applications. Nevertheless, the mathematical analysis of the method was carried out much later; the relationships of the assumed stresses hybrid method with other non-standard variational principles were studied in [2, 3] and the asymptotic error estimates were proved in [4] for second-order problems and in [5, 6] for fourth-order problems. Afterwards, in [7], the methods itself was reinterpreted from a different point of view, namely as a 'displacement method with projection'. Based on this idea, a slight modification of the method was introduced in [8] in order to have a simpler treatment of the non-linearity in the von Kármán plate equations, and the convergence of the method was proved in a neighbourhood of an isolated solution.

The aim of this paper is to extend the results of [8] to the case of critical points such as normal limit points and simple bifurcation points. For this we shall make a wide use of the general theory of [9 to 11]; we shall show that such a theory applies, and estimate the abstract error bounds in the present case. As a byproduct we show that our modified method provides optimal error bounds for the linear eigenvalue problem (bifurcation from the trivial branch) even in the case of a piecewise linear approximation of the stress field (while the classical one requires at least piecewise quadratics [12]).

An outline of the paper is the following. In Section 14.2 we introduce the modified assumed stresses hybrid method on a linear model problem and we prove optimal error bounds; the method is then applied to the non-linear von Kármán plate equations. In Section 14.3 we recall the general abstract results of [9 to 11]. Finally, in Section 14.4 we evaluate the terms appearing in the abstract error bounds when applied to the present situation; we derive optimal error bounds for non-degenerated turning points and simple bifurcation points.

For the sake of simplicity, we shall deal here with the 'lowest degree' case, i.e. cubic interelement displacements, linear interelement normal derivatives, and linear stresses within each element. However, the 'higher degree' cases (see [6] for linear problems) so not seem to be a straightforward generalization of the present analysis.

Throughout the paper the following notations and conventions will be used. The symbol $\| \|_{s,D}$ will denote the usual Sobolev norm of order s over the domain D [13]; the same symbol will also be used for norms defined on pairs of functions or symmetric tensors. When unnecessary, the indication D of the domain will be dropped and $\| \|$ will be used instead of $\| \|_0$. The symbol $(,)$ will denote an L^2-type inner product for functions, pairs of functions, or symmetric tensors. The duality pairing between a distributional space and its dual will also be denoted by $(,)$ or, sometimes, by an integral. The convention of summation of repeated indices will constantly be used. We also set:

$$\phi_{/i} = \frac{\partial \phi}{\partial x_i}; \qquad \phi_{/n} = \frac{\partial \phi}{\partial n}; \qquad \phi_{/ij} = \frac{\partial^2 \phi}{\partial x_i \, \partial x_j} \qquad i, j = 1, 2$$

$$D_2\phi = \begin{pmatrix} \phi_{/11} & \phi_{/12} \\ \phi_{/21} & \phi_{/22} \end{pmatrix} \tag{14.1}$$

The letter γ will always denote a constant which depends only on the minimum angle in the decomposition; its optimum value will obviously change from one place to another.

14.2 MODIFIED HYBRID APPROXIMATIONS OF THE VON KÁRMÁN EQUATIONS

Let Ω be (for the sake of simplicity) a convex polygon that we assume to be the portion of the (x_1, x_2) plane occupied by a homogeneous, isothropic thin plate when no forces are acting on it. Let $w^1(x)$ be the Airy function of the inplane displacements and $w^2(x)$ the transversal displacement. The von Kármán non-linear plate-bending equations can be written as follows:

$$\begin{gathered} \Delta^2 w + N(w, \lambda, \mu) = 0 \\ \text{Boundary conditions} \end{gathered} \tag{14.2}$$

where $w = (w^1, w^2)$ and the non-linear term $N(w, \lambda, \mu) = (N^1, N^2)$ is given by

$$\begin{aligned} N^1(w, \lambda, \mu) &= \tfrac{1}{2}[w^2, w^2] \\ N^2(w, \lambda, \mu) &= -[w^1 + \mu \bar{w}^1, w^2] + \lambda p \end{aligned} \tag{14.3}$$

with $[\phi, \psi]$ defined by

$$[\phi, \psi] = \phi_{/11}\psi_{/22} + \phi_{/22}\psi_{/11} - 2\phi_{/12}\psi_{/12} \tag{14.4}$$

Here λ, μ are real parameters and $\lambda p(x)$ is the assumed transversal load; $\bar{w}^1$ is a given function which takes care of the forces acting on the lateral surface of the plate and of the boundary conditions (we assume, for the sake of simplicity, that the plate is clamped along the entire boundary) so that we may also assume that

$$w^i = \frac{\partial w^i}{\partial n} = 0 \text{ on } \partial\Omega, \qquad i = 1, 2 \tag{14.5}$$

Let now $\{\mathcal{T}_h\}_h$ be a regular family of triangulations of Ω (see, for example, [13]). For any given $\mathcal{T}_h$ we consider the space W_h of C^1 functions ϕ which: (a) are cubic along each side; (b) have linear normal derivatives on each side; and (c) satisfy the boundary conditions $\phi = \phi_{/n} = 0$ on $\partial\Omega$. In order to make W_h finite dimensional we also assume: (d) the functions ϕ of W_h satisfy $\Delta^2\phi = 0$ in each triangle (although such a condition will never be used in practice, since all the computations will be based on the value of ϕ and grad ϕ on the interelement boundaries).

We also introduce the space S_h of symmetric tensors σ_{ij} $(i, j = 1, 2)$ which are linear inside each element (and indepently assumed on each element).

Let Π_h be the projection operator on S_h, defined by

$$\sigma_h = \Pi_h \sigma \qquad \text{iff } (\Pi_h\sigma, \tau) = (\sigma, \tau)\ \forall\, \tau \in S_h \tag{14.6}$$

and let D_2 be defined by (14.1). The following formula can easily be checked, integrating by parts:

$$(D_2\phi, \tau) = \sum_{K\in\mathcal{T}_h} \left[\int_{\partial K} (\phi_{/i}\tau_{ij}n_j - \phi\tau_{ij/j}n_i)\, \mathrm{ds} + \int_K \tau_{ij/ij}\phi\, \mathrm{dx} \right] \tag{14.7}$$

so that $\Pi_h\phi$ can be computed from the knowledge of ϕ and grad ϕ at the interelement boundaries only.

The following lemma is proved in [5].

Lemma 14.1 *There exists a positive constant $\gamma > 0$, independently of h, such that*

$$\|\Pi_h D_2\phi\| \geq \gamma\, \|D_2\phi\| \qquad \text{for all } \phi \in W_h. \tag{14.8}$$

For the sake of simplicity we define first our modified hybrid method on the linear model problem

$$\begin{aligned} \Delta^2\psi &= f \text{ in } \Omega \\ \psi &= \psi_{/n} = 0 \text{ on } \partial\Omega \end{aligned} \tag{14.9}$$

More precisely we define, for a given f in $H^{-1}(\Omega)$, $T_hf = (\psi^h, \sigma^h)$ as follows:

$$\begin{aligned} &\psi^h \text{ is the unique solution in } W_h \text{ of} \\ &(\Pi_h D_2\psi^h, \Pi_h D_2\phi = (f, \phi^I)\ \forall\ \phi \in W_h \end{aligned} \tag{14.10}$$

$$\sigma^h = \Pi_h D_2 \psi^h \tag{14.11}$$

where ϕ^I is the piecewise linear continuous function which interpolates ϕ at the vertices of the decomposition.

Remark 14.1 It is easy to check that the stiffness matrix originated by (14.10) and (14.11) coincides exactly with the one originated by the classical assumed stresses hybrid method. However, the right-hand side is different in the two cases and the computation of the one in (14.10), (14.11) seems easier; in fact our formulation does not need an *a priori* knowledge of a particular solution $\bar{\sigma}$ of the equilibrium equation $\bar{\sigma}_{ij/ij}=f$ which can be a source of difficulties in the original formulation of [1] (see also [14, 15, and 2]; for a different technique in order to avoid such difficulties within the original formulation see [16, 17]).

The following lemma will be used in the sequel.

Lemma 14.2 *Let $\psi \in H_0^2(\Omega)$ with $\Delta^2\psi \in H^{-1}(\Omega)$ and let $\psi(h)$ be the unique function in $H_0^2(\Omega)$ that satisfies*

$$\begin{aligned} &\psi(h)=\psi, \qquad \psi_{/i}(h)=\psi_{/i}\ (i=1,2)\ \text{on}\ \partial K \\ &\Delta^2\psi(h)=0\ \text{in}\ K \end{aligned} \tag{14.12}$$

for each $K \in \mathcal{T}_h$. Then:

$$\|D_2(\psi-\psi(h))\| \leqslant \gamma\,|h|\,\|\Delta^2\psi\|_{-1} \tag{14.13}$$

where $|h|$ is the biggest diameter of the elements of $\mathcal{T}_h$.

Proof We have, for any K in $\mathcal{T}_h$,

$$\begin{aligned} \|D_2\psi - D_2\psi(h)\|_{0,K}^2 &= \int_K [\psi-\omega(h)]\Delta^2[\psi-\psi(h)]\,\mathrm{d}x \\ &= \int_K [\psi-\psi(h)]\Delta^2\psi\,\mathrm{d}x \end{aligned} \tag{14.14}$$

On the other hand, using the Poincaré–Friedrichs inequality (compare, for example, [13]) on the master element $\hat{K}$, we have

$$\begin{aligned} \|\psi-\psi(h)\|_{1,K} &\leqslant \gamma\,\|\hat{\psi}-\hat{\psi}(h)\|_{1,\hat{K}} \\ &\leqslant \gamma\,\|D_2\hat{\psi}-D_2\hat{\psi}(h)\|_{0,\hat{K}} \\ &\leqslant \gamma\,|h|\,\|D_2\psi-D_2\psi(h)\|_{0,K} \end{aligned} \tag{14.15}$$

so that

$$\|\psi-\psi(h)\|_{1,\Omega} \leqslant \gamma\,|h|\,\|D_2\psi-D_2\psi(h)\|_{0,\Omega} \tag{14.16}$$

Combining (14.14) and (14.16) we get

$$\begin{aligned}\|D_2\psi-D_2\psi(h)\|_{0,\Omega}^2&=\sum_K\int_K[\psi-\psi(h)]\Delta^2\psi\,\mathrm{d}x\\&=\int_\Omega[\psi-\psi(h)]\Delta^2\psi\,\mathrm{d}x\\&\leqslant\|\psi-\psi(h)\|_1\,\|\Delta^2\psi\|_{-1}\\&\leqslant\gamma\,|h|\,\|D_2\psi-D\psi(h)\|_{0,\Omega}\,\|\Delta^2\psi^{\cdot}\|_{-1}\end{aligned}\tag{14.17}$$

and (14.13) follows.

Remark 14.2 If $\Delta^2\psi\in L^2(\Omega)$ we would have, with the same proof,

$$\|D_2\psi-D_2\psi(h)\|\leqslant\gamma\,|h|^2\,\|\Delta^2\psi\|_0\tag{14.18}$$

Lemma 14.3 *If* $\psi\in H^3(\Omega)\cap H_0^2(\Omega)$ *then:*

$$\operatorname*{Inf}_{\phi\in W_h}\|D_2\psi-D_2\phi\|\leqslant|h|\,\|\psi\|_3\gamma\tag{14.19}$$

Proof Let ϕ_h be the interpolant of ψ in W_h and let ϕ_c be the interpolant of ψ by means of the reduced Hsieh–Clough–Tocher element (see, for instance, [13]). Clearly ϕ_h and ϕ_c coincides (with their first derivatives) at the interelement boundaries. The minimum potential energy principle, applied to each K, yields

$$\|D_2\psi(h)-D_2\phi_h\|_{0,K}\leqslant\|D_2\psi-D_2\phi_c\|_{0,K}^2\tag{14.20}$$

(remember that $\Delta^2(\psi(h)-\phi_h)=0$ in K!) so that:

$$\|D_2\psi-D_2\phi_h\|\leqslant\|D_2\psi-D_2\psi(h)\|+\|D_2\psi-D_2\phi_c\|\leqslant\gamma\,|h|\,\|\psi\|_3\tag{14.21}$$

Theorem 14.1 *Let* ψ *be the solution of* (14.9), *let* $\sigma=D_2\psi$, *and let* $(\psi^h,\sigma^h)=T_hf$ *as defined in* (14.10), (14.11). *We have*

$$\|D_2\psi-D_2\psi^h\|+\|\sigma-\sigma^h\|\leqslant\gamma\,|h|\,\|f\|_{-1}\tag{14.22}$$

Proof Let $W=H_0^2(\Omega)$, $S=(L^2(\Omega))_s^4=\{\tau\mid\tau_{ij}\in L^2(\Omega),\ \tau_{12}=\tau_{21}\}$. We remark that (ψ,σ) and (ψ^h,σ^h) are the solutions of the problem

$$\begin{aligned}(\sigma,\tau)&=(D_2\psi,\tau)\qquad\text{for all }\tau\in S\\(\sigma,D_2\phi)&=(f,\phi)\qquad\text{for all }\phi\in W\end{aligned}\tag{14.23}$$

and of the problem

$$\begin{aligned}(\sigma^h,\tau)&=(D_2\psi^h,\tau)\qquad\text{for all }\tau\in S_h\\(\sigma^h,D_2\phi)&=(f,\phi^I)\qquad\text{for all }\phi\in W_h\end{aligned}\tag{14.24}$$

respectively. Lemma 14.1 enables us to apply the general theory of [18], which gives

$$\|D_2 \quad - D_2\psi^h\| + \|\sigma - \sigma^h\| \leqslant \gamma \left\{ \operatorname*{Inf}_{\phi \in W_h} \|D_2\psi - D_2\phi\| + \operatorname*{Inf}_{\tau \in S_h} \|\sigma - \tau\| \right.$$

$$\left. + \sup_{\phi \in W_h} \|D_2\phi\|^{-1} (f, \phi - \phi^I) \right\}$$

$$\leqslant \gamma\,|h|\,\|\psi\|_3 + \gamma\,|h|\,\|\sigma\|_1 + \gamma\,|h|\,\|f\|_{-1} \leqslant \gamma\,|h|\,\|f\|_\infty \tag{14.25}$$

We may now introduce the modified hybrid approximation to the von Kármán equations.

We define first the space $V = [H_0^2(\Omega) \times (L^2(\Omega))_s^4]^2$; the current element of V will be denoted by $v = [\{(\psi^1, \tau^1), (\psi^2, \tau^2)\}$ with $\psi^i \in H_0^2(\Omega)$ and $\tau^i \in [L^2(\Omega)]_s^4$. We then set $X = (L^1(\Omega))^2$ and define, for $v \in V$ and λ, μ real:

$$G(v, \lambda, \mu) = (\tfrac{1}{2}[\tau^2, \tau^2], -[\tau^1 + \mu\bar{\tau}^1, \tau^2] + \lambda p) \tag{14.26}$$

where $\bar{\tau}^1 = D_2\bar{w}^1$ is a given tensor and $[\sigma, \tau] = \sigma_{11}\tau_{22} + \sigma_{22}\tau_{11} - 2\sigma_{12}\tau_{12}$. It is easy to check that G is a C^∞ mapping from $V \times R \times R$ into X. We finally set, for a general $f = (f^1, f^2) \in X : v = Tf$ iff $\Delta^2\psi^i = f^i$ and $\tau^i = D_2\psi^i$ $(i = 1, 2)$. In this framework the von Kármán equations (14.2) can be written

$$u + TG(u, \lambda, \mu) = 0 \tag{14.27}$$

for $u = \{(w^1, \sigma^1), (w^2, \sigma^2)\}$ and $\sigma^i = D_2w^i$. We set now, for $f = (f^1, f^2) \in X$,

$$T_hf = v_h \equiv \{(\psi_h^1, \tau_h^1), (\psi_h^2, \tau_h^2)\} \qquad \text{iff } (\psi_h^i, \tau_h^i) = T_hf^i \ (i = 1, 2) \tag{14.28}$$

where T_h is defined by (14.9), (14.10). The discrete problem will now be written, in compact form,

$$u_h + T_hG(u_h, \lambda, \mu) = 0 \tag{14.29}$$

with $u_h = \{(w_h^1, \sigma_h^1), (w_h^2, \sigma_h^2)$ and $\sigma_h^i = \Pi_hD_2w_h^i$ $(i = 1, 2)$.

In the following, we shall use, as much as possible, the compact formulation (14.29). However, it is better, for the sake of clarity, to write the discrete equations once in the expanded form; we shall do that now, summarizing, at the same time, the whole discretization.

14.1.1 Summary of the discrete formulation

We look for $w_h^1(x)$, $w_h^2(x)$ in W_h ($\equiv$cubic at the interelement boundaries with linear normal derivatives and vanishing on $\partial\Omega$ with their gradients) and for $\sigma_h^1(x)$, $\sigma_h^2(x)$ in S_h ($\equiv$piecewise linear discontinuous symmetric tensors) such that

$$\int_{\Omega} (\sigma_h^1)_{ij}\varphi_{/ij}\, dx = -\tfrac{1}{2}\int_{\Omega} [\sigma_h^2, \sigma_h^2]\phi^I\, dx \qquad \text{for all } \phi \in W_h \tag{14.30}$$

$$\int \tau_{ij}(w_h^1)_{/ij}\, dx = \int (\sigma_h^1)_{ij}\tau_{ij}\, dx \qquad \text{for all } \tau \in S_h \tag{14.31}$$

$$\int_{\Omega} (\sigma_h^2)_{ij}\phi_{/ij}\, dx = \int_{\Omega} ([\sigma_h^1 + \mu\bar{\tau}^1, \sigma_h^2] + \lambda p)\phi^I\, dx \qquad \text{for all } \phi \in W_h \tag{14.32}$$

$$\int_{\Omega} \tau_{ij}(w_h^2)_{/ij}\, dx = \int_{\Omega} (\sigma_h^2)_{ij}\tau_{ij}\, dx \qquad \text{for all } \tau \in S_h \tag{14.33}$$

We recall that ϕ^I is the piecewise linear continuous interpolant of ϕ and that the integrals on the left-hand side of (14.30) to (14.33) are computed as boundary integrals on ∂K via formula (14.7). We also recall that $\bar{\tau}^1$ and p are given, that λ and μ are real parameters, and that $[\sigma, \tau] = \sigma_{11}\tau_{22} + \sigma_{22}\tau_{11} - 2\sigma_{12}\tau_{12}$.

14.3 ABSTRACT ERROR BOUNDS

We recall here the abstract error bounds of [9, 10, 11] for finite dimensional approximations of non-linear problems. In view of the application to (14.27) and (14.29), we shall present such results in a simplified form, which will be enough to deal with our case. The precise definitions of 'regular branch' 'non-degenerated turning point', and 'simple bifurcation point' can be found, for instance, in [9, 10, 11] together with the proofs of the following statements.

Assume that we are given two Banach spaces V and X, a C^∞ mapping $G(v, \lambda, \mu)$ from $V \times R \times R$ into X, and a linear compact operator T from X back into V. Assume in addition that we are given a family $\{T_h\}_h$ of linear operators from X into V such that

$$\lim_{h\to 0} \|T - T_h\|_{\mathscr{L}(X,V)} = 0 \tag{14.34}$$

and consider the continuous abstract problem

$$u + TG(u, \lambda, \mu) = 0 \tag{14.35}$$

and its approximation

$$u + T_hG(u, \lambda, \mu) = 0 \tag{14.36}$$

We remark that, *a priori*, the solution of (14.36) is sought in V, although any solution of (14.36) must belong to the range of T_h.

Theorem 14.2 *Let, for fixed $\mu = \bar{\mu}$, $u = u(\lambda)$ be a regular branch of solutions of (14.35) for $\lambda \in [\lambda_1, \lambda_2]$. Then there exist $h_0 > 0$ and $\rho > 0$ such that, for any $h \leqslant h_0$, (14.36) has a unique branch $u_h = u_h(\lambda)$, $\lambda \in [\lambda_1, \lambda_2]$, which belongs to the 'tube': $\{v \in V$ s.t. $\|v - u(\lambda)\|_V \leqslant \rho$ for some $\lambda \in [\lambda_1, \lambda_2]\}$. Moreover, $u_h(\lambda)$ satisfies, for m integer $\geqslant 0$,*

$$\left\| \frac{\mathrm{d}^m}{\mathrm{d}\lambda} (u(\lambda) - u_h(\lambda)) \right\|_V \leqslant \gamma_m \sum_{l=0}^{m} \left\| \frac{\mathrm{d}^l}{\mathrm{d}\lambda^l} (T - T_h) G(u(\lambda), \lambda, \bar{\mu}) \right\|_V \tag{14.37}$$

uniformly in $\lambda \in [\lambda_1, \lambda_2]$.

Remark 14.3 A similar statement obviously holds, interchanging the roles of λ and μ.

Theorem 14.3 *Let, for fixed $\mu = \bar{\mu}$, (u_0, λ_0) be a non-degenerated turning point of the problem*

$$u + TG(u, \lambda, \bar{\mu}) = 0 \tag{14.38}$$

in the variables (u, λ). Let $(u(t), \lambda(t), \bar{\mu})$, $|t| \leqslant t_0$ be a parametrization of the unique branch of solutions of (14.38) through (u_0, λ_0). There exist $h_0 > 0$ and $\rho > 0$ such that for all $h \leqslant h_0$ there is a unique branch $(u_h(t), \lambda_h(t))$, $|t| \leqslant t_0$, of solutions of

$$u_h + T_h G(u_h, \lambda_h, \bar{\mu}) = 0 \tag{14.39}$$

in the tube $\{(v, \lambda) \in V \times R$ s.t. $\|v - u(t)\|_V + |\lambda - \lambda(t)| < \rho$ for some $t \in [-t_0, t_0]\}$. Moreover, we have, for any integer $m \geqslant 0$:

$$\left\| \frac{\mathrm{d}^m}{\mathrm{d}t^m} (u(t) - u_h(t)) \right\|_V + \left| \frac{\mathrm{d}^m}{\mathrm{d}t^m} (\lambda(t) - \lambda_h(t)) \right|$$
$$\leqslant \gamma_m \sum_{l=0}^{m} \left\| \frac{\mathrm{d}^l}{\mathrm{d}t^l} (T - T_h) G(u(t), \lambda(t), \bar{\mu}) \right\|_V \tag{14.40}$$

uniformly in $|t| \leq t_0$. Finally, on the branch $(u_h(t), \lambda_h(t))$ there is a unique non-degenerated turning point u_h^c, λ_h^c and we have:

$$\|u_h^c - u_0\|_V \leqslant \gamma \sum_{l=0}^{1} \left\| \frac{\mathrm{d}^l}{\mathrm{d}t} (T - T_h) G(u(t), \lambda(t), \bar{\mu})_{|t=0} \right\|_V \tag{14.41}$$

$$|\lambda_h^c - \lambda_0| \leqslant \gamma\{\|u_h^c - u_0\|_V^2 + \langle (T - T_h) G(u_0, \lambda_0, \bar{\mu}), \phi^* \rangle$$
$$+ \|[(T - T_h) D_u G(u_0, \lambda_0, \bar{\mu})]^* \phi^*\|_{V^*} \|u_h^c - u_0\|_V\} \tag{14.42}$$

where $\phi^* \in V^*$ is the eigenvector of $(I + TD_u G(u_0, \lambda_0, \bar{\mu}))^*$ corresponding to the eigenvalue zero.

Theorem 14.4 *Let, for fixed $\lambda = \bar{\lambda}$, (u_0, μ_0) be a simple bifurcation point of the problem*

$$u + TG(u, \bar{\lambda}, \mu) = 0 \tag{14.43}$$

in the variables (u, μ) *and let* $(u_1(t), \mu_1(t))$ *and* $(u_2(t), \mu_2(t))$ *be the parametrizations, for* $|t| \leq t_0$, *of the two branches of solutions of* (14.43) *crossing at* (u_0, μ_0). *There exist* $h_0 > 0$ *and* $\rho > 0$ *such that for all* $h \leq h_0$, *the set* $\mathcal{S}_h^\rho$ *of solutions* (u_h, μ_h) *of*

$$u_h + T_h G(u_h, \bar{\lambda}, \mu_h) = 0 \tag{14.44}$$

in the sphere $S_\rho = \{(v, \mu) \in V \times R \text{ s.t. } \|v - u_0\|_V + |\mu - \mu_0| \leq \rho\}$ *is composed of two* C^∞ *branches. These two branches, in general, do not intersect. the distance between* $\mathcal{S}_h^\rho$ *and* $\mathcal{S}^\rho = \{\text{set of solutions of (14.43) in } S_\rho\}$ *is bounded by*

$$\mathcal{D}(\mathcal{S}_h^\rho, \mathcal{S}^\rho) \leq \gamma \sup_{|t| \leq t_0} \sum_{i=1}^{2} \sum_{l=0}^{1} \left\| \frac{\mathrm{d}^l}{\mathrm{d}t^l} (T - T_h) G(u_i(t), \bar{\lambda}, \mu_i(t)) \right\|_V^2 + d(h) \tag{14.45}$$

where, in its turn, $d(h)$ *is bounded by*

$$|d(h)| \leq \gamma \{ |\langle (T - T_h) G(u_0, \bar{\lambda}, \mu_0), \phi^* \rangle| \, \|(T - T_h) G(u_0, \bar{\lambda}, \mu_0)\|_V \times \|[(T - T_h) D_u G(u_0, \bar{\lambda}, \mu_0)]^* \phi^*\|_{V^*} \} \tag{14.46}$$

Theorem 14.5 *In the same hypotheses of Theorem* 14.4, *assume that* $u_0 = 0$ *and that problem* (14.43) *has a trivial branch of solutions* $u_1(t) \equiv 0$. *Then the approximate problem* (14.44) *has a trivial branch of solutions* $u_1^h(t) \equiv 0$ *and simple bifurcation point* $(0, \mu_h^c)$ *with*

$$|\mu_0 - \mu_h^c| \leq \gamma \{ |\langle (T - T_h) D_u G(0, \bar{\lambda}, \mu_0) \phi, \phi^* \rangle + \|(T - T_h) G(0, \bar{\lambda}, \mu_0)\|_V \, \|[(T - T_h) D_u G(0, \bar{\lambda}, \mu_0)]^* \phi^*\|_{V^*} \tag{14.47}$$

where $\phi \in V$ *is the eigenvector of the operator* $I + D_u G(0, \bar{\lambda}, \mu_0)$ *corresponding to the eigenvalue zero. Moreover, the other branch* $u_2^h(t)$, $\mu_2^h(t)$ *of solutions of* (14.44) *satisfies*

$$\left\| \frac{\mathrm{d}^m}{\mathrm{d}t^m} (u_2(t) - u_2^h(t)) \right\|_V + \left| \frac{\mathrm{d}^m}{\mathrm{d}t^m} (\mu_2(t) - \mu_2^h(t)) \right| \leq \gamma_m \sum_{l=0}^{m+1} \left\| \frac{\mathrm{d}^l}{\mathrm{d}t^l} (T - T_h) G(u_2(t), \bar{\lambda}, \mu_2(t)) \right\|_V \tag{14.48}$$

for any integer $m \geq 0$, *uniformly in* $|t| \leq t_0$.

14.4 ERROR BOUNDS FOR THE MODIFIED HYBRID APPROXIMATIONS OF THE VON KÁRMÁN EQUATIONS

We want now to apply the abstract error bounds of Section 14.3 to our example (14.27), (14.29) defined in Section 14.2. We remark first that Theorem 14.1 implies

$$\|(T - T_h) f\|_V \leq \gamma \, |h| \, \|Tf\|_v \tag{14.49}$$

for any $f \in X = (L^1(\Omega))^2$ such that $Tf \in v = [H^3(\Omega) \times (H^1(\Omega))^4_s]^2$. We recall that for $v = \{((\psi^1, \tau^1), (\psi^2, \tau^2)\}$ we have

$$\|v\|_V = (\|D_2\psi^1\|_0^2 + \|\tau^1\|_0^2 + \|D_2\psi^2\|_0^2 + \|\tau^2\|^2)^{1/2}$$

We remark now that if $f = G(u_0, \lambda_0, \mu_0)$ and (u_0, λ_0, μ_0) is a solution of (14.27), then $Tf = -u_0$; similarly, if $u(t), \lambda(t), \mu(t)$ is a branch of solutions of (14.27), and if $f = (d^l/dt^l)G(u(t), \lambda(t), \mu(t))$, then $Tf = -(d^l/dt^l)u(t)$. Therefore, formulae (14.37), (14.40), (14.41), and (14.48) provide at once (using (14.49)) error estimates of optimal type (that is $O(|h|)$) for regular branches, non-degenerated turning points, critical solutions of snap-through type, and bifurcation from the trivial branch.

On the other hand, formulae (14.42), (14.45), and (14.47) need some extra work. More precisely we are going to prove that, in our case:

$$|\langle (T - T_h)G(u_0, \lambda_0, \mu_0, \phi^*\rangle| \leq \gamma\, |h|^2 \tag{14.50}$$

$$\|[(T - T_h)D_uG(u_0, \lambda_0, \mu_0)]\phi^*\|_{V^*} \leq \gamma\, |h| \tag{14.51}$$

and

$$|\langle (T - T_h)D_uG(u_0, \lambda_0, \mu_0)\phi, \phi^*\rangle| \leq \gamma\, |h|^2 \tag{14.52}$$

This will provide: (a) a 'double order' estimate $O(|h|^2)$ for the critical load parameter at a non-degenerated turning point (snap-through point); (b) an optimal order estimate $O(|h|)$ in the case of a general simple bifurcation point; (c) a 'double order' estimate $O(|h|^2)$ for the critical parameter μ (of the boundary forces) at a simple bifurcation from the trivial branch (which implies an optimal $O(|h|^2)$ estimate for the eigenvalue problem). Therefore our aim is now to prove (14.50) to (14.52). In order to do that, we need some more information about the eigenvalues ϕ, ϕ^*. Assume that (u_0, λ_0, μ_0) is a critical point of (14.27) with

$$u_0 = \{(w_0^1, \sigma_0^1), (w_0^2, \sigma_0^2)\} \tag{14.53}$$

Following [11] we denote by $\chi \in (H_0^2(\Omega))^2$ an eigenfunction of the linearized von Kármán operator

$$\chi = (\chi^1, \chi^2) \to (\Delta^2\chi^1 + [w_0^2, \chi^2], \Delta^2\chi^2 - [w_0^2, \chi^1] - [w_0^1 + \mu_0\bar{w}^1, \chi^2]) \tag{14.54}$$

corresponding to the zero eigenvalue. Similarly, we denote by $\chi_* \in (H_0^2(\Omega))^2$ an eigenfunction of the formal adjoint operator corresponding to the zero eigenvalue; this means that

$$\Delta^2\chi_* + D_2^*\Lambda(\chi_*) = 0 \tag{14.55}$$

where

$$\Lambda(\zeta) = (\zeta^2 A\sigma_0^2, \chi^1 A\sigma_0^2 - \chi^2 A(\sigma_0^1 + \mu_0\bar{\tau}^1)) \tag{14.56}$$

$$A\tau = \begin{pmatrix} \tau_{22} & -\tau_{12} \\ -\tau_{12} & \tau_{11} \end{pmatrix} \tag{14.57}$$

and

$$D_2^*\tau = (\tau^1_{ij/ij}, \tau^2_{ij/ij}) \tag{14.58}$$

The following lemma is proved in [11].

Lemma 14.4 *We have (up to a normalizing factor):*

$$\phi = \{(\chi^1, \omega^2), (\chi^2, \omega^2)\} \tag{14.59}$$

with χ *given by* (14.54) *and* $\omega^i = D_2\chi^i$ $(i = 1, 2)$;

$$\phi^* = \{(0, \eta^1), (0, \eta^2)\} \tag{14.60}$$

with $\eta = -\Lambda(\chi_*)$ *and* χ_* *given by* (14.55).

We remark that, as a consequence,

$$D_2^*\eta = \Delta^2\chi_* \tag{14.61}$$

Theorem 14.6 *If the solution* w_0 *and the eigenfunctions* χ *and* χ_* *(defined by* (14.54) *and* (14.55) *respectively) belong to* $(H^3(\Omega))^2$, *if in addition* η *defined by* (14.60) *belongs to* $[(H^1(\Omega))_s^4]^2$, *then* (14.50) *to* (14.52) *hold.*

Proof Let us consider (14.50) first. We set

$$\bar{u}_h \equiv \{(\bar{w}_h^1, \bar{\sigma}_h^1), \qquad (\bar{w}_h^1, \bar{\sigma}_h^1)\} = T_h G(u_0, \lambda_0, \mu_0) \tag{14.62}$$

and also

$$\chi_h^i = \text{interpolant of } \chi_*^i \text{ in } W_h \qquad (i = 1, 2) \tag{14.63}$$

Then we have

$$\begin{aligned}
\langle (T - T_h)G(u_0, \lambda_0, \mu_0), \phi^* \rangle &= (\sigma_0^i - \bar{\sigma}_h^i, \eta^i) \\
&= (D_2 w_0^i - \Pi_h D_2 \bar{w}_h^i, \eta^i - \Pi_h \eta^i) \\
&\quad + (D_1 w_0^i - \Pi_h D_2 \bar{w}_h^i, \Pi_h \eta^i) \\
&\leq \gamma\, |h|^2\, \|w_0\|_3\, \|\eta\|_1 + (D_2 w_0^i - D_2 \bar{w}_h^i, \Pi_h \eta^i) \\
&= \gamma\, |h|^2 + (D_2 w_0^i - D_2 \bar{w}_h^i, \Pi_h \eta^i - \eta^i) \\
&\quad + (D_2 w_0^i - D_2 \bar{w}_h^i, \eta^i) \\
&\leq \gamma\, |h|^2 + (w_0^i - \bar{w}_h^i, D_2^* \eta^i) \\
&= \gamma\, |h|^2 + (w_0^i - \bar{w}_h^i, \Delta^2 \chi_*^i) \\
&= \gamma\, |h|^2 + (D_2 w_0^i - D_2 \bar{w}_h^i, D_2 \chi_*^i) \\
&= \gamma\, |h|^2 + (D_2 w_0^i - D_2 \bar{w}_h^i, D_2 \chi_*^i - D_2 \chi_h^i) \\
&\quad + (D_2 w_0^i - D_2 \bar{w}_h^i, D_2 \chi_h^i) \\
&\leq \gamma\, |h|^2 + (D_2 w_0^i - D_2 \bar{w}_h^i, D_2 \chi_h^i) \\
&= \gamma\, |h|^2 + (D_2 w_0^i, D_2 \chi_h^i) - (D_2 \bar{w}_h^i, \Pi_h D_2 \chi_h^i) \\
&\quad + (D_2 \bar{w}_h^i, (\Pi_h - I) D_2 \chi_h^i) \\
&= \gamma\, |h|^2 + (\Delta^2 w_0^i, \chi_h^i - (\chi_h^i)^I) \\
&\quad + ((I - \Pi_h) D_2 \bar{w}_h^i, (\Pi_h - I) D_2 \chi_h^i) \\
&\leq \gamma\, |h|^2 + (\Delta^2 w_0^i, \chi_h^i - (\chi_h^i)^I \\
&\quad + \|(I - \Pi_h) D_2 \bar{w}_h^i\|\, \|(I - \Pi_h) D_2 \chi_h^i\|
\end{aligned} \tag{14.64}$$

We finally remark that

$$\|(I-\Pi_h)D_2 w_h^i\| \leq \|(I-\Pi_h)D_2(w_h^i - w_0^i)\| + \|(I-\Pi_h)D_2 w_0^i\| \leq \gamma\,|h|\,\|W_0\|_3$$
$$|(I-\Pi_h)D_2 \chi_h^i\| \leq \gamma\,|h|\,\|\chi_*\|_3 \tag{14.65}$$

Similarly, $(\Delta^2 w_0^i, \chi_h^i - (\chi_h^i)^I) \leq \gamma\,\|\Delta^2 w_0\|_{-1}\,\|\chi_*\|_3$ so that:

$$|\langle (T-T_h)G(u_0, \lambda_0, \mu_0), \phi^*\rangle| \leq \gamma\,|h|^2\,(\|w_0\|_3^2 + \|\chi_*\|_3^2 + \|\eta\|_1^2) \tag{14.66}$$

In order to check (14.51) we observe that

$$\|[T-T_h)D_uG(u_0, \lambda_0, \mu_0)]^*\phi^*\|_{V^*} = \sup_{\|v\|_V=1} \langle (T-T_h)D_uG(u_0, \lambda_0, \mu_0)v, \phi^*\rangle \tag{14.67}$$

and we proceed as in the previous case. However, here for a general $v \in V$ we cannot ensure that $\|(T-T_h)D_uG(u_0, \lambda_0, \mu_0)v\|_V \leq \gamma\,|h|$ (due to the lack of regularity) so that, in the end, we get

$$|\langle (T-T_h)D_uG(u_0, \lambda_0, \mu_0)v, \phi^*\rangle| \leq \gamma\,|h|\,(\|v\|_V^2 + \|\eta\|_1^2 + \|\chi_*\|_3^2) \tag{14.68}$$

which implies (14.51). Finally, (14.52) follows from the fact that $TD_uG(u_0, \lambda_0, \mu_0)\phi = -\phi$, with the same procedure used to prove (14.50).

14.4.1 Summary of the error bounds

We may roughly summarize the results of this section as follows.

If $(u(\lambda), \lambda, \bar{\mu})$ is a regular branch of solutions of (14.27) and $u(\lambda)$ is smooth enough, we have for $m \geq 0$:

$$\left\| \frac{\mathrm{d}^m}{\mathrm{d}\lambda^m}(u(\lambda) - u_h(\lambda)) \right\|_V \leq \gamma_m |h| \tag{14.69}$$

If $(u(t), \lambda(t), \bar{\mu})$ is a branch of solutions in the neighbourhood of a 'snap-through' point (u_0, λ_0) and $u(t)$ is smooth enough, we have for $m \geq 0$:

$$\left\| \frac{\mathrm{d}^m}{\mathrm{d}t^m}(u(t) - u_h(t)) \right\|_V + \left| \frac{\mathrm{d}^m}{\mathrm{d}t^m}(\lambda(t) - \lambda_h(t)) \right| \leq \gamma_m\,|h| \tag{14.70}$$

Moreover, the approximate branch shows a 'snap-through' point u_h^c, λ_h^c with

$$|h|\,\|u_0 - u_h^c\|_V + |\lambda_0 - \lambda_h^c| \leq \gamma\,|h|^2 \tag{14.71}$$

If, for $\lambda = \bar{\lambda}$ fixed, $(u_i(t), \mu_i(t))$ $(i = 1, 2)$ are two smooth branches crossing at the simple bifurcation point (u_0, μ_0), then the set of discrete solutions consists of two smooth branches, and in a fixed neighbourhood of (u_0, μ_0) the distance of the two sets of continuous and (respectively) discrete solutions is bounded by $\gamma\,|h|$.

If finally the continuous problem has a bifurcation from the trivial branch (say $u=0$, $\lambda=0$, $\mu \in R$) for $\mu=\mu_0$, then the discrete problem bifurcates from the trivial branch at $\mu=\mu_h^c$ with

$$|\mu_0-\mu_h^c| \leqslant \gamma\,|h|^2 \tag{14.72}$$

and the distance between the continuous bifurcating branch $(u(t), \mu(t))$ and the discrete bifurcating branch $(u_h(t), \mu_h(t))$ is bounded by

$$\left\|\frac{\mathrm{d}^m}{\mathrm{d}t^m}(u(t)-u_h(t))\right\|_V + \left|\frac{\mathrm{d}^m}{\mathrm{d}t^m}(\mu(t)-\mu_h(t))\right| \leqslant \gamma\,|h| \tag{14.73}$$

for any integer $m \geqslant 0$.

We recall that here

$$u=\{w^1,\sigma^1),(w^2,\sigma^2)\} \tag{14.74}$$

with w^1 = Airy function, w^2 = transversal displacement, and $\sigma^i_{rs}=w^i_{/rs}$ $(i=1,2)$. We also recall that

$$\|u\|_V^2=\left(\sum_{i=1}^2\sum_{r=1}^2\sum_{s=1}^2 \|w_{/rs}\|_{L^2}^2+\|\sigma_{rs}\|_{L^2}^2\right)^{1/2} \tag{14.75}$$

(i.e. some kind of *energy norm*).

Remark 14.4 For a more detailed analysis of the von Kármán equations from the mathematical point of view we refer to [19]; for more references on non-standard approximations of fourth-order problems see [20].

REFERENCES

1. T. H. H. Pian, 'Element stiffness matrices for boundary compatibility and for prescribed boundary stresses', *Proc. First Conf. on Matrix Methods in Structural Mechanics*, pp. 457–477, Wright-Patterson Air Force Base, 1965, AFDL-TR-66-80, 1966.
2. B. Fraeijs de Veubeke, 'Variational principles and the patch test', *Int. J. Num. Meth. in Eng.*, **8,** 783–801 (1974).
3. G. Sander and P. Beckers, 'The influence of the choice of connectors in the finite element method', *Mathematical Aspects of Finite Element Methods*, Lecture Notes in Mathematics, No. 606, pp. 316–342, Springer, 1977.
4. F. Brezzi, 'Sur une méthode hybride pour l'approximation du problème de la torsion d'une barre élastique', *Rend. First Lombardo Sci. Lett. A.*, **108,** 274–300 (1974).
5. F. Brezzi, 'Sur la méthode des éléments finis hybrides pour le problème biharmonique', *Num. Math.*, **24,** 103–131 (1975).
6. F. Brezzi and L. D. Marini, 'On the numerical solution of plate bending problems by hybrid methods', *RAIRO*, **R3,** 5–50 (1975).
7. F. Brezzi, 'Hybrid method for fourth order elliptic equations', in *Mathematical Aspects of Finite Element Methods*, Lecture Notes in Mathematics, No. 606, pp. 35–46, Springer, 1977.
8. A. Quarteroni, 'Hybrid finite element methods for the von Kármán equations', *Calcolo*, **XVI,** 271–288 (1979).

9. F. Brezzi, J. Rappaz, and P. A. Raviart, 'Finite dimensional approximations of nonlinear problems. Part I: Branches of nonsingular solutions', *Num. Math.*, **36,** 1–25 (1980).
10. F. Brezzi, J. Rappaz, and P. A. Raviart, 'Finite dimensional approximations of nonlinear problems. Part II: Limit points', *Num. Math.* (to appear).
11. F. Brezzi , J. Rappaz, and P. A. Raviart, 'Finite dimensional approximations of nonlinear problems. Part III: Simple bifurcation points', Submitted to *Num. Math.*
12. C. Canuto, 'A hybrid FEM to compute the free vibration frequencies of a clamped plate', RAIRO (to appear).
13. P. G. Ciarlet, *The Finite Element Method for Elliptic Problems*, North-Holland, Amsterdam, 1978.
14. T. H. H. Pian and P. Tong, 'Basis of finite element methods for solid continua', *Int. J. Num. Meth. in Eng.*, **1,** 3–68 (1969).
15. T. H. H. Pian, 'Finite element formulation by variational principles with relaxed continuity requirements', in *The Mathematical Foundations of the F.E.M. with Applications to P.D.E.* (Ed. A. K. Aziz), pp. 671–687, Academic Press, New York, 1973.
16. L. D. Marini, *Implementation of Hybrid Finite Element Methods and Associated Numerical Problems, Part I,* Pubblicazione No. 136, IAN-CNR, Pavia, 1976.
17. L. D. Marini, *Implementation of Hybrid Finite Element Methods and Associated Numerical Problems, Part II,* Pubblicazione No. 182, IAN-CNR, Pavia, 1978.
18. F. Brezzi, 'On the existence uniqueness and approximation of saddle point problems arising from Lagrangian multipliers', *RAIRO,* **8-R2,** 129–151 (1974).
19. P. G. Ciarlet and P. Rabier, *Les Equations de von Kármán,* Lecture Notes in Mathematics, No. 806, Springer, Berlin, 1980.
20. F. Brezzi, 'Non standard finite elements for fourth order elliptic problems', in *Energy Methods in Finite Element Analysis* (Eds. R. Glowinsky, E. Y. Rodin, and O. C. Zienkiewicz), Chap. 10, John Wiley, Chichester.

Hybrid and Mixed Finite Element Methods
Edited by S. N. Atluri, R. H. Gallagher and O. C. Zienkiewicz

Chapter 15

Optimal Design for Torsional Rigidity

Robert Kohn and Gilbert Strang

15.1 INTRODUCTION

Mixed and hybrid methods have a special relationship to the ideas of duality and the method of Lagrange multipliers. In the classical problems of linear elasticity, duality appears in the connection between potential energy and complementary energy; one is a function of strains and the other of stresses, and both are minimized by the solution. The hybrid method came from Pian's suggestion of a subtle way to work simultaneously with stress and displacement fields, an idea which continues to be fruitful.

I believe these methods play a particularly important role in problems with constraints. Above all, that means the equations of an incompressible fluid or of incompressible elasticity; the constraint is $\operatorname{div} u = 0$ and its Lagrange multiplier is the pressure. It is a central problem in computational mathematics to find the right way to impose that constraint. However, there are other constraints of importance—from yield conditions in plasticity and from friction or non-penetration in contact problems—and those are *inequality constraints*. In this note we study still another class of constrained problems, those of optimal design, and we work out the solution for an example of some engineering interest. This is part of a more comprehensive study of *shape optimization* (or optimal redesign) which is described in [1].

Our study is still at the analytical stage, concentrating on individual problems in order to understand the characteristics of their solutions. Perhaps the most striking property is this: in addition to regions where the available material is fully used, and regions where no material is required, the optimal structure can include an area where it is a certain *fibre density* that gives maximum strength at minimum cost—certainly not minimum cost of manufacture, but minimum cost of material. This phenomenon has been noticed before; the fibres form a new kind of material with different properties from the original, and completely anisotropic. This intermediate material arises directly from the optimality conditions (the Kuhn–Tucker conditions) that identify the design of least weight.

We certainly expect these fibres to reappear in more complicated problems, where numerical methods are needed to study the solution. And we believe that duality theory will be central to the numerical analysis, as it already is in the mixed method for problems with equality constraints. The literature shows very substantial activity in structural optimization, with direct minimization and direct solution of the optimality conditions originally viewed as the chief contenders. More recently, the work of Schmit [2], Fleury [3], Pedersen [4], and very many others has brought these techniques more towards cooperation than contention. Pedersen suggests that an optimal design now requires about the same computer time as ten analyses, which represents important progress—due to a careful choice of design variables and local linearization, and to the possibility of solving exactly the dual of key subproblems [5].

15.2 THE TORSION PROBLEM

We come to a specific problem: the design of a long cylindrical rod to achieve a given torsional rigidity with minimum material. There will be only one constraint on the geometry: *the cross-section of the rod must lie within a given square.* For convenience we pick the unit square Ω. There must also be a condition on the prescribed torsional rigidity D; it cannot exceed the value $D_{\max}$ which is achieved for a fully plastic solid rod.

The computation of $D_{\max}$ is a classical problem. For an elastic rod there is a membrane analogy, since Prandtl's stress function for torsion shares the Poisson equation $\Delta\psi = -2$ that governs a membrane under uniform load. For perfect plasticity, the yield condition imposes an additional constraint on the stress $\sigma = (\sigma_{xz}, \sigma_{yz}) = (\psi_y, -\psi_x)$; note that $\operatorname{div} \sigma = 0$ from equilibrium, and with a suitable normalization of units plasticity requires $|\sigma| = (\sigma_{xz}^2 + \sigma_{yz}^2)^{1/2} \leqslant 1$, or $|\nabla\psi| \leqslant 1$. The graph of ψ in fully plastic torsion becomes a 'sandhill' with base Ω and slope 1 or a membrane forced into its extreme deflection—in our case, a square pyramid.

This solution can be determined either from σ or ψ; the two are linked by

$$\iint (x\sigma_{yz} - y\sigma_{xz})\, dx\, dy = -\iint (x\psi_x + y\psi_y)\, dx\, dy = \iint 2\psi\, dx\, dy \tag{15.1}$$

In this application of Green's formula we used the boundary condition $\sigma n = 0$ which implies that ψ is constant on the boundary Γ of the square; its tangential derivative is

$$\frac{\partial\psi}{\partial t} = \nabla\psi t = (-\sigma_2, \sigma_1)(-n_2, n_1) = \sigma n = 0$$

and we can choose $\psi = 0$ on Γ. The torsional rigidity is determined by either of the maximum principles:

$$D_{\max} = \max_{\substack{\psi=0 \text{ on } \Gamma \\ |\nabla\psi|\leq 1}} 2\iint \psi \, dx \, dy$$

$$D_{\max} = \max_{\substack{\operatorname{div} \sigma=0 \\ \sigma n=0 \\ |\sigma|\leq 1}} \iint (x\sigma_{yz} - y\sigma_{xz}) \, dx \, dy$$

From the first it is easy to see that ψ should climb as rapidly as possible as we move to the interior of the square. The rate of increase is restricted by $|\nabla\psi|\leq 1$, so that the optimal ψ at any point is exactly the distance to the boundary—and the graph of this function is a pyramid.

Now we come to the question of optimal design *to achieve a given* $D\leq D_{\max}$ *with a minimum amount of material.* In other words, we remove as much of the rod as possible; it is a problem of *redesign.* The natural way to begin is to bore a hole, or several holes, down the rod. The loss of rigidity is expected to be least when the hole is near the centre, and for a circular rod this leads directly to the optimal solution. In that case the circular pyramid (or cone) which represents the graph of $\psi_{\max}$ is truncated by a plane, so that the resulting ψ satisfies $2\iint \psi \, dx \, dy = D$. If Ω were the unit circle, and the truncation occurred at the radius $R<1$, then the optimal design is given by the hollow cylinder $R\leq r\leq 1$. We note that $\sigma n = 0$ on the unloaded boundary of the hole, implying as before that $\psi = \text{constant}$ and that formula (15.1) is correct—but we are not at liberty to set $\psi = 0$ on this boundary too. Instead the constant is automatically chosen by the maximization of $2\iint \psi \, dx \, dy$. This multiply-connected case is beautifully illustrated by the photographs of sandhills which are reproduced by Nadai [6], and the analytical solution for several holes was given by Ting [7].

In our case Ω is a square, and the shape of the hole is not at all clear. In fact, a single hole is not optimal. The best design includes not only a hole at the centre but also removal of the corners of the square, making the outer boundary more nearly (but not exactly) circular. Worse than this, even the right choice of these five holes does not give the minimal weight. It becomes possible, although more and more delicate, to maintain the rigidity D with less material by removing thin holes that are roughly quarter-circles running parallel to and inside the rounded corners. In short, the whole design is becoming unmanageable.

Our chief problem is to find the right formulation of our original question; in other words, to restate it as a conventional minimum principle. What we learn from constructing more and more holes is that the final design will partly resemble a 'Michell truss'. It will be a *continuum of fibres,* with the density varying from one point to another in Ω. Where the material is completely removed ($\sigma=0$ and $\psi=$constant) the density is zero, and where there are no holes the density is unity. Elsewhere we keep only a fraction of the material, *according to the fraction of the constraint* $|\sigma|\leqslant 1$ *which is actually needed in the final design.* In other words, it is $|\sigma|$ itself which reveals the fibre density, and the volume of material is the integral of $|\sigma|$ (or $|\nabla\psi|$). Therefore our problem becomes, first in terms of ψ and then equivalently in terms of σ,

$$\text{Minimize} \iint_\Omega |\nabla\psi| \text{ subject to } \psi=0 \text{ on } \Gamma,\ |\nabla\psi|\geqslant 1,\ \iint 2\psi\leqslant D. \tag{15.2}$$

$$\text{Minimize} \iint_\Omega |\sigma| \quad \text{subject to } \operatorname{div}\sigma=0,\ \sigma n=0 \text{ on } \Gamma,\ |\sigma|\leqslant 1,\ -\iint \sigma\mathbf{v}\geqslant D. \tag{15.3}$$

It is this minimum that we compute Ω is a square. In the statement (15.3) we introduced the vector $\mathbf{v}=(y,-x)$ in order to write more concisely the left-hand side of formula (15.1). (For an arbitrary $\mathbf{v}$ it would be $\operatorname{div}(-\mathbf{v})^\perp$ which appears in (15.2); here rotation through $\pi/2$ gives $(-\mathbf{v})^\perp=(x, y)$, and its divergence is just our constant factor 2.) In the case of a circular Ω, it should be possible to verify optimality when $\psi=1-r$ in an annulus $R\leqslant r\leqslant 1$ and $\psi=1-R$ in the hole, corresponding to a unit circumferential vector $\sigma=u_\theta$ in the annulus and $\sigma=0$ in the centre. In this case $|\sigma|=1$ or $|\sigma|=0$, and the intermediate case of a fibre density happens to be unnecessary. That is extremely exceptional, however, and for a square we must expect a region where $0<|\sigma|<1$.

To justify the formulations (15.2) and (15.3), we prove in [1] that they give the greatest lower bound on the area which remains after removing genuine holes from the rod. The optimal cylinder is the limiting case of a sequence of ordinary but increasingly complicated designs.

15.3 DUALITY AND OPTIMALITY

For a circle we could somehow establish that the hollow tube is the correct solution. For other domains Ω, the way to confirm that a proposed σ or ψ is minimal is to verify the Kuhn–Tucker conditions for optimality. Therefore our next step is to construct the dual problems, introducing a Lagrange multiplier w, and to recognize the conditions which link the minimizing σ to the maximizing w. In applications w is the *warping function.* Proceeding formally, and integrating by parts,

$$\min_{\substack{\operatorname{div}\sigma=0\\ \sigma n=1\\ |\sigma|\leqslant 1\\ -\iint\sigma\mathbf{v}\geqslant D}} = \min_{\substack{\sigma n=0\\ |\sigma|\leqslant 1}} \max_{\substack{w\\ \lambda\geqslant 0}} \iint (|\sigma|+w\operatorname{div}\sigma)+\lambda\left(D+\iint\sigma\mathbf{v}\right)$$

$$= \max_{\substack{w\\ \lambda\geqslant 0}} \min_{|\sigma|\leqslant 1} \iint (|\sigma|-\sigma\nabla w+\lambda\sigma\mathbf{v})+\lambda D$$

We can replace w by λw, since it leaves the maximization unchanged, and then the whole key is to concentrate on the inner minimization:

$$m = \min_{|\sigma|\leq 1} |\sigma| - \sigma\lambda(\nabla w - \mathbf{v})$$

The minimum occurs when the vector $\boldsymbol{\sigma}$ has the same direction as $\nabla w - \mathbf{v}$ and there are three possibilities for the magnitude of σ:

If $|\nabla w - \mathbf{v}| < 1/\lambda$ then $\sigma = 0$ and $m = 0$.

If $|\nabla w - \mathbf{v}| = 1/\lambda$ then $0 \leq |\sigma| \leq 1$ and $m = 0$.

If $|\nabla w - \mathbf{v}| > 1/\lambda$ then $|\sigma| = 1$ and $m = 1 - \lambda\,|\nabla w - \mathbf{v}|$.

These are exactly the Kuhn–Tucker conditions, distinguishing regions that are empty, fibred, or solid. Substituting the minimum value m into the maximization, we reach the dual:

$$\text{Maximize} \quad \iint (1 - \lambda\,|\nabla w - \mathbf{v}|)_{-}\,dx\,dy + \lambda D \quad \text{subject to} \quad \lambda \geq 0. \tag{15.4}$$

The subscript "$-$" indicates that we have zero when the quantity in parentheses is positive, as in the first two cases above. For large λ the contribution from the integral is more and more negative, eventually striking the right balance with the positive term λD—unless $D > D_{\max}$, in which case the maximum in the dual is $+\infty$ and the constraints in the primal were not feasible.

15.4 THE OPTIMAL DESIGN IN A SQUARE

We come now to the explicit construction of the optimal σ. There is one key observation which tells us what to look for in the fibred region, where $|\nabla w - v| = 1/\lambda = \text{constant}$. Remembering that σ is parallel to $\nabla w - \mathbf{v}$, we shall show that *the fibres lie along circular arcs of fixed radius* $r = 1/2\lambda$. These are the curves traced out by the vector field σ and coincide with the level curves $\psi = \text{constant}$. The argument is this: since $\lambda(\nabla w - \mathbf{v})$ is a unit vector $(\cos\theta, \sin\theta)$, we have

$$\lambda\left(\frac{\partial w}{\partial x} - y\right) = \cos\theta, \qquad \lambda\left(\frac{\partial w}{\partial y} + x\right) = \sin\theta$$

Differentiating with respect to y and x respectively and subtracting the first from the second, w disappears to leave

$$2\lambda = \cos\theta\,\frac{d\theta}{dx} + \sin\theta\,\frac{d\theta}{dy} = \text{curvature} \tag{15.5}$$

Since σ is everywhere parallel to $(\cos\theta, \sin\theta)$, it lies along circular arcs of

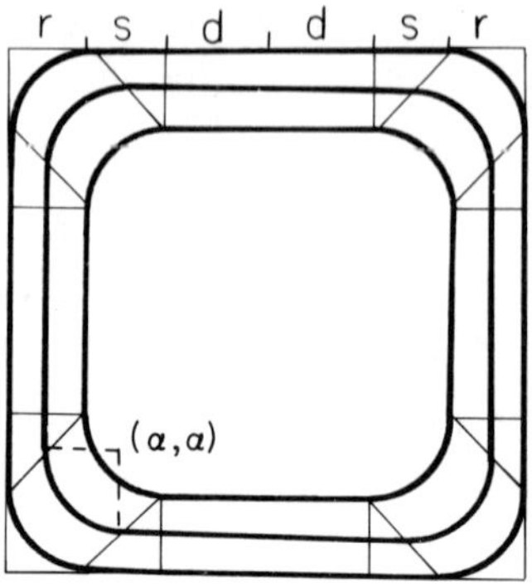

Figure 15.1 The optimal design: level curves of ψ

radius r. We can make this explicit by guessing that these arcs begin by rounding off the corners of the square and that their centres lie along the diagonal of the square (Figure 15.1). In this case the circles in the lower left corner, with centre at (α, α), are

$$(x-\alpha)^2+(y-\alpha)^2=r^2 \tag{15.6}$$

and on this circle the stress function has the constant value $\psi=\alpha-r$. The corresponding σ, not yet proved optimal, is

$$\sigma=(\psi_y, -\psi_x)=(\alpha_y, -\alpha_x) \tag{15.7}$$

Differentiating (15.6) with respect to x and also y,

$$2(x-\alpha)(1-\alpha_x)+2(y-\alpha)(-\alpha_x)=0=2(x-\alpha)(-\alpha_y)+2(y-\alpha)(1-\alpha_y)$$

Solving for α_x and α_y, (15.7) gives σ on this arc:

$$\sigma=\left(\frac{y-\alpha}{x+y-2\alpha}, \frac{\alpha-x}{x+y-2\alpha}\right) \tag{15.8}$$

Note that σ is perpendicular to the vector $(\alpha-x, \alpha-y)$ which points to the centre of the circle, and it is immediate to check that $|\sigma|\leqslant 1$. At the points $x=\alpha-r$, $y=\alpha$ and $x=\alpha$, $y=\alpha-r$, the vector $\boldsymbol{\sigma}$ becomes $(0, -1)$ and $(1, 0)$, which signals the end of the fibre and the beginning of the solid region where $|\sigma|=1$. These points are separated by exactly a quarter of a circle, as in Figure 15.1.

In the solid region we have exactly the same solution as for fully plastic torsion. The vectors σ run parallel to the boundaries of the square, the stress function ψ is the distance from the boundary, and its graph over this part of the square is the same pyramid as before. The key step is to determine how far into the centre of the square this picture can extend. In other words, we have to determine the relation of the dimensions r and d which describe the inner hole in Figure 15.1.

The key is to introduce the vector $\boldsymbol{\tau}$, produced by rotating $\nabla w - \mathbf{v}$ through an angle $\pi/2$:

$$\nabla w - \mathbf{v} = (w_x - y, w_y + x), \qquad \text{so} \quad \tau = (-w_y - x, w_x - y)$$

The divergence of τ equals -2, by direct calculation. Furthermore, the second of our three optimality conditions (the one for fibres) implies that $|\nabla w - \mathbf{v}| = 1/\lambda$, or $|\tau| = 2r$, at the boundary of any hole. Since σ runs along the edge of the hole, τ is perpendicular to it; with our construction it points inwards. Therefore the divergence theorem $\iint \operatorname{div} \tau = \iint \tau n$ in the central hole yields

$$-2(\text{area of hole}) = -2r(\text{perimeter of hole}) \tag{15.9}$$

Now we can determine the relation of r to d. The perimeter consists of four quarter-circles plus four straight edges, $2\pi r + 8d$. Computing the area in a similar way, (15.9) reduces to $d = (\sqrt{\pi}/2)r$. Together with the fact that $2r + 2d + 2s = 1$, on the outer edge of the unit square this determines d and s in terms of r. With this one remaining parameter we satisfy the original constraint of given torsional rigidity D:

$$\iint 2\psi = -\iint \sigma \mathbf{v} = D$$

The picture becomes clearer as we recognize the extreme cases in Figure 15.2:

If $D = D_{\max}$, then $d = r = 0$ and we have fully plastic torsion; the graph of ψ is a square pyramid.
If $D \to 0$, then $s \to 0$ and the optimal design is concentrated on a thin curve that runs along the centre of each side and in an arc around each corner.

There remains one final step: to show that the proposed σ (or ψ) is

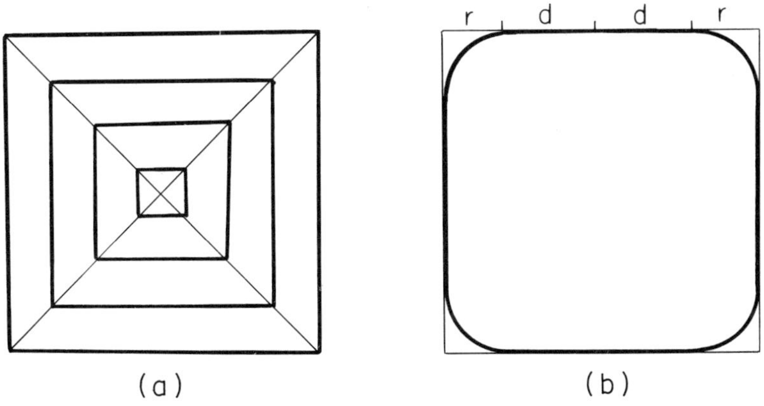

Figure 15.2 (a) Fully plastic: $D = D_{\max}$. (b) Hinge line as $D \to 0$, $s \to 0$

optimal by constructing the corresponding warping function w. For fully plastic torsion, which is the limiting case of our design, this was carried out by Mosolov [8]. In our case the construction is simple whenever there is material; σ is non-zero and $\nabla w - v$ is parallel to it. The more delicate problem is to show that w can be continued into the central hole and into the corners of the square in such a way that $|\nabla w - v| < 1/\lambda = 2/r$. By pure good fortune, this coincides with the limit analysis problem solved in [9]. In terms of $\tau = (\nabla w - v)^{\perp}$, we are required to satisfy $\operatorname{div} \tau = -2$, and the flattened circle in the centre of our design is exactly the hinge line discovered in our earlier paper. We are right at the limit load and the duality theorem of that paper establishes the existence of τ (although we have never succeeded in computing it). A similar reasoning applies to the extension into the corners, where again $\sigma = 0$ and we must achieve $|\tau| < 2/r$.

This completes our specific example, but the underlying ideas extend to a wide class of optimal designs. A second problem, for a cylinder in shear, is worked out in [10]. Beyond that there is a tremendous range of current activity in shape optimization, and we can pay tribute here only to two men, A. G. M. Mitchell and William Prager, whose contributions were sufficient to create and develop a whole subject of scientific research.

ACKNOWLEDGMENT

The authors are grateful for the support of the National Science Foundation through a postdoctoral fellowship and contract MCS 78-12363. and of the Army Research Office (DAAG29-K0033).

REFERENCES

1. R. Kohn and G. Strang, *Optimal Design and Convex Analysis* (to appear).
2. L. A. Schmit and C. Fleury, 'Structural synthesis by combining approximation concepts and dual methods', *AIAA J.*, **18,** 1252–1260 (1980).
3. C. Fleury, *Reconciliation of Mathematical Programming and Optimality Criteria Approaches to Structural Optimization*, NATO Inst. on Modern Structural Optimization, Liège, 1980 (to appear).
4. P. Pedersen, 'The integrated approach of FEM–SLP for solving problems in optimal design', DCAMM Report 182, Copenhagen, 1980.
5. C. Fleury and L. A. Schmit, 'Primal and dual methods in structural optimization', *J. Struct. Div., ASCE,* **106,** 1117–1133 (1980).
6. A. Nadai, *Theory of Flow and Fracture in Solids*, McGraw-Hill, New York, 1950.
7. T. W. Ting, 'Torsional rigidities for bars under fully plastic torsion', *SIAM J. Appl. Math.*, **25,** 54–68 (1973).
8. P. P. Mosolov, 'On the torsion of a rigid-plastic cylinder', *PMM,* **41,** 344–353 (1977).
9. G. Strang, *A Minimax Problem in Plasticity Theory*, Lecture Notes on Mathematics No. 701, Springer, 1980.
10. G. Strang and R. Kohn, 'Optimal design of cylinders in shear', *MAFELAP Conf.*, Brunel University, 1981.

Hybrid and Mixed Finite Element Methods
Edited by S. N. Atluri, R. H. Gallagher, and O. C. Zienkiewicz

Chapter 16

The Structural Stability Criterion for Mixed Principles

Roger Valid

16.1 INTRODUCTION

One of the most promising tendencies in modern methods of structural analysis consists in working mixed or mixed hybrid variational principles, which are characterized by multiple unknown fields. In effect, in the static or dynamic cases, the calculation of rather accurate stresses involves some difficulties, which are alleviated in general at the cost of expensive computations. This requirement seems to be particularly severe in all stability problems and the calculation of critical loads.

The parameters of cost and accuracy would in the future be optimized by some techniques consisting of associating mixed principles with more economical ones, say the primal principle, in some coupled or postponed methods [1 to 3]. It should be noted incidentally that some studies, both theoretical and numerical, of the various principles, in which equilibrium equations are introduced in a weak and selective form, should allow for a judicious choice.

In this perspective, it appeared necessary to extend the static stability criterion for elastic structures to the case of mixed principles. In fact, the classical structural stability criterion of Lagrange–Dirichlet has to be applied to the direct potential energy of the structure and takes a different form in the case where mixed principles are worked up with other forms of the energy.

Let us recall that the classical structural stability criterion, proposed by Lagrange in 1788 and named since that time the energy criterion, states that the necessary and sufficient condition for a discrete conservative system to be in a stable equilibrium is that its total potential energy be a local minimum in that state. If this energy is a maximum, the equilibrium is unstable.

Based on the linearization, and therefore relative to infinitely small displacements, this criterion was criticized by Dirichlet (1846)—for an asymptotic expansion anticipated on the stability—who extended the criterion to finite perturbations. Its generalization was the extension to elastic bodies. It states that the necessary and sufficient condition for an elastic body submitted to a conservative loading, in isothermal conditions, to be in a stable equilibrium state is that its total potential energy presents a weak relative minimum for virtual kinematically admissible displacements.

Even restricted to these specified conditions, the Lagrange–Dirichlet criterion remains open to numerous criticisms, the most severe of which being that it is in fact not relevant to any precise definition of the stability. (This objection was suppressed in some way, by Hadamard, who adapted to this criterion a definition *ad hoc*). Nevertheless, one will find in reference [4] a rigorous definition of the Lagrange stability as a more restrictive case of the Liapounov stability. We noted then in [5]: 'As a static test, restricted to conservative loads, and to virtual kinematically admissible displacements, it applies neither to dynamic cases, nor to the case of non-conservative loadings, or of the actual displacements, nor also to finite displacements. (It supposed besides, as a reminder, that the equilibrium is not neutral, that is to say non-isolate.)'

Admitted in the majority of the practical applications as a necessary and sufficient condition of stability of continuous systems, the necessary and sufficient condition for the energy W to reach at the equilibrium a local weak minimum is easily shown, for the discrete systems, and is expressed by the condition $\delta^2 W > 0$, where $\delta^2 W$ is the second variation of W for virtual kinematically admissible displacements $\delta U \neq 0$. But this condition, actually written in terms of Frechet's differentials, is not sufficient for arbitrary continuous systems, as can be shown on some litigious cases [6]. It is then necessary to make use of conditions of the coercivity type:

$$\delta^2 W \geqslant c\rho, \qquad c > 0$$

where ρ is a measure of the stability. We shall adopt it anyway in the sequel for discretized systems.

Let us remark at this point that, for the discretized systems, the adoption of norm as mathematical measures makes them all equivalent, but that, for a discretized continuous system, even if the measures are equivalent, they are practically different each from one another and thus introduce appreciable numerical differences, which are so much larger that the mesh is more refined and the solution converges more to the exact one [7].

16.2 NOTATIONS

It is worth noting that in the following we shall make use, for the sake of simplicity and of clarity, of a general intrinsic formulation, in which all the symbols will be precisely defined in due time. In general we shall often make a difference between a function and its value, but without any change of letter, so that if a quantity y (scalar, vector, mapping, tensor, etc.) is a function of a quantity x which belongs to a set X, we shall note

$$y = \mathbf{y}(x), \qquad x \in X$$

where $\mathbf{y}$ is the function and y its value at the point x of X. Moreover, in the case where X is a normed space and $\mathbf{y}$ a mapping which applies X in a normed space, then:

(a) $(\partial y/\partial x)$ will be the first derivative of y at a point $x \in \sigma$, an open set of X.
(b) $dy = (\partial y/\partial x)\, dx$ will be the differential vector dy relative to a differential vector field dx defined in an open set σ of X.
(c) $(\partial^2 y/\partial x\, \partial x)$ will be the second derivative of y at point x, and d^2y the second variation of y due to the first and second variations dx and d^2x of x defined in an open set σ of X, such that if

$$dx = f(x), \qquad x \in \boldsymbol{\sigma} \subset \mathbf{X}, \ \boldsymbol{f} \textbf{ differentiable in } \boldsymbol{\sigma}$$

$$d^2x = \frac{\partial f(x)}{\partial x}\, dx$$

16.3 THE STATISTIC STABILITY CRITERION FOR CONSTRAINED SYSTEMS

In the following, all the quantities in consideration will be supposed to be differentiable as many times as necessary, but the extension of the formulae in terms of distribution derivatives is straightforwrd.

We shall recall in this section the stability conditions relative to constrained systems. These conditions will be shown from free systems and in order to make a self-contained statement.

We want to search for the minimum of a scalar quantity y, a function of a variable $x \in X$, where X is a linear normed space, so:

$$\min_{x \in X} y = \tilde{y}(x) \tag{16.1}$$

The necessary stationary condition is written

$$\delta y = 0, \qquad \forall\, \delta x \in \mathbf{X} \quad \text{or} \quad \frac{\partial y}{\partial x} = 0 \tag{16.2}$$

and the necessary stability condition is

$$\delta^2 y \geq 0, \qquad \forall\, \delta x \neq 0, \delta^2 x \in \mathbf{X} \tag{16.3}$$

In fact the conditions (16.3) are written:

$$\delta^2 y = \left[\frac{\partial^2 y}{\partial x\, \partial x}\right]_0 (\delta x)(\delta x) + \left[\frac{\partial y}{\partial x}\right]_0 (\delta^2 x) \geq 0$$

where the derivatives $[\partial^2 y/\partial x\, \partial x]_0$ and $[\partial y/\partial x]_0$ are calculated at the point x_0 given by the resolution of (16.2), which gives again

$$\left[\frac{\partial^2 y}{\partial x\, \partial x}\right](\delta x)(\delta x) \geqslant 0, \qquad \forall\, \delta x \in \mathbf{X} \tag{16.4}$$

It could also be supposed, in (16.3) that $\delta x = C^{\underline{\text{te}}}$ or $\delta^2 x = 0$.

In this case of a constrained minimum, one will look in the minimum of y with the constraint

$$z = \underset{\sim}{z}(x) = 0 \in \mathbf{E}$$

$\mathbf{E}$ being a normed vectorial space. Then (16.1) becomes

$$min\; y = y(x), \qquad \text{when } x \in \{x \in X, \text{s.t.* } z = \underset{\sim}{z}(x) = 0\} \tag{16.5}$$

The necessary condition of stationarity becomes here

$$\left.\begin{aligned} z &= 0 \\ \delta y &= 0 \end{aligned}\right\} \qquad \forall\, \delta x \text{ s.t. } \delta z = 0 \tag{16.6}$$

while the stability condition (16.3) becomes

$$\delta^2 y \geqslant 0, \qquad \forall\, \delta x \neq 0, \qquad \delta^2 x \in \mathbf{X}, \text{ s.t. } \delta z = \delta^2 z = 0 \tag{16.7}$$

Before transforming the conditions (16.6) and (16.7) to make them more easily workable, it is worth noting that one could tentatively write the condition (16.4) in the form

$$\left[\frac{\partial^2 y}{\partial x\, \partial x}\right]_0 (\delta x)(\delta x) \geqslant 0, \qquad \forall\, \delta x \neq 0 \in \mathbf{X} \text{ and s.t. } \delta z = 0$$

without taking account of $\delta^2 x$ and therefore of $\delta^2 z = 0$, but this condition would be false. In fact, δx does not only have to verify that $\delta z = 0$, for, as an example, if δx meets

$$\left[\frac{\partial z}{\partial x}\right]_0 (\delta x) = 0$$

while δx is supposed constant, that is to say such that

$$\delta^2 x = 0$$

as

$$z = 0 \Rightarrow \delta z = \delta^2 z = 0, \qquad \forall\, \delta x, \delta^2 x$$

there exists between δx and $\delta^2 x$ the constraint

$$\left[\frac{\partial^2 z}{\partial x\, \partial x}\right]_0 (\delta x)(\delta x) + \left[\frac{\partial z}{\partial x}\right]_0 (\delta^2 x) = 0$$

* The abbreviation s.t. stands for 'such that'.

which then becomes

$$\left[\frac{\partial^2 z}{\partial x\,\partial x}\right]_0 (\delta x)(\delta x) = 0$$

In other words, δx has to belong only to the set of vectors which cancel this quadratic form and which are in the plane

$$\left[\frac{\partial z}{\partial x}\right]_0 (\delta x) = 0$$

Using (16.6) and (16.7), the stationary condition can be transformed classically as folows: $\exists$ a Lagrange multiplier $\lambda \in \mathbf{E}'$ such that

$$\frac{\partial y}{\partial x}\,\delta x - \lambda \frac{\partial z}{\partial x}\,\delta x = 0, \qquad \forall\, \delta x \in \mathbf{X} \qquad z = 0 \tag{16.8}$$

or

$$\frac{\partial y}{\partial x} - \lambda \frac{\partial z}{\partial x} = 0 \qquad z = 0 \tag{16.9}$$

Thus, (16.9), which replaces (16.6), provides the values $x = x_0$, $\lambda = \lambda_0$. The stability condition (16.7) is written again:

$$\delta^2 y = \left[\frac{\partial^2 y}{\partial x\,\partial x}\right]_0 (\delta x)(\delta x) + \left[\frac{\partial y}{\partial x}\right]_0 (\delta^2 x) \geq 0$$

$$\forall\, \delta x \neq 0 \in \mathbf{X}, \text{ s.t. } \left[\frac{\partial z}{\partial x}\right]_0 (\delta x) = 0,$$

$$\forall\, \delta x \neq 0 \text{ and } \delta^2 x \in \mathbf{X}, \text{ s.t. } \left[\frac{\partial^2 z}{\partial x\,\partial x}\right]_0 (\delta x)(\delta x) + \left[\frac{\partial z}{\partial x}\right]_0 (\delta^2 x) = 0 \tag{16.10}$$

Now at point $x = x_0$, the first equation of (16.9) holds. Therefore

$$\left[\frac{\partial y}{\partial x}\right]_0 (\delta^2 x) = \lambda_0 \left[\frac{\partial z}{\partial x}\right]_0 (\delta^2 x)$$

Hence (16.10) can be written

$$\left[\frac{\partial^2 y}{\partial x\,\partial x}\right]_0 (\delta x)(\delta x) + \lambda_0 \left[\frac{\partial z}{\partial x}\right]_0 (\delta^2 x) \geq 0$$

$$\forall\, \delta x \neq 0 \in \mathbf{X} \text{ and s.t. } \left[\frac{\partial z}{\partial x}\right]_0 (\delta x) = 0$$

$$\forall\, \delta x \neq 0 \text{ and } \delta^2 x \in \mathbf{X}, \text{ s.t. } \left[\frac{\partial^2 z}{\partial x\,\partial x}\right]_0 (\delta x)(\delta x) + \left[\frac{\partial z}{\partial x}\right]_0 (\delta^2 x) = 0 \tag{16.11}$$

Taking account of the first part of (16.11) in the last part to eliminate $\delta^2 x$, the equation is written

$$\left[\frac{\partial^2 y}{\partial x\, \partial x}\right]_0 (\delta x)(\delta x) - \lambda_0 \left[\frac{\partial^2 z}{\partial x\, \partial x}\right]_0 (\delta x)(\delta x) \geqslant 0$$
$$\forall\ \delta x \neq 0 \in \mathbf{X} \text{ and s.t. } \left[\frac{\partial z}{\partial x}\right]_0 \delta x = 0 \tag{16.12}$$

To recapitulate, to solve the problem (16.5), conditions (16.6) and (16.7) become

$$z = 0$$
$$\frac{\partial y}{\partial x} - \lambda \frac{\partial z}{\partial x} = 0$$
$$\left[\frac{\partial^2 y}{\partial x\, \partial x}\right]_0 (\delta x)(\delta x) - \lambda_0 \left[\frac{\partial^2 z}{\partial x\, \partial x}\right]_0 (\delta x)(\delta x) \geqslant 0 \tag{16.13}$$
$$\forall\ \delta x \neq 0 \in \mathbf{X} \text{ and s.t. } \left[\frac{\partial z}{\partial x}\right]_0 \delta x = 0$$

16.3.1 Other writing

It is possible to transform (16.13) if the multiplier λ is taken as an independence variable, a way which appears to be particularly useful in the numerical computations by discretized methods. It is only necessary to write classically:

$$\delta(y - \lambda z) = 0, \qquad \forall\ \delta x \in \mathbf{X},\ \delta\lambda \in \mathbf{E}'$$
$$\delta^2(y - \lambda z) \geqslant 0 \qquad \text{at point } x_0, \lambda_0 \tag{16.14}$$
$$\forall\ \delta x \neq 0 \in \mathbf{X} \text{ and s.t. } \delta z = 0 \text{ at point } x_0$$

In fact, the first equation of (16.14)

$$\delta y - \lambda\, \delta z - \delta\lambda z = 0, \qquad \forall\ \delta x \in \mathbf{X},\ \delta\lambda \in \mathbf{E}'$$

or

$$\frac{\partial y}{\partial x} - \lambda \frac{\partial z}{\partial x} = 0$$
$$z = 0$$

Then the second statement of (16.14) is written:

$$\delta^2(y-\lambda z)=\left[\frac{\partial^2 y}{\partial x\,\partial x}-\lambda\frac{\partial^2 z}{\partial x\,\partial x}\right]_0(\delta x)(\delta x)+$$

$$+\left[\frac{\partial y}{\partial x}-\lambda\frac{\partial z}{\partial x}\right]_0\delta^2 x-2\,\delta\lambda\left[\frac{\partial z}{\partial x}\right]_0\delta x-\delta^2\lambda z_0\geqslant 0,$$

$$\forall\ \delta x\neq 0\in\mathbf{X}\text{ and s.t. }\left[\frac{\partial z}{\partial x}\right]_0\delta x=0 \quad (16.15)$$

So the third statement of (16.13) is found again.

16.3.2 Remarks

(a) The conditions of the first part of (16.12) (respectively the second part of 16.14)) have to take account of the second part of (16.12) (respectively the third part of (16.14)), that is to say $\delta z=0$, and not of $\delta[y-\lambda z]=0$, the latter being only a stationary condition which provides the point (x_0, λ_0).

(b) It may be supposed at the onset, in the computation of the second part of (16.14), that δx and $\delta\lambda$ are constant fields.

(c) The condition (16.14) gives a weak formulation of the problem and of the criterion.

16.4 APPLICATION TO MIXED PRINCIPLES

The problems of elasticity can be treated, as we recalled, by means of mixed principles, the most representative of which is the two-field principle of Hellinger–Reisener, where the unknown fields are the stress S and the displacement U. It can be obtained from the primal principle, in terms of the displacement, in introducing the deformation as an independent unknown by means of a Lagrange multiplier, the constraint between deformation and displacement being thus taken into account in the principle. Through this process, the energy criterion applied to the total potential energy is transformed into a criterion with constrained variables, thus allowing one to apply the preceding results.

Let us consider a hyperelastic structure which occupies, in its natural state ξ_0 [5], a domain Ω_0 of boundary Σ_0 of generic point M_0, and submitted to a volume density f of given external forces and a surface density F of forces given on a part Σ_{0F} of Σ_0. One supposes that f and F are square summable in Ω_0 and Σ_0 respectively. A displacement field $U=U_g$ on the complementary part Σ_{0U} of Σ_0 is also given.

If M is transformed point of M_0 in the deformation and D the deformation at M referred to Ω_0, one has

$$D=\frac{1}{2}\left[\overline{\frac{\partial M}{\partial M_0}}\frac{\partial M}{\partial M_0}-1_{E_3}\right]=\bar{D} \tag{16.16}$$

where the bar designates the transposition in the three-dimensional Euclidian space $\mathbf{E}_3$.* Putting, then,

$$M=M_0+U$$

one has

$$D=\underset{\sim}{D}(U)=\frac{1}{2}\left[\frac{\partial U}{\partial M_0}+d\overline{\frac{\partial U}{\partial M_0}}+\overline{\frac{\partial U}{\partial M_0}}\frac{\partial U}{\partial M_0}\right] \tag{16.17}$$

Let α be the volume density of strain energy, function of D, and possibly of M which will be implied. The primal principle is written as follows:

U_0 makes stationary the functional

$$y=y(\underset{\sim}{U})=\int_{\Omega_0}\alpha\,\mathrm{d}\Omega_0-\int_{\Omega_0}\bar{f}U\,\mathrm{d}\Omega_0-\int_{\Sigma_{0F}}\bar{F}U\,\mathrm{d}\Sigma_0,\quad \forall\,\underset{\sim}{U}\in\mathcal{U}$$

with

$\mathcal{U}=\{\underset{\sim}{U}\in[W^{1,4}(\Omega_0)]^3,\ U=U_g \text{ on } \Sigma_{0U}\}\Leftrightarrow$

U kinematically admissible (KA)

and

$$\alpha=\underset{\sim}{\alpha}(\underset{\sim}{D}(U)) \tag{16.18}$$

Let us then bring the constraint (16.17) into (16.18) by means of a multiplier $t=\bar{t}\in\mathcal{L}(\mathbf{E}_3,\mathbf{E}_3)$. To simplify the writing, we shall make use of the following scalar product:

$$T_r(D_1D_2)=\tilde{D}_1D_2$$

for all D_1, D_2, Hermitian mappings, $\in\mathcal{L}(\mathbf{E}_3,\mathbf{E}_3)\Leftrightarrow\forall\,\underset{\sim}{D_1},\,D_2\in\mathbf{E}_6$, where $\mathbf{E}_6$ is an Euclidian vector space of dimension 6, and where $\tilde{D}$ is the transpose of D according to this scalar product. This notation is extended without any difficulty to the associated functional spaces.

* So that the inner product of two vectors V_1, $V_2\in\mathbf{E}_3$ will be $\bar{V}_1V_2$.

The functional then becomes:*

$$y+\underset{\sim}{t}[\underset{\sim}{D}-D]=\int_{\Omega_0}[\alpha+\tilde{t}[\underset{\sim}{D}(U)-D]]\,d\Omega_0$$

$$-\int_{\Omega_0}\bar{f}U\,d\Omega_0-\int_{\Sigma_{0F}}\bar{F}U\,d\Sigma_0$$

$$\forall\begin{cases} U \quad \text{KA} & \\ D=\underset{\sim}{D}(M_0), & D\in[L^2(\Omega_0)]^6=\zeta \\ t=\underset{\sim}{t}(M_0), & \underset{\sim}{t}\in[L^2(\Omega_0)]^6=\zeta \end{cases}$$

$$\text{with } \alpha=\underset{\sim}{\alpha}(D) \qquad (16.19)$$

The condition of stationarity (16.14) is written here:

$$\delta[y+\underset{\sim}{t}[\underset{\sim}{D}-\underset{\sim}{D}]]$$

$$=\int_{\Omega_0}\left[\left[\frac{\partial\alpha}{\partial D}-\tilde{t}\right]\delta D+\delta\tilde{t}[\underset{\sim}{D}(U)-D]+\tilde{t}\,\delta[\underset{\sim}{D}(U)]\right.$$

$$\left.-\bar{f}\,\delta U\right]d\Omega_0-\int_{\Sigma_{0F}}\bar{F}\,\delta U\,d\Sigma_0=0$$

$$\forall\,\delta U\ \text{KA},\qquad \forall\,\delta\underset{\sim}{D}\in\zeta,\qquad \forall\,\delta\underset{\sim}{t}\in\zeta \qquad (16.20)$$

In (16.20), one has used the expression δU KA for the following definition:

$$\delta\underset{\sim}{U}\in\{\delta\underset{\sim}{U}\in[W^{1,4}(\Omega_0)]^3,\ \delta U=0 \text{ on } \Sigma_{0U}\}\Leftrightarrow\delta U\ \text{KA}$$

In the sequel, we will not always express the various functional spaces fully where there is no ambiguity.

The Euler equations of principle (16.20) are then classically [5]:

$$\left.\begin{aligned}\frac{\partial\alpha}{\partial D}-\tilde{t}=0\\ \underset{\sim}{D}(U)-D=0\end{aligned}\right\}\text{ in }\Omega_0$$

$$t \text{ statistically admissible (SA)} \qquad (16.21)$$

These equations allow theoretical calculation of the solution (U_0, D_0) and t_0, which was called primarily x_0 and λ_0 in Section 16.3.

The first equation of (16.21) shows that the multiplier is assimilable to the stress S, which results itself from the constitutive law, that is to say,

$$\tilde{S}=\frac{\partial\alpha}{\partial D}=\tilde{t}$$

* Let us recall here, for instance, that $\tilde{t}D$ is expressed by $t^{ij}\varepsilon_{ij}$ in conventional notations, which is a duality product in fact.

If we suppose that this equation is strictly verified, that is to say, if

$$\frac{\partial}{\partial \underset{\sim}{D}}[y+\tilde{t}[\underset{\sim}{D}-\underset{\sim}{D}]]=0$$

the independent variable D has to be eliminated, precisely by the first equation of (16.21), or assuming the existence of the inverse of the constutitive law (convexity of the strain energy),

$$D=B(t) \tag{16.22}$$

Equation (16.20) then becomes

$$\delta\left[\int_{\Omega_0}[\tilde{t}\underset{\sim}{D}(U)-\beta-\bar{f}U]\,\mathrm{d}\Omega_0-\int_{\Sigma_{0F}}\bar{F}U\,\mathrm{d}\Sigma_0\right]=0$$

$$\forall\, U\ \mathrm{KA},\ t\in\zeta$$

$$\text{with}\quad \beta=\underset{\sim}{\beta}(t)=\int_0^t B(\zeta)\,\mathrm{d}\zeta \tag{16.23}$$

This is the classical two-field principle of Hellinger–Reissner [5].

Let us note that the Legendre transformation consists of a consideration of the complementary energy

$$\beta=\underset{\sim}{\beta}(t)=\tilde{t}D-\underset{\sim}{\alpha}(D),\qquad \text{where } D=\underset{\sim}{D}(t)\text{ (see 16.22)} \tag{16.24}$$

which gives for the first integral of (16.19):

$$\int_{\Omega_0}[\tilde{t}\underset{\sim}{D}(U)-\alpha-\tilde{t}D]\,\mathrm{d}\Omega_0=\int_{\Omega_0}[\tilde{t}\underset{\sim}{D}(U)-\beta]\,\mathrm{d}\Omega.$$

It remains to calculate the first member of the second statement of (16.14) to obtain the stability condition, that is to say:

$$\delta^2[y+\tilde{t}[\underset{\sim}{D}-\underset{\sim}{D}]]$$

It should be noted that it is out of the question to take account of Euler's equation (the first equation of 16.21), which was used to state the principle (16.23) in its standard form, for the first variation $\delta[y+\tilde{t}[\underset{\sim}{D}-\underset{\sim}{D}]]$ would cut off some terms which could be of primary importance in the computation of the second variation. With (16.20) one has, with constant δD, δU, and δt:

$$\begin{aligned}&\delta^2[y+\tilde{t}[\underset{\sim}{D}-\underset{\sim}{D}]]\\&=\int_{\Omega_0}\left[\frac{\partial^2\alpha}{\partial D\,\partial D}(\delta D)(\delta D)+2\,\delta\tilde{t}\left[\frac{\partial D}{\partial U}\delta U-\delta D\right]\right.\\&\left.+\tilde{t}\frac{\partial^2 D}{\partial U\,\partial U}(\delta U)(\delta U)\right]\mathrm{d}\Omega_0\geqslant 0\end{aligned}$$

$$\forall\,\delta D,\ \delta U\ \mathrm{KA},\ \text{s.t.}\ \frac{\partial D}{\partial U}\delta U-\delta D=0 \tag{16.25}$$

or again

$$\int_{\Omega_0}\left[\frac{\partial^2\alpha}{\partial D\,\partial D}(\delta D)(\delta D)+\tilde{t}_0\frac{\partial^2 D}{\partial U\,\partial U}(\delta U)(\delta U)\right]\mathrm{d}\Omega_0\geqslant 0$$

$$\forall\ \delta D_0,\ \delta U\ \text{KA, s.t.}\ \frac{\partial D}{\partial U}\delta U-\delta D=0 \qquad (16.26)$$

In that expression, all the derivatives have to be calculated at the point D_0, U_0, t_0 solution of (16.23). It has been also indicated that t also had to take its value for the solution t_0. Let us note also that (see 16.17):

$$\delta \underset{\sim}{D}(U)=\frac{\partial D}{\partial U}\delta U=\frac{1}{2}\left[\frac{\partial\delta U}{\partial M_0}+\overline{\frac{\partial\delta U}{\partial M_0}}+\overline{\frac{\partial U}{\partial M_0}}\frac{\partial\delta U}{\partial M_0}+\overline{\frac{\partial\delta U}{\partial M_0}}\frac{\partial U}{\partial M_0}\right]$$

$$\frac{\partial^2 D}{\partial U\,\partial U}(\delta U)(\delta U)=\overline{\frac{\partial\delta U}{\partial M_0}}\frac{\partial\delta U}{\partial M_0}$$

16.4.1 Verification

The preceding stability condition can easily be verified. In effect, the stability condition coming from the direct principle (16.19) would be written, in applying the criterion (16.3) without any constraint:

$$\delta^2 y=\delta^2\int_{\Omega_0}\alpha\,\mathrm{d}\Omega_0,\qquad \text{where } \alpha=\underset{\sim}{\alpha}(\underset{\sim}{D}(U))$$

Now

$$\delta y=\int_{\Omega_0}\frac{\partial\alpha}{\partial D}\delta[\underset{\sim}{D}(U)]\,\mathrm{d}\Omega_0=\int_{\Omega_0}\frac{\partial\alpha}{\partial D}\frac{\partial D}{\partial U}\delta U\,\mathrm{d}\Omega_0$$

$$\delta^2 y=\int_{\Omega_0}\left[\frac{\partial^2\alpha}{\partial D\,\partial D}(\delta\underset{\sim}{D}(U))(\delta\underset{\sim}{D}(U))+\frac{\partial\alpha}{\partial D}(\delta^2\underset{\sim}{D}(U))\right]\mathrm{d}\Omega_0\qquad\text{with constant }\delta U$$

Hence

$$\delta^2 y=\int_{\Omega_0}\left[\frac{\partial^2\alpha}{\partial D\,\partial D}\left(\frac{\partial D}{\partial U}\delta U\right)\left(\frac{\partial D}{\partial U}\delta U\right)+\frac{\partial\alpha}{\partial D}\frac{\partial^2 D}{\partial U\,\partial U}(\delta U)(\delta U)\right]\mathrm{d}\Omega_0\geqslant 0 \qquad (16.27)$$

If in (16.26) one takes account of the constraint, on the one hand, and if one eliminates D, on the other hand, in taking account of the first equation of (16.21) one again arrives at formula (16.27).

16.4.2 Complete weak form of the stability criterion

The totally weak form of the criterion (16.26) is easily written from (16.20), which also gives the constraint

$$\int_{\Omega_0}\left[\frac{\partial^2\alpha}{\partial D\,\partial D}(\delta D)(\delta D)+\tilde{t}_0\frac{\partial^2 D}{\partial U\,\partial U}(\delta U)(\delta U)\right]\mathrm{d}\Omega_0\geqslant 0,\quad \forall\;\delta D,\delta U\text{ KA}$$

such that

$$\int_{\Omega_0}\delta\tilde{t}\left[\frac{\partial D}{\partial U}\delta U-\delta D\right]\mathrm{d}\Omega_0=0,\qquad \forall\;\delta t\in\zeta \tag{16.28}$$

This form will be useful for the discretization.

16.4.3 Application to the principle of Hellinger–Reissner

The criterion (16.27), or (16.28), cannot be applied in such a form to the principle of Hellinger–Reissner, but rather to the three-field principle (16.19), the so-called principle of Hu–Washizu, before the elimination of D [5]. However, in the formulation (16.27), there is nothing to prevent account being taken of the stationarity condition, or at least some of these conditions according to the chosen principle (the derivative in (16.27) being calculated at the solution point), e.g. the first equation in (16.19). Thus:

$$\frac{\partial\alpha}{\partial D}=\tilde{t},\qquad \frac{\partial\beta}{\partial t}=\tilde{D}$$

Now (16.24) gives

$$\beta=\tilde{t}D-\alpha,\qquad \text{where }\tilde{D}=\frac{\partial\beta}{\partial t}\text{ and }\beta=\underset{\sim}{\beta}(t)$$

$$\delta\beta=\delta\tilde{t}D+\tilde{t}\,\delta D-\delta\alpha$$

$$=\delta\tilde{t}D+\tilde{t}\,\delta D-\frac{\partial\alpha}{\partial D}\delta D=\delta\tilde{t}D$$

With constant δt and δD, this becomes:

$$\delta^2\beta=\delta\tilde{t}\,\delta D=\frac{\partial^2\alpha}{\partial D\,\partial D}(\delta D)(\delta D)=\delta^2\alpha \tag{16.29}$$

Hence

$$\int_{\Omega_0}\left[\frac{\partial^2\beta}{\partial t\,\partial t}(\delta t)(\delta t)+t_0\frac{\partial^2 D}{\partial U\,\partial U}(\delta U)(\delta U)\right]\mathrm{d}\Omega_0\geqslant 0$$

$$\forall\;\delta t\in\zeta,\ \delta U\text{ KA, s.t.}\left[\int_{\Omega_0}\delta\tilde{t}\left[\frac{\partial D}{\partial U}\delta U-\frac{\partial^2\beta}{\partial t\,\partial t}\delta t\right]\mathrm{d}\Omega_0=0,\ \forall\underline{\delta t}\in\zeta\right] \tag{16.30}$$

As in the preceding, the verification is straightforward and again gives (16.27).

16.5 DISCRETIZATION

The discretization of the unknown fields t and U, for instance by the finite element method, applied to the principal of Hellinger–Reissner, (16.23), provides particular matrix equations for (16.30). If U and t are discretized by columns q_U and q_t respectively, after assemblage (16.30) is expressed by

$$\begin{aligned} &\overline{\delta q_t}\mathscr{B}\,\delta q_t+\overline{\delta q_U}\mathscr{D}\,\delta q_U \geqslant 0, \qquad \mathscr{B}=\bar{\mathscr{B}},\, \mathscr{D}=\bar{\mathscr{D}}, \\ &\forall\ \delta q_U,\, \delta q_t,\ \text{s.t. } \mathscr{A}\,\delta q_U-\mathscr{B}\,\delta q_t=0; \qquad \delta q_t \in \mathbb{R}^{N_t},\, \delta q_U \in \mathbb{R}^{N_U} \end{aligned} \tag{16.31}$$

In (16.31), the symmetric matrices $\mathscr{B}$ and $\mathscr{D}$ represent quadratic forms. The matrix $\mathscr{B}$, representing the Hessian of the volume density of the complementary energy, evidently admits an inverse in our case; hence:

$$\delta q_t=\mathscr{B}^{-1}\mathscr{A}\,\delta q_U$$

and (16.31) is reduced to

$$\overline{\delta q_U}\,[\bar{\mathscr{A}}\mathscr{B}^{-1}\mathscr{A}+\mathscr{D}]\,\delta q_U \geqslant 0, \qquad \forall\ \delta q_U \in \mathbb{R}^{N_U} \tag{16.32}$$

The matrices $\mathscr{A}$ and $\mathscr{B}$ are naturally functions of the solution t_0 and U_0 in the actual state.

Remark 16.1 The classical notions of matrices of elastic stiffness and of geometrical stiffness appear clearly in (16.32), that is to say $\bar{\mathscr{A}}\mathscr{B}^{-1}\mathscr{A}$ and $\mathscr{D}$ respectively, the matrix $\mathscr{D}$ being, as it must be, linear in function of the stress t_0. These notions remain valid therefore in structural stability, even in the case of non-linear hyperelasticity and for the mixed principle too, due to the fact that the structural stability needs only to use the quadratic forms of the second variation (subject to the examination in the critical state of the derivatives of larger orders).

Remark 16.2 The result which appears in (16.30), or in (16.32), seems to be obvious at first, since it can be found by a direct demonstration which would consist of use of the expression of the direct potential energy for the stability criterion and that of (16.27) and (16.29), and knowing, as is well known, that principle (16.23) leads to a saddle-point. It must be observed, however, that the constraint betwen δt and δU, which is fundamental, as seen from (16.32), must not be forgotten.

Remark 16.3 The given result takes its true interest in the utilization of the mixed principle of Hellinger–Reissner, the aim of which is to provide stresses and displacements directly. It is observed that (16.32) is expressed indirectly through the results.

Remark 16.4 The proposed method is general in the sense that it allows consideration of arbitrary non-holonomic constraints. This is also the case for non-separate variables, contrary to the case considered above, where δD and δU are fortunately separate. One thinks in particular of cases where the structure contains incompressible materials.

In the case of non-separate variables the discretized criterion appears in the following general form:

$$\begin{aligned} \bar{q}Hq \geqslant 0, \qquad & \forall\, q \neq 0 \in \mathbb{R}^n, \text{ s.t.} \\ \bar{T}_1 q = 0 & \end{aligned} \tag{16.33}$$

where H is a symmetric matrix, due to (16.13), and T_1 is a row of vectors of $\mathbb{R}^n$, with $p<n$, which is supposed of course to be independent:

$$T_1 = [T_{11} \quad T_{12} \cdots T_{1p}]$$

These p vectors subtend a subspace of $\mathbb{R}^n$, the second equation of (16.33) indicating that q must be in the supplementary orthogonal subspace. Let

$$\pi' = T_1[\bar{T}_1 \quad T_1]^{-1}\bar{T}_1 = \bar{\pi}'$$

be the orthogonal projector upon the subspace subtended by T_1 and let

$$\pi = 1_{\mathbb{R}^n} - \pi' = \bar{\pi}$$

be the orthogonal projector on the supplementary subspace. Then consider a basis T_2 of the range of π and the basis

$$T = [T_1 \quad T_2]$$

of $\mathbb{R}^n$. If one makes the change of basis

$$q = Tq' = [T_1 \quad T_2]\begin{bmatrix} {}^1q' \\ {}^2q' \end{bmatrix}$$

(16.33) becomes

$$\bar{T}\pi H \pi T = \begin{bmatrix} \bar{T}_1\pi H \pi T_1 & \bar{T}_1 \pi H \pi T_2 \\ \bar{T}_2 \pi H \pi T_1 & \bar{T}_2 \pi H \pi T_2 \end{bmatrix} = \begin{bmatrix} 0 & 0 \\ 0 & \bar{T}_2 H T_2 \end{bmatrix}$$

since

$$\pi T_2 = T_2, \qquad \pi T_1 = 0, \qquad H = \bar{H}$$

This gives

$$\overline{{}^2q'}\, \bar{T}_2 H T_2 {}^2q' \geqslant 0, \qquad \forall\, {}^2q' \neq 0 \in \mathbb{R}^{n-p} \tag{16.34}$$

It therefore remains to find $T_2 \perp T_1$. For that purpose, it seems that the simplest method is to utilize the orthogonalization process of Schmidt.

In an applied procedure of the non-linear mixed principle by small increments of loading, one could then apply this process by linearization [8].

16.6 THE CASE OF LINEAR BUCKLING: STEP-BY-STEP METHOD

Application of the principle of Hellinger–Reissner is carried out in general by load increments through a linearized formulation which takes account of the preceding loading at each step. It is therefore interesting to take a prestressed state ζ_0 as the reference state and to check at each step the stability of the state ζ. If it is supposed then that the increments D of the deformation are small enough so that the strain volume density α is a function limited to the quadratic terms in D, one has

$$\alpha = \left[\frac{\partial \alpha}{\partial D}\right]_0 D + \tfrac{1}{2}\tilde{D}AD = \tilde{S}_0 D + \tfrac{1}{2}\tilde{D}AD, \qquad \text{where } A = \tilde{A} \in \mathscr{L}(\mathbf{E}_6, \mathbf{E}_6) \quad (16.35)$$

In (16.35), S_0 is the prestress which is reached in state ξ_0.

It is supposed that, in the reference state ξ_0, the body occupies a volume Ω_0 of boundary Σ_0 and that it is in equilibrium under the action of the volume density f_0 of gives forces and of the surface density F_0 of given forces on the part Σ_{0F} of Σ_0. As before,

$$U = U_g \quad \text{on} \quad \Sigma_{0U}$$

Then, a supplementary volume density f_1 in Ω_0 and a supplementary surface density F_1 are applied, the latter still on Σ_{0F}. The consequence is an extra stress S_1. The point M_0 comes to M and the deformations D is such that

$$D = \frac{1}{2}\left[\overline{\frac{\partial M}{\partial M_0}}\frac{\partial M}{\partial M_0} - 1_{E_3}\right], \qquad \text{with } M = M_0 + U \qquad (16.36)$$

The stationary condition (16.19) is then written successively with (16.35):

$$\delta\left[\int_{\Omega_0} [\underset{\sim}{\alpha}(D) + \tilde{t}[\underset{\sim}{D}(U) - D] - \bar{f}U]\, d\Omega_0 - \int_{\Sigma_{0F}} \bar{F}U\, d\Sigma_0\right]$$
$$= 0, \qquad \forall\, \delta t, \delta D, \delta U \text{ KA}$$

$$= \int_{\Omega_0} \left[\frac{\partial \alpha}{\partial D}\delta D + \tilde{t}[\delta \underset{\sim}{D}(U) - \delta D] + \delta t[\underset{\sim}{D}(U) - D] - \bar{f}\,\delta U\right] d\Omega_0 - \int_{\Sigma_{0F}} \bar{F}\,\delta U\, d\Sigma_0$$
$$= 0 \qquad \forall\, \delta t, \delta D, \delta U \text{ KA}$$

with $S_1 = AD$

Let us call t_0 the value of t when $D=U=0$ in state ξ_0 and t_1 its increment in the deformed state ξ. It becomes

$$\int_{\Omega_0} [[\tilde{S}_0+\tilde{S}_1-\tilde{t}_0-\tilde{t}_1]\,\delta D+[\tilde{t}_0+\tilde{t}_1]\,\delta \underset{\sim}{D}(U)+\delta t[\underset{\sim}{D}(U)-D]-[\bar{f}_0+\bar{f}_1]\,\delta U]\,d\Omega_0$$

$$-\int_{\Sigma_{0F}} [\bar{F}_0+\bar{F}_1]\,\delta U\,d\Sigma_0=0, \qquad \forall\, \delta t, \delta D, \delta U \text{ KA}$$

Now (16.17) gives

$$\delta \underset{\sim}{D}(U)=\frac{1}{2}\left[\frac{\partial \delta U}{\partial M_0}+\overline{\frac{\partial \delta U}{\partial M_0}}+\overline{\frac{\partial U}{\partial M_0}}\frac{\partial \delta U}{\partial M_0}+\overline{\frac{\partial \delta U}{\partial M_0}}\frac{\partial U}{\partial M_0}\right]$$

and for $U=0$,

$$\delta \underset{\sim}{D}(U)\,|_{U=0}=\frac{1}{2}\left[\frac{\partial \delta U}{\partial M_0}+\overline{\frac{\partial \delta U}{\partial M_0}}\right]$$

As the equilibrium is supposed to be realized in state ξ_0, it becomes

$$\int_{\Omega_0}\left[[\tilde{S}_0-\tilde{t}_0]\,\delta D+\tilde{t}_0\frac{\partial \delta U}{\partial M_0}-\bar{f}_0\,\delta U\right]d\Omega_0$$

$$-\int_{\Sigma_{0F}} \bar{F}_0\,\delta U\,d\Sigma_0=0, \qquad \forall\, \delta D, \delta U \text{ KA}$$

The component t_0 of t is equal to the prestress S_0 (of Cauchy).

This remains for the stationary condition, and keeping only terms of the first order in t_1, U, and D:

$$\int_{\Omega_0}\left[[\tilde{S}_1-\tilde{t}_1]\,\delta D+\tilde{S}_0\overline{\frac{\partial U}{\partial M_0}}\frac{\partial \delta U}{\partial M_0}+\tilde{t}_1\frac{\partial \delta U}{\partial M_0}+\delta t_1\left[\frac{\partial U}{\partial M_0}-D\right]-\bar{f}_1\,\delta U\right]d\Omega_0$$

$$-\int_{\Sigma_{0F}} \bar{F}_1\,\delta U\,d\Sigma_0=0, \qquad \forall\, \delta D, \delta t_1=\overline{\delta t_1}, \delta U \text{ KA} \quad (16.37)$$

Incidentally, the terms which have been left aside are

$$\int_{\Omega_0}\left[\tilde{t}_1\overline{\frac{\partial U}{\partial M_0}}\frac{\partial \delta U}{\partial M_0}+\delta t_1\overline{\frac{\partial U}{\partial M_0}}\frac{\partial U}{\partial M_0}\right]d\Omega_0$$

If we suppose that the component t_1 of the multiplier is equal to the stress S_1, that is to say, if the following Euler equation strictly holds:

$$S_1-t_1=0 \qquad \text{in } \Omega_0$$

the linearized principle of Hellinger–Reissner, with a prestress, is now written:

$$\int_{\Omega_0}\left[\tilde{S}_0\frac{\overline{\partial U}}{\partial M_0}\frac{\partial \delta U}{\partial M_0}+\tilde{t}_1\frac{\partial \delta U}{\partial M_0}+\delta\tilde{t}_1\left[\frac{\partial U}{\partial M_0}-D\right]-\bar{f}_1\,\delta U\right]\mathrm{d}\Omega_0$$

$$-\int_{\Sigma_{0F}}\bar{F}_1\,\delta U\,\mathrm{d}\Sigma_0=0 \qquad \forall\,\delta D,\ \delta t_1=\overline{\delta t_1},\ \delta U\ \mathrm{KA} \quad (16.38)$$

where D has to be eliminated by means of the constitutive equation linearized from the state ξ_0, which is (see 16.35)

$$S_1=t_1=AD \qquad \text{or} \qquad D=A^{-1}t_1=Bt_1,\ \text{with } B=\tilde{B}\in\mathscr{L}(\mathbf{E}_6,\mathbf{E}_6) \quad (16.39)$$

Finally, one has the stationarity condition:

$$\delta\left[\int_{\Omega_0}\left[\tfrac{1}{2}\tilde{S}_0\frac{\overline{\partial U}}{\partial M_0}\frac{\partial U}{\partial M_0}+\tilde{t}_1\frac{\partial U}{\partial M_0}-\tfrac{1}{2}\tilde{t}_1Bt_1-\bar{f}_1U\right]\mathrm{d}\Omega_0-\int_{\Sigma_{0F}}\bar{F}_1U\,\mathrm{d}\Sigma_0\right]=0$$

$$\forall\,\delta t_1=\overline{\delta t_1},\ \delta U\ \mathrm{KA} \quad (16.40)$$

As in Section 16.4, the stability criterion must be calculated from (16.37), say with constant δt_1, δU, δD, and (to be complete) in keeping the non-linear terms in t_1 and U:

$$\delta\left[\int_{\Omega_0}\left[[\widetilde{AD}-\tilde{t}_1]\,\delta D+\tilde{S}_0\frac{\overline{\partial U}}{\partial M_0}\frac{\partial\delta U}{\partial M_0}+\tilde{t}_1\left[\frac{\partial\delta U}{\partial M_0}+\frac{\overline{\partial U}}{\partial M_0}\frac{\partial\delta U}{\partial M_0}\right]\right.\right.$$

$$\left.\left.+\,\delta t_1[\underset{\sim}{D}(U)-D]-\bar{f}_1\,\delta U\right]\mathrm{d}\Omega_0-\int_{\Sigma_{0F}}\bar{F}_1\,\delta U\,\mathrm{d}\Sigma_0\right]\geqslant 0$$

$$\forall\,\delta D,\ \delta U\ \mathrm{KA},\ \text{s.t.}\int_{\Omega_0}\underline{\delta t_1}[\delta\underset{\sim}{D}(U)-\delta D]\,\mathrm{d}\Omega_0=0$$

which becomes with (16.39):

$$\int_{\Omega_0}\left[\delta t_1B\,\delta t_1+[\tilde{S}_0+\tilde{t}_1]\frac{\overline{\partial\delta U}}{\partial M_0}\frac{\partial\delta U}{\partial M_0}\right]\mathrm{d}\Omega_0\geqslant 0$$

$$\forall\,\delta t_1\in\zeta,\ \delta U\ \mathrm{KA},\ \text{s.t.} \quad (16.41)$$

$$\left[\int_{\Omega_0}\underline{\delta t_1}\left[\left[1_{E_3}+\frac{\overline{\partial U}}{\partial M_0}\right]\frac{\partial\delta U}{\partial M_0}-B\,\delta t_1\right]\mathrm{d}\Omega_0=0, \qquad \forall\,\underline{\delta t_1}=\overline{\underline{\delta t_1}}\in\zeta\right]$$

In that weak formulation of the criterion, similar to (16.30), the non-linear terms have been kept inside to be complete, but to study the stability of state ξ_0 itself, it is necessary to cancel t_1 in the first equation of (16.41) and U is the second.

Let us rcmark then that, in that case, the formulation (16.41), which is relative to the linear buckling, strictly gives again the notion of Timoshenko's quotient, such as it appears in [9]. The formulation (16.30) therefore gives a generalization of it in the non-linear case.

16.7 MODIFIED PRINCIPLE OF HELLINGER–REISSNER AND HYBRID DUAL PRINCIPLE

Let us consider the stationarity condition of the principle of Hellinger–Reissner given by (16.37) in the linearized prestressed case:

$$\int_{\Omega_0}\left[\left[\frac{\partial\alpha}{\partial D}-\tilde{t}_1\right]\delta D+\tilde{S}_0\,\overline{\frac{\partial U}{\partial M_0}}\,\frac{\partial\delta U}{\partial M_0}+\tilde{t}_1\,\frac{\partial\delta U}{\partial M_0}+\widetilde{\delta t_1}\left[\frac{\partial U}{\partial M_0}-D\right]-\bar{f}_1\,\delta U\right]\mathrm{d}\Omega_0$$
$$-\int_{\Sigma_{0F}}\bar{F}_1\,\delta U\,\mathrm{d}\Sigma_0=0$$

$$\forall\,\delta D=\overline{\delta D},\ \delta t_1=\overline{\delta t_1},\ \delta U\ \mathrm{KA} \tag{16.42}$$

In (16.42), one has willingly kept the term in δD. That becomes, by applying Stokes' formula and calling n_0 the unit normal of Σ_0 oriented outwards:

$$\left[\int_{\Omega_0}\left[\left[\frac{\partial\alpha}{\partial D}-\tilde{t}_1\right]\delta D-\left[\mathrm{div}\left[S_0\,\overline{\frac{\partial U}{\partial M_0}}+t_1\right]+\bar{f}_1\right]\right.\right.$$
$$\left.\delta U-\mathrm{div}\,\delta t_1\cdot U-\widetilde{\delta t_1}D\right]\mathrm{d}\Omega_0$$

$$+\int_{\Sigma_{0U}}\left[\tilde{n}_0 S_0\,\overline{\frac{\partial U}{\partial M_0}}\,\delta U+\tilde{n}_0\,\delta t_1\,U_g\right]\mathrm{d}\Sigma_0+\int_{\Sigma_{0F}}\left[\left[\tilde{n}_0 S_0\,\overline{\frac{\partial U}{\partial M_0}}+\tilde{n}_0 t_1-\bar{F}_1\right]\right.$$
$$\left.\delta U+\tilde{n}_0\,\delta t_1\,U\right]\mathrm{d}\Sigma_0=0$$

$$\forall\,\delta D=\overline{\delta D},\ \delta t_1=\overline{\delta t_1},\ \delta U\ \mathrm{KA},$$

and where

$$\alpha=\tfrac{1}{2}\tilde{D}AD,\qquad \mathrm{A}=\tilde{\mathrm{A}}\in\mathscr{L}(\mathbf{E}_6,\mathbf{E}_6) \tag{16.43}$$

If we take account in (16.43) of the equations

$$\frac{\partial\alpha}{\partial D}=\tilde{t}_1$$
$$D=Bt_1,\qquad B=\tilde{B}=\mathrm{A}^{-1}$$

equation (16.43) becomes formally:

$$\int_{\Omega_0} \left[-\left[\operatorname{div}\left[S_0 \overline{\frac{\partial U}{\partial M_0}} + t_1 \right] + \bar{f}_1 \right] \delta U - \operatorname{div} \delta t_1 \cdot U - \widetilde{\delta t_1} B t_1 \right] \mathrm{d}\Omega_0$$

$$+ \int_{\Sigma_{0U}} \bar{n}_0\, \delta[t_1 U]\, \mathrm{d}\Sigma_0 + \int_{\Sigma_{0F}} \left[\left[\bar{n}_0 S_0 \overline{\frac{\partial U}{\partial M_0}} + \bar{n}_0 t_1 - \bar{F}_1 \right] \mathrm{d}U + \bar{n}_0\, \delta t_1 U \right] \mathrm{d}\Sigma_0$$

$$= 0$$

$$\forall\, \delta t_1 = \overline{\delta t_1},\ \delta U \text{ KA} \tag{16.44}$$

This equation provides immediately the modified principle of Hellinger–Reissner, that is to say:

$$\delta\left[\int_{\Omega_0} \left[\tfrac{1}{2} \tilde{S}_0 \overline{\frac{\partial U}{\partial M_0}} \frac{\partial U}{\partial M_0} - [\operatorname{div} t_1 + \bar{f}_1] U - \tfrac{1}{2} \tilde{t}_1 B t_1 \right] \mathrm{d}\Omega_0 \right.$$

$$\left. + \int_{\Sigma_{0U}} \bar{n}_0 t_1 U\, \mathrm{d}\Sigma_0 + \int_{\Sigma_{0F}} [\bar{n}_0 t_1 - \bar{F}_1] U\, \mathrm{d}\Sigma_0 \right] = 0$$

$$\forall\, t_1 \in \{H(\operatorname{div}, \Omega_0),\ t_1 = \bar{t}_1\}, \qquad U \text{ KA} \tag{16.45}$$

This formulation can be discretized by the finite element methods in discretizing the space

$$\zeta_1 = \{H(\operatorname{div}, \Omega_0),\ t_1 = \bar{t}\}$$

where t_1 and div t_1 are square-summable in Ω_0, as explained in [1, 10] but in a non-prestressed linear case.

However, it is necessary here to suppose that U is kinematically admissible still in the preceding meaning because of the presence of the prestress term. Therefore the formulation (16.45) does not appear to be very interesting for applications. Nevertheless, the stability criterion is calculated as before from formula (16.43) before doing any assumptions. Once the calculus has been carried out for the second variation and for the weakened constraint, it is possible to apply the assumptions which are specific to the problem under consideration. One can then easily find for the studied state ξ_0:

$$\int_{\Omega_0} \left[\tilde{\delta t_1} B\, \delta t_1 + \tilde{S}_0 \overline{\frac{\partial \delta U}{\partial M_0}} \frac{\partial \delta U}{\partial M_0} \right] \mathrm{d}\Omega_0 \geqslant 0$$

$$\forall\, \delta t_1 \in \zeta_1 \quad \text{and} \quad \delta U \text{ KA, s.t.}$$

$$\int_{\Omega_0} \underline{\delta t_1} \left[\frac{\partial \delta U}{\partial M_0} - B t_1 \right] \mathrm{d}\Omega_0 = 0, \qquad \forall\, \underline{\delta t_1} \in \zeta_1 \tag{16.46}$$

Let us consider now the non-linear hybrid dual principle [11, 12], which consists of the assumption that the stress belongs to a class such that (quasi-equilibrium in the interior of Ω_0):

$$\operatorname{div} t_1 + \bar{f}_1 = 0 \qquad \text{in } \Omega_0 \tag{16.47}$$

More precisely, in the applications by the finite element method, one supposes that (16.47) holds in each element and that one takes account of the equality of the fluxes $t_1 n_0$ in a weak form at the interfaces, as could be seen in (16.45).

It is known that condition (16.47) can be verified by the use of a stress function, and also in weakening the stress symmetry [1, 5, 12, 13].

The formulation (16.45) then becomes

$$\delta\left[\int_{\Omega_0}\left[\tfrac{1}{2}\tilde{S}_0\overline{\frac{\partial U}{\partial M_0}\frac{\partial U}{\partial M_0}} - \tfrac{1}{2}\bar{t}_1 B t_1\right]\mathrm{d}\Omega_0 + \int_{\Sigma_{0U}} \bar{n}_0 t_1 U\,\mathrm{d}\Sigma_0 + \int_{\Sigma_{0F}} [\bar{n}_0 t_1 - \bar{F}_1] U\,\mathrm{d}\Sigma_0\right] = 0$$

$$\forall\, U \text{ KA} \quad \text{and} \quad t_1 \in \zeta_2$$

where $$\zeta_2 = \{t_1 = \bar{t}_1 \in [L^2(\Omega_0)]^6;\ \operatorname{div} t_1 + \bar{f}_1 = 0\} \tag{16.48}$$

The stability condition of principle (16.47) can be formally calculated from (16.42), i.e.

$$\int_{\Omega_0}\left[\delta^2\alpha + 2\tilde{t}_1\left[\frac{\partial \delta U}{\partial M_0} - \delta D\right] + \tilde{S}_0\overline{\frac{\partial \delta U}{\partial M_0}\frac{\partial \delta U}{\partial M_0}}\right]\mathrm{d}\Omega_0 \geqslant 0$$

$$\forall\, \delta U \text{ KA}, \delta D, \delta t_1 \in \zeta, \text{ s.t. } \left[\int_{\Omega_0} \underline{\delta t_1}\left[\frac{\partial \delta U}{\partial M_0} - \delta D\right]\mathrm{d}\Omega_0 = 0, \ \forall\, \delta t_1 \in \zeta\right] \tag{16.49}$$

One may apply to (16.49) all the constraints and hypotheses which are specific to (16.48). Then

$$\delta^2\alpha = \delta^2\beta = \delta t_1 B\,\delta t_1, \qquad \forall\, \delta t_1 \in \zeta_{20}$$

with

$$\zeta_{20} = \{\delta t_1 = \overline{\delta t_1} \in [L^2(\Omega_0)]^6;\ \operatorname{div} \delta t_1 = 0\}$$

The criterion then becomes

$$\int_{\Omega_0}\left[\delta t_1 B\,\delta t_1 + \tilde{S}_0\overline{\frac{\partial \delta U}{\partial M_0}\frac{\partial \delta U}{\partial M_0}}\right]\mathrm{d}\Omega_0 \geqslant 0$$

$$\forall\, \delta U \text{ KA} \quad \text{and} \quad \delta t_1 \in \zeta_{20}, \text{ s.t. } \left[\int_{\Omega_0} \underline{\delta t_1}\left[\frac{\partial \delta U}{\partial M_0} - B\,\delta t_1\right]\mathrm{d}\Omega_0 = 0, \ \forall\, \delta t_1 \in \zeta_{20}\right] \tag{16.50}$$

Criterion (16.50) is formally similar to the preceding ones, but the functional space relative to δt_1 is different. The discretization leads again to express the first statement of (16.50) as a function of the variable which discretizes the displacement, but still through the stress solution. Experience shows that principle (16.48) is very efficient; hence the interst of (16.50).

It is now proposed to illustrate the preceding results for the case of beams and shells, where the problems of buckling are stated almost exclusively. However, the study will be limited to examples of formulae relative to linear buckling.

16.8 LINEAR BUCKLING OF BEAMS IN BENDING

Let us consider the very simple case of straight beams in bending. We shall call W the deflection, E the Young modulus, and I the section inertia, which are functions of the abscissa x of the mean line of the beam of length L. We pose:

$$\alpha = \tfrac{1}{2}EID^2 \qquad D = \frac{\mathrm{d}^2 w}{\mathrm{d}x^2} \tag{16.51}$$

Let F be the given compressive force; the preceding criterion (16.37) can be found again in that particular case with the functional.

$$\int_0^L \tfrac{1}{2}EID^2\,\mathrm{d}x + \tfrac{1}{2}F\int_0^L \left[\frac{\mathrm{d}w}{\mathrm{d}x}\right]^2 \mathrm{d}x + \int_0^L t\left[\frac{\mathrm{d}^2 w}{\mathrm{d}x^2} - D\right]\mathrm{d}x, \qquad \forall\, w \text{ KA} \tag{16.52}$$

where all the expressions corresponding to those of (16.37) can be identified for the properly so-called prestressed state.

The stationarity condition is written:

$$\int_0^L EID\,\delta D\,\mathrm{d}x + F\int_0^L \frac{\mathrm{d}w}{\mathrm{d}x}\frac{\mathrm{d}\delta w}{\mathrm{d}x}\mathrm{d}x + \int_0^L \left[t\left[\frac{\mathrm{d}^2\delta w}{\mathrm{d}x^2} - \delta D\right] + \delta t\left[\frac{\mathrm{d}^2 w}{\mathrm{d}x^2} - D\right]\mathrm{d}x\right] = 0$$

$$\forall\, \delta D, \delta t, \delta w \text{ KA} \tag{16.53}$$

The stability criterion is then written, with constant δt, δD, δw:

$$\int_0^L EI[\delta D]^2\,\mathrm{d}x + F\int_0^L \left[\frac{\mathrm{d}\delta w}{\mathrm{d}x}\right]^2 \mathrm{d}x \geqslant 0$$

$$\forall\, \delta D, \delta w \text{ KA, s.t.} \int_0^L \delta t\left[\frac{\mathrm{d}^2\delta w}{\mathrm{d}x^2} - \delta D\right]\mathrm{d}x = 0, \qquad \forall\, \delta t$$

If D and δD are eliminated thanks to (16.53):

$$t = EID, \qquad D = \frac{t}{EI}, \qquad \delta D = \frac{\delta t}{EI}$$

it becomes

$$\int_0^L \frac{[\delta t]^2}{EI}\, dx + F \int_0^L \left[\frac{d\delta w}{dx}\right]^2 dx \geqslant 0$$

$$\forall\ \delta t,\ \delta w \text{ KA, s.t. } \int_0^L \underline{\delta t}\left[\frac{d^2\, \delta w}{dx^2} - \frac{\delta t}{EI}\right] dx = 0,$$

$$\forall\ \underline{\delta t} \tag{16.54}$$

If the second part of (16.54) is taken in a strong form, the elimination of δt easily gives, in that particularly simple case the classical criterion of linear buckling of beams.

16.9 LINEAR BUCKLING OF SHELLS IN KIRCHHOFF–LOVE HYPOTHESES

In Kirchhoff–Love hypotheses, according to which the geometrical normal to the middle surface always remains identical to the material normal during the deformation process and the normal strain energy is disregarded, the virtual strain energy is written [12]:

$$\delta w = \int_{\Sigma_{m_0}} T_r(\mathsf{n}\ \delta\Gamma + \mathsf{m}\ \delta K)\, d\Sigma_0 \tag{16.55}$$

where Σ_{m_0} (of boundary C_{m_0}) represents the middle surface of generic point m_0 in a reference state ξ_0, Γ represents the tangential deformation, K the curvature variation, and n and m the dual surface stresses of these surface deformations respectively. These four last quantities are Hermitian mappings of the tangent plane $\mathbf{E}_{20}$ at m_0 of Σ_{m_0}.

Let N_0 be the unit normal to Σ_{m_0} at m_0, m the point of the deformed middle surface Σ_m corresponding to m_0, N the unit normal to Σ_m at m, and $\mathbf{E}_2$ the tangent plane of Σ_m at m. Let also $\partial N/\partial m$ be the curvature operator of Σ_m which defines the second fundamental form of Σ_m [5]. We have

$$\frac{\partial N}{\partial m} = \overline{\frac{\partial N}{\partial m}} \in \mathscr{L}(\mathbf{E}_2, \mathbf{E}_2), \qquad \frac{\partial N_0}{\partial m_0} = \overline{\frac{\partial N_0}{\partial m_0}} \in \mathscr{L}(\mathbf{E}_{20}, \mathbf{E}_{20})$$

where the bar denotes the transposition in space $\mathbf{E}_3$.

Finally, let V be the displacement vector of m_0. We have [5, 12]

$$\Gamma = \frac{1}{2}\left[\overline{\frac{\partial m}{\partial m_0}}\frac{\partial m}{\partial m_0} - 1_{E_{20}}\right] = \boldsymbol{\Gamma}(V) = \bar{\Gamma} \in \mathscr{L}(\mathbf{E}_{20}, \mathbf{E}_{20})$$

$$K = \overline{\frac{\partial m}{\partial m_0}}\frac{\partial N}{\partial m}\frac{\partial m}{\partial m_0} - \frac{\partial N_0}{\partial m_0} = \underset{\sim}{K}(V) = \bar{K} \in \mathscr{L}(\mathbf{E}_{20}, \mathbf{E}_{20}) \qquad (16.56)$$

with

$$V = m - m_0 \in \mathbf{E}_3$$

If we call α the surface elastic strain energy, we have

$$\alpha = \underset{\sim}{\alpha}(\Gamma, K) \qquad \text{with } \Gamma = \underset{\sim}{\Gamma}(V),\ K = \underset{\sim}{K}(V) \qquad (16.57)$$

If we are given the surface density f of forces on Σ_{m_0} and the line density F of forces on the part C_{m_0F} of the boundary C_{m_0}, and if the displacement V_g is also given on the complementary part C_{m_0V} of this boundary, the functional of the primal principle is written:

$$\int_{\Sigma_{m_0}} \alpha \, d\Sigma_{m_0} - \int_{\Sigma_{m_0}} \bar{f}V \, d\Sigma_{m_0} - \int_{C_{m_0F}} \bar{F}V \, d\xi$$
$$\text{with (16.56) and (16.57), and } V \text{ KA} \qquad (16.58)$$

where ξ is the line abscissa on C_{m_0}(Σ_{m_0}, C_{m_0}, N_0 are canonically oriented with respect to $\mathbf{E}_3$).

If from now on the two surface deformations Γ and K are taken as independent variables, the two constraints (16.56) have to be introduced into the functional by means of two multipliers called t and s respectively. The stationarity condition is then written:

$$\delta\left[\int_{\Sigma_{m_0}} \left[\underset{\sim}{\alpha}(\Gamma, K) + \tilde{t}[\underset{\sim}{\Gamma}(V) - \Gamma] + \tilde{s}[\underset{\sim}{K}(V) - K] - \bar{f}V\right] d\Sigma_{m_0} - \int_{C_{m_0F}} \bar{F}V \, d\xi\right] = 0$$

$$\forall t, s, \Gamma, K, V \text{ KA} \qquad (16.59)$$

The multipliers t and s, as Γ and K, are Hermitian operators on $\mathbf{E}_{20}$, and so one has posed the inner product

$$T_r(\Gamma_1\Gamma_2) = \tilde{\Gamma}_1\Gamma_2$$

if $\Gamma_1, \Gamma_2 \in \mathbf{E}_{30}$, an Euclidian space of Hermitian mappings of $\mathbf{E}_{20}$, where $\tilde{\Gamma}$ is the transpose of Γ according to this inner product. Condition (16.59) immediately gives

$$\int_{\Sigma_{m_0}} \left[\left[\frac{\partial \alpha}{\partial \Gamma} - \tilde{t}\right]\delta\Gamma + \left[\frac{\partial \alpha}{\partial K} - \tilde{s}\right]\delta K + \delta[\tilde{t}\underset{\sim}{\Gamma}(V) + \tilde{s}\underset{\sim}{K}(V)]\right.$$

$$\left. - \delta t\Gamma - \delta s K - \bar{f}\,\delta V\right] \mathrm{d}\Sigma_{m_0} - \int_{C_{m_0F}} \bar{F}\,\delta V\,\mathrm{d}\xi = 0$$

$$\forall\, \delta\Gamma, \delta K, \delta t, \delta s \in \mathbf{E}_{30}, \delta V \text{ KA} \tag{16.60}$$

which is an expression strictly similar to (16.20) with the following changes:

$$\begin{bmatrix} \mathsf{in} \\ \mathsf{im} \end{bmatrix} \Rightarrow S, \quad \begin{bmatrix} \Gamma \\ K \end{bmatrix} \Rightarrow D, \quad \begin{bmatrix} t \\ s \end{bmatrix} \Rightarrow t, \quad V \Rightarrow U, \quad S = \frac{\partial \alpha}{\partial D} \tag{16.61}$$

The stationary condition of the principle of Hellinger–Reissner is written here:

$$\delta\left[\int_{\Sigma_{m_0}} [\widetilde{\mathsf{in}}\underset{\sim}{\Gamma}(V) + \widetilde{\mathsf{im}}\underset{\sim}{K}(V) - \beta - \bar{f}V]\,\mathrm{d}\Sigma_{m_0} - \int_{C_{m_0F}} \bar{F}V\,\mathrm{d}\xi\right] = 0$$

$$\forall\, \mathsf{in}, \mathsf{im}, V \text{ KA}$$

with

$\beta = \underset{\sim}{\beta}(\mathsf{in}, \mathsf{im}) = \widetilde{\mathsf{in}}\Gamma + \widetilde{\mathsf{im}}K - \underset{\sim}{\alpha}(\Gamma, K)$, where Γ and K are functions of in and im by inversion of the constitutive equation (16.62)

Then, working as in Section 16.6 and calling in_0, im_0 the prestresses which in state ξ_0 equilibrate the given force densities f_0 and F_0 in that state, (16.59) successively gives

$$\int_{\Sigma_{m^0}} \left[(\widetilde{\mathsf{in}}_0 + \widetilde{\mathsf{im}}]\,\delta\Gamma + [\widetilde{\mathsf{im}}_0 + \widetilde{\mathsf{im}}_1]\,\delta K + \tilde{t}[\delta\underset{\sim}{\Gamma}(V) - \delta\Gamma] + \tilde{s}[\delta\underset{\sim}{K}(V) - \delta K]\right.$$

$$\left. + \delta t[\underset{\sim}{\Gamma}(V) - \Gamma] + \delta s[\underset{\sim}{K}(V) - K] - \bar{f}\,\delta V\right] \mathrm{d}\Sigma_{m_0} - \int_{C_{m_0F}} \bar{F}\,\delta V\,\mathrm{d}\xi = 0$$

$$\forall\, \delta\Gamma, \delta K, \delta t, \delta s \in \mathbf{E}_{30}, \delta V \text{ KA}$$

with the linear law

$$\begin{bmatrix} \mathsf{in}_1 \\ \mathsf{im}_1 \end{bmatrix} = \begin{bmatrix} {}^1\mathrm{A}_1 & {}^1\mathrm{A}_2 \\ \widetilde{{}^1\mathrm{A}_2} & {}^2\mathrm{A}_2 \end{bmatrix} \begin{bmatrix} \Gamma \\ K \end{bmatrix} \quad \text{or} \quad S = AD, \text{ with } \mathrm{A} = \tilde{\mathrm{A}} \in \mathscr{L}(\mathbf{E}_{30}, \mathbf{E}_{30}) \tag{16.63}$$

Let us decompose, as in Section 6, the multipliers into premultipliers with a subscript zero and increments with a subscript one, as also for the given external densities of force. Condition (16.63) becomes

$$\int_{\Sigma_{m_0}} [[\widetilde{\mathsf{in}}_0+\widetilde{\mathsf{in}}_1-\tilde{t}_0-\tilde{t}_1]\,\delta\Gamma+[\widetilde{\mathsf{im}}_0+\widetilde{\mathsf{im}}_1-\tilde{s}_0-\tilde{s}_1]\,\delta K$$

$$+(\tilde{t}_0+\tilde{t}_1)\,\delta\underset{\sim}{\Gamma}(V)+(\tilde{s}_0+\tilde{s}_1)\,\delta\underset{\sim}{K}(V)+\widetilde{\delta t}_1[\underset{\sim}{\Gamma}(V)-\Gamma]+\widetilde{\delta s}_1[\underset{\sim}{K}(V)-K]$$

$$-[\bar{f}_0+\bar{f}_1]\,\delta V]\,\mathrm{d}\Sigma_{m_0}-\int_{C_{m_0F}}[\bar{F}_0+\bar{F}_1]\,\delta V\,\mathrm{d}\xi=0$$

$$\forall\,\delta\Gamma,\,\delta K,\,\delta t,\,\delta s\in\mathbf{E}_{30}\text{ and }\delta V\text{ KA} \tag{16.64}$$

One now finds [12, 14]

$$\underset{\sim}{\Gamma}(V)=\frac{1}{2}\left[\pi_0\frac{\partial V}{\partial m_0}+\overline{\pi_0\frac{\partial V}{\partial m_0}}\right]+\frac{1}{2}\left[\overline{\frac{\partial V}{\partial m_0}}\,\pi_0\frac{\partial V}{\partial m_0}+2\overline{\frac{\partial V}{\partial m_0}}\,N_0\bar{N}_0\frac{\partial V}{\partial m_0}\right]+\cdots$$

$$\underset{\sim}{K}(V)=\overline{\pi_0\frac{\partial V}{\partial m_0}\frac{\partial N_0}{\partial m_0}}-\pi_0\frac{\partial}{\partial m_0}\left[\overline{\frac{\partial V}{\partial m_0}}N_0\right]+\pi_0\frac{\partial}{\partial m_0}\left[\overline{\pi_0\frac{\partial V}{\partial m_0}}\,\pi_0\overline{\frac{\partial V}{\partial m_0}}N_0\right]$$

$$-\overline{\frac{\partial V}{\partial m_0}}\frac{\partial}{\partial m_0}\left[\overline{\frac{\partial V}{\partial m_0}}N_0\right]+\cdots \tag{16.65}$$

and

$$\delta\Gamma=\frac{1}{2}\left[\pi_0\frac{\partial\delta V}{\partial m_0}+\overline{\pi_0\frac{\partial\delta V}{\partial m_0}}\right]+\frac{1}{2}\overline{\frac{\partial V}{\partial m_0}}\frac{\partial\delta V}{\partial m_0}+\overline{\frac{\partial\delta V}{\partial m_0}}\frac{\partial V}{\partial m_0}+\cdots$$

$$\delta K=\overline{\pi_0\frac{\partial\delta V}{\partial m_0}\frac{\partial N_0}{\partial m_0}}-\pi_0\frac{\partial}{\partial m_0}\left[\overline{\frac{\partial\delta V}{\partial m_0}}N_0\right]+\cdots \tag{16.66}$$

In (16.66), π_0 is the Hermitian (or orthogonal) projector from $\mathbf{E}_3$ upon the tangent planes $\mathbf{E}_{20}$ of Σ_{m_0}, which is supposed to be Riemannian. Moreover, the preceding developments rigorously take account of the Riemannian connection.

Supposing that the equilibrium is realized in the prestressed state ξ_0 and neglecting the energies due to the bending prestress im_0 and to the premultiplier s_0, we have first (for $V=0$):

$$\int_{\Sigma_{m_0}}\left[[\widetilde{\mathsf{in}}_0-\tilde{t}_0]\,\delta\Gamma+\tfrac{1}{2}\widetilde{\mathsf{in}}_0\left[\pi_0\frac{\partial\delta V}{\partial m_0}+\overline{\pi_0\frac{\partial\delta V}{\partial m_0}}\right]-\bar{f}_0\,\delta V\right]\mathrm{d}\Sigma_{m_0}$$

$$-\int_{C_{m_0F}}\bar{F}_0\,\delta V\,\mathrm{d}\xi=0\qquad\forall\,\delta\Gamma,\,\delta V,\text{ KA}$$

Then, in keeping only the terms of the first order in in_1, im_1, t_1, s_1, and V:

$$\int_{\Sigma_{m_0}} \Bigg[[\widetilde{\mathsf{in}}_1 - \tilde{t}_1]\,\delta\Gamma + [\widetilde{\mathsf{im}}_1 - \tilde{s}_1]\,\delta K$$

$$+\tfrac{1}{2}\widetilde{\mathsf{in}}_0\left[\overline{\frac{\partial V}{\partial m_0}}\frac{\partial \delta V}{\partial m_0} + \overline{\frac{\partial \delta V}{\partial m_0}}\frac{\partial V}{\partial m_0}\right]$$

$$+\tfrac{1}{2}\tilde{t}_1\left[\pi_0\frac{\partial \delta V}{\partial m_0} + \overline{\pi_0\frac{\partial \delta V}{\partial m_0}}\right]$$

$$+\tilde{s}_1\left[\overline{\pi_0\frac{\partial \delta V}{\partial m_0}\frac{\partial N_0}{\partial m_0}} - \pi_0\frac{\partial}{\partial m_0}\left[\overline{\frac{\partial \delta V}{\partial m_0}}N_0\right]\right]$$

$$+\delta t_1\left[\frac{1}{2}\left[\pi_0\frac{\partial V}{\partial m_0} + \overline{\pi_0\frac{\partial V}{\partial m_0}}\right] - \Gamma\right]$$

$$+\delta s_1\left[\overline{\pi_0\frac{\partial V}{\partial m_0}\frac{\partial N_0}{\partial m_0}} - \pi_0\frac{\partial}{\partial m_0}\left[\overline{\frac{\partial V}{\partial m_0}}N_0\right] - K\right]$$

$$-\bar{f}_1\,\delta V\Bigg]\,\mathrm{d}\Sigma_{m_0} - \int_{C_{m_0F}} \bar{F}_1\,\delta V\,\mathrm{d}\xi = 0$$

$$\forall\ \delta\Gamma,\ \delta K,\ \delta t_1,\ \delta s_1 \in \mathbf{E}_{30},\ \delta V \text{ KA} \tag{16.67}$$

The stability criterion may then be calculated from (16.67), in the linear case, and still with constant δt_1, δs_1, $\delta\Gamma$, δK, and δV; i.e.

$$\int_{\Sigma_{m_0}} \Bigg[\widetilde{\delta\mathsf{in}}_1\,\delta\Gamma + \widetilde{\delta\mathsf{im}}_1\,\delta K + \widetilde{\mathsf{in}}_0 \overline{\frac{\partial \delta V}{\partial m_0}}\frac{\partial \delta V}{\partial m_0}$$

$$+2\delta t_1\left[\frac{1}{2}\left[\pi_0\frac{\partial \delta V}{\partial m_0} + \overline{\pi_0\frac{\partial \delta V}{\partial m_0}}\right] - \Gamma\right] + 2\delta s_1$$

$$\left[\overline{\pi_0\frac{\partial \delta V}{\partial m_0}\frac{\partial N_0}{\partial m_0}} - \pi_0\frac{\partial}{\partial m_0}\left[\overline{\frac{\partial \delta V}{\partial m_0}}N_0\right] - K\right]\mathrm{d}\Sigma_{m_0} \geqslant 0 \tag{16.68}$$

$\forall\ \delta\Gamma, \delta K, \delta V$ KA, such that

$$\int_{\Sigma_{m0}} \widetilde{\delta\mathsf{in}}_1\left[\frac{1}{2}\left[\pi_0\frac{\partial \delta V}{\partial m_0} + \overline{\pi_0\frac{\partial \delta V}{\partial m_0}}\right] - \delta\Gamma\right]\mathrm{d}\Sigma_{m_0} = 0, \qquad \forall\ \underline{\delta\mathsf{in}_1}$$

$$\int_{\Sigma_{m_0}} \widetilde{\delta\mathsf{im}}_1\left[\overline{\pi_0\frac{\partial \delta V}{\partial m_0}\frac{\partial N_0}{\partial m_0}} - \pi_0\frac{\partial}{\partial m_0}\left[\overline{\frac{\partial \delta V}{\partial m_0}}N_0\right] - K\right]\mathrm{d}\Sigma_{m_0} = 0, \qquad \forall\ \underline{\delta\mathsf{im}_1}$$

Finally, in utilizing the Legendre transformation

$$\beta = \beta(\mathsf{in}_1, \mathsf{im}_1) = \widetilde{\mathsf{in}}_1\Gamma + \widetilde{\mathsf{im}}_1 K - \underset{\sim}{\alpha}(\Gamma, K)$$

where

$$\begin{bmatrix}\Gamma\\ K\end{bmatrix} = \begin{bmatrix}{}^1\mathrm{A}_1 & {}^1\mathrm{A}_2\\ \widetilde{{}^1\mathrm{A}_2} & {}^2\mathrm{A}_2\end{bmatrix}^{-1}\begin{bmatrix}\mathsf{in}_1\\ \mathsf{im}_1\end{bmatrix}$$

one finds the following criterion in the case of linear buckling:

$$\int_{\Sigma_{m_0}} \left[\tfrac{1}{2}[\widetilde{\delta \mathrm{in}_1}, \widetilde{\delta \mathrm{im}_1}] \begin{bmatrix} {}^1A_1 & {}^1A_2 \\ \widetilde{{}^1A_2} & {}^2A_2 \end{bmatrix}^{-1} \begin{bmatrix} \delta \mathrm{in}_1 \\ \delta \mathrm{im}_1 \end{bmatrix} + \widetilde{\mathrm{in}_0}\, \overline{\frac{\partial \delta V}{\partial m_0} \frac{\partial \delta V}{\partial m_0}} \right] d\Sigma_{m_0} \geqslant 0$$

$\forall\ \delta \mathrm{in}, \delta \mathrm{im}_1$, and δV KA, such that:

$$\int_{\Sigma_{m_0}} [\underline{\widetilde{\delta \mathrm{in}_1}}, \underline{\widetilde{\delta \mathrm{im}_1}}] \left\{ \begin{bmatrix} \tfrac{1}{2}\left[\pi_0 \dfrac{\partial \delta V}{\partial m_0} + \overline{\pi_0 \dfrac{\partial \delta V}{\partial m_0}}\right] \\ \overline{\pi_0 \dfrac{\partial \delta V}{\partial m_0} \dfrac{\partial N_0}{\partial m_0}} - \pi_0 \dfrac{\partial}{\partial m_0}\left[\overline{\dfrac{\partial \delta V}{\partial m_0}} N_0\right] \end{bmatrix} - \begin{bmatrix} {}^1A_1 & {}^1A_2 \\ \widetilde{{}^1A_2} & {}^2A_2 \end{bmatrix}^{-1} \begin{bmatrix} \delta \mathrm{in}_1 \\ \delta \mathrm{im}_1 \end{bmatrix} \right\}$$

$$d\Sigma_{m_0} = 0 \qquad \forall\ \underline{\delta \mathrm{in}}, \underline{\delta \mathrm{im}_1} \tag{16.69}$$

It should be noted that in the prestress term the derivative $\partial \delta V/\partial m_0$ appears, without any further approximation, with all its components (tangential and normal), and that on the other hand the formulation (16.69) of the criterion is strictly similar to that of the three-dimensional case.

It is quite possible to apply this criterion, subject to obvious modifications of functional spaces, to the case of the hybrid dual principle, as it is presented in [12] in the case of non-linear shells.

16.10 THE STABILITY CRITERION IN THE CASE OF THE COMPLEMENTARY ENERGY PRINCIPLE WITH ROTATION

A so-called principle of complementary energy was proposed in 1972 by Fraeijs de Veubeke, in the case of non-linear deformations [13]. This principle received some numerical applications, in particular in the case of deformations and buckling of plates [15, 16]. It has been extended to the case of shells of Kirchhoff and Love [17].

We propose in the following to apply the preceding method to the research of the static stability criterion for that principle whose statement lies, as for the preceding ones, in the introduction to the primal principle of a particular constraint. We shall briefly recall the statement of the principle in the simple three-dimensional case, and even the way to obtain it, as this is necessary for our purpose. Then we shall apply the general method which was described in Section 16.3 to come to the particular stability criterion relative to this principle. Taking again the notations of Section 16.4, the polar decomposition of the Jacobian of the transformation is introduced:

$$M = \underset{\sim}{M}(M_0)$$

that is:

$$\frac{\partial M}{\partial M_0} = R[1_{E_3} + h], \qquad R\bar{R} = \bar{R}R = 1_{E_3}, \qquad h = \bar{h} \in \mathscr{L}(\mathbf{E}_3, \mathbf{E}_3) \tag{16.70}$$

where, given the Jacobian $\partial M/\partial M_0$, R and h are unique and represent respectively the local rotation and the local extension functions of M_0.

The constraint (16.70) is then brought into the primal principle (16.18) by means of a Lagrange multiplier $\bar{t} \in \mathscr{L}(\mathbf{E}_3, \mathbf{E}_3')$, which gives the stationary condition:

$$\delta\left[\int_{\Omega_0}\left[\alpha + T_r\left(t\left[\frac{\partial M}{\partial M_0} - R(1_{E_3}+h)\right]\right) \mathrm{d}\Omega_0\right]\right.$$
$$\left. - \int_{\Omega_0} \bar{f}U \,\mathrm{d}\Omega_0 - \int_{\Sigma_{0F}} \bar{F}U \,\mathrm{d}\Sigma_0\right] = 0$$
$$\forall\ \delta U \text{ KA}, \delta t, \delta R \mid R\bar{R} = \bar{R}R = 1_{E_3}, \delta h = \delta h \qquad (16.71)$$

with $\alpha = \underset{\sim}{\alpha}(h)$.

We have posed $\alpha = \underset{\sim}{\alpha}(h)$, for (16.70) gives

$$D = \frac{1}{2}\left[\frac{\partial \bar{M}}{\partial M_0}\frac{\partial M}{\partial M_0} - 1_{E_3}\right] = \tfrac{1}{2}[[1_{E_3}+h]^2 - 1_{E_3}] = h + \frac{h^2}{2} \qquad (16.72)$$

and as the medium is supposed to be hyperelastic, α was initially a function of D.

Equation (16.71) then provided the following results [13, 17]:

$$\frac{\partial \alpha}{\partial h} = r = \frac{tR + \overline{tR}}{2}$$
$$R[1_{E_3}+h]t = \overline{R(1_{E_3}+h)t} \qquad (16.73)$$
$$t = S[1_{E_3}+h]\bar{R} \qquad \left(\text{or } t = S\overline{\frac{\partial M}{\partial M_0}}\right)$$

In the first equation of (16.73), r, which is the dual quantity of the extension h, represents the Jauman stress and $\tilde{r}$, its transpose in $\mathbf{E}_g$, represents a so-called space of the mappings of $\mathbf{E}_3$, such that

$$\forall\ t_1, t_2 \in \mathscr{L}(\mathbf{E}_3, \mathbf{E}_3), \qquad T_r(t_1 t_2) = \tilde{t}_1 \bar{t}_2{}^*$$

The first relation of (16.73) is provided by a variation of δh in (16.71) and the second one of (16.73) by a variation δR. The elimination of h, in taking

* It should be noted that here t_1, t_2 are non-symmetric mappings of $\mathbf{E}_3$ or vectors of $\mathbf{E}_g$, and still $\tilde{t}_1 \bar{t}_2 \Rightarrow t_1^{ij} t_{2ij}$ in conventional notations.

account of the first relation of (16.73) which also provides

$$h = \underset{\sim}{h}(r)$$

then gives the following mixed principle:

$$\delta\left[\int_{\Omega_0}\left[T_r\left(t\frac{\partial U}{\partial M_0}+t-tR\right)-\beta-\bar{f}U\right]\mathrm{d}\Omega_0-\int_{\Sigma_{0F}}\bar{F}U\,\mathrm{d}\Sigma_0\right]=0$$

$$\forall\, t\in\mathscr{L}(\mathbf{E}_3,\mathbf{E}_3),\ U \text{ KA},\ R \mid R\bar{R}=1_{E_3}$$

$$\text{with } \beta=\underset{\sim}{\beta}(r)=\int_0^r \underset{\sim}{h}(\rho)\,\mathrm{d}\rho \text{ and } r=\frac{tR+\overline{tR}}{2} \qquad (16.74)$$

For statically admissible stresses, (16.14) gives the principle of complementary energy with rotation of Fraeijs de Veubeke. One may also remark that the complementary energy β is given by the Legendre transformation:

$$\beta=\underset{\sim}{\beta}(r)=\bar{r}\bar{h}-\underset{\sim}{\alpha}(h), \qquad \text{where } h=\underset{\sim}{h}(r) \qquad (16.75)$$

To find the stability criterion relative to principle (16.74), it is important to emphasize that this criterion can neither be deduced from (16.74), as we have already said, nor even from (16.71), for (16.71) does not include explicitly all the constraints. In fact, the three variables U, R, and h are constrained by (16.70), that is to say, mainly by

$$\frac{\partial M}{\partial M_0}=R[1_{E_3}+h], \qquad h=\bar{h}$$
$$\bar{R}R=1_{E_3} \qquad (16.76)$$

while the constraint $R\bar{R}=1_{E_3}$ results from the second equation of (16.76). It is proper to introduce them into (16.18) with the exception of condition $h=\bar{h}$, which will be taken account of 'externally' since it does not add any supplementary term to the criterion.

One will introduce the second equation of (16.76) by means of a supplementary multiplier

$$s=\bar{s}\in\mathscr{L}(\mathbf{E}_3,\mathbf{E}_3)$$

giving the principle

$$\delta\left[\int_{\Omega_0}\left[\alpha+T_r\left(t\left[\frac{\partial M}{\partial M_0}-R[1_{E_3}+h]\right]\right.\right.\right.$$
$$\left.\left.+s(\bar{R}R-1_{E_3})-\bar{f}U\right]\mathrm{d}\Omega_0-\int_{\Sigma_{0F}}\bar{F}U\,\mathrm{d}\Sigma_0\right]=0$$

$$\forall\,\delta U \text{ KA},\ \delta R,\ \delta t,\ \delta s,\ \delta h-\overline{\delta h}$$

$$\text{with } \alpha=\underset{\sim}{\alpha}(h) \qquad (16.77)$$

This easily gives, besides the preceding relation:

$$s+\bar{s}=[1_{E_3}+h]tR=\overline{[1_{E_3}+h]tR}=2s^* \tag{16.78}$$

An immediate application of the criterion (16.13), for instance, then gives the following stability criterion:

$$\int_{\Omega_0}[\delta^2\underset{\sim}{\alpha}(h)+T_r(-2t\,\delta R\,\delta h+[1_{E_3}+h]tR\,\overline{\delta R}\,\delta R)]\,\mathrm{d}\Omega_0\geqslant 0$$

$$\text{with}\quad h=\underset{\sim}{h}(r),\ r=\frac{tR+\overline{tR}}{2},\ \delta h=C^{t_e}_{=}$$

$$\forall\,\delta R\mid \bar{R}R=1_{E_3}\quad\text{and}\quad \delta h=\frac{\partial h}{\partial r}\,\delta r \tag{16.79}$$

Using a similar argument to that which gave (16.29) and utilizing (16.75), it is seen that, with constant δr

$$\delta^2\alpha=\delta^2\beta=\frac{\partial^2\beta}{\partial r\,\partial r}(\delta r)(\delta r),\qquad \left(\tilde{h}=\frac{\partial\beta}{\partial r}\text{ supposed to exist}\right) \tag{16.80}$$

Criterion (16.79) can then be written in the following way (successively):

$$\int_{\Omega_0}\left[\frac{\partial^2\beta}{\partial r\,\partial r}(\delta r)(\delta r)+T_r(-2t\,\delta R\bar{R}R\,\delta h\right.$$
$$\left.+[1_{E_3}+h]tR\,\overline{\delta R}R\bar{R}\,\delta R)\right]\mathrm{d}\Omega_0\geqslant 0 \tag{16.81}$$

$$\int_{\Omega_0}\left[\frac{\partial^2\beta}{\partial r\,\partial r}(\delta r)(\delta r)+T_r(2tR\,\overline{\delta RR}\,\delta h\right.$$
$$\left.-[1_{E_3}+h]tR\,\overline{\delta RR}\,\overline{\delta RR})\right]\mathrm{d}\Omega_0\geqslant 0 \tag{16.82}$$

$$\int_{\Omega_0}\left[\frac{\partial^2\beta}{\partial r\,\partial r}(\delta r)(\delta r)+2\frac{\partial^2\beta}{\partial r\,\partial r}(\delta r)(tR\,\overline{\delta RR})-T_r([1_{E_3}+h]tR[\overline{\delta RR}]^2)\right]\mathrm{d}\Omega_0\geqslant 0$$
$$\forall\,\overline{\delta RR}=-\bar{R}\,\delta R\quad\text{and}\quad \delta r=\overline{\delta r} \tag{16.83}$$

Condition (16.83) is the stability criterion relative to principle (16.74). It should be noted that, due to the particular form of the constraint between U, h, and R, this criterion is independent of δU, and only depends, as normally expected, on the variables which exist in the principle. If the complementary principle itself is considered, the functional space of the stresses is properly modified.

* The meaning of this stress s is to constrain the mapping R of $\mathbf{E}_3$ to be a pure unitary rotation, skew-symmetric when linearized.

16.10.1 Verification

The field U being given by (16.72), h and δh may be considered as functions of U and δU only. Thus, with constant δU:

$$\begin{aligned} \delta^2 \underset{\sim}{\alpha}(U) &= \frac{\partial^2 \alpha}{\partial h\, \partial h}[\delta \underset{\sim}{h}(U)](\delta \underset{\sim}{h}(U)) + \frac{\partial \alpha}{\partial h}\, \delta^2 \underset{\sim}{h}(U) \\ \delta^2 \underset{\sim}{\alpha}(U) &= \frac{\partial^2 \alpha}{\partial h\, \partial h}(\delta h(U))(\delta \underset{\sim}{h}(U)) + \tilde{r}\, \delta^2 \underset{\sim}{h}(U) \end{aligned} \tag{16.84}$$

Now (16.72) gives

$$\delta D = \delta h + h\, \delta h$$

$$\delta^2 D = \delta^2 h[1_{E_3} + h] + [\delta h]^2$$

Hence

$$\delta^2 h[1_{E_3} + h] = \delta^2 D - [\delta h]^2 \tag{16.85}$$

Let us consider the linearized constraint

$$\frac{\partial \delta U}{\partial M_0} = \delta R[1_{E_3} + h] + R\, \delta h \tag{16.86}$$

We have seen, in Section 16.4, that with constant δU:

$$\delta^2 D = \overline{\frac{\partial \delta U}{\partial M_0}}\, \frac{\partial \delta U}{\partial M_0}$$

which can be written with (16.86):

$$\delta^2 D = [[1_{E_3} + h]\, \overline{\delta R} + \delta h \bar{R}][\delta R[1_{E_3} + h] + R\, \delta h]$$

or

$$\begin{aligned} \delta^2 D = {} & [1_{E_3} + h]\, \overline{\delta R}\, \delta R[1_{E_3} + h] + [1_{E_3} + h]\, \overline{\delta R}\, R\, \delta h \\ & + \delta h\, \bar{R}\, \delta R[1_{E_3} + h] + [\delta h]^2 \end{aligned}$$

Equation (16.85) then gives

$$\begin{aligned} \delta^2 h[1_{E_3} + h] = {} & [1_{E_3} + h]\, \overline{\delta R}\, \delta R[1_{E_3} + h] \\ & + [1_{E_3} + h]\, \overline{\delta R}\, R\, \delta h + \delta h\, \bar{R}\, \delta R[1_{E_3} + h] \end{aligned} \tag{16.87}$$

Let us consider, for the sake of simplification, the case where the extension h is negligible compared with 1_{E_3} (what is incidentally a reasonable practical assumption). Equation (16.87) then gives in that approximation:

$$\delta^2 h \simeq \overline{\delta R}\, \delta R + \overline{\delta R}\, R\, \delta h + \delta h\, \bar{R}\, \delta R \tag{16.88}$$

However, the third equation of (16.73) also gives in that approximation:

$$tR \simeq S = \bar{S} = \frac{\widetilde{\partial \alpha}}{\partial D}$$

and (16.72) gives

$$D \simeq h$$

Thus:

$$tR \simeq S = \bar{S} \simeq \frac{\widetilde{\partial \alpha}}{\partial h} = r \qquad (16.89)$$

Equation (16.85) becomes, taking account of (16.89) and (16.88):

$$\frac{\partial^2 \alpha}{\partial h\, \partial h}(\delta h)(\delta h) + T_r(tR[\overline{\delta R}\, \delta R + \overline{\delta R}\, R\, \delta h + \delta h\, \bar{R}\, \delta R])$$

which is precisely the integrand of (16.79) in that approximation, or also of (16.82).

Notice that it is quite possible to refine this approximation by putting

$$\frac{\partial M}{\partial M_0} = R[1_{E_3} + \varepsilon h], \qquad \frac{\partial \delta U}{\partial M_0} = \delta R[1_{E_3} + \varepsilon h] + \varepsilon R\, \delta h, \qquad \varepsilon \ll 1$$

and neglecting the terms of the same order (16.82) and (16.27).

16.10.2 Another form of the stability criterion

Let us consider the criterion (16.83) relative to the so-called principle of complementary energy and let us pose:

$$\int_{\Omega_0} \frac{\partial^2 \beta}{\partial r\, \partial r}(\delta r)(\delta r)\, d\Omega_0 = \widetilde{\delta r}\, K_1\, \delta r, \qquad K_1 = \tilde{K}_1$$

$$\int_{\Omega_0} \frac{\partial^2 \beta}{\partial r\, \partial r}(\delta r)(tR\, \overline{\delta R}\, R)\, d\Omega_0 = \widetilde{\delta r}\, K_1[k(\overline{\delta R}\, R)] = \widetilde{\delta r}\, K_1 k\, \overline{\delta R}\, R \qquad (16.90)$$

$$\int_{\Omega_0} T_r([1_{E_3} + h]tR[\overline{\delta R}\, R]^2)\, d\Omega_0 = \widetilde{\overline{\delta R}\, R} K_2\, \overline{\delta R}\, R, \qquad K_2 = \tilde{K}_2$$

where we have extended, with obvious notations, the transposition and inner product in the functional space.

Then, the criterion (16.83) becomes, with (16.90):

$$\widetilde{\delta r}\, K_1\, \delta r + 2\, \widetilde{\delta r}\, K_1 k\, \overline{\delta R}\, R + \widetilde{\overline{\delta R}\, R}\, K_2\, \overline{\delta R}\, R \geqslant 0$$

$$\forall\ \delta r = \widetilde{\delta r} \quad \text{and} \quad \overline{\delta R}\, R = -\bar{R}\, \delta R \in \mathscr{L}(\mathbf{E}_3, \mathbf{E}_3) = \mathbf{E}_g$$
$$\text{with} \quad K_1 = \tilde{K}_1,\ k, \quad \text{and} \quad K_2 = \tilde{K}_2 \in \mathscr{L}(\mathbf{E}_g, \mathbf{E}_g) \qquad (16.91)$$

This can also be written in matrix form:

$$[\widetilde{\delta r}\ \widetilde{\overline{\delta R}\, R}]\begin{bmatrix} K_1 & \widetilde{K_1 k} \\ K_1 k & K_2 \end{bmatrix}\begin{bmatrix} \delta r \\ \overline{\delta R}\, R \end{bmatrix} \geqslant 0$$

$$\forall\ \delta r = \overline{\delta r},\ \overline{\delta R}\, R = -\bar{R}\,\delta R \in \mathscr{L}(\mathbf{E}_3, \mathbf{E}_3) \qquad (16.92)$$

The matrices of (16.92) being Hermitian in $\mathbf{E}_g \times \mathbf{E}_g$, it is seen that, going from a stable state, the critical point will be obtained for solutions t_0 and R_0, which depends on the terms of this matrix such that

$$K_1\ \delta r + \widetilde{K_1 k}\ \overline{\delta R}\, R = 0 \qquad (16.93)$$

$$K_1 k\ \delta r + K_2\ \overline{\delta R}\, R = 0$$

From the first equation of (16.90), K_1 is positive definite; therefore the first equation of (16.93) gives

$$\delta r = -K_1^{-1}\ \widetilde{K_1 k}\ \overline{\delta R}\, R$$

Hence

$$[-K_1 k K_1^{-1} \widetilde{K_1 k} + K_2]\ \overline{\delta R}\, R = 0, \qquad \overline{\delta R}\, R \neq 0 \qquad (16.94)$$

The critical solution makes the operator singular:

$$K_2 - K_1 k K_1^{-1} \widetilde{K_1 k}$$

The transformation of these formulae in matrix form, through the discretization of the unknown fields, is immediate, at least in its principle.

It can also be seen from (16.90) that k and K_2 are linear with respect to tR, i.e. the solution (practically with respect to the Jauman stress which has been reached). It is again easy to find the classical notions of elastic and geometrical stiffnesses in (16.94). With regards to linear buckling one would have, for instance,

$$\left.\begin{aligned} K_1 k &= \lambda [K_1 k]^* \\ K_2 &= \lambda K_2^* \end{aligned}\right\}, \qquad \lambda \in [0, +\infty[$$

Conditions (16.93) then become

$$K_1\ \delta r + \lambda [\widetilde{K_1 k}]^*\ \overline{\delta R}\, R = 0$$

$$(K_1 k)^* + K_2^*\ \overline{\delta R}\, R = 0$$

while (16.94) becomes

$$[-\lambda [K_1 k]^* K_1^{-1} [\widetilde{K_1 k}]^* + K_2^*]\ \overline{\delta R}\, R = 0$$

If K_2 admitted an inverse, one would again find

$$[K_1 - \lambda[\widetilde{K_1 k}]^* K_2^{*-1} K_1 k]\,\delta r = 0, \qquad \delta r \neq 0$$

a formulae where λ occupies the usual place, in problems of linear buckling, as the coefficient of the geometrical stiffness.

16.11 CONCLUSION

As we have seen, the static stability criterion of elastic structures, in the case of mixed or mixed hybrid principles, cannot be deduced immediately from the principles themselves. It is proper to go back deliberately to the formulation of these principles from the primal one of the total potential energy.

The utilization of a very general theorem relative to the constrained systems allows us to find the formulation of the criterion in all cases. The result becomes very simple and almost obvious in most mixed principles, but may appear much less obvious in the cases of the so-called principle of complementary energy in the non-linear case, and also in the case of non-holonomic constraints.

We have insisted on the necessity of taking some elementary but essential precautions when applying the general theorem, where all the constraints have to be taken into account *a priori* and to be brought in a weak form into the principle in consideration.

It is of interest that the notions of elastic and geometrical stiffnesses remain valid in all cases.

The application relative to the linear buckling of shells can be complicated, rather notably, as we saw, if one takes account of the variation of the Riemannian connection, which must be done rigorously. Let us note incidentally that the extension to shells of the stability criterion in the case of complementary energy can be done without any particular difficulty, by means of the extension which has been made of such a principle to the case of thin shells by Kirchhoff and Love [17].

REFERENCES

1. S. Chryssafi, *Principes Variationnels Mixtes en Statique. Couplage de ces Principes avec le Principe des Déplacements*, Doctoral thesis, 3rd ed., N. T. ONERA No. 1980–2, Paris, 1980.
2. R. Ohayon and R. Valid, *Principe Duaux en Vibrations Harmoniques. Formulation Symétrique Couplée et Sous-structuration* (to be published).
3. Ph. Gibert, *Les Principes Variationnels Mixtes en Mécanique des Vibrations: Vibrations Harmoniques de Structures Présentant des Liaisons non Holonomes—Encadrement Numérique de Fréquences Propres de Structures Complexes*, Docteur-Ingénieur thesis, Paris, 1980. Publication ONERA (to appear).

4. R. J. Knops and E. W. Wilkes, *Theory of Elastic Stability*, Handbook of Physics, Vol. Vl, a/3, Mechanics of Solids III, 1973.
5. R. Valid, *La Mécaniques des Milieux Continus et le Calcul des Structures*, Eyrolles, Paris, 1977; North-Holland, 1981.
6. W. T. Koiter, 'Stability of Equilibrium of continuous bodies', Technical Report No. 79, Brown University, April 1962.
7. R. Valid, 'La théorie et les méthodes de calcul des structures en vibration', *Cycle de Conférences CEA-EDF: Vibrations des Structures dans le Domaine Industriel*, Jouy-en-Josas, France, 15–19 Oct. 1979. Publication ONERA No. 1980–1.
8. R. Valid, 'Calcul sélectif de l'influence de grandes modifications structurales sur les modes propres d'une structure linéaire', *La Rech. Aérosp.*, **6,** 359–364 (1971).
9. Y. Mesière, 'Calcul des charges critiques en élasticité sous chargement conservatif', *Third Congrès Français de Mécanique*, Grenoble, France, 6–9 Sept. 1977.
10. P. A. Raviart, *Cours de D. E. A. d'Analyse Numérique*, Vol. VI, Université Paris, 1979–80.
11. P. L. Boland and T. H. H. Pian, 'Large deflection analysis of these elastic structures by the assumed stress hybrid elastic structures by the assumed stress hybrid element method', *Computers and Structures*, **7,** 1–12 Feb. 1977.
12. R. Valid, 'An intrinsic formulation for the non-linear theory of shells and some approximations', *Trends in Computerized Structural Analysis and Synthesis*, 30 Oct.–1 Nov. 1978, Washington, D. C.; *Computers and Structures*, **10,** 183–194 (1979); T. P. ONERA No. 1979–2.
13. B. Fraeijs de Veubeke, 'A new variational principle for finite elastic displacements', *Int. J. Eng. Sc.*, **10,** 745–763 (1972).
14. R. Valid, *Déformations Non Lineaires et Flambage Statique*, Publication ONERA No. 1977–2.
15. C. Sanders and E. Carnoy, 'Hybrid stress finite element model for elastic-plastic analysis', *Int. Conf. on Finite Elements in Non-Linear Solid and Structural Mechanics*, Gello, Norway, 29 Aug.–1 Sept. 1977, A 04, 1–19.
16. S. N. Atluri and H. Murakawa, 'On hybrid finite element models in non-linear solid mechanics', *Int. Conf. on Finite Elements in Non-Linear Solid and Structural Mechanics*, Gello, Norway, 29 Aug.–1 Sept. 1977, A 01, 1–38.
17. R. Valid, 'Le principe complementaire en théorie non lineaire des coques', *CRA Sc.* t. 289, December 1979, p. 293; 'The principle of complementary energy in the non-linear theory of shells', *Com. Fifteenth ICTAM*, Toronto, 17–23 Aug. 1980, T. P. ONERA 1980–86.

Hybrid and Mixed Finite Element Methods
Edited by S. N. Atluri, R. H. Gallagher, and O. C. Zienkiewicz

Chapter 17

Finite Element Analysis of Stress and Strain Singularity Eigenstate in Inhomogeneous Media or Composite Materials

Y. Yamada and H. Okumura

17.1 INTRODUCTION

Inverse square root, $1/\sqrt{r}$, singularity is well established for the stress and strain distribution at the crack tip of homogeneous isotropic [1 to 4] and anisotropic [5] elastic materials. In the finite element analysis this property is utilized for the development of the quarter-point element [6, 7] and the stress intensity factors are determined for two- and three-dimensional specimens of elastic fracture mechanics. The $1/\sqrt{r}$ singularity at the tip of the crack does not hold for the inhomogeneous media, a typical example of which is composite material. Therefore, in general, the state of stress and strain singularities should be determined for each case of interest or concern by solving the pertinent eigenvalue problem. Analyses of a two-material wedge by Hein and Erdogan [8] and of the stress at the crack normal to the interface of two materials by Cook and Erdogan [9] are examples of such study. Reconsiderations were given later to similar problems by Lin and Mar [10] who derived a more compact analysis procedure and discussed in detail the crack lying along as well as normal to the bimaterial interface. Lin and Mar were also the forerunners in the application of the finite element method to this type of singular boundary value problem. Their method is characterized by the use of the hybrid crack element as was the case in an earlier work of Tong, Pian, and Lasry [11] and contrasts with the formulation of Fujitani [12] who solved the two-dimensional crack problem by a method based essentially on the virtual work principle of conventional type. The extensions of the finite element and/or finite difference methods to the three-dimensional cases are due to Bažant and Estenssoro [13], Fujitani [14], Benthem [15], and Benthem and Douma [16].

The present paper introduces a simpler approach and illustrates its competence by a couple of examples, furnishing satisfactory predictions for the stress and strain eigenstate at the crack tip. It features the use of the singularity transformation proposed by Yamada *et al.* [17] and can be easily incorporated in the standard or conventional finite element programmes. The derivation of relevant equations or formulae is described first with respect to the case of plane stress and strain distribution and then extended to three dimensions. Numerical examples cover elementary problems in elastic fracture mechanics, characteristic stress eigenstate at the plane crack in composite materials, and the three-dimensional stress singularity near the terminal point of the crack front at the surface of a semi-infinite body. The solutions have been verified by comparison with the analytical and/or precedent numerical solutions, when available.

17.2 BASIC FORMULATION

The finite element used in the present analysis is a triangular or sector element for the plane stress and strain cases and is placed around the crack tip as shown in Figure 17.1. In conventional non-singular applications, the mapping of the element configuration from the parametric ξ, η coordinates plane to the physical one is expressed as follows. Referring to Figure 17.2 and using the polar coordinates r, θ in the physical plane,

$$r=\frac{1+\xi}{2}r_0 \qquad \text{or} \qquad \rho=\frac{r}{r_0}=\frac{1+\xi}{2} \tag{17.1}$$

$$\theta=H_1\theta_1+H_2\theta_2+H_3\theta_3 \tag{17.2}$$

where

$$H_1=\tfrac{1}{2}(-\eta+\eta^2), \qquad H_2=1-\eta^2, \qquad H_3=\tfrac{1}{2}(\eta+\eta^2) \tag{17.3}$$

r and θ represent the polar coordinate radius and the angle of a generic

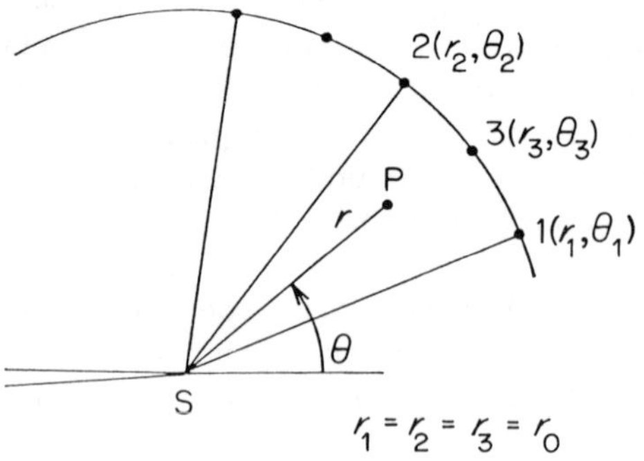

Figure 17.1 Triangular sector element at the crack tip

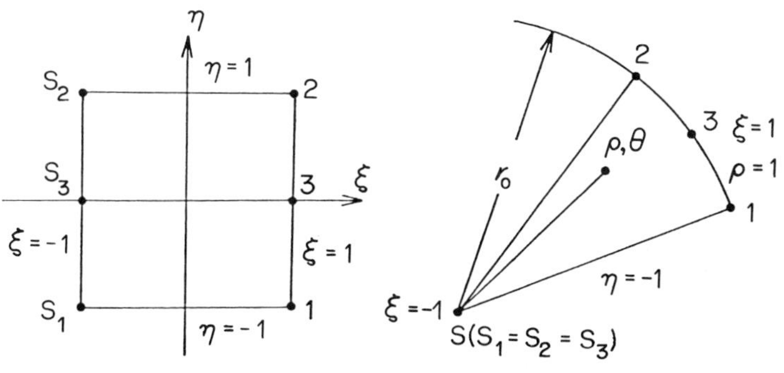

Figure 17.2 Mapping of element from parametric coordinates ξ, η plane

point P, and r_0 is the radius of the common circular arc on which the nodes 1, 2 and 3 are placed.

Equations (17.1) and (17.2) are consistent with the isoparametric formulation which assumes the displacement field as the following:

$$\mathbf{u}-\mathbf{u}_s=[H_1(\mathbf{u}_1-\mathbf{u}_s)+H_2(\mathbf{u}_2-\mathbf{u}_s)+H_3(\mathbf{u}_3-\mathbf{u}_s)]\frac{1+\xi}{2} \tag{17.4}$$

$\mathbf{u}_s$ represents the displacement vector at the vertex S of the sector and $\mathbf{u}_i$ ($i=1, 2, 3$) are the displacement vectors at the nodes on the circular arc of radius $r=r_0$. Introducing the relative displacement $\bar{\mathbf{u}}=\mathbf{u}-\mathbf{u}_s$ and $\bar{\mathbf{u}}_i=\mathbf{u}_i-\mathbf{u}_s$, we can rewrite equation (17.4) as

$$\bar{\mathbf{u}}=(H_1\bar{\mathbf{u}}_1+H_2\bar{\mathbf{u}}_2+H_3\bar{\mathbf{u}}_3)\frac{1+\xi}{2} \tag{17.5}$$

As a consequence of the use of equation (17.5), written in terms of the relative displacement, the characteristic equation derived in the sequel will not contain the eigenvalue corresponding to the rigid body mode. It is noted that the variable-number nodes concept is applicable to the above finite element routine by the method of Bathe and Wilson [18] or the procedure proposed by Okabe, Yamada, and Nishiguchi [19, 20].

In the singular transformation of Yamada *et al.* [17], the following mapping is applied in place of equation (17.1):

$$r=\left(\frac{1+\xi}{2}\right)^{1/\lambda} r_0 \quad \text{or} \quad \rho=\frac{r}{r_0}=\left(\frac{1+\xi}{2}\right)^{1/\lambda} \tag{17.6}$$

This transformation gives rise to a term ρ^λ in the expression of the relative displacement field of equation (17.5) as

$$\bar{\mathbf{u}}=(H_1\bar{\mathbf{u}}_1+H_2\bar{\mathbf{u}}_2+H_3\bar{\mathbf{u}}_3)\rho^\lambda \tag{17.7}$$

The singularity of stress and strain thus introduced is in the order of $\rho^{\lambda-1}$.

From equations (17.6), (17.2), and (17.3), we have

$$\frac{\partial r}{\partial \xi}=\frac{r_0}{2\lambda}\rho^{1-\lambda}, \qquad \frac{\partial r}{\partial \eta}=0$$
$$\frac{\partial \theta}{\partial \xi}=0, \qquad \frac{\partial \theta}{\partial \eta}=\tfrac{1}{2}(\theta_3-\theta_1) \tag{17.8}$$

Note that it is assumed that $\theta_2=(\theta_1+\theta_3)/2$. Inversion of the equations (17.8) yield

$$\frac{\partial \xi}{\partial r}=\frac{1}{2\,|J|}(\theta_3-\theta_1), \qquad \frac{\partial \eta}{\partial r}=0$$
$$\frac{\partial \xi}{\partial \theta}=0, \qquad \frac{\partial \eta}{\partial \theta}=\frac{1}{2\,|J|}\frac{r_0}{\lambda}\rho^{1-\lambda} \tag{17.9}$$

where $|J|$ is the determinant of the Jacobian matrix:

$$|J|=\begin{vmatrix}\partial r/\partial \xi & \partial r/\partial \eta\\ \partial \theta/\partial \xi & \partial \theta/\partial \eta\end{vmatrix}=\frac{r_0}{4\lambda}\rho^{1-\lambda}(\theta_3-\theta_1) \tag{17.10}$$

The strain components ε_r, ε_θ, and $\gamma_{r\theta}$ in polar coordinates are given as

$$\varepsilon_r=\frac{\partial u_r}{\partial r}$$
$$\varepsilon_\theta=\frac{1}{r}\frac{\partial v_\theta}{\partial \theta}+\frac{u_r}{r} \tag{17.11}$$
$$\gamma_{r\theta}=\frac{1}{r}\frac{\partial u_r}{\partial \theta}+\frac{\partial v_\theta}{\partial r}-\frac{v_\theta}{r}$$

where u_r and v_θ represent the radial and circumferential components of the displacement vector $\mathbf{u}$. Strains are dependent merely on the relative displacement, so we can use equation (17.5) for the evaluation of ε_r, ε_θ, and $\gamma_{r\theta}$ of equation (17.11). Thus, considering equations (17.9) and (17.10), we have, for ε_r,

$$\varepsilon_r=\frac{\partial \bar{u}_r}{\partial r}=\frac{\partial \bar{u}_r}{\partial \xi}\frac{\partial \xi}{\partial r}+\frac{\partial \bar{u}_r}{\partial \eta}\frac{\partial \eta}{\partial r}$$
$$=\frac{\lambda}{r_0}\rho^{\lambda-1}(H_1\bar{u}_{r1}+H_2\bar{u}_{r2}+H_3\bar{u}_{r3})$$

Similarly, for ε_θ and $\gamma_{r\theta}$,

$$\varepsilon_\theta=\frac{1}{r_0}\rho^{\lambda-1}\left[(H_1\bar{u}_{r1}+H_2\bar{u}_{r2}+H_3\bar{u}_{r3})+\frac{2}{\theta_s}\left(\frac{\partial H_1}{\partial \eta}\bar{v}_{\theta 1}+\frac{\partial H_2}{\partial \eta}\bar{v}_{\theta 2}+\frac{\partial H_3}{\partial \eta}\bar{v}_{\theta 3}\right)\right]$$

$$\gamma_{r\theta}=\frac{1}{r_0}\rho^{\lambda-1}\left[\frac{2}{\theta_s}\left(\frac{\partial H_1}{\partial \eta}\bar{u}_{r1}+\frac{\partial H_2}{\partial \eta}\bar{u}_{r2}+\frac{\partial H_3}{\partial \eta}\bar{u}_{r3}\right)+(\lambda-1)\right.$$
$$\left.\times(H_1\bar{v}_{\theta 1}+H_2\bar{v}_{\theta 2}+H_3\bar{v}_{\theta 3})\right]$$

where θ_s denotes the including angle $\theta_3 - \theta_1$ of the sector element at the vertex S.

The above relations can be summarized in the following matrix form:

$$\{\varepsilon\} = \begin{Bmatrix} \rho_r \\ \varepsilon_\theta \\ \gamma_{r\theta} \end{Bmatrix} = \sum_i [B_i]\{\bar{u}_i\} = [B]\{\bar{u}\}, \qquad \{\bar{u}_i\} = \begin{Bmatrix} \bar{u}_{ri} \\ \bar{u}_{\theta i} \end{Bmatrix} \tag{17.12}$$

where

$$[B] = \frac{1}{r_0} \rho^{\lambda-1} (\lambda [B_a] + [B_b]) \tag{17.13}$$

The generic components of the matrices $[B_a]$ and $[B_b]$ are respectively

$$[B_{ia}] = \begin{bmatrix} H_i & 0 \\ 0 & 0 \\ 0 & H_i \end{bmatrix} \quad [B_{ib}] = \begin{bmatrix} 0 & 0 \\ H_i & \dfrac{2}{\theta_s} \dfrac{\partial H_i}{\partial \eta} \\ \dfrac{2}{\theta_s} \dfrac{\partial H_i}{\partial \eta} & -H_i \end{bmatrix} \quad \text{for } i = 1, 2, 3 \tag{17.14}$$

The exponent $(\lambda - 1)$ to the non-dimensional radius ρ in equation (17.13) represents the strain singularity in the present analysis. It is noted that the modification is only pertinent to the term multiplied by λ in the strain–nodal displacement matrix $[B]$, compared to the conventional non-singular case where λ equals unity.

17.3 DERIVATION OF CHARACTERISTIC EQUATION

The characteristic equation for the determination of eigenstate can be obtained from the virtual work principle. For plane problems the principle is expressed as follows. Considering the region having the outer radius r_0 depicted in Figure 17.3,

$$\int_0^{r_0} \int_{-\pi}^{\pi} (\sigma_r \delta\varepsilon_r + \sigma_\theta \, \delta\varepsilon_\theta + \tau_{r\theta} \, \delta\gamma_{r\theta}) tr \, \mathrm{d}r \, \mathrm{d}\theta = r_0 \int_{-\pi}^{\pi} (T_r \, \delta\bar{u}_r + T_\theta \, \delta\bar{v}_\theta) t \, \mathrm{d}\theta \tag{17.15}$$

where t represents the thickness of the element. Each finite element or sector contributes to the left-hand side of equation (17.15) by an amount given as the following. Writing in matrix form and transforming to the parametric coordinates ξ, η:

$$r_0 \iint_{-1}^{1} \lfloor \delta\varepsilon \rfloor \{\sigma\} \rho \, |J| \, t \, \mathrm{d}\xi \, \mathrm{d}\eta \tag{17.16}$$

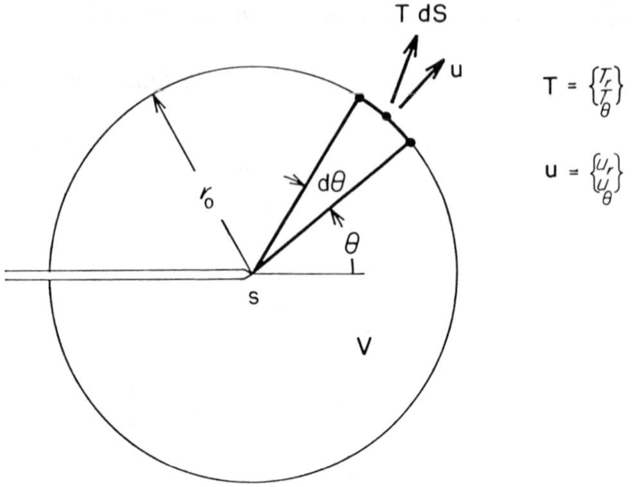

Figure 17.3 Circular region with outer radius r_0 around the stress-free crack tip

Substituting $|J|$ from equation (17.10) and using equation (17.6), the above relation becomes

$$\frac{r_0^2}{2} t\theta_s \int_{-1}^{1} \int_{0}^{1} \lfloor \delta\varepsilon \rfloor \{\sigma\} \rho \, d\rho \, d\eta, \qquad \theta_s = \theta_3 - \theta_1 \tag{17.17}$$

In view of equations (17.12) and (17.13) and from the constitutive relation $\{\sigma\} = [D]\{\varepsilon\}$, we conclude the contribution from each element as

$$\frac{1}{4\lambda} t\theta_s \lfloor \delta\bar{u} \rfloor \int_{-1}^{1} (\lambda[B_a]^T + [B_b]^T)[D](\lambda[B_a] + [B_b]) \, d\eta \{\bar{u}\} \tag{17.18}$$

Considering next the outer surface r_0 of the element, we note that the components of traction are $T_r = \sigma_r$ and $T_\theta = \tau_{r\theta}$ there. As $\xi = 1$ on the surface $r = r_0$, the displacement is given, from equation (17.5),

$$\mathbf{u} = H_1\bar{\mathbf{u}}_1 + H_2\bar{\mathbf{u}}_2 + H_3\bar{\mathbf{u}}_3 = [H]\{\bar{u}\} \tag{17.19}$$

Therefore the contribution of each element to the right-hand side of equation (17.15) comes to

$$r_0 t \int_{\theta_1}^{\theta_3} \lfloor \delta u \rfloor \{\sigma\} \, d\theta = \frac{r_0}{2} t\theta_s \lfloor \delta\bar{u} \rfloor \int_{-1}^{1} [H]^T [d][B] \, d\eta \{\bar{u}\} \tag{17.20}$$

Relation $d\theta = \theta_s \, d\eta/2$ has been used and $[d]$ is the matrix comprising two rows of the stress–strain matrix $[D]$ which are connected with the stress components σ_r and $\tau_{r\theta}$. Substitution of $[B]$ from equation (17.13) into equation (17.20) yields, as $\rho = 1$ for $r = r_0$,

$$\tfrac{1}{2} t\theta_s \lfloor \delta\bar{u} \rfloor \int_{-1}^{1} [H]^T [d](\lambda [B_a] + [B_b]) \, d\eta \{\bar{u}\} \tag{17.21}$$

Synthesizing the element contributions given by equations (17.18) and (17.21) to the left- and right-hand sides of equation (17.15) respectively, we have the following characteristic equation due to the arbitrariness of the relative nodal displacement vector $\{\delta\bar{U}\}$:

$$(\lambda^2[A] + \lambda[B] + [C])\{\bar{U}\} = 0 \tag{17.22}$$

where

$$[A] = \sum ([k_a] - [k_{sa}]), \qquad [B] = \sum ([k_b] - [k_{sb}]), \qquad [C] = \sum [k_c] \tag{17.23}$$

Summation extends over the whole finite elements covering the circular region V of Figure 17.3. Definition of terms in equation (17.23) are

$$\begin{aligned} [k_a] &= \int_{-1}^{1} [B_a]^T [D][B_a] \, d\eta \\ [k_c] &= \int_{-1}^{1} [B_b]^T [D][B_b] \, d\eta \\ [k_b] &= \int_{-1}^{1} ([B_a]^T [D][B_b] + [B_b]^T [D][B_a]) \, d\eta \end{aligned} \tag{17.24}$$

and

$$\begin{aligned} [k_{sa}] &= 2 \int_{-1}^{1} [H]^T [d][B_a] \, d\eta \\ [k_{sb}] &= 2 \int_{-1}^{1} [H]^T [d][B_b] \, d\eta \end{aligned} \tag{17.25}$$

The characteristic equation (17.22), which is quadratic in λ, can be transformed to the standard eigenvalue problem as follows:

$$[S]\begin{Bmatrix} \bar{V} \\ \bar{U} \end{Bmatrix} = \frac{1}{\lambda}\begin{Bmatrix} \bar{V} \\ \bar{U} \end{Bmatrix}, \qquad [S] = \begin{bmatrix} 0 & I \\ -C^{-1}A & -C^{-1}B \end{bmatrix} \tag{17.26}$$

Equation (17.26) is solved by the double *QR* method [21] to have the eigenvalue λ which includes the complex roots of the conjugate pair in general. After the determination of λ, the eigenmode $\{\bar{U}\}$ or the nodal displacement corresponding to each λ can be obtained by solving equation (17.26) algebraically.

17.4 EXTENSION TO THREE DIMENSIONS

The formulation of the preceding section can be extended to three dimensions by the application of the pyramidal element shown in Figure 17.4 by incorporating pertinent singlar transformation. Polar coordinates r, θ, ϕ are conveniently adopted in the analysis of the three-dimensional case of fracture mechanics. In these coordinates, the virtual work principle corresponding to equation (17.15) of the plane problems is written as

$$\int_0^{r_0}\int^{\theta}\int^{\phi}\lfloor\delta\varepsilon\rfloor\{\sigma\}r^2\sin\theta\,\mathrm{d}r\,\mathrm{d}\theta\,\mathrm{d}\phi = r_0^2\int^{\theta}\int^{\phi}\lfloor\delta\bar{u}\rfloor\{T\}\sin\theta\,\mathrm{d}\theta\,\mathrm{d}\phi \tag{17.27}$$

Related formulas giving the components $[B_{ia}]$ and $[B_{ib}]$ of the strain–nodal displacement matrix $[B]$ and the components of matrices $[A]$, $[B]$, and $[C]$ of the characteristic equation in three-dimensional cases have been given elsewhere [22]. Therefore, for three-dimensional cases, further details of solutions are merely appended in the next section discussing the numerical examples.

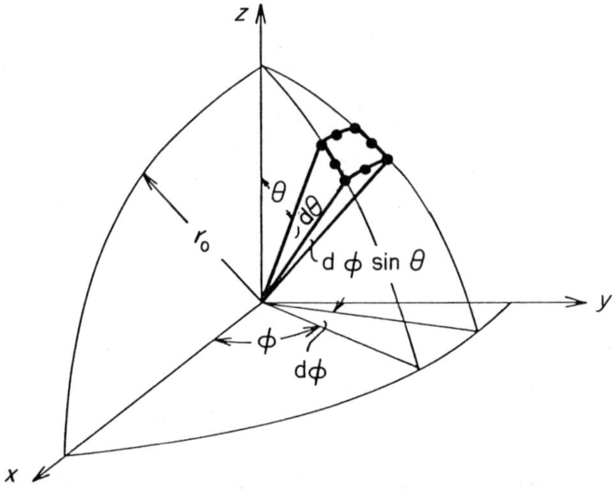

Figure 17.4 Pyramidal element for three-dimensional singularity analysis

17.4 NUMERICAL EXAMPLES

17.4.1 Verification through solutions of elementary problems

Verification of the present formulation has been carried out through a study of two elementary problems, i.e. stress singularities at the tip of the

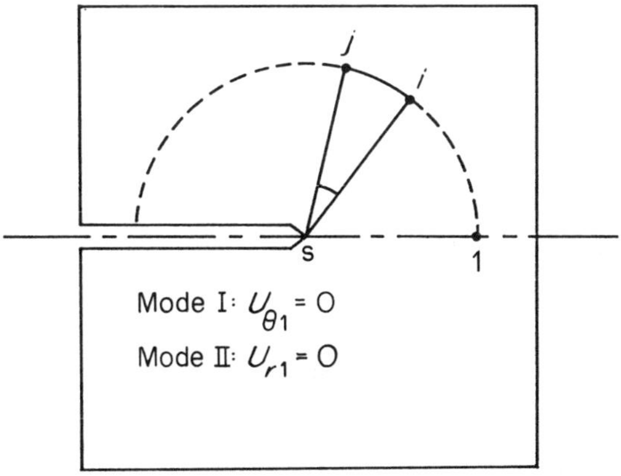

Figure 17.5 Stress singularities at the tip of symmetric crack in homogeneous material

stress-free symmetric crack of Figure 17.5 and at the angular corner of the plate of Figure 17.6 under the clamped-free boundary condition. The plane stress state and the homogeneity of material properties are assumed in these examples. Unless otherwise specified, Young's modulus E, Poisson's ratio ν, and thickness t used in the numerical analysis are

$$E = 1.0, \qquad \nu = 0.3, \qquad \text{and} \qquad t = 1.0 \tag{17.28}$$

It is noted that the eigensolution is not dependent on E in the examples concerned here.

Table 17.1 compares the lowest eigenvalues obtained by various elements and mesh division for the stress-free plane crack of Figure 17.5. Results are

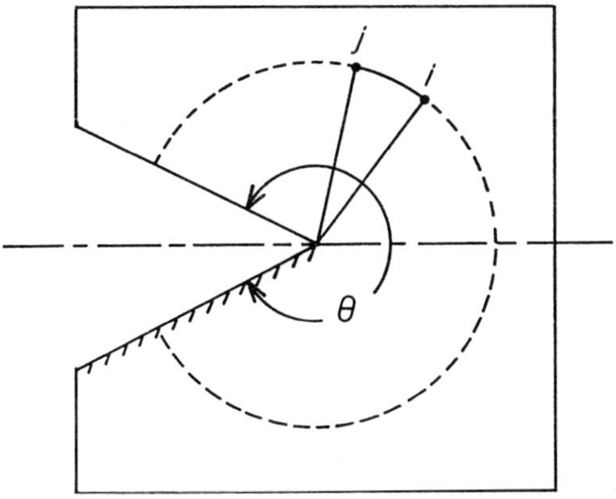

Figure 17.6 Angular corner of plate under clamped–free boundary condition

Table 17.1 Comparison of the lowest eigenvalue for the stress-free plane crack of Figure 17.5

Element identification	Linear		Quadratic	Exact
Number of element	10	20	10	
Mode I	0.5115	0.5031	0.5010	0.5
Mode II	0.5361	0.5090	0.5001	0.5

given for the so-called mode I and mode II of displacement discontinuities. Identification of an element by a quadratic implies the element of Figure 17.1 with interpolations expressed by equations (17.4) and (17.3) for displacement $\bar{\mathbf{u}}$, while a linear element assumes

$$\bar{\mathbf{u}} = H_1\bar{\mathbf{u}}_1 + H_2\bar{\mathbf{u}}_2, \qquad H_1 = \tfrac{1}{2}(1-\eta), \qquad H_2 = \tfrac{1}{2}(1+\eta) \tag{17.29}$$

The number of elements applies to the upper half region analysed as

Table 17.2 Comparison of eigenvalues λ, opening node I deformation of the crack of Figure 17.5

Order of λ	Present solution*	Exact values	Percentage error
1	0.503	0.5	+0.6
2	1.017	1.0	+1.6
3	1.495	1.5	−0.3
4	2.043	2.0	+2.2
5	2.483	2.5	−0.6
6	3.054	3.0	+1.8
7	3.529	3.5	+0.8

* Obtained by linear element, number of divisions = 20.

indicated in Figure 17.5. It can be seen that twenty linear or ten quadratic elements lead to a satisfactory estimate of the eigenvalues. Table 17.2 confirms the conclusion by summarizing seven lowest eigenvalues determined by twenty linear elements. Table 17.3 shows a confirmation that the eigenvalue of the problem of Figure 17.5 is not dependent on the value of Poisson's ratio ν.

Table 17.3 Lowest eigenvalue estimates for materials with different Poisson's ratio, mode I deformation of the crack of Figure 17.5

Poisson's ratio	0.0	0.15	0.3	0.45
Eigenvalue*	0.5028	0.5029	0.5031	0.5033

* As in Table 17.2.

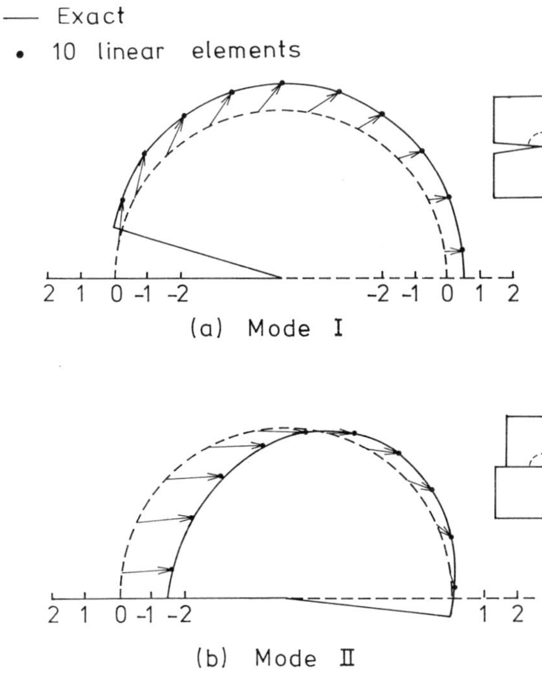

Figure 17.7 Displacement at the crack root as the function of angle θ

Figure 17.7(a) and (b) portray respectively the displacement at the root of the crack as the function of angle θ for mode I and mode II displacement discontinuities. As stated above, the solution concerns the upper half region of the crack tip. The displacement constraints in the finite element analysis for mode I and II deformations are respectively $v_\theta = 0$ and $u_r = 0$ at node 1 of Figure 17.5. Coincidence of the solutions with exact ones is almost complete, even though only ten linear elements are used in the numerical analysis.

As for the eigenstate at the angular corner of the plate of Figure 17.6, the real parts Re (λ) of the first two lowest eigenvalues λ_1 and λ_2 are shown in Figure 17.8 as the function of the vertex angle θ. It is known that λ_1 is complex for the range of included angles $131° < \theta \leq 360°$. The exact solution for Re (λ_1) in Figure 17.8 is read from Figure 1 of [1] which shows the root of the following transcendental equation for the plane stress case

$$\sin^2 (\lambda\theta)^2 = \frac{4}{(3-\nu)(1+\nu)} - \frac{1+\nu}{3-\nu}\frac{\sin^2\theta}{\theta^2}(\lambda\theta)^2 \tag{17.30}$$

Solutions of Eq. (17.30) have been discussed in detail by England [23].

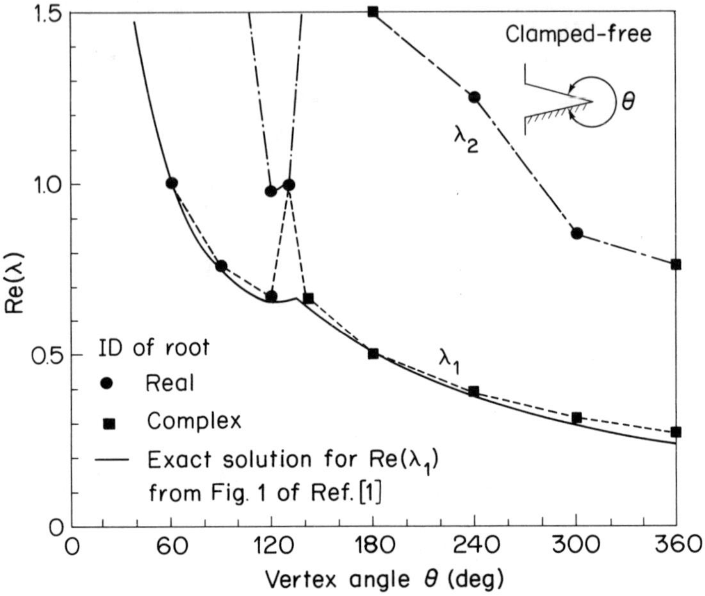

Figure 17.8 Real part of first two lowest eigenvalues λ_1 and λ_2 for the angular corner under clamped–free condition

17.4.2 Plane crack in bimaterial or composite

Important applications of the present formulation are stress singularities at the tip of the crack normal to and along the bimaterial interface of Figure 17.9 and the two-material joint of Figure 17.10.

Exact solution for the crack normal to the interface is given by the solution λ of the following equations for both mode I and II deformations [9, 10]:

$$f(\lambda) = \sin \lambda\pi = 0 \tag{17.31}$$

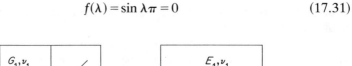

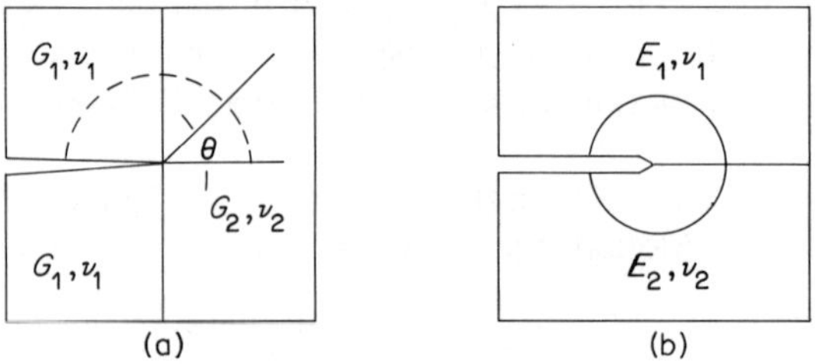

Figure 17.9 Crack normal to and along the two-material interface

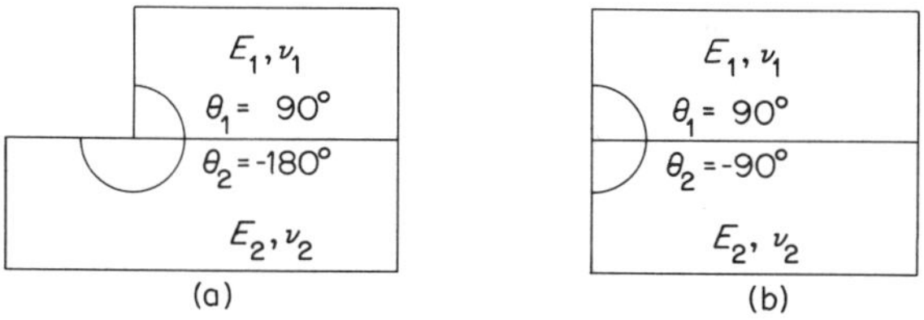

Figure 17.10 Two-material joint or laminate corner

and

$$g(\lambda) = \lambda^2(-4\alpha^2 + 4\alpha\beta) + 2\alpha^2 - 2\alpha\beta + 2\alpha - \beta + 1 + (-2\alpha^2 + 2\alpha\beta - 2\alpha + 2\beta)\cos\lambda\pi = 0 \tag{17.32}$$

where

$$\alpha = \frac{G_1/G_2 - 1}{\kappa_1 + 1}, \qquad \beta = \frac{G_1(\kappa_2 + 1)}{G_2(\kappa_1 + 1)} \tag{17.33}$$

$$\kappa = \begin{cases} 3 - 4\nu & \text{for plane strain} \\ (3-\nu)/(1+\nu) & \text{for plane stress} \end{cases} \tag{17.34}$$

where G stands for the shear modulus and the subscript 1 refers to the material or region where the crack is lying.

Table 17.4 shows a comparison of numerical results with exact solutions for the lowest eigenvalue. Due to symmetry, only the upper half of the region is considered and fifty quadratic elements are used. Eigensolutions are dependent on the ratio m of shear moduli G_1 and G_2 and Poisson's ratios ν_1 and ν_2. Assumed values of Poisson's ratios for epoxy and aluminium are 0.35 and 0.30 respectively. It can be seen from Table 17.4 that the singularity becomes more severe when the crack is contained in the harder material side. Table 17.5 summarizes the lowest ten eigenvalues for the case of the crack lying in the softer material, that is $m = G_1/G_2 = 0.043$. The coincidence of numerical predictions with the exact solutions is satisfactory.

As for the crack along the bimaterial interface, the exact solution for the

Table 17.4 Lowest eigenvalue at the crack tip normal to bimaterial interface

Material pair			Plane stress		Plane strain	
1	2	$m = G_1/G_2$	Numerical*	Exact†	Numerical*	Exact†
Epoxy	Al	0.043	0.7133	0.7110	0.6636	0.6619
Al	Epoxy	23.08	0.1728	0.1758	0.1759	0.1752

* Obtained by linear element, 50 divisions.
† References [9, 10].

Table 17.5 Comparison of eigenvalues at the crack tip normal to the Epoxy–Al interface, plane stress condition

Order of λ	Present analysis*	Exact $f(\lambda)=0$	Exact† $f(\lambda)=0$
1	0.7133	—	0.7110
2	1.004	1.0	—
3	(1.733, 0.477)	—	(1.732, 0.473)
4	1.985	2.0	—
5	3.201	3.0	—
6	3.727	4.0	—
7	(3.825, 1.065)	—	(3.814, 1.061)
8	5.037	5.0	—
9	5.888	6.0	—
10	(5.896, 1.311)	—	(5.852, 1.337)

*,† As in Table 17.4.
Numbers in parenthesis represent real Re (λ) and imaginary Im (λ) eigenvalues.

singular eigenstate is known to be [8]

$$\lambda = \tfrac{1}{2} - i \ln\left[\left(\frac{G_1}{G_2}\kappa_2 + 1\right) \Big/ \left(\frac{G_1}{G_2} + \kappa_1\right)\right] \tag{17.35}$$

or alternatively [10]

$$\lambda = \tfrac{1}{2} \pm i \frac{1}{2\pi} \cosh^{-1} \frac{(1+\alpha)^2 + (\alpha-\beta)^2}{2(1+\alpha)(\beta-\alpha)} \tag{17.36}$$

α and β being as given by equation (17.33). Eigenstate is again defined uniquely by the ratio of m of the shear moduli G_1 and G_2 and Poisson's ratios ν_1 and ν_2 of two materials. In the numerical example, we assume that $\nu_1 = \nu_2 = 0.20$. Therefore, the results are expressible equivalently in terms of E_1/E_2. In Figure 17.11, finite element solutions are compared with equations (17.35) or (17.36) for a wide range of values of E_1/E_2. Eigenvalues are complex, except for the case of a homogeneous material. As the problem is non-symmetric, the whole region spanning 360° should be analysed by dividing into finite elements.

The last examples of a plane crack are concerned with the singularities at the stepped corner and the flush surface of a two-material joint which correspond respectively to the cases of Figure 17.10(a) and (b). Solutions are shown in Figures 17.12 and 17.13 and are compared with the results read from Figures 14 and 8 of [8].

It is seen from Figure 17.12 that the lowest eigenvalue in the case of a stepped corner is real for $E_1/E_2 < 10$ and imaginary for $E_1/E_2 > 10$. Moreover, two real eigenvalues can be less than unity for some range of E_1–E_2 combination in $E_1/E_2 < 10$. Therefore, terms up to two are to be

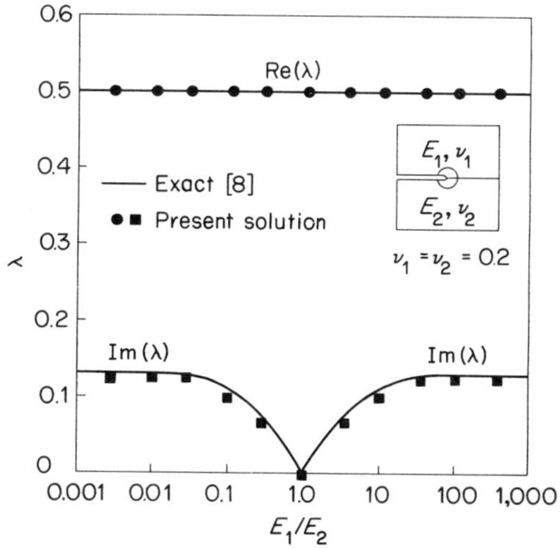

Figure 17.11 Eigenvalues for crack lying along the two-material interface as the function of E_1/E_2

associated with the stress singularities for that range of E_1–E_2 combination. In the case of the flush joint surface of Figure 17.13, eigenvalues are imaginary except for the homogeneous material or $E_1/E_2 = 1.0$. As the real part Re (λ) is greater than 0.5, stress singularity is not so severe in this case, compared to that relevant to the crack tip of a bimaterial interface.

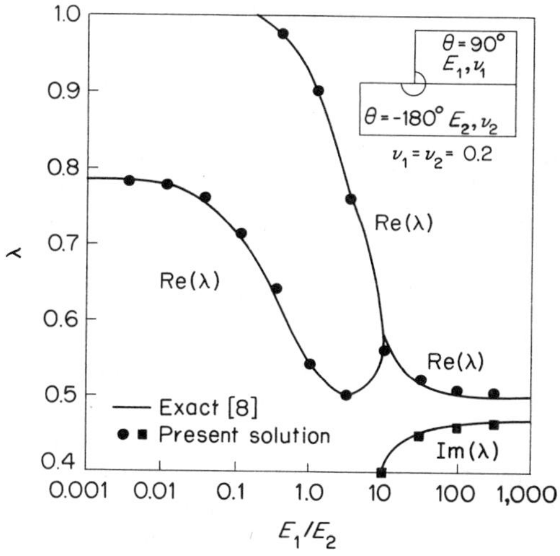

Figure 17.12 Eigenvalues for the stepped corner of two-material joint

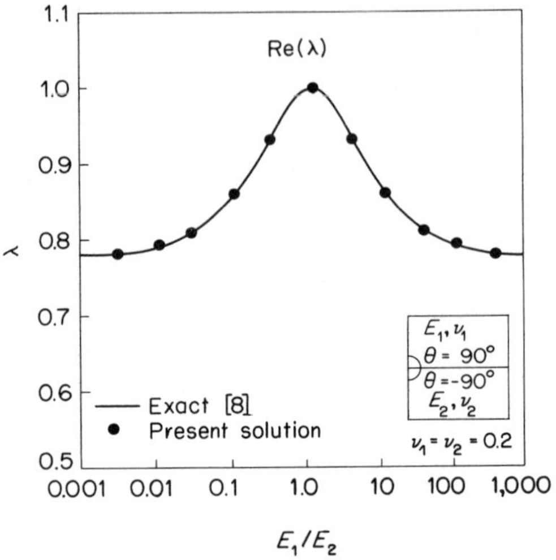

Figure 17.13 Real part of eigenvalues for flush surface of two-material joint

17.4.3 A three-dimensional problem

The pyramidal element of Figure 17.4 can be used conveniently for the analysis of the eigenstate near the terminal point of the crack front at the surface of the semi-infinite body. This problem has been solved by Bažant and Estenssoro [13], Fujitani [14], Benthem [15], and Benthem and Douma [16], and the result of an earlier computation by the authors has been reported elsewhere [22]. Figure 17.14 depicts the mesh division and Gauss integration points adopted in the computation.

Figure 17.15 shows the displacement around the crack root on the surface of the semi-infinite body for mode I deformation. Poisson's ratio is assumed to be $\nu = 0.3$ and the lowest eigenvalue of $\lambda = 0.5465$ is obtained by the finite element model of Figure 17.14, which is compared to $\lambda = 0.5477$ of the exact solution of Benthem. In case of mode II deformation, the displacement changes as shown in Figure 17.16. The finite element solution of the lowest eigenvalue for mode II deformation is found to be $\lambda = 0.3910$. The above results for mode I as well as mode II deformations agree satisfactorily with those from exact solutions of Benthem, with the use of fewer elements than Bažant and Estenssoro. Moreover, departure of the method of authors from the standard finite element analysis routine is minimal; contrast with the recent work of Fujitani [24] who has employed the Rayleigh–Ritz method for obtaining the solutions to almost the same degree of accuracy.

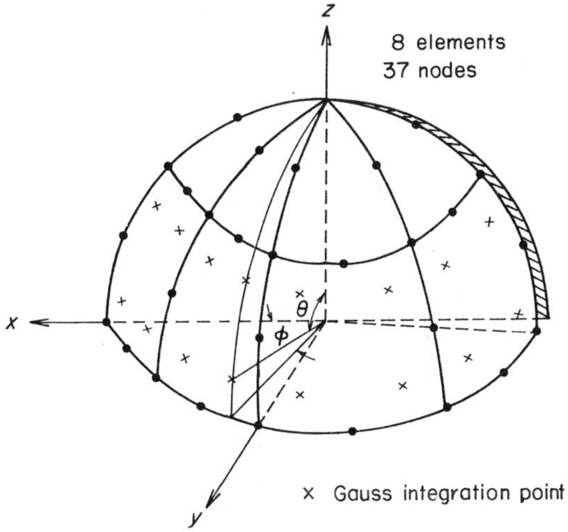

Figure 17.14 Mesh division for analysis of eigen state at the terminal point of the crack front on the surface of semi-infinite body

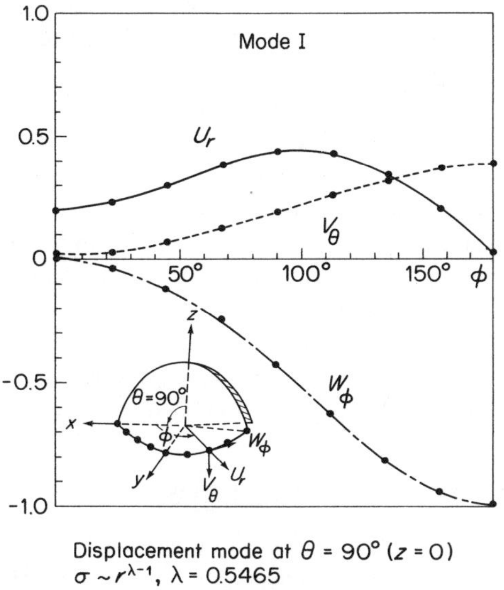

Figure 17.15 Displacement on $z = 0$ plane where plane crack runs into the surface of a semi-infinite body in the case of mode I deformation

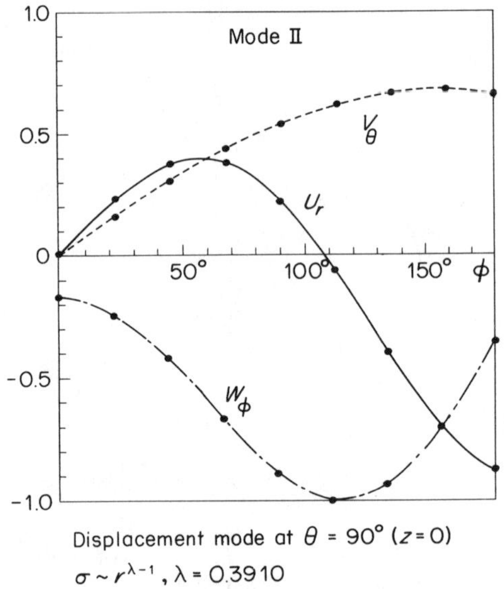

Figure 17.16 Displacement on $z = 0$ plane in the case of mode II deformation

17.5 CONCLUDING REMARKS

A compact finite element formulation is presented for analysis of the eigenstate of stress and strain singularities in continua. It incorporates the singular transformation and is applicable to a number of important problems of fracture mechanics in a consistent way. Elementary examples of stress singularities in homogeneous media are first studied and then extended to the plane crack at the bimaterial interface and a case of three-dimensional crack. Numerical results show good correlation with the existing analytical and numerical solutions, which have been brought out by tremendous analytical efforts or sophisticated numerical procedures.

It can be concluded that the formulation of the proposed method and the associated computational routines have been verified by the examples studied. Further extension of the application is expected by virtue of the simplicity and generality of the basic ideas of the method.

REFERENCES

1. M. L. Williams, 'Stress singularities resulting from various boundary conditions in angular corners of plates in extension', *J. Appl. Mech.*, **19,** 526–528 (1952).
2. M. L. Williams, 'On the stress distribution at the base of a stationary crack', *J. Appl. Mech.*, **24,** 109–114 (1957).
3. G. C. Sih and H. Liebowicz, *Mathematical Theories of Brittle Fracture*, Vol. II, Chap. 2, Academic Press, New York, 1968.

4. T. H. H. Pian, '*Crack elements*', Paper presented at *World Congress on Finite Element Methods in Structural Mechanics*, Dorset, England, 12–17 Oct. 1975.
5. G. C. Sih, P. C. Paris and G. R. Irwin, 'On cracks in rectilinearly anisotropic bodies', *Int. J. Fracture Mechanics*, **3,** 189–203 (1965).
6. R. D. Henshell and K. G. Shaw, 'Crack tip finite elements are unnecessary', *Int. J. Num. Meth. in Eng.*, **9,** 495–507 (1975).
7. R. S. Barsoum, 'On the use of isoparametric finite elements in linear fracture mechanics', *Int. J. Num. Meth. in Eng.*, **10,** 25–37 (1976).
8. V. L. Hein and F. Erdogan, 'Stress singularities in a two material wedge', *Int. J. Fracture Mechanics*, **7,** 317–330 (1971).
9. T. S. Cook and F. Erdogan, 'Stresses in bonded materials with a crack perpendicular to the interface', *Int. J. Engineering Science*, **10,** 667–697 (1972).
10. K. Y. Lin and J. W. Mar, 'Finite element analysis of stress intensity factors for cracks at a bi-material interface', *Int. J. of Fracture*, **12,** 521–531 (1976).
11. P. Tong, T. H. H. Pian and S. J. Lasry, 'A hybrid-element approach to crack problems in plane elasticity', *Int. J. Num. Meth. in Eng.*, **7,** 297–308 (1973).
12. Y. Fujitani, 'Finite element analysis of the singular solution in crack problems, I. Two dimensional crack problems', *Seisan Kenkyu, Monthly J. of Inst. of Ind. Sci., Univ. of Tokyo*, **29,** 459–462 (1977).
13. Z. P. Bažant and L. F. Estenssoro, 'Surface singularity and crack propagation', *Int. J. Solids and Structures*, **15,** 405–426 (1979).
14. Y. Fujitani, 'Finite element analysis of the singular solution in crack problems, II. Formulation in the case of three dimensional crack problems', *Seisan Kenkyu, Monthly J. of Inst. of Ind. Sci. Univ. of Tokyo*, **29,** 515–518 (1977).
15. J. P. Benthem, 'The quarter-infinite crack in a half space; Alternative and additional solutions', *Int. J. Solids and Structures*, **16,** 119–130 (1980).
16. J. P. Benthem and Th. Douma, 'Graphs of the Three-Dimensional State of Stress at the Vertex of a Quarter-Infinite Crack in a half-space', WTHD Report No. 123, Delft University of Technology, 1980.
17. Y. Yamada, Y. Ezawa, I. Nishiguchi and M. Okabe, 'Reconsiderations on singularity or crack tip elements', *Int. J. Num. Meth. in Eng.*, **14,** 1525–1544 (1979).
18. K. J. Bathe and E. L. Wilson, *Numerical Methods in Finite Element Analysis*, Prentice-Hall, Englewood Cliffs, 1976.
19. M. Okabe, Y. Yamada and I. Nishiguchi, 'Basis transformation of trial function space in Lagrange interpolation', *Comp. Meth. in Appl. Mech. Eng.*, **23,** 85–99 (1980).
20. M. Okabe, Y. Yamada and I. Nishiguchi, 'Reconsideration of rectangular Lagrange families with hierarchy ranking basis', *Comp. Meth. in Appl. Mech. Eng.*, **23,** 369–390 (1980).
21. J. G. F. Francis, 'The QR transformation, a unitary analogue to the LR transformation, Parts I and II', *Computer J.*, **4,** 265–271 and 332–345 (1961 and 1962).
22. Y. Yamada and H. Okumura, 'Analysis of local stress in composite materials by the 3D finite element', Paper submitted to Japan–US Symposium on Composite Materials, Tokyo, January 1981.
23. A. H. England, 'On stress singularities in linear elasticity', *Int. J. Eng. Sci.*, **9,** 571–585 (1971).
24. Y. Fujitani, 'Analysis of the singular solution in the three dimensional surface crack problem by Rayleigh–Ritz method', Vol. 28, pp. 129–137. Report of Faculty of Engineering, Hiroshima University, 1980.

Hybrid and Mixed Finite Element Methods
Edited by S. N. Atluri, R. H. Gallagher, and O. C. Zienkiewicz

Chapter 18

Some Fracture Mechanics Analyses Using Finite Element Method With Penalty Function

Genki Yagawa and Tatsuhiko Aizawa

18.1 INTRODUCTION

The importance of knowledge of cracks in structures is well recognized in assessing the integrity of structural components based on fracture mechanics. Nevertheless, few exact solutions are available even for simple crack problems due to mathematical intractabilities encountered. As for the numerical methods, the finite element method has proved to be a very useful tool for solving relevant boundary value problems and more than a decade has passed since the first application of the finite element method to the fracture mechanics analysis appeared in the literature [1, 2]. As is well known, the main concern for the finite element method is about computer cost, especially when one applies the method to such problems as fracture mechanics which involves singularity. To overcome this disadvantage of the finite element fracture mechanics analysis, several techniques have been reported for the accurate calculations of the stress intensity factor, such as the direct method [3], virtual crack extension method [4, 5], superposition method [6 to 12], distorted element method [13, 14], special singular element method [15 to 17], and so on.

The second problem concerning fracture mechanics analysis is the consideration of material non-linearity because the stress around the crack tip is beyond the yield point of material in practical situations. According to Hutchinson [18] and Rice and Rosengren [19], the so-called HRR (Hutchinson–Rice–Rosengren) singularity is found to dominate at the vicinity of the crack tip, assuming that the material obeys the power law type of constitutive equation and that the deformation theory of plasticity is valid. Taking the HRR singularity into account, non-linear crack analyses have been made to obtain such non-linear fracture parameters as the J-integral value or crack opening displacement both in plane stress and antiplane shear [20, 21]. For the case of the incompressible material and the plane strain at the same time, it is well recognized that special care must be taken in the finite element algorithm, since there exists no rigorous correspondence between stress and strain components [22 to 24].

The present authors proposed a superposition method [9 to 12], in which the nodal displacements of the conventional finite element method together with the unknown constants of the analytical solution are determined in a modified displacement-type variational principle. The authors studied another superposition method with the use of a penalty function formulation applied to linear and non-linear fracture mechanics problems both under compressible and incompressible material conditions [25, 26]. This paper gives a review on the method and some new results obtained recently as well.

18.2 THEORY

Let us discuss the theoretical background of the present method. For simplicity, the first part of this section is limited to the linear fracture analysis.

Consider the two-dimensional crack problem. As shown in Figure 18.1, the whole region is fictitiously divided into the near crack tip region $V^{(1)}$ and the surrounding region $V^{(0)}$. In $V^{(0)}$, the regular finite elements are employed since the crack tip singularity is dominant only near the crack tip, whereas the analytical functions are used in $V^{(1)}$, which are superposed on

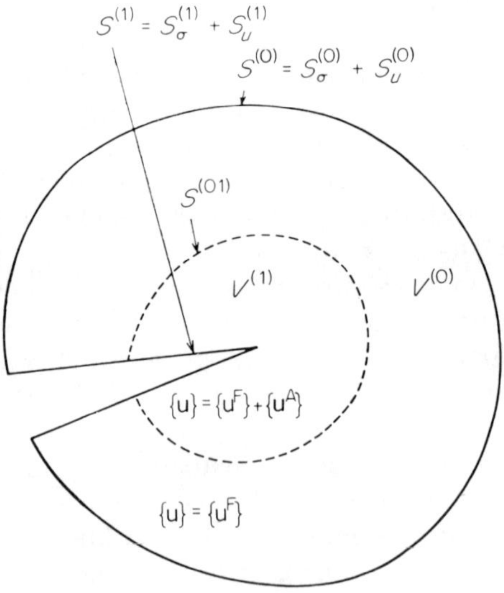

Figure 18.1 A superposition method for a cracked body

the finite element functions, i.e.

$$\mathbf{u}^{(0)} = \mathbf{u}^F \qquad \text{in } V^{(0)} \tag{18.1}$$

$$\mathbf{u}^{(1)} = \mathbf{u}^F + \mathbf{u}^A \qquad \text{in } V^{(1)} \tag{18.2}$$

where $\mathbf{u}^{(0)}$ and $\mathbf{u}^{(1)}$ are the total displacements, $\mathbf{u}^F$ the contribution by the finite elements and $\mathbf{u}^A$ the analytical displacement corresponding to the particular crack tip singularity. The fictitious boundary between these two regions is designated as $S^{(01)}$, where the appropriate continuity conditions are to be satisfied. The traction forces $\bar{\mathbf{T}}^{(0)}$ and $\bar{\mathbf{T}}^{(1)}$ are prescribed on the boundaries $S_\sigma^{(0)}$ of $V^{(0)}$ and $S_\sigma^{(1)}$ of $V^{(1)}$ respectively, and the boundary displacements $\mathbf{u}^{(0)}$ and $\mathbf{u}^{(1)}$ are prescribed on $S_u^{(0)}$ of $V^{(0)}$ and $S_u^{(1)}$ of $V^{(1)}$ respectively.

The functional to be minimized for the above case is given by [27]

$$\pi_1 = \sum_{i=0}^{1} \int_{V^{(i)}} A(\mathbf{u}^{(i)})\,\mathrm{d}V - \int_{S_\sigma^{(i)}} \bar{\mathbf{T}}^{(i)}\mathbf{u}^{(i)}\,\mathrm{d}S \tag{18.3}$$

with the additional equality constraint

$$\mathbf{u}^{(1)} - \mathbf{u}^{(0)} = 0 \qquad \text{on } S^{(01)} \tag{18.4}$$

where $A(\mathbf{u}^{(i)})$ is the strain energy density. Equations (18.3) with (18.4) can be reduced to the following functional without constraint:

$$\pi_2 = \sum_{i=0}^{1} \int_{V^{(i)}} A(\mathbf{u}^{(i)})\,\mathrm{d}V + \int_{S^{(01)}} \boldsymbol{\lambda}(\mathbf{u}^{(1)} - \mathbf{u}^{(0)})\,\mathrm{d}S - \int_{S_\sigma^{(i)}} \bar{\mathbf{T}}^{(i)}\mathbf{u}^{(i)}\,\mathrm{d}S \tag{18.5}$$

It is noted here that $\boldsymbol{\lambda}$ in equation (18.5) has an implicit relation with $(\mathbf{u}^{(1)} - \mathbf{u}^{(0)})$ as [28]

$$\boldsymbol{\lambda} = \frac{\alpha}{2}(\mathbf{u}^{(1)} - \mathbf{u}^{(0)}) \tag{18.6}$$

where α is the stiffness on $S^{(01)}$, which is called the penalty number.

Substituting equation (18.6) into equation (18.5), we have the following functional:

$$\pi_3 = \sum_{i=0}^{1} \int_{V^{(i)}} A(\mathbf{u}^{(i)})\,\mathrm{d}V + \frac{\alpha}{2}\int_{S^{(01)}} (\mathbf{u}^{(1)} - \mathbf{u}^{(0)})^2\,\mathrm{d}S - \int_{S_\sigma^{(i)}} \bar{\mathbf{T}}^{(i)}\mathbf{u}^{(i)}\,\mathrm{d}S \tag{18.7}$$

or in matrix form

$$\pi_3 = \tfrac{1}{2}\lfloor d^{\mathrm{F}}, d^{\mathrm{A}} \rfloor \begin{bmatrix} K^{\mathrm{FF}} & K^{\mathrm{AF}} \\ K^{\mathrm{FA}} & K^{\mathrm{AA}} \end{bmatrix} \begin{Bmatrix} d^{\mathrm{F}} \\ d^{\mathrm{A}} \end{Bmatrix} - \lfloor d^{\mathrm{F}}, d^{\mathrm{A}} \rfloor \begin{Bmatrix} f^{\mathrm{F}} \\ f^{\mathrm{A}} \end{Bmatrix}$$
$$+ \frac{\alpha}{2}\lfloor d^{\mathrm{F}}, d^{\mathrm{A}} \rfloor [\mathrm{A}]^t[\mathrm{A}] \begin{Bmatrix} d^{\mathrm{F}} \\ d^{\mathrm{A}} \end{Bmatrix} \tag{18.8}$$

where $\{d^F\}$ is the nodal displacement vector of the finite element method, $\{d^A\}$ the coefficient vector of the singular function, and

$$[K^{FF}]=\int_{V^{(0)}+V^{(1)}}[B]^t[D][B]\,dV \tag{18.9a}$$

$$[K^{AA}]=\int_{V^{(1)}}[Q]^t[D][Q]\,dV \tag{18.9b}$$

$$[K^{AF}]=[K^{FA}]=\int_{V^{(1)}}[B]^t[D][Q]\,dV \tag{18.9c}$$

$$\{f^F\}=\int_{S_\sigma^{(1)}+S_\sigma^{(0)}}[N_F]\{\bar{T}\}\,dS \tag{18.9d}$$

$$\{f^A\}=\int_{S_\sigma^{(1)}}[N_A]\{\bar{T}\}\,dS \tag{18.9e}$$

Here, $[B]$, $[Q]$, $[N_F]$, and $[N_A]$ are respectively defined as

$$\begin{aligned}\{\varepsilon^F\}&=[B]\{d^F\}, & \{\varepsilon^A\}&=[Q]\{d^A\}\\ \{u^F\}&=[N_F]\{d^F\}, & \{u^A\}&=[N_A]\{d^A\}\end{aligned} \tag{18.10}$$

with $\{\varepsilon^F\}$ and $\{\varepsilon^A\}$ being the strains pertinent to the finite element and the analytical displacements respectively. The matrix $[D]$ in equations (18.9a, b, c) is the usual elastic stress–strain matrix. The subsidiary condition corresponding to equation (18.4) may be formally written as follows:

$$[A]\begin{Bmatrix} d^F \\ d^A \end{Bmatrix}=0 \tag{18.11}$$

where $[A]$ is a proper matrix.

Though the above formulation is useful for crack analyses in the plane stress condition, some modification is necessary when one uses it for the case when the Poisson's ratio becomes around 0.5 in the plane strain or three-dimensional state. The modification taken here is to add the term

$$\sum_{i=0}^{1}\int_{V^{(i)}}\frac{\alpha_1}{2}\left(\frac{1}{V_e}\int_{V_e}\varepsilon_{ii}^F\,dV\right)^2 \tag{18.12}$$

to the original functional. Here V_e is the volume of each element, α_1 the penalty number, and it is assumed that the analytical displacement $\mathbf{u}^A$ is so chosen as to satisfy the condition $\varepsilon_{ii}^A=0$. The mean stress σ_m in this method is given in each element by [28]

$$\sigma_m=\frac{\alpha_1}{V_e}\int_{V_e}\varepsilon_{ii}^F\,dV \tag{18.13}$$

Consider a rectangular plate with a centre crack under uniform tension

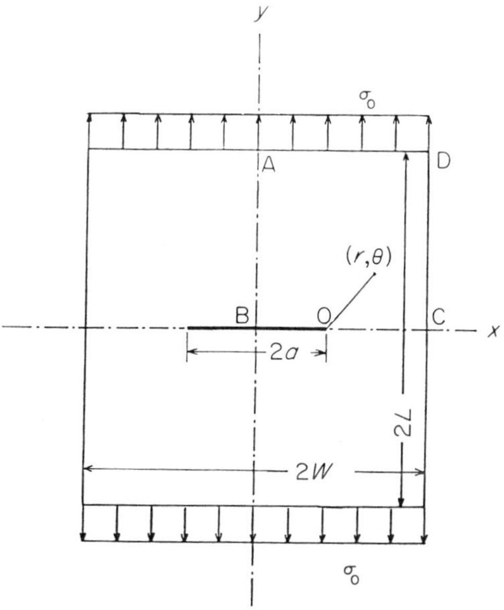

Figure 18.2 A rectangular plate with a centre crack

(Figure 18.2). For simplicity, the whole region of the plate is taken to be $V^{(1)}$, i.e. the analytical displacements are superposed on the finite elements throughout the plate. As for the analytical displacements $\mathbf{u} = (u^{\mathrm{A}}, v^{\mathrm{A}})$, we employ the expansions

$$u^{\mathrm{A}} = \sqrt{r} \sum_{m=1}^{N_1} a_m F_u^m(\theta), \qquad v^{\mathrm{A}} = \sqrt{r} \sum_{m=1}^{N_2} b_m F_v^m(\theta) \tag{18.14}$$

where r and θ denote the polar coordinates with the origin being the crack tip, N_1 and N_2 are the numbers of terms adopted in the calculation, a_m and b_m $(m = 1, 2, \ldots)$ are the unknown coefficients, and $F_u^m(\theta)$ and $F_v^m(\theta)$ $(m = 1, 2, \ldots)$ are the functions of θ. In the case of the plane stress and the antiplane shear, a_m and b_m are linearly independent and $F_u^m(\theta)$ and $F_v^m(\theta)$ are given by

$$F_u^m(\theta) = \cos\frac{m}{2}\theta, \qquad F_v^m(\theta) = \sin\frac{m}{2}\theta \tag{18.15}$$

In the case of the incompressible materials, these coefficients are *a priori* arranged so that $\varepsilon_{ii}^{\mathrm{A}}$ becomes null.

With the intention of analysing one-quarter of the plate (that is ABCD in Figure 18.2) due to the symmetrical condition, we take into account the following boundary conditions as the penalty terms in the functional:

$$u = 0 \qquad \text{on AB} \tag{18.16}$$

$$\sigma_y = \sigma_{xy} = 0 \qquad \text{on BO} \tag{18.17}$$

where σ_{ij} are the stress tensors. In addition, the incompressibility condition of the finite element displacement is considered in terms of another penalty term in the functional.

Thus, the functional in this case can be written as follows:

$$\pi_5 = \sum_{\text{element}} \left\{ \int_{V_e} \tfrac{1}{2}\bar{\sigma}\bar{\varepsilon}\, dV + \frac{\alpha_1}{2}\left[\frac{1}{V_e}\int_{V_e} (\varepsilon_x^F + \varepsilon_y^F)\, dV\right]^2 \right\}$$

$$- \int_{S_\sigma} \bar{\mathbf{T}}^t \mathbf{u}\, dS + \frac{\alpha_2}{2}\int_{S_{AB}} \sigma_x^2\, dS + \frac{\alpha_3}{2}\int_{S_{BO}} \sigma_y^2\, dS + \frac{\alpha_4}{2}\int_{S_{BO}} \sigma_{xy}^2\, dS \tag{18.18}$$

where $\bar{\sigma}$ and $\bar{\varepsilon}$ are respectively the equivalent stress and strain, and $\alpha_1 \sim \alpha_4$ are the penalty numbers. The second term on the right-hand side of equation (18.18) is necessary for the incompressibility condition of the finite element displacement, and the last three terms are for the boundary conditions, equations (18.16) and (18.17). It is noted that the stress-free requirement on BO, which is very important to obtain accurate results in the crack analyses, is enforced by means of the last two terms on the right-hand side of equation (18.18), although this is satisfied *a priori* as the usual natural boundary condition. The other boundary conditions are the natural boundary conditions or are satisfied with the use of equations (18.14) as the analytical displacements.

Next, we extend the present method to the non-linear crack analyses, in which the HRR singularity [18, 19] is taken into account and the constitutive equation is given as

$$\bar{\varepsilon} = \kappa \bar{\sigma}^n \tag{18.19}$$

where n is the strain-hardening exponent and κ the proportional coefficient. According to the J_2 deformation plasticity theory, the multiaxial stress–strain law can be written as

$$\boldsymbol{\varepsilon} = \tfrac{3}{2}\kappa \bar{\sigma}^{n-1} \boldsymbol{\sigma}' \tag{18.20}$$

where $\boldsymbol{\sigma}'$ is the stress deviation vector. Based on the HRR singularity, the distributions of displacement, strain, and stress near the crack tip are given as follows:

$$\mathbf{u} = \kappa \left(\frac{J}{\kappa I_n}\right)^{n/(n+1)} r^{1/(n+1)} \tilde{\mathbf{u}}(\theta)$$

$$\boldsymbol{\varepsilon} = \kappa \left(\frac{J}{\kappa I_n}\right)^{n/(n+1)} r^{-1/(n+1)} \tilde{\boldsymbol{\varepsilon}}(\theta) \tag{18.21}$$

$$\boldsymbol{\sigma} = \kappa \left(\frac{J}{\kappa I_n}\right)^{1/(n+1)} r^{-1/(n+1)} \tilde{\boldsymbol{\sigma}}(\theta)$$

where I_n is a function of n and J is the J-contour integral value. $\tilde{\mathbf{u}}(\theta)$, $\tilde{\boldsymbol{\varepsilon}}(\theta)$, and $\tilde{\boldsymbol{\sigma}}(\theta)$ are the characteristic functions of θ. To apply the present method to these non-linear crack problems, we use the following analytical displacements:

$$u^{\mathrm{A}}=r^{1/(n+1)}\sum_{m=1}^{N_1} a_m \cos\frac{m}{2}\theta, \qquad v^{\mathrm{A}}=r^{1/(n+1)}\sum_{m=1}^{N_2} b_m \sin\frac{m}{2}\theta \quad (18.22)$$

In the plane stress condition, the coefficients a_m and b_m in equation (18.22) are linearly independent, while in the plane strain condition they are so arranged as to satisfy the incompressibility condition, $\varepsilon_x^{\mathrm{A}}+\varepsilon_y^{\mathrm{A}}=0$.

Taking the plate as shown in Figure 18.2, the functional in this case can be written as follows:

$$\pi_6=\sum_{\text{element}}\left\{\int_{V_e}\frac{n}{n+1}\bar{\sigma}\bar{\varepsilon}\,\mathrm{d}V+\frac{\alpha_1}{2}\left[\frac{1}{V_e}\int_{V_e}(\varepsilon_x^{\mathrm{F}}+\varepsilon_y^{\mathrm{F}})\,\mathrm{d}V\right]^2\right\}$$

$$-\int_{S_\sigma}\bar{\mathbf{T}}^t\mathbf{u}\,\mathrm{d}S+\frac{\alpha_2}{2}\int_{S_{\mathrm{AB}}}\sigma_x^2\,\mathrm{d}S+\frac{\alpha_3}{2}\int_{S_{\mathrm{BO}}}\sigma_y^2\,\mathrm{d}S+\frac{\alpha_4}{2}\int_{S_{\mathrm{BO}}}\sigma_{xy}^2\,\mathrm{d}S \quad (18.23)$$

Using equations (18.10) as well as equation (18.20) leads equation (18.23) to a non-linear function of the vectors $\mathbf{d}^{\mathrm{F}}$ and $\mathbf{d}^{\mathrm{A}}$. Extremizing the functional with respect to $\mathbf{d}^{\mathrm{F}}$ and $\mathbf{d}^{\mathrm{A}}$, we have a set of non-linear equations which can be solved with the use of proper solution techniques such as the Newton–Raphson iteration.

18.3 SOME TESTS ON ACCURACY AND CONVERGENCE

In the first place, we study the convergence in the linear crack analysis (Figure 18.2). Assuming the compressibility of material, i.e. the Poisson's ration $\nu\neq 0.5$, the following analytical displacements are employed:

$$\begin{aligned} u^{\mathrm{A}}&=a_1\sqrt{r}\left[\left(\frac{3-\nu}{1+\nu}-\frac{1}{2}\right)\cos\frac{\theta}{2}-\tfrac{1}{2}\cos\tfrac{3}{2}\theta\right]\\ v^{\mathrm{A}}&=a_1\sqrt{r}\left[\left(\frac{3-\nu}{1+\nu}+\frac{1}{2}\right)\sin\frac{\theta}{2}-\tfrac{1}{2}\sin\tfrac{3}{2}\theta\right]\end{aligned} \quad (18.24)$$

It is noted here that coefficient a_1 in equations (18.24) is directly related to the stress intensity factor K_I as

$$a_1=\frac{1+\nu}{E}\frac{K_I}{\sqrt{(2\pi)}} \quad (18.25)$$

where E is Young's modulus.

Table 18.1 Convergence test with increasing penalty number α_2

	Single precision (32 bits/word)		Double precision (64 bits/word)	
α_2	f	$\lvert u\rvert/(L\sigma_0/E)$	f	$\lvert u\rvert/(L\sigma_0/E)$
10^2	1.884	1.12	1.884	1.12
10^3	1.418	0.18	1.418	0.18
10^4	1.339	0.03	1.341	0.03
10^5	1.346	3.0×10^{-3}	1.332	3.0×10^{-3}
10^6	1.344	3.0×10^{-4}	1.331	3.0×10^{-4}
10^7	1.729	3.0×10^{-5}	1.331	3.0×10^{-5}
10^8	−1.751	3.2×10^{-6}	1.331	3.0×10^{-6}
10^9	0.030	2.8×10^{-7}	1.331	3.0×10^{-7}

In this test, the terms with α_1, α_3, and α_4 in equation (18.18) are set to be zero, while the penalty number α_2 is varied from 10^2 to 10^9. Table 18.1 shows how the non-dimensional stress intensity factor f $[=K_I/\sigma_0\sqrt{(\pi a)}]$ converges to the exact value ($f=1.334$) [29] and how the displacement u on AB in Figure 18.2 which should be zero in the exact solution diminishes with an increasing value of α_2 in equation (18.18). In this calculation, Young's modulus and Poisson's ratio are assumed to be 1.0 and 0.3 respectively, and $L=W=16$, $a/W=0.5$. The finite element mesh of one-quarter of the plate is 4×4. It is seen from the table that the double precision word length should be used in order to get better solutions in the case of higher values of the penalty number, which is required for the rigorous satisfaction of the constraint.

Next, consider the case of incompressible material in the plane strain condition ($\nu=0.5$). The analytical displacements in this problem are assumed as follows.

$$u^{\mathrm{A}}=a_1\sqrt{r}\,\sin^2\frac{\theta}{2}\cos\frac{\theta}{2}$$
$$v^{\mathrm{A}}=a_1\sqrt{r}\,\sin^3\frac{\theta}{2} \tag{18.26}$$

where the unknown a_1 has the relation with K_{I} as

$$a_1=\frac{3}{\sqrt{(2\pi)}}\frac{K_{\mathrm{I}}}{E} \tag{18.27}$$

Table 18.2 shows the effects of the mesh arrangement and the penalty number α_1 in equation (18.18) on the solution. It is noted that the penalty numbers α_2, α_3, and α_4 are always set to be 10^5, and the 64 bits word length is used in this calculation. As can be seen from the table, even with such a coarse mesh as 4×4, the value of f converges to the exact one within 2 per cent error if α_1 is increased over 10^5. On the other hand, keeping α_1 to be

Table 18.2 The effect of the number of freedom and penalty number α_1 on the non-dimensional stress intensity factor f

Non-dimensional stress intensity factor f					
		Penalty number α_1			
Mesh arrangement	Number of freedoms	10^0	10^2	10^5	10^7
2×2	40			1.450	1.450
4×4	126	1.618	1.364	1.360	1.360
6×6	260			1.340	
8×8	442			1.332	
Ref. 29			1.334		

10^5 and varying the mesh from 2×2 to 8×8, the discrepancy between exact and present values is seen to diminish monotonically into less than 0.2 per cent.

Finally, the convergence in the Newton-type interation is studied for the non-linear crack analysis. The compressible material is used in this case. The geometry of the plate and the material data are assumed as $L = W = 16$, $a/W = 0.5$, $n = 3$, and $\kappa = 1.0$. Six iterations were required until the convergence solution was obtained, in which the criterion of convergence was assumed as $(\pi_i - \pi_{i-1})/\pi_i < 0.01$ with π_i and π_{i-1} being the values of functionals at ith and $(i-1)$th iteration steps respectively.

18.4 NON-LINEAR FRACTURE MECHANICS ANALYSES

The J integral value, J, the crack opening displacement, δ, and the residual load point displacement, Δ, are known to be important fracture parameters in the cracks of work hardening materials [30], and can be determined as the function of geometry of structure, applied load, and material constants. Assuming that the material obeys the power law as defined in equation (18.19), it seems better to take into account the HRR singularity which is given by equation (18.21) in the forming the trial functions for the calculations of fracture parameters with enough accuracy.

Through the dimensional analysis, the following non-dimensional forms of J, δ, and Δ are known to be useful [21]:

$$\begin{aligned} J &= a\kappa(1-\lambda)^{-n} g_1(\lambda, n)\sigma_0^{n+1} \\ \delta &= a\kappa(1-\lambda)^{-n} g_2(\lambda, n)\sigma_0^{n} \\ \Delta &= a\kappa(1-\lambda)^{-n} g_3(\lambda, n)\sigma_0^{n} \end{aligned} \tag{18.28}$$

where g_1, g_2, and g_3 are the non-dimensional values of the J integral value, and crack opening displacement, and the residual load point displacement respectively. In this analysis, the J value is calculated from the contour integral, the crack opening displacement from the vertical displacement at the centre of the crack (i.e. point B in Figure 18.2), and the load point displacement is determined by

$$\Delta = \Delta_{\text{crack}} - \Delta_{\text{no crack}} \tag{18.29}$$

where Δ_{crack} and $\Delta_{\text{no crack}}$ are the load point displacements of the cracked plate and uncracked one respectively, which are defined as

$$\Delta_{\text{crack}} = \frac{1}{2W}\int_{-W}^{W} (v|_{y=L} - v|_{y=-L})\,dx = \frac{2}{W}\int_{0}^{W} v|_{y=L}\,dx$$
$$\Delta_{\text{no crack}} = 2L\kappa\sigma_0^n \tag{18.30}$$

At first, in order to study the effect of λ as well as n on the non-linear fracture parameters g_1, g_2, and g_3, we take the rectangular plate of $L/W = 0.2$ under uniform tension with a centre crack of length $2a$ in the plane stress condition (see Figure 18.2). In Table 18.3, we show the parameters g_1 and g_2 calculated for the same plate but with $\lambda = \frac{1}{2}$ and n being up to 7. As seen from the table, the differences between the present results and those of Shih and Hutchinson [21] are very small, whereas the solutions by Ranaweera and Leckie [23] underestimate the parameters.

In the case of a plate with a side edge crack under an inplane bending moment M, the same non-dimensional form as equations (18.28) are written as

$$J = \kappa(W-a)h_1(\lambda, n)\left(\frac{M}{M_0}\right)^{n+1}$$
$$\delta = \kappa a h_2(\lambda, n)\left(\frac{M}{M_0}\right)^{n} \tag{18.31}$$
$$\psi = \kappa a h_3(\lambda, n)\left(\frac{M}{M_0}\right)^{n}$$

Table 18.3 Non-dimensional non-linear fracture parameters g_1 and g_2 with the values of the strain hardening exponent n for centre cracked plane ($\lambda = \frac{1}{2}$) under uniform tension (plane stress)

		$n=1.0$	$n=3.0$	$n=5.0$	$n=7.0$
Present	g_1	2.223	2.046	1.817	1.597
	g_2	2.365	1.724	1.291	1.097
Ref. 21	g_1	2.212	2.056	1.812	1.644
	g_2	2.382	1.703	1.307	1.084
Ref 23 (coarse mesh)	g_1	2.192	1.994	1.608	1.222
	g_2	2.355	1.668	1.166	0.864
Ref 23 (fine mesh)	g_1	2.198	2.016	1.750	1.382
	g_2	2.367	1.682	1.263	0.904

Table 18.4 Non-dimensional non-linear fracture parameters h_1, h_2, and h_3 with the values of the strain hardening exponent n for side edge cracked plate ($\lambda = \frac{1}{2}$) under inplane bending (plane stress)

		$n=1.0$	$n=2.0$	$n=3.0$
Present	h_1	1.088	0.921	0.806
	h_2	5.111	3.711	2.993
	h_3	2.698	2.403	2.081
Ref 21	h_1	1.104	0.957	0.851
	h_2	5.129	3.640	2.947
	h_3	2.749	2.359	2.032

where ψ is the residual rotation at the end of the applied moment and is defined in the same way as in equations (18.29) and (18.30). $M_0[= 0.2679(W-a)^2]$ is the limit moment. In Table 18.4, we show the comparison between the present results and Shih and Hutchinson's ones [21]. The difference is within 3 per cent, which is worse than in centre cracked plate analysis.

Consider the non-linear crack analysis under antiplane shear, for which Shih [31] has obtained the non-linear fracture parameters. Taking a rectangular plate of $L/W = 2.0$ under antiplane shear with a centre crack, we show in Table 18.5 the comparison of the present non-dimensional parameters with Shih's results for various values of λ and n. The difference between both results is under 0.5 per cent for almost the whole range.

Table 18.5 Non-dimensional non-linear fracture parameters g_1, g_2, and g_3 with the values of the strain hardening exponent n and the crack length to plate ratio λ for centre cracked plate under antiplane shear

		$n=1.0$		$n=2.0$		$n=3.0$	
		Present	Ref 31	Present	Ref. 31	Present	Ref. 31
$\lambda=\frac{1}{8}$	g_1	1.383	1.389	1.785	1.791	2.016	2.023
	g_2	1.766	1.746	2.069	2.054	2.241	2.226
	g_3	0.170	0.171	0.327	0.328	0.488	0.489
$\lambda=\frac{1}{4}$	g_1	1.234	1.242	1.438	1.446	1.497	1.506
	g_2	1.542	1.526	1.610	1.601	1.589	1.581
	g_3	0.301	0.300	0.508	0.506	0.674	0.670
$\lambda=\frac{1}{2}$	g_1	0.993	1.000	0.979	0.983	0.919	0.925
	g_2	1.124	1.111	0.932	0.926	0.783	0.779
	g_3	0.441	0.438	0.559	0.556	0.586	0.583
$\lambda=\frac{3}{4}$	g_1	0.800	0.804	0.699	0.703	0.629	0.632
	g_2	0.685	0.681	0.424	0.422	0.303	0.302
	g_3	0.407	0.404	0.352	0.350	0.284	0.283

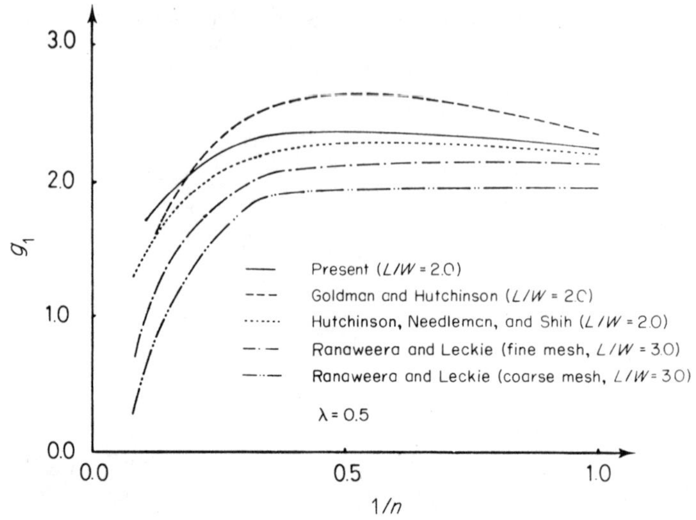

Figure 18.3 Comparison of the non-dimensional J integral value g_1 between several numerical methods for centre cracked plate under uniform tension (plane strain)

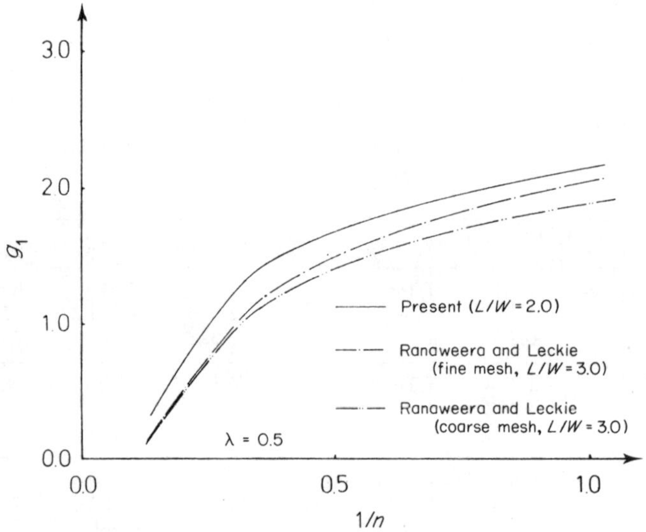

Figure 18.4 Comparison of the non-dimensional J integral value g_1 between several numerical methods for double-edged cracked plate under uniform tension (plane strain)

Table 18.6 Comparison of the non-linear fracture parameters, g_1, g_2, and g_3 between three methods for centre cracked plate under uniform tension (plane strain)

		$n=1.0$	$n=1.5$	$n=2.0$	$n=3.0$
$\lambda=\frac{1}{4}$	g_1	2.580* (2.58†, 2.57‡)	2.194 (3.01, 2.90)	3.159 (3.21, 3.11)	3.468 (3.24, 3.55)
	g_2	2.895 (2.615, 2.79)	3.050 (2.737, 2.99)	3.131 (2.789, 3.09)	3.182 (2.719, 3.14)
	g_3	0.559 (0.515, 0.548)	0.762 (0.692, 0.752)	0.954 (0.856, 0.942)	1.291 (1.101, 1.27)
$\lambda=\frac{1}{2}$	g_1	2.239 (2.34, 2.19)	2.323 (2.58, 2.25)	2.349 (2.63, 2.27)	2.319 (2.50, 2.18)
	g_2	2.121 (1.992, 2.09)	2.020 (1.905, 1.99)	1.899 (1.784, 1.87)	1.665 (1.510, 1.61)
	g_3	0.838 (0.789, 0.789)	1.006 (0.939, 0.949)	1.118 (1.035, 1.06)	1.222 (1.086, 1.15)
$\lambda=\frac{3}{4}$	g_1	2.087 (2.05, 2.10)	1.924 (2.04, 1.85)	1.780 (1.88, 1.80)	1.593 (1.54, 1.57)
	g_2	1.410 (1.263, 1.40)	1.117 (0.998, 1.08)	0.910 (0.786, 0.899)	0.663 (0.523, 0.637)
	g_3	0.843 (0.754, 0.814)	0.822 (0.725, 0.763)	0.761 (0.645, 0.733)	0.626 (0.487, 0.593)

* Present.
† Ref. [22].
‡ Ref. [24].

For the centre cracked and double edge cracked plate in the plane strain condition under uniform tension, the same fracture parameters are calculated with the assumptions as $L/2 = W = 16$, $\kappa = \sigma_0 = 1.0$ and $N_1 + N_2 = 7$, and are shown in Figures 18.3 and 18.4, respectively. In the case of the centre cracked plate, Table 18.6 shows the comparison between the present results and the other ones for g_1, g_2, and g_3 in equations (18.28). It can be seen from the table that the present solutions agree well with those in [24] and the results in [22] seem to have some discrepancies from those in [24] and in this paper.

18.5 CONCLUSION

An analysis method using the superposition technique with the penalty function method is proposed with applications to crack problems both in linear and non-linear fracture mechanics. It is found through several numerical examples that the present method may be useful to calculate various fracture mechanics parameters in compressible as well as incompressible materials.

REFERENCES

1. R. H. Gallagher, 'A survey of finite element fracture mechanics analysis', *Proc. First Int. Conf. on Structural Mechanics in Reactor Technology*, **5,** L6/9 (1972).
2. M. C. Apostal, S. Jordan, and P. V. Marcal, *Finite Element Techniques for Postulated Flaws in Shell Structures*, EPRI SR-22, Electric Power Research Institute, 1975.
3. A. S. Kobayashi, D. Maiden, B. Simon, and S. Iida, 'Application of the method of finite element analysis to two-dimensional problems in fracture mechanics', *Paper* 69*WA PVP*-12, ASME Winter Annual Meeting, 1969.
4. D. M. Parks, 'A stiffness derivative finite element technique for determination of elastic crack tip stress intensity factors, *Int. J. of Fracture*, **10,** 487–502 (1974).

5. T. K. Hellen, 'On the method of virtual crack extensions', *Int. J. Num. Meth. in Eng.*, **9,** 187–208 (1975).
6. Y. Yamamoto, 'Finite element approaches with the aid of analytical solutions', in *Recent Advances on Matrix Method of Structural Analysis and Design* (Eds. R. H. Gallagher, Y. Yamada, and J. T. Oden), pp. 85–103, University of Alabama Press, 1971.
7. Y. Yamamoto, N. Tokuda, and Y. Sumi, 'Finite element treatment of singularities of boundary value problems and its application to analysis of stress intensity factors', in *Theory and Practice of Finite Element Methods for Structural Analysis* (Eds. Y. Yamada and R. H. Gallagher), pp. 75–90, University of Tokyo Press, 1973.
8. Y. Yamamoto and N. Tokuda, 'Determination of stress intensity factors in cracked plates by the finite element method', *Int. J. Num. Meth. in Eng.*, **6,** 427–439 (1973).
9. G. Yagawa, T. Nishioka, Y. Ando, and N. Ogura, 'The finite element calculation of stress intensity factors using superposition', in *Computational Fracture Mechanics* (Eds. E. Rybicki and S. Benzley), pp. 21–34, ASME, 1975.
10. G. Yagawa, N. Miyazaki, and Y. Ando, 'Superposition method of finite element and analytical solutions for transient creep analysis', *Int. J. Num. Meth. in Eng.*, **11,** 1107–1115 (1977).
11. G. Yagawa and T. Nishioka, 'Finite element analysis of stress intensity factors for plane extension and plate bending', *Int. J. Num. Meth. in Eng.*, **14,** 727–740 (1979).
12. G. Yagawa and T. Nishioka, 'Superposition method for semi-circular surface crack', *Int. J. Solids and Structures*, **16,** 585–595 (1980).
13. R. D. Henshell and K. G. Shaw, 'Crack tip elements are unnecessary', *Int. J. Num. Meth. in Eng.*, **9,** 495–507 (1975).
14. R. S. Barsoum, 'On the use of isoparametric finite elements in linear fracture mechanics', *Int. J. Num. Meth. in Eng.*, **10,** 25–37 (1976).
15. E. Byskov, 'The calculation of stress intensity factors using the finite element method with cracked elements', *Int. J. Fracture Mechanics*, **6,** 159–167 (1970).
16. T. H. H. Pian, P. Tong, and C. H. Luk, 'Elastic crack analysis by a finite element hybrid method', *Proc. Third Conf. on Matrix Methods in Structural Mechanics*, pp. 661–682, AFFDL-Tr-71-160, 1973.
17. S. N. Atluri, A. S. Kobayashi, and M. Nakagaki, 'An assumed displacement hybrid finite element model for linear fracture mechanics', *Int. J. Fracture*, **11,** 257–271 (1975).
18. J. W. Hutchinson, 'Singular behavior at the end of a tensile crack in a hardening material', *J. Mechanics and Physics of Solids*, **16,** 13–31 (1968); also 'Plastic stress and strain fields at a crack tip', *J. Mechanics and Physics of Solids*, **16,** 337–347 (1968).
19. J. R. Rice and G. F. Rosengren, 'Plane strain deformation near crack tip in power-law hardening material', *J. Mechanics and Physics of Solids*, **16,** 1–12 (1968).
20. P. D. Hilton and J. W. Hutchinson, 'Plastic intensity factors for cracked plates', *Engineering Fracture Mechanics*, **3,** 435–451 (1971).
21. C. F. Shih and J. W. Hutchinson, 'Fully plastic solutions and large scale yielding estimates for plane stress crack problems', *ASME J. Engineering Materials and Technology*, **98,** 289–295 (1976).
22. N. L. Goldman and J. W. Hutchinson, 'Fully plastic crack problems: the center-cracked strip under plane strain', *Int. J. Solids and Structures*, **11,** 575–591 (1975).

23. M. P. Ranaweera and F. A. Leckie, 'Solution of nonlinear elastic fracture problems by direct optimization', in *Numerical Methods in Fracture Mechanics* (Eds. A. R. Luxmoore and D. R. J. Owen), pp. 450–463, Swansea, 1978.
24. J. W. Hutchinson, A. Needleman, and C. F. Shih, *Fully Plastic Problems in Bending and Tension*, MECH-6, Harvard University, 1978.
25. G. Yagawa, T. Aizawa, and Y. Ando, 'Crack analysis of power hardening materials using a penalty function and superposition method', *ASTM STP*, **700,** 439–452 (1980).
26. G. Yagawa, T. Aizawa, and Y. Ando, 'Linear and nonlinear elastic analysis of cracked plate: Application of a penalty function and superposition method', *Int. J. Num. Meth. in Eng.*, **17,** 719–733 (1981).
27. K. Washizu, *Variational Methods in Elasticity and Plasticity*, 2nd ed., Pergamon Press, 1975.
28. O. C. Zienkiewicz, *The Finite Element Method*, 3rd ed., McGraw-Hill, 1977.
29. M. Ishida, 'Methods of Laurent expansion for internal crack Problems', *Methods of Analysis and Solutions of Crack Problens* (Ed, G. C. Sih), pp. 56–130, Noordhoff, Leiden (1973).
30. C. E. Turner, 'Methods for post-yield fracture safety assessment', *Post-yield Fracture Mechanics* (Ed. D. G. H. Latzko), pp. 23–210, Applied Science Publishers, London, 1979.
31. C. F. Shih, '*J*-integral estimates for strain hardening materials in antiplane shear using fully plastic solution', *ASTM STP*, **590,** 3–26, 1976.

Hybrid and Mixed Finite Element Methods
Edited by S. N. Atluri, R. H. Gallagher, and O. C. Zienkiewicz

Chapter 19

Accuracy Considerations for Finite Element Calculations of the Stress Intensity Factor by the Method of Superposition

Yoshiyuki Yamamoto, Naoaki Tokuda, and Yoichi Sumi

19.1 INTRODUCTION

The method of solution based on the superposition of analytical and finite element solutions has been developed for determining the stress concentration factor on notched surfaces and the stress intensity factor of cracks; various versions have been developed [1 to 15]. In the method proposed by the present authors [1 to 7], the stress concentration factor or the stress intensity factor is determined by certain collocation conditions. This method is easy to apply because the scheme for numerical computation is the same for any analytical solutions, even if they have singularities of a higher order. Sumi, Nemat-Nasser, and Keer [6, 7] applied the method of superposition to calculate the derivative of the stress intensity factor with respect to the crack length. The accuracy of this method has so far been discussed empirically by solving simple problems whose solution is given in analytical form [2]; it will be investigated from the theoretical point of view in the present paper.

The accuracy of solutions will be investigated herein by considering the behaviours of the residual solution and the difference between the actual and analytical solutions. Errors arising from the residual solution may be called the *inherent error*, and decrease with the residual stress values at the collocation points. As a result, a guidance can be derived for choosing the analytical solutions with singularities and the collocation points.

19.2 ELASTIC STRESS ANALYSIS IN A TWO-DIMENSIONAL ELASTIC BODY

Consider a two-dimensional elastic problem defined in a cracked domain D

by a set of equations for the displacement u:

$$-L[u]=b \qquad \text{in } D \tag{19.1}$$

$$B[u]=f \qquad \text{on } S_\sigma \tag{19.2}$$

$$u=v \qquad \text{on } S_u \tag{19.3}$$

where S_σ and S_u form the boundary of D, and the body force b, the surface traction f, and the displacement v are prescribed in D, on S_σ or on S_u respectively. The problem under consideration can be written in a weak form. Let u be an arbitrary regular function satisfying the condition

$$\delta u=0 \qquad \text{on } S_u \tag{19.4}$$

Since the problem is self-adjoint, the inner product of $-L[u]$ and δu in D can be expressed as

$$(-L[u], \delta u)=\langle u, \delta u\rangle-((B[u], \delta u)) \tag{19.5}$$

where

$$(-L, \delta u)=\int (-L)^t\, \delta u \, \mathrm{d}D \tag{19.6}$$

$$\langle u, \delta u\rangle=\int \varepsilon[u]^t E \varepsilon[\delta u]\, \mathrm{d}D \tag{19.7}$$

$$((B, \delta u))=\int B^t\, \delta u \, \mathrm{d}S_\sigma \tag{19.8}$$

Here ε is a linear vector operator giving a strain and E is the elastic modulus tensor. Introducing equations (19.1) and (19.2) into equation (19.5) leads to the principle of virtual work

$$\langle u, \delta u\rangle-(b, \delta u)-((f, \delta u))=0 \tag{19.9}$$

or equivalently to the minimum principle of potential energy

$$\tfrac{1}{2}\langle u, u\rangle-(b, u)-((f, u))=\text{minimum} \tag{19.10}$$

with the subsidiary condition given by equation (19.3). The finite element method can be formulated on the basis of equations (19.9) or (19.10). The solution of the boundary value problem defined by equations (19.1) and (19.3) is uniquely determined within trivial solutions, which will be assumed to vanish for the sake of simplicity.

Assume that the boundary S_σ has a singular point O at a crack tip. Introduce a rectangular coordinate system (x, y) and a polar coordinate system (r, θ) with the origin O so that the x axis is in contact with the crack surface (see Figure 19.1). Then the solution u may be singular at the origin O. Let $[D]$ be a neighbourhood of the crack tip O and let $[S_\sigma]$ be the corresponding part of S_σ. For the sake of simplicity, assume that b and f

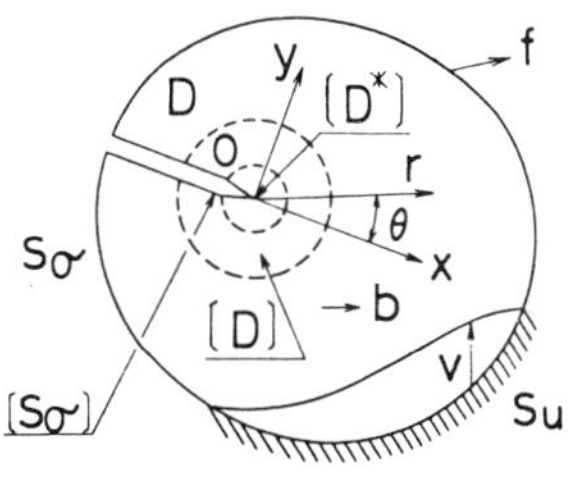

Figure 19.1 Cracked domain D

vanish in $[D]$ or on $[S_\sigma]$ respectively:

$$b=0 \qquad \text{in } [D]$$
$$f=0 \qquad \text{on } [S_\sigma]$$

Then the characteristics of the analytical solution u^{a} can be investigated by introducing $u^{\mathrm{a}}=r^{\lambda}g(\theta)$ into the following homogeneous equations:

$$\begin{aligned} -L[u^{\mathrm{a}}]&=0 \qquad \text{in } [D] \\ B[u^{\mathrm{a}}]&=0 \qquad \text{on } [S_\sigma] \end{aligned} \tag{19.11}$$

Then λ and g can be determined as eigenpairs, which give a set of analytical functions u_j^{a} $(j=1,2,\ldots)$. It will be assumed, by disregarding the governing equations, that the domain of definition for the u_j^{a} can be extended analytically throughout D. Now the analytical solution u^{a} is expressed as a linear combination of the u_j^{a}:

$$u^{\mathrm{a}}=\sum_{j=1}^{N} m_j u_j^{\mathrm{a}} \tag{19.12}$$

Here u_j^a $(j=1,\ldots,N)$ have singularities and u_j^{a} $(j=N+1,\ldots)$ are regular.

Assume that the solution of the original problem is expressed as a sum of the analytical solution and a regular residual solution u^{r}:

$$u=u^{\mathrm{a}}+u^{\mathrm{r}} \tag{19.13}$$

where

$$\begin{aligned} -L[u^{\mathrm{r}}]&=b+L[u^{\mathrm{a}}] \qquad \text{in } D \\ B[u^{\mathrm{r}}]&=f-B[u^{\mathrm{a}}] \qquad \text{on } S_\sigma \\ u^{\mathrm{r}}&=v-u^{\mathrm{a}} \qquad \text{on } S_{\mathrm{u}} \end{aligned} \tag{19.14}$$

Since u^r is regular, it can be obtained numerically by a conventional finite element scheme and is expressed in the form:

$$u^{\mathrm{r}}=u^{\mathrm{e}}+\sum_{j=1}^{N} m_j u_j^{\mathrm{re}} \tag{19.15}$$

where the u^e and u_j^{re} are numerical solutions of the equations

$$\begin{aligned} -L[u^e] &= b && \text{in } D \\ B[u^e] &= f && \text{on } S_\sigma \\ u^e &= v && \text{on } S_u \\ -L[u_j^{re}] &= L[u_j^a] && \text{in } D \\ B[u_j^{re}] &= -B[u_j^a] && \text{on } S_\sigma \\ u_j^{re} &= -u_j^a && \text{on } S_u \end{aligned}$$

Then the solution of the original problem is expressed as

$$u = u^e + \sum_{j=1}^{N} m_j(u_j^a + u_j^{re}) \tag{19.16}$$

where the m_j are constants to be determined. Introducing u from equation (19.16) into equation (19.9) leads to the equation

$$\left\langle u^e + \sum_{k=1}^{N} m_k(u_k^a + u_k^{re}),\ u_j^a + u_j^{re} \right\rangle - (b,\ u_j^a + u_j^{re}) - ((f,\ u_j^a + u_j^{re})) = 0$$

$$j = 1, \ldots, N \tag{19.17}$$

Then the m_j can be determined by equation (19.17). Pian, Tong and Luk [8] and Yagawa *et al.* [9 to 13] proposed a direct application of equation (19.17). In this formulation, fairly complicated integrals should be evaluated for determining inner products. The present authors derived a set of collocation equations equivalent to equation (19.17).

Introducing

$$u = u^a \qquad \text{and} \qquad \delta u = u_j^a + u_j^{re}$$

into the equality given by equation (19.5) leads to

$$(-L[u^a],\ u_j^a + u_j^{re}) = \langle u^a,\ u_j^a + u_j^{re} \rangle - ((B[u^a],\ u_j^a + u_j^{re})) \tag{19.18}$$

From equations (19.17) and (19.18), it follows that

$$\langle u^r,\ u_j^a + u_j^{re} \rangle - (-L[u^r],\ u_j^a + u_j^{re}) - ((B[u^r],\ u_j^a + u_j^{re})) = 0 \tag{19.19}$$

This relation will be simplified by considering the characteristics of numerical solutions. If equation (19.14) can be solved rigorously, $u_j^a + u_j^{re}$ vanishes throughout the domain. In reality, u_j^{re} is a solution by a conventional finite element method, and therefore $u_j^a + u_j^{re}$ does not vanish near the crack tip or in $[D]$:

$$u_j^a + u_j^{re} = 0, \qquad \varepsilon[u_j^a + u_j^{re}] = 0 \qquad \text{in } D - [D] \tag{19.20}$$

Since

$$-L[u^{r}]=b+L[u^{a}]=0 \qquad \text{in } [D]$$
$$B[u^{r}]=f-B[u^{a}]=0 \qquad \text{on } [S]$$

the following relation holds approximately:

$$(-L[u^{r}],\, u_j^{a}+u_j^{re})+((B[u^{a}],\, u_j^{a}+u_j^{re}))=0$$

Now equation (19.18) can be written in a simple form:

$$\langle u^{r},\, u_j^{a}+u_j^{re}\rangle=0 \qquad j=1,\ldots,N \tag{19.21}$$

or approximately

$$\int_{[D]}\left(\sum_{k=1}^{N} m_k\varepsilon[u_k^{re}]+\varepsilon[u^{e}]\right)^{t} E\varepsilon[u_j^{a}+u_j^{re}]\,\mathrm{d}D=0 \qquad j=1,\ldots,N \tag{19.22}$$

From equation (19.20), $\varepsilon[u_j^{a}+u_j^{re}]$ is significant only in a small domain $[D^*]$ in $[D]$, and there holds an approximate relation given by

$$\varepsilon[u_j^{a}+u_j^{re}]=\varepsilon[u_j^{a}] \qquad \text{in } [D^*]$$

Here $[D^*]$ may be a small assemblage of elements near the crack tip. Let σ be the stress caused by the strain $\varepsilon[u]$:

$$\sigma=E\varepsilon[u]$$

It then follows that

$$\int_{[D]}\left(\sum_{k=1}^{N} m_k\sigma_k^{re}+\sigma^{e}\right)^{t}\varepsilon[u_j^{a}]\,\mathrm{d}D=0 \qquad j=1,\ldots,N \tag{19.23}$$

or approximately

$$\sum_{A_i\subset[D^*]}\left(\sum_{k=1}^{N} m_k\sigma_k^{re}+\sigma^{e}\right)^{t}\varepsilon[u_j^{a}]A_i=0 \qquad j=1,\ldots,N \tag{19.24}$$

Here A_i is the area of the finite element i in $[D^*]$ and its coefficients are evaluated at the representative point of each element. In general, $\varepsilon[u_j^{a}]$, $j=1,\ldots,N$, have the common significant components, say $(\varepsilon[u_j^{a}])_{\mu k}$, $j=1,\ldots,N$, $k=1,\ldots,$ and then equation (19.24) can be simplified as

$$(\sigma^{r})_{\mu_k}=\left(\sum_{j=1}^{N} m_j\sigma_j^{re}+\sigma^{e}\right)_{\mu k}=0 \qquad k=1,\ldots \tag{19.25}$$

which are evaluated at certain collocation points chosen in $[D^*]$.

The same collocation relations can be derived from physical considerations. The stress-free condition on $[S_\sigma]$ can be expressed as

$$(\sigma)_{\mu_k}=0 \qquad (k=1,2) \qquad \text{or} \qquad \sigma_{yy}=\sigma_{xy}=0$$

Since σ^{a} satisfies this condition exactly, the following relations hold:

$$(\sigma^{r})_{\mu_k} = 0 \qquad \text{on } [S_\sigma] \qquad (k = 1, 2)$$

Because of the regularity of σ^{r}, this condition can be replaced by a collocation relation at a certain number of points chosen near the crack tip:

$$\left(\sum_{j=1}^{N} m_j \sigma_j^{re} + \sigma^{e}\right)_{\mu_k} = 0 \qquad k = 1, 2 \tag{19.26}$$

In the case of finite element calculations, accurate evaluation of stresses is difficult on free surfaces. Edge or nodal averaging gives satisfactory results for inner points, even in the case of conventional finite elements [16]. Therefore, centres of element edges and nodal points other than the crack tip can be chosen for the collocation points for evaluating stress values in equation (19.26). In reality, the residual stress components may take non-vanishing values, a fact which causes inherent errors for the present method: this will be discussed hereafter.

19.3 STRESSES AROUND A CRACK TIP

According to Williams [17], the stress is given in the following form around the crack tip:

$$\sigma = \sum_{k=1,2} \frac{K_k}{\sqrt{(2\pi r)}} f_k + s + O(\sqrt{r}) \tag{19.27}$$

where

$$\begin{aligned} f_{1xx} = f_{1yy} = f_{2xy} &= \cos\frac{\theta}{2}\left(1 \mp \sin\frac{\theta}{2}\sin\frac{3\theta}{2}\right) \\ f_{1xy} = f_{2yy} &= \sin\frac{\theta}{2}\cos\frac{\theta}{2}\cos\frac{3\theta}{2} \\ f_{2xx} &= -f_{2yy} - 2\sin\frac{\theta}{2} \end{aligned} \tag{19.28}$$

Here K_1 and K_2 are the stress intensity factors for the opening and inplane shear modes for the stress field σ and s is a constant stress satisfying the stress-free condition such that

$$s_{yy} = s_{xy} = 0$$

According to the present theory, $(2\pi r)^{-1/2} f_k$ $(k = 1, 2)$ can be used as the analytical solution with singularities. As can be seen from equation (19.27), the stress has terms of order $r^{1/2}$ which cannot be expressed by shape functions of conventional finite elements; therefore such terms of order $r^{1/2}$

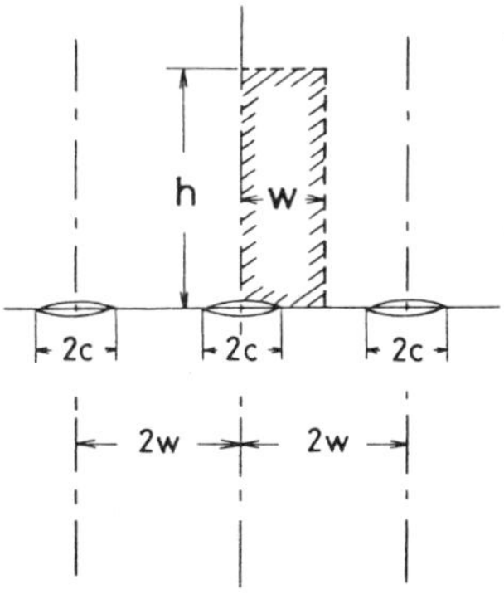

Figure 19.2 Collinear cracks

should be included into the category of the analytical solution. Analytical solutions can also be obtained as solutions of similar problems. These procedures will be exemplified by a simple example, an infinite plate with collinear cracks of length $2c$ as shown in Figure 19.2. Westergaard [18] obtained analytical solutions of the plate under uniform tensile or shear stress σ^0 at infinity. The corresponding stress intensity factors are given by

$$K_1 = \left(\frac{2w}{\pi c}\tan\frac{\pi c}{2w}\right)^{1/2} K_0, \qquad K_2 = 0 \qquad \text{for a stretched plate}$$

$$K_1 = 0, \qquad K_2 = \left(\frac{2W}{\pi c}\tan\frac{\pi c}{2w}\right)^{1/2} K_0 \quad \text{for a sheared plate} \tag{19.29}$$

where $K_0 = \sigma^0\sqrt{\pi c}$ is the stress intensity factor for a stretched infinite plate with a crack of length $2c$. Finite element analyses will be performed for obtaining the residual stress only on the shaded domain; on the boundary of the domain, displacements and stresses calculated by Westergaard's solution will be applied.

The following three sets of the analytical solutions will be used for expressing singular stress distributions around crack tips:

(a) Two-term approximation:

$$\sigma_j^{\mathrm{a}} = \frac{K_0}{\sqrt{(2\pi r)}} f_i \qquad (m_j = K_j/K_0),\ j = 1, 2 \tag{19.30}$$

(b) Four-term approximation:

$$\begin{aligned} \sigma_j^{\mathrm{a}} &= \frac{K_0}{\sqrt{(2\pi r)}} f_i \qquad (m_j = K_j/K_0),\ j = 1, 2 \\ \sigma_j^{\mathrm{a}} &= \sqrt{r} f_i \qquad\qquad j = 3, 4 \end{aligned} \tag{19.31}$$

where

$$\left.\begin{matrix} f_{jxx} \\ f_{jyy} \end{matrix}\right\} = \frac{5}{4}\left(2+\left\{\begin{matrix} \sin^2\theta \\ \cos^2\theta \end{matrix}\right.\right)P_j \pm \tfrac{3}{2}\sin(2\theta)\frac{dP_j}{d\theta}+\left\{\begin{matrix} \cos^2\theta \\ \sin^2\theta \end{matrix}\right. \frac{d^2p_j}{d\theta^2}$$

$$f_{jxy} = -\tfrac{5}{8}\sin(2\theta)P_j - \tfrac{3}{2}\cos(2\theta)\frac{dP_j}{d\theta}+\tfrac{1}{2}\sin(2\theta)\frac{d^2P_j}{d\theta^2}$$

$$P_j = \left\{\begin{matrix} \cos \\ -\sin \end{matrix}\right. (\theta/2) + \left\{\begin{matrix} -(1/5)\cos \\ \sin \end{matrix}\right. (5\theta/2) \qquad j=3,4$$

(c) Approximation by the use of Westergaard's solutions for an infinite plate with a crack of length $2c$, which are derived from the stress functions F_j $(j=1,2)$ such that

$$F_1 = \operatorname{Re}\bar{Z} + y \operatorname{Im}\bar{Z} - \tfrac{1}{2}\sigma^0 y^2, \qquad F_2 = -y \operatorname{Re}\bar{Z} \tag{19.32}$$

where

$$Z = \frac{\sigma^0}{\sqrt{[1-(c/\zeta)^2]}}, \qquad \zeta = x+iy, \qquad \frac{dZ}{d\zeta} = \bar{Z},$$

The mesh pattern used for the present analysis is shown in Figure 19.3 and the finite element is the constant strain triangle. Patterns A and B are

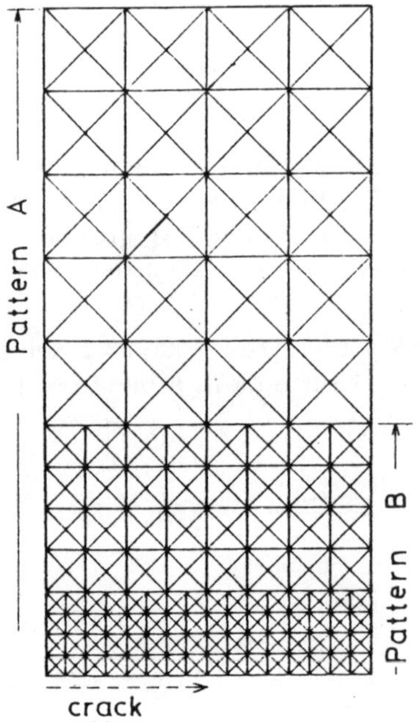

Figure 19.3 Mesh subdivision

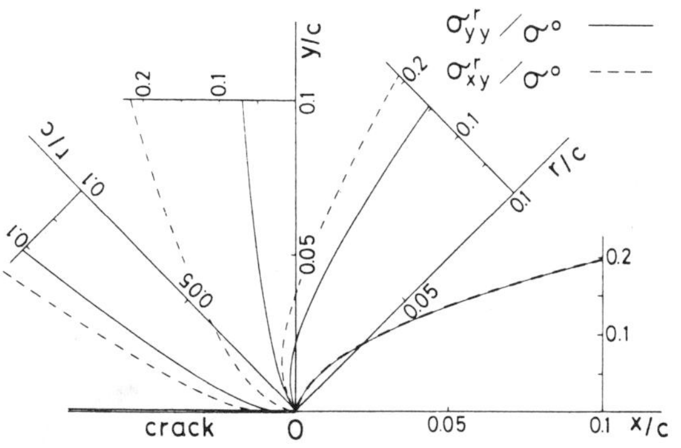

Figure 19.4 Residual stress for two-term approximation ($c/w = 8/16$)

for the stretched or sheared plate respectively. In the case of a stretched plate, the stress around the crack tip can be expressed only by f_1 and f_3, and in the case of a sheared plate, it can be expressed by f_2 and f_4.

Assume that the rigorous solution is given in an analytical form. Then the stress σ is determined exactly and can be resolved into a sum of the analytical and residual solutions in the following form:

$$\sigma = \sigma^{\mathrm{r}} + \sum m_j^0 \sigma_j^{\mathrm{a}} \tag{19.33}$$

where the σ^{r} and m_j^0 are given exactly. Figures 19.4 to 19.6 show the residual stresses along radial vectors for the infinite stretched and sheared plates with collinear cracks corresponding to three sets of analytical solutions. Here σ^0 is the uniform tensile or shear stress at infinity. The residual stress components $\sigma^{\mathrm{r}}_{\mathrm{yy}}$ and $\sigma^{\mathrm{r}}_{\mathrm{xy}}$ tend to zero at the crack tip. In Figure 19.4,

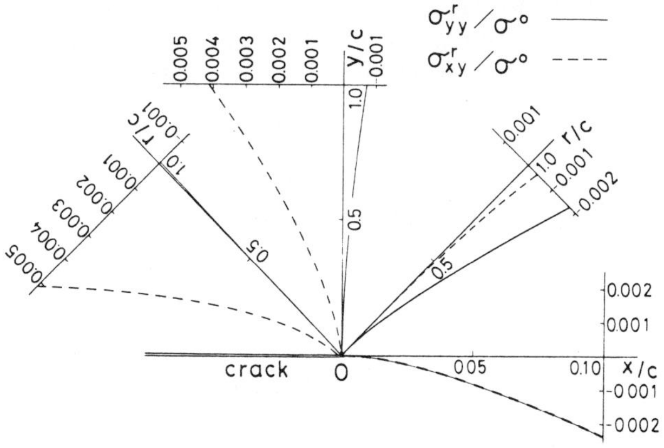

Figure 19.5 Residual stress for four-term approximation ($c/w = 8/16$)

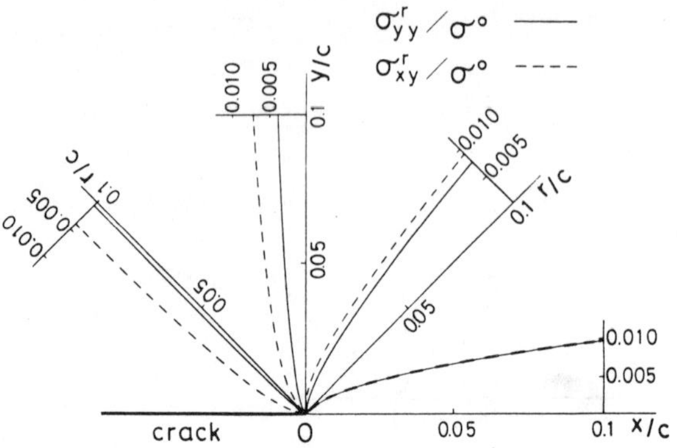

Figure 19.6 Residual stress for approximation by Westergaard's solution of an infinite plate with a crack ($c/w = 8/16$)

the effect of the term of order $r^{1/2}$ can be seen, and in Figures 19.5 and 19.6, four-term approximation and approximation with Westergaard's solution for an infinite plate with a crack give extremely small residual stresses around the crack tip.

19.4 INHERENT ERRORS OF THE METHOD OF SUPERPOSITION

For finite element calculations, accurate evaluations for stresses at the crack tip are difficult, and therefore equation (19.26) has to be evaluated at certain points near the crack tip. Since the edge or nodal averaging is effective for this purpose, the edge centres and the nodes near the crack tip are conveniently chosen as the collocation points for evaluating stresses in equation (19.26). For the mesh pattern shown in Figure 19.7, some of the following

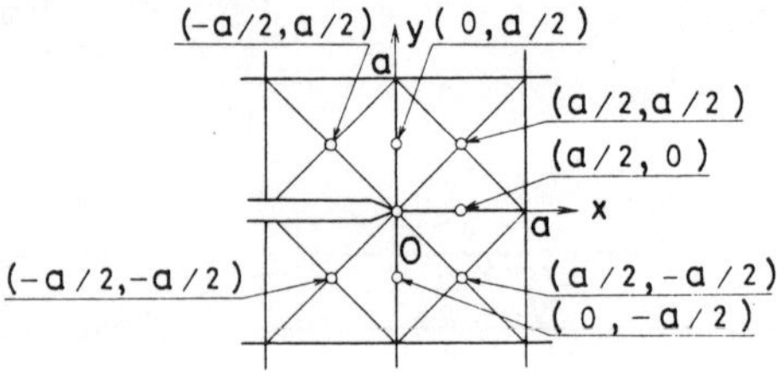

Figure 19.7 Collocation points and meshes around a crack tip

are used as the collocation points:

$$(a/2, 0), (\pm a/2, \pm a/2), (0, \pm a/2)$$

where a is the diameter of meshes or the mesh size.

By virtue of equation (19.26), the m_j are determined from N equations obtained by manipulating the conditions that the residual stress components $(\sigma^r)_{\mu_k}$, $k = 1, 2$, vanish at N^* ($\leq N$) collocation points chosen properly [2, 5]:

$$\left(\sigma - \sum_{j=1}^{N} m_j \sigma_j^a\right)_{\mu_k} = 0, \qquad k = 1, 2 \text{ at } N^* \text{ collocation points} \tag{19.34}$$

The exact values of the m_j^0 satisfy equation (19.33); i.e.

$$\left(\sigma - \sum_{j=1}^{N} m_j^0 \sigma_j^a\right)_{\mu_k} = (\sigma^r)_k, \qquad k = 1, 2 \text{ at the collocation points} \tag{19.35}$$

It then follows that

$$\sum_{j=1}^{N} (m_j - m_j^0)(\sigma_j^a)_{\mu_k} = (\sigma^r)_{\mu_k} \qquad k = 1, 2 \text{ at } N \text{ collocation points} \tag{19.36}$$

which gives the errors for m_j. Errors thus obtained can be called *inherent errors* because they are independent of the accuracy of the numerical stress analysis. They depend not only upon the collocation points but also upon the set of analytical solutions adopted. Therefore, the analysis of the inherent error gives a guidance for the choice of the collocation points and the set of analytical solutions.

As an example, consider a stretched infinite plate with collinear cracks. Take two-term approximation of analytical solutions. From symmetry of the stress field, it follows that

$$\sigma = \sigma^r + m_1^0 \sigma_1^a \qquad (m_2^0 = 0) \tag{19.37}$$

Since σ_{xy} vanishes on the crack surface, m_1^0 and m_1 are determined by

$$\sigma_{yy} - m_1^0 \sigma_{1yy}^a = \sigma_{yy}^r \qquad \text{at a collocation point}$$

$$\sigma_{yy} - m_1 \sigma_{1yy}^a = 0 \qquad \text{at the collocation point}$$

Then the inherent error is given by

$$\frac{m_1 - m_1^0}{m_1^0} = \frac{K_1 - K_1^0}{K_1^0} = \frac{\sigma_{yy}^r}{\sigma_{yy} - \sigma_{yy}^r} \qquad \text{at the collocation point} \tag{19.38}$$

where K_1^0 is the exact value of the stress intensity factor. Results obtained are shown in Figure 19.8. The inherent error is approximately in proportion

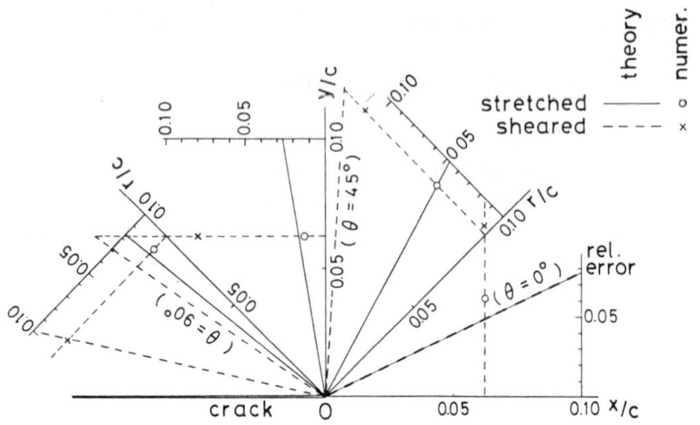

Figure 19.8 Inherent error for two-term approximation ($c/w = 8/16$)

to the radial distance of the collocation point from the origin. For this analytical solution, the preferable direction for the collocation point is $|\theta| = 90° \sim 135°$. Figure 19.8 shows errors of the numerical solutions by the finite element method with the use of the collocation point $(a/2, 0)$, $(a/2, a/2)$, $(0, a/2)$, or $(-a/2, a/2)$, where $a = w/16$; they are in good accordance with the theoretical values.

Equation (19.36) means that the set of analytical solutions should be chosen so that the residual stress components σ_{yy}^{r} and σ_{xy}^{r} become small in comparison to the stress value at the collocation points. In the case where Westergaard's solution for the infinite plate with a crack is used as the analytical solution, the residual stresses are very small, as shown in Figure 19.6, and therefore the inherent error for m_1 is much smaller than the above case (see Figure 19.9). As can be expected, four-term approximation gives the best solutions from the viewpoint of the inherent error. In this case, two collocation points are used for determining m_1. In the case where $(a/2, 0)$ and $(0, a/2)$ are adopted as the collocation points, the theoretical and numerical inherent errors become as follows:

$$\left.\begin{aligned}&\text{Theoretical relative error} = 0.00041\\&\text{Numerical relative error} = 0.00027\end{aligned}\right\} \quad (\text{for } c/w = 8/16)$$

which show satisfactory coincidence.

Similar results are obtained for the sheared infinite plate with collinear cracks, the inherent errors being shown in Figures 19.8 and 19.9 together with numerical results. Figure 19.8 shows significant discrepancies between theoretical and numerical results; it must be due to the fact that the finite element approximation used here might be inadequate for calculating shear stresses.

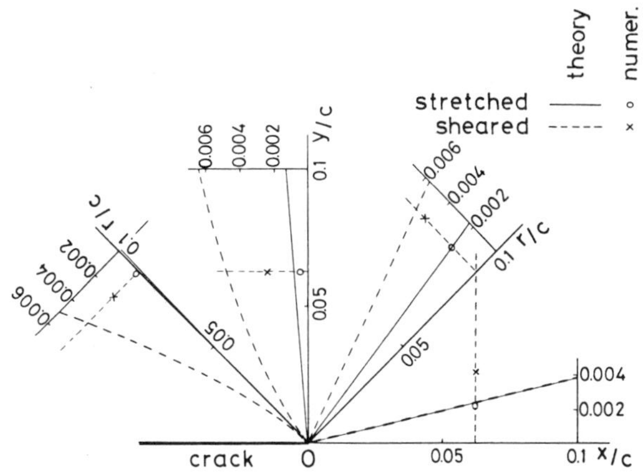

Figure 19.9 Inherent error for approximation by Westergaard's solution of an infinite plate with a crack ($c/w = 8/16$)

19.5 MODIFICATION OF FINITE ELEMENT MODELLING

Figures 19.4 to 19.6 indicate that the residual stress components σ_{yy}^{r} and σ_{xy}^{r} take very small values around the crack tip and satisfy the stress-free conditions approximately on the x axis near the origin. This fact suggests a possibility of modification of the finite element modelling for numerical calculation of the residual stress; the modelled crack length, $2c^*$, may not be identical with the actual one, $2c$. For example, assume that the modelled crack surface is short in comparison to the actual crack, as shown in Figure 19.10. The analytical solutions are determined in connection with the actual crack length $2c$ and the numerical solutions σ^{e} and σ_{i}^{re} are determined so as to satisfy the stress-free conditions on the modelled crack surface.

As for the choice of collocation points, there are many alternatives; two special cases will be considered in the following. The collocation points will be chosen near the actual or modelled crack tip. In the case where the crack length is elongated for modelling, collocation points will not be taken near the actual crack tip because it is meaningless. The resulting residual stress

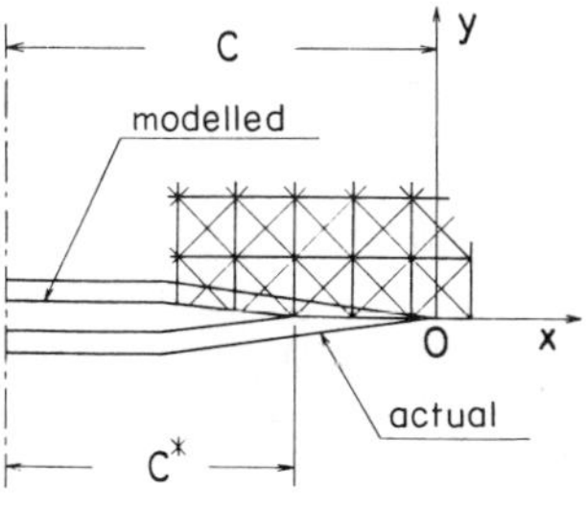

Figure 19.10 Actual and modelled cracks

Table 19.1 Stress intensity factor calculated by the use of modified crack ($c/w = 8/16$, $K_1^0/K_0 = K_2^0/K_0 = 1.12837$)

		K_1/K_0 for a stretched plate				K_2/K_0 for a sheared plate			
		Collocation points near modelled crack tips		Collocation points near actual crack tips		Collocation points near modelled crack tips		Collocation points near actual crack tips	
c^*/w	$(c^*-c)/c$	Four-term approximation	Two-term approximation	Four-term approximation	Two-term approximation	Four-term approximation	Two-term approximation	Four-term approximation	Two-term approximation
4/16	−4/8	1.1439 (0.014)	0.8890	1.1465 (0.016)	1.4366	1.1402 (0.012)	0.9901	1.1608 (0.03)	1.4828
5/16	−3/8	1.353 (0.006)	0.9602	1.1381 (0.009)	1.3478	1.1357 (0.007)	1.0551		
6/16	−2/8	1.1307 (0.002)	1.0258 (−0.09)	1.1327 (0.004)	1.2670 (0.12)	1.1337 (0.005)	1.1142 (−0.013)	1.1411 (0.011)	1.3222
7/16	−1/8	1.1288 (0.0004)	1.0865 (−0.04)	1.1298 (0.001)	1.1982 (0.06)	1.334 (0.004)	1.1681 (0.035)	1.1370 (0.008)	1.2611 (0.12)
8/16	0	1.1287 (0.0003)	1.1430 (0.013)			1.1340 (0.005)	1.2170 (0.08)		
9/16	1/8	1.1296 (0.001)	1.1954 (0.06)			1.1348 (0.006)	1.2616		
10/16	2/8	1.1309 (0.002)	1.2439			1.1352 (0.006)	1.3016		
11/16	3/8	1.1319 (0.003)	1.2882			1.1345 (0.005)	1.3369		
12/16	4/8	1.1319 (0.003)	1.3280			1.1320 (0.004)	1.3674		

Numerals in parentheses indicate the relative errors.
Collocation points are taken along $\theta = 0°$ and $90°$ for four-term approximation and $\theta = 90°$ for two-term approximation with respect to the origin at the modelled or actual crack tip.

Table 19.2 Stress intensity factor calculated by the use of modified crack modelling and four-term approximation

	K_1/K_0 for a stretched plate			K_2/K_0 for a sheared plate			
c/w	Consistent modelling $c=c^*$	Shortened crack model $c=c^*+w/16$	Elongated crack model $c=c^*-w/16$	Consistent modelling $c=c^*$	Shortened crack model $c=c^*+w/16$	Elongated crack model $c=c^*-w/16$	Exact value $K_1^0/K_0 = K_2^0/K_0$
3/16	1.0197 (0.005)	1.0208 (0.006)	1.0285 (0.014)				1.0149
4/16	1.0300 (0.003)	1.0305 (0.003)	1.0352 (0.008)	1.0563 (0.029)	1.0535 (0.027)	1.0630 (0.036)	1.0270
6/16	1.0662 (0.001)	1.0665 (0.001)	1.0686 (0.004)	1.0780 (0.013)	1.0761 (0.011)	1.0815 (0.016)	1.0651
8/16	1.1287 (0.0003)	1.1288 (0.0004)	1.1296 (0.001)	1.1340 (0.005)	1.1334 (0.005)	1.1348 (0.006)	1.1284
10/16	1.2345 (−0.0002)	1.2344 (−0.0003)	1.2338 (−0.0007)	1.2332 (−0.001)	1.2347 (0.0)	1.2298 (−0.004)	1.2347
12/16	1.4308 (−0.0005)	1.4304 (−0.0008)	1.4252 (−0.004)	1.4130 (−0.013)	1.4206 (−0.008)	1.3958 (−0.025)	1.4315
13/16	1.6060 (−0.0007)	1.6054 (−0.001)	1.5917 (−0.009)	1.5622 (−0.027)	1.5802 (−0.016)	1.5219 (−0.051)	1.6072
14/16	1.9111 (−0.0007)	1.9097 (−0.001)	1.8581 (−0.027)	1.7834 (−0.065)			1.9125

Numerals in parentheses indicate the relative errors.
Collocation points are taken along $\theta = 0°$ and $90°$ with respect to the origin at the modelled crack tip.

will satisfy the stress-free condition on the crack surface in the sense of numerical computations. If this modification gives satisfactory results, the stress intensity factors for a continuously growing crack can be analysed with the use of a mesh pattern chosen properly.

As the first example, the stress intensity factor is analysed for collinear cracks of length $2c = w$ with the use of finite element models with various crack lengths $2c^*$. Results obtained are given in Table 19.1, and four-term approximation shows satisfactory accuracy for a wide range of the model's crack length. In general, errors decrease with the absolute value of the difference of length of both cracks. As for collocation points, they should be chosen in connection to the modelled crack tip (see Figure 19.10) but there are no significant differences.

As the second example, the stress intensity factor of a plate with collinear cracks of various lengths $2c$ is analysed with the use of models with collinear cracks of lengths longer or shorter by one mesh size $a = w/16$. Results obtained are given in Table 19.2, and four-term approximation shows splendid accuracy. Here collocation points are taken in connection to the modelled crack tip.

19.6 AXIALLY SYMMETRIC THREE-DIMENSIONAL PROBLEMS

So far two-dimensional problems have been discussed. Hereafter a simple three-dimensional example will be considered. Figure 19.11 shows an infinite body with a penny-shaped crack of radius c subjected to a uniform tension σ^0 at infinity in which shear-type deformations do not appear in front of the crack tip line. The displacement and stress of this problem can be expressed by a potential function F:

$$\begin{aligned} u_\rho &= z\frac{\partial^2 F}{\partial\rho\,\partial z} + (1-2\nu)\frac{\partial F}{\partial\rho} \\ u_z &= z\frac{\partial^2 F}{\partial z^2} - 2(1-\nu)\frac{\partial F}{\partial z} \qquad (19.39) \\ u_\phi &= 0 \end{aligned}$$

$$\left.\begin{matrix}\sigma_{\rho\rho}\\ \sigma_{\phi\phi}\end{matrix}\right\} = 2G\left[\nu\left(1+\frac{1}{1-2\nu}z\frac{\partial}{\partial z}\right)\nabla^2 F + (1-2\nu)\left\{\begin{matrix}\dfrac{\partial^2 F}{\partial\rho^2}\\ \dfrac{\partial F}{\rho\,\partial\rho}\end{matrix}\right. + \left\{\begin{matrix}z\dfrac{\partial^3 F}{\partial\rho^2\,\partial z} - 2\dfrac{\partial^2 F}{\partial z^2}\\ \dfrac{z}{\rho}\dfrac{\partial^2 F}{\partial\rho\,\partial z}\end{matrix}\right.\right]$$

$$\sigma_{zz} = 2G\left[\nu\left(1+\frac{1}{1-2\nu}z\frac{\partial}{\partial z}\right)\nabla^2 F + z\frac{\partial^3 F}{\partial z^3} - \frac{\partial^2 F}{\partial z^2}\right] \qquad (19.40)$$

$$\sigma_{\rho z} = 2Gz\frac{\partial^3 F}{\partial\rho\,\partial z^2}, \qquad \sigma_{\rho z} = \sigma_{\phi z} = 0$$

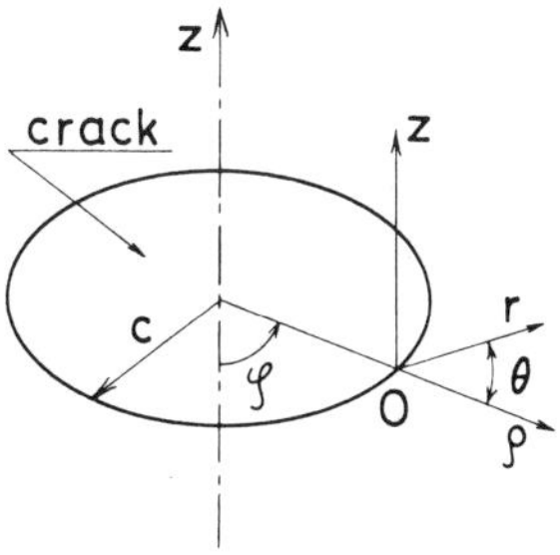

Figure 19.11 Infinite body with a penny-shaped crack

where G and ν are the shear modulus and Poisson's ratio and (ρ, ϕ, z) is the cylindrical coordinate system shown in Figure 19.11. The exact solution of this problem is given by the potential

$$F^0 = \frac{\sigma^0 c^2}{4\pi G}\left[\frac{\sqrt{\lambda}}{c}\left(1+\frac{3\mu}{c^2}\right)-\left(1-\frac{\lambda+\mu}{c^2}-\frac{3\lambda\mu}{c^4}\right)\tan^{-1}\frac{\sqrt{\lambda}}{c}\right]$$

where (λ, μ) is an elliptical coordinate system defined by

$$c^2\rho^2 = (c^2+\lambda)(c^2+\mu), \qquad c^2z^2 = -\lambda\mu$$

Introducing this potential into equation (19.40) leads to the exact stress distribution given by

$$\begin{aligned}\sigma_{zz} &= \frac{2\sigma^0}{\pi}\left[\frac{\sqrt{\lambda}/c}{(\lambda-\mu)^3}(-5\lambda\mu^2+\mu^3-5c^2\lambda\mu+c^2\lambda^2)+\tan^{-1}\frac{\sqrt{\lambda}}{c}\right]\\ \sigma_{\rho z} &= -\frac{2\sigma^0}{\pi}\frac{\lambda\sqrt{-\mu}\sqrt{(c^2+\mu)}}{c(\lambda-\mu)^3\sqrt{c^2+\lambda}}(\mu^2-5\lambda\mu-3c^2\mu-c^2\lambda)\end{aligned} \tag{19.41}$$

The stress distribution around the crack tip line can be expressed in the following form:

$$\sigma = \sum_{j=1,3} m_j\sigma_j^{a} + s + O(r^{3/2}) \tag{19.42}$$

where the σ_j^{a} $(j = 1, 3)$ are derived from the displacement potentials

$$\begin{aligned}F_1 &= \tfrac{3}{2}K_0Gc^{-3}\left(\frac{r}{c}\right)^{3/2}\left(\cos\frac{3\theta}{2}-\frac{1}{4}\frac{r}{c}\cos\frac{\theta}{2}\right)\\ F_3 &= \left(\frac{r}{c}\right)^{5/2}\cos\frac{5\theta}{2}\end{aligned} \tag{19.43}$$

Here (r, θ) is the local polar coordinate such that

$$\rho = c + r\cos\theta, \qquad z = r\sin\theta$$

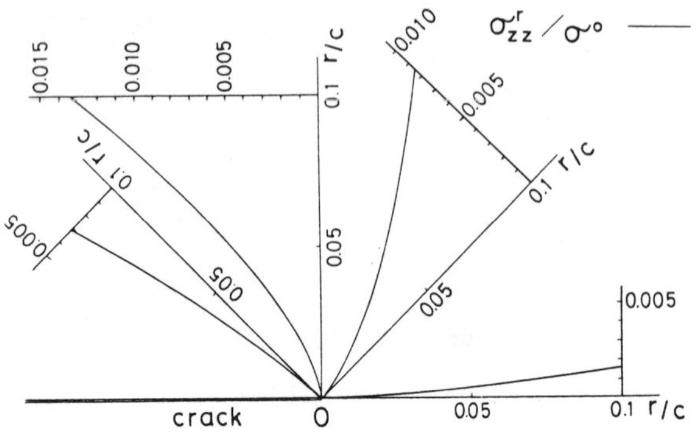

Figure 19.12 Residual stress in an infinite body with a penny-shaped crack

The constant m_1 in equation (19.42) corresponds to the stress intensity factor of the opening mode:

$$m_1 = \frac{K_1}{K_0}$$

By the same way as before, the residual stress around the crack tip line can be obtained in an analytical form and is given in Figure 19.12, which shows that it takes extremely small values around the crack tip line. Therefore, the previous results hold in the three-dimensional problems.

19.7 CONCLUSION

In the present paper, the inherent error of the method of superposition of analytical and finite element solutions has been investigated in relation to the residual stress. From the viewpoint of the inherent error, the analytical solutions should be chosen so that certain components of the residual stress become small at the collocation points for the method. If the analytical solutions are taken properly, modelling of the crack surface is not sensible for calculating the stress intensity factor by the finite element method. Therefore, the stress intensity factor of a continuously growing crack can be obtained by the use of one mesh pattern. Results obtained can be applied for other boundary value problems defined on a cracked body.

REFERENCES

1. Y. Yamamoto, 'Finite element approaches with the aid of analytical solutions', *Recent Advances on Matrix Methods of Structural Analysis and Design*, pp. 85–103, University of Alabama Press, 1971.
2. Y. Yamamoto and N. Tokuda, 'Determination of stress intensity factors in

cracked plates by the finite element method', *Int. J. Num. Meth. in Eng.*, **6,** No. 3, 427–439 (1973).

3. Y. Yamamoto, N. Tokuda, and Y. Sumi, 'Finite element treatment of singularities of boundary value problems and its application to analysis of stress intensity factors', *Theory and Practice of Finite Element Methods for Structural Analysis*, pp. 75–90, University of Tokyo Press, 1973.
4. Y. Yamamoto and Y. Sumi, 'Stress intensity factor for three-dimensional cracks', *Int. J. Fracture,* **14,** No. 1, 17–38 (1978).
5. Y. Yamamoto and Y. Sumi, 'Finite element analysis for stress intensity factors', Lecture Notes in Math. No. 704, *Comput. Meth. in Appl. Sci. and Eng.*, **1977,** 1, Springer, pp. 154–168, 1979.
6. Y. Sumi, S. Nemat-Nasser, and L. M. Keer, 'A new combined analytical and finite-element solution method for stability analysis of the growth of interacting tension cracks in brittle solids', *Int. J. Eng. Sci.*, **18,** No. 1, 211–224 (1980).
7. S. Nemat-Nasser, Y. Sumi, and L. M. Keer, 'Unstable growth of tension cracks in brittle solids: Stable and unstable bifurcations, snap-through, and imperfection sensitivity', *Int. J. Solids and Structures,* **16,** No. 11, 1017–1035 (1980).
8. T. H. H. Pian, P. Tong, and C. H. Luk, 'Elastic crack analysis by a finite element hybrid method', *Proc. Third Conf. on Matrix Meth. of Struct. Mech.*, pp. 690–711, AFFDL-TR-71-160, 1971.
9. G. Yagawa, T. Nishioka, Y. Ando, and N. Ogura, 'The finite element calculation of stress intensity factors using superposition', *Computational Fracture Mech.*, pp. 21–34, ASME Special Publ., 1975.
10. G. Yagawa, T. Nishioka, and Y. Ando, 'Elastic-plastic finite element analysis using superposition', *Nucl. Eng. Design,* **34,** 247–254 (1976).
11. G. Yagawa, N. Miyazaki, and Y. Ando, 'Superposition method of finite element and analytical solutions for transient creep analysis', *Int. J. Num. Meth. in Eng.*, **11,** No. 7, 1107–1115 (1977).
12. G. Yagawa and N. Miyazaki, 'Steady state creep analysis of a cracked body using the superposition method', *Nucl. Eng. Design,* **54,** 79–89 (1979).
13. G. Yagawa and T. Nishioka, 'Finite element analysis of stress intensity factors for plane extension and plate bending', *Int. J. Num. Meth. in Eng.*, **14,** No. 5, 727–740 (1979).
14. T. H. H. Pian, 'Cracked elements', Presented at the *World Cong. on Finite Element Meth. in Struc. Mech.*, Bournemouth, Dorset, England, 1975.
15. S. B. Dong, 'Global–local finite element methods', *State-of-the-Art in Finite Element Methods,* ASME (to be published).
16. Y. Yamamoto and N. Tokuda, 'A note on convergence of finite element solutions', *Int. J. Num. Meth. in Eng.*, **3,** No. 4, 485–493 (1971).
17. M. L. Williams, 'Stress singularities resulting from various boundary conditions in angular corners of plates in extension', *J. Appl. Mech., Trans. ASME,* **74,** No. 4, 526–528 (1952).
18. H. M. Westergaard, 'Bearing pressure and cracks', *J. Appl. Mech.*, **31,** No. 2, A49–A53 (1939).

Hybrid and Mixed Finite Element Methods
Edited by S. N. Atluri, R. H. Gallagher, and O. C. Zienkiewicz

Chapter 20

Incompatible Plate Elements Based Upon the Generalized Variational Principles

Wei-Zang Chien

20.1 INTRODUCTION

In finite element calculation for a plate, the continuity or equilibrium conditions along interelement boundaries can be satisfied by compatible plate elements at the expense of higher degrees of freedom or large amounts of calculation. Pian [1], Tong [2], and Herrmann [3, 4] pointed out in a series of papers that the simple incompatible plate elements can be used if these interelement boundary conditions are to be relaxed to the extent that they are satisfied in an integral sense, and hence will be completely satisfied when the element size becomes infinitesimally small. This formulation thus calls for modified (or generalized) variational principles, for which the interelement conditions are introduced as conditions of constraint and appropriate boundary variables are used as the corresponding Lagrange multipliers. In this paper, the above-mentioned boundary variables are further identified in terms of the original variable and its derivatives, so that no new variables are needed in the generalized variational principles. Thus further simplification in finite element computations for incompatible plate elements has been satisfied.

20.2 PRINCIPLE OF MINIMUM POTENTIAL ENERGY AND ITS GENERALIZATION FOR INCOMPATIBLE PLATE ELEMENTS

For a thin plate of bending rigidity D and Poisson's ratio μ under the action of lateral loads $f(x, y)$, the lateral deflection $w(x, y)$ is given by

$$\nabla^2\nabla^2 w = \frac{f}{D} \qquad (\text{in } A) \tag{20.1}$$

where

$$\nabla^2 = \frac{\partial^2}{\partial x^2} + \frac{\partial^2}{\partial y^2} \tag{20.2}$$

There are various kinds of boundary conditions:

(a) Equivalent shearing force (or edge force) $H_\nu = Q_\nu + M_{\nu s,s}$ is given or lateral deflection w is given:

$$H_\nu = \bar{H} \qquad (\text{in } S_{\sigma_1}) \tag{20.3a}$$

$$w = \bar{w} \qquad (\text{in } S_{\omega_1}) \tag{20.3b}$$

(b) Edge moment M is given or normal slope of deflections in the outward normal direction w is given:

$$M_\nu = \bar{M} \qquad (\text{in } S_{\sigma_2}) \tag{20.4a}$$

$$w_{,\nu} = \bar{w}_{,\nu} \qquad (\text{in } S_{\omega_2}) \tag{20.4b}$$

where

$$S_{\sigma_1} + S_{\omega_1} = S_{\sigma_2} + S_{\omega_2} = \text{entire boundary} \tag{20.5}$$

There are also various kinds of conditions at edge corners. Corner forces normal to the plate are given or the corner deflections are given:

$$P_{k_1} = \bar{P}_{k_1} \qquad (\text{at corners } k_1 = 1, 2, \ldots, k_\sigma) \tag{20.6a}$$

$$w_{k_2} = \bar{w}_{k_2} \qquad (\text{at corners } k_2 = 1, 2, \ldots, k) \tag{20.6b}$$

where we take i as the total number of corners on the boundary

$$k_\sigma + k_w = i \tag{20.7}$$

The principle of minimum potential energy may be stated as follows. Among all possible $w(x, y)$ which satisfy deflection boundary conditions (20.3b), (20.4b), (20.6b) in the boundaries S_{w_1}, S_{w_2} and corners k_w, the one which minimizes the functional Π gives the solution of (20.1) in A under the actions of boundary forces (20.3a), (20.4a), (20.6a) in the boundaries S_{σ_1}, S_{σ_2} and the corners k_σ, and the distributed load $f(x, y)$ in A.

In this case, the functional Π may be written as

$$\Pi = \Pi_0 - \iint_A fw \, dA - \int_{S_{\sigma_1}} \bar{H}w \, ds - \int_{S_{\sigma_2}} \bar{M}w_{,\nu} \, ds - \sum_{k_1=1}^{k_\sigma} \bar{P}_{k_1}\omega_{k_1} \tag{20.8}$$

in which Π_0 is the bending energy in the plate

$$\Pi_0 = \iint_A \frac{D}{2}\left\{\left(\frac{\partial^2 w}{\partial x^2} + \frac{\partial^2 w}{\partial y^2}\right)^2 - 2(1-\mu)\left(\frac{\partial^2 w}{\partial x^2}\frac{\partial^2 w}{\partial y^2} - \frac{\partial^2 w}{\partial x\,\partial y}\frac{\partial^2 w}{\partial x\,\partial y}\right)\right\} dA \tag{20.9}$$

This principle can be easily proved by taking variation of Π, in which [5]

$$\delta\Pi_0=\iint_A D\nabla^2\nabla^2 w\,dA-\int_S M_\nu(w)\frac{\partial\delta w}{\partial\nu}\,ds+\int_S H_\nu(w)\,\delta w\,ds+\sum_{k=1}^{i}P_k(w)\,\delta w_k \tag{20.10}$$

where $M_\nu(w)$, $H_\nu(w)$, $P_k(w)$ are the following linear functions of w and its derivatives on the boundaries or at edge corners:

$$M_\nu(w)=-D\left\{\mu\nabla^2 w+(1-\mu)\frac{\partial^2 w}{\partial\nu^2}\right\}\qquad\text{(on boundary } S) \tag{20.11a}$$

$$M_\nu(w)=-D\left\{\frac{\partial}{\partial\nu}\left[\nabla^2 w+(1-\mu)\frac{\partial^2 w}{\partial s^2}\right]-(1-\mu)\frac{\partial}{\partial s}\frac{1}{\rho_s}\frac{\partial w}{\partial s}\right\}$$

$$\text{(on boundary } S) \tag{20.11b}$$

$$P_k(w)=-(1-\mu)D\Delta\left\{\frac{\partial^2 w}{\partial\nu\,\partial s}-\frac{1}{\rho_s}\frac{\partial w}{\partial s}\right\}_k\qquad\text{(at corners } k) \tag{20.11c}$$

in which $\Delta\left\{\frac{\partial^2 w}{\partial\nu\,\partial s}-\frac{1}{\rho_s}\frac{\partial w}{\partial s}\right\}$ represents the increment of $\frac{\partial^2 w}{\partial\nu\,\partial s}-\frac{1}{\rho_s}\frac{\partial w}{\partial s}$ at the kth corner on the boundary curves S, and in general we assume that there are i edge corners on S. ρ_s is the radius of curvature of boundary curve S and is positive when the boundary curve is convex at that point. For straight edges, $1/\rho_s$ vanishes.

Minimization of Π in (20.8) gives not only the field equation (20.1) in A but also the force boundary conditions (20.3a), (20.4a), and the force corner conditions (20.6a). In fact, (20.11a, b, c) are the expressions of edge bending moment, edge shearing force, and corner force respectively in the plate edges in terms of the derivatives of the deflection.

The minimum potential energy variational principle can be generalized by considering the deflection boundary conditions (20.3b), (20.4b), (20.6b) as conditions of constraint and using corresponding Lagrange multipliers. Thus we have the following generalized variational principle based upon the principle of minimum potential energy.

Among all possible $w(x, y)$, the one which makes the following functional $\tilde{\Pi}$ stationary gives the solution of the field equation (20.1) under the boundary conditions (20.3a, b), (20.4a, b), and the corner conditions (20.6a, b). This functional $\tilde{\Pi}$ can be written in terms of undetermined Lagrange multipliers $\lambda_{(1)}$, $\lambda_{(2)}$, and $\lambda_{(3)k_2}$ (where $k_2=1, 2, \ldots, k_w$) as follows:

$$\tilde{\Pi}=\Pi_0-\iint_A fw\,dA-\int_{S_{\sigma_1}}\bar{H}w\,ds+\int_{S_{\sigma_2}}\bar{M}w_{,\nu}\,ds-\sum_{k_1=1}^{k_\sigma}\bar{P}_{k_1}w_{k_1}$$

$$+\int_{S_{\omega_1}}\lambda_{(1)}(s)(w-\bar{w})\,ds-\int_{S_{\omega_2}}\lambda_{(2)}(s)(w_{,\nu}-\bar{w}_{,\nu})\,ds$$

$$+\sum_{k_2=1}^{k_w}\lambda_{(3)k_2}(w_{k_2}-\bar{w}_{k_2}) \tag{20.12}$$

where $\lambda_{(1)}, \lambda_{(2)}$ are functions of boundary arc length coordinates and $\lambda_{(3)k}$ $(k_2=1,2,\ldots,k_w)$ are undetermined constants.

By means of (20.10), the variation of $\tilde{\Pi}$ gives

$$\delta\tilde{\Pi}=\iint_A (D\nabla^2\nabla^2 w-f)\delta w\,\mathrm{d}A+\int_{S_{\sigma_1}}(H_\nu-\bar{H})\delta w\,\mathrm{d}s-\int_{S_{\sigma_2}}(M_\nu-\bar{M})\frac{\partial\delta w}{\partial\nu}\mathrm{d}s$$
$$+\sum_{k_1=1}^{k_\sigma}(P_{k_1}-\bar{P}_{k_1})\delta w_{k_1}+\int_{S_{w_1}}[\lambda_{(1)}(s)+H_\nu(w)]\delta w\,\mathrm{d}s$$
$$+\int_{S_{w_2}}[\lambda_{(2)}(s)-M_\nu(w)]\frac{\partial\delta w}{\partial\nu}\,\mathrm{d}s$$
$$+\sum_{k_2=1}^{k_w}\left\{\lambda_{(3)k_2}+P_{k_2}\right\}\delta w_{k_2}+\int_{S_{w_1}}(w-\bar{w})\delta\lambda_{(1)}\,\mathrm{d}s+\int_{S_{w_2}}(w_{,\nu}-\bar{w}_{,\nu})\delta\lambda_{(2)}\,\mathrm{d}s$$
$$+\sum_{k_2=1}^{k_w}(w_{k_2}-\bar{w}_{k_2})\delta\lambda_{(3)k_2} \tag{20.13}$$

The stationary condition of this functional

$$\delta\tilde{\Pi}=0 \tag{20.14}$$

gives not only the field equation (20.1), the boundary conditions (20.3a, b), (20.4a, b), and the corner conditions (20.5a, b) but also the definitions of $\lambda_{(1)}(s)$, $\lambda_{(2)}(s)$, and $\lambda_{(3)k_2}$. They are

$$\lambda_{(1)}(s)=-H_\nu(w)=D\left\{\frac{\partial}{\partial\nu}\left[\nabla^2 w+(1-\mu)\frac{\partial^2 w}{\partial s^2}\right]-(1-\mu)\frac{\partial}{\partial s}\frac{1}{\rho_s}\frac{\partial w}{\partial s}\right\} \tag{20.15a}$$

$$\lambda_{(2)}(s)=M_\nu(w)=-D\left\{\mu\nabla^2 w+(1-\mu)\frac{\partial^2 w}{\partial\nu^2}\right\} \tag{20.15b}$$

$$\lambda_{(3)k_2}=-P_{k_2}(w)=(1-\mu)D\Delta\left\{\frac{\partial^2 w}{\partial\nu\,\partial s}-\frac{1}{\rho_s}\frac{\partial w}{\partial s}\right\}_{k_2} \tag{20.15c}$$

Substitution of $\lambda_{(1)}(s)$, $\lambda_{(2)}(s)$, $\lambda_{(3)k_2}$ from (20.15a, b, c) into (20.13), we obtain the expression of the functional $\tilde{\Pi}$ of our generalized variational principle:

$$\tilde{\Pi}=\Pi_0-\iint_A fw\,\mathrm{d}A-\int_{S_{\sigma_1}}\bar{H}w\,\mathrm{d}s+\int_{S_{\sigma_2}}\bar{M}\frac{\partial w}{\partial\nu}\mathrm{d}s-\sum_{k_1=1}^{k_\sigma}\bar{P}_{k_1}w_{k_1}$$
$$-\int_{S_{w_1}}H_\nu(w)(w-\bar{w})\,\mathrm{d}s+\int_{S_{w_2}}M_\nu(w)(w_{,\nu}-\bar{w}_{,\nu})\,\mathrm{d}s-\sum_{k_2=1}^{k_w}P_{k_2}(w)$$
$$\times(w_{k_2}-\bar{w}_{k_2}) \tag{20.16}$$

If the field variable $w(x, y)$ in this variational principle is chosen so that it satisfies various deflection boundary conditions (20.3b), (20.4b), and (20.6b), then the generalized functional $\tilde{\Pi}$ in (20.16) reduces to the form Π in (20.8).

Now we turn to discuss various variational principles for finite element calculation. Let us suppose that the region of the plate is subdivided into a finite number of discrete elements, say N elements altogether, and each element possesses r continuous segments and r corners, where r in general is equal to or greater than 2 for a curved segment and is equal to or greater than 3 for a straight segment. Let us suppose that we formulate the finite element field functions with the compatible model, i.e. the deflections and normal slopes of deflection are continuous along all the interelement boundaries of any two neighbouring elements and also the deflection of any corner of an element is equal to the deflections of common corners of all neighbouring elements. Let us suppose that there are c_F points (x_{c_1}, y_{c_1}) in the interior of the plate domain A loaded by given concentrated loads $\bar{F}_{c_1}$, and c_w points (x_{c_2}, y_{c_2}) in A supported with given deflections $\bar{w}_{c_2}$. That is,

$$F_{c_1} = \bar{F}_{c_1} \qquad \text{(at points } x_{c_1}, y_{c_1}, \quad \text{where} \quad c_1 = 1, 2, \ldots, c_F) \tag{20.17a}$$

$$w_{c_2} = \bar{w}_{c_2} \qquad \text{(at supporting points } x_{c_2}, y_{c_2}, \text{ where } c_2 = 1, 2, \ldots, c_w) \tag{20.17b}$$

Let us further suppose that all $c_F + c_w$ points are at the same time the common nodal corner points of the discrete elements.

Thus the generalized variational principles of this plate for the finite element field function with the compatible model can be stated as follows. Among all possible sets of compatible finite element field functions $w^{(n)}(x, y)$, $n = 1, 2, \ldots, N$, the set which makes the following functional $\tilde{\Pi}_f$ stationary gives the solution of the field equation (20.1) with the boundary conditions (20.3b), (20.4b), plate corner conditions (20.6b), and the point supported conditions (20.17b) under the actions of the distributed load $f(x, y)$, concentrated loads $\bar{F}_{c_1}$ of (20.17a), boundary bending moments $\bar{M}$ of (20.4a), boundary shearing forces $\bar{H}$ of (20.3a), and the boundary corner forces $\bar{P}_{k_1}$ of (20.6a):

$$\begin{aligned}\tilde{\Pi}_f = \sum_{n=1}^{N} \Pi_f^{(n)} - \sum_{k_1=1}^{k_\sigma} \bar{P}_{k_1} w_{k_1} - \sum_{k_2=1}^{k_w} \sum_{j=1}^{r1} p_{k_2}^{(j)}(w)(w_{k_2}^{(j)} - \bar{w}_{k_2}) \\ - \sum_{c_1=1}^{c_F} \bar{F}_{c_1} w_{c_1} - \sum_{c_2=1}^{c_w} \sum_{j=1}^{r} p_{c_2}^{(j)}(w)(w_{c_2}^{(j)} - \bar{w}_{c_2})\end{aligned} \tag{20.18a}$$

$$\delta \tilde{\Pi}_f = 0 \tag{20.18b}$$

in which

$$\Pi_f^{(n)} = \Pi_{0f}^{(n)} - \iint_{A^{(n)}} f w^{(n)}\, dA^{(n)} - \int_{S_{\sigma_1}^{(n)}} \bar{H} w^{(n)}\, ds^{(n)} + \int_{S_{\sigma_2}^{(n)}} \bar{M} \frac{\partial w^{(n)}}{\partial \nu^{(n)}}\, ds^{(n)}$$

$$- \int_{S_{w_1}^{(n)}} H_\nu(w^{(n)})(w^{(n)} - \bar{w})\, ds^{(n)} + \int_{S_{w_2}^{(n)}} M\nu(w^{(n)}) \left(\frac{\partial w^{(n)}}{\partial v^{(n)}} - \frac{\partial \bar{w}}{\partial v}\right) ds^{(n)} \tag{20.19a}$$

$$\Pi_{0f}^{(n)} = \iint_{A^{(n)}} \frac{D}{2} \left\{ \left(\frac{\partial^2 w^{(n)}}{\partial x^2} + \frac{\partial^2 w^{(n)}}{\partial y^2} \right)^2 \right.$$

$$\left. -2(1-\mu) \left(\frac{\partial^2 w^{(n)}}{\partial x^2} \frac{\partial^2 w^{(n)}}{\partial y^2} - \frac{\partial^2 w^{(n)}}{\partial x\, \partial y} \frac{\partial^2 w^{(n)}}{\partial x\, \partial y} \right) \right\} dA^{(n)} \tag{20.19b}$$

$$p_{k_2}^{(j)}(w) = -(1-\mu) D \Delta_{k_2} \left\{ \frac{\partial^2 w^{(j)}}{\partial \nu^{(j)}\, \partial s^{(j)}} - \frac{1}{\rho_s^{(j)}} \frac{\partial w^{(j)}}{\partial s^{(j)}} \right\} \tag{20.19c}$$

$$p_{c_2}^{(j)}(w) = -(1-\mu) D \Delta_{c_2} \left\{ \frac{\partial^2 w^{(j)}}{\partial \nu^{(j)}\, \partial s^{(j)}} - \frac{1}{\rho_s^{(j)}} \frac{\partial w^{(j)}}{\partial s^{(j)}} \right\} \tag{20.19d}$$

r' = number of elements with common corner k_2 along the plate boundary, where deflection w_{k_2} is given (20.19e)

r = number of elements with common corner in the interior of plate domain at c_2, where the deflection w_{c_2} is given (20.19f)

$$w_{k_1} = w_{k_1}^{(j)} \qquad \text{for } j = 1, 2, \ldots, r'' \tag{20.19g}$$

$$w_{c_1} = w_{c_1}^{(j)} \qquad \text{for } j = 1, 2, \ldots, r''' \tag{20.19h}$$

r'' = number of elements with common corner k_1 along the plate boundary, where corner force P_{k_1} is given (20.19i)

r''' = number of elements with common corner in the interior of plate domain at c_1, where the concentrated load F_{c_1} is given (20.19j)

In (20.18a, b), $w^{(n)}$ of the nth element satisfies the following interelement boundary conditions, i.e. if (n), (n') are two neighbouring elements, then we have

$$w^{(n)} - w^{(n')} = 0 \qquad \text{(on } n-n' \text{ interelement boundary)} \tag{20.20a}$$

$$\frac{\partial w^{(n)}}{\partial \nu^{(n)}} + \frac{\partial w^{(n')}}{\partial \nu^{(n')}} = 0 \qquad \text{(on } n-n' \text{ interelement boundary)} \tag{20.20b}$$

in which

$$d\nu^{(n)} = -d\nu^{(n')} \qquad \text{(on } n-n' \text{ interelement boundary)} \tag{20.20c}$$

It is easily seen that (20.19g, h) are continuity conditions at common corners of the elements for the deflection. Furthermore, $S_{\sigma_1}^{(n)}, S_{\sigma_2}^{(n)}, S_{w_1}^{(n)}, S_{w_2}^{(n)}$ are respectively the plate boundary segments with prescribed forces and deflections in the nth element.

We can further generalize the variational principles (20.18a, b) by considering interelement boundary conditions (20.20a,b) and common corner conditions (20.19g, h) as conditions of constraint and using corresponding Lagrange multipliers. Thus we have the following generalized variational principles for the incompatible model of finite elements.

Among all possible sets of finite element field functions $w^{(n)}(x, y), n = 1, 2, \ldots, N$, which are not necessarily compatible with each other, the set which makes the following functional $\tilde{\Pi}_f^*$ stationary gives the solution of the field equation (20.1) with boundary conditions (20.3b), (20.4b), (20.6b), point supported conditions (20.18b), and interelement continuity conditions (20.20a, b), (20.19g, h) under the actions of the distributed load $f(x, y)$, concentrated loads $\bar{F}_{c_1}$ of (20.18a), boundary bending moment $\bar{M}$ of (20.4a), boundary shearing force $\bar{H}$ of (20.3a), and the boundary corner forces $\bar{P}_{k_1}$ of (20.6a):

$$\begin{aligned}\tilde{\Pi}_f^* = &\sum_{n=1}^{N} \Pi_f^{(n)} - \sum_{k_1=1}^{k_\sigma} \bar{P}_{k_1} w_{k_1}^{(1'')} + \sum_{k_1=1}^{k_\sigma} \sum_{n''=2''}^{r''} \Lambda_{3k_1}^{(1''n'')}(w_{k_1}^{(1'')} - w_{k_1}^{(n'')})\\ &- \sum_{c_1=1}^{c_F} \bar{F}_{c_1} w_{c_1}^{(1''')} + \sum_{c_1=1}^{c_F} \sum_{n'''=2'''}^{r'''} \Lambda_{3c_1}^{(1'''n''')}(w_{c_1}^{(1''')} - w_{c_1}^{(n''')})\\ &- \sum_{k_2=1}^{k_w} \sum_{j=1}^{r'} p_{k_2}^{(j)}(w)(w_{k_2}^{(j)} - \bar{w}_{k_2}) - \sum_{S_2=1}^{S_w} \sum_{j=1}^{\gamma} p_{s_2}^{(j)}(w)(w_{s_2}^{(j)} - \bar{w}_{s_2})\\ &+ \sum_{\text{All }(nn')} \int \Lambda_1^{(nn')}(s)(w^{(n)} - w^{(n')})\, ds^{(n)}\\ &+ \sum_{\text{All }(nn')} \int \Lambda^{(nn')}(s)\left(\frac{\partial w^{(n)}}{\partial \nu^{(n)}} + \frac{\partial w^{(n')}}{\partial \nu^{(n')}}\right) ds^{(n)}\end{aligned} \tag{20.21a}$$

$$\delta \tilde{\Pi}_f^* = 0 \tag{20.21b}$$

where $\Lambda_1^{(nn')}(s), \Lambda_2^{(nn')}(s)$ are Lagrange multipliers defined as functions of arc length s along an interelement boundary of element (n) and (n'). $\Lambda_{3k_1}^{(1''n'')}$ $(n'' = 2'', 3'', \ldots, r'')$ are Lagrange multipliers defined as constant at each common corner k_1 $(=1, 2, \ldots, k)$ related to a pair of neighbouring elements $(1'')$ and (n'') on the plate boundary. Similarly, $\Lambda_{3c}^{(1'''n''')}$ $(n''' = 2''', 3''', \ldots, r''')$ are Lagrange multipliers defined as constant at each common corner c_1 $(=1, 2, \ldots c_F)$ related to a pair of neighbouring elements $(1''')$ and (n''') in the interior of the plate.

The generalized variational principle for the incompatible model of finite element is the stationary condition (20.21b) of the functional $\tilde{\Pi}_f^*$, where

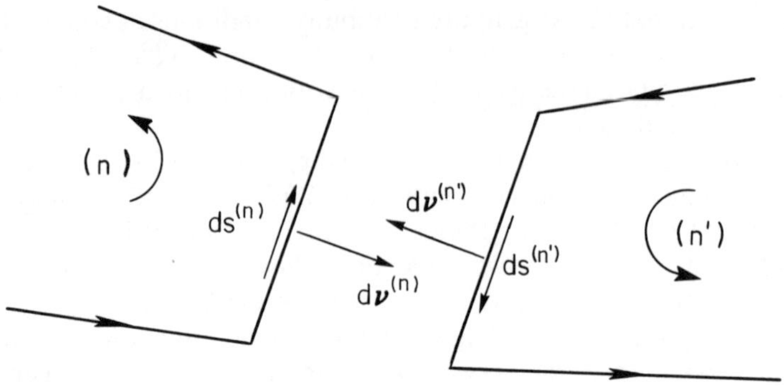

Figure 20.1 The positive direction of ds and dω in neighbouring elements

$w^{(n)}$, $\Lambda_1^{(nn')}(s)$, $\Lambda_2^{(nn')}(s)$, $\Lambda_{3k_1}^{(1''n'')}$, $\Lambda_{3c_1}^{(1'''n''')}$ are taken to be independent variables. It should be noted that if we take the counterclockwise direction of d$s^{(n)}$ for the boundary of element (n) as the positive direction, then from Figure 20.1, for the interelement boundary d$s^{(n)}$ and d$s^{(n')}$, we have

$$\mathrm{d}s^{(n)} = -\mathrm{d}s^{(n')}, \qquad \mathrm{d}\nu^{(n)} = -\mathrm{d}\nu^{(n')} \tag{20.22}$$

If we taken

$$\Lambda_1^{(nn')} = \Lambda_1^{(n'n)} \tag{20.23}$$

then we have

$$\int_{(nn')} \Lambda_1^{(nn')}(w^{(n)} - w^{(n')})\,\mathrm{d}s^{(n)} = \int_{(nn')} \Lambda_1^{(nn')} w^{(n)}\,\mathrm{d}s^{(n)} + \int_{(n'n)} \Lambda^{(n'n)} w^{(n')}\,\mathrm{d}s^{(n')} \tag{20.24}$$

or

$$\sum_{\substack{\text{All}\\(nn')}} \int_{(nn')} \Lambda_1^{(nn')}(w^{(n)} - w^{(n')})\,\mathrm{d}s^{(n)} = \sum_{\text{All } n} \sum_{\substack{\text{All } n'\\ \text{neigh.}\\ \text{to } n}} \int_{(nn')} \Lambda_1^{(nn')} w^{(n)}\,\mathrm{d}s^{(n)} \tag{20.25}$$

Similarly, if we take

$$\Lambda_2^{(nn')} = -\Lambda_2^{(n'n)} \tag{20.26}$$

we have

$$\int_{(nn')} \Lambda_2^{(nn')} \left\{ \frac{\partial w^{(n)}}{\partial \nu^{(n)}} + \frac{\partial w^{(n')}}{\partial \nu^{(n')}} \right\} \mathrm{d}s^{(n)} = \int_{(nn')} \Lambda_2^{(nn')} \frac{\partial w^{(n)}}{\partial \nu^{(n)}}\,\mathrm{d}s^{(n)} + \int_{(n'n)} \Lambda_2^{(n'n)} \frac{\partial w^{(n')}}{\partial \nu^{(n')}}\,\mathrm{d}s^{(n')} \tag{20.27}$$

or

$$\sum_{\substack{\text{All}\\(nn')}} \int_{(nn')} \Lambda_2^{(nn')} \left\{ \frac{\partial w^{(n)}}{\partial \nu^{(n)}} + \frac{\partial w^{(n')}}{\partial \nu^{(n')}} \right\} \mathrm{d}s^{(n)}$$

$$= \sum_{\text{All } n} \sum_{\substack{\text{All } n'\\ \text{neigh.}\\ \text{to } n}} \int_{(nn')} \Lambda_2^{(nn')} \frac{\partial w^{(n)}}{\partial \nu^{(n)}} \mathrm{d}s^{(n)} \tag{20.28}$$

Substituting (20.25), (20.28) into (20.21a), this functional can be further simplified into the form

$$\begin{aligned}\tilde{\Pi}_f^* = {} & \sum_{n=1}^{N} \Pi_f^{*(n)} - \sum_{k_1=1}^{k_\sigma} P_{k_1} w_{k_1}^{(1'')} + \sum_{k_1=1}^{k_\sigma} \sum_{n''=2''}^{\gamma''} \Lambda_{3k_1}^{(1''n'')} (w_{k_1}^{(1'')} - w_{k_1}^{(n'')}) \\ & - \sum_{k_2=1}^{k_w} \sum_{j=1}^{r'} p_{k_2}^{(j)}(w)(w_{k_2}^{(j)} - \bar{w}_{k_2}) - \sum_{c_2=1}^{c_w} \sum_{j=1}^{r} p_{c_2}^{(j)}(w)(w_{c_2}^{(j)} - \bar{w}_{c_2}) \\ & - \sum_{c_1=1}^{c_F} \bar{F}_{c_1} w_{c_1}^{(1''')} + \sum_{c_1=1}^{c_F} \sum_{n'''=2'''}^{\gamma'''} \Lambda_{3c_1}^{(1'''n''')} (w_{c_1}^{(1''')} - w_{c_1}^{(n''')}) \end{aligned} \tag{20.29a}$$

where

$$\Pi_f^{*(n)} = \Pi_f^{(n)} + \sum_{\substack{\text{All } n'\\ \text{neigh.}\\ \text{to } n}} \left\{ \int_{(nn')} \Lambda_1^{(nn')} w^{(n)} \, \mathrm{d}s^{(n)} + \int_{(nn')} \Lambda_2^{(nn')} \frac{\partial w^{(n)}}{\partial \nu^{(n)}} \mathrm{d}s^{(n)} \right\} \tag{20.29b}$$

This functional is similar to the functional introduced by Tong [2] and Pian [1] for the same problems, except that the corner conditions at k_1, k_2, c_1, c_2 are taken into consideration. This kind of formulation has the disadvantage of introducing too many unknowns in the finite element calculation and consequently a very large matrix is used even for a simple problem.

It can be shown through the variation of (20.29a) that $\Lambda_1^{(nn')}, \Lambda_2^{(nn')}, \Lambda_{3k_1}^{(1''n'')}, \Lambda_{3c_1}^{(1'''n''')}$ (where $n'' = 2'', 3'', \ldots, r''$; $n''' = 2''', 3''', \ldots, r'''$) can be expressed in terms of functions of w and its derivatives.

The variations of (20.29a) can be written as

$$\begin{aligned}\delta\tilde{\Pi}_f^* = {} & \sum_{n=1}^{N} \delta\tilde{\Pi}_f^{*(n)} + \sum_{(nn')} \delta\tilde{\Pi}^{(nn')} + \sum_{k_1=1}^{k_\sigma} \delta\tilde{\Pi}_{k_1} + \sum_{k_2=1}^{k_w} \delta\tilde{\Pi}_{k_2} \\ & + \sum_{c_1=1}^{c_F} \delta\tilde{\Pi}_{c_1} + \sum_{c_2=1}^{c_w} \delta\tilde{\Pi}_2 \end{aligned} \tag{20.30}$$

in which

$\delta\tilde{\Pi}_f^{*(n)}$ = variations of $\tilde{\Pi}_f^*$ for nth element (20.30a)

$\delta\tilde{\Pi}^{(nn')}$ = variations of $\tilde{\Pi}_f^*$ for interelement bounaary (n,n') (20.30b)

$\delta\tilde{\Pi}_{k_1}$ = variations of $\tilde{\Pi}_f^*$ for boundary corners with prescribed corner force $\bar{P}_{k_1}$ (20.30c)

$\delta\tilde{\Pi}_{k_2}$ = variations of $\tilde{\Pi}_f^*$ for boundary corners with prescribed deflection $\bar{w}_{k_2}$ (20.30d)

$\delta\tilde{\Pi}_{c_1}$ = variations of $\tilde{\Pi}_f^*$ for interelement corners in the interior of plate with prescribed concentrated load $\bar{F}_{c_1}$ (20.30e)

$\delta\tilde{\Pi}_{c_2}$ = variations of $\tilde{\Pi}_f^*$ for interelement corners in the interior of plate with prescribed deflection $\bar{w}_{c_1}$ (20.30f)

They are

$$\begin{aligned}\delta\tilde{\Pi}_f^{*(n)} = & \iint_{A^{(n)}} (D\nabla^2\nabla^2 w - f)\delta w^{(n)}\,dA^{(n)} + \int_{S_{\sigma_1}^{(n)}} [H_\nu(w^{(n)}) - \bar{H}]\delta w^{(n)}\,ds^{(n)} \\ & - \int_{S_{\sigma_2}^{(n)}} (w^{(n)} - \bar{w})\delta H_\nu(w^{(n)})\,ds^{(n)} - \int_{S_{\sigma_2}^{(n)}} [M_\nu(w^{(n)}) - \bar{M}]\frac{\partial \delta w^{(n)}}{\partial \nu^{(n)}}\,ds^{(n)} \\ & + \int_{S_{w_2}^{(n)}} \left(\frac{\partial w^{(n)}}{\partial \nu^{(n)}} - \frac{\partial w}{\partial \nu}\right)\delta M_\nu(w^{(n)})\,ds^{(n)}\end{aligned} \tag{20.31a}$$

$$\begin{aligned}\delta\tilde{\Pi}^{(nn')} = & \int_{(nn')} \{\Lambda_1^{(nn')} + H_\nu(w^{(n)})\}\delta w^{(n)}\,ds^{(n)} + \int_{(n'n)} \{\Lambda_1^{(n'n)} + H_\nu(w^{(n')})\} \\ & \times \delta w^{(n')}\,ds^{(n')} \\ & + \int_{(nn')} \{\Lambda_2^{(nn')} - M_\nu(w^{(n)})\}\frac{\partial \delta w^{(n)}}{\partial \nu^{(n)}}\,ds^{(n)} \\ & + \int_{(n'n)} \{\Lambda_2^{(n'n)} - M_\nu(w^{(n')})\}\frac{\partial \delta w^{(n')}}{\partial \nu^{(n')}}\,ds^{(n')} \\ & + \int_{(nn')} (w^{(n)} - w^{(n')})\delta\Lambda_1^{(nn')}\,ds^{(n)} + \int_{(n'n)} \left(\frac{\partial w^{(n)}}{\partial \nu^{(n)}} + \frac{\partial w^{(n')}}{\partial \nu^{(n')}}\right)\delta\Lambda_2^{(nn')}\,ds^{(n')}.\end{aligned} \tag{20.31b}$$

$$\begin{aligned}\delta\tilde{\Pi}_{k_1} = & \sum_{n''=2''}^{\gamma''} (w_{k_1}^{(1'')} - w_{k_1}^{(n'')})\delta\Lambda_{3k_1}^{(1''n'')} + \left[\sum_{n''=2''}^{\gamma''} \Lambda_{3k_1}^{(1''n'')} - \bar{P}_{k_1} + p_{k_1}^{(1'')}(w)\right]\delta w_{k_1}^{(1'')} \\ & + \sum_{n''=2''}^{\gamma''} [-\Lambda_{3k_1}^{(1''n'')} + p_{k_1}^{(n'')}(w)]\delta w_{k_1}^{(n'')}\end{aligned} \tag{30.31c}$$

$$\delta\tilde{\Pi}_{k_2} = -\sum_{j=1}^{\gamma'} [w_{k_2}^{(j)} - \bar{w}_{k_2}]\delta p_{k_2}^{(j)}(w) \tag{20.31d}$$

$$\begin{aligned}\delta\tilde{\Pi}_{c_1} = & \sum_{n'''=2'''}^{\gamma'''} [w_{c_1}^{(1''')} - w_{c_1}^{(n''')}]\delta\Lambda_{3c_1}^{(1'''n''')} + \left[\sum_{n'''=2'''}^{\gamma'''} \Lambda_{3c_1}^{(1'''n''')} - \bar{F}_{c_1} + p_{c_1}^{(1''')}(w)\right]\delta w_{c_1}^{(1''')} \\ & + \sum_{n'''=2'''}^{\gamma''} [-\Lambda_{3c_1}^{(1'''n''')} + p_{c_1}^{(n''')}(w)]\delta w_{c_1}^{(n''')}\end{aligned} \tag{20.31e}$$

$$\delta\tilde{\Pi}_{c_2} = -\sum_{j=1}^{\gamma} [w_{c_2}^{(j)} - \bar{w}_{c_2}]\delta p_{c_2}^{(j)}(w) \tag{20.31f}$$

where $H_\nu(w^{(n)})$, $M_\nu(w^{(n)})$, $P_{k_1}(w^{(n)})$ are given in (20.11a, b, c); $p_{k_2}^{(i)}(w)$, $p_{c_2}^{(j)}(w)$ are given in (20.19c, d); and $p_{k_1}^{(n'')}(w)$, $p_{c_1}^{(n''')}(w)$ are given similarly as follows:

$$p_{k_1}^{(n'')}(w) = -(1-\mu)D\Delta_{k_1}\left\{\frac{\partial^2 w}{\partial\nu\,\partial s} - \frac{1}{\rho s}\frac{\partial w}{\partial s}\right\}^{(n'')} \quad \begin{Bmatrix} n'' = 1'', 2'', \ldots, \gamma'' \\ k_1 = 1, 2, \ldots, k_\sigma \end{Bmatrix} \tag{20.32}$$

$$p_{c_1}^{(n''')}(w) = -(1-\mu)D\Delta_{c_1}\left\{\frac{\partial^2 w}{\partial\nu\,\partial s} - \frac{1}{\rho_s}\frac{\partial w}{\partial s}\right\}^{(n''')} \quad \begin{Bmatrix} n''' = 1''', 2''', \ldots, \gamma''' \\ c_1 = 1, 2, \ldots, c_F \end{Bmatrix} \tag{20.33}$$

The stationary condition of variation of $\bar{\Pi}_f^*$ in (20.30) gives not only the field equation (20.1) and the boundary conditions (20.3a, b), (20.4a, b), and edge corner conditions (20.6a, b) for each element, but also the interelement boundary conditions (20.20a, b) and common corner conditions (20.19g, h). Furthermore, the Lagrange multipliers are also determined by the following relations:

$$\begin{aligned}\Lambda_1^{(nn')} &= D\left\{\frac{\partial}{\partial\nu}\left[\nabla^2 w + (1-\mu)\frac{\partial^2 w}{\partial s^2}\right] + (1-\mu)\frac{\partial}{\partial s}\frac{1}{\rho_s}\frac{\partial w}{\partial s}\right\}^{(n)} = -H_\nu(w^{(n)}) \\ &= D\left\{\frac{\partial}{\partial\nu}\left[\nabla^2 w + (1-\mu)\frac{\partial^2 w}{\partial s^2}\right] + (1-\mu)\frac{\partial}{\partial s}\frac{1}{\rho_s}\frac{\partial w}{\partial s}\right\}^{(n')} = -H_\nu(w^{(n')})\end{aligned} \tag{20.34a}$$

$$\begin{aligned}\Lambda_2^{(nn')} &= -D\left\{\mu\nabla^2 w + (1-\mu)\frac{\partial^2 w}{\partial\nu^2}\right\}^{(n)} = M_\nu(w^{(n)}) \\ &= D\left\{\mu\nabla^2 w + (1-\mu)\frac{\partial^2 w}{\partial\nu^2}\right\}^{(n')} = -M_\nu(w^{(n')})\end{aligned} \tag{20.34b}$$

$$\begin{aligned}\Lambda_{3k_1}^{(1''n'')} &= -(1-\mu)D\Delta_{k_1}\left\{\frac{\partial^2 w}{\partial\nu\,\partial s} - \frac{1}{\rho_s}\frac{\partial w}{\partial s}\right\}^{(n'')} \\ &= p_{k_1}^{(n'')}(w) \quad \begin{Bmatrix} n'' = 2'', 3'', \ldots, r'' \\ k_1 = 1, 2, \ldots, k_\sigma \end{Bmatrix}\end{aligned} \tag{20.35a}$$

$$\sum_{n''=2''}^{\gamma''} \Lambda_{3k_1}^{(1''n'')} + p_{k_1}^{(1'')}(w) = \sum_{n''=1''}^{\gamma''} p_{k_1}^{(n'')}(w) = \bar{P}_{k_1} \tag{20.35b}$$

$$\begin{aligned}\Lambda_{3c_1}^{(1'''n''')} &= -(1-\mu)D\Delta_{c_1}\left\{\frac{\partial^2 w}{\partial\nu\,\partial s} - \frac{1}{\rho_s}\frac{\partial w}{\partial s}\right\}^{(n''')} \\ &= p_{c_1}^{(n''')}(w) \quad \begin{Bmatrix} n''' = 2''', 3''', \ldots, \gamma''' \\ c_1 = 1, 2, \ldots, c_F \end{Bmatrix}\end{aligned} \tag{20.36a}$$

$$\sum_{n'''=2'''}^{\gamma'''} \Lambda_{3c_1}^{(1'''n''')} + p_{c_1}^{(1''')}(w) = \sum_{n'''=1'''}^{\gamma'''} p_{c_1}^{(n''')}(w) = \bar{F}_{c_1} \tag{20.36b}$$

The relations (20.34a, b) represent the continuity conditions of the equivalent shearing force and the bending moment along the interelement boundaries. The relations (20.35a, b), (20.36a, b) represent the fact that the resultant shearing force contributed from all elements to their common corners is equal to the applied loads. Thus we have

$$\int_{(nn')} \Lambda_1^{(nn')}[w^{(n)} - w^{(n')}]\,ds^{(n)}$$

$$= -\int_{(nn')} H_\nu(w^{(n)})w^{(n)}\,ds^{(n)} - \int_{(nn')} H_\nu(w^{(n')})w^{(n')}\,ds^{(n')} \quad (20.37a)$$

$$\int_{(nn')} \Lambda_2^{(nn')}\left(\frac{\partial w^{(n)}}{\partial \nu^{(n)}} + \frac{\partial w^{(n')}}{\partial \nu^{(n')}}\right) ds^{(n)}$$

$$= \int_{(nn')} M_\nu(w^{(n)})\frac{\partial w^{(n)}}{\partial \nu^{(n)}}\,ds^{(n)} + \int_{(nn')} M_\nu(w^{(n')})\frac{\partial w^{(n')}}{\partial \nu^{(n')}}\,ds^{(n')} \quad (20.37b)$$

$$\sum_{n''=2''}^{\gamma''} \Lambda_{3k_1}^{(1''n'')}[w_{k_1}^{(1'')} - w_{k_1}^{(n'')}] - \bar{P}_{k_1}w_{k_1}^{(1'')}$$

$$= \left\{\sum_{n''=2''}^{\gamma''} p_{k_1}^{(n'')}(w) - \bar{P}_{k_1}\right\}w_{k_1}^{(1'')} - \sum_{n''=2''}^{\gamma''} p_{k_1}^{(n'')}(w)w_{k_1}^{(n'')} \quad (20.38a)$$

$$\sum_{n'''=2'''}^{\gamma'''} \Lambda_{3c_1}^{(1'''n''')}[w_{c_1}^{(1''')} - w_{c_1}^{(n''')}] - \bar{F}_{c_1}w_{c_1}^{(1''')}$$

$$= \left\{\sum_{n'''=2'''}^{\gamma'''} p_{c_1}^{(n''')}(w) - \bar{F}_{c_1}\right\}w_{c_1}^{(1''')} - \sum_{n'''=2'''}^{\gamma'''} p_{c_1}^{(n''')}(w)w_{c_1}^{(n''')} \quad (20.38b)$$

The relations (20.37a, b) can be written as

$$\int_{(nn')} \Lambda_1^{(nn')}[w^{(n)} - w^{(n')}]\,ds^{(n)}$$

$$= -\int_{(nn')} \tfrac{1}{2}[H_\nu(w^{(n)}) + H_\nu(w^{(n')})]w^{(n)}\,ds^{(n)}$$

$$-\int_{(nn')} \tfrac{1}{2}[H_\nu(w^{(n)}) + H_\nu(w^{(n')})]w^{(n')}\,ds^{(n')} \quad (20.39a)$$

$$\int_{(nn')} \Lambda_2^{(nn')}\left[\frac{\partial w^{(n)}}{\partial \nu^{(n)}} + \frac{\partial w^{(n')}}{\partial \nu^{(n')}}\right] ds^{(n)}$$

$$= \int_{(nn')} \tfrac{1}{2}[M_\nu(w^{(n)}) + M_\nu(w^{(n')})]\frac{\partial w^{(n)}}{\partial \nu^{(n)}}\,ds^{(n)}$$

$$+\int_{(nn')} \tfrac{1}{2}[M_\nu(w^{(n)}) + M_\nu(w^{(n')})]\frac{\partial w^{(n')}}{\partial \nu^{(n')}}\,ds^{(n')} \quad (20.39b)$$

Therefore, we may finally write (20.29a,b) as

$$\Pi_f^{**} = \sum_{n=1}^{N} \Pi_f^{**(n)} + \sum_{k_1=1}^{k_\sigma} \left\{ \sum_{n''=2''}^{\gamma''} p_{k_1}^{(n'')}(w)[w_{k_1}^{(1'')} - w_{k_1}^{(n'')}] - \bar{P}_{k_1} w_{k_1}^{(1'')} \right\}$$
$$+ \sum_{c_1=1}^{C_F} \left\{ \sum_{n'''=2'''}^{\gamma'''} p_{c_1}^{(n''')}(w)[w_{c_1}^{(1''')} - w_{c_1}^{(n''')}] - \bar{F}_{c_1} w_{c_1}^{(1''')} \right\}$$
$$- \sum_{k_2=1}^{k_w} \sum_{j=1}^{\gamma'} p_{k_2}^{(j)}(w)[w_{k_2}^{(j)} - \bar{w}_{k_2}] - \sum_{c_2=1}^{c_w} \sum_{j=1}^{\gamma} p_{c_2}^{(j)}[w_{c_2}^{(j)} - \bar{w}_{c_2}] \tag{20.40a}$$

$$\Pi_f^{**(n)} = \Pi_f^{(n)} + \sum_{\substack{\text{All } n' \\ \text{neigh.} \\ \text{to } n}} \int_{(nn')} \tfrac{1}{2}[M_\nu(w^{(n)}) + M_\nu(w^{(n')})] \frac{\partial w^{(n)}}{\partial \nu^{(n)}} \, ds^{(n)}$$
$$- \sum_{\substack{\text{All } n' \\ \text{neigh.} \\ \text{to } n}} \int_{(nn')} \tfrac{1}{2}[H_\nu(w^{(n)}) + H_\nu(w^{(n')})] w^{(n)} \, ds^{(n)} \tag{20.40b}$$

in which $\Pi_f^{(n)}$ represents (20.19a).

The superindices $1'', 2'', \ldots, r''$ represent r'' elements having a common corner at the plate boundary point k_1, and we can take any element among $1'', 2'', 3'', \ldots, r''$ as the $1''$th element. Similarly, the superindices $1''', 2''', \ldots, r'''$ represent r''' elements having a common interelement corner c_1 in the interior of the plate, and we can take any one of them as $1'''$th element.

It is easily proved that the stationary condition of variation of Π_f^{**}, i.e.

$$\delta \Pi_f^{**} = 0 \tag{20.41}$$

gives not only the field equation (20.1), the boundary conditions (20.3a, b), (20.4a, b), and the edge corner conditions (20.6a, b) for each element, but also the interelement boundary conditions (20.20a, b) and the common corner conditions (20.19a, b). Furthermore, (20.41) also gives the conditions of continuity of the equivalent shearing force (20.34a) and the bending moment (20.34b) along the interelement boundaries, and the conditions that the resultant shearing force contributed from all the neighbouring elements to their common corners equal to the applied concentrated loads respectively.

Hence, the stationary condition of the variation of the functional (20.40a) truly represents the original plate-bending problem, whose solution is continuous everywhere in the given domain.

20.3 INCOMPATIBLE FINITE ELEMENTS FOR PLATE BENDING

Let us consider following 6 or 9 degrees of freedom interpolation functions for incompatible triangular serendipity elements (Figure 20.2). The field

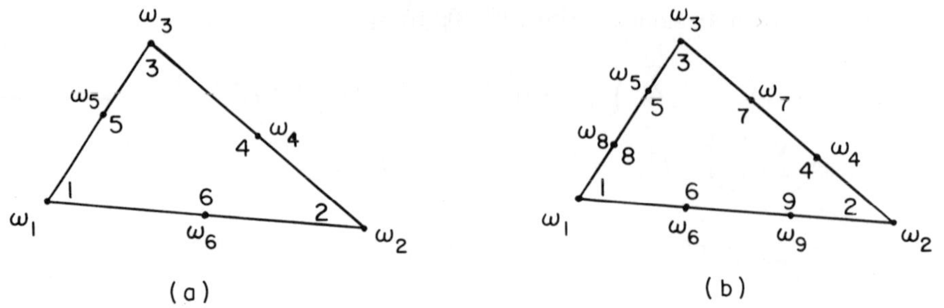

Figure 20.2 Incompatible triangular serendipity elements. (a) 6-degrees of freedom. (b) 9-degrees of freedom

function in the nth element may be written as

$$w^{(n)} = \sum_{i=1}^{6 \text{ or } 9} N_i^{(n)} w_i^{(n)} = \mathbf{N}^{(n)} \mathbf{w}^{(n)} \tag{20.42}$$

For 6 degrees of freedom interpolation functions, we have

$$\begin{aligned} N_i^{(n)} &= 2L_i^{(n)}(L_i^{(n)} - \tfrac{1}{2}) && \text{for } i = 1, 2, 3 \\ N_i^{(n)} &= 4L_j L_k && \text{for } i = 4, 5, 6 \ (i-3, j, k \text{ are cyclic of } 1, 2, 3) \end{aligned} \tag{20.43a}$$

For 9 degrees of freedom interpolation functions, we have

$$\begin{aligned} N_i^{(n)} &= \tfrac{9}{2} L_i^{(n)}(L_i^{(n)} - \tfrac{1}{3})(L_i^{(n)} - \tfrac{2}{3}) && \text{for } i = 1, 2, 3 \\ N_i^{(n)} &= \tfrac{27}{2} L_j^{(n)} L_k^{(n)}(L_j^{(n)} - \tfrac{1}{3}) && \text{for } i = 4, 5, 6 \ (i-3, j, k \text{ are cyclic of } 1, 2, 3) \\ N_i^{(n)} &= \tfrac{27}{2} L_j^{(n)} L_k^{(n)}(L_k^{(n)} - \tfrac{1}{3}) && \text{for } i = 7, 8, 9 (i-6, j, k \text{ are cyclic of } 1, 2, 3) \end{aligned} \tag{20.43b}$$

where

$$\mathbf{N}^{(n)} = [N_1, N_2, \ldots, N_{60\text{rq}}]^{(n)} \tag{20.44}$$

$$\mathbf{w}^{(n)T} = [w_1, w_2, \ldots, w_{60\text{rq}}]^{(n)} \tag{20.45}$$

$L_1^{(n)}, L_2^{(n)}, L_3^{(n)}$ are area coordinates of the nth triangular element and $w_i^{(n)} (i = 1, 2, \ldots, 6 \text{ or } 9)$ are the deflections of nodal points.

With these interpolation functions, if we take the deflection of boundary nodal points of neighbouring elements as being equal to each other (Figure 20.3), the interelement boundary conditions of (20.20a) and (20.19g, h) are automatically satisfied. Hence the functional $\tilde{\Pi}_f^{**}$ of (20.40a, b) can be

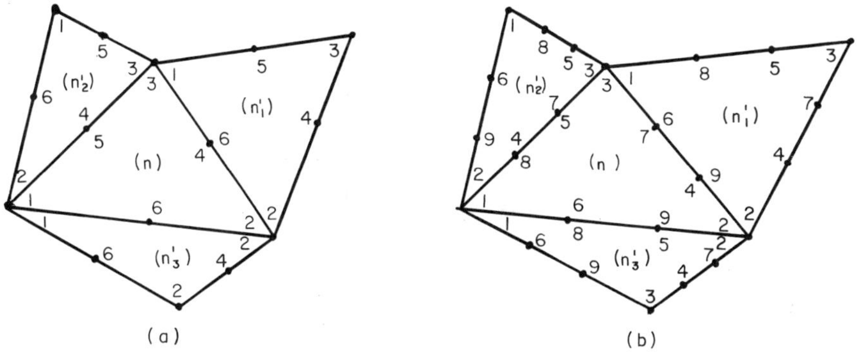

Figure 20.3 The position of nodal points of neighbouring elements. (a) 6-degrees of freedom. (b) 9-degrees of freedom

simplified to the form:

$$\tilde{\Pi}_f^{**} = \sum_{n=1}^{N} \tilde{\Pi}_f^{**(n)} - \sum_{k_1=1}^{k_\sigma} \bar{P}_{k_1} w_{k_1} - \sum_{k_2=1}^{k_w} (w_{k_2} - \bar{w}_{k_2}) \sum_{j=1}^{r'} p_{k_2}^{(j)}(w)$$
$$- \sum_{c_1=1}^{c_F} \bar{F}_{c_1} w_{c_1} - \sum_{c_2=1}^{c_w} (w_{c_2} - \bar{w}_{c_2}) \sum_{j=1}^{\gamma} p_{c_2}^{(j)}(w) \tag{20.46}$$

$$\tilde{\Pi}_f^{**(n)} = \Pi_f^{(n)} + \sum_{\substack{\text{All } n' \\ \text{neigh. to } n}} \int_{(nn')} \tfrac{1}{2}[M_\nu(w^{(n)}) + M_\nu(w^{(n')})] \frac{\partial w^{(n)}}{\partial \nu^{(n)}} \, ds^{(n)} \tag{20.47}$$

where $\Pi_f^{(n)}$ is given in (20.19a, b) and

$$M_\nu(w^{(n)}) = -D\left\{\mu \nabla^2 w + (1-\mu)\frac{\partial^2 w}{\partial \nu^2}\right\}^{(n)} \tag{20.48a}$$

$$M_\nu(w^{(n')}) = -D\left\{\mu \nabla^2 w + (1-\mu)\frac{\partial^2 w}{\partial \nu^2}\right\}^{(n')} \tag{20.48b}$$

$$p_{k_2}^{(j)}(w) = -(1-\mu) D \Delta_{k_2}\left(\frac{\partial^2 w}{\partial \nu \, \partial s}\right)^{(j)} \tag{20.48c}$$

$$p_{c_2}^{(j)}(w) = -(1-\mu) D \Delta_{c_2}\left(\frac{\partial^2 w}{\partial \nu \, \partial s}\right)^{(j)} \tag{20.48d}$$

It should be noted that, in $\Pi_f^{(n)}$, there are terms involving $H_\nu(w^{(n)})$. In this interpolation function,

$$H_\nu(w^{(n)}) = 0 \qquad \text{(for 6 degrees of freedom)} \tag{20.49a}$$

$$H_\nu(w^{(n)}) = \text{constant} \qquad \text{(for 9 degrees of freedom)} \tag{20.49b}$$

In the case of 6 degrees of freedom interpolation functions, $H_\nu(w^{(n)})$

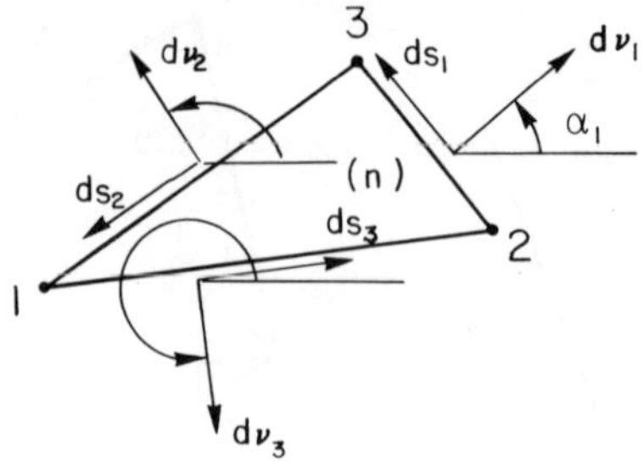

Figure 20.4 The angle of inclination for various edges of triangular element

vanishes everywhere; thus

$$\int_{S_{w_1}^{(n)}} H_\nu(w^{(n)})(w^{(n)} - \bar{w})\, ds^{(n)} = 0 \tag{20.49c}$$

We should retain this condition of prescribed deflection by putting

$$w_j^{(n)} - \bar{w}_j^{(n)} = 0 \tag{20.49d}$$

where j represents the number of nodal points in this part of the plate boundary $S_{w_1}^{(n)}$.

In order to calculate $\partial^2 w/\partial \nu^2$, $\partial^2 w/\partial \nu\, \partial s$, $\partial^2 w/\partial s^2$, let us use the inclination angle of normal ν of the triangle edges to the r axis α as shown in Figure 20.4. Consider first a rotation of the coordinate axis (x, y) through an angle α about the origin. Let the new axes be denoted by ν and s. The relationship between the two systems may be expressed as

$$\nu = x \cos \alpha + y \sin \alpha, \qquad s = -x \sin \alpha + y \cos x \tag{20.50}$$

or

$$x = \nu \cos \alpha - s \sin \alpha, \qquad y = \nu \sin \alpha + s \cos \alpha \tag{20.51}$$

Differentiation with respect to ν and s gives

$$\frac{\partial x}{\partial \nu} = \cos \alpha, \qquad \frac{\partial y}{\partial \nu} = \sin \alpha, \qquad \frac{\partial x}{\partial s} = -\sin \alpha \qquad \frac{\partial y}{\partial s} = \cos \alpha \tag{20.52}$$

Hence we have

$$\begin{bmatrix} \dfrac{\partial w}{\partial \nu} \\ \dfrac{\partial w}{\partial s} \end{bmatrix} = \begin{bmatrix} \cos & \sin \alpha \\ -\sin \alpha & \cos \alpha \end{bmatrix} \begin{bmatrix} \dfrac{\partial w}{\partial x} \\ \dfrac{\partial w}{\partial y} \end{bmatrix} \tag{20.53}$$

and

$$\begin{bmatrix} \dfrac{\partial^2 w}{\partial \nu^2} \\ \dfrac{\partial^2 w}{\partial \nu\, \partial s} \\ \dfrac{\partial^2 w}{\partial s^2} \end{bmatrix} = \begin{bmatrix} \cos^2\alpha & 2 \sin \alpha \cos \alpha & \sin^2\alpha \\ -\sin \alpha \cos \alpha & \cos^2\alpha - \sin^2\alpha & \sin \alpha \cos \alpha \\ \sin^2\alpha & -2 \sin \alpha \cos \alpha & \cos^2\alpha \end{bmatrix} \begin{bmatrix} \dfrac{\partial^2 w}{\partial x^2} \\ \dfrac{\partial^2 w}{\partial x\, \partial y} \\ \dfrac{\partial^2 w}{\partial y^2} \end{bmatrix} \tag{20.54}$$

Thus, from (20.48a,b), we find

$$M_\nu = -D[1 \;\; 0 \;\; \mu] \begin{bmatrix} \dfrac{\partial^2 w}{\partial \nu^2} \\ \dfrac{\partial^2 w}{\partial \nu\, \partial s} \\ \dfrac{\partial^2 w}{\partial s^2} \end{bmatrix} = -D\boldsymbol{\tau} \begin{bmatrix} \dfrac{\partial^2 w}{\partial x^2} \\ \dfrac{\partial^2 w}{\partial x\, \partial y} \\ \dfrac{\partial^2 w}{\partial y^2} \end{bmatrix} \tag{20.55}$$

where

$$\boldsymbol{\tau} = [\cos^2\alpha - \mu \sin^2\alpha,\ 2(1-\mu) \sin \alpha \cos \alpha,\ \sin^2\alpha - \mu \cos^2\alpha] \tag{20.56}$$

From (20.42), we find

$$\begin{bmatrix} \dfrac{\partial^2 w}{\partial x^2} \\ \dfrac{\partial^2 w}{\partial x\, \partial y} \\ \dfrac{\partial^2 w}{\partial y^2} \end{bmatrix} = \mathbf{G}\mathbf{w} \tag{20.57}$$

where **G** is a matrix in this element. For the case of 6 degrees of freedom, all the elements of this matrix are constants, while for the case of 9 degrees of freedom, all the elements of this matrix are linear functions of x and y:

$$\mathbf{G} = \begin{bmatrix} \dfrac{\partial^2 N_1}{\partial x^2} & \dfrac{\partial^2 N_2}{\partial x^2} & \cdots\cdots & \dfrac{\partial^2 N_6}{\partial x^2} \left(\text{or } \dfrac{\partial^2 N_q}{\partial x^2}\right) \\ \dfrac{\partial^2 N_1}{\partial x\, \partial y} & \dfrac{\partial^2 N_2}{\partial x\, \partial y} & \cdots\cdots & \dfrac{\partial^2 N_6}{\partial x\, \partial y} \left(\text{or } \dfrac{\partial^2 N_q}{\partial x\, \partial y}\right) \\ \dfrac{\partial^2 N_1}{\partial y^2} & \dfrac{\partial^2 N_2}{\partial y^2} & \cdots\cdots & \dfrac{\partial^2 N_6}{\partial y^2} \left(\text{or } \dfrac{\partial^2 N_q}{\partial y^2}\right) \end{bmatrix} \tag{20.58}$$

We obtain finally from (20.48a)

$$M_\nu(w) = -D\boldsymbol{\tau}\mathbf{G}\mathbf{w} = M_\nu(N)\mathbf{w} \tag{20.59}$$

where

$$M_\nu(N) = [M_\nu(N_1), M_\nu(N_2), \ldots\ldots, M_\nu(N_6) \text{ or } M_\nu(N_q)] \tag{20.60}$$

$$M_\nu(N_i) = -D\boldsymbol{\tau}\begin{bmatrix} \dfrac{\partial^2 N_i}{\partial x^2} \\ \dfrac{\partial^2 N_i}{\partial x\,\partial y} \\ \dfrac{\partial^2 N_i}{\partial y^2} \end{bmatrix} \tag{20.61}$$

This formula for bending moment (20.59) can be used for various interelement boundaries of element (n). For example, let us take the boundary edge 1 (see Figure 20.4) of this element, then the matrices $\mathbf{G}^{(n)}$ and $\mathbf{w}^{(n)}$ are calculated for the element (n), while $\boldsymbol{\tau}^{(n)}$ is calculated from (20.56) by means of α_1. We may denote this $\boldsymbol{\tau}^{(n)}$ as $\boldsymbol{\tau}_1^{(n)}$. In general, we find for various triangle edges

$$\boldsymbol{\tau}_j^{(n)} = [\cos^2\alpha_j - \mu \sin^2\alpha_j, 2(1-\mu)\sin\alpha_j \cos\alpha_j, \sin^2\alpha_j - \mu\cos^2\alpha_j] \tag{20.62a}$$

$$M_{\nu j}(w^{(n)}) = -D\boldsymbol{\tau}_j^{(n)}\mathbf{G}^{(n)}\mathbf{w}^{(n)} \qquad \text{for } j = 1, 2, 3 \tag{20.62b}$$

Similarly we find for the element (n') neighbouring to element (n),

$$M_{\nu_j}(w^{(n')}) = -D\boldsymbol{\tau}_j^{(n')}\mathbf{G}^{(n')}\mathbf{w}^{(n')} \qquad \text{for } j = 1, 2, 3 \tag{20.63}$$

According to Figure 20.3, there are three neighbouring elements, that is $n'_i = n'_1, n'_2, n'_3$. The bending moment acting on the interelement boundary (n, n'_i) of the elements n and n'_i are respectively

$$M_{\nu_3}(w^{(n'_1)}) = -D\boldsymbol{\tau}_3^{(n'_1)}\mathbf{G}^{(n'_1)}\mathbf{w}^{(n_1)} \tag{20.64a}$$

$$M_{\nu_1}(w^{(n'_2)}) = -D\boldsymbol{\tau}_1^{(n'_2)}\mathbf{G}^{(n'_2)}\mathbf{w}^{(n_2)} \tag{20.64b}$$

$$M_{\nu_2}(w^{(n'_3)}) = -D\boldsymbol{\tau}_2^{(n'_3)}\mathbf{G}^{(n'_3)}\mathbf{w}^{(n'_3)} \tag{20.64c}$$

Furthermore, we have

$$\frac{\partial w^{(n)}}{\partial \nu_i^{(n)}} = [\cos\alpha_i, \sin\alpha_i]\begin{bmatrix} \dfrac{\partial w}{\partial x} \\ \dfrac{\partial x}{\partial y} \end{bmatrix}_i^{(n)} = \mathbf{R}_i^{(n)}\mathbf{w}^{(n)} \tag{20.65}$$

where

$$\mathbf{R}_i^{(n)} = [\cos\alpha_i, \sin\alpha_i] \begin{bmatrix} \dfrac{\partial N_1}{\partial x}, \dfrac{\partial N_2}{\partial x}, & \cdots\cdots, & \dfrac{\partial N_6}{\partial x} \left(\text{or } \dfrac{\partial N_q}{\partial x}\right) \\ \dfrac{\partial N_1}{\partial y}, \dfrac{\partial N_2}{\partial y}, & \cdots\cdots, & \dfrac{\partial N_6}{\partial y} \left(\text{or } \dfrac{\partial N_q}{\partial y}\right) \end{bmatrix}^{(n)} \tag{20.65a}$$

We are now in the position to calculate the integrals in (20.47):

$$\begin{aligned} &\sum_{n_1' n_2' n_3'} \int_{(nn')} \tfrac{1}{2}[M_\nu(w^{(n)}) + M_\nu(w^{(n')})] \frac{\partial w^{(n)}}{\partial \nu^{(n)}} \, \mathrm{d}s^{(n)} \\ &\quad = -\int_{(nn_1')} \frac{D}{2} [\boldsymbol{\tau}_1^{(n)} \mathbf{G}^{(n)} \mathbf{w}^{(n)} + \boldsymbol{\tau}_3^{(n_1')} \mathbf{G}^{(n_1')} \mathbf{w}^{(n_1')}] \mathbf{R}_1^{(n)} \mathbf{w}^{(n)} \, \mathrm{d}s_1^{(n)} \\ &\qquad - \int_{(nn_2')} \frac{D}{2} [\boldsymbol{\tau}_2^{(n)} \mathbf{G}^{(n)} \mathbf{w}^{(n)} + \boldsymbol{\tau}_1^{(n_2')} \mathbf{G}^{(n_2')} \mathbf{w}^{(n_2')}] \mathbf{R}_2^{(n)} \mathbf{w}^{(n)} \, \mathrm{d}s_2^{(n)} \\ &\qquad - \int_{(nn_3')} \frac{D}{2} [\boldsymbol{\tau}_3^{(n)} \mathbf{G}^{(n)} \mathbf{w}^{(n)} + \boldsymbol{\tau}_2^{(n_3')} \mathbf{G}^{(n_3')} \mathbf{w}^{(n_3')}] \mathbf{R}_3^{(n)} \mathbf{w}^{(n)} \, \mathrm{d}s_3^{(n)} \\ &\quad = \tfrac{1}{2} \mathbf{w}^{(n)T} \mathbf{K}_0^{(n)} \mathbf{w}^{(n)} + \tfrac{1}{2} \sum_{i=1}^{3} \mathbf{w}^{(n)T} \mathbf{K}_{(i)}^{(nn_i')} \mathbf{w}^{(n_i')} \end{aligned} \tag{20.66}$$

where $$\mathbf{K}_0^{(n)} = -D \sum_{i=1}^{3} \int_{(nn')} \mathbf{R}^{(n)T} \boldsymbol{\tau}_i^{(n)} \mathbf{G}^{(n)} \, \mathrm{d}s_i^{(n)} \tag{20.67a}$$

$$\mathbf{K}_{(1)}^{(nn_1')} = -D \int_{(nn_1')} \mathbf{R}_1^{(n)T} \boldsymbol{\tau}_3^{(n_1')} \mathbf{G}^{(n_1')} \, \mathrm{d}s_1^{(n)} \tag{20.67b}$$

$$\mathbf{K}_2^{(nn_2')} = -D \int_{(nn_2')} \mathbf{R}_2^{(n)T} \boldsymbol{\tau}_1^{(n_2')} \mathbf{G}^{(n_2')} \, \mathrm{d}s_2^{(n)} \tag{20.67c}$$

$$\mathbf{K}_3^{(nn_3')} = -D \int_{(nn_3')} \mathbf{R}_3^{(n)T} \boldsymbol{\tau}^{(n_3')T} \boldsymbol{\tau}_2^{(n_3')} \mathbf{G}^{(n_3')} \, \mathrm{d}s_3^{(n)} \tag{20.67d}$$

By means of (20.42), $\Pi_{0f}^{(n)}$ of (20.18b) may be calculated from

$$\Pi_{0f}^{(n)} = \iint_{A^{(n)}} \frac{D}{2} \mathbf{w}^{(n)T} \mathbf{G}^{(n)T} \boldsymbol{\mu} \mathbf{G}^{(n)} \mathbf{w}^{(n)} \, \mathrm{d}A^{(n)} = \tfrac{1}{2} \mathbf{w}^{(n)T} \mathbf{K}_{0f}^{(n)} \mathbf{w}^{(n)} \tag{20.68}$$

in which

$$\mathbf{K}_{0f}^{(n)} = \iint_{A^{(m)}} D \mathbf{G}^{(n)T} \boldsymbol{\mu} \mathbf{G}^{(n)} \, \mathrm{d}A^{(n)} \tag{20.69a}$$

$$\boldsymbol{\mu} = \begin{bmatrix} 1 & \cdots & \mu \\ & 2(1-\mu) & \\ \mu & \cdots & 1 \end{bmatrix} \tag{20.69b}$$

Similarly, we have on S_w the integral

$$\int_{S_{w_2}^{(n)}} M_\nu(w^{(n)}) \frac{\partial w^{(n)}}{\partial \nu^{(n)}} \, ds^{(n)} = \mathbf{w}^{(n)T} \mathbf{K}_{S_{w_2}}^{(n)} \mathbf{w}^{(n)} \tag{20.70}$$

where

$$\mathbf{K}_{S_{w_2}}^{(n)} = -\int_{S_{w_2}^{(n)}} D\mathbf{R}_{S_{w_2}}^{(n)} \boldsymbol{\tau}_{S_{w_2}}^{(n)} \mathbf{G}^{(n)} \, ds_{w_2}^{(n)} \tag{20.70a}$$

Let us now calculate the equivalent shearing force $H_\nu(w)$ from (20.11b). In the present case, $1/\rho_s = 0$; thus

$$H_\nu(w) = -D \frac{\partial}{\partial \nu} \left[\frac{\partial^2 w}{\partial \nu^2} + (2-\mu) \frac{\partial^2 w}{\partial s^2} \right] \tag{20.71}$$

Using (20.53), (20.54), we have

$$\frac{\partial}{\partial \nu} = \frac{\partial x}{\partial \nu} \frac{\partial}{\partial x} + \frac{\partial y}{\partial \nu} \frac{\partial}{\partial y} = \cos\alpha \frac{\partial}{\partial x} + \sin\alpha \frac{\partial}{\partial y} \tag{20.72a}$$

$$\begin{aligned} \frac{\partial^2 w}{\partial \nu^2} + (2-\mu) \frac{\partial^2 w}{\partial s^2} &= [\cos^2\alpha + (2-\mu) \sin^2\alpha] \frac{\partial^2 w}{\partial x^2} - 2(1-\mu) \sin\alpha \cos\alpha \frac{\partial^2 w}{\partial x \, \partial y} \\ &\quad + [\sin^2\alpha + (2-\mu)\cos^2\alpha] \frac{\partial^2 w}{\partial y^2} \end{aligned} \tag{20.72b}$$

Hence $M(w)$ in (20.71) can be regrouped into matrix form:

$$H_\nu(w) = -D\boldsymbol{\sigma} \begin{bmatrix} \dfrac{\partial^3 w}{\partial x^3} \\ \dfrac{\partial^3 w}{\partial x^2 \, \partial y} \\ \dfrac{\partial^3 w}{\partial x \, \partial y^2} \\ \dfrac{\partial^3 w}{\partial y^2} \end{bmatrix} = -D\boldsymbol{\sigma}\mathbf{J}\mathbf{w} \tag{20.73}$$

where

$$\boldsymbol{\sigma} = [\sigma_{(1)}, \sigma_{(2)}, \sigma_{(3)}, \sigma_{(4)}] \tag{20.74}$$

$$\sigma_{(1)} = [\cos^2\alpha + (2-\mu)\sin^2\alpha] \cos\alpha \tag{20.74a}$$

$$\sigma_{(2)} = [(2-\mu)\sin^2\alpha - (1-2\mu)\cos^2\alpha] \sin\alpha \tag{20.74b}$$

$$\sigma_{(3)} = [(2-\mu)\cos^2\alpha - (1-2\mu)\sin^2\alpha] \cos\alpha \tag{20.74c}$$

$$\sigma_{(4)} = [\sin^2\alpha + (2-\mu)\cos^2\alpha] \sin\alpha \tag{20.74d}$$

and

$$\begin{bmatrix} \dfrac{\partial^3 w}{\partial x^3} \\ \dfrac{\partial^3 w}{\partial x^2\,\partial y} \\ \dfrac{\partial^3 w}{\partial x\,\partial y^2} \\ \dfrac{\partial^3 w}{\partial y^3} \end{bmatrix} = \mathbf{J}\mathbf{w} \tag{20.75}$$

$$\mathbf{J} = \begin{bmatrix} \dfrac{\partial^3 N_1}{\partial x^3} & \dfrac{\partial^3 N_2}{\partial x^3} & \cdots\cdots & \dfrac{\partial^3 N_6}{\partial x^3}\left(\text{or } \dfrac{\partial^3 N_q}{\partial x^3}\right) \\ \dfrac{\partial^3 N_1}{\partial x^2\,\partial y} & \dfrac{\partial^3 N_2}{\partial x^2\,\partial y} & \cdots\cdots & \dfrac{\partial^3 N_6}{\partial x^2\,\partial y}\left(\text{or } \dfrac{\partial^3 N_q}{\partial x^2\,\partial y}\right) \\ \dfrac{\partial^3 N_1}{\partial x\,\partial y^2} & \dfrac{\partial^3 N_2}{\partial x\,\partial y^2} & \cdots\cdots & \dfrac{\partial^3 N_6}{\partial x\,\partial y^2}\left(\text{or } \dfrac{\partial^3 N_q}{\partial x\,\partial y^2}\right) \\ \dfrac{\partial^3 N_1}{\partial y^3} & \dfrac{\partial^3 N_2}{\partial y^3} & \cdots\cdots & \dfrac{\partial^3 N_6}{\partial y^3}\left(\text{or } \dfrac{\partial^3 N_q}{\partial x^2\,\partial y}\right) \end{bmatrix} \tag{20.76}$$

From (20.59), (20.72), we can calculate the following integral:

$$\int_{s_{w_1}^{(n)}} H_\nu(w^{(n)})w^{(n)}\,\mathrm{d}s^{(n)} = \mathbf{w}^{(n)T}\mathbf{K}_{s_{w_1}}^{(n)}\mathbf{w}^{(n)} \tag{20.77a}$$

$$\int_{s_{w_2}^{(n)}} M_\nu(w^{(n)})\frac{\partial w^{(n)}}{\partial \nu^{(n)}}\,\mathrm{d}s^{(n)} = \mathbf{w}^{(n)T}\mathbf{K}_{s_{w_2}}^{(n)}\mathbf{w}^{(n)} \tag{20.77b}$$

where $\mathbf{K}_{s_{w_2}}^{(n)}$ is given in (20.70a) and

$$\mathbf{K}_{s_{w_1}}^{(n)} = -\int_{s_{w_1}^{(n)}} D\mathbf{N}_{s_{w_1}}^{(n)T}\boldsymbol{\sigma}_{s_{w_1}}^{(n)}\mathbf{J}^{(n)}\,\mathrm{d}s_{w_1}^{(n)} \tag{20.78}$$

$\mathbf{N}_{s_{w_1}}^{(n)}, \boldsymbol{\sigma}_{s_{w_1}}^{(n)}, \mathbf{J}^{(n)}$ are calculated according to (20.44), (20.74), and (20.76) respectively.

Using (20.68), (20.77a, b), we may write (20.19a) similarly in terms of matrix notation:

$$\begin{aligned}\Pi_f^{(n)} = {}& \tfrac{1}{2}\mathbf{w}^{(n)T}\mathbf{K}_{0f}^{(n)}\mathbf{w}^{(n)} - \bar{\mathbf{f}}^{(n)}\mathbf{w}^{(n)} - \bar{H}^{(n)}\mathbf{w}^{(n)} + \bar{M}^{(n)}\mathbf{w}^{(n)} \\ & - \mathbf{w}^{(n)T}\mathbf{K}_{s_{w1}}^{(n)}\mathbf{w}^{(n)} - \mathbf{w}^{(n)T}\mathbf{K}_{s_{w2}}^{(n)}\mathbf{w}^{(n)} + \bar{\mathbf{w}}^{(n)}\mathbf{w}^{(n)} - \bar{\mathbf{w}}_{s\nu}^{(n)}\mathbf{w}^{(n)}\end{aligned} \tag{20.79}$$

where

$$\bar{\mathbf{f}}^{(n)} = \iint_{A^{(n)}} \mathbf{f}\mathbf{N}^{(n)}\, dA^{(n)} \tag{20.80a}$$

$$\bar{\mathbf{H}}^{(n)} = \int_{s_{\sigma_1}^{(n)}} \bar{H}\mathbf{N}^{(n)}\, ds_{\sigma_1}^{(n)} \tag{20.80b}$$

$$\bar{\mathbf{M}}^{(n)} = \int_{s_{\sigma_2}^{(n)}} \bar{M}\mathbf{R}_{s_{\sigma_2}}^{(n)}\, ds_{\sigma_2}^{(n)} \tag{20.80c}$$

$$\bar{\mathbf{w}}^{(n)} = -\int_{s_{w_1}^{(n)}} D\bar{w}\boldsymbol{\tau}_{s_{w_1}}^{(n)}\mathbf{J}^{(n)}\, ds_{w_1}^{(n)} \tag{20.80d}$$

$$\bar{\mathbf{w}}_{,\nu}^{(n)} = -\int_{s_{w_2}^{(n)}} D\bar{w}_{,\nu}\boldsymbol{\tau}_{s_{w_2}}^{(n)}\mathbf{G}^{(n)}\, ds_{w_2}^{(n)} \tag{20.80e}$$

Equation (20.79) can be further simplified:

$$\Pi_f^{(n)} = \tfrac{1}{2}\mathbf{w}^{(n)T}\mathbf{K}_f^{(n)}\mathbf{w}^{(n)} - \bar{\mathbf{F}}_f^{(n)}\mathbf{w}^{(n)} \tag{20.81}$$

where

$$\mathbf{K}_f^{(n)} = \mathbf{K}_{0f}^{(n)} - 2\mathbf{K}_{s_{w_1}}^{(n)} - 2\mathbf{K}_{s_{w_2}}^{(n)} \tag{20.81a}$$

$$\bar{\mathbf{F}}_f^{(n)} = \bar{\mathbf{f}}^{(n)} + \bar{\mathbf{H}}^{(n)} - \bar{\mathbf{M}}^{(n)} - \bar{\mathbf{w}}^{(n)} + \bar{\mathbf{w}}_{,\nu}^{(n)} \tag{20.81b}$$

$\bar{F}_f^{(n)}$ represents the resultant influence of the prescribed loads and deflections on the element area and its plate boundary.

The summation of (20.81) and (20.66) gives

$$\tilde{\Pi}^{**(n)} = \tfrac{1}{2}\mathbf{w}^{(n)T}\mathbf{K}^{**(n)}\mathbf{w}^{(n)} + \tfrac{1}{2}\sum_{i=1}^{3}\mathbf{w}^{(n)T}\mathbf{K}_{(i)}^{(nn'_i)}\mathbf{w}^{(n'_i)} - \bar{\mathbf{F}}_f^{(n)}\mathbf{w}^{(n)} \tag{20.82}$$

where

$$\mathbf{K}^{**(n)} = \mathbf{K}_0^{(n)} + \mathbf{K}_f^{(n)} \tag{20.83}$$

Let us now consider the matrix representation of $p_{k_2}^{(j)}(w)$ from (20.19c). For the case of straight boundaries, we have

$$p_{k_2}^{(j)}(w) = -(1-\mu)D\Delta_{k_2}\left(\frac{\partial^2 w}{\partial\nu\,\partial s}\right)^{(j)} \tag{20.84}$$

From (20.54), we have

$$\frac{\partial^2 w}{\partial\nu\,\partial s} = [-\sin\alpha\cos\alpha,\ \cos^2\alpha - \sin^2\alpha,\ \sin\alpha\cos\alpha]\begin{bmatrix}\dfrac{\partial w}{\partial x^2}\\[2mm] \dfrac{\partial^2 w}{\partial x\,\partial y}\\[2mm] \dfrac{\partial^2 w}{\partial y^2}\end{bmatrix} = \boldsymbol{\pi}\mathbf{G}\mathbf{w} \tag{20.85}$$

where

$$\boldsymbol{\pi} = [-\sin\alpha\cos\alpha, \cos^2\alpha - \sin^2\alpha, \sin\alpha\cos\alpha] \tag{20.85a}$$

Thus, (20.84) can be written as

$$p_{k_2}^{(j)}(w) = -(1-\mu)D\Delta_{k_2}(\boldsymbol{\pi}\mathbf{G})^{(j)}\mathbf{w}^{(j)} = \mathbf{p}_{k_2}^{(j)}\mathbf{w}^{(j)} \tag{20.86}$$

where

$$p_{k_2}^{(j)} = -(1-\mu)D\Delta_{k_2}(\boldsymbol{\pi}\mathbf{G})^{(j)} \tag{20.86a}$$

Therefore, (20.46) is

$$\begin{aligned}\tilde{\Pi}_f^{**} = \sum_{n=1}^{N} &\left\{\tfrac{1}{2}\mathbf{w}^{(n)T}\mathbf{K}^{**(n)}\mathbf{w}^{(n)} + \tfrac{1}{2}\sum_{i=1}^{3}\mathbf{w}^{(n)T}\mathbf{K}_{(i)}^{(nn_i')}\mathbf{w}^{(n_i')} - \bar{\mathbf{F}}_f^{(n)}\mathbf{w}^{(n)}\right\} \\ &- \sum_{k_2=1}^{k_w}(w_{k_2} - \bar{w}_{k_2})\sum_{j=1}^{\gamma'}\mathbf{p}_{k_2}^{(j)}\mathbf{w}^{(j)} - \sum_{k_1=1}^{k_\sigma}\bar{P}_{k_1}w_{k_1} \\ &- \sum_{c_2=1}^{c_w}(w_{c_2} - \bar{w}_{c_2})\sum_{j=1}^{\gamma}\mathbf{p}_{c_2}^{(j)}\mathbf{w}^{(j)} - \sum_{c_1=1}^{c_F}\bar{F}_{c_1}w_{c_1}\end{aligned} \tag{20.87}$$

or in assembled form

$$\tilde{\Pi}_f^{**} = \tfrac{1}{2}\mathbf{w}^T\mathbf{K}^{**}\mathbf{w} - \bar{\mathbf{F}}\mathbf{w} \tag{20.88}$$

where $\mathbf{K}^{**}$ is the assembled rigidity matrix, which is defined by

$$\begin{aligned}\tfrac{1}{2}\mathbf{w}^T\mathbf{K}^{**}\mathbf{w} = \sum_{n=1}^{N} &\left\{\tfrac{1}{2}\mathbf{w}^{(n)T}\mathbf{K}^{**(n)}\mathbf{w}^{(n)} + \tfrac{1}{2}\sum_{i=1}^{3}\mathbf{w}^{(n)T}\mathbf{K}_{(i)}^{(nn_i)}\mathbf{w}^{(i')}\right\} \\ &- \sum_{c_2=1}^{c_w} w_{c_2}\sum_{j=1}^{\gamma}\mathbf{p}_{c_2}^{(j)}\mathbf{w}^{(j)} - \sum_{k_2=1}^{k_w} w_{k_2}\sum_{j=1}^{\gamma'}\mathbf{p}_{k_2}^{(j)}\mathbf{w}^{(j)}\end{aligned} \tag{20.89a}$$

$$\begin{aligned}\bar{\mathbf{F}}\mathbf{w} = \sum_{n=1}^{N}\bar{\mathbf{F}}_f^{(n)}\mathbf{w}^{(n)} &- \sum_{k_2=1}^{k_w}\bar{w}_{k_2}\sum_{j=1}^{\gamma'}\mathbf{p}_{k_2}^{(j)}\mathbf{w}^{(j)} \\ &- \sum_{c_2=1}^{c_w}\bar{w}_{c_2}\sum_{j=1}^{\gamma}\mathbf{p}_{c_2}^{(j)}\mathbf{w}^{(j)} + \sum_{k_1=1}^{k_\sigma}\bar{P}_{k_1}w_{k_1} + \sum_{c_1=1}^{c_F}\bar{F}_{c_1}w_{c_1}\end{aligned} \tag{20.89b}$$

The variation of (20.88) gives the equations for the determination of $\tilde{\Pi}_f^{**}$, that is

$$\delta\tilde{\Pi}_f^{**} = 0, \qquad \mathbf{K}\mathbf{w} = \bar{\mathbf{F}} \tag{20.90}$$

where in general

$$\mathbf{K} = \tfrac{1}{2}\mathbf{K}^{**} + \tfrac{1}{2}\mathbf{K}^{**T} \tag{20.91}$$

For a simply supported square plate under a uniformly distributed load, the present calculation can be greatly simplified. If we assume that the deflection of the plate edge is satisfied in the final calculation, then (20.46), (20.47), (20.19a, b) can be written as

$$\tilde{\Pi}_f^{**(n)} = \Pi_f^{(n)} + \sum_{\substack{\text{All } n' \\ \text{neigh. to } n}} \int_{(nn')} \tfrac{1}{2}[M_\nu(w^{(n)}) + M_\nu(w^{(n')})] \frac{\partial w^{(n)}}{\partial \nu^{(n)}} \, ds^{(n)} \tag{20.92}$$

$$\Pi_f^{(n)} = \iint_{A^{(n)}} \frac{D}{2} \left\{ \left(\frac{\partial^2 w}{\partial x^2} + \frac{\partial^2 w}{\partial y^2} \right)^2 - 2(1-\mu) \left(\frac{\partial^2 w}{\partial x^2} \frac{\partial^2 w}{\partial y^2} - \frac{\partial^2 w}{\partial x \, \partial y} \frac{\partial^2 w}{\partial x \, \partial y} \right) \right\}^{(n)} dA^{(n)} - \iint_{A^{(n)}} f w^{(n)} \, dA^{(n)} \tag{20.93}$$

Hence, the assembled form is the same as (20.88), and $\mathbf{K}^{**}$, $\bar{\mathbf{F}}$ are defined by

$$\tfrac{1}{2}\mathbf{w}^T \mathbf{K}^{**} \mathbf{w} = \sum_{n=1}^{N} \left\{ \tfrac{1}{2} \mathbf{w}^{(n)T} \mathbf{K}^{**(n)} \mathbf{w}^{(n)} + \tfrac{1}{2} \sum_{i=1}^{3} \mathbf{w}^{(n)T} \mathbf{K}_{(i)}^{(nn_i')} \mathbf{w}^{(n)} \right\} \tag{20.94a}$$

$$\bar{\mathbf{F}} \mathbf{w} = \sum_{n=1}^{N} \bar{\mathbf{f}}^{(n)} \mathbf{w}^{(n)} \tag{20.94b}$$

where $\mathbf{K}^{**(n)}$ is given by (20.83), (20.81a), (20.67a), and $\mathbf{K}_{(i)}^{(nn_i')}$ are given in (20.67b, c, d).

REFERENCES

1. Theodore H. H. Pian and Pin Tong, 'Finite element methods in continuum mechanics', in *Advances in Applied Mechanics* (Ed. Chia-shun Yih), Vol. 12, pp. 1–58, Academic Press, 1972.
2. Pin Tong, 'New displacement hybrid finite element model for solid continua', *Int. J. Num. Meth. in Eng.*, **2,** 78–83 (1970).
3. L. R. Herrmann, 'A bending analysis for plates', *Proc. First Conf. Matrix Methods in Struct. Mech.*, 1965, pp. 1165–1194, AFFDL-TR-68-150, 1966.
4. L. R. Herrmann, 'Finite element bending analysis for plates', *J. Eng. Mech. Div., ASCE,* **98,** No. EM5, 13–26 (1967).
5. Wei-Zang Chien, *Variational Principles and Finite Elements* (in Chinese), Science Publications, Peking, 1980.

Hybrid and Mixed Finite Element Methods
Edited by S. N. Atluri, R. H. Gallagher, and O. C. Zienkiewicz

Chapter 21

Mixed and Irreducible Formulations in Finite Element Analysis

Some general comments and applications to the incompressibility problem

O. C. Zienkiewicz, R. L. Taylor, and J. M. W. Baynham

21.1 INTRODUCTION

In the early days of the finite element method, terminology was introduced such as 'potential energy formulation' or 'complementary energy formulation' in order to differentiate between the procedures of elastic stress analysis in which the discretization was applied to the direct minimization of corresponding variational principles. Later, discretizations requiring the stationarity of mixed variational principles (such as that of Hellinger–Reissner) were termed 'mixed methods', and a particular form of this, in which different variables were discretized in separate parts of the domain (and which was developed by Pian), was given the name of 'hybrid'.

As the application of finite elements has now progressed to the solution of many diverse forms of physical system, ranging from electromagnetics through to fluid dynamics and biological systems, the terminology should be defined in a more general manner. This is particularly important as in most situations the discretization is made either at the differential equation level by the use of weighted residual methods or at the level of the weak (integral) forms—neither method relies upon the existence of a variational principle for which stationarity would be sought. What then are the definitions which can be coined in such a context?

21.2 MIXED AND IRREDUCIBLE FORMS

A physical problem to be solved can invariably be defined by a set of 'm' differential equations, in which $\mathbf{\Phi}=[\phi_1, \phi_2, \ldots, \phi_n]^T$ are the independent variables, and which can be written as

$$\mathbf{L\Phi}+\mathbf{q}=0 \qquad \text{(in thc domain } \Omega) \tag{21.1}$$

Here the operator $\mathbf{L}$ is such that m differential equations result.

The appropriate boundary conditions on Γ can be similarly written as

$$\mathbf{B\Phi}+\mathbf{g}=0 \qquad (\text{on } \Gamma) \tag{21.2}$$

In the numerical solution, the approximation to the independent variable is made in terms of known shape functions $\mathbf{N}_i$ and a set of unknown parameters $\mathbf{a}_i$, so that

$$\mathbf{\Phi}\approx\hat{\mathbf{\Phi}}=\sum_1^n \mathbf{N}_i\mathbf{a}_i \tag{21.3}$$

The discrete (algebraic) equation set for which an approximate solution is desired is cast in a form of 'weighted residuals' as

$$\int_\Omega \mathbf{W}_j^T(\mathbf{L}\hat{\mathbf{\Phi}}+\mathbf{q})\,\mathrm{d}\Omega+\int_\Gamma \bar{\mathbf{W}}_j^T(\mathbf{B}\hat{\mathbf{\Phi}}+\mathbf{g})\,\mathrm{d}\Gamma=0 \qquad j=1\to n \tag{21.4}$$

where $\mathbf{W}_j$ and $\bar{\mathbf{W}}_j$ are independently chosen sets of weighting functions.

Whilst the accuracy of a numerical solution, and the effort required to find it, depend very much upon the particular form of the shape and weighting functions used, it is of even greater importance to consider the set of differential equations to which the discretization is to be applied. This set can be large—consisting of low-order differential equations—or smaller—when direct elimination of variables with generally higher order derivates is involved.

Such differing sets of equations can be obtained by direct elimination of variables, and immediately a possible definition suggests itself.

A set of differential equations (and of corresponding variables) is *irreducible* if, by a simple process of elimination, no further reduction can be made in the number of variables still leaving a solvable problem.* Any other form of governing differential equations and of corresponding variables is termed *mixed.*

If now discretization is applied to a problem, in the manner of equation (21.3), then irreducible and mixed finite element methods arise.

Examples of such problems abound in physical situations, and some are given below.

Example 21.1: Beam bending Here the basic problem can be stated by

* In a (say) two-dimensional elasticity analysis with two displacement variables it is of course possible to eliminate one of the displacements. This problem is, however, not solvable due to boundary conditions.

writing four equations, which in usual notation are

$$\begin{aligned} \frac{\mathrm{d}S}{\mathrm{d}x} - q &= 0 \\ \frac{\mathrm{d}M}{\mathrm{d}x} - s &= 0 \\ M - EI\frac{\mathrm{d}\theta}{\mathrm{d}x} &= 0 \\ \theta - \frac{\mathrm{d}w}{\mathrm{d}x} &= 0 \end{aligned} \tag{21.5}$$

The four independent variables are

$$\mathbf{\Phi}^T = [S, M, \theta, w] \tag{21.6}$$

which correspond respectively to shear force, bending moment, rotation, and lateral displacement. In the above, q is the specified loading intensity.

Clearly a 'mixed' formulation will result if all the variables are independently approximated (as indeed was done by Nemat-Nasser and Lee [1], who used an appropriate variational theorem).

Alternatively, other mixed forms can be derived by successive elimination of variables until a finally irreducible form is reached in terms of the single displacement variable w:

$$\frac{\mathrm{d}^2}{\mathrm{d}x^2}\left[EI\frac{\mathrm{d}^2 w}{\mathrm{d}x^2}\right] - q = 0 \tag{21.7}$$

Equivalent equations can be derived describing the problem of plate bending in a similar manner.

Example 21.2: Linear elasticity Here we consider a uniaxial bar with stress σ and displacement u giving

$$\begin{aligned} \frac{\mathrm{d}\sigma}{\mathrm{d}x} + b &= 0 \\ \sigma - E\frac{\mathrm{d}u}{\mathrm{d}x} &= 0 \end{aligned} \tag{21.8}$$

The above set represents a 'mixed' form, with

$$\mathbf{\Phi}^T = [\sigma, u] \tag{21.9}$$

A simple elimination leads to the irreducible equation:

$$\frac{\mathrm{d}}{\mathrm{d}x}\left[E\frac{\mathrm{d}u}{\mathrm{d}x}\right] + b = 0 \tag{21.10}$$

from which the conventional 'displacement' formulation results.

Clearly, elasticity equations in a three-dimensional form are but an extension of the above example. Now the most general, 'mixed', form of equations can be written as

$$\begin{aligned} \sigma_{ij,j} + b_i &= 0 && \text{(equilibrium)} \\ \sigma_{ij} &= D_{ijkl}\varepsilon_{kl} && \text{(constitutive)} \\ \Sigma_{kl} &= \tfrac{1}{2}[u_{k,l} + u_{l,k}] && \text{(strain definition)} \end{aligned} \qquad (21.11)$$

for $i, j, k, l = 1, 2, 3$

Various formulations are possible—from an independent σ_{ij}, Σ_{kl}, u_k form to the standard displacement approach in terms of u_k.

Example 21.3: Steady state heat conduction Here the one-dimensional problem—that of a lagged bar with longitudinal heat flow q with temperature at a cross-section T and distributed heat sources Q—can be described by the equations:

$$\begin{aligned} \frac{\mathrm{d}q}{\mathrm{d}x} - Q &= 0 \\ q + k\frac{\mathrm{d}T}{\mathrm{d}x} &= 0 \end{aligned} \qquad (21.12)$$

The above set (cf. Example 21.1) represents a 'mixed' form with variables:

$$\mathbf{\Phi}^T = [q, T] \qquad (21.13)$$

Simple elimination leads to the irreducible equation:

$$\frac{\mathrm{d}}{\mathrm{d}x}\left[k\frac{\mathrm{d}T}{\mathrm{d}x}\right] + Q = 0 \qquad (21.14)$$

The corresponding three-dimensional problem can obviously be similarly treated.

Example 21.4: Wave equations In wave-type problems, typified by the propagation of a shock through an elastic bar, it is often the practice to use the primitive problem variables, i.e. the velocity u and the stress σ. Now dynamic and continuity considerations give (on neglect of convective terms) the equation set:

$$\begin{aligned} \rho\frac{\partial u}{\partial t} + \frac{\partial \sigma}{\partial x} &= 0 \\ \frac{\partial u}{\partial x} - \frac{1}{E}\frac{\partial \sigma}{\partial t} &= 0 \end{aligned} \qquad (21.15)$$

This mixed form can result in one or other 'irreducible' sets, depending on which variable is chosen for elimination. Thus we can arrive at

$$\frac{\partial^2 u}{\partial x^2} = \frac{1}{c^2}\frac{\partial^2 u}{\partial t^2} \tag{21.16}$$

or at the equation:

$$\frac{\partial^2 \sigma}{\partial x^2} = \frac{1}{c^2}\frac{\partial^2 \sigma}{\partial t^2} \tag{21.17}$$

where $c^2 = E/\rho$, giving the well-known alternatives to the wave equation.

Again similar expressions can be written in the two- or three-dimensional cases.

21.3 SPECIAL MIXED AND IRREDUCIBLE PENALTY FORMS

In some sets of equations it appears possible to eliminate only a limited number of variables by using all the equations. For instance, in the case of one-dimensional elasticity given by equations (21.8) we found that σ could be readily eliminated, leading to the displacement form. But the elimination of u to give the formulation in terms of σ did not appear possible, as one of the equation contained only σ. Such problems are known as *constrained.* In this example the first, equilibrium, equation is acting as the 'constraint'.

Examples of this kind abound in many physical situations, a classic one being, for instance, that of incompressible elasticity. Here we write, in place of equations (21.11), the set:

$$\begin{aligned} \sigma_{ij} + b_i &= 0 \\ \sigma_{ij} &= 2\mu\varepsilon_{ij} + \delta_{ij}p \\ \varepsilon_{ij} &= \tfrac{1}{2}[u_{i,j} + u_{j,i}] \end{aligned} \tag{21.18}$$

and having introduced a new variable p (pressure), an additional (constraint) equation has to be added. This is simply the incompressibility condition:

$$\varepsilon_{ii} = 0 \tag{21.19}$$

In the above equation set apparently no further reduction is possible beyond that which retains u and p as variables, and the problem seems to have been contracted to an 'irreducible' set, written for brevity as:

$$\begin{aligned} L_{ij}u_j + p_{,i} + b_i &= 0 \\ u_{i,j} &= 0 \end{aligned} \qquad \text{for } i, j = 1, 2, 3 \tag{21.20}$$

or as

$$\begin{aligned} \mathbf{Lu} + \nabla p + \mathbf{b} &= 0 \\ \nabla^{\mathrm{T}}\mathbf{u} &= 0 \end{aligned}$$

Now, however, we can proceed by a penalty substitution [2] and write the second equation as

$$u_{i,i} = \frac{p}{\alpha} \tag{21.21}$$

where $\alpha \to \infty$.

With this substitution, p can be eliminated and a finally irreducible form is obtained, which can be written as

$$L_{ij}u_j + \alpha u_{k,ki} + b_i = 0 \tag{21.22}$$

This equation can now be used for the numerical solution of incompressibility problems, as has frequently been demonstrated [3, 4, 5]. We thus have to slightly modify the definition of 'mixed' forms, to be 'Those forms in which further reduction is impossible, even by the introduction of a penalty parameter'.

Indeed, the use of such a penalty parameter makes different forms of 'irreducible' formulations possible in otherwise familiar situations. For instance, if we consider the equations describing the one-dimensional elasticity problem in Example 21.2 (equation 21.8), it would appear that the only irreducible set is that known as the displacement form (equations 21.10). However, if we write equation (21.8) as the nearly equivalent set:

$$\frac{\mathrm{d}\sigma}{\mathrm{d}x} + b = \frac{u}{\alpha}$$

$$\sigma - E\frac{\mathrm{d}u}{\mathrm{d}x} = 0 \tag{21.23}$$

where $\alpha \to \infty$, then elimination of u is possible and we obtain

$$\sigma - E\alpha\frac{\mathrm{d}}{\mathrm{d}x}\left[\frac{\mathrm{d}\sigma}{\mathrm{d}x} + b\right] = 0 \tag{21.24}$$

This is a primary form which could be solved with stress as the single variable and approximates the so-called complementary energy forms. Such a formulation has in fact been used with success for the solution of two-dimensional elasticity problems by Taylor and Zienkiewicz [6].

21.4 MIXED VERSUS IRREDUCIBLE FORMS

In recent literature many papers can be found extolling the virtues of mixed as opposed to irreducible (displacement) approximations. To what extent are such claims justified?

It might appear, on the face of it, that mixed approximations would in general be more costly, requiring a larger number of free parameters to determine a given solution field. However, the latter proposition may not always be true if, as in the case of beams, plates, and shells, the irreducible form requires a higher order of continuity and hence more discretization parameters.

As with most such general statements, their truth may at times be only partial. If we examine the simple case of one-dimensional elasticity, defined in equations (21.8), we will find that the mixed formulation may yield inferior, identical, or indeed superior results, depending upon the type of trial functions **N** and weighting functions **W** used for each variable. This particular problem has been examined by Osborne [7] and later by Slater and Taylor [8], and below we discuss some of the possibilities of solution.

Let us consider for instance the general case, with the shape functions:

$$u = \mathbf{N}^u \mathbf{u} \qquad \text{and} \qquad \sigma = \mathbf{N}^\sigma \boldsymbol{\sigma} \tag{21.25}$$

where $\mathbf{u}$ and $\boldsymbol{\sigma}$ stand for the nodal values.

We can immediately write the general approximation to equations (21.8), using the typical weighted forms (equation 21.4) as

$$\begin{bmatrix} 0 & \mathbf{Q}_1 \\ \mathbf{Q}_2 & \mathbf{K} \end{bmatrix} \begin{bmatrix} \mathbf{u} \\ \boldsymbol{\sigma} \end{bmatrix} = \begin{bmatrix} \mathbf{f} \\ 0 \end{bmatrix} \tag{21.26}$$

where

$$\mathbf{K} = \int \mathbf{W}_2^T \mathbf{N}^\sigma \, \mathrm{d}x + \text{b.t.}$$

$$\mathbf{Q}_1 = \int \mathbf{W}_1^T \frac{\mathrm{d}}{\mathrm{d}x} [\mathbf{N}^\sigma] \, \mathrm{d}x + \text{b.t.}$$

$$\mathbf{Q}_2 = -\int E\mathbf{W}_2^T \frac{\mathrm{d}}{\mathrm{d}x} [\mathbf{N}^u] \, dx + \text{b.t.} \equiv \int E\left[\frac{\mathrm{d}}{\mathrm{d}x}[\mathbf{W}_2]\right]^T \mathbf{N}^u \, \mathrm{d}x + \text{b.t.}$$

b.t. standing for boundary terms not detailed here.

In the above the weighting functions $\mathbf{W}_1$ and $\mathbf{W}_2$ can be arbitrarily chosen, and we note that with $\mathbf{W}_1 = \mathbf{N}^u$ and $\mathbf{W}_2 = \mathbf{N}^\sigma$ the resulting form is non-symmetric (because of the multiplier E) and no variational theorem exists for the problem.

If the second of equations (21.8) is scaled before weighting, then the equation set becomes

$$\begin{bmatrix} 0 & \mathbf{Q}_3 \\ \mathbf{Q}_4 & \hat{\mathbf{K}} \end{bmatrix} \begin{bmatrix} \mathbf{u} \\ \boldsymbol{\sigma} \end{bmatrix} = \begin{bmatrix} \mathbf{f} \\ 0 \end{bmatrix} \tag{21.27}$$

where

$$\hat{\mathbf{K}} = \int \frac{1}{E} \mathbf{W}_2^T \mathbf{N}^\sigma \, dx + \text{b.t.}$$

$$\mathbf{Q}_3 = \int \mathbf{W}_1^T \frac{d}{dx} [\mathbf{N}^\sigma] \, du + \text{b.t.}$$

$$\mathbf{Q}_4 = -\int E \mathbf{w}_2^T \frac{d}{dx} \left[\mathbf{N}^v \right] dx + \text{b.t.} \equiv \int E \left[\frac{d}{dx} [\mathbf{w}_2] \right]^T \mathbf{N}^u \, dx + \text{b.t.}$$

and now if $\mathbf{W}_1 = \mathbf{N}^u$ and $\mathbf{W}_2 = \mathbf{N}^\sigma$, the equation set is symmetric.

Various solutions will now be considered for the example of the one-dimensional stretching under constant stress of a bar of constant cross-section but variable Young's modulus E (Figure 21.1A). The displacement form (equation 21.10) gives displacements which are on average correct (Figure 21.1A), and discontinuous stresses in the element which include the step change of E. Solution of equations 21.27 with σ piecewise constant and u linear results in displacements which are nodally exact (Figure 21.1B) and stresses which are everywhere exact. Thus the mixed form is better in this case both for displacement and stresses.

This form of mixed solution can be extended to all stress analysis and will show improvement over the displacement forms if E is variable or if other non-linearities, e.g. plasticity, exist. With independent, element-based interpolation for stresses elimination can be done at the element level leaving the final form of analysis similar to that derived from displacements alone. Such a formulation could therefore be termed one of 'generalized displacements'. In practice such methods can be cost-effective if the number of stress parameters within an element is not excessive, thus resulting only in small matrix inversions.

We note that a C^0 continuous choice for both $\mathbf{N}^u$ and $\mathbf{N}^\sigma$ will now give results which are decidedly inferior to those obtained using the displacement forms, while if piecewise constant $\mathbf{N}^u$ and linear $\mathbf{N}^\sigma$ are used, a different approximation again will be obtained. Some such approximations may indeed be an improvement on the irreducible form.

Obviously much depends upon the type of mixed formulation chosen and on the discretization made. It is here that variations between the different published procedures arise (usually distinguished by the different variational principles used) and at this stage the success or otherwise of the approximation will be decided.

We could now state generally, without detailed proof, that:

(a) If expansions *consistent* with an irreducible form are used in the mixed formulations (e.g. strain, or stress, functions being given as derivatives

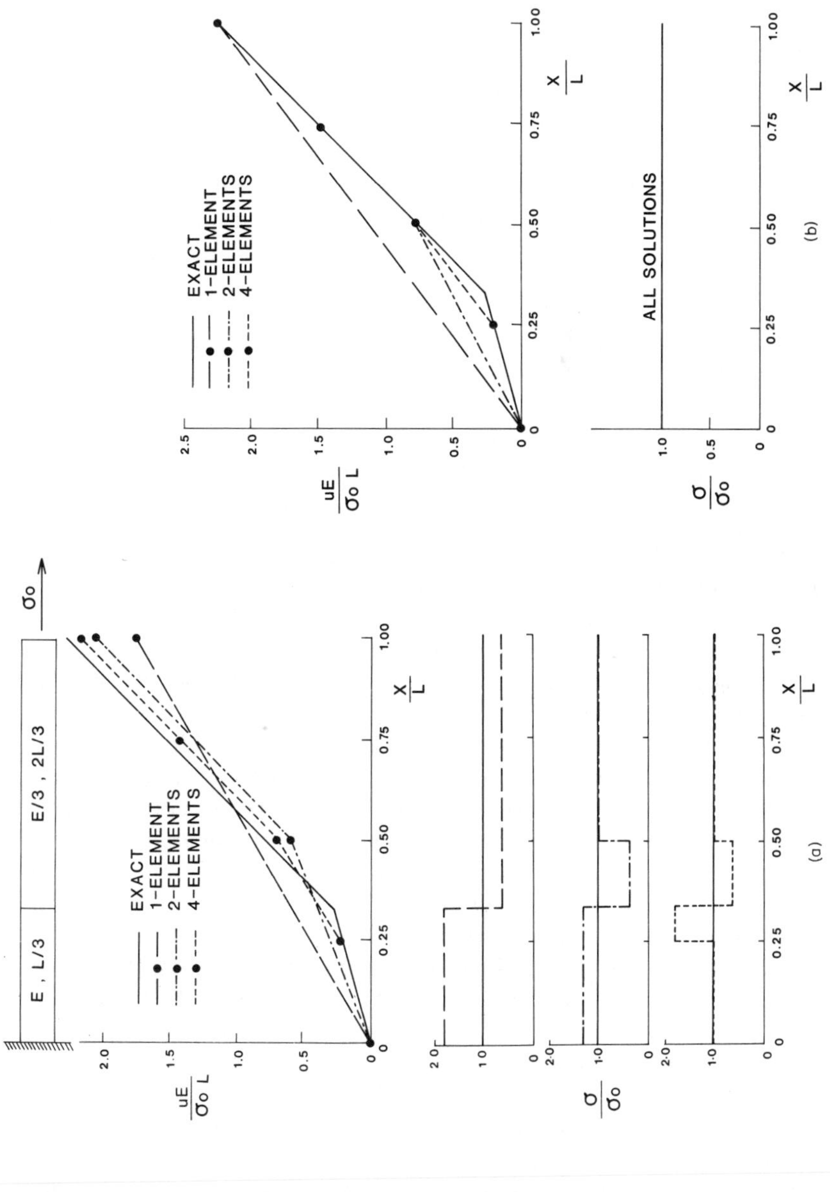

Figure 21.1 One-dimensional elasticity problem. (a) Results of displacement analysis. (b) Results of mixed analysis

of displacements), then the mixed formulation will be at least as good as that obtained by the irreducible form.

(b) If imposition of excessive continuity between elements is implied in the shape functions of the problem, then the approximation will suffer viz à viz the corresponding irreducible form—in which such continuity is not implied.

In the example of one-dimensional elasticity just quoted, such excessive continuity is generated by requiring the C^0 continuity of σ *and* u simultaneously.

An interesting example of such deterioration of performance has been observed recently in wave problems (such as that characterized by equation (21.15) or by the slightly more involved variant used for meteorological or shallow water wave problems:

$$\frac{\partial u}{\partial t}+U\frac{du}{dx}-fv+g\frac{\partial h}{\partial x}=0$$

$$\frac{\partial v}{\partial t}+U\frac{\partial v}{\partial x}+fu=0 \qquad (21.28)$$

$$\frac{\partial h}{\partial t}+U\frac{\partial h}{\partial x}+H\frac{\partial u}{\partial x}=0$$

In the above u, v, and h are the dependent variables, standing for velocity components and surface elevation, and U, H, f, and g are stream velocity, stream depth, and Coriolis and gravity accelerations respectively.

Here, the linearized form allows reduction to an irreducible problem, but computational convenience (evident in problems with non-linear convective terms) suggests a mixed treatment of the u, v, and h variables.

Attempts at using a linear, C^0, interpolation for all variables fail computationally, leading to inaccurate solutions [9], a phenomenon observed earlier in standard finite difference approximations. Through the use of C^0 interpolation for the velocities and piecewise constant approximation for the h variable much improved accuracy is obtained. This indeed is almost equivalent to the use of interweaving finite difference meshes for which such improved performance has been earlier observed [10].

In the solution of two-dimensional wave-type equations, which arise for example in problems of estuary and tidal flow, similar difficulties have been earlier observed in mixed formulations which use the complete variable set. This has led to the derivation of rather complex irreducible forms of equation, in the single h variable [11]. The proper use of a mixed formulation once again has been found to lead to excellent results [12]. In Figure 21.2 [9] we compare the results obtained in the steady state solution of equations (21.28), by the use of C^0/C^0 and C^0/discontinuous approximations.

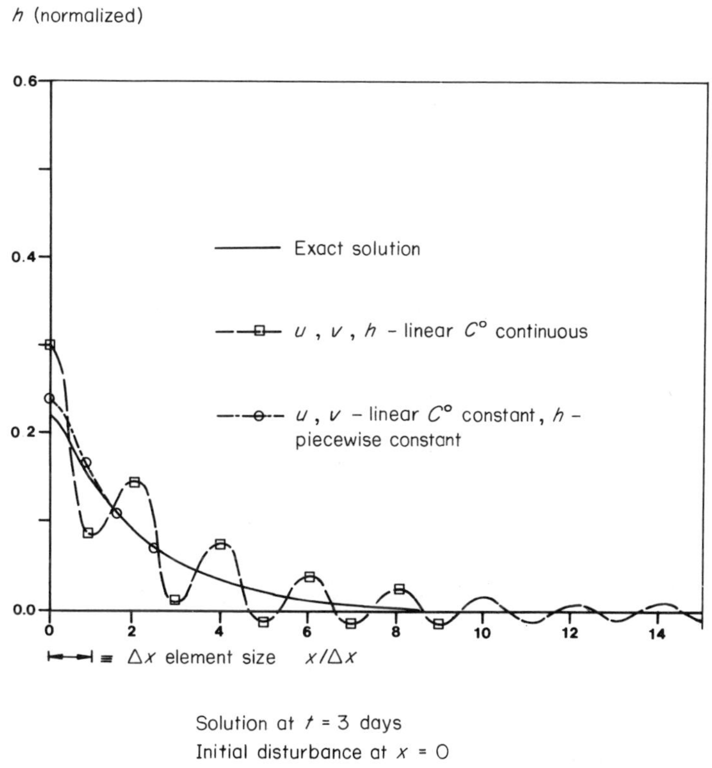

Figure 21.2 Problem of geostropic adjustment

At this stage the question may well be raised as to the advisability of the use of mixed forms which at best appear to give only comparatively small additional accuracy over the direct solution of the irreducible form. However, some points in favour of mixed forms have already been made:

(a) The mixed form may relax continuity requirements on interpolation (plates, shells, etc.).
(b) It frequently permits a more direct and efficient treatment of non-linearity.

Finally, and this will be referred to in the next section:

(c) It allows appropriate emphasis to be placed on individual variables, by varying the number of degrees of freedom attached to each in the expansions.

The second listed advantage in using the mixed form is well illustrated in a problem of non-linear pressure waves in a cavitating fluid, recently solved by Newton [13]. Here the notion of a displacement potential, ψ, is used to define the 'mass displacement' as

$$\psi_{,i} = \rho u_i \tag{21.29}$$

The 'mass displacement' divergence thus becomes

$$\nabla^2 \psi \equiv \psi_{,ii} = s \tag{21.30}$$

and the equation

$$\ddot{\psi} = p \tag{21.31}$$

gives the relationship between ψ and the pressure p.

Equations (21.30) and (21.31) are supplemented by a constitutive relationship:

$$p = p(s) \tag{21.32}$$

which suggests the possibility that treating the set of equations (21.30) to (21.32) as a mixed problem may be advantageous. For a linear relationship such as

$$p = c^2 s \tag{21.33}$$

(where c is the velocity of sound), the 'irreducible' form is simply obtained as

$$\nabla^2 \psi = \frac{1}{c^2} \ddot{\psi} \tag{21.34}$$

the standard and well-tested Helmholz equation. However, if the relationship is non-linear, elimination of the variable s leads to a complex equation system, while a mixed formulation permits an easy time-marching procedure to be developed. In this problem a simple collocation weighting is conveniently used on equation (21.31), and this procedure for discretization has been used successfully by Newton [13].

The third advantage of the mixed form can perhaps be illustrated by again referring to the example of the one-dimensional elasticity problem of equations (21.8). Here we can decide, *a priori*, that only linear terms will be required for the expansion of σ, as a nearly constant stress may occur. We can, however, use a larger number of parameters to describe u, which may vary abruptly if discontinuities in the value of the elastic modulus E are likely. Such a decision could be manifested by the use of different expressions for each variable within an element, and would reflect the importance of the individual equations.

On the debit side of the use of the mixed formulation we should add the possibility of the singularity of the approximation matrix—a phenomenon which sometime occurs.

General rules have not yet been formulated for the avoidance of such a singularity, and the investigation has to proceed cautiously.

21.5 EQUIVALENCE OF MIXED PENALTY FORM AND OF REDUCED INTEGRATION

Any general statement regarding the occasional equivalence of approximations made by mixed and irreducible formulations is a difficult one to prove. Intuitively it appears that if, in the mixed form, we specify expansions which are of identical form:

(a) to those describing corresponding variables in the irreducible form and
(b) to expansions which correspond to the variation (in the irreducible form) of the eliminated variable,

and further that if we use a consistent set of weighting functions, then equivalence of the two methods should result. Thus, for instance, in the particular case of approximation to the one-dimensional elasticity problem, we have indicated that the use of, say, a linear expansion for displacement u and a piecewise constant one for stresses σ leads to the equivalence (identity) of results when weighting functions were suitably chosen.

For the particular case of penalty formulations, the equivalence has been demonstrated by Malkus and Hughes [14], and due to very general interest in this class of problems we devote this section to its general discussion.

As a model we take the set of typical, penalty constrained equations written in a general form:

$$\begin{aligned}\mathbf{Lu}+\mathbf{C}p+\mathbf{b}&=0\\ \bar{\mathbf{C}}\mathbf{u}-\frac{p}{\alpha}&=0\end{aligned} \qquad (21.35)$$

With particular choice of operators L, C, and $\bar{C}$ the above equations become identical to equations (21.20) and (21.21), representing incompressible elasticity (or indeed incompressible viscous flow). With suitable meaning attached to the variables $\mathbf{u}$ and p and appropriate operators, the above set will also represent such problems as the bending of plates and shells, in which independent displacement and rotation expansions have been used, and indeed other problems.

In all of the above cases the operators C and $\bar{C}$ contain only first derivatives and therefore, in an approximation by trial functions of the form:

$$\hat{u}=\hat{\mathbf{N}}^u\mathbf{u} \quad \text{and} \quad \hat{p}=\mathbf{N}^p\mathbf{p} \qquad (21.36)$$

element discontinuous forms of $\hat{\mathbf{N}}^p$ are permissible. The irreducible form of equations (21.35) includes only consideration of variables u, and involves solution of the following equation set:

$$[\mathbf{L}+\mathbf{C}\alpha\bar{\mathbf{C}}]\mathbf{u}+\mathbf{b}=0 \qquad (21.37)$$

If a discretization of the mixed formulation of equation (21.35) is made using weighting functions $\mathbf{W}^u$ and $\mathbf{W}^p$ respectively we have a set of algebraic equations:

$$\begin{bmatrix} \mathbf{K} & \mathbf{Q} \\ \bar{\mathbf{Q}} & -\mathbf{H} \end{bmatrix} \begin{bmatrix} \mathbf{u} \\ p \end{bmatrix} + \begin{bmatrix} \mathbf{f} \\ 0 \end{bmatrix} = 0 \tag{21.38}$$

where

$$\begin{aligned} \mathbf{K} &= \int \mathbf{W}_u^T[\mathbf{L}\mathbf{N}^u]\,\mathrm{d}\Omega + \text{b.t.} \\ \mathbf{Q} &= \int \mathbf{W}_u^T[\mathbf{C}\mathbf{N}^p]\,\mathrm{d}\Omega + \text{b.t.} \\ \bar{\mathbf{Q}} &= \int \mathbf{W}_p^T[\bar{\mathbf{C}}\mathbf{N}^u]\,\mathrm{d}\Omega + \text{b.t.} \\ \mathbf{H} &= \frac{1}{\alpha}\int \mathbf{W}_p^T N^p\,\mathrm{d}\Omega + \text{b.t.} \\ \mathbf{f} &= \int \mathbf{W}_u^T \mathbf{b}\,\mathrm{d}\Omega \end{aligned} \tag{21.39}$$

On elimination of p via the second equation, the final algebric system becomes

$$[\mathbf{K} + \alpha\bar{\mathbf{K}}]\mathbf{u} + \mathbf{f} = 0 \tag{21.40}$$

where

$$\bar{\mathbf{K}} = [\mathbf{Q}\mathbf{H}^{-1}\bar{\mathbf{Q}}] \tag{21.41}$$

On the other hand, the discretization of the irreducible form (equation 21.37) by weighting with $\mathbf{W}^u$ results directly in

$$[\mathbf{K} + \alpha\bar{\bar{\mathbf{K}}}]\mathbf{u} + \mathbf{f} = 0 \tag{21.42}$$

where $\mathbf{K}$ and $\mathbf{f}$ are exactly as in the mixed form, but

$$\bar{\bar{\mathbf{K}}} = \int \mathbf{W}_u^T[\mathbf{C}\bar{\mathbf{C}}\mathbf{N}^u]\,\mathrm{d}\Omega \tag{21.43}$$

The two formulations (21.40) and (21.42) will yield identical results if the equality of $\bar{\mathbf{K}}$ and $\bar{\bar{\mathbf{K}}}$ is assured.

As shown by Malkus and Hughes [14] this equality will in many cases be assured,* providing that in the mixed form discontinuous p approximations are used so that the elimination can be achieved at the element level as in

* The equivalence is assured for linear plane stress or strain problems, for instance, but fails for axisymmetric situations. Here it appears that the mixed form gives superior results.

equations (21.40) and (21.41), and that suitable quadrature rules are used for the evaluation of the integrals in the irreducible form. Indeed, with so-called 'reduced' quadrature of matrix $\bar{\bar{K}}$ (corresponding to the least order of integration to ensure convergence), we find that the number of quadrature points corresponds precisely to the number of free parameters defining p at the element level.

Such reduced quadrature has over the years become increasingly popular [15, 16], and its success variously explained [17, 18].

21.6 DEGREE OF CONSTRAINT AND PERFORMANCE OF PENALTY FORMS

Whenever a constrained system of equations exists (e.g. the limiting case of equation 21.35 with $\alpha = \infty$) then every degree of freedom of p imposed on the algebraic equation tends to restrict the ability of the variables u to satisfy the first of equations (21.38). If then m stands for the number of parameters of p, and n for the number of parameters of u, we must have

$$m > n \tag{21.44}$$

if non-trivial (non-zero) answers are to be obtained from the approximation equations.

The ratio of the two numbers must hence be greater than 1, if meaningful results are to be obtained, that is

$$\beta = \frac{m}{n} > 1 \tag{21.45}$$

As the exact count of the degrees of freedom imposed on an assumed problem is tedious, we require a simple guide to the efficiency of elements in solving fully constrained problems (and hence penalty forms with large α values). Perhaps the best way to implement such a guide is to make a count of the changes Δm and Δn when an additional element is added to a problem field [19]. Here we take

$$\beta = \frac{\Delta m}{\Delta n} \tag{21.46}$$

and element efficiency can be assured by giving this ratio an optimal value which compromises between ensuring a sufficient number of degrees of freedom for $\mathbf{u}$ and imposing a sufficient number of constraints by means of $\mathbf{p}$.

In Figure 21.3 we compare the performances of several well-known elements and define the number of reduced integration points or equivalent degrees of freedom of p, for the problem of incompressible behaviour of an elastic solid. (Performances in incompressible fluid mechanics are similar.)

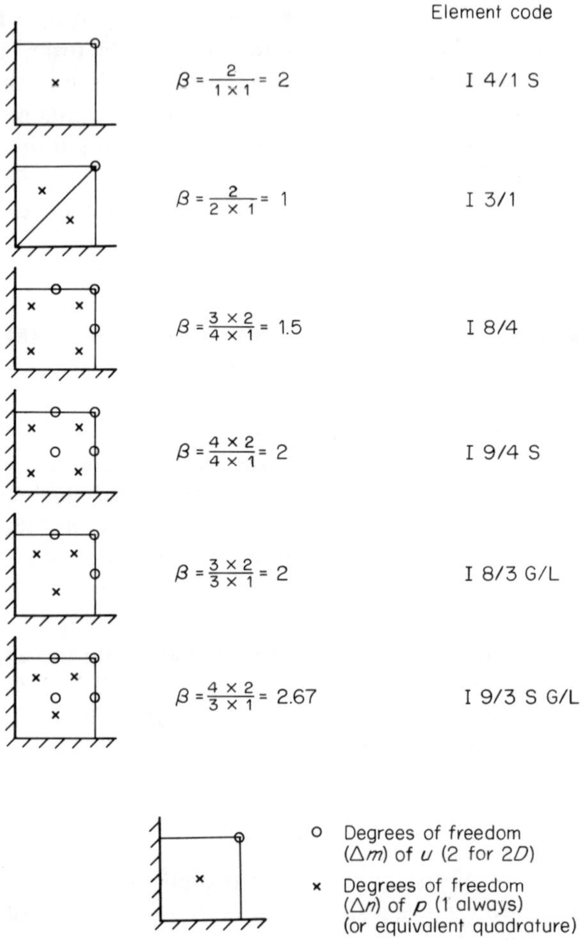

Figure 21.3 The freedom (β) numbers of various elements when incompressible analysis is carried out using one of the mixed u/p formulations or equivalent 'reduced' integration

The first of these elements is the two-dimensional Herrmann triangle [20], well known to be very deficient in performance. The second is the isoparametric bilinear quadrilateral derived by Hughes, Taylor, and Levy [4]. The third is the serendipity quadratic element, which was first used in incompressible situations by Naylor [21]. The fourth is the Lagrangian-type quadratic element used with success in many (flow-type) metal-forming situations [22, 23]. The final two are obvious developments of the serendipity and Lagrangian quadratic elements, in which a local linear p variation is assumed. The first suggestion of such an element is implied in the work of

Nagtegaal, Parks, and Rice [24], but it does not appear to have been implemented so far in actual computation.

For this reason (and indeed because some recent mathematics has suggested that certain of the elements devised will not work [25, 26]), we outline their performances in a few examples, and look at the evaluation of the velocity field as well as that of the pressures.

The elements in Figure 21.3 are given code numbers, such as I 9/4 S, which indicates an element which is isoparametric (I), nine noded, and has four integration points (or pressure parameters). S in the above code means that selective integration needs to be used to avoid singularity of the (overall) K matrix, which could result from the use of four quadrature points in the integration of the matrix K of equation (21.42), for example.

For the elements to which mixed p interpolation is applied directly as in equation (21.40), there are two possible ways of expanding the pressure variable—either in terms of global coordinates x and y (two-dimensional) or in terms of local ones Σ and η. Hence the additional coding 'G' or 'L'. At this stage we can state without giving results that G-type elements give a slightly better performance when the mesh is irregular, and this element will be used henceforth here. (Naturally, for rectangular grids, no differences are evident between the performances of G or L-type elements).

We now investigate the performances of the four elements—I 8/4, I 8/3 G, I 9/4 S, and I 9/3 G—in various problems involving incompressible Newtonian flow.

Test 21.1 The first, almost trivial example, is that of uniform fully developed plane Newtonian flow, with a regular mesh (see Figure 21.4a). All boundary velocities are specified, the fully developed velocity profile giving parabolic variation of axial velocity across the section and zero non-axial velocity. The velocity fields for all the elements are similar to that shown in Figure 21.4(b), with no significant non-axial velocity being evident.

Various pressure fields are shown in the figure. That for the I 8/4 element (Figure 21.4d) exhibits 'overconstrained' behaviour, while those for the other elements all reproduce the expected linear pressure variation in the direction of flow (Figure 21.4c).

Test 21.2 This example (Figure 21.5) is again that of uniform fully developed plane Newtonian flow, but here the mesh consists of irregularly shaped elements. Boundary conditions are the same as in Test 21.1.

The use of an irregular mesh dramatically affects the results. The velocity field for the I 8/4 element (Figure 21.5b) is significantly in error, while those for the I 8/3 G, I 9/4 S, and I 9/3 G elements are similar to that shown in Figure 21.4(b).

The pressure field for the I 8/4 element is highly oscillatory, with pressure within individual elements varying between plus and minus 40 Δp (where Δp

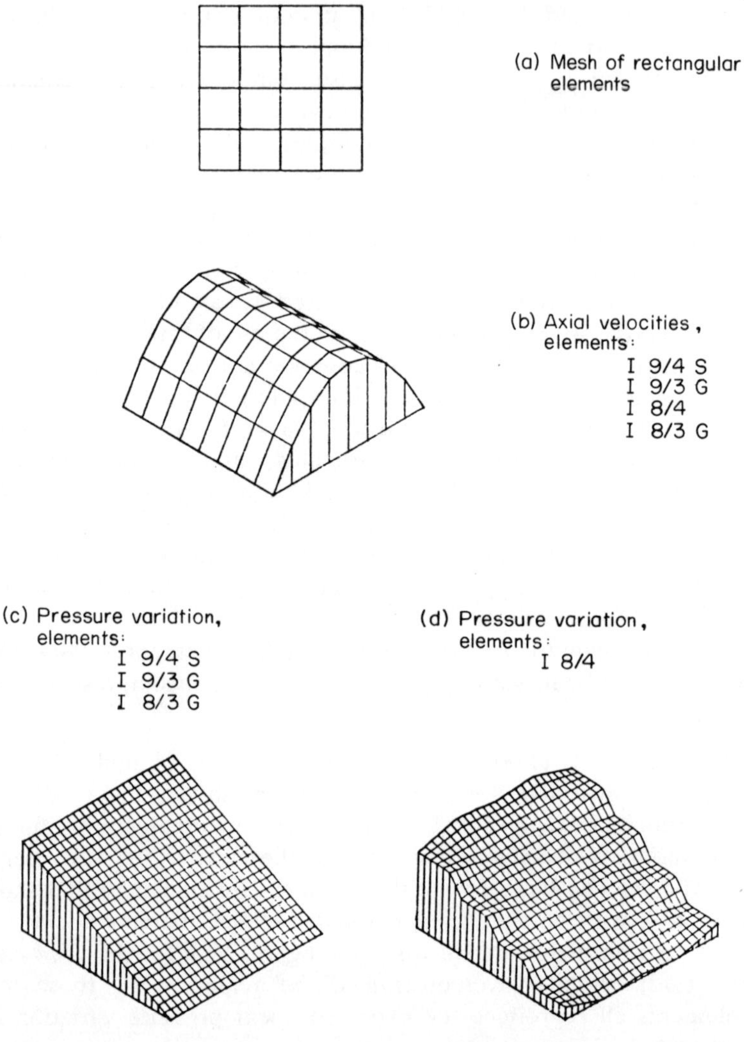

Figure 21.4 Test 1: uniform fully developed plane Newtonian flow, with mesh of rectangular elements

is the total expected pressure drop in the mesh); this pressure field is not shown in the figure. The field for the I 8/3 G element (Figure 21.5c) shows a marked improvement over that for the I 8/4 element. The I 9/4 S and I 9/3 G elements both give pressure fields similar to that shown in Figure 21.4(c).

Test 21.3 This example is that of steady plane Newtonian flow (at a low Reynolds number) through a constriction, using a mesh of rectangular

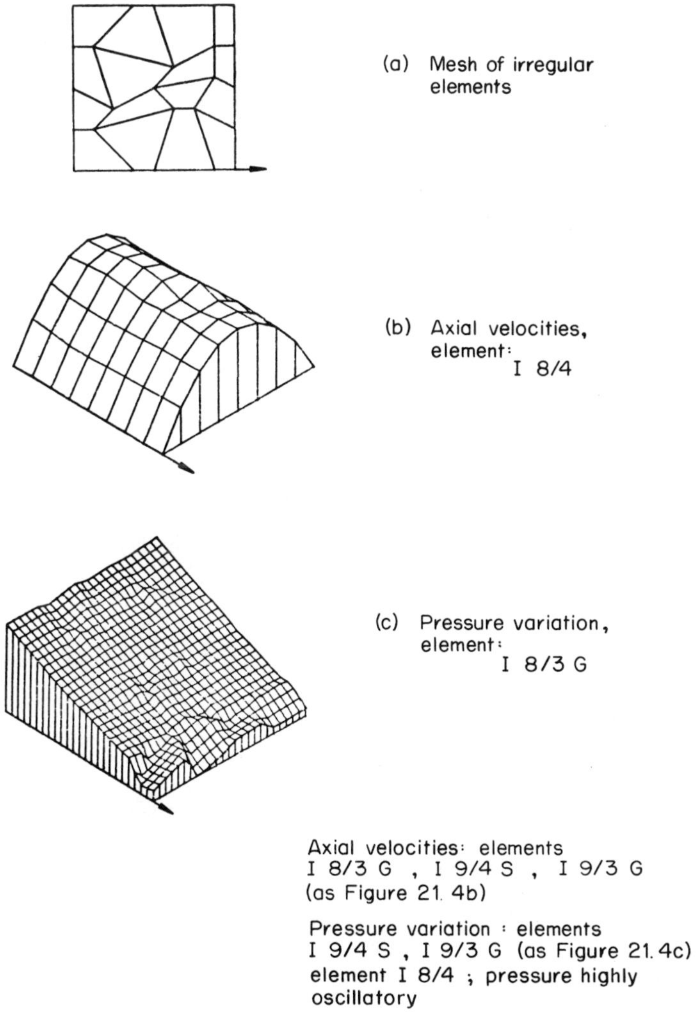

Figure 21.5 Test 2: uniform fully developed plane Newtonian flow, with mesh of irregular elements

elements (Figure 21.6a). Boundary velocities at the inlet are assumed to be fully developed, and at the outlet the condition of zero normal traction is applied, while non-axial velocity is specified to be zero.

The pressure field for the I 8/4 element is highly oscillatory and is not shown in the figure. That for the I 8/3 G element (Figure 21.6b) is again a marked improvement over that for the I 8/4 element. Also shown are the fields for the I 9/4 S (Figure 21.6c) and I 9/3 G (Figure 21.6d) elements, the former producing slight oscillations, the latter being relatively well behaved.

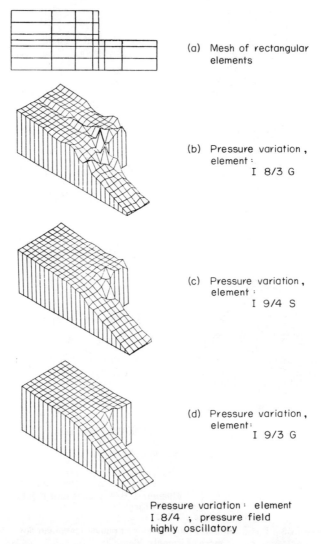

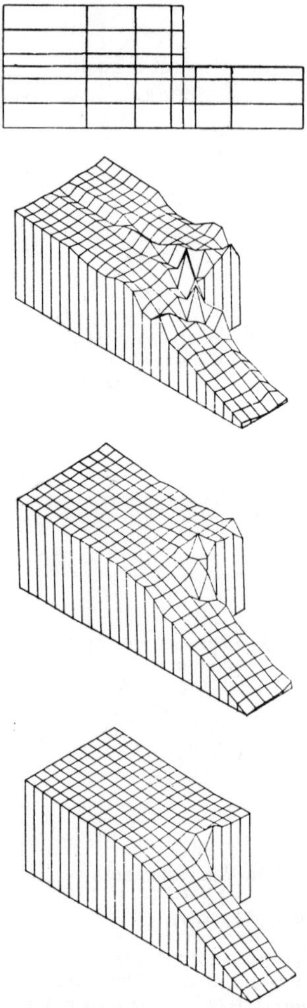

(a) Mesh of rectangular elements

(b) Pressure variation, element: I 8/3 G

(c) Pressure variation, element: I 9/4 S

(d) Pressure variation, element: I 9/3 G

Pressure variation: element I 8/4 ; pressure field highly oscillatory

Figure 21.6 Test 3: steady plane, low Reynolds number flow through a constriction

The velocity fields (Figure 21.6e and f) obtained with the I 8/3 G, I 9/4 S, and I 9/3 G elements are all similar, while that for the element I 8/4 is not smooth (Figure 21.6g and h).

Test 21.4 This example is the same as that of Test 21.3, except that here the mesh includes non-rectangular elements near the corner of the constriction (Figure 21.7a).

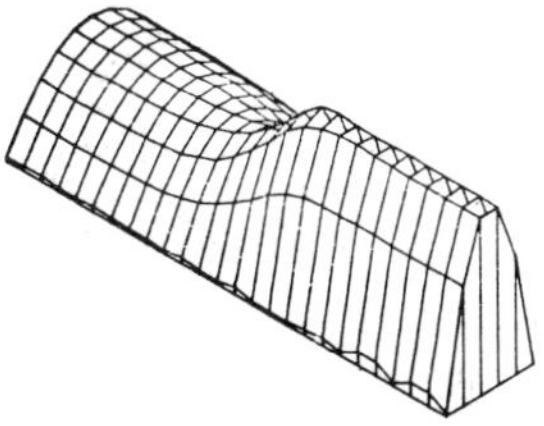

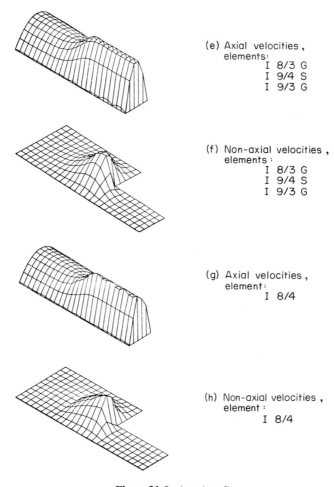

(e) Axial velocities, elements:
I 8/3 G
I 9/4 S
I 9/3 G

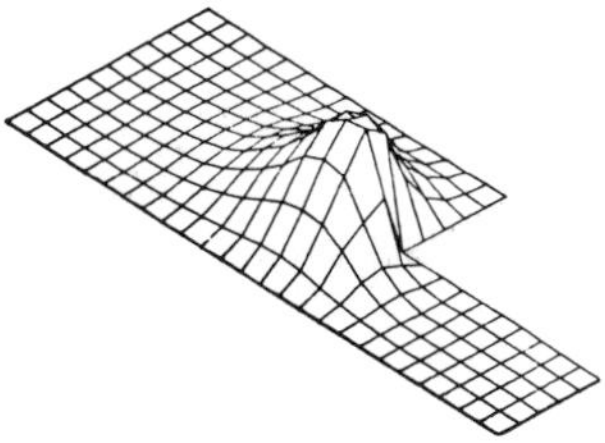

(f) Non-axial velocities, elements:
I 8/3 G
I 9/4 S
I 9/3 G

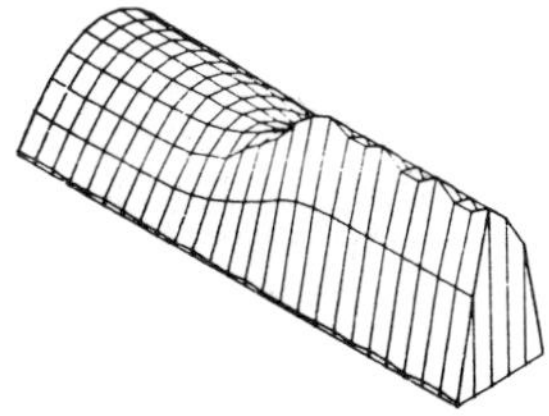

(g) Axial velocities, element:
I 8/4

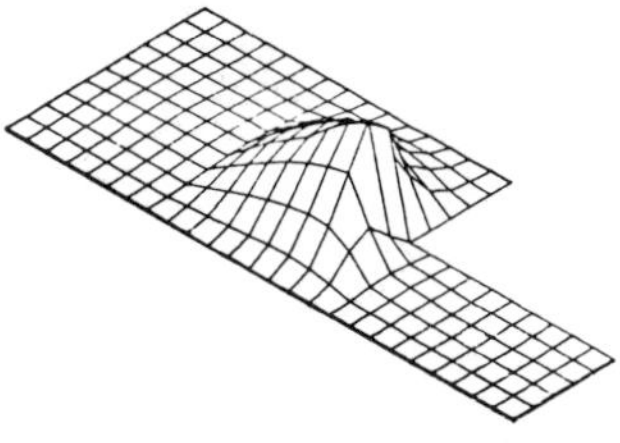

(h) Non-axial velocities, element:
I 8/4

Figure 21.6 (*continued*)

The pressure field for the I 8/4 element is again highly oscillatory, and is not shown. The field for the I 8/3 G element (Figure 21.7b) is slightly more oscillatory than that for the same element in Test 21.3 (Figure 21.6b). Fields for the I 9/4 S and I 9/3 G elements are shown, the former (Figure 21.7c) being more oscillatory than that for the same element in Test 21.3 (Figure 21.6c). The latter (Figure 21.7d) is marginally better than that for the same element in Test 21.3 (Figure 21.6d).

Velocity fields for the I 8/4 and I 8/3 G elements are shown in Figure 21.7. The former (Figure 21.7e and f) gives a non-smooth and slightly oscillatory velocity field. The latter (Figure 21.7g and h) shows some non-smoothness, particularly for the non-axial velocities.

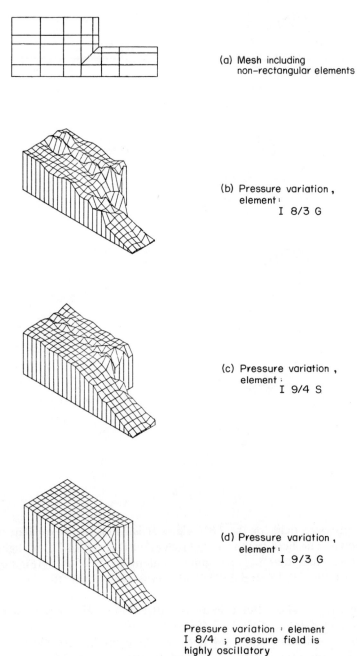

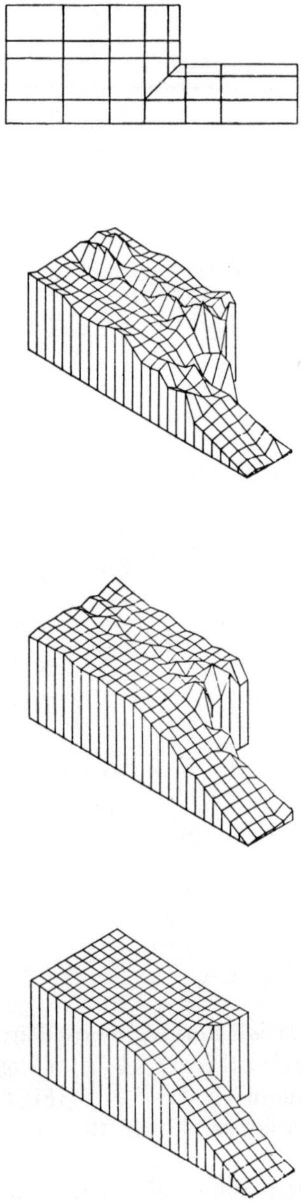

Figure 21.7 Test 4: steady plane, low Reynolds number, Newtonian flow through a constriction

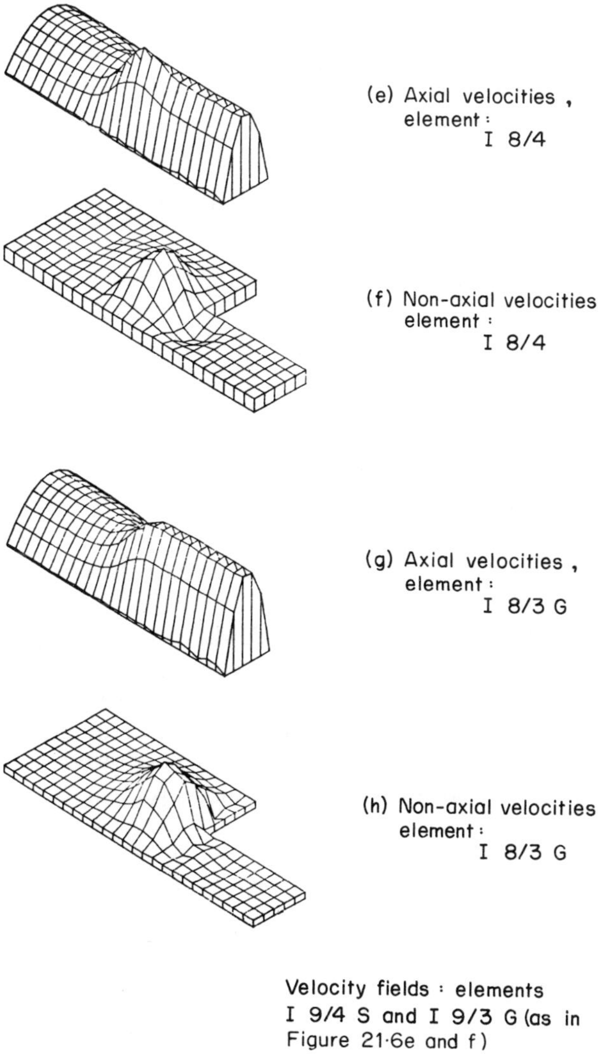

Figure 21.7 (*continued*)

21.6.1 Comments on results

The tests indicate that the I 8/4 element is overconstrained, producing inaccurate or oscillatory pressure fields, and possibly even inaccurate velocity fields, in particular when non-rectangular elements are used.

The I 8/3 G elements, however, gives vastly improved results. Pressure fields are now very good when elements are rectangular and the velocity field is uniform, but tend to deteriorate significantly otherwise. Velocity fields are markedly improved, although in the case where non-rectangular elements are used *and* the velocity field is non-uniform, the field may become slightly oscillatory

Both the I 9/4 S and the I 9/3 G element give very good velocity fields in the tests imposed here.

The pressure fields obtained with the I 9/4 S element are excellent when the velocity field is uniform. In cases of non-uniform velocity fields, the pressures become non-smooth, particularly so when non-rectangular elements are used.

The I 9/3 G element gives excellent pressure fields when velocities are uniform. In other cases the pressure fields are very good. In regions of rapid pressure variation local mesh design (naturally) is important, and use of non-rectangular elements to produce a better mesh has here *improved* the pressure field.

21.7 CONCLUDING REMARKS

In the preceding pages we have outlined what we believe is a reasonable definition of 'mixed' and 'irreducible' procedures. Further we have demonstrated some merits of mixed formulations in various solutions, including the much discussed penalty forms. Many problems still remain, to which attention should be given if the behaviour of mixed processes is to be fully exploited. Here we could list such outstanding problems as:

(a) that of ensuring non-singularity of a numerical formulation;
(b) the determination of suitable patch tests for mixed finite elements, and the study of their convergence;
(c) the conditions for equivalence with irreducible forms including such general areas as reduced integration, etc.

In this paper attention was focused on the 'constraint' imposed in penalty forms for incompressible formulations, with a suggestion that the 'freedom number β' (equation 21.46) should be greater than unity and that the performance of elements would be optimal at a certain predetermined value of this number. It has been stated earlier that for the two-dimensional situation this number should be 2 (corresponding to the ratio of the velocity to pressure variables), but numerical experiments show that in fact the best element was one where this ratio was exceeded. Again this can be considered as a problem area in which further research is needed.

With regard to definitions, we have only specified the differences between mixed and irreducible forms without introducing definitions of so-called hybrid methods or more general coupled problems. Perhaps we should state these briefly.

Coupled problems are mixed forms in which the domains of the governing equations differ (or may differ). Indeed the coupling often occurs only through an interface between two domains—e.g. problems of structure—fluid interaction. Moreover, in some problems it arises in an intersection of two domains—e.g. problems of soil mechanics where pore pressure of the fluid may be non-zero only in part of the total solid skeleton.

At time the two domains can overlap completely, as for instance in thermally coupled strain analysis, and here the differences between mixed and coupled forms are less clear.

Coupling of the first kind, in which interfaces only need to be considered, has led to many techniques for numerical formulation. Occasionally these involve the introduction of additional variables (Lagrangian multipliers) which are defined elsewhere.

This leads to the definition of *hybrid processes.* These are problems of the single-field type in which subdivision into separate domains is made and a solution attempted by imposing coupling conditions on interfaces.

Such a definition leads to the various forms introduced early in the finite element game by Pian [27, 28] (see also Chap. 12 of [15]), and which have in many cases proved efficient in practice.

REFERENCES

1. S. Nemat-Nasser and K. N. Lee, 'Finite element formulations for elastic plates by general variational statements with discontinuous fields', *Comp. Meth. Appl. Mech. Eng.*, **2,** 33–41 (1973).
2. O. C. Zienkiewicz, 'Constrained variational principles and penalty analysis function methods in finite elements', *Conf. Numerical Solution of Differential Equations,* Dundee, 1973; Lecture Notes on Mathematics, Springer, 1973.
3. O. C. Zienkiewicz and P. N. Godbole, 'Viscous incompressible flow with special reference to non Newtonian (plastic) flow', in *Finite Elements in Fluids* (Eds. J. T. Oden, O. C. Zienkiewicz, R. H. Gallagher, and C. Taylor), Vol. 1, Chap. 2, pp. 25–71, John Wiley and Sons, 1975.
4. T. J. R. Hughes, R. L. Taylor, and J. F. Levy, 'A finite element method for incompressible viscous flows', *Proc. Second Int. Symp. on Finite Elements in Fluid Problems ICCAD,* pp. 1–16, St. Margharita Ligure, Italy, 1976.
5. O. C. Zienkiewicz and P. N. Godbole, 'Penalty function approach to problems of plastic flow of metals with large surface deformations', *J. Strain Analysis,* **10,** 180–183 (1975).
6. R. L. Taylor and O. C. Zienkiewicz, 'Complementary energy with penalty functions in finite element analysis', in *Energy Methods in Finite Element*

Analysis (Eds. R. Glowinski, E. Y. Rodin, and O. C. Zienkiewicz), Chap. 8, John Wiley and Sons, 1979.
7. J. Osborne, Private communication, University of Maryland, 1980.
8. J. M. Slater and R. L. Taylor, 'Mixed model finite element analysis of inelastic plane frames', Report CE299, Department of Civil Engineering, University of California, Berkeley, 1980.
9. A. Schoenstadt, 'A transfer function analysis of numerical schemes used to simulate geostropic adjustment', *Monthly Weather Rev.*, **1980,** 108 (1980).
10. R. T. Williams and O. C. Zienkiewicz, 'Improved finite element forms for the shallow water equations', *Int. J. Num. Meth. Flow Problems* (to be published 1981).
11. D. R. Lynch and W. G. Gray, 'A wave equation model for finite element tidal computations', *Comp. Fluids*, **7,** 207–228 (1979).
12. S. Nakazawa, Private communication, University of Wales, Swansea, 1981.
13. R. E. Newton, 'Finite element study of induced cavitation', Preprint 80–110, ASCE Convention, Oregon, 14–18 April 1980.
14. D. S. Malkus and T. J. R. Hughes, 'Mixed finite element methods: Reduced and selective integration techniques: A unification of concepts', *Comp. Meth. in Appl. Mech. Eng.*, **15,** 63–81 (1978).
15. O. C. Zienkiewicz, *The Finite Element Method*, 3rd ed., Chap. 11, McGraw-Hill, 1977.
16. O. C. Zienkiewicz, R. L. Taylor, and J. M. Too, 'Reduced integration techniques in general analysis of plates and shells', *Int. J. Num. Meth. in Eng.*, **3,** 275–290 (1971).
17. S. F. Pawsey and R. W. Clough, 'Improved numerical integration of thick shell finite elements', *Int. J. Num. Meth. in Eng.*, **3,** 545–586 (1971).
18. G. J. Fix, 'On the effect of quadrature errors in the finite element method', in *Advances in Computational Methods in Structural Mechanics and Design* (Eds. J. T. Oden, R. W. Clough, and Y. Yamamoto), pp. 55–68, University of Alabama Press, 1972. (Also see pp. 25–56, *The Mathematical Foundations of the Finite Element Method with Applications to Differential Equations* (Ed. A. K. Aziz), Academic Press, 1972.).
19. O. C. Zienkiewicz and E. Hinton, 'Reduced integration, function smoothing and non-conformity in finite element analysis', *J. Franklin Inst.*, **302,** 443–461 (1976).
20. L. R. Herrman, 'Elasticity equations for incompressible, or nearly incompressible materials by a variational theorem', *AIAA J.*, **3,** 1896 (1965).
21. D. J. Naylor, 'Stresses in nearly incompressible materials for finite elements with application to the calculation of excess pore pressures', *Int. J. Num. Meth. in Eng.*, **8,** 443–460 (1974).
22. O. C. Zienkiewicz, P. C. Jain, and E. Onate, 'Flow of solids during forming and extrusion: Some aspects of numerical solutions', *Int. J. Solids and Structures,* **14,** 15–38 (1978).
23. O. C. Zienkiewicz, E. Onate, and J. C. Heinrich, 'Applications of numerical methods to forming processes', Winter Annual Meeting of ASME, San Francisco, California, 1978.
24. J. C. Nagtegaal, D. M. Parks, and J. R. Rice, 'On numerically accurate finite element solutions in the fully plastic range', *Comp. Meth. Appl. Mech. Eng.*, **4,** 153–178 (1974).
25. J. T. Oden, 'R.I.P. Method for Stokesian flows', *Proc. Third Int. Conf. Finite Elements in Flow Problems*, Vol. 2, Banff, Alberta, Canada, 10–13 June 1980.

26. J. T. Oden, 'R.I.P. method for Stokesian flows', TICOM Report 80-11, University of Texas, Austin, August 1980.
27. T. H. H. Pian, 'Derivation of element stiffness matrices by assumed stress distributions', *AIAA J.*, **2,** 1333–1335 (1964).
28. T. H. H. Pian, 'Hybrid Models', in *Numerical and Computer Methods in Applied Mechanics* (Eds. S. J. Fenves *et al.*), Academic Press, 1971.

Hybrid and Mixed Finite Element Methods
Edited by S. N. Atluri, R. H. Gallagher, and O. C. Zienkiewicz

Chapter 22

Selection of Finite Element Methods

D. N. Arnold, I. Babuška, and J. Osborn

22.1 INTRODUCTION

The goal of engineering computations is to obtain quantitative information about engineering problems. This goal is usually achieved by the approximate solution of a mathematically formulated problem. Although a relevant mathematical formulation of the problem and its approximate solution are closely related (see, for example, [1, 2]), here we shall suppose that a mathematical formulation has already been determined and is amenable to an approximate treatment. We shall discuss a broad class of approaches based on variational methods of discretization which allow one to find the approximate solution within a desired range of accuracy.

Let H denote the linear space of possible solutions and $u \in H$ the exact solution of the problem. A (*linear, consistent*) *variational method of discretization* consists of a finite dimensional linear subspace $S \subset H$ called the *trial space* in which the approximate solution is sought, a *test space* V (of the same dimension as the trial space S), and a *bilinear form* $B(u, v)$ defined on $H \times V$. The approximate solution, denoted by Pu, is then determined by the conditions

$$Pu \in S \tag{22.1a}$$

$$B(Pu, v) = B(u, v) \qquad \text{for all } v \in V \tag{22.1b}$$

In order that Pu be computable, the following two conditions should be satisfied:

For any $v \in V$, $B(u, v)$ is computable from the data of the problem (without knowing u). (22.2a)

For any $s \in S$ there is some $v \in V$ such that $B(s, v) \neq 0$. (22.2b)

It follows that (22.1) leads to a system of linear equations which is uniquely solvable. It is obvious that $Pu = u$ for any $u \in S$. The approximate solution Pu obviously depends on the selection of S, V, and B.

The acceptability of the approximate solution is stated in terms of a norm $\|\cdot\|$ of the difference of u and Pu, i.e. we accept Pu if

$$\|Pu - u\| \leq \tau \, \|u\| \tag{22.3}$$

where τ is an *a priori* given tolerance. (An absolute error criterion or other variant is equally possible.) Thus, given $\|\cdot\|$ and τ, the goal is to select S, V, and B so that (22.3) is achieved in the most effective way. (We do not give here an exact meaning to the word 'effective'.)

For each mathematical problem *there exists a wide variety of possible variational methods of discretization.* In this paper we shall discuss properties of these methods which enable us to distinguish among them and which aid in the selection or design of a method which is effective in achieving the given goals of the computation. In Section 22.2 some general considerations are discussed. The remainder of the paper is devoted to specific illustrative results. We conclude this introduction with a very simple example in terms of which some of the main ideas will be explained.

Let us consider a longitudinally loaded bar on an elastic support. For $0<x<l$ denote by $u(x)$ and $\sigma(x)$ the longitudinal displacement and normal stress respectively. A classical formulation consists of the boundary value problem

$$E(x)u'(x)=\sigma(x) \tag{22.4a}$$

$$-(F(x)\sigma(x))'+c(x)u(x)=p(x) \tag{22.4b}$$

$$u(0)=0, \qquad u(l)=0 \tag{22.4c}$$

Here $E(x)$ denotes the modulus of elasticity, $F(x)$ the cross-section, $c(x)$ the spring constant of the elastic support, and $p(x)$ the longitudinal load. We can cast this problem in a variational form in various ways. For example, define the bilinear form

$$B_1(u, \sigma; v, \rho)=\int_0^l (Eu'\rho-\sigma\rho+F\sigma v'+cuv)\,\mathrm{d}x \tag{22.5}$$

Then

$$B_1(u, \sigma; v, \rho)=\int_0^l pv\,\mathrm{d}x \tag{22.6}$$

for any $v\in \mathring{H}^1$ and $\rho\in H^0$, where

$$\mathring{H}^1=\left\{v\,\middle|\,\int_0^l [v^2+(v')^2]\,\mathrm{d}x<\infty,\ v(0)=v(l)=0\right\}$$

and

$$H^0=\left\{\rho\,\middle|\,\int_0^l \rho^2\,\mathrm{d}x<\infty\right\}$$

and so $B_1(u, \sigma; v, \rho)$ is computable without explicitly knowing the exact solution (u, σ). Note that $B_1(u, \sigma; v, \rho)$ is defined for all $(u, \sigma)\in \mathring{H}^1\times H^0=H$ and all $(v, \rho)\in H$.

There are many other bilinear forms which could be used in a variational formulation of (22.4). For example, let

$$B_2(u, \sigma; v, \rho) = \int_0^l \left(u'\rho - \frac{\sigma\rho}{E} + F\sigma v' + cuv \right) \mathrm{d}x \tag{22.7}$$

Then

$$B_2(u, \sigma; v, \rho) = \int_0^l pv \, \mathrm{d}x$$

for (u, σ), $(v, \rho) \in H$ (with H defined as above). Integrating by parts in (22.7) we get

$$B_3(u, \sigma; v, \rho) = \int_0^l \left[-u\rho' - \frac{\sigma\rho}{E} - (F\sigma)'v + cuv \right] \mathrm{d}x \tag{22.8}$$

so

$$B_3(u, \sigma, v, \rho) = \int_0^l pv \, \mathrm{d}x$$

Here we assume $u, \ v \in H^0$ and

$$\sigma, \rho \in H^1 = \left\{ v \,\middle|\, \int_0^l [v^2 + (v')^2] \, \mathrm{d}x < \infty \right\}$$

In both cases the bilinear form is obviously computable from data. For a final example set

$$B_4(u, v) = \int_0^l (EFu'v' + cuv) \, \mathrm{d}x \tag{22.9}$$

By eliminating σ from (22.4) we see that

$$B_4(u, v) = \int_0^l pv \, \mathrm{d}x$$

for $u, v \in \mathring{H}^1$. This is the usual form used in displacement finite element methods. Of course many other variational formulations of (22.4) are possible (in fact an infinite number).

To complete the specification of a discretization method we must select, in addition to the bilinear form, the trial and test spaces S and V. For example, in the case of B_1 we may select any finite dimensional subspaces S and V of H which are of the same dimension and satisfy (22.2b).

Let us now define some of the norms which we will consider. For k a non-negative integer set $u^{[k]} = \dfrac{d^k u}{dx^k}$ and let

$$\|u\|_{H^k} = \left[\int_0^l \sum_{j=0}^{k} u^{[j]2}(x)\,dx\right]^{1/2}$$

$$\|u\|_{H^k} = \sum_{j=0}^{k} \operatorname{ess}\sup_{0\leq x\leq l} |u^{[j]}(x)|$$

We also define analogous norms for k a negative integer. For such k define $v = u^{[k]}$ by $u(x) = v^{[-k]}(x)$, choosing the constants of integration so that $\int_0^l v^{[2m]}\,dx = 0$, $0 \leq m \leq (-k-1)/2$; $v^{[2m+1]}(0) = v^{[2m+1]}(l) = 0$, $0 \leq m \leq (-k-2)/2$. Then we will write

$$\|u\|_{H^k} = \|v\|_{H^0}.$$

These negatively indexed norms emphasize the effect of oscillations of u less than do the positively indexed norms. Analogous norms can be defined when u depends on two variables.

22.2 GENERAL CONSIDERATIONS

To achieve the acceptance criterion (22.3) it is certainly necessary that

$$Z(u, S) = \inf_{s\in S} \|u - s\| \leq \tau \|u\| \tag{22.10}$$

The quantity $Z(u, S)$ measures the error in best possible approximation of u by elements of S with respect to the chosen norm $\|\cdot\|$, i.e. the *approximability* of u by S.

The choice of S is clearly essential to the effectiveness of the discretization method. The solution u is unknown *a priori*, and often only the information that $u \in H$ is available. In such cases S has to be selected so that *every* element in H can be approximated well. More information about u allows more effective choice of S. Such information can be achieved through a learning process during the computation and thus S can be selected adaptively (see, for example, [3, 4]).

That the trial functions approximate the solution well, i.e. that the magnitude of $Z(u, S)$ is small, does not alone insure that the approximate solution Pu is close to the exact solution u. Therefore it is reasonable to ask that the method be *quasi-optimal*. This means that

$$\|u - Pu\| \leq KZ(u, S) = K \inf_{s\in S} \|u - s\| \qquad \text{for all } u \in H \tag{22.11}$$

where K is a constant which is not too large. The smallest value of K for which (22.11) holds is called the *quasi-optimality constant*.

Condition (22.11) is equivalent to another condition called the *stability condition.* This states that

$$\|Pu\| \leqslant K^* \|u\| \qquad \text{for all } u \in H \tag{22.12}$$

The smallest value of K^* for which (22.12) holds is called the *stability constant.* To see that quasi-optimality and stability are equivalent, assume that (22.12) holds. Now, if $s \in S$, then

$$\begin{aligned}\|u - Pu\| &= \|(u-s) - P(u-s)\| \leqslant \|u-s\| + \|P(u-s)\| \\ &\leqslant \|u-s\| + K^* \|u-s\| \leqslant (K^*+1) \|u-s\|\end{aligned}$$

(Here we used the fact that $Ps = s$, as mentioned earlier.) Thus (22.11) holds with $K \leqslant K^* + 1$. On the other hand, assuming (22.11), we have that

$$\|Pu\| \leqslant \|Pu - u\| + \|u\| \leqslant KZ(u, S) + \|u\| \leqslant (K+1) \|u\|$$

and so (22.12) holds with $K^* \leqslant K+1$. The importance of (22.12) is that it is often easier to verify than (22.11).

Note that while approximability is affected only by the choice of the trial space S, stability (or quasi-optimality) depends on the interplay between B, H, S, and V. Because the test functions are not needed for approximation purposes, *the main goal in the selection of V is to achieve stability* with the smallest possible constant K. Let us remark that for certain bilinear forms and certain norms, the choice $V = S$ leads to the stability constant 1. In such cases, the performance of the method depends solely on the selection of S.

Note, further, that both approximability and stability depend heavily on the norm under consideration. Changing the norm can violate quasi-optimality although the computational algorithm remains the same. Because of this, the method must be investigated in close relation to the given acceptance criterion.

Although approximability and stability are essential and of primary interest for the method, there are other important features to be considered in the rational selection of discretization procedures.

22.2.1 Robustness

The bilinear form B and the solution u may depend on various parameters, e.g. in the above example of the bar problem, E, F, and c may play a significant role. Both approximability and stability will depend on such parameters. A method is called robust when its performance is relatively uninfluenced by the variation of the parameters within a large range.

22.2.2 A *posteriori* estimates and adaptive approaches

A typical acceptance criterion, as mentioned above, is

$$\|u - Pu\| \leqslant \tau \|u\|$$

where τ is a given tolerance. Although we have

$$\|u-Pu\| \leqslant KZ(u, s)$$

this estimate may have no direct practical importance. In the first place we will in general not know precise values for the quasi-optimality constant K or for $Z(u, S)$. Moreover, even when these are known the resulting estimate may be very pessimistic. The reason is that the quasi-optimality constant K is based on the worst case (since 22.11 must hold for all $u \in H$), while the true solution may have special properties unknown to us. The only general ways to implement the acceptance criterion reliably are based on *a posteriori* analysis of the approximate solution Pu. Thus a computable error estimator ε is introduced, which depends solely on input data and Pu and satisfies

$$\varepsilon \sim \|u-Pu\|$$

in the sense that

$$C_1\varepsilon \leqslant \|u-Pu\| \leqslant C_2\varepsilon$$

and

$$\theta = \frac{\varepsilon}{\|u-Pu\|} \to 1 \qquad \text{as } \varepsilon \to 0$$

This can be achieved (see, for example, [3 to 6]), but not every selection of H, S, V, B, and $\|\cdot\|$ allows for estimators with the same effectiveness and reliability. Feasibility of adaptive selection of test and trial spaces may also be an important feature to be considered in the selection of the form B.

22.3 ILLUSTRATIVE RESULTS

In this section we discuss some concrete mathematical results illustrating the ideas introduced above.

22.3.1 Approximability

First we consider some questions related to approximation. In engineering computations the solution we are interested in approximating usually has special properties. For example, it may be smooth except for some singular behaviour in the neighbourhood of a known point such as a crack tip, corner, or concentrated load. Moreover, the qualitative nature of such singularities is known.

22.3.1.1 The one-dimensional case

The one-dimensional analogue of 'corner' behaviour in two and three dimensions is given by functions $u_\gamma(x) = x^\gamma$, γ a real number. This function has the property that

$$\sum_{l=0}^{k} \int_0^1 x^{2l-2k+\alpha}\left(\frac{\mathrm{d}^l u}{\mathrm{d}x^l}\right)^2 \mathrm{d}x < \infty \tag{22.13}$$

for any integer $k>0$ and any real number $\alpha>2k-1-2\gamma$. Suppose we are interested in approximating in the H^0 norm a function satisfying (22.13). It can be shown that there exists a sequence of subspaces $S^{(n)}$ of H^0 such that $S^{(n)}$ has dimension n and the $S^{(n)}$ satisfy the following approximation property: if $u \in H^0$ is any function satisfying (22.13) for a non-negative integer k and any real number α such that $2k>\alpha\geq 0$, or if $u\in H^k$, then

$$Z(u, S^{(n)}) \leq C(k, \alpha)n^{-k} \tag{22.14}$$

Moreover, (22.14) exhibits the best rate of convergence achievable by any subspaces $S^{(n)}$ of dimension n (see [7]). This is a very robust approximation property. In particular, all the functions u_γ, with $\gamma>-\frac{1}{2}$, can be approximated with this rate. (For $\gamma\leq-\frac{1}{2}$, $\mathbf{u}_\gamma \notin H^0$ so such a result cannot apply.)

In fact, since the functions u_γ have additional properties, even better approximation than indicated by (22.14), namely an exponential rate of convergence, may be obtained for them by another choice of spaces. Thus there exists a sequence of subspaces $\bar{S}^{(n)}$ of dimension n such that if $\gamma>-\frac{1}{2}$, then

$$Z(u_\gamma, \bar{S}^{(n)}) \leq C\exp(-\beta\sqrt{n}) \tag{22.15}$$

for some $\beta>0$ (see [8] for details).

The two results quoted above relate to the existence of a sequence of subspaces of H^0 with good approximation properties. We now consider the quality of approximation achieved by some concrete choices of the sequences $S^{(n)}$ suitable for computation. First let $P^{(n)}$ be the space of polynomials of degree less than n. Then any function satisfying (22.13) can be approximated with the error

$$Z(u, P^{(n)}) = \inf_{S\in P^{(n)}} \|u-s\| \leq C(\alpha, k)n^{-\min(k,2k-\alpha)} \tag{22.16}$$

Applying (22.6) to the functions u_γ $(\gamma>-\frac{1}{2})$ we get the estimate

$$Z(u_\gamma, P^{(n)}) \leq C(\gamma, \varepsilon)^{-(1+2\gamma)+\varepsilon} \tag{22.17}$$

with $\varepsilon>0$ arbitrarily small. It can also be shown that the estimate (22.17) is essentially the best possible one.

Let us now select $S^{(n)}=S_p^{(n)}$, the space of all piecewise polynomials of degree less than p on a quasi-uniform partition [0, 1] into n elements. This space has dimensions roughly proportional to n. The results analogous to

(22.16) and (22.17) in this case are

$$Z(u, S_p^{(n)}) \leqslant C(k, \alpha, p) n^{-\min(p, k-\alpha/2)} \tag{22.18}$$

and

$$Z(u_\gamma, S_p^{(n)}) \leqslant C(\gamma, \varepsilon) n^{-\min[p, (1+2\gamma)/2]+\varepsilon} \tag{22.19}$$

and these rates are essentially unimprovable. Comparing (22.16) and (22.18) we see that for functions u only assumed to satisfy (22.13) the rate of approximation achieved by the polynomials is certainly not worse and may be better than that achieved by the piecewise polynomials. For the functions u_γ, (22.17) and (22.19) show that the rate achieved by the polynomials is at least twice that achieved by the piecewise polynomials (for more details, see [9]).

The estimate (22.18) is in essence the classical estimate

$$Z(u, S_p^{(n)}) \leqslant C(k, p) n^{-k} \|u\|_{H^k}$$

when $p > k$ and $\alpha = 0$ (see, for example, [10]). The question arises whether under these conditions an expression for $C(k, p)$ can be given which explicitly characterizes the behaviour with respect to p. In [11] such an expression is given in both the one- and two-dimensional cases (and for approximation in H^1). It is shown there that on the right-hand side we can have $\hat{C}(k) n^{-k+\varepsilon}$ with $\hat{C}$ independent of p.

Neither the piecewise polynomial spaces nor the polynomial spaces achieve the optimal rate of convergence characterized by (22.14). For example, for $\gamma > -\frac{1}{2}$ sufficiently small, the function u_γ is not approximated at the optimal rate of n^{-k} for either of these cases. Such a rate can be achieved in the first case by a proper refinement of the mesh in a neighbourhood of the origin and in the second case by changing the polynomials to some other system of functions (see [11, 12]). The importance of this observation is that for engineering computations it appears likely that approximation spaces can be created which yield a rate of convergence which is better than polynomial and is probably exponential.

Thus far we have considered approximation in the H^0 norm. Similar results are available for all the H^l norms, l both positive and negative. An interesting fact is that when l decreases the rate of convergence furnished by either $p^{(n)}$ or $S_p^{(n)}$ increases linearly with l (for more details, for example, see [13]).

22.3.1.2 The multidimensional case

So far we have discussed only the one-dimensional case. Analogous results exist in more than one dimension, but these are far from complete. We will not go into details here, but refer the reader, for example, to [9, 11, 14].

22.3.1.3 The h, p, and h–p versions of the finite element method

As was stated above, there are important cases when selecting the same trial and test spaces leads to a stability constant of 1; thus approximability by the trial space determines the performance of the method. The classical finite element method uses piecewise polynomials of fixed degree p on meshes which are refined to achieve accuracy. Because the size of the elements is usually denoted by h, this method is called the *h version* and the approximability properties (22.18) and (22.19) for such spaces over quasi-uniform meshes are used. The *p version* achieves accuracy by fixing the mesh and increasing the degree p of the polynomial. In this case the approximation results (22.16) and (22.17) are applicable. The p version has been implemented in the program COMET X. We refer the reader to [9] and [15] and references therein for detailed information. Finally, the *h–p version* combines both of these approaches. The exponential convergence rate given in (22.15) can be realized in the h–p version.

22.3.2 Finite element methods

We turn now to a discussion of finite element methods. As discussed in Section 22.2, the quality of approximation yielded by such a method is assured by stability in conjunction with approximability. The stability of a method depends on the interplay between the spaces S and V, the bilinear form B, and the norm $\|\cdot\|$. This is illustrated in the first example.

22.3.2.1 An example illustrating the role of the trial and test spaces in stability

First we consider a one-dimensional problem with the simplest possible bilinear form. Setting $l=1$, $EF=1$, and $c=0$ in (22.9), we get the form

$$B(u, v) = \int_0^1 u'v' \, dx \qquad (22.20)$$

The solution $u \in \mathring{H}^1$ satisfies $B(u, v) = \int_0^1 pv \, dx$ for all $v \in \mathring{H}^1$, and the related two-point boundary value problem is

$$-u'' = p$$
$$u(0) = u(1) = 0$$

For discretization we define spaces of smooth splines. Let $\Delta = \{0 = x_0 < x_1 < \cdots < x_n = 1\}$ be a mesh of $[0, 1]$ and set $h_i = x_i - x_{i-1}$. For $\gamma \geqslant 1$, the mesh is called γ *quasi-uniform* if

$$\max_{i,j} \frac{h_i}{h_j} \leqslant \gamma$$

A weaker restriction is that the mesh be γ *locally quasi-uniform*, i.e. that

$$\max_{i} \frac{h_i}{h_{i\pm 1}} \leqslant \gamma$$

Given any mesh Δ we define for $r = 0, 1, 2, \ldots$ the *smooth splines of degree r subordinate to* Δ to be the piecewise polynomials of degree r with $r-1$ continuous derivatives. The space of all such splines is denoted $M^r(\Delta)$. In particular $M^0(\Delta)$ is the space of piecewise constant functions and $M^1(\Delta)$ is the space of continuous piecewise linear functions. We also denote by $\dot{M}^1(\Delta)$ the space of piecewise linear functions with zero boundary values and by $\dot{M}^3(\Delta)$ the space of natural cubic splines, that is $\dot{M}^3(\Delta) = \{v \in M^3(\Delta) \mid v = v'' = 0 \text{ when } x = 0 \text{ or } 1\}$.

We consider the use of such spline spaces for S and V in conjunction with the bilinear form B defined in (22.20). It is possible to show that if S and V are taken to be spaces of smooth splines of degree r_1 and r_2 respectively (with appropriate boundary conditions), and if r_1 and r_2 are either both odd or both even, then condition (22.2b) is satisfied. Hence $Pu \in S$ is uniquely defined by (22.1).

The most standard case occurs with $S = V = \dot{M}^1(\Delta)$. The stability properties of this method in several norms are summarized in the following theorem.

Theorem 22.1 *Let $S = V = \dot{M}^1(\Delta)$ for an arbitrary partition Δ. Then*

$$\|Pu - u\|_{H^1} \leqslant K \inf_{s\in S} \|u - s\|_{H^1} \qquad (22.21a)$$

$$\|Pu - u\|_{H^1_\infty} \leqslant K \inf_{s\in S} \|u - s\|_{H^1_\infty} \qquad (2r.21b)$$

$$\|\mathrm{Pu} - \mathrm{u}\|_{H^0_\infty} \leqslant K \inf_{s\in S} \|u - s\|_{H^0_\infty} \qquad (22.21c)$$

for all $u \in \mathring{H}^1$, with K independent of Δ. However, for any $C > 0$ and any mesh Δ there exists $u \in H^1$ such that

$$\|Pu - u\|_{H^0} \geqslant C \inf_{s\in S} \|u - s\|_{H^0}$$

For more details see [16].

The case where $S = \dot{M}^1(\Delta)$, $V = \dot{M}^3(\Delta)$ is less familiar and more involved.

Theorem 22.2 *Let* $S = \dot{M}^1(\Delta)$, $V = \dot{M}^3(\Delta)$. *Then*:

(*a*) *For an arbitrary partition* Δ,

$$\|Pu - u\|_{H^0} \leq K \inf_{s \in S} \|u - s\|_{H^0}$$

with K *independent of* Δ *and* $u \in \mathring{H}^1$.

(*b*) *For any* $\gamma \geq 1$ *there exists a constant* $K(\gamma)$ *such that for all* γ *quasi-uniform partitions and all* $u \in \mathring{H}^1$,

$$\|Pu - u\|_{H^1} \leq K(\gamma) \inf_{s \in S} \|u - s\|_{H^1}$$

and

$$\|Pu - u\|_{H^{-1}} \leq K(\gamma) \inf_{s \in S} \|u - s\|_{H^{-1}}$$

(*c*) *However, for any* $C > 0$ *and any partition* Δ, *there exists* $u \in H^1$ *such that*

$$\|Pu - u\|_{H^{-2}} \geq C \inf_{s \in S} \|u - s\|_{H^{-2}}$$

(*d*) *If* $1 \leq \gamma < \gamma_0 = 1 + \sqrt{3} + \sqrt{(3 + 2\sqrt{3})} = 5.2745\ldots$, *there exists a constant* $\bar{K}(\gamma)$ *such that*

$$\|Pu - u\|_{H^1} \leq \bar{K}(\gamma) \inf_{s \in s} \|u - s\|_{H^1}$$

and

$$\|Pu - u\|_{H^{-1}} \leq \bar{K}(\gamma) \inf_{s \in S} \|u - s\|_{H^{-1}}$$

for any γ *locally quasi-uniform partition* Δ *and* u. *However,* $\lim_{\gamma \to \gamma 0} \bar{K}(\gamma) = \infty$.

Thus we see that there is a very fine interplay between the trial and test spaces, even for the simplest bilinear form.

So far we have analysed the form (22.20). We now consider the form

$$B(u, v) = \int_0^1 Eu'v' \,\mathrm{d}x \tag{22.22}$$

where $0 < e_0 \leq E(x) \leq e_1 < \infty$. The question arises whether Theorems 22.1 and 22.2 remain true as stated. It is possible to show that if $E(x)$ is sufficiently smooth, then Theorems 22.1 and 22.2 hold without change. The requirement of smoothness means that K may also depend on the maximum of the first few derivatives of E as well as on e_0 and e_1. It can be shown that (22.21a) holds with K depending only on e_0 and e_1, but (22.21c) is not true when no differentiability restrictions are made on E. Thus we may say that the performance of that method is more robust with respect to the coefficient E in the norm $\|\cdot\|_{H^1}$ than it is in the norm $\|\cdot\|_{H^0_\infty}$.

22.3.2. The bilinear form and robustness

We continue to consider the simple problem (22.4) but now consider the effect of the choice of the bilinear form on the robustness of the method. For simplicity assume that $F=1$ and $c=0$. Let $E(x)$ be given satisfying $0<e_0\leqslant E(x)\leqslant e_1$ and consider the bilinear forms (22.5) and (22.7). The bilinear form (22.5) clearly stems from the system of equations

$$\begin{aligned} Eu' &= \sigma \\ \sigma' &= p \end{aligned} \tag{22.23}$$

while (22.7) comes from

$$\begin{aligned} u' &= \frac{\sigma}{E} \\ \sigma' &= p \end{aligned} \tag{22.24}$$

Assume now that we take

$$S = V = \dot{M}^1(\Delta)\times M^0(\Delta) \tag{22.25}$$

in both cases. It is easy to see that σ can be eliminated from the system of linear equations arising from (22.5) with the choice of spaces given in (22.13), and we then get the same method as when (22.10) is used with $S=V=\dot{M}^1(\Delta)$. The properties of this method were summarized in Theorem 22.1. The form (22.7), however, gives different results, which we now consider in detail.

Letting $(\bar{u}, \bar{\sigma}) = P(u, \sigma)$ we get:

Theorem 22.3 *Let Δ be an arbitrary mesh. Define P by the bilinear form (22.7) with S and V defined by (22.25). Then there exists a constant K depending only on e_0 and e_1 such that*

(a) $$\|u-\bar{u}\|_{H^1}+\|\sigma-\bar{\sigma}\|_{H^0}\leqslant K \inf_{(\chi,\psi)\in S} [\|u-\chi\|_{H^1}+\|\sigma-\psi\|_{H^0}]$$

(b) $$\|u-\bar{u}\|_{H_\infty}+\|\sigma-\bar{\sigma}\|_{H_\infty}\leqslant K\left[\inf_{(\chi,\psi)\in S} \|u-\chi\|_{H^0_\infty}+\|\sigma-\psi\|_{H^0_\infty}\right]$$

(c) $$\|\sigma-\bar{\sigma}\|_{H^0}\leqslant K \inf_{\psi\in M^0(\Delta)} \|\sigma-\psi\|_{H^0}$$

(d) $$\|\sigma-\bar{\sigma}\|_{H^0_\infty}\leqslant K \inf_{\psi\in M^0(\Delta)} \|\sigma-\psi\|_{H^0_\infty}$$

The statements analogous to (c) and (d) for the error $\|u-\bar{u}\|$ are not true. In order to elaborate this point let us introduce a further notation. For $\chi \in \dot{M}^1(\Delta)$ let $\tilde{\chi}$ be defined by

$$\begin{aligned} \tilde{\chi}(x_j) &= \chi(x_j) && \text{for } j=0,1,\ldots,n \\ (E\tilde{\chi}')' &= 0 \text{ on } (x_{j-1},x_j) && \text{for } j=1,2,\ldots,n \end{aligned}$$

Then we have:

Theorem 22.4 *Let $\bar{u}$ be defined as in Theorem 22.3 (i.e. using the form (22.7) and spaces of (22.25). Then*

$$\|u-\bar{u}\|_{H^0_\infty} \leq K(e_0, e_1, V(E)) \inf_{\chi \in \dot{M}^1(\Delta)} [\|u-\chi\|_{H^0_\infty} + \|u-\tilde{\chi}\|_{H^0_\infty}]$$

and

$$\max_j |u(x_j)-\bar{u}(x_j)| \leq K(e_0, e_1, V(E)) \inf_{\chi \in \dot{M}^1(\Delta)} \|u-\tilde{\chi}\|_{H^0_\infty}$$

with K depending on e_0, e_1, and the variation $V(E)$ of the function E.

We remark that this theorem is not valid when the dependence of K on $V(E)$ is suppressed. Moreover, while the term $\inf \|u-\tilde{\chi}\|_{H^0_\infty}$ in the second estimate is necessary, it is usually smaller than the first term. Comparing Theorem 22.4 with the previous results we see that the form (22.7) is much more robust than (22.5) with respect to all the norms we have considered except the H^1 norm, and should be preferred in most situations.

Let us comment on the system of linear equations which the approximate solution led to when $\bar{\sigma}$ is eliminated in either of the two methods discussed in this section. In both cases $Pu \in \dot{M}^1(\Delta)$ is defined by a system of the form

$$\int_0^1 E_\Delta (Pu)'v' \, dx = \int_0^1 pv \, dx \qquad \text{for all } v \in \dot{M}^1(\Delta)$$

When the form (22.5) is used

$$E_\Delta = \frac{1}{h_i}\int_{x_{i-1}}^{x_i} E(x)\, dx \qquad \text{on } (x_{i-1}, x_i)$$

while when (22.7) is used

$$E_\Delta = \left[\frac{1}{h_i}\int_{x_{i-1}}^{x_i} \frac{1}{E(x)}\, dx\right]^{-1} \qquad \text{on } (x_{i-1}, x_i)$$

$i=1,2,\ldots,n$. Thus in the former case E is replaced by its piecewise average and in the latter by its piecewise harmonic average. The above-stated results show that the usual finite element method does not have as good stability properties when $E(x)$ changes significantly over an element, and so should not be used. The change can be measured by the ratio of the average to the harmonic average.

22.3.2.3 Changing the dependent variable to improve approximability

Now we turn to the analysis of the form defined in (22.8) where for simplicity we take $F=1$ and $c=0$. If we choose $S=V=M^0(\Delta)\times M^1(\Delta)$ and set $(\bar{u}, \bar{\sigma})=P(u, \sigma)$, it is easy to prove that

$$\|u-\bar{u}\|_{H^0}+\|\sigma-\bar{\sigma}\|_{H^1}\leqslant C \inf_{(\chi,\psi)\in S} [\|u-\chi\|_{H^0}+\|\sigma-\psi\|_{H^1}] \tag{22.26}$$

with C independent of Δ but depending on E. Now for any $g\in H^1$ consider the variational formulation

$$B_3(u_g, \sigma_g; v, \rho)=\int_0^l (p-g')v\,dx-\int_0^l \frac{\rho g}{E}\,dx \tag{22.27}$$

instead of the method just considered:

$$B_3(u, \sigma; v, \rho)=\int_0^l pv\,dx$$

The new variables are related to the old by

$$\sigma_g=\sigma-g, \qquad u_g=u$$

Thus we may compute σ_g and then take $\sigma=\sigma_g+g$ for the stress component of the approximate solution. Because the same bilinear form occurs in both these approaches, the stability is unchanged and we get:

Theorem 22.5 *For the method associated with (22.27)*

$$\|u-\bar{u}\|_{H^0}+\|\sigma_g-\bar{\sigma}_g\|_{H^1}\leqslant C \inf_{(\chi,\rho)\in S}\left[\|u-\chi\|_{H^0}+\|\sigma_g-\rho\|_{H^1}\right] \tag{22.28}$$

where the C is the same constant which appears in (22.26) (and therefore is independent of g). Moreover, $\sigma-\bar{\sigma}=\sigma_g-\bar{\sigma}_g$.

Now we note that the proper choice of g can increase the smoothness of σ_g, thereby increasing its approximability and so improving the accuracy of the method. In this simple one-dimensional case the best choice is $g=\int p\,dx$ so σ_g is constant. (In some situations it is as easy or easier for the user to input g as p.) In this case the last term in (22.28) will disappear and $\bar{\sigma}$ will exactly equal σ. A similar idea may be fruitfully applied to related mixed methods in more than one dimension.

22.3.2.4 A robust method for a parameter-dependent problem

In Subsection 22.3.2.2 we discussed a simple case in which changing the bilinear form significantly increased the robustness of the method. We now discuss another example in which a parameter enters in a direct fashion. The problem to be considered models the deflection of a beam allowing for the effect of shear stress. In the simplest case this model can be described by the system of equations

$$\begin{aligned} -\phi_d'' + d^{-2}(\phi_d - \omega_d') &= 0 \qquad 0<x<1 \\ d^{-2}(\phi_d - \omega_d')' &= g \qquad 0<x<1 \end{aligned} \tag{22.29}$$

with the boundary conditions

$$\phi_d(0) = \phi_d(1) = \omega_d(0) = \omega_d(1) = 0$$

Physically $\phi_d(x)$ represents the displacement, $\omega_d(x)$ the rotation of the cross-section, and $g(x)$ the transverse load. The solution depends on the beam thickness d. We associate to the problem (22.29) the bilinear form

$$B_d(\phi, \omega; \psi, \nu) = \int_0^1 [\phi'\psi' + d^{-2}(\phi - \omega')(\psi - \nu')]\,\mathrm{d}x$$

Let $S = V = \dot{M}^1(\Delta) \times \dot{M}^1(\Delta)$. Then we have

$$B_d(\phi_d, \omega_d; \psi, \nu) = \int_0^1 g\nu\,\mathrm{d}x \tag{22.30}$$

and so the bilinear form is computable. Denoting $P(\phi_d, \omega_d)$ by $(\bar{\phi}_d, \bar{\omega}_d)$ we get

Theorem 22.6

$$\|\phi_d - \bar{\phi}_d\|_{H^1} + \|\omega_d - \bar{\omega}_d\|_{H^1} \leq C(d) \inf_{(\chi,\rho)\in S} [\|\phi_d - \chi\|_{H^1} + \|\omega_d - \rho\|_{H^1}]$$

The constant $C(d)$ is independent of ϕ, $\omega \in \mathring{H}^1$, but $C(d) \to \infty$ as $d \to 0$.

A corollary of this theorem is:

Theorem 22.7 *Let any non-zero load g be given and let $0<\sigma<1$ be arbitrary. Then for any partition Δ there exists a value of d depending on Δ such that*

$$\|\phi_d - \bar{\phi}_d\|_{H^1} \geq \sigma\, \|\phi\|_{H^1}$$
$$\|\omega_d - \bar{\omega}\|_{H^1} \geq \sigma\, \|\omega_d\|_{H^1}$$

Theorem 22.7 shows that for small d the method based on (22.30) is virtually useless.

We now associate to our problem another bilinear form, in which we introduce a new variable ξ, representing the shear stress.

$$\hat{B}_d(\phi, \omega, \xi; \psi, \nu, \eta) = \int_0^1 [\phi'\psi' + \xi(\psi - \nu') + \eta(\phi - \omega') - d^2\xi\eta]\, dx \tag{22.31}$$

The functions ϕ_d, ω_d, and $\xi_d = d^{-2}(\phi_d - \omega_d')$ satisfy

$$\hat{B}_d(\phi_d, \omega_d, \xi_d; \psi, \nu, \eta) = \int_0^1 g\nu\, dx \tag{22.32}$$

for any $\psi \in \mathring{H}^1$, $\nu \in \mathring{H}^1$, $\eta \in H^0$. Select now $S = V = \dot{M}^1(\Delta) \times \dot{M}^1(\Delta) \times M^0(\Delta)$, and let $(\hat{\phi}_d, \hat{\omega}_d, \hat{\xi}_d) = P(\phi_d, \omega_d, \xi_d)$. The robustness of the new method with respect to the parameter d is evidenced by the following result.

Theorem 22.8 *For the method associated with (22.32)*

$$\|\phi_d - \hat{\phi}_d\|_{H^1} + \|\omega_d - \hat{\omega}_d\|_{H^1} + \|\xi_d - \hat{\xi}_d\|_{H^0}$$
$$\leq C \inf_{(\chi, \rho, \lambda) \in S} [\|\phi_d - \chi\|_{H^1} + \|\omega_d - \rho\|_{H^1} + \|\xi_d - \lambda\|_{H^0}]$$

with C independent of Δ *and* d.

When g in (22.29) is smooth, then ϕ_d, ω_d, and ξ_d are smooth also, and may well be approximated independently of d. It follows that computations based on (22.31) and (22.32) give very good results while we have seen that computations based on (22.30) yield extremely poor results for small d. This difference in the robustness of the two methods with respect to d is very striking in practice. It is also worth noting that the additional variable $\hat{\xi}_d$ may be eliminated from (22.32). The resulting method is identical with the method based on (22.30) except that the integrals are calculated by the composite mid-point rule. By employing this reduced integration implementation, the mixed method entails no extra expense whatever (for more details, see [17]).

22.3.2.5 Robust methods in two dimensions

So far we have discussed various ideas concerning the proper selection of S, V, and B in the context of simple one-dimensional examples. Analogous ideas can be used in more dimensions also, although much less is known at present. Nevertheless we will briefly consider some examples. Consider the problem

$$\frac{\partial}{\partial x} a \frac{\partial u}{\partial x} + \frac{\partial}{\partial y} a \frac{\partial u}{\partial y} = f$$

on

$$\Omega = \{(x, y) \mid |x| < 1, |y| < 1\}$$

with the condition $u = 0$ on $\partial\Omega$. Assume that $a = a_0$ for $x < 0$ and $a = a_1$ for $x \geq 0$, where a_0 and a_1 are distinct positive constants. Having selected a triangulation $\mathcal{T}$ (with minimal angle condition), the usual method employs the bilinear form

$$B(u, v) = \int_\Omega \left(a \frac{\partial u}{\partial x} \frac{\partial v}{\partial x} + a \frac{\partial u}{\partial y} \frac{\partial v}{\partial y} \right) dx\, dy \tag{22.33}$$

and equal trial and test spaces consisting of functions which are continuous, linear on every triangle, and zero on $\partial\Omega$. If the interface $x = 0$ does not coincide with the boundaries of the triangles, then the solution, which is not smooth at the interface, will be approximated less accurately than for a problem with a smooth coefficient. We shall show how to proceed in a slightly different fashion which will avoid this problem.

We select for V the space of continuous piecewise linear functions. Thus the restriction of a test function to a triangle is a linear combination of the three functions 1, x, and y and each test function is continuous at the nodes. The trial functions are taken to restrict on each triangle to a linear combination of the functions 1, $\int_0^x dt\, a$, and y and to be continuous at the nodes. The trial functions, unlike the test functions, need not be continuous on element boundaries except at the nodes (and so are 'non-conforming'). Now interpret (22.33) as a sum of integrals over the individual triangles and replace the norm $\|\cdot\|_{H^1}$ by the norm $\|\cdot\|_{H^1(\mathcal{T})}$ defined as the square root of a sum of integrals over the triangles. This is a common approach in non-conforming finite element methods.

Theorem 22.9 *For this method*

$$\|u - Pu\|_{H^1(\mathcal{T})} \leq C \inf_{\chi \in S} \|u - \chi\|_{H^1(\mathcal{T})}$$

with C independent of $\mathcal{T}$ and u. The value of the unusual trial space S used here is that while stability still holds (as stated in Theorem 22.9), these trial functions mimic the behaviour of the solution u and thus greatly improve the approximability. That is, $Z(u, S)$ is generally much smaller for this choice of S than if S is taken equal to V (the usual choice).

The resulting method therefore gives superior results. An important observation to be made here is that the difficulties encountered with non-conforming methods generally arise from the non-conformity of the test space. Non-conforming trial functions cause no such problems.

A similar idea can be applied to corner problems. Consider solving Laplace's equation on a domain with a corner angle of $\frac{3}{2}\pi$, and zero boundary conditions. The solution then has a singular component of the type $r^{2/3} \sin 2\pi\theta/3$. For test functions we use the usual linear elements but for trial functions we use elements based on the functions 1, $r^{2/3} \sin 2\pi\theta/3$, and $r^{2/3} \cos 2\pi\theta/3$, instead of 1, x, and y. Using this approach, the loss of accuracy due to the singular behaviour of the solution in a neighbourhood of the corner is prevented. We remark that this procedure need not entail any computational difficulties because it can be implemented in a way which preserves the symmetry of the linear equations, and one may work in the usual way with the microstiffness matrices and nodal variables.

22.4 CONCLUSIONS

We summarize here the main ideas we have presented. As we have seen, there is virtually an unlimited variety of possible variational discretization methods. Such a method is characterized by the bilinear form and the trial and test space. In selecting a method it is of paramount importance to consider the goals of the computation, in particular the norm with which the error is to be assessed. The goals of the computations are best achieved by a method which has good approximability and stability properties with respect to the desired norm. The method should be robust in the sense that these properties apply uniformly over the relevant class of problems. We note that often the obvious method is not best and various variations can lead to strikingly improved results.

REFERENCES

1. M. Vogelius and I. Babuška, 'On a dimensional reduction method: Parts I, II, III', *Math. of Comput.*, **37,** 31–46 (Part I), 47–68 (Part II), to appear (Part III) (1981).
2. I. Babuška and W. Rheinboldt, 'Computational aspects of the finite element method', in *Math. Software III*, pp. 225–255, Academic Press, 1977.
3. I. Babuška and W. Rheinboldt, 'Reliable error estimation and mesh adaptation for the finite element method', in *Computational Methods in Nonlinear Mechanics* (Ed. J. T. Oden), pp. 67–108, North Holland, (1980).
4. I. Babuška, 'Analysis of optimal finite element meshes in R^1', *Math. of Comp.*, **1979,** 435–463 (1979).
5. I. Babuška and W. Rheinboldt, '*A-posteriori* error analysis of finite element solutions for one dimensional problems', *SIAM J. Num. Anal.*, **1981,** 365–389 (1981).
6. I. Babuška and A. Miller, '*A-posteriori* error estimates and adaptive techniques for a finite element method', Inst. for Phys. Sci. & Tech., Lab. for Num. Anal., University of Maryland Tech. Note BN-968, June 1981.
7. H. Triebel, *Interpolation Theory Function Spaces, Differential Operators,* North Holland, Amsterdam, 1978.
8. R. De Vore and K. Scherer, 'Variable knot variable degree spline approximation to x quantitative approximation', *Proc. Bonn Conference,* North Holland, 1978.
9. I. Babuška, B. A. Szabo, and I. N. Katz, 'The P-version of the finite element method', *SIAM J. Num. Anal.*, **1981,** 515–545 (1981).

10. P. G. Ciarlet, *The Finite Element Methods for Elliptic Problems*, North Holland, Amsterdam, 1978.
11. I. Babuška, and M. R. Dorr, 'Error estimates for the combined h and p versions of the finite element method', *Num. Math.*, **1981,** 257–277 (1981).
12. M. R. Dorr, Dissertation (in preparation).
13. I. Babuška and A. K. Aziz, 'Survey Lectures on the mathematical foundations of the finite element method', in *The Mathematical Foundations of the Finite Element Method with Applications to Partial Differential Equations* (Ed. A. K. Aziz), Academic Press, 1972.
14. I. Babuška, R. B. Kellogg, and J. Pitkäranta, 'Direct and inverse estimates for finite elements with mesh refinement', *Num. Math.*, **1979,** 447–471 (1979).
15. I. Babuška and B. A. Szabo, 'On the rates of convergence of the finite element method', Center for Comp. Mechanics, Washington Univ., St. Louis, Rept. WU/CCM-80/2. To appear in *Int. J. Num. Meth. in Eng.*, 1981.
16. I. Babuška and J. Osborn, 'Analysis of finite element methods for second order boundary value problems using mesh dependent norms', *Num. Math.*, **34,** 41–62 (1980).
17. D. N. Arnold, 'Discretization by finite elements of a model parameter dependent problem', *Num. Math.*, **37,** 405–421 (1981).

Hybrid and Mixed Finite Element Methods
Edited by S. N. Atluri, R. H. Gallagher, and O. C. Zienkiewicz

Chapter 23

Complementary Energy Revisited

R. H. Gallagher, J. C. Heinrich, and N. Sarigul

23.1 INTRODUCTION

When the finite element method as we know it today was introduced and attracted wide attention twenty-five years ago, considerable attention was given to the possibility of 'bracketing' the exact solution through calculation of upper-bound and lower-bound finite element solutions [1]. These alternatives were represented by the force and displacement methods. It was soon found that bracketing of a solution applied straightforwardly to only certain circumstances, and even for those it was necessary to exercise care in the selection of basis functions and in the construction of all contributions to the relevant energy expressions.

Subsequently, the force method vanished from practice. Stiffness equations, primarily associated with assumed displacement formulations, also derived the hybrid formulations and were themselves no longer associated with solution bounds. Indeed, this symposium attests to the popularity of hybrid formulations and, by implication, to the limited interest that prevails today in the bracketing of solutions.

Although broad operational capabilities for solution bracketing, in our opinion, lie well into the future, there is strong motivation for a stress-based, or complementary, formulation of the finite element problem. For example, it is desirable to calculate directly the values of equilibrium stress fields which evidence continuity across element interfaces.

Fraejis de Veubeke and Zienkiewicz [2] proposed an approach to this objective fourteen years ago. Founded in the use of stress functions, it was attractive on many counts. Principally, it represented a complete algebraic dual to the assumed displacement method. Within a short period of time, a large number of developments of the assumed stress function approach appeared [3 to 11]. It is fair to say, however, that interest in this avenue has flagged within the past decade. Although contributions on stress-based methods have appeared (e.g. [12, 13]), the overall concept has not made its way into today's widely distributed, popular finite element programmes.

In our view, the fundamental limitation on use of stress-function-based finite element formulations results from an awkwardness in dealing with boundary conditions. The force vector that derives from the boundary integral term of the assumed displacement potential energy expression is easily calculated and is customarily populated with one or more non-zero terms. The vector obtained from the boundary integral term of the complementary energy expression is customarily zero, however. Other means must be sought for representation of boundary conditions.

Numerous proposals have been advanced for treatment of boundary conditions in an assumed stress function, complementary energy finite element formulation. Some of these lead to algebraic conditions of constraint, which can be handled by algebraic manipulation by Lagrange multipliers, or by penalty function methods. Another possibility in the treatment of boundary conditions is a 'direct' manner, where the applied edge tractions are identified with element nodal point parameters. This is hardly a new or innovative approach, but has not found use on account of the large number of nodal parameters that must apparently be introduced when it is employed.

In this paper we propose the use of special boundary elements which enable the direct specification of boundary conditions on the edges that correspond to the boundaries but which do not carry the nodal parameters, used for this purpose, into the interior. This approach requires the definition of special basis functions. The concepts of blending interpolants [14] are used to advantage for this purpose.

23.2 COMPLEMENTARY ENERGY

The principle of stationary complementary energy furnishes a variational basis for the direct formulation of element flexibility equations, i.e. expressions for element displacement parameters in terms of force parameters. The complementary energy (Π_c) of a structure is given by the sum of the complementary strain energy (U^*) and the potential of boundary forces acting through prescribed displacements (V^*), i.e.

$$\Pi_c = U^* + V^* \tag{23.1}$$

The principle of stationary complementary energy can be stated as follows: 'Among all states of stress which satisfy the equilibrium conditions in the interior of the body and the prescribed surface stress conditions, the state of stress which also satisfies the stress–displacement relations in the interior and all prescribed displacement boundary conditions makes the complementary energy assume a stationary value'. Thus:

$$\delta\Pi_c = \delta U^* + \delta V^* = 0 \tag{23.2}$$

For stable equilibrium $\delta\Pi_c$ is a minimum:

$$\delta^2\Pi_c = \delta^2 U^* + \delta^2 V^* > 0 \tag{23.3}$$

The complementary strain energy is defined as (we neglect initial strains)

$$U^* = \frac{1}{2}\int_{\text{vol}} \boldsymbol{\sigma}[E]^{-1}\boldsymbol{\sigma}\,\mathrm{d(vol)} \tag{23.4}$$

where $\boldsymbol{\sigma}$ is the vector of stress components. In the presence of prescribed displacements $\bar{\mathbf{u}}$

$$V^* = -\int_{S_u} \bar{\mathbf{u}}\cdot\mathbf{T}\,\mathrm{d}S \tag{23.5}$$

where S_u is the surface on which $\bar{\mathbf{u}}$ is prescribed and $\mathbf{T}$ is the surface load.

23.3 PLANE STRESS

In the case of plane stress $\boldsymbol{\sigma} = [\sigma_x \sigma_y \tau_{xy}]^T$ is the vector of stress components and $\boldsymbol{\varepsilon} = [\varepsilon_x \varepsilon_y \gamma_{xy}]^T$ is the vector of strain components.

We now express the stress vector $\boldsymbol{\sigma}$ in terms of the Airy stress function Φ where, by the usual definition, $\sigma_x = \Phi_{yy}$, etc., and the subscripts on Φ denote differentiation with respect to the indicated variables. Thus

$$\boldsymbol{\sigma} = [\Phi_{yy}\Phi_{xx} - \Phi_{xy}] = [\Phi''] \tag{23.6}$$

and, by substitution of this into (23.4),

$$\Pi_c = \frac{1}{2}\int_{\text{vol}} [\Phi''][E]^{-1}[\Phi'']^T\,\mathrm{d(vol)} - \int_{S_u} \bar{\mathbf{u}}\cdot\mathbf{T}\,\mathrm{d}S = U^* + V^* \tag{23.7}$$

We will exclude from this discussion the possibility of non-zero prescribed displacements so that $V^* = 0$ and $\Pi_c = U^*$. Also, with respect to force boundary conditions, we consider only distributed edge loadings. The condition of zero prescribed displacement is the usual practical circumstance and, as we shall see, it results in a special difficulty.

To illustrate finite element formulative concepts the constant thickness rectangle of dimension $a \times b$ (Figure 23.1) can be studied. The necessary interelement equilibrium conditions, requiring continuity of both ϕ and

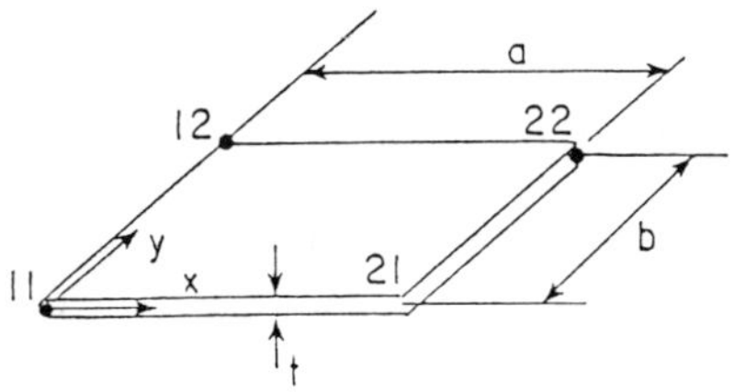

Figure 23.1 Rectangular element

$\partial\phi/\partial n$ on each boundary, can be met by the bivariate Hermite polynomial interpolation of third order. We have for this sixteen-term function

$$\Phi = [N]\{\Phi^e\} \tag{23.8}$$

where

$$[N] = [(N_1(x)N_1(y)) \cdots (N_{x1}(x)N_{y2}(y))] \tag{23.9}$$

$$\{\Phi^e\} = [\Phi_{11} \cdots \Phi_{xy_{12}}]^T \tag{23.10}$$

$[N]$ and $\{\Phi^e\}^T$ are 16×1 matrices.

The shape functions $N_1(x), \ldots, N_{x1}(x)$ are as follows (with $\xi = x/a$):

$$\begin{aligned} N_1(x) &= 1 - 3\xi^2 + 2\xi^3, & N_2(x) &= 3\xi^2 - 2\xi^3 \\ N_{x1}(x) &= a(\xi - 2\xi^2 + \xi^3), & N_{x2}(x) &= a(\xi^3 - \xi^2) \end{aligned} \tag{23.11}$$

where $\xi = x/a$.

$N_1(y) \cdots N_{y2}(y)$ are similarly defined by replacing a with b and $\xi = x/a$ by $\eta = y/b$. We shall refer to the *stress function parameters* $\{\Phi^e\}$ as *d.o.f.* (degrees of freedom) in the following, even though the latter term has a displacement connotation.

Establishing now the stress vector by appropriate differentiation we have, symbolically

$$\boldsymbol{\sigma} = \begin{Bmatrix} \Phi_{yy} \\ \Phi_{xx} \\ -\Phi_{xy} \end{Bmatrix} = \begin{bmatrix} [N_{yy}] \\ [N_{xx}] \\ [N_{xy}] \end{bmatrix} \{\bar{\Phi}^e\} = [G]\{\bar{\Phi}^e\} \tag{23.12}$$

By substitution of equation (23.12) into equation (23.7), the element complementary strain energy (U^{e*}) becomes

$$U^{e*} = \frac{[\bar{\Phi}^e]}{2}[f^e]\{\bar{\Phi}^e\} \tag{23.13}$$

where

$$[f^e] = \left[\iint_A [G]^T[E]^{-1}[G]t\,\mathrm{d}A\right] \tag{23.14}$$

$[f^e]$ denoting the *element flexibility matrix* and t the plate thickness.

Because of the analogy between the homogeneous forms of the equilibrium governing differential equation for plate flexure (in terms of transverse displacement w) and the compatibility differential equation for plate stretching (in terms of the Airy stress function Φ), the existing stiffness formulations for plate flexure give the flexibility coefficients defined by equation (23.14). Due account must be taken of the differences in the constants and signs between terms of the material compliance ($[E]^{-1}$) and stiffness $[E]$ matrices.

We next examine the applied loads on the element boundary, which lead

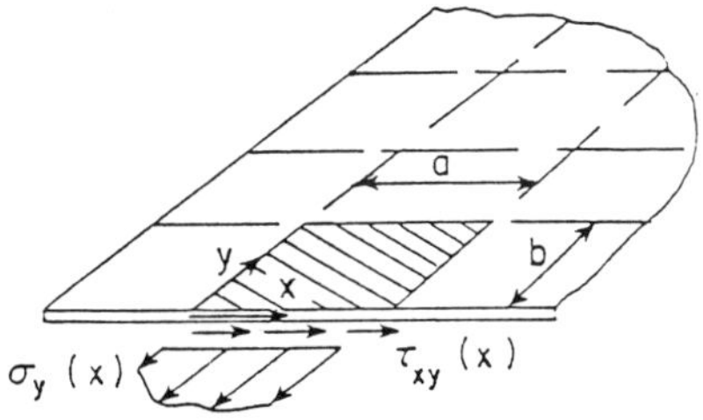

Figure 23.2 Distributed edge stresses

to constraint conditions on the degrees of freedom. Consider an element with an edge parallel to the x axis and acted upon by distributed stresses $\sigma_y(x)$ (Figure 23.2). At any point along the edge:

$$\frac{\partial^2 \Phi}{\partial x^2} = \sigma_y(x) \tag{23.15}$$

Similarly, for shear,

$$\frac{\partial^2 \Phi}{\partial x \, \partial y} = -\tau_{xy}(x) \tag{23.16}$$

If the edge is parallel to the y direction, then

$$\frac{\partial^2 \Phi}{\partial y^2} = \sigma_x(y) \tag{23.17}$$

23.4 GLOBAL EQUATIONS

The synthesis of the global equations begins with the determination of the complementary strain energy (U^*), using as a basis the component element complementary energies (U^{e*}). In plane stress the requirements to be met in interconnection of elements to retain a valid stationary complementary energy formulation are those of continuity of the normal and shear stresses across the element boundaries. Use of the bivariate Hermitian polynomial interpolation function for Φ meets such requirements. Thus, the required continuity is achieved when, at each joint in the system, the degrees of freedom $\{\bar{\Phi}^e\}$ of the elements meeting at each joint are equated to each other.

It follows from the well-known concepts of direct stiffness analysis that the system flexibility matrix is formed by simple addition of all element flexibility coefficients with like subscripts. Thus

$$U^* = \frac{\lfloor \bar{\Phi} \rfloor}{2} [f]\{\bar{\Phi}\} \tag{23.18}$$

where $[\Phi]=[\Phi_j \Phi_{x_j} \Phi_{y_j} \Phi_{xy_j}]$ lists all joint degrees of freedom in the system and $[f]$ is the system flexibility matrix, established as

$$f_{ij} = \sum_{e=1}^{e} f_{ij}^{e} \tag{23.19}$$

for the elements at the joint i. The global system complementary energy is therefore

$$\Pi_c = U_c = \frac{[\bar{\Phi}]}{2}[f]\{\bar{\Phi}\} \tag{23.20}$$

For a system with n degrees of freedom the matrix $[f]$ is of order $n \times n$ and $[\bar{\Phi}]$ is of order $1 \times n$.

Suppose the boundary conditions are represented by constraint equations. The element boundary conditions assemble to form the m global constraint equations:

$$[C]\{\bar{\Phi}\} - \{B\} = 0 \tag{23.21}$$

Before the condition of stationary complementary energy can be invoked account must be taken of the constraint equations. Using the method of Lagrange multipliers we have

$$\Pi_c = \frac{[\bar{\Phi}]}{2}[f]\{\bar{\Phi}\} - [\lambda]\{B\} + [\bar{\Phi}][\Gamma]^T\{\lambda\} \tag{23.22}$$

The function $\bar{\Pi}_c$ is subjected to the stationary condition so that by variation of all parameters in $\{\bar{\Phi}\}$ and $\{\lambda\}$

$$\begin{matrix} n \downarrow \\ \uparrow \\ m \downarrow \end{matrix} \left[\begin{array}{c|c} d & x^T \\ \hline c & 0 \end{array}\right] \left\{\begin{array}{c} \bar{\Phi} \\ \hline \lambda \end{array}\right\} = \left\{\begin{array}{c} 0 \\ \hline \{B\} \end{array}\right\} \tag{23.23}$$

$$|\leftarrow n \rightarrow|\leftarrow m \rightarrow| \qquad (n+m)$$

Equation (23.23) can now be solved for all $\{\Phi\}$ and $\{\lambda\}$. The stress function parameters, upon substitution into the differential relationships between the stresses and stress functions, yield the element stresses.

23.5 CONSTRUCTION OF BOUNDARY TRACTION ELEMENTS

As equations (23.15) to (23.17) show, stress function formulations of plane elasticity have traction boundary conditions which involve second derivatives of the stress function. These cannot be implemented directly with the bicubic Hermite element, which possesses only first derivatives. However, it is possible to modify such an element using the technique of blending function interpolation [14]. This allows us to introduce additional degrees of freedom along the side of such elements which lie on the boundary so as to incorporate the boundary condition directly while keeping the conformal character of the elements as well as its compatibility with the internal 16 degree of fredom bicubic Hermite elements.

This procedure follows the lines described in [15] and is outlined below for the case of a unit element, defined over the square $[0, 1]\times[0, 1]$ where the x axis coincides with the domain boundary which is subject to a boundary condition of the form of equation (23.15).

An appropriate element will carry the usual 16 degrees of freedom, plus it should be able to interpolate σ_y along $y=0$, as illustrated in Figure 23.3.

Consider now the cubic polynomials defined in $0\leqslant s\leqslant 1$:

$$\begin{aligned} C^1(s) &= 1-3s^2+2s^3 \\ C^2(s) &= 3s^2-2s^3 \\ C^3(s) &= s-2s^2+s^3 \\ C^4(s) &= -s^2+s^3 \end{aligned} \qquad 0\leqslant s\leqslant 1 \tag{23.24}$$

These are the shape functions for a one-dimensional Hermite cubic element in one dimension defined in the interval $0\leqslant s\leqslant 1$, i.e. the expression

$$C[\mu(s)]\equiv C^1(s)\mu(0)+C^2(s)\mu(1)+C^3(s)\mu_s(0)+C^4(s)\mu_s(1) \tag{23.25}$$

interpolates the function $\mu(s)$ and its first derivative at the points $s=0$ and $s=1$.

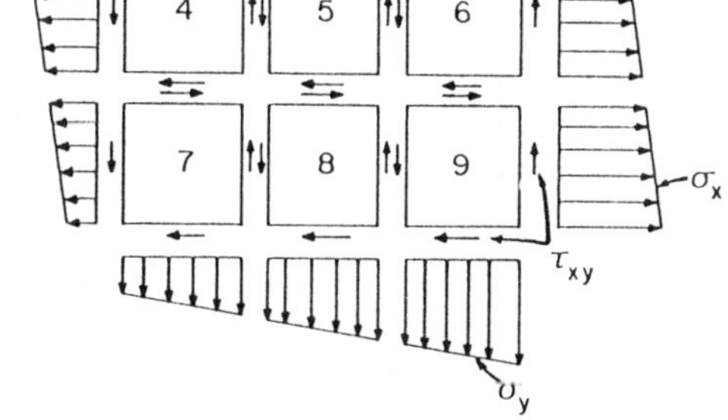

Element Type	Degrees of Freedom
1	20
2	18
3	20
4	18
5	16
6	18
7	20
8	18
9	20

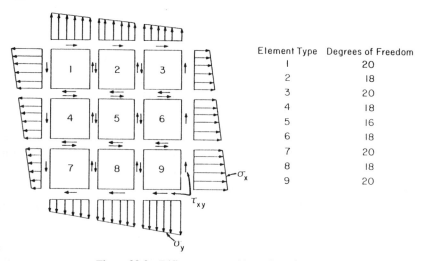

Figure 23.3 Different types of boundary elements

With these functions, we define the two projection operators:

$$P_x[\phi(x, y)] = C^1(x)\phi(0, y) + C^2(x)\phi(1, y) + C^3(x)\phi_x(0, y) + C^4(x)\phi_x(1, y)$$
$$P_y[\phi(x, y)] = C^1(y)\phi(x, 0) + C^2(y)\phi(x, 1) + C^3(y)\phi_y(x, 0) + C^4(y)\phi_y(1, 1) \tag{23.26}$$

These are two bivariate functions with the property that P_x takes the value ϕ and $\partial P_x/\partial x$ the value ϕ_x along the boundaries $y = 0$ and $y = 1$, while P_y equals ϕ and $\partial P_y/\partial y$ equals ϕ_y along $x = 0$ and $x = 1$.

If we now define an operator P as the Boolean sum of P_x and P_y, i.e.

$$\begin{aligned} P[\phi] &\equiv Px[\phi] + Py[\phi) - Px[Py[\phi]] \\ &= Px[\phi] + Py[\phi] - Py[Px[\phi]] \end{aligned} \tag{23.27}$$

the resulting bivariate function is such that it assumes the value (x, y) along the element boundary and $\partial P/\partial n = \phi_n$ there too, where n denotes the outward normal derivative to the boundary of the unit square.

We can now choose the particular form of $\phi(x, y)$ and $\phi_n(x, y)$ that we wish to have along each boundary almost arbitrarily, the only restriction coming from the commutativity of the compound operator required by (23.27).

We now introduce the quintic polynomials for $0 \leqslant s \leqslant 1$:

$$\begin{aligned} Q^1(s) &= 1 - 10s^3 + 15s^4 - 6s^5 \\ Q^2(s) &= 10s^3 - 15s^4 + 6s^5 \\ Q^3(s) &= s - 6s^3 + 8s^4 - 3s^5 \\ Q^4(s) &= -4s^3 + 7s^4 - 3s^5 \\ Q^5(s) &= \tfrac{1}{2}(s^2 - 3s^3 + 3s^4 - s^5) \\ Q^6(s) &= \tfrac{1}{2}(s^3 - 2s^4 + s^5) \end{aligned} \tag{23.28}$$

and define

$$Q[\mu(s)] \equiv Q^1(s)\mu(0) + Q^2(s)\mu(1) + Q^3(s)\mu_s(0) + Q^4(s)\mu_s(1) + Q_5(s)\mu_{ss}(0) + Q_6(s)\mu_{ss}(1)$$

which is the quintic Hermite interpolant of $\mu(s)$ for the function and its first and second derivatives at $s = 0$ and $s = 1$.

The required element is now given by (23.27) with Px and Py defined as

$$Px[\phi] = C^1(x)C[\phi(0, y)] + C^2(x)C[\phi(1, y)] + C^3(x)C[\phi_x(0, y)] + C^4(x)C[\phi_x(1, y)]$$
$$Py[\phi] = C^1(y)Q[\phi(x, 0)] + C^2(y)C[\phi(x, 1)] + C^3(y)C[\phi_y(x, 0)] + C^4(x)C[\phi_y(x, 1)] \tag{23.30}$$

Table 23.1 Shape functions for boundary rectangles

	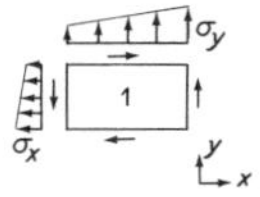	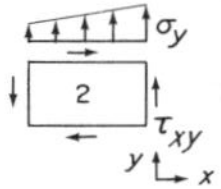	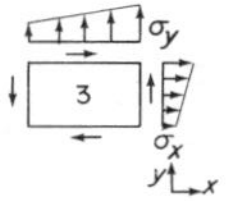	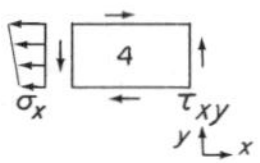	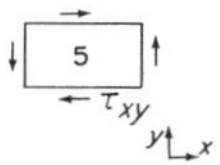	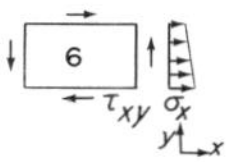	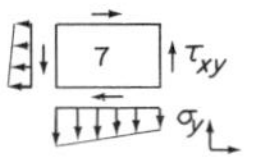	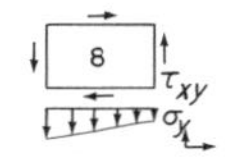	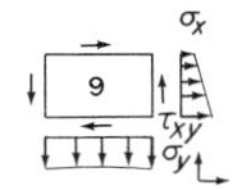
$N_1(\xi,\eta)$	$C^1(\xi)Q^1(\eta)$	$C^1(\xi)C^1(\eta)$	$C^1(\xi)C^1(\eta)$	$C^1(\xi)Q^1(\eta)$	$C^1(\xi)C^1(\eta)$	$C^1(\xi)C^1(\eta)$	$C^1(\xi)Q^1(\xi)+$ $Q^1(\xi)C^1(\eta)-$ $C^1(\xi)C^1(\eta)$	$Q^1(\xi)C^1(\eta)$	$Q^1(\xi)C^1(\eta)$
$N_2(\xi,\eta)$	$C^3(\xi)C^1(\eta)$	$C^3(\xi)C^1(\eta)$	$C^3(\xi)C^1(\eta)$	$C^3(\xi)C^1(\eta)$	$C^3(\xi)C^1(\eta)$	$C^3(\xi)C^1(\eta)$	$Q^3(\xi)C^1(\eta)$	$Q^3(\xi)C^1(\eta)$	$Q^3(\xi)C^1(\eta)$
$N_3(\xi,\eta)$	$C^1(\xi)Q^3(\eta)$	$C^1(\xi)C^3(\eta)$	$C^1(\xi)C^3(\eta)$	$C^1(\xi)C^3(\eta)$	$C^1(\xi)C^3(\eta)$	$C^1(\xi)C^3(\eta)$	$C^1(\xi)Q^3(\eta)$	$C^1(\xi)C^3(\eta)$	$C^1(\xi)C^3(\eta)$
$N_4(\xi,\eta)$	$C^3(\xi)C^3(\eta)$	$C^3(\xi)C^3(\eta)$	$C^3(\xi)C^3(\eta)$	$C^3(\xi)C^3(\eta)$	$C^3(\xi)C^3(\eta)$	$C^3(\xi)C^3(\eta)$	$C^3(\xi)C^3(\eta)$	$C^3(\xi)C^3(\eta)$	$C^3(\xi)C^3(\eta)$
$N_5(\xi,\eta)$	$C^1(\xi)Q^5(\eta)$	$C^2(\xi)C^1(\eta)$	$C^2(\xi)Q^1(\eta)$	$C^1(\xi)Q^5(\eta)$	$C^3(\xi)C^1(\eta)$	$C^2(\xi)Q^1(\eta)$	$Q^5(\xi)C^1(\eta)$	$Q^5(\xi)C^1(\eta)$	
$N_6(\xi,\eta)$	$C^2(\xi)C^1(\eta)$	$C^4(\xi)C^1(\eta)$	$C^4(\xi)C^1(\eta)$	$C^2(\xi)C^1(\eta)$	$C^4(\xi)C^1(\eta)$	$C^4(\xi)C^1(\eta)$	$C^1(\xi)Q^5(\eta)$	$Q^2(\xi)C^1(\eta)$	$Q^1(\eta)C^2(\xi)+$ $Q^3(\xi)C^1(\eta)$ $-C^1(\xi)C^1(\eta)$
$N_7(\xi,\eta)$	$C^4(\xi)C^1(\eta)$	$C^2(\xi)C^3(\eta)$	$C^2(\xi)Q^3(\eta)$	$C^4(\xi)C^1(\eta)$	$C^2(\xi)C^3(\eta)$	$C^2(\xi)Q^3(\eta)$	$Q^2(\xi)C^1(\eta)$	$Q^4(\xi)C^1(\eta)$	$Q^4(\xi)C^1(\eta)$
$N_8(\xi,\eta)$	$C^2(\xi)C^3(\eta)$	$C^4(\xi)C^3(\eta)$	$C^4(\xi)C^3(\eta)$	$C^2(\xi)C^3(\eta)$	$C^4(\xi)C^3(\eta)$	$C^4(\xi)C^3(\eta)$	$Q^4(\xi)C^1(\eta)$	$C^2(\xi)C^3(\eta)$	$C^2(\xi)Q^3(\eta)$
$N_9(\xi,\eta)$	$C^4(\xi)C^3(\eta)$	$Q^2(\xi)C^2(\eta)$	$C^2(\xi)Q^5(\eta)$	$C^3(\xi)C^3(\eta)$	$C^2(\xi)C^2(\eta)$	$C^2(\xi)Q^5(\eta)$	$C^2(\xi)C^3(\eta)$	$C^4(\xi)C^3(\eta)$	$C^4(\xi)C^3(\eta)$
$N_{10}(\xi,\eta)$	$Q^2(\xi)C^2(\eta)$	$Q^4(\xi)C^2(\eta)$	$C^2(\xi)Q^2(\eta)+$ $Q^2(\xi)C^2(\eta)-$ $C^2(\xi)C^2(\eta)$	$C^2(\xi)C^1(\eta)$	$C^4(\xi)C^2(\eta)$	$C^2(\xi)Q^2(\eta)$	$C^4(\xi)C^3(\eta)$	$Q^6(\xi)C^1(\eta)$	$Q^6(\xi)C^1(\eta)$
$N_{11}(\xi,\eta)$	$Q^4(\xi)C^2(\eta)$	$C^2(\xi)C^4(\eta)$	$Q^4(\xi)C^2(\eta)$	$C^4(\xi)C^2(\eta)$	$C^2(\xi)C^3(\eta)$	$C^4(\xi)C^2(\eta)$	$Q^6(\xi)C^1(\eta)$	$C^2(\xi)C^2(\eta)$	$C^2(\xi)Q^5(\eta)$
$N_{12}(\xi,\eta)$	$C^2(\xi)C^4(\eta)$	$C^4(\xi)C^4(\eta)$	$C^2(\xi)Q^4(\eta)$	$C^2(\xi)C^4(\eta)$	$C^4(\xi)C^4(\eta)$	$C^2(\xi)Q^4(\eta)$	$C^2(\xi)C^2(\eta)$	$C^4(\xi)C^2(\eta)$	$C^2(\xi)Q^2(\eta)$
$N_{13}(\xi,\eta)$	$C^4(\xi)C^4(\eta)$	$Q^6(\xi)C^2(\eta)$	$C^4(\xi)C^4(\eta)$	$C^4(\xi)C^4(\eta)$	$C^1(\xi)C^2(\eta)$	$C^4(\xi)C^4(\eta)$	$C^4(\xi)C^2(\eta)$	$C^2(\xi)C^4(\eta)$	$C^4(\xi)C^2(\eta)$
$N_{14}(\xi,\eta)$	$Q^6(\xi)C^2(\eta)$	$Q^1(\xi)C^2(\eta)$	$Q^6(\xi)C^2(\eta)$	$C^1(\xi)Q^2(\eta)$	$C^3(\xi)C^2(\eta)$	$C^2(\xi)Q^6(\eta)$	$C^2(\xi)C^4(\eta)$	$C^4(\xi)C^4(\eta)$	$C^2(\xi)Q^4(\eta)$
$N_{15}(\xi,\eta)$	$Q^1(\xi)C^2(\eta)+$ $C^2(\xi)Q^2(\eta)-$ $C^1(\xi)C^2(\eta)$	$Q^3(\xi)C^2(\eta)$	$C^2(\xi)Q^6(\eta)$	$C^3(\xi)C^2(\eta)$	$C^1(\xi)C^4(\eta)$	$C^1(\xi)C^2(\eta)$	$C^4(\xi)C^4(\eta)$	$C^1(\xi)C^2(\eta)$	$C^4(\xi)C^4(\eta)$
$N_{16}(\xi,\eta)$	$Q^3(\xi)C^2(\eta)$	$C^1(\xi)C^4(\eta)$	$Q^1(\xi)C^2(\eta)$	$C^1(\xi)Q^4(\eta)$	$C^3(\xi)C^4(\eta)$	$C^3(\xi)C^2(\eta)$	$C^1(\xi)C^2(\eta)$	$C^3(\xi)C^2(\eta)$	$C^2(\xi)Q^6(\eta)$
$N_{17}(\xi,\eta)$	$C^1(\xi)Q^4(\eta)$	$C^3(\xi)C^4(\eta)$	$Q^3(\xi)C^2(\eta)$	$C^3(\xi)C^4(\eta)$		$C^1(\xi)C^4(\eta)$	$C^3(\xi)C^2(\eta)$	$C^1(\xi)C^4(\eta)$	$C^1(\xi)C^2(\eta)$
$N_{18}(\xi,\eta)$	$C^3(\xi)C^4(\eta)$	$Q^5(\xi)C^2(\eta)$	$C^1(\xi)C^4(\eta)$	$C^1(\xi)Q^6(\eta)$		$C^3(\xi)C^4(\eta)$	$C^1(\xi)Q^4(\eta)$	$C^3(\xi)C^4(\eta)$	$C^3(\xi)C^2(\eta)$
$N_{19}(\xi,\eta)$	$Q^5(\xi)C^2(\eta)$		$C^3(\xi)C^4(\eta)$				$C^3(\xi)C^4(\eta)$		$C^1(\xi)C^4(\eta)$
$N_{20}(\xi,\eta)$	$C^1(\xi)Q^6(\eta)$		$Q^5(\xi)C^2(\eta)$				$C^1(\xi)Q^6(\eta)$		$C^3(\xi)C^4(\eta)$

It should be clear that if the boundary coincides with any of the other sides of the element, we simply need to replace the cubic interpolant over that boundary by the quintic interpolant; this is done in the two intersecting sides in the case of a corner where stresses are applied on two sides of the element. In any case, at least two elements need to be constructed, one for the corner case and the one just described; the rest can be obtained by rotations of these typical elements.

For the case of the element defined by equations (23.27) and (23.30) with degrees of freedom ordered as shown in Figure 23.1, the shape functions $N^i(x, y)$ are given by

$$
\begin{array}{lll}
N^1(x,y)=Q^1(x)C^1(y) & N^7(x,y)=C^4(x)C^2(y) & N^{13}(x,y)=C^3(x)C^3(y)\\
N^2(x,y)=Q^2(x)C^1(y) & N^8(x,y)=C^3(x)C^2(y) & N^{14}(x,y)=C^4(x)C^3(y)\\
N^3(x,y)=C^2(x)C^2(y) & N^9(x,y)=C^1(x)C^3(y) & N^{15}(x,y)=C^4(x)C^4(y)\\
N^4(x,y)=C^1(x)C^2(y) & N^{10}(x,y)=C^2(x)C^3(y) & N^{16}(x,y)=C^3(x)C^4(y)\\
N^5(x,y)=Q^3(x)C^1(y) & N^{11}(x,y)=C^2(x)C^4(y) & N^{17}(x,y)=Q^5(x)C^1(y)\\
N^6(x,y)=Q^4(x)C^1(y) & N^{12}(x,y)=C^1(x)C^4(y) & N^{18}(x,y)=Q^6(x)C^1(y)
\end{array}
\tag{23.31}
$$

It is easily verified that the compatibility condition of (23.27) is satisfied as well as for a corner element. Basically the only difficulties introduced by these new boundary elements are the variable number of degrees of freedom, which can now be 16, 18, or 20, depending on whether the element is interior, side, or corner, and the need of a higher order Gaussian quadrature to numerically integrate elements in the boundary, where quintic polynomials appear. On the other hand, stress boundary conditions can now be directly applied.

There are nine different types of boundary elements (see Figure 23.3). Table 23.1 gives the shape functions for these different types of boundary rectangles.

For each type of boundary element a flexibility matrix is programmed, using the 4×4-point Gauss–Legendre interpolation formula as a reasonable compromise between computational speed and numerical accuracy.

23.6 NUMERICAL EXAMPLES

The problems studied are shown in Figures 23.4 and 23.5.

In the first example, shown in Figure 23.4, an elastic analysis of a rectangular plate subject to parabolically distributed stresses σ_y along opposite faces is performed. For this example a 'classical' solution is available [16] for the stresses at the centre of the plate. Uniform meshes with four and sixteen elements are used to discretize one quadrant of the plate after taking

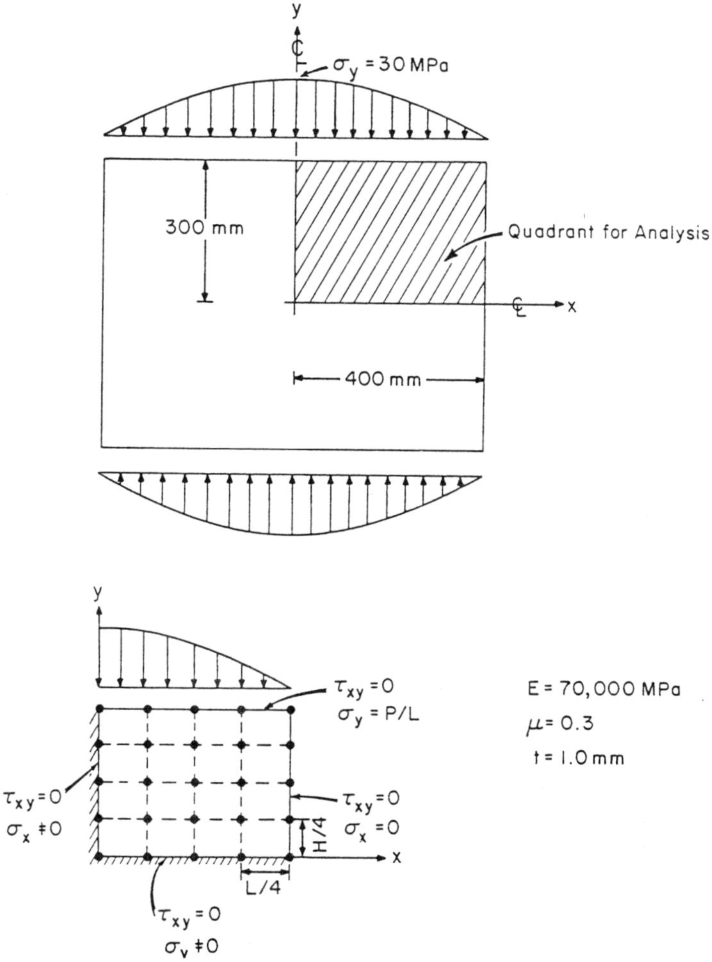

Figure 23.4 Rectangular plate under parabolically varying edge stress

advantage of the symmetry of the problem about both coordinate axes. In this problem we are free to use the standard 16 degrees of freedom bicubic Hermite elements along $x = 0$ and $y = 0$ or to obtain the stresses σ_x along $y = 0$ and σ_y along $x = 0$ directly using the proposed elements.

The second example, as shown in Figure 23.5, consists of a simply supported deep beam subject to a uniform load. Symmetry about the vertical centreline allows the discretization of half the problem where an eighteen-element uniform mesh is used. Displacement boundary conditions not present in the previous example are applied over part of the boundary.

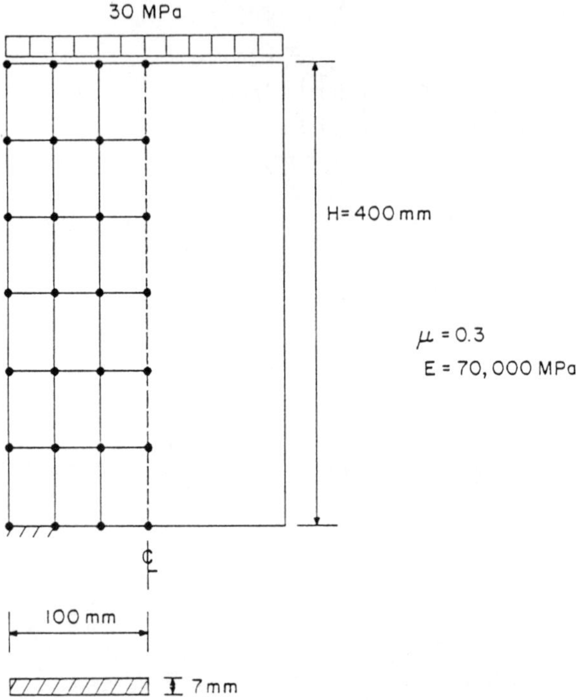

Figure 23.5 Uniformly loaded, simply supported deep beam

23.7 CONCLUDING REMARKS

A robust direct flexibility alternative to the existing predominant direct stiffness method requires more than an effective scheme for resolution of boundary conditions. It requires a convenient approach to conditions where the equations of equilibrium are not homogeneous, a simple representation of stiffener elements, a means for the efficient analysis of three-dimensional elasticity problems, and numerous other extensions. Concepts for dynamic [17] and buckling [18] analysis have been proposed and must be integrated with these extensions.

REFERENCES

1. J. H. Argyris and S. Kelsey, *Energy Theorems and Structural Analysis*, Butterworths, London, 1960.
2. B. Fraejis de Veubeke and O. C. Zienkiewicz, 'Strain-energy bounds in finite element analysis by slab-analogy', *J. of Strain Analysis*, **2,** No. 4 (1967).
3. R. H. Gallagher, *Finite Element Analysis: Fundamentals*, Prentice-Hall, Englewood Cliffs, New Jersey, 1975.

4. R. G. Charlwood, 'Dual formulations of linear elasticity using finite elements', *Computer Aided Engineering* (Ed. G. L. M. Gladwell), University of Waterloo, 1971.
5. R. H. Gallagher and A. K. Dhalla, 'Direct flexibility finite element elastoplastic analysis', *Proc. First Int. Conf. on Struct. Mechanics in Reactor Technology*, **6,** Part M, 444–462 (1972).
6. E. F. Rybicki and L. A. Schmit, 'An incremental complementary energy method of nonlinear stress analysis', *AIAA J.*, **8,** 10, 119 (1969).
7. V. B. Watwood and B. J. Hartz, 'An equilibrium stress field model for finite element solutions of two dimensional elastostatic problems', *Int. J. Solids and Structures*, **4,** 857–873 (1968).
8. L. S. D. Morley, 'The triangular equilibrium element in the solution of plate bending problems', *The Aero. Quarterly*, **19,** 149–169 (1968).
9. L. S. D. Morley, 'A triangular equilibrium element with linearly varying bending moments for plate bending problems', *J. Royal Aeronautical Society*, **October 1967,** 71 (1967).
10. B. M. Fraejis de Veubeke and A. Millard, 'Discretization of stress fields in the finite element method', *J. Franklin Inst.*, **302,** No. 5/6, 389–412 (November/December 1976).
11. Z. M. Elias, 'Duality in finite element methods', *Proc. ASCE, J. Eng. Mech. Div.*, **94,** No. EM4, 931–946 (1968).
12. R. L. Taylor and O. C. Zienkiewicz, 'Complementary energy with penalty functions in finite element analysis', in *Energy Methods in Finite Element Analysis* (Ed. R. Glowinski), pp. 153–174, John Wiley & Sons, 1979.
13. R. Lawther and A. P. Kabaila, 'Single domain equilibrium approximations in two dimensional elasticity', *Proc. Third Int. Conf. in Australia on Finite Element Methods*, pp. 101–113, July 1979.
14. W. J. Gordon and C. A. Hall, 'Transfinite element method: Blending function interpolation over arbitrary curved element domains', *Num. Math.*, **21,** 109–129 (1973).
15. D. S. Watkins, 'On the construction of conforming rectangular plate elements', *Int. J. Num. Meth. in Eng.*, **10,** 925–933 (1976).
16. S. Timoshenko and J. N. Goodier, *Theory of Elasticity*, 3rd ed., McGraw-Hill, 1970.
17. B. Tabarrok and D. Sodhi, 'On the generalization of stress function procedure for the dynamic analysis of plates', *Int. J. Num. Meth. in Eng.*, **5,** 523–542 (1972).
18. C. Sundararajan, 'Stability analysis of plates by a complementary energy method', *Int. J. Num. Meth. in Eng.*, **15,** No. 3, 343–350 (1981).

Hybrid and Mixed Finite Element Methods
Edited by S. N. Atluri, R. H. Gallagher, and O. C. Zienkiewicz

Chapter 24

Mixed Finite Element Approximations via Interior and Exterior Penalties for Contact Problems in Elasticity

J. T. Oden

24.1 INTRODUCTION

The use of exterior penalty formulations of boundary value problems as a basis for the development of finite element methods has gained much popularity in recent times. Such methods frequently lead to fewer unknowns than multiplier methods; they sometimes provide regularity that admits the use of numerical schemes that might not be otherwise applicable, and they can produce one-parameter families of approximate solutions, depending upon the penalty parameter ε, all members of which have a useful physical or mathematical significance. For a discussion of some aspects of finite element methods based on exterior penalty ideas, see, for instance, [1] or [2].

There are two important aspects of penalty methods that motivate the present investigation. First of all, penalty methods are widely used in optimization problems in $\mathbb{R}^n$, and in the optimization literature one finds not only exterior penalty methods in frequent use, but also so-called interior penalty methods, hybrid (exterior/interior) methods, exponential penalties, etc. Are any of these alternative penalty approaches useful in variational formulation and approximation of boundary value problems? Second, the penalty finite element methods are closely related to mixed finite element methods and a brief examination of some of their properties seems appropriate for a symposium devoted to this general class of methods.

We shall be concerned here with penalty-type formulations and finite element approximations of a class of contact problems in elasticity of the form

$$\begin{aligned} (\mathbf{u}, \sigma) \in V \times W : a(\mathbf{u}, \mathbf{v}) + [\sigma, v_n] &= f(\mathbf{v}) \qquad \text{for all} \quad v \in V \\ [\tau - \sigma, u_n - s] &\geq 0 \qquad \text{for all} \quad \tau \in M \end{aligned} \tag{24.1}$$

wherein the following notation and conventions are used:

$V=\{\mathbf{u}=(u_1, u_2, \ldots, u_N)\in(H^1(\Omega))^N \mid \mathbf{u}=0$ on $\Gamma_D\}$ = space of admissible displacements
(Here Ω is a smooth, open, bounded domain in $\mathbb{R}^N$ with boundary $\partial\Omega$, $H^1(\Omega)$ is the usual Sobolev space order 1, and Γ_D is the portion of the boundary of Ω on which displacements are prescribed.)

$W=H^{1/2}(\Gamma_C)$ = Sobolev space of functions defined on the contact surface Γ_C ($\bar{\Gamma}_C\cap\bar{\Gamma}_D=\phi$)

$a(\mathbf{u},\mathbf{v})=\int_\Omega E_{ijkl}u_{k,l}v_{i,j}\,dx$ = the virtual work done by the stress

$\sigma_{ij}(\mathbf{u})=E_{ijkl}u_{k,l}$ on the strain $\varepsilon_{ij}(\mathbf{v})=\frac{1}{2}(v_{i,j}+v_{j,i})$

$v_{i,j}=\partial v_i/\partial x_j$, etc., $dx=dx_1\,dx_2\cdots dx_N$

$[\cdot,\cdot]$ = duality pairing ('scalar product') on $W'\times W$

$f(\mathbf{v})=\int_\Omega \mathbf{f}\cdot\mathbf{v}\,dx+\int_{\Gamma_F}\mathbf{t}\cdot\mathbf{v}\,ds$ = virtual work of external forces:

$\mathbf{f}$ = body forces $\in L^2(\Omega)$

$\mathbf{t}$ = surface tractions $\in L^2(\Gamma_F)$, Γ_F being the portion of the boundary on which tractions are applied ($f\in V'$)

u_n = normal trace of $\mathbf{u}\in V$; if $\mathbf{n}$ is a unit exterior normal to Γ and $\gamma: V\to W$ is the trace operator, then $u_n\overset{\text{def}}{=}\gamma(\mathbf{u})\cdot\mathbf{n}$

σ = the contact pressure $[\sigma=\sigma_n=\sigma_{ij}(\mathbf{u})n_in_j]$

$M=\{\tau\in W' \mid \tau\leqslant 0\}$ = the negative cone in W' corresponding to the ordering, $\phi\leqslant 0$, $\phi\in W\Rightarrow\phi\leqslant 0$ a.e. in Γ_C

Problem (24.1) is merely a variational statement of Signorini's problem in elastostatics: the problem of unilateral contact of an elastic body with a rigid frictionless foundation. When $\mathbf{u}$ is sufficiently smooth, (24.1) is equivalent to the classical elastostatics problem:

$$
\begin{aligned}
&\left.\begin{aligned}\sigma_{ij}(\mathbf{u})_{,j}+f_i&=0\\ \sigma_{ij}(\mathbf{u})&=E_{ijkl}u_{k,l}\end{aligned}\right\}\quad(\text{in }\Omega)\\
&u_i=0\ \text{ on }\ \Gamma_D;\qquad \sigma_{ij}(\mathbf{u})n_j=t_i\qquad(\text{on }\Gamma_F)\\
&\left.\begin{aligned}\sigma=\sigma_{ij}(\mathbf{u})n_in_j&\leqslant 0\\ u_n&\leqslant s\\ \sigma(u_n-s)&=0\\ \sigma_{Ti}=\sigma_{ij}n_j-\sigma n_i&=0\end{aligned}\right\}\quad(\text{on }\Gamma_C)\\
&\qquad i,j,k,l=1,2,\ldots,N
\end{aligned}
\tag{24.2}
$$

The elasticities E_{ijkl} are assumed to have the usual symmetries and positive definitiveness so that the bilinear form $a(\cdot,\cdot)$ is symmetric, continuous, and

V-elliptic, i.e. constants $m, M > 0$ exist such that for all $\mathbf{u}, \mathbf{v} \in V$,

$$a(\mathbf{u}, \mathbf{v}) \leqslant M\,|\mathbf{u}|_1\,|\mathbf{v}|_1, \qquad a(\mathbf{u}, \mathbf{u}) \geqslant M\,|\mathbf{u}|_1^2 \tag{24.3}$$

where $|\cdot|_1$ is the norm on V;

$$|\mathbf{v}|_1^2 = \int_\Omega v_{i,j} v_{i,j}\,\mathrm{d}x \qquad \text{for} \quad i, j = 1, 2, \ldots, N \tag{24.4}$$

Under these conditions, (24.1) is equivalent to the problem of minimizing the total potential energy functional

$$F(\mathbf{v}) = \tfrac{1}{2} a(\mathbf{v}, \mathbf{v}) - f(\mathbf{v}) \qquad \mathbf{v} \in V \tag{24.5}$$

subject to the constraint $u_n \leqslant s$, s being a measure of the initial 'gap' between the body and the foundation prior to deformation given in W. Thus, the contact pressure σ enters as a Lagrange multiplier corresponding to this constraint.

We shall explore two distinct penalty approaches to problem (24.1) which are best described within the context of the constrained minimization problem:

$$\text{Find } \mathbf{u} \in K \text{ such that } \min_{\mathbf{v} \in K} F(\mathbf{v}) = F(\mathbf{u}) \tag{24.6}$$

where K is the constraint set

$$K = \{\mathbf{v} \in V \mid v_n \leqslant s \qquad (\text{on } \Gamma_C)\} \tag{24.7}$$

In this paper, we discuss several finite element methods based on these penalty formulations; we comment on their relationship with certain mixed methods; we describe the results of several numerical experiments; and we comment on advantages and disadvantages of each approach.

24.2 EXTERIOR AND INTERIOR PENALTIES

The functional F in (24.5) and the constraint set K in (24.7) have several properties convenient in standard optimization problems:

F is

(1) strictly convex
(2) Gâteaux-differentiable; indeed,

$$\langle DF(\mathbf{u}), \mathbf{v}\rangle = a(\mathbf{u}, \mathbf{v}) - f(\mathbf{v}); \qquad \mathbf{u}, \mathbf{v} \in K \tag{24.8}$$

(3) coercive (i.e. $F(\mathbf{v}) \to +\infty$ as $|\mathbf{v}|_1 \to \infty$)

K is non-empty, convex, and closed in V

Under conditions (24.8), we are always guaranteed the existence of a unique solution u to the minimization problem (24.6). Moreover, this solution is characterized as the solution of the variational inequality

$$a(\mathbf{u}, \mathbf{v}-\mathbf{u}) \geq f(\mathbf{v}-\mathbf{u}) \qquad \text{for all } v \in K \tag{24.9}$$

We remark that such constrained minimization problems can also be approached using Lagrange multipliers and this involves seeking saddle points $(\mathbf{u}, \sigma)$ of the functional

$$L: V \times M \to \mathbb{R}; \qquad L(\mathbf{v}, \tau) = F(\mathbf{v}) - [\tau, v_n - s] \tag{24.10}$$

which satisfy (24.1).

Neither the variational inequality (24.9) nor the saddle-point formulation using (24.10) provide particularly convenient bases for the construction of finite element methods. In (24.9) we have the problem of constructing approximations of the constraint set K and this is often difficult. In (24.10) the set K does not appear, but we find explicitly in the formulation the contact pressure τ (or σ); hence, (24.10) leads to a mixed formulation in which an additional unknown (the multiplier) appears.

It appears that the use of penalty methods may overcome both of these disadvantages: the constraint set then enters only in the construction of the penalty functional, only one unknown field appears in the formulation, and the contact pressures can be calculated with little effort after the unconstrained problem has been solved.

Penalty methods correspond to perturbations of the constrained minimization problem (24.6) in which a minimizing sequence $\{\mathbf{u}_\varepsilon\}$ is produced ($\varepsilon > 0$, $\varepsilon \to 0$), the entries $\mathbf{u}_\varepsilon$ being solutions to the perturbed minimization problem.

We must emphasize that the optimization problem is used as a basis for discussing penalty ideas only for convenience and simplicity; penalty methods of the type we discuss here are also applicable to cases in which $a(\cdot, \cdot)$ is unsymmetric and, therefore, no energy function F exists.

Penalty methods in optimization theory typically fall into one of two general categories:

(a) *Exterior penalty methods*, in which the perturbed problems are constructed so that the sequence $\mathbf{u}_\varepsilon$ lies exterior to the constraint set K and
(b) *Interior penalty methods*, in which the perturbed problems are constructed so that the sequence $\mathbf{u}_\varepsilon$ lie interior to the constraint set K.

There are, of course, 'hybrid' penalty methods which combine features of both exterior and interior penalty methods, and we shall make a brief comment on the role of one such hybrid in use in our work on finite elasticity.

24.2.1 Exterior penalty methods

In the exterior penalty method, we introduce an *exterior penalty functional* $P: V \to \mathbb{R}$ with the properties

Property 24.1 P is weakly lower semicontinuous on V (it is sufficient to replace this requirement by the conditions that P be convex and Gâteaux-differentiable on V).

Property 24.2 P is positive semidifinite on K:

$$P(\mathbf{v}) \geqslant 0, \qquad P(\mathbf{v}) = 0 \qquad \text{iff } \mathbf{v} \in K$$

Given a P with these properties, we construct a perturbation F_ε of the functional F of (24.5) according to

$$\varepsilon > 0, \qquad F_\varepsilon(\mathbf{v}) = F(\mathbf{v}) + \frac{1}{\varepsilon} P(\mathbf{v}), \qquad \mathbf{v} \in V \tag{24.11}$$

We can then easily establish the following result:

Theorem 24.1 *Let conditions* (24.1) *hold and let P be given which satisfies Properties* 24.1 *and* 24.2. *Then, for each* $\varepsilon > 0$, *there exists a unique minimizer* u_ε *of the functional* (24.4). *Moreover, each minimizer is characterized as a solution of the variational boundary value problem,*

$$a(\mathbf{u}_\varepsilon, \mathbf{v}) + \frac{1}{\varepsilon} \langle DP(\mathbf{u}_\varepsilon), \mathbf{v} \rangle = f(\mathbf{v}) \qquad \text{for all } \mathbf{v} \in V \tag{24.12}$$

where $DP: V \to V'$ *is the Gâteaux differential of P. In addition, there exists a subsequence* $\varepsilon' \to 0$ *such that* $\mathbf{u}_{\varepsilon'}$ *converges weakly to the unique minimizer* $\mathbf{u} \in K$ *of F (i.e. u is the solution of* 24.6).

For the contact problem, we may take

$$P(v) = \frac{1}{2} \int_{\Gamma_C} (v_n - s)_+^2 \, ds \tag{24.13}$$

where $(\ )_+$ denotes the positive part of the function (for a complete discussion of these ideas, see [2]). Then (24.11) becomes

$$a(\mathbf{u}_\varepsilon, \mathbf{v}) + \varepsilon^{-1}(((\mathbf{u}_\varepsilon)_n - s)_+, v_n) = f(\mathbf{v}) \qquad \text{for all } \mathbf{v} \in V \tag{24.14}$$

where $(\cdot, \cdot)$ is the L^2 inner product on $L^2(\Gamma_C)$. An approximation σ_ε of the contact pressure σ can also be constructed according to

$$\sigma_\varepsilon = -\varepsilon^{-1}((\mathbf{u}_\varepsilon)_n - s)_+ \tag{24.15}$$

However, an additional condition must be imposed if we are to have convergence of σ_ε to σ in W'. In particular, we demand that the following Babuska–Brezzi condition holds [cf 9–12].

There exists $\beta > 0$ such that

$$\beta \|\tau\|_{W'} \leq \sup_{\mathbf{v} \in V} \frac{\|[\tau, v_n]\|}{|\mathbf{v}|_1} \qquad \text{for all } \tau \in w \tag{24.16}$$

We can then establish the following result (for a proof, see [2]).

Theorem 24.2 *If, in addition to the conditions of Theorem* 24.1, *condition* (24.6) *holds, then the solutions* $\mathbf{u}_\varepsilon$ *of* (24.14) *and the functions* σ_ε *of* (24.15) *converge strongly in V and W respectively to the solution* $(\mathbf{u}, \sigma)$ *of* (24.1). *Moreover,*

$$|\mathbf{u} - \mathbf{u}_\varepsilon|_1 + \|\sigma - \sigma_\varepsilon\|_{W'} \leq C\sqrt{\varepsilon} \tag{24.17}$$

24.2.2 Interior penalty methods

While general interior penalty theories can be devised (see, for example [3]), their use in variational boundary value problems appears, at this time, to involve more specialized considerations than the exterior penalty methods. For an interior penalty formulation of problem (24.5), we introduce an *interior penalty* or *barrier functional* $Q: V \to \bar{\mathbb{R}}$ satisfying the following conditions:

Property 24.3 Q is proper, convex, and lower semicontinuous.

Property 24.4

$$Q(v) = \begin{cases} < +\infty & \text{if } \mathbf{v} \in \mathring{K} \equiv \text{interior of } K \\ = +\infty & \text{if } \mathbf{v} \notin \mathring{K} \end{cases}$$

Property 24.5 $F \equiv F + \varepsilon Q$, $\varepsilon > 0$, is coercive, $F_\varepsilon : V \to \bar{\mathbb{R}}$.

Property 24.6 For every $\mathbf{v}_\varepsilon \in$ interior of K which converges weakly to $\mathbf{v} \in K$ as $\varepsilon \to 0$, we have $\liminf_{\varepsilon \to 0} F(\mathbf{v}_\varepsilon) \geq 0$.

These conditions were employed by Oden [4] to prove the following result:

Theorem 24.4 *Let Properties* 24.3 *to* 24.6 *and* (24.8) *hold. Then, for all* $\varepsilon > 0$ *there exists a unique minimizer* $\mathbf{u}_\varepsilon$ *of the functional* $F_\varepsilon = F + \varepsilon Q$. *Moreover,* $\mathbf{u}_\varepsilon$ *lies in the interior of the constraint set K and there exists a subsequence* $\varepsilon' \to 0$ *such that* $\mathbf{u}_{\varepsilon'}$ *converges weakly in V to the solution* $\mathbf{u}$ *of the minimization problem* (24.6). *If, in addition, the barrier functional Q is Gâteaux-differentiable, then the minimizers* $\mathbf{u}_\varepsilon$ *of* F_ε *are characterized by*

$$a(\mathbf{u}_\varepsilon, \mathbf{v}) + \varepsilon \langle DQ(\mathbf{u}_\varepsilon), \mathbf{v} \rangle = f(\mathbf{v}) \qquad \text{for all } \mathbf{v} \in V \tag{24.18}$$

For our contact problem, we can take, the following barrier functionals:

(a) *The inverse barrier functional*:

$$Q_1(\mathbf{v}) = \begin{cases} -\int_{\Gamma_C} \frac{ds}{v_n - s} & \text{if} \quad v \in \mathring{K} \\ +\infty & \text{if} \quad v \notin \mathring{K} \end{cases} \tag{24.19}$$

(b) *The logarithmic barrier functional*:

$$Q_2(\mathbf{v}) = \begin{cases} -\int_{\Gamma_C} \ln(s - v_n)\, ds & \text{if } v \in \mathring{K} \\ +\infty & \text{if } \mathbf{v} \in \mathring{K} \end{cases} \tag{24.20}$$

It can be shown [4] that both of these functionals satisfy properties 24.3 to 24.6 for F and K given in (24.8). Moreover, we can compute

$$\begin{aligned} \langle DQ_1(\mathbf{u}), \mathbf{v}\rangle &= \int_{\Gamma_C} v_n (u_n - s)^{-2}\, ds \qquad \text{for } \mathbf{u}, \mathbf{v} \in \mathring{K} \\ \langle DQ_2(\mathbf{u}), \mathbf{v}\rangle &= \int_{\Gamma_C} v_n (s - u_n)^{-1}\, ds \qquad \text{for } \mathbf{u}, \mathbf{v} \in \mathring{K} \end{aligned} \tag{24.21}$$

Approximate contact pressures are then

$$\sigma_\varepsilon = \varepsilon((\mathbf{u}_\varepsilon)_n - s)^2 \quad \text{for} \quad Q_1; \qquad \sigma_\varepsilon = \varepsilon((\mathbf{u}_\varepsilon)_n - s)^{-1} \; \textit{for } Q_2 \tag{24.22}$$

When (24.16) holds, one can again show that a subsequence of perturbed pressures $\{\sigma_{\varepsilon'}\}$ exists, for either choice in (24.22), that converges weakly in W' to an element σ, where $(\mathbf{u}, \sigma)$ is the solution of (24.1).

Under stronger conditions, one can guarantee the strong convergence of $(\mathbf{u}_\varepsilon, \sigma_\varepsilon)$ to $(\mathbf{u}, \sigma)$; e.g. if $(-p)^{-1/2}, (-p_\varepsilon)^{-1/2} \in W$, then for Q_1 we can show that

$$|\mathbf{u} - \mathbf{u}_\varepsilon| \leqslant C\sqrt{\varepsilon} \tag{24.23}$$

whereas if $(-p)^{-1}, (-p_\varepsilon)^{-1} \in W$ and Q_2 is used we have the higher order result

$$|\mathbf{u} - \mathbf{u}_\varepsilon|_1 < C\varepsilon \tag{24.24}$$

24.2.3 Hybrid penalty methods

There are other types of penalty methods which combine features of the exterior and interior penalty methods discussed above. These are sometimes referred to as hybrid methods. As an example, let P be an exterior penalty functional for the constraint set K in (24.7) and let Q be a barrier functional for this set. Then we can consider the problem of seeking minimizers of the perturbed functional

$$F_\varepsilon(\mathbf{v}) = F(\mathbf{v}) + \frac{1}{\varepsilon} P(\mathbf{v}) + \varepsilon Q(\mathbf{v}) \qquad \text{for } \mathbf{v} \in V \tag{24.25}$$

While we have not analysed such formulations in detail, such formulations have been employed in numerical experiments in solving finite elasticity problems in which the total potential energy is of the form

$$F(\mathbf{v}) = \int_\Omega W(\nabla\mathbf{v})\,\mathrm{d}x - E_0 \int_\Omega \ln(\det \nabla\mathbf{v})\,\mathrm{d}x$$
$$+ \tfrac{1}{2}E_1 \int_\Omega (\det \nabla\mathbf{v} - 1)^2\,\mathrm{d}x \tag{24.26}$$

where W is a polynomial in invariants of $\nabla\mathbf{v}^T \cdot \nabla v$ ($\mathbf{v}$ being the motion) and E_0 and E_1 are 'material' constants. Clearly, F represents the energy in a compressible elastic material (since $\det \nabla\mathbf{v} \neq 1$). However, if E_0 is regarded as an interior penalty parameter and E_1^{-1} as an exterior penalty parameter then (24.26) represents a hybrid penalty functional for minimization problems in incompressible finite elasticity. Some numerical results obtained using this strategy are to be discussed in a forthcoming paper.

24.3 FINITE ELEMENT APPROXIMATIONS

Formally, the construction of finite element approximations of penalized variational problems is straightforward. We partition Ω into a collection $\{\Omega_e\}_{e=1}^E$ of finite elements ($\bar{\Omega} = \cup_{e=1}^E \bar{\Omega}_e$) over which continuous piecewise polynomial approximations of the components v_i of the virtual displacements are constructed. In this way we obtain a basis for a finite dimensional subspace V_h of V, h being the usual mesh parameter. By regular refinements of the mesh, we produce a family $\{V_h\}_{0<h\leq 1}$ of subspaces of V with special interpolation properties. Here we shall assume that the spaces V_h are such that (at least as $h \to 0$) interpolation estimates of the following type hold:

> For any $\mathbf{v} \in (H^m(\Omega))^N \cap V$, constants $C_1, C_2 > 0$, independent $\mathbf{v}$ and h, and an element $\tilde{\mathbf{v}}^h \in V_h$ exists such that

$$|\mathbf{v} - \tilde{\mathbf{v}}^h|_1 \leq C_1 h^\mu \|\mathbf{v}\|_{m,\Omega}$$
$$\mu = \min(k, m-1) \tag{24.27}$$
$$\|\gamma(\mathbf{v}) - \gamma(\tilde{\mathbf{v}}^h)\|_{0,p,\Gamma_C} \leq C_2 h^{r-p} \|\gamma(\mathbf{v})\|_{r,\Gamma_C}$$

Here k is the order of the complete polynomial used in the construction of V_h (that is $P_k(\Omega) \subset V_h$, $k \geq 1$), γ is the trace operator mapping V onto W, m and r are non-negative real numbers, but p may be a negative real number.

Typical elements of the space V_h are, of course, of the form

$$v_i^h(\mathbf{x}) = \sum_{\alpha=1}^{M} v_i^\alpha \phi_\alpha(\mathbf{x}), \qquad \mathbf{v}^h = (v_1^h, v_2^h, \ldots, v_N^h) \tag{24.28}$$

where $\mathbf{x} = (x_1, x_2, \ldots, x_N) \in \bar{\Omega}$, $v_i^\alpha = v_k^h(\mathbf{x}^\alpha)$, $\mathbf{x}^\alpha$ being a nodal point in the mesh and ϕ_α the global basis functions normalized so that $\phi_\alpha(\mathbf{x}^\beta) = \delta_\alpha^\beta$; $\alpha, \beta = 1, 2, \ldots, M$.

24.3.1 Exterior penalty finite element approximations

Exterior penalty finite element approximations of (24.12) with P given by (24.14) involve seeking $\mathbf{u}_\varepsilon^h \in V_h$ such that

$$a(\mathbf{u}_\varepsilon^h, \mathbf{v}^h) + \varepsilon^{-1} I[((\mathbf{u}_\varepsilon^h)_n - s)_+ v_n^h] = f(\mathbf{v}^h) \qquad \text{for all } \mathbf{v}^h \in V_h \tag{24.29}$$

where $I[\cdot]$ indicates a numerical quadrature rule of the type

$$I(f) = \sum_{e=1}^{E} \sum_{j=1}^{G_e} W_j^e f(\boldsymbol{\xi}_j^e) \qquad \text{for } f \in C^0(\bar{\Omega}) \tag{24.30}$$

Here W_j^e are the quadrature weights at the jth integration point in element e and ξ_j^e denotes the quadrature point j in element e. Of course, numerical integration is also used, in practice, to evaluate the stiffness matrix $a(\phi_\alpha, \phi_\beta)$, but the order G_e of the quadrature rule used in this calculation can be taken to be large enough to yield exact values of these integrals. The penalty term $I(((\mathbf{u}_\varepsilon^h)_n - s)_+ v_n^h)$, on the other hand, may or may not be integrated exactly (see [5, 6] or [2]).

For each choice of the space V_h (equivalently for each choice of polynomials basis ϕ_α) and for each integration rule $I(\cdot)$, there is defined a corresponding finite dimensional space W_h approximating W'. For example, if C^0 piecewise quadratics are used to form V_h and exact integration is used, W_h will consist of (traces of) C^0 piecewise polynomials defined on the approximation Γ_C^h of Γ_C. On the other hand, if quadratics are used for V_h and the trapezoid rule is used for $I(\cdot)$, W_h will be spanned by C^0 piecewise linear functions. We then have defined W_h so that

$$I(\tau^h \hat{\tau}^h) = [\tau^h, \hat{\tau}^h] \qquad \text{for all } \tau^h, \hat{\tau}^h \in W_h \tag{24.31}$$

and W_h is such that the exterior penalty approximation of the contact pressure is then uniquely defined by

$$\sigma_\varepsilon^h \in W_h : \sigma_\varepsilon^h(\boldsymbol{\xi}_j^e) = -\varepsilon^{-1}((\mathbf{u}_\varepsilon^h)_n - s)_+(\boldsymbol{\xi}_j^e) \tag{24.32}$$

In other words, *in the discrete problem we impose condition* (24.15) *only at the quadrature points corresponding to the rule* $I(\cdot)$.

Conditions (24.30) and (24.31) are insufficient to guarantee the stability of the scheme (24.29), (24.31). We must also impose a discrete version of the

Babuska–Brezzi condition (24.16). In particular, for stability of the approximation (24.31) in $L^2(\Gamma_C)$, we introduce the discrete Babuska–Brezzi condition:

There exists $\beta_h > 0$ such that

$$\beta_h \|\tau^h\|_{0,\Gamma_C} \leq \sup_{v^h \in V_h} \frac{|I(\tau^h v_n^h)|}{|\mathbf{v}^h|_1} \qquad \text{for all } \tau^h \in W_h \tag{24.33}$$

The following result establishes the relationship between the exterior penalty method (24.31), (24.32) and a form of mixed methods for this class of problems.

Theorem 24.4 *Let the conditions* (24.8) and (24.27) hold and let $(\mathbf{u}, \sigma)$ be the solution of (24.1). In addition, let conditions (24.31), (24.32), and (24.33) hold. Then there exists a unique solution $\mathbf{u}_\varepsilon^h$ of (24.29) for every $\varepsilon > 0$. Moreover, for fixed h, $(\mathbf{u}_\varepsilon^h, \sigma_\varepsilon^h)$ converges to a solution $(\mathbf{u}^h, \sigma^h) \in V_h \times W_h$ of the mixed finite element approximation,

$$\begin{aligned} a(\mathbf{u}^h, \mathbf{v}^h) - I(\sigma^h v_n^h) &= f(\mathbf{v}^h) \qquad &&\text{for all } \mathbf{v}^h \in V_h \\ I[(\tau^h - \sigma^h)(u^h - s)] &\geq 0 &&\text{for all } \tau^h \in M_h \end{aligned} \tag{24.34}$$

where

$$M_h = \{\tau^h \in W_h \mid \tau^h(\boldsymbol{\xi}_j^e) \leq 0, \quad 1 \leq e \leq E, \quad 1 \leq j \leq G_e\} \tag{24.35}$$

Moreover,

$$\begin{aligned} |\mathbf{u} - \mathbf{u}^h|_1 \leq C\{&|\mathbf{u} - \mathbf{v}^h|_1 + \beta_h^{-1} \|u_n - v_n^h\|_{0,\Gamma_C} + \|\sigma - \tau^h\|_{W'} + |[\tau - \sigma^h, u_n - s]|^{1/2} \\ &+ |[\tau^h - \sigma, u_n - s]|^{1/2} + E_1 \end{aligned} \tag{24.36}$$

$$\|\sigma - \sigma^h\|_{0,\Gamma_C} \leq \|\sigma - \tau^h\|_{0,\Gamma_C} + \beta_h^{-1}\{\|\sigma - \tau^h\|_{W'} + M\,|\mathbf{u} - \mathbf{u}^h|_1 + E_2\}$$

for all $\mathbf{v}^h \in V_h$, $\tau^h \in M_h$, $\tau \in W'$, E_1 and E_2 representing quadrature errors.

For proofs of results of this type, see [2]. The important point we wish to make here is that the convergence of the method depends upon the interpolation properties (24.27) of V_h, the stability parameter β_h and whether or not it depends upon h, and the integration errors E_1, E_2. Also, the relationship between the exterior penalty methods apparent in (24.34) is worth noting. Similar equivalences have been pointed out by Malkus and Hughes in [13]; see also Becouvier in [8] and the author's papers [2, 4, 5, 7]. We comment further on the implication of these results in Section 24.4 of this paper.

24.3.2 Interior penalty finite element methods

An exterior penalty method for problem (24.1) consists of seeking $w_\varepsilon^h \in V_h$ such that

$$a(\mathbf{w}_\varepsilon^h, \mathbf{v}^h) + \varepsilon I[(s - \mathbf{w}_\varepsilon^h \cdot \mathbf{n})^\alpha \mathbf{v}^h \cdot \mathbf{n}] = f(\mathbf{v}^h) \qquad \text{for all } \mathbf{v}^h \in V_h \tag{24.37}$$

where $\alpha = 2$ if the barrier function Q_1 is used while $\alpha = 1$ if Q_2 is employed. Again, V_h and $I(\cdot)$ determine a space W_h of approximate multipliers (contact pressure), but in the present case the members of W_h are complicated rational polynomials. The interpolation properties of such spaces are not known; nevertheless, several features of the exterior penalty method carry over to the interior penalty case. There is, for example, also a form of mixed method directly related to (24.34). The following properties of these methods were established in [14];

(a) Under the conditions laid down thus far, there exists a unique solution $\mathbf{w}_\varepsilon^h$ to (24.34) for each $\varepsilon > 0$.

(b) If $\{W_h\}$ is a family of approximation spaces corresponding to V_h and $I(\cdot)$ (as noted above) and $\bar{\cup}_h W_h = W'$, if the approximate contact pressure p_ε^h is now defined by

$$p_\varepsilon^h \in W_h;\, I(p_\varepsilon^h v_n^h) = \varepsilon I[(s - \mathbf{w}_\varepsilon^h \cdot \mathbf{n}) + v_n^h) + v_n^h] \qquad \text{for all } v^h \in V_h \tag{24.38}$$

and (most importantly) if the discrete Babuska–Brezzi condition (24.33) holds, then there exists $\varepsilon_k' \to 0$ such that $(w_{\varepsilon_k'}^h, p_{\varepsilon_k'}^h)$ converges strongly in $V_h \times W_h$ to a pair $(\mathbf{w}^h, p^h)$ which represents the interior penalty mixed finite element approximation

$$\begin{aligned} &a(\mathbf{w}^h, \mathbf{v}^h) - I(p^h v_n) = f(\mathbf{v}^h) && \text{for all } \mathbf{v}^h \in V_h \\ &I[(\tau^h - p^h)(w_n^h - s)] \geq 0 && \text{for all } \tau^h \in M_h \end{aligned} \tag{24.39}$$

Here s is assumed to be continuous and $M_h = \{\tau^h \in W_h \mid \tau_n \leq 0\}$.

(c) If, in addition to the assumptions in (b), we have

$$(-p_\varepsilon^h)^{-1/2}, (-p^h)^{-1/2} \in W_h \text{ for } Q = Q_1$$

or

$$(-p_\varepsilon^h)^{-1}, (-p^h)^{-1} \in W_h \text{ for } Q = Q_2$$

and if the spaces W_h exhibit interpolation properties of the type

$$\left.\begin{aligned} \|\sqrt{-p} - \sqrt{-p^h}\|_{0,\Gamma_C} &\leq Ch^\sigma \text{ for } Q = Q_1 \\ \|(-p)^{-1} - (-p^h)^{-1}\|_{0,\Gamma_C} &\leq Ch^\sigma \text{ for } Q = Q_2 \end{aligned}\right\} \sigma > 0 \tag{24.40}$$

then constants C_1', C_2', independent of h and ε, exist such that (as $h \to 0$)

$$|\mathbf{w}^h - \mathbf{w}_\varepsilon^h|_1 \leq \begin{cases} C(h^\sigma\sqrt{\varepsilon} + \sqrt{\varepsilon}) & \text{for the inverse barrier functional} \\ Ch^\sigma\varepsilon & \text{for the logarithmic barrier function} \end{cases}$$

(d) If the conditions of item (c) hold and $(\mathbf{u}, \sigma)$ is the solution of (24.1), then

$$
\begin{aligned}
|\mathbf{u}-\mathbf{u}^h|_1 \leqslant{} & \bar{C}_1[|\mathbf{u}-\mathbf{v}^h|_1+\|\sigma-\tau^h\|_{W'}+\beta_h^{-1}\,\|(\mathbf{u}-\mathbf{v}^h)\cdot\mathbf{n}\|_{0,\Gamma_C}] \\
& +\bar{C}_2[\|\sigma-\tau^h\|_{0,\Gamma_C}\,\|u_n-v_n^h\|_{0,\Gamma_C}+[\tau-p^h+q\tau^h-\sigma,u_n-s]^{1/2}] \\
\|\sigma-p^h\|_{0,\Gamma_C} \leqslant{} & \bar{C}_3\beta_h^{-1}(\|p-\tau^h\|_{W'}+|\mathbf{u}-\mathbf{w}^h|_1)+\|\sigma-\tau^h\|_{0,\Gamma_C}
\end{aligned}
\tag{24.42}
$$

for all $\mathbf{v}^h \in V_h$, $\tau^h \in W_h$, $\bar{C}_\alpha$ being constants independent of h. To reduce these estimates further, we must identify the interpolation properties of the rational polynomials in W_h. As noted earlier, this project remains to be completed.

It is clear that the interior penalty methods may behave somewhat differently from the exterior methods. Both, however, produce special types of mixed methods as $\varepsilon \to 0$ and the stability of both depend upon the satisfaction of a discrete Babuska–Brezzi condition of the form (24.33).

24.4 ALGORITHMS AND NUMERICAL RESULTS

24.4.1 The exterior method

The success of the exterior penalty method or its closely associated mixed methods for the finite element analysis of contact problems depends upon the existence of a stability parameter β_h in (24.33) which is independent of h. An analysis of this condition in [2] yielded the following conclusions:

(a) The use of mixed methods of the type in (24.34) which employ non-conforming finite element approximations of the multipliers generally lead to a stability parameter β_h which is zero or dependent on h and are therefore unstable. This, of course, corresponds to the use of integration rules $I(\cdot)$ in the penalty approach which do not involve integration points which coincide with boundary nodes.
(b) Conforming mixed approximations of σ which employ polynomials of degrees less than or equal to those of the degree k of traces of elements in V_h are typically stable (in $L^2(\Gamma_C)$) (β_h is independent of h), but suboptimal rates of convergence are obtained if $r<h$. Thus, it appears that one should use exact integration of the penalty terms whenever feasible, except possibly in large problems in which an accurate description of the contact pressure is not needed.

As specific examples, we cite for a two-dimensional contact problem with regular boundary and a uniform mesh:

Comparison of exterior penalty methods		
V_h	$I(\cdot)(W_h)$	Stability (β_h)/convergence rate
C^0 piecewise biquadratics	Simpton's rule/C^0 piecewise quadratics	Stable/optimal ($\lvert\mathbf{u}-\mathbf{u}^h\rvert_1 = O(h^2)$)
C^0 piecewise biquadratics	Trapezoid rule/C^0 piecewise linear	Stable/optimal ($\lvert\mathbf{u}-\mathbf{u}^h\rvert_1 = O(h)$)
C^0 piecewise biquadratics	One-point gauss/piecewise constant	Unstable
C^0 piecewise biquadratics	Two-point Gauss/discontinuous piecewise linear	Unstable
C^0 piecewise bilinear	Trapezoid rule/C^0 piecewise linear	Stable/optimal
C^0 piecewise bilinear	Two-point Gauss/discontinuous piecewise linear	Unstable

Having selected a stable and convergent exterior penalty scheme, the questions of solving the systems of non-linear equations produced by (24.29) arises. There are many choices available, but we have found that a simple relaxation algorithm with projections is very effective for problems of this type. The basic steps in this algorithm are listed as follows:

(a) Set the first iterate $\mathbf{u}_\varepsilon^{h(1)} = 0$.

(b) Define

$$\alpha_t(\mathbf{u}_\varepsilon^{h(t)}) = \mathbf{u}_\varepsilon^{u(t)} \cdot \mathbf{n} - s$$

$$\beta_{t+1}(\mathbf{u}_\varepsilon^{h(t)}) = \begin{cases} \alpha_t(\mathbf{u}_\varepsilon^{h(t)}) & \text{if } \alpha_t(\mathbf{u}_\varepsilon^{h(t)}) > 0 \\ 0 & \text{if } \alpha_t(\mathbf{u}_\varepsilon^{h(t)}) \leqslant 0 \end{cases}$$

(c) Obtain the tth iterate $\mathbf{u}_\varepsilon^{h(t)}$ by solving the linear system

$$a(\mathbf{u}_\varepsilon^{h(t)}, \mathbf{v}^h) + \varepsilon^{-1} I[\alpha_t(\mathbf{u}_\varepsilon^{h(t)}) v_n^h] = f(\mathbf{v}^h) \qquad \text{for all } \mathbf{v}^h \in V_h$$

(d) Repeat this process until the relative error

$$e_R^{t+1} = \frac{\lvert \mathbf{u}_\varepsilon^{h(t+1)} - \mathbf{u}_\varepsilon^{h(t)} \rvert_1}{\lvert \mathbf{u}_\varepsilon^{h,(t+1)} \rvert_1}$$

is less that a preassigned tolerance ε_R.

Typically, the identifications $\alpha_t = \mathbf{u}^{h,t} \cdot \mathbf{n} - s$ are made only at nodes on Γ_C^h. The numerical results described below were obtained using this algorithm.

24.4.2 The interior method

The performance of the interior penalty methods depend strongly on the form of the barrier functional chosen and, of course, the spacc V_h and integration rule $I(\cdot)$. We have performed numerical experiments using the inverse and logarithmic barrier functionals Q_1 and Q_2 described earlier, and we have employed C^0 piecewise biquadratic and C^0 piecewise bilinear (nine-node quad and four-node quad) elements in the displacement approximation. We have also chosen the integration order G to be quite high so as to avoid effects of inaccurate numerical integration of the penalty terms. For this rather limited number of cases we have experienced no numerical instabilities of the type sometimes encountered with exterior methods, and we conjecture that for these methods the stability parameter β_h is a positive constant, independent of h.

The interior penalty finite element approximation (24.37) leads to a system of non-linear algebraic equations of the form

$$\mathbf{K}\mathbf{u} + \varepsilon \mathbf{Q}(\mathbf{u}) = \mathbf{f}$$

where $\mathbf{K}$ is the usual stiffness matrix, $\mathbf{u}$ the vector of unknown nodal values of $\mathbf{u}_\varepsilon^h$, $\mathbf{Q}(\mathbf{u})$ the non-linear term arising from the barrier functional, and $\mathbf{f}$ the load vector. In the numerical experiments described below, we solve this system using a standard Newton–Raphson scheme which employs a Gauss elimination profile solver for treating linear equations in each iterate.

24.4.3 Numerical results

We consider as a numerical example a problem that has become a benchmark in much of our numerical work on contact problems (see, for example, [2, 3, 5, 7]): the problem of indentation of a rigid cylindrical punch into a rectangular slab of homogeneous, isotropic, linearly elastic material. The problem is one of plane strain. The dimensions of the punch and slab and the mechanical properties of the material are indicated in Figure 24.1(a). The slab is allowed to displace tangentially to the rigid supports, as indicated. The indentation δ of the sphere into the cylinder is prescribed. The problem approximates the classical Hertz problem of the indentation of an elastic half-space by a rigid sphere.

As a first test, we consider an analysis of this problem using the exterior penalty methods with V_h constructed using the mesh of nine-node biquadratic elements indicated in Figure 24.1(b) and an indentation $\delta = 0.8$. We employ first Simpson's rule and then two-point Gaussian quadrature in the quadrature formula $I(\cdot)$, meaning that the multipliers τ^h are C^0 piecewise quadratics or discontinuous piecewise linears. Computed contact pressures for each case are shown in Figure 24.2. As predicted, the conforming

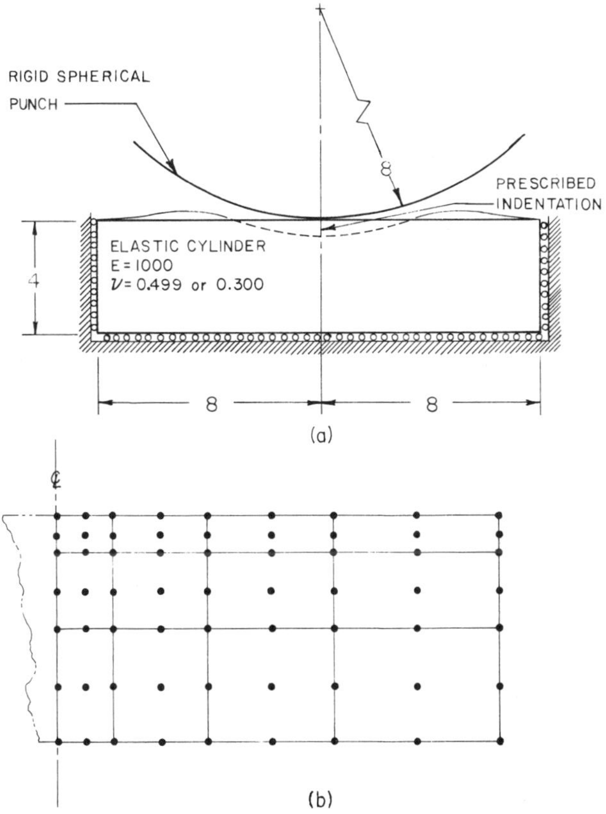

Figure 24.1 (a) An elastic slab indented by a rigid cylindrical punch. (b) A discretization of half a plane of the slab using nine-node biquadratic elements

contact pressure approximation corresponding to Simpson's rule is stable whereas the method employing discontinuous pressures is unstable. Indeed, in this latter method, severe oscillations are observed in the calculated contact pressures and these grow in magnitude and frequency as the mesh is refined. The calculations illustrated in Figure 24.2 correspond to an almost incompressible material with Poisson's rates $\nu = 0.3$. Similar behaviour was observed for materials with $\nu = 0.3$.

Another test case was the problem of an elastic half-cylinder pushed into a frictionless rigid foundation. The finite element model and other details of this problem are given in Figure 24.3(a). Simpson's rule is employed to integrate the exterior penalty term and good results were again obtained using only three or four iterations of the algorithm described earlier. The computed deformed shape of the body is shown in Figure 24.3(a) and the

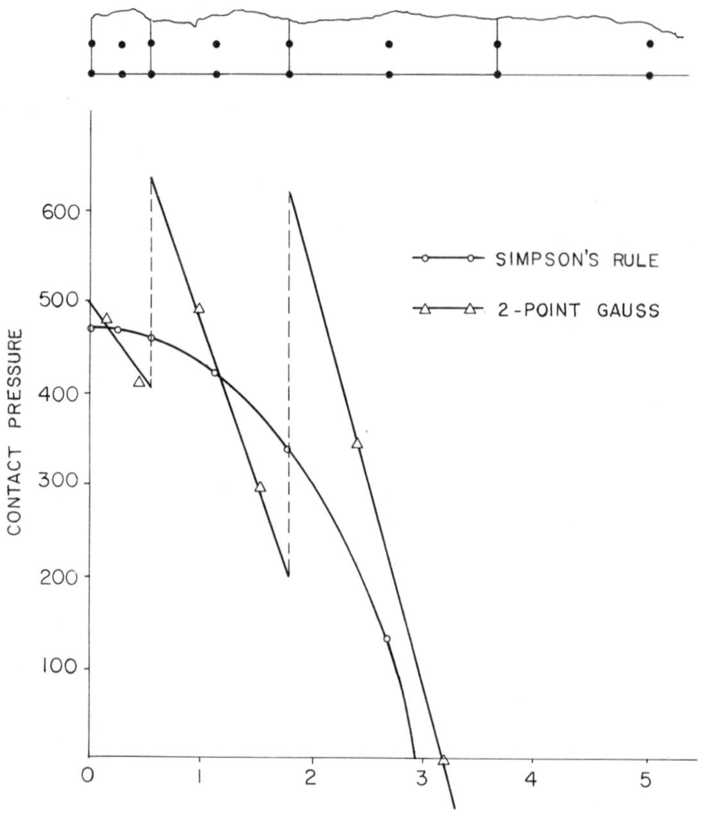

Figure 24.2 Contact pressure profiles computed using exterior penalty methods with the penalty functional integrated using Simpson's rule and two-point Gaussian quadrature

computed contact pressure profile is indicated in Figure 24.3(b). The results compare very well with the classical Hertz solution for a cylinder pushed into a rigid half-space. For example, the maximum contact pressure predicted by the Hertz theory is $\sigma_{max} = 291$ compared to the calculated $\sigma_{max} = 286$. Likewise, the length b of the contact zone is 3.496 according to the Hertz theory whereas we calculated a value of $b = 3.6$.

Next we analyse a similar class of problems using interior penalty finite element methods. Again the problem indicated in Figure 24.1(a) is treated, except that now a Poisson's ratio of $\nu = 0.3$ with $\delta = 0.5$ is employed and the uniform mesh of nine-node biquadratic elements shown in Figure 24.4(a) is utilized. Both inverse and logarithmic barrier functionals were employed,

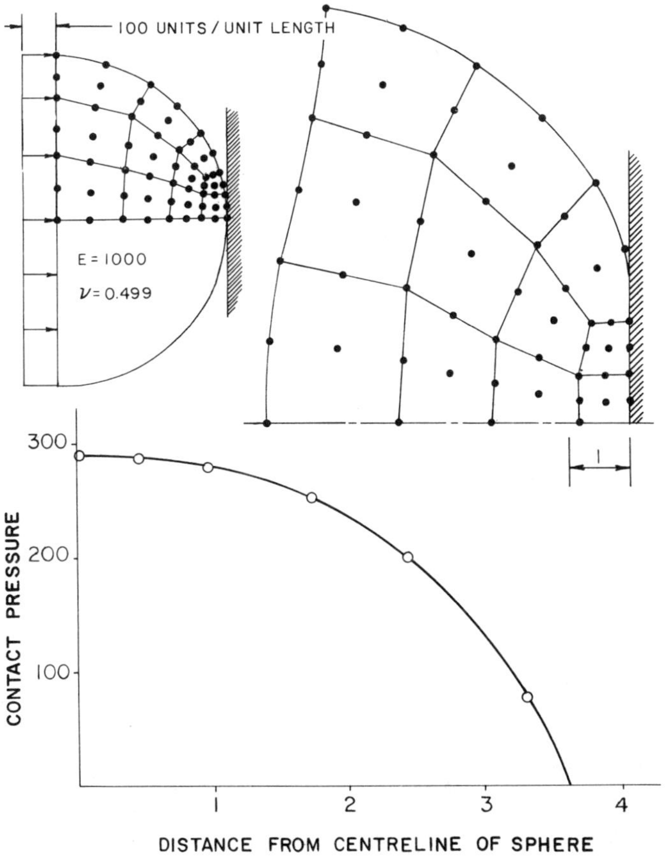

Figure 24.3 Computed deformed shape and contact pressure of the problem of compressing an elastic half-cylinder into a rigid half-space using exterior penalty methods

computed deformed contact areas and contact pressures being plotted in Figure 24.4(b) and (c). Results obtained by using the inverse barrier functional and the logarithmic barriers were virtually indistinguishable after eight to ten iterations, but the behaviour of the algorithm for each choice of barrier was quite different. The inverse barrier functional, for example, leads to more slowly convergent schemes than did the logarithmic. This same problem was also solved using the exterior penalty algorithm and essentially the same numerical results were obtained as the interior methods with only four iterations.

For a comparison of the interior inverse, interior logarithmic, and exterior penalty methods for this problem, we examine the computed values of the

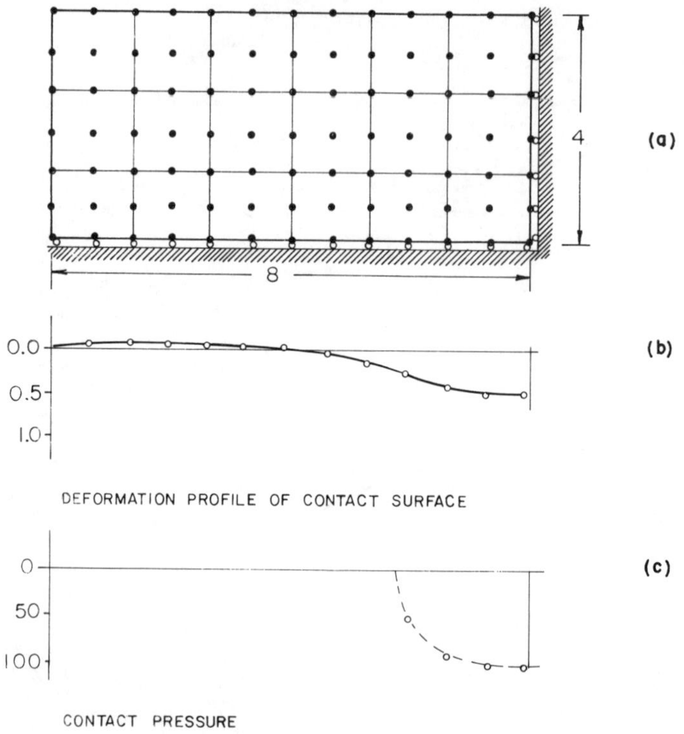

Figure 24.4 (a) Uniform mesh of nine-node biquadratic elements used to solve the problem in Figure 24.1 by interior penalty methods for the case $\nu = 0.3$, (b) computed deformed contact surface, and (c) contact pressure

contact pressure at various nodes along the contact surface in the table below.

These results are typical of our experiments with interior methods: while they appeared to be stable and to ultimately produce good approximations of the contact area and pressures, they are generally 'stiff' and require many iterations to bring them within reasonable limits of the actual solution. The logarithmic interior penalty method generally proved to be superior to the inverse interior method, requiring consistently fewer Newton–Raphson iterations.

We also observed that the behaviour of our interior methods (particularly the rate of convergence of the iterative scheme) was very sensitive to the choice of starting values: an initial displacement too far from the correct contact area sometimes leads to a divergent scheme, whereas many iterations are sometimes needed to produce a small improvement when profiles close to the actual solution are reached.

Computed nodal contact pressures						
Method	Penalty parameter ε	Contact pressure (lb/in^2) Node number 1	2	3	4	Number of iterations
Exterior penalty	10^{-3}	209.3	196.2	179.1	112.5	4
Interior penalty	10^{-5}	221.0	255.2	176.0	134.6	2
logarithmic barrier	10^{-5}	208.8	195.7	179.5	112.5	9
Interior penalty	10^{-8}	216.0	207.9	178.6	118.8	2
inverse barrier	10^{-8}	208.4	195.8	179.1	112.5	8

The exterior penalty methods, of course, may be unstable, but we feel that these methods are now sufficiently well understood that these unstable cases can be avoided. Hence, at this time, the exterior methods seem to have some significant advantages over interior penalty methods for the types of problems considered here. Nevertheless, there may be certain classes of non-linear problems, such as the finite elasticity problems characterized by (24.26), for which interior methods or hybrid interior/exterior may have significant advantages over other methods.

ACKNOWLEDGEMENT

The work reported here was completed during the course of a research project supported by the U.S. Air Force of Scientific Research under Contract F-49620-78-C-0083. It is also a pleasure to acknowledge many fruitful discussions of the material discussed here with Professor N. Kikuchi, Dr. Y. J. Song, and Mr. S. J. Kim.

REFERENCES

1. M. Bercouvier, 'Perturbation of mixed variational problems', *RAIRO*, **12,** No. 3, 1978.
2. J. T. Oden, N. Kikuchi and Y. J. Song, 'Reduced integration and exterior penalty methods for finite element approximations of contact problems in incompressible elasticity', *TICOM Report*, No. 80-2, Austin, 1980.
3. Y. J. Song, J. T. Oden and N. Kikuchi, 'Discrete LBB-conditions for RIP-finite element methods', *TICOM Report*, No. 80-7, Austin, 1980.
4. J. T. Oden, 'RIP methods for Stokesian flows', in *Finite Elements in Fluids* (Eds. R. H. Gallagher *et al.*, Vol. IV, John Wiley and Sons, London, 1981 (to appear).
5. J. T. Oden, 'Exterior penalty methods for contact problems in elasticity', *Lecture Notes in Mathematics*, Springer-Verlag, Berlin, 1981 (to appear).

6. J. T. Oden, 'Penalty-finite element approximations of unilateral problems in elasticity', *Approximation Theory*, (Ed. G. Lorenz), Vol. III, pp. 1–5, Academic Press, New York, 1981.
7. J. T. Oden and N. Kikuchi, 'Finite element methods for constrained problems in elasticity', *Int. J. Num. Meth. in Eng.* (to appear).
8. N. Kikuchi and J. T. Oden, *Contact Problems Elasticity*, SIAM Publications, Philadelphia (to appear).
9. I. Babuska, 'Error bounds for finite element method', *Num. Math.*, **16,** 322–333 (1971).
10. I. Babuska, 'The finite element method for elliptic equations with discontinuous coefficients', *Computing*, **5,** 207–213 (1970).
11. I. Babuska and A. K. Aziz 'Survey lectures on the mathematical foundations of the finite element method', in *Mathematical Foundations of the Finite Element Method with Applications to Partial Differential Equations* (Ed. A. K. Aziz), pp. 1–354, Academic Press, New York, 1972.
12. F. Brezzi, 'On the existence, uniqueness, and approximation of saddle point problems arising from Lagrangian multipliers', *RAIRO*, **8,** 129–151, 1974.
13. D. S. Malkus and T. J. R. Hughes, 'Mixed finite element methods—reduced and selective integration techniques: A unification of concepts', *Comp. Meth. Appl. Mech. Eng*:, **15,** 63–81, 1978.
14. J. T. Oden and S. J. Kim, 'Interior penalty methods for the Signorini problem in elastostatics', *Computers and Mathematics with Applications*, **8,** 1, 35–56 (1981).

Hybrid and Mixed Finite Element Methods
Edited by S. N. Atluri, R. H. Gallagher, and O. C. Zienkiewicz

Chapter 25

A General Penalty/Mixed Equivalence Theorem For Anisotropic, Incompressible Finite Elements

Thomas J. R. Hughes and David S. Malkus

25.1 BACKGROUND

The work presented here represents the resolution of two seemingly different questions raised by the equivalence theorem for reduced/selective integration methods [1]. In [1] it is shown that the use of numerical integration in a prescribed fashion leads to the equivalence of penalty and perturbed Lagrangian matrices and solutions obtained by the finite element discretization of a variety of constrained media problems. In consequence of heuristic arguments presented in [1] and later made quite rigorous in [2, 3], the integration formula applied to the volumetric terms must be a 'reduced' formula. The equivalence theorem states that the penalty method is equivalent to a *numerically integrated* perturbed Lagrangian. In a linear, constant coefficient elasticity problem in Cartesian coordinates, it turns out that the numerical integration formula often exactly integrates the perturbed Lagrangian. This is the case, for example, with four-node bilinear elements with one-point quadrature or nine-node biquadratic elements with two-by-two quadrature. Unfortunately, there is ample evidence that these elements do not satisfy the LBB stability condition required to guarantee velocity–displacement and pressure convergence in the usual norms [4 to 7]. Although it seems that the LBB condition is not necessary in some cases if convergence in other norms is acceptable [4 to 7], alternative elements have been proposed which use reduced integration formulae of even lower accuracy than those mentioned above. An example would be the nine-node biquadratic with three-point, or even one-point quadrature. These elements seem to satisfy the LBB condition [4], but if the reduced/selective integration approach of [1] is employed—even in Cartesian coordinates—the equivalent perturbed Lagrangian may not be integrated exactly.

Thus the reduced/selective integration techniques may introduce an integration error term whose stability and accuracy must be investigated [8, 9].

This seems to pose some particularly thorny questions in the non-linear case when the tensor of elastic coefficients in the relevant integrands are incremental quantities based on a finite element solution at a previous iterate in a sequence of linearizations. So the first question addressed here is that of uncoupling the integration error term from the choice of pressure interpolation. The equivalence theorem will establish when this can be done while still retaining an exact—or sometimes only approximate—equivalence between penalty and perturbed Lagrangian formulations.

The second question addressed here concerns the interaction between the reduced/selective approach and the form of the tensor of material constants. In [1], only isotropic materials were considered, though the reduced/selective approach can be generalized to the anisotropic case when, as in isotropy, there is a natural splitting of the tensor of elastic constants into a deviatoric part and a dilatational part. The dilatational part must appear multiplied by a bulk modulus factor which is large compared to the elastic constants of the deviatoric term. When this splitting can be accomplished, reduced integration can be applied selectively to the dilatational term, and there is an equivalence to a perturbed Lagrangian formulation when the appropriate conditions are met [1]. In the general anisotropic case, making such a splitting may be computationally inconvenient. This seems to be particularly the case when the tensor of material constants arise from an incremental scheme in non-linear elasticity. The second question addressed here is how to treat the linearized constraint of near incompressiblity by splitting the strain components rather than the tensor of material constants. This leads to a treatment of nearly incompressible materials which has corresponding penalty and perturbed Lagrangian forms, but which can be implemented without requiring *a priori* segregation of the volumetric components of the tensor of material constants.

25.2 VARIATIONAL PRINCIPLES

The theorem proved here applies rigorously to linear analysis when a penalty method and corresponding perturbed Lagrangian can be identified for the continuous problem.

25.2 Notation and preliminaries

Following the work of Key [10], we use the following strain deviator:

$$e'_{ij} = e_{ij} - \frac{\beta}{3}\,\theta\,\delta_{ij}$$

$$\theta \equiv e_{ii} \tag{25.1}$$

where e_{ij} is the infinitesimal strain tensor. In [10], Key speculates that it might be of interest to consider a continuous variation of the parameter β. For the purposes at hand, we will argue below that $\beta=0$ and $\beta=1$ are the only useful values. In [11], Hughes uses equation (25.1) with $\beta=1$ to develop his 'B matrix method' for improvement of the FEM dilatational approximation. The B matrix method is similar to, and often coincides with, reduced/selective integration techniques. The work presented here may be regarded as the establishment of an equivalence theorem for the B matrix method.

In general the techniques discussed here are applied to a constitutive equation

$$\sigma_{ij}=C_{ijkl}e_{kl} \tag{25.2}$$

for a compressible material. At the operational level, we make no assumption on C_{ijkl} other than that it has the required symmetries [12] which reduce the number of possible independent components of 21. Such a material may be nearly incompressible in the sense that its compliance to isotropic stress is very small, and this would be reflected in the values of C_{ijkl} in some unspecified manner. For *theoretical purposes only*, we assume a deviatoric/dilatational splitting of C_{ijkl} which allows identification of a bulk modulus. First we observe that the following decomposition of C_{ijkl} can always be accomplished:

$$\begin{aligned}
C_{ijkl} &= C^{d}_{ijkl}+\tfrac{1}{3}C^{dv}_{kl}\,\delta_{ij}+\tfrac{1}{3}C^{vd}_{ij}\,\delta_{kl}+\kappa\,\delta_{ij}\,\delta_{kl}\\
C^{dv}_{kl} &\equiv C_{nnkl}-\tfrac{1}{3}C_{nnmm}\,\delta_{kl}\\
C^{vd}_{ij} &\equiv C_{ijnn}-\tfrac{1}{3}C_{mmnn}\,\delta_{ij}\\
\kappa &\equiv C_{mmnn}/9\\
C^{d}_{ijkl} &\equiv C_{ijkl}-\tfrac{1}{3}C^{dv}_{kl}\,\delta_{ij}-\tfrac{1}{3}C^{vd}_{ij}\,\delta_{kl}-\kappa\,\delta_{ij}\,\delta_{kl}
\end{aligned} \tag{25.3}$$

Our assumption will be that there is a convenient compressibility parameter, γ, that C^{d}_{ijkl} and C^{dv}_{ij} $(=C^{vd}_{ij})$ stay bounded as $\gamma\to 0$ while $\kappa(\gamma)\to+\infty$. These assumptions seem reasonable for a continuum theory in which pressure and shear can be uncoupled. We also assume $\kappa(\gamma)$ is spatially constant.

25.2 The Variational Principles

The basic variational principle discussed here is the familiar

$$J(u_m)=\frac{1}{2}\int_{\Omega} e_{ij}C_{ijkl}e_{kl}\,\mathrm{d}\Omega+\hat{F}(u_m) \tag{25.4}$$

where u_m is the displacement vector associated with e_{ij} and $\hat{F}$ is a linear functional representing the energy of applied loads. We assume boundary conditions which imply positive definiteness of the strain energy. A major point of this paper is that we can deal directly with equation (25.4) by the strain-splitting method introduced in [11] and amplified below. The correct handling of the volumetric terms will be automatic. For theoretical purposes, we consider the segregation of the tensor of material constants into terms which are bounded as $\gamma \to 0$ and the penalty term identified by $\kappa(\gamma)$. To this end we define

$$C'_{ijkl} = C^{d}_{ijkl} + \tfrac{1}{3}C^{dr}_{kl}\,\delta_{ij} + \tfrac{1}{3}C^{vd}_{ij}\,\delta_{kl} \tag{25.5}$$

The segregated variational principle of equation (25.5) is

$$J(u_m) = \frac{1}{2}\int_{\Omega} \{e_{ij}C'_{ijkl}e_{kl} + \kappa(\gamma)\theta^2\}\,\mathrm{d}\Omega + \hat{F}(u_m) \tag{25.6}$$

On the space of all admissible displacements, an equivalent perturbed Lagrangian can be constructed, following the spirit of [13], by appending the dilatation equation

$$p = \kappa(\gamma)\theta + \tfrac{1}{3}C^{dv}_{ij}e'_{ij} \tag{25.7}$$

This is the same definition of the pressure function used by Key in [10]. The associated perturbed Lagrangian is [10]:

$$\Lambda(u_m, p) = \frac{1}{2}\int_{\Omega} \{W(e'_{ij}) + 2p\theta - \kappa^{-1}(\gamma)(p - \tfrac{1}{3}C^{dv}_{ij}e'_{ij})^2\}\,\mathrm{d}\Omega + \hat{F}(u_m) \tag{25.8}$$

$W('_{ij})$ will henceforth be used to denote $e'_{ij}C^{d}_{ijkl}e'_{kl}$. As $\gamma \to 0$, equation (25.7) is the condition of incompressibility and equation (25.8) becomes the associated Lagrange multiplier method.

25.3 IMPROVED DILATATIONAL APPROXIMATION

25.3.1 Strain operator splitting

We define the following operator which behaves formally like a tensor of rank 3 and operates by contraction of the third index:

$$L_{ijk} \equiv \frac{1}{2}\left(\frac{\partial}{\partial x_j}\,\delta_{ki} + \frac{\partial}{\partial x_i}\,\delta_{kj}\right) \tag{25.9}$$

It thus follows that

$$e_{ij} = L_{ijk}u_k \tag{25.10}$$

Applying the strain splitting of equation (25.1) we define

$$\begin{aligned} L^{v}_{ijk} &= \frac{\beta}{3}\,\delta_{ij}\,\frac{\partial}{\partial x_k} \\ L^{d}_{ijk} &= L_{ijk} - L^{v}_{ijk} \end{aligned} \tag{25.11}$$

and thus

$$e'_{ij} = L^d_{ijk} u_k \tag{25.12}$$

as defined in equation (25.1).

25.3.2 Finite element approximation

From now on we shall be concerned with solving the minimality and stationarity problems of equations (25.6) and (25.8) respectively on a finite element displacement trial space $S^h \subset H^1$. H^1 is the space of admissible displacements for the continuous problem. H^1 is assumed to satisfy appropriate essential boundary conditions. Whether dealing with the penalty or perturbed Lagrangian formulations, we will need a pressure trial space $T^h \subset H^0$. H^0 is the space of admissible pressures for the continuous problem. Let $Q^h \subset H^0$ be the smallest subspace containing all the scalar functions which are partial derivatives of components of S^h and also containing T^h. The key to what follows is a linear smoothing operator

$$P_h : Q^h \to T^h \tag{25.13}$$

which will be constructed to improve the dilatational approximation. A typical choice is the best approximation or projection operator of H^0. Since Q^h contains piecewise continuous functions, P_h can also be an interpolation operator at points in the interior of elements, and even on element boundaries, if T^h is an 'unconnected' FEM trial space (that is T^h allows interelement discontinuity).

25.3.3 Strain operator improvement

In equations (25.11), we replace L^v_{ijk} by

$$\bar{L}^v_{ijk} = \frac{\beta}{3} \delta_{ij} P_h \frac{\partial}{\partial x_k} \tag{25.14}$$

which is the improved volumetric approximation. It is only defined when applied to $u_k \in S^h$. For such u_k we can now define the total improved strain operator

$$\bar{L}_{ijk} = L^d_{ijk} + \bar{L}^v_{ijk} = \frac{1}{2}\left(\frac{\partial}{\partial x_j}\delta_{ki} + \frac{\partial}{\partial x_i}\delta_{kj} - \frac{2}{3}\beta\,\delta_{ij}\frac{\partial}{\partial x_k}\right) + \frac{\beta}{3}\delta_{ij} P_h \frac{\partial}{\partial x_k} \tag{25.15}$$

The improved variational principle in unsegregated form is then

$$J(u_m) = \int_\Omega \bar{L}_{ijn} u_n C_{ijkl} \bar{L}_{klm} u_m \, d\Omega + \hat{F}(u_m) \tag{25.16}$$

Remarks 25.1 The most important aspect of equation (25.16) is that the method is implemented exactly as given, without segregation of volumetric terms.

Below it will be shown that the improvement operator P_h can be constructed and β chosen so that all implicit penalty terms are subjected to improved dilatational approximation.

This method is essentially the B matrix method. The B matrix can be obtained by taking $u_k^h \in S^h$ expanded in local or global shape functions, N_k^Δ,

$$u_k^h = N_{(k)}^\Delta u_{\Delta(k)} \qquad \text{(no sum on } k\text{)}$$

where $u_{\Delta k}$ is the kth vector component of the nodal-value vector, with local or global numbering given by Δ. The B matrix is then the usual matrix arrangement [11] of

$$\bar{L}_{ijk} N_k^\Delta \qquad \text{(sum on } k\text{)}$$

(with $\beta = 1$).

25.3.4 The choice of the parameter β

The parameter β has yet to be determined. To deal with the implied volumetric terms in equation (25.16) without segregating them, we have two choices: first, we could use uniformly reduced numerical integration; second, we could improve the volumetric terms using P_h and employ exact or highly accurate integration in equation (25.16). The latter would require that all penalty terms in equation (25.16) consist only of improved volumetric terms. Using explicit segregation, we find that

$$\begin{aligned} \bar{L}_{ijn} u_n C_{ijkl} \bar{L}_{klm} u_m = {} & W(e'_{ij}) + 2 L^d_{ijn} u_n C'_{ijkl} \bar{L}^v_{klm} u_m \\ & + \underbrace{\bar{L}^v_{ijn} u_n C'_{ijkl} \bar{L}^v_{klm} u_m}_{\to 0} + \kappa(\gamma)[LC^d_{iin} u_n)^2 \\ & + 2(L^d_{iin} u_n)(\bar{L}^v_{kkm} u_m) + (\bar{L}^v_{iin} u_n)^2] \end{aligned} \tag{25.17}$$

The penalty terms involving

$$L^d_{iin} u_n = (1-\beta)\theta \tag{25.18}$$

determine β; they could lead to an unimproved θ factor, multiplied by $\kappa(\gamma)$. Therefore, the only resolution is that $\beta = 1$. In the $\beta = 1$ case all penalty terms are improved, since

$$\bar{L}_{ijn} u_n C_{ijkl} \bar{L}_{klm} u_m = \hat{W}(e_{ij}) + \kappa(\gamma)[P_h \theta]^2 \tag{25.19}$$

where $\hat{W}(e_{ij})$ gives the energy density of all non-penalty terms in equation (25.17).

Any other choice of β would lead to unimproved volumetric penalty terms. This would require the use of uniform reduced integration if segregation of volumetric terms is to be avoided. In that case, improvement is not required, so $\beta = 0$ is the other meaningful choice. The authors do not recommend the use of uniform reduced integration, so we take the $\beta = 1$ alternative hereafter.

25.4 THE EQUIVALENCE THEOREM

The remainder of this paper will be devoted to determination of which smoothing operators P_h will lead to equivalence of the penalty method of equation (25.6) and the perturbed Lagrangian of equation (25.8) when solutions are sought on the trial space $S^h \times T^h$.

25.4.1 The role of numerical integration

In the method described here, now that we have assumed that $\beta = 1$, the only role of numerical integration is to accurately and stably evaluate the energy expressions in equations (25.4), (25.6), and (25.8). All improvement of volumetric approximation is carried out via the choice of P_h. To that end, we denote by a slash through the integral sign the fact that we may be employing a numerical integration formula:

$$\fint_\Omega F(\mathbf{x})\, d\Omega = \sum_{k=1}^{m} \omega_k F(\boldsymbol{\xi}_k) \tag{25.20}$$

However the slash does not rule out the possibility that *exact* integration may be employed.

25.4.2 The equivalence theorem

Our theorem follows from choosing a P_h which is a *projection* in the inner product induced by $\fint_\Omega \cdot\, d\Omega$ on the volumetric terms. Note that this may be a discrete inner product. Such a P_h would satisfy:

$$P_h Q^h = T^h \tag{25.21a}$$

$$P_h^2 = P_h \tag{25.21b}$$

$$\fint_\Omega (P_h\theta)\sigma\, d\Omega = \fint_\Omega \theta(P_h\sigma)\, d\Omega \qquad \text{for all } \theta, \sigma \in Q^h \tag{25.21c}$$

$$\fint_\Omega (\theta - P_h\theta) g\, d\Omega = 0 \qquad \text{for all } g \in T^h \tag{25.21d}$$

It is important to note that P_h is not an exact least squares projection unless $\fint_\Omega \cdot\, d\Omega$ is exact. If P_h satisfies the requirements of equations (25.21), it is not hard to deduce the following:

Theorem 25.1 *If P_h is the projection in the volumetric inner product induced by $\fint_\Omega \cdot\, d\Omega$ then the following variational principles produce the same finite* *element solutions* $(u_m^h, p^h) \in S^h \times T^h$:

$$J(u_m) = \frac{1}{2} ⨏_\Omega \bar{L}_{ijn} u_n C_{ijkl} \bar{L}_{klm} u_m \, d\Omega + \hat{F}(u_m)$$

$$p \equiv \kappa(\gamma) P_h \theta + \tfrac{1}{3} P_h C_{ij}^{dv} e'_{ij} \tag{25.22a}$$

$$\Lambda(u_m, p) = \tfrac{1}{2} ⨏_\Omega \{W(e'_{ij}) + 2p\theta - \kappa^{-1}(\gamma) \times (p - \tfrac{1}{3} P_h C_{ij}^{dv} \bar{L}_{ijk}^d u_k)^2\} \, d\Omega + \hat{F}(u_m) \tag{25.22b}$$

Furthermore, the discrete equations of (25.22a) can be obtained by eliminating the pressures of (25.22b) in the manner of [1]. If in addition T^h is unconnected, then the element matrices of (25.22a) can be obtained from those of (25.22b) by eliminating the pressures on the element level.

Remarks 25.2 Note that Key's variational principle [10] differs from (25.22b) in that no P_h acts on the C_{ij}^{dv} term. Thus when $C_{ij}^{dv} \neq 0$, our theorem may be interpreted as an exact equivalence to a modified mixed principle with a projected volume coupling term. Alternatively, it may be viewed as an approxiamte equivalence to Key's principle.

As was the case with reduced/selective techniques, the theorem shows that mixed method solutions may be obtained from the minimum principle (25.22a).

When $⨏_\Omega \cdot dt$ is exact and P_h is the least squares projection, the method is a generalization of the 'mean dilatation' method of Nagtegaal [11, 14]; it is precisely that method when T^h is composed of piecewise constants.

The method suggested by Hughes [11] in which P_h is taken to be an interpolation operator does not lead to an equivalence in general, because P_h is not then a true projection (except in special cases). One might expect that good results can be obtained nevertheless, because of an approximate equivalence when the projection is replaced by nodal interpolation in T^h.

It appears that the assumption that $\kappa(\gamma)$ is spatially constant can be relaxed whenever $\kappa(\gamma) P_h \theta = P_h \kappa(\gamma) \theta$ at the $\boldsymbol{\xi}_i$. This occurs, for example, in the reduced/selective case and also when $\kappa(\gamma)$ is a (different) constant in each element and T^h is unconnected.

Finally we point out that equivalence can easily be obtained when $⨏_\Omega \cdot d\Omega$ is not exact. In [11] it is observed that when $⨏_\Omega \cdot d\Omega$ is exact, the least squares projection required to get equivalence involves the mass matrix of T^h. When $⨏_\Omega \cdot d\Omega$ is not exact, use of the mass matrix integrated with $⨏_\Omega \cdot d\Omega$ leads to the conclusion that the P_h so constructed is the projection with respect to $⨏_\Omega \cdot d\Omega$. It should be pointed out that the appropriate interpretation of $p = \kappa(\gamma) P_h \theta + \tfrac{1}{3} P_h C_{ij}^{dv} e'_{ij}$ is that

$$⨏_\Omega pg \, d\Omega = ⨏_\Omega (\kappa(\gamma)\theta + \tfrac{1}{3} C_{ij}^{dv} e'_{ij}) g \, d\Omega \qquad \text{for all } g \in T^h \tag{25.23}$$

and p must be computed in this way for the equivalence to hold. These facts are also the keys to the proof of the isotropic reduced/selective case [1].

25.5 CONCLUSIONS

The major consequences of the method and equivalence theorem given here are two-fold: first, mixed method solutions can be obtained by a minimum principle in which volumetric terms need not be segregated in the implementation; second non-constant elastic coefficients in C_{ijkl}, geometric, and/or isoparametric coefficients in the integrand are integrated by $\int_\Omega \cdot \, d\Omega$. This formula can be exact or chosen solely on the basis of stability and accuracy requirements, because the nodes of T^h and volumetric integration points need not coincide.

REFERENCES

1. D. S. Malkus and T. J. R. Hughes, 'Mixed finite element methods—reduced and selective integration techniques' A unification of concepts', *Comp. Meth. Appl. Mech. Eng.*, **15,** 63–81 (1978).
2. M. Bercovier, 'Perturbation of mixed variational problems. Application to mixed finite element methods', *RAIRO Analyse Numerique*, **12,** 211–236 (1978).
3. G. F. Carey and J. T. Oden, *Finite Elements*: *A Second Course*, Vol. II, Prentice-Hall, Englewood Cliffs, N.J., 1981.
4. J. T. Oden and N. Kikuchi, 'Penalty methods for constrained problems in elasticity', *Int. J. Num. Meth. in Eng.* (to appear 1981).
5. D. S. Malkus, 'Eigenproblems associated with the discrete LBB condition for incompressible finite elements', *Int. J. Eng. Sci.* (to appear 1981).
6. D. S. Malkus and E. T. Olsen, 'A convergence theorem for incompressible finite elements based on a constrained approximation condition', Illinois Institute of Technology Research Report No. 81–3, for NSF Grant No. CME 80–17549 (Dept. of Mathematics), April 1981.
7. C. Johnson and J. Pitkaranta, 'Analysis of some mixed finite element methods related to reduced integration', Chalmers Institute of Technology Research Report 80.02 (Dept. of Computer Sciences), 1980.
8. F. Brezzi, 'On the existence, uniqueness, and approximation of saddle point problems arising from Lagrangian multipliers', *RAIRO Analyse Numerique*, **8,** 129–151 (1974).
9. G. Fix, 'On the effects of quadrature errors in the finite element method', in *Advances in Computational Methods in Structural Mechanics and Design* (Eds. J. T. Oden, R. W. Clough, and Y. Yamamoto), University of Alabama (Huntsville) Press, 1972.
10. S. W. Key, 'A variational principle for incompressible and near-incompressible anisotropic elasticity', *Int. J. Solids and Structures*, **5,** 951–964 (1969).
11. T. J. R. Hughes, 'Generalization of selective integration procedures to anisotropic and non-linear media', *Int. J. Num. Meth. in Eng.*, **15,** No. 9, 1413–1418 (1980).
12. Y. C. Fung, *Foundations of Solid Mechanics*, Prentice-Hall, Englewood Cliffs, N.J., 1965.

13. R. L. Taylor, K. Pister, and L. R. Hermann, 'A variational principle for incompressible and nearly-incompressible orthotropic elasticity', *Int. J. Solids and Structures,* **4,** 875–883 (1968).
14. J. C. Nagtegaal and J. E. De Jong, 'Some computational aspects of elastic plastic large strain analysis', Research Report M.S. 10.910, Technical University, Eindhoven, The Netherlands, 1979.

Hybrid and Mixed Finite Element Methods
Edited by S. N. Atluri, R. H. Gallagher, and O. C. Zienkiewicz

Chapter 26

On a Finite Element Procedure for Non-Linear Incompressible Elasticity

M. Bercovier, Y. Hasbani, Y. Gilon, and K. J. Bathe

26.1 INTRODUCTION

We consider an ideally incompressible body that fills a domain Ω at rest, $\Omega \subset R^N$, $N=2,3$. We assume that the energy function W is known and that W depends on the strains only.

As a model problem we shall suppose that the body is fixed along part of the boundary $\Gamma_0 \subset \Gamma$ of Ω and that we apply a traction $\mathbf{f}$ on $\Gamma_1 \subset \Gamma, \Gamma_1 \cap \Gamma_0 = \emptyset$.

The resulting deformation field $\mathbf{u}$ is a stationary point of the functional

$$I(\mathbf{u}) = \int_\Omega W(I_1, I_2)\, \mathrm{d}\Omega + \int_{\Gamma_{g_1}} \mathbf{u}^T \mathbf{f}\, \mathrm{d}\Gamma \tag{26.1}$$

satisfying the incompressibility constraint

$$|\mathbf{F}(\mathbf{u})| = 1 \tag{26.2}$$

where $\mathbf{F}(\mathbf{u})$ is the deformation gradient $(f_{ij} = \delta_{ij} + u_{i,j})$, $|\mathbf{F}(\mathbf{u})| = \det \mathbf{F}(\mathbf{u})$, and I_1, I_2 are the first and second strain invariants [1, 2].

Our aim is to define an effective finite element formulation to compute approximate solutions of (26.1), (26.2). By 'effective' we mean that it must be easily included in standard finite element codes and efficient enough to allow the analysis of complex problems with large displacements and large strains. As also recognized recently by other researchers, the standard selective–reduced integration schemes cannot necessarily be applied directly to these analyses conditions (see, for example, [3]).

In the development of our formulation we face two difficulties: a highly non-linear problem and a highly non-linear constraint. In the following we develop a penalty function formulation for the constraint (26.2). This approach is justified both by its ease of implementation in existing finite element programmes and by the fact that it includes the analysis of quasi-incompressible materials.

Using a mixed variational formulation we first show that the so-called reduced integration approach is not always equivalent to a consistent 'approximate constraint' formulation, even in the case of a kinematically linear constraint (div $\mathbf{u}=0$).

We first devise a penalty function formulation with 'approximate constraints' in such a way as to try to take advantage of the properties of multidimensional Gauss–Legendre integration rules. Although inspired by the 'reduced integration' method, this formulation is not in general equivalent to the reduced integration procedure.

On bilinear (respectively trilinear) elements for two- (respectively three-) dimensional analysis we define as the 'approximate constraint', a constraint based on the mean volume. Such a construction was considered by Malkus [4] for the trilinear element. In order to avoid 'locking' or poor pressure predictions we must generalize our approach to quadratic *curved* isoparametric elements. This generalization is not straightforward.

Identifying the problems that we encounter leads to the general concept of an 'approximate constraint' formulation. The basic step of this formulation is to use a projection of the original constraint on a simple function space, element by element. Such a function space can be defined in natural element coordinates or in global coordinates. Taking as an example three-dimensional second-order elements, we construct the penalty function corresponding to an 'approximate constraint' in global coordinates. We then discuss briefly how the formulation can be implemented effectively in a Newton–Raphson (total Lagrangian) formulation.

26.2 THE PENALTY FUNCTION APPROACH

In his study of rubber-like materials, Rivlin used the third invariant $I_3(=|\mathbf{F}|^2)$ instead of (25.2); that is

$$I_3=1$$

and defined the mixed variational formulation as:

Find a stationary point $(\mathbf{u}, p)$ of

$$\mathscr{L}(\mathbf{u}, p)=I(\mathbf{u})+\int_\Omega p(I_3-1)\,\mathrm{d}\Omega \tag{26.3}$$

Finite element applications based on (26.3) can be found in [1]. More recently finite elements based on assumed stress formulations were also devised for (26.3) by Pian and Lee [5].

The finite element implementation of (26.3) necessitates judiciously related choices of trial spaces for $\mathbf{u}$ and p (e.g. [6]). We shall use that fact later and show that it amounts to a *weakening* of condition (26.2). For instance, instead of (26.2) being taken pointwise, it will be taken in the mean over each element; hence the notion of 'approximate constraints'.

Let us recall that for small deformations and displacements, neglecting all terms in $u_{i,j}$ of higher order than one, (26.2) is replaced by

$$\operatorname{div} \mathbf{u} = 0 \tag{26.4}$$

This is the source of major difficulties in the finite element analysis of incompressible viscous fluid flow. The mathematical justification of the choices of trial spaces for the approximation of (26.4) in Stokes' equations were given by Fortin [7] and Crouzeix and Raviart [8].

To avoid the use of hydrostatic pressure degrees of freedom, we want to replace (26.3) by a penalty function approach:

Find a stationary point of

$$I_\varepsilon(\mathbf{u}) = I(\mathbf{u}) + \frac{1}{\varepsilon} \int_\Omega (|\mathbf{F}| - 1)^2 \, d\Omega \tag{26.5}$$

A more general formulation would use

$$I_\varepsilon(\mathbf{u}) = I(\mathbf{u}) + \frac{1}{\varepsilon} \int_\Omega g(|\mathbf{F}| - 1) \, d\Omega \tag{26.6}$$

where $g(x)$ is a penalty function satisfying $g(x) = 0$ if $x = 0$.

Since (26.2) is not a linear constraint, and since we have no nice convexity properties, there is no reason, in general, that the solution of (26.3) and the limit solution of (26.5) be the same when $\varepsilon \to 0$. Le Tallec [9] has studied this question and Oden [10] has given some results as to when (26.5) converges to (26.1), (26.2).

Still it is clear that for the finite element formulation of (26.5) to converge to a corresponding solution of (26.1), (26.2) the penalty function will have to be weakened in some way analogous to the weakening carried out for problem (26.3). We can find in [11] an example in which the solution 'blows up' as $\varepsilon \to 0$ for Stokes' equations when the proper conditions on the trial space are not met. The standard weakening procedure is the so-called penalty and (selective) reduced integration approach which has been used successfully in flow problems. (For details see [11 to 15]). A study of this approach and its relation to (26.3) was carried out by Malkus [4] who reported some interesting results. However, as we shall show in the following, the reduced integration method is under certain conditions an erroneous approach, even in the linear case (26.4). Instead, a proper extension of the mixed finite element method of [7] and [8] leads to a method of 'approximate constraints' that reduces in simple cases to reduced integration methods.

26.2.1 Quasi-incompressible materials

Considering actual rubber elasticity problems it is appropriate also to treat the material as not completely incompressible. For instance, Tavarra and Rimondi [16] remark that the ratio K/E_0 between the bulk modulus K and the initial (for small displacements) Young's modulus E_0 varies from 10^{-3} for pure rubber to 10^{-2} for carbon loaded rubbers. Hence, one possibility is to replace in (26.1) the function $W(I_1, I_2)$ by one that includes I_3 or $|\mathbf{F}|$ also, namely,

$$I(\mathbf{u}) = \int_{\Omega} W(I_1, I_2)\, \mathrm{d}\Omega + \int_{\Omega} h(|\mathbf{F}|)\, \mathrm{d}\Omega + \int_{\Gamma_1} \mathbf{u} \cdot \mathbf{f}\, \mathrm{d}\Gamma \tag{26.7}$$

Now $h(|\mathbf{F}|)$ is a function that will be 'stiff' in $(|\mathbf{F}|-1)$ since we have a nearly incompressible material. Hence in [16] we find, for example:

$$\begin{aligned} W(I_1, I_2) + h(|\mathbf{F}|) &= C_{10}(I_1-3) + C_{01}(I_2-3) + C_{11}(I_1-3)(I_2-3) + \cdots \\ &+ C_{20}(I_1-3)^2 + C_{02}(I_2-3)^2 + C_{30}(I_1-3)^3 + \alpha_1(|F|^2-1) + \alpha_2(|F|^2-1)^2 \end{aligned} \tag{26.8}$$

where

$$|C_{ij}| \leq 5 \cdot 10^{-2}$$
$$\alpha_1 = 7 \cdot 10^{-2}$$

and

$$\alpha_2 = 0(10)$$

Hence α_2 is actually a penalty term of order 10^{+3}.

It is clear that to approach stiff problems of the type given in (26.7) we must use a finite element that is stable and convergent for (26.5) when ε tends to zero, and this independently of the mesh. Thus, we have an additional motivation for studying the penalty function approach, namely the problems of type (26.7).

Before explaining our approach in the linear case let us mention that a related method, the so-called method of augmented Lagrangian, has been successfully used by Le Tallec [9] for rubber materials. One aspect of this method (dealing with the non-linear equations) is not related to our problem. The second aspect of this method, approaching (26.2) by an augmented Lagrangian method, is related to our method. This approach, as given for a Q_1 element, is equivalent to the reduced integration method for the penalty function part. The results obtained by Le Tallec are very interesting; nevertheless implementation of this method to higher order elements appears difficult.

26.3 REDUCED CONSTRAINTS AND PENALTY METHOD FOR STOKES' EQUATIONS

The linear analogue of (26.1), (26.2), is:

Minimize

$$E(\mathbf{u}) = \mu \int_\Omega \sum_{ij} [\tfrac{1}{2}(u_{i,j} + u_{j,i})^2] \, d\Omega + \int_\Omega \mathbf{u} \cdot \mathbf{f} \, d\Gamma \tag{26.9}$$

subject to the constraint

$$\text{div}\, \mathbf{u} = 0 \tag{26.10}$$

The mixed formulation corresponding to (26.3) is:

Find a saddle-point to

$$\mathscr{L}(\mathbf{u}, p) = E(\mathbf{u}) + \int_\Omega p \, \text{div}\, \mathbf{u} \, d\Omega \tag{26.11}$$

and the penalty approach is given by

$$E_\varepsilon(\mathbf{u}) = E(\mathbf{u}) + \frac{1}{\varepsilon} \int_\Omega (\text{div}\, \mathbf{u})^2 \, d\Omega \tag{26.12}$$

For a theoretical study of the finite element method for (26.11) and (26.12) we refer to Bercovier [13]. An effective way to establish a solution scheme for (26.11) and (26.12) is to define a trial space W for the Lagrange multiplier p of a lower order than the one, say V, of the displacements and to define W elements per element (i.e. discontinuously). This corresponds to a weakening of (26.10) which is now replaced by

$$\text{div}_h\, \mathbf{u} = 0 \tag{26.13}$$

where div_h is a 'local' operator, introduced at the element level. The mathematical background of this method is given by the following result due to Brezzi and Bauska:

Introduce the following hypothesis:

$$\sup_{\mathbf{v} \in V - \{0\}} \frac{\left| \int_\Omega p \, \text{div}_h \, \mathbf{v} \, d\Omega \right|}{\|\mathbf{v}\|_v} \geqslant k \, \|p\|_W \tag{26.14}$$

where $\|\cdot\|_V$ is the (energy) norm of the displacement field trial functions space and $\|\cdot\|_W$ the norm of the hydrostatic pressure trial function space. If (26.14) is satisfied, then (26.11) has a unique solution; moreover, we can show [13] that the problem:

Minimize

$$E_\varepsilon(\mathbf{u}) = E(\mathbf{u}) + \frac{1}{\varepsilon}\int_\Omega (\mathrm{div}_h\, \mathbf{u})^2\, \mathrm{d}\Omega \tag{26.15}$$

has a unique solution $\mathbf{u}_\varepsilon$ and

$$\|\mathbf{u} - \mathbf{u}_\varepsilon\|_V + \left\| p - \frac{1}{\varepsilon}\,\mathrm{div}\, \mathbf{u}_\varepsilon \right\|_W \leqslant c\varepsilon \tag{26.16}$$

where $(\mathbf{u}, p)$ is the unique solution of (26.11).

26.3.1 A simple example

Let Ω be in R^2 and let (T_h) be a quadrangulation of Ω. For V we choose bilinear isoparametric elements. Then for W a natural choice is the space of constant pressure per element $K \in (T_h)$. The corresponding approximate divergence is then defined by

$\mathrm{div}_h\, \mathbf{u}$ is a *constant* on k, such that

$$\int_K p\, \mathrm{div}_h\, \mathbf{u}\, \mathrm{d}K = \int_K p\, \mathrm{div}\, \mathbf{u}\, \mathrm{d}K \qquad \text{for all } p \in W$$

Since p is constant on K, we obtain:

$$\mathrm{div}_h\, \mathbf{u}|_K = [\mathrm{meas}\,(K)]^{-1} \int_K \mathrm{div}\, \mathbf{u}\, \mathrm{d}K \tag{26.17}$$

so that (26.10) has been weakened to

$$\int_K \mathrm{div}\, \mathbf{u}\, \mathrm{d}K = 0$$

and the corresponding penalty function is

$$\frac{1}{\varepsilon}\int_\Omega (\mathrm{div}_h\, \mathbf{u})^2\, \mathrm{d}\Omega = \frac{1}{\varepsilon} \sum_{K \in (T_h)} [\mathrm{meas}\,(K)]^{-1} \left(\int_K \mathrm{div}\, \mathbf{u}\, \mathrm{d}K \right)^2 \tag{26.18}$$

This is, in essence, the function used by Nagtegaal, Parks, and Rice [6]. Note that for any quadrilateral element

$$\int_K \mathrm{div}\, \mathbf{u}\, \mathrm{d}K = \mathrm{div}\, \mathbf{u}\,(c_K) \times [\mathrm{meas}\,(K)] \tag{26.19}$$

where c_K is the centre of element K, so that

$$\left(\int_K \mathrm{div}\, \mathbf{u}\, \mathrm{d}K \right)^2 = \mathrm{div}^2\, \mathbf{u}(c_K) \times [\mathrm{meas}\,(K)]^2 \tag{26.20}$$

Hence (26.18) is equivalent to the 'evaluation' of $\int_K (\text{div}\,\mathbf{u})^2\,dK$ by means of a one-point formula (for details see, for example, [17]). This is the essence of the reduced integration method.

This 'nice' integration property is, however, limited to the four-node quadrilateral element for two-dimensional plane strain analysis. It does not carry over to axisymmetric or three-dimensional analysis. Also, for curved elements, the standard numerical error estimates such as those given by Ciarlet and Raviart [see Theorem 6, p. 444 of [18]] do not hold.

The above analysis can also be extended to the Q_2 nine-node element with the pressure in Q_1. Here again the reduced integration 'trick' is equivalent to a Q_1 approximate constraint on straight-sided elements, but we have been unsuccessful in generalizing the method to higher order, variable-number nodes and three-dimensional elements. In the following section we apply and discuss this approach to non-linear large strain analysis.

26.4 PENALTY FUNCTION, NON-LINEAR CASE

We now turn to the problem of defining a consistent approximation to

$$\frac{1}{\varepsilon}\int_\Omega (|\mathbf{F}|-1)^2\,d\Omega$$

We shall proceed in the following way:

(a) Find a weaker condition corresponding to (26.2), namely

$$(|\mathbf{F}|-1)_h = 0 \qquad \text{on each element } K$$

(Note that if our procedure is invariant for the constants over K it is equivalent to $|\mathbf{F}|_h = 1$.)

(b) Compute

$$\frac{1}{\varepsilon}\int_K (|\mathbf{F}|-1)_h^2\,dK$$

on each element K.

It is natural to look for a definition that is consistent with the one given for the linear case; i.e. we want to obtain results similar to those in [14] when the displacements and deformations are very small.

26.4.1 Bilinear element

Let us consider the case for which the displacements are given by conforming 'bilinear' isoparametric elements. By analogy with the linear case we define an approximate ratio of deformation $(|\mathbf{F}|-1)_h$, constant on each element. More precisely

$$\int_K (|\mathbf{F}|-1)_h \, \mathrm{d}K = \int_K (|\mathbf{F}|-1) \, \mathrm{d}K \tag{26.21}$$

Thus, on each element K,

$$(|\mathbf{F}|-1)_h|_K = [\text{meas}\,(K)]^{-1} \int_K (|\mathbf{F}|-1) \, \mathrm{d}K \tag{26.22}$$

In order to compute the right-hand side of (26.22), we go back to the reference element $\hat{K}$:

$$(|\mathbf{F}|-1)_h|_K = [\text{meas}\,(K)]^{-1} \int_K (|\mathbf{F}(x(\mathbf{r}))-1) \det \mathbf{J}_K(\mathbf{r}) \, \mathrm{d}\hat{K} \tag{26.23}$$

where $\mathbf{r}=(r_i)$, $i=1, \ldots, N$, are the element natural coordinates and $\mathbf{J}_K(\mathbf{r})$ is the Jacobian of the isoparametric transformation. Note that the integration in (26.23) cannot in general be computed exactly by a one-point integration rule!

Our problem is now to find a stationary point of the functional:

$$I_\varepsilon(\mathbf{u}) = \int_\Omega W(I_1, I_2) \, \mathrm{d}\Omega + \frac{1}{\varepsilon} \int_\Omega (|\mathbf{F}|-1)_h^2 \, \mathrm{d}\Omega + \int_{\Gamma_1} \mathbf{u}^T \mathbf{f} \, \mathrm{d}\Gamma \tag{26.24}$$

where

$$\frac{1}{\varepsilon} \int_\Omega (|\mathbf{F}|-1)_h^2 \, \mathrm{d}\Omega = \frac{1}{\varepsilon} \sum_{K \in (T_h)} \int_K \left\{ \left(\int_K |\mathbf{F}| \, \mathrm{d}K \right) \Big/ \text{meas}\,(K) - 1 \right\}^2 \mathrm{d}K \tag{26.25}$$

Hence our approximate constraint penalty function is defined by

$$\text{meas}\,(K) \left\{ \left(\int_K |\mathbf{F}| \, \mathrm{d}K \right) \Big/ \text{meas}\,(K) - 1 \right\}^2$$

and not $\int_K (|\mathbf{F}|-1)^2 \, \mathrm{d}K$, whether with or without reduced integration. A version of this element together with a logarithmic penalty function has been used successfully in [19]. Malkus and Hughes [17], using reduced integration, have suggested as a weakened constraint

$$|\mathbf{F}\,(c_K)| = 1 \tag{26.26}$$

where c_K denotes the element centre point. This amounts to introducing the following penalty term:

$$\frac{\text{meas}\,(K)}{\varepsilon} (|\mathbf{F}(c_K)|-1)^2$$

The same function was used by Le Tallec [9] for the augmented Lagrangian method. Now it is clear that except for rectangles (26.25) and (26.26) give rise to different penalty functions. In the two-dimensional case (26.26) introduces an error $o(h)$ compared to (26.25) (h being the 'size' of the element). As stated before, (26.26) will lead to meaningless results for a general eight-node isoparametric brick! (Let us note that the $o(h)$ error in (26.25) for parallepipedic three-dimensional elements can be the source of serious difficulties, as shown by Malkus [4]).

26.4.2 Q_2–Q_1 element

Consider a nine-node isoparametric element K. We want to define a Q_1 ('bilinear') approximate dilatation operator $(|\mathbf{F}|-1)_h$. First, let us suppose that K is a straight-sided quadrilateral (subparametric case). We define $(|\mathbf{F}|-1)_h$ as the 'bilinear' function over K satisfying:

$$\int_K q(|\mathbf{F}|-1)_h \, \mathrm{d}K = \int_K q(|\mathbf{F}|-1) \, \mathrm{d}K \qquad \text{for all } q \in Q_1(K) \tag{26.27}$$

We next choose as a basis for $Q_1(K)$ the four shape functions based on the four Gaussian nodes of the 2×2 quadrature rule, i.e.

$$\psi_i \in Q_1(K); \qquad \psi_i(g_j) = \delta_{ij} \qquad \text{for } i, j = 1, \ldots, 4$$

Then, by definition:

$$(|\mathbf{F}|-1)_h = \sum_{i=1}^{4} (|\mathbf{F}|\,(g_i) - 1)_h \psi_i$$

To compute the left-hand side of (26.27) we go back to the reference element $\hat{K}$:

$$\int_K q(|\mathbf{F}|-1)_h \, \mathrm{d}K = \int_{\hat{K}} \hat{q}(|\mathbf{F}|-1)_h \det \mathbf{J}_K \, \mathbf{dr}$$

The right-hand side of this equality is a polynomial of degree at most 3 and it can be computed exactly by the 2×2 rule. Taking $\hat{q} = \hat{\psi}_i$, $i = 1, \ldots, 4$, we obtain

$$[\det \mathbf{J}_K(g_i) \operatorname{meas}(K)] \cdot (|\mathbf{F}|-1)_h(g_i) = \int_{\mathbf{K}} \hat{\psi}_i(|\mathbf{F}|-1)_h \det \mathbf{J}_K \, \mathbf{dr} \tag{26.28}$$

Now, defining

$$c_i = \operatorname{meas}(K) \cdot \det \mathbf{J}_K(g_i)$$

we thus obtain

$$(|\mathbf{F}|-1)_h(g_i) = c_1^{-1} \int_K \psi_i(|\mathbf{F}|-1) \, \mathrm{d}K \tag{26.29}$$

Going back to the reference element, we have

$$\int_K \psi_i(|\mathbf{F}|-1)\,\mathrm{d}K = \int_{\hat{K}} \hat{\psi}_i(|\mathbf{F}|-1)(\mathbf{x}(\mathbf{r}))\det \mathbf{J}_K(\mathbf{r})\,\mathrm{d}\mathbf{r} \tag{26.30}$$

The quantity $|\mathbf{F}| \det \mathbf{J}_K$ is in general a rational function on the reference element. Hence we have to compute (26.30) by means of a 3×3 Gaussian rule. Consider now the evaluation of the penalty functional

$$\frac{1}{\varepsilon}\sum_{K\in(T_h)}\int_K (|\mathbf{F}|-1)_h^2\,\mathrm{d}K$$

Since, by construction, $(|\mathbf{F}|-1)_h$ and $\det \mathbf{J}_K$ are both bilinear on the reference element we can use the 2×2 Gaussian rule to evaluate exactly these integrals:

$$\int_K (|\mathbf{F}|-1)_h^2\,\mathrm{d}K = \sum_{i=1}^{4} c_i(|\mathbf{F}|-1)_h^2(g_i) \tag{26.31}$$

and from (26.29) we obtain

$$\int_K (|\mathbf{F}|-1)_h^2\,\mathrm{d}K = \sum_{i=1}^{4} c_i^{-1}\left[\int_K \psi_i(|\mathbf{F}|-1)\,\mathrm{d}K\right]^2 \tag{26.32}$$

Now we observe that (26.32) could not be obtained by the reduced integration technique.

Let us consider how we would extend our construction to a higher order three-dimensional brick element based on an eight-node subparametric transformation. The quantity $\det \mathbf{J}_K$ is then a polynomial of degree 2 in each variable; thus (26.28) and (26.31) do not hold any more. The 'trick' (inspired initially by the reduced integration method) of choosing as a basis for $(|\mathbf{F}|-1)_h$ the shape functions based on the nodes of a Gaussian quadrature rule is of no help here!

In the two-dimensional case things also get more involved if we suppose now that K has curved sides. In this case we must ask how to represent $(|\mathbf{F}|-1)_h \in Q_1(K)$ and what we mean by $Q_1(K)$.

Originally we wanted to preserve the attractive properties of selective–reduced integration and introduced the following constructions. Let $\tilde{K}$ be the straight-sided quadrilateral defined by the property that its four Gaussian nodes corresponding to the 2×2 integration are the same as those of the curved element $\tilde{K}$ (see Figure 26.1). Note that

$$\text{meas}\,(K) = \text{meas}\,(\tilde{K})$$

Defining

$$(|\mathbf{F}|-1)_h \in Q_1(\tilde{K})$$

then

$$\int_{\tilde{K}} (|\mathbf{F}|-1)_h q\,\mathrm{d}\tilde{K} = \int_K (|\mathbf{F}|-1)q\,\mathrm{d}\tilde{K} \qquad \text{for all } q\in Q_1(\tilde{K}) \tag{26.33}$$

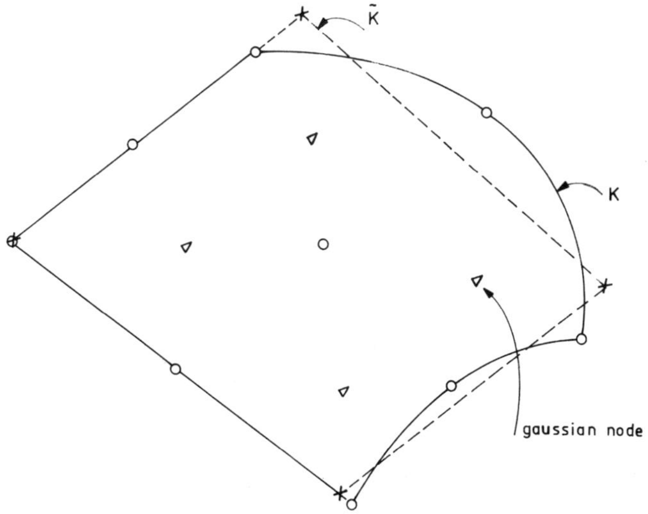

Figure 26.1 Construction of element $\tilde{K}$

The numerical integration is performed over the straight-sided element $\tilde{K}$ so that (26.28) to (26.31) hold with $\det \mathbf{J}_{\tilde{K}}$ instead of $\det \mathbf{J}_K$.

A particular difficulty of this approach is that the right-hand side of (26.33) is not easy to compute. Since $|\mathbf{F}|$ is defined by means of shape functions over K and not $\tilde{K}$, we may consider replacing (26.33) by

$$\int_{\tilde{K}} (|\mathbf{F}|-1)_h q \, d\tilde{K} = \int_K (|\mathbf{F}|-1) q \, dK \qquad (26.33a)$$

where we now have $q \in Q_1(K)$ which is still difficult to work with. Trying to stay close to what made the reduced integration attractive does not appear to be effective. Hence, we must go back to the original formulation that allowed us to justify the reduced integration trick in the first place.

26.5 CONSTRAINT APPROXIMATION

The following formulation represents a natural extension of selective integration techniques and is closely related to ideas presented earlier (e.g. [3 to 13]).

26.5.1 Abstract formulation

Given a trial space $T(K)$ different from the original one (say $Q_2(K)$), we define the projection G_h on $T(K)$ of any function $G(\mathbf{x})$ over K by

$$\int_K G_h q \, dK = \int_K G q \, dK \qquad \text{for all } q \in T(K);\ G_h \in T(K) \qquad (26.34)$$

Let ψ_i $(i=1,\ldots,s)$ be a basis of $T(K)$, that is:

$$G_h(x)=\sum_{i=1}^{s}\alpha_i\psi_i(\mathbf{x})$$

where α_i is the 'nodal parameter' corresponding to $\psi_i(\mathbf{x})$. The expression in (26.34) is equivalent to

$$\int_K \left(\sum \alpha_i\psi_i(\mathbf{x})\right)\psi_j(\mathbf{x})\,\mathrm{d}K=\int_K G(\mathbf{x})\psi_j(\mathbf{x})\,\mathrm{d}K \qquad \text{for } j=1,\ldots,s \tag{26.34a}$$

Let us introduce the 'weight matrix'

$$\mathbf{W}=(w_{ij});\qquad w_{ij}=\int_K \psi_i(\mathbf{x})\psi_j(\mathbf{x})\,\mathrm{d}K \tag{26.35}$$

Its inverse $\mathbf{W}^{-1}=(w_{ij})^{-1}$ and the vector

$$\mathbf{G}=(G_i);\qquad G_i=\int_K G\psi_i\,\mathrm{d}K$$

The relation in (26.34a) can be rewritten as

$$(\alpha_1,\ldots,\alpha_s)^T=\mathbf{W}^{-1}\mathbf{G} \tag{26.36}$$

Suppose now that G is a function giving rise to an 'energy' term

$$E=\int_K G^2(\mathbf{x})\,\mathrm{d}K$$

(in our case it will be the penalty functional) and that we want to replace this energy term by its projection

$$E_h=\int_K G_h^2(\mathbf{x})\,\mathrm{d}K$$

or equivalently

$$E_h=(\alpha_1,\ldots,\alpha_s)\mathbf{W}(\alpha_1,\ldots,\alpha_s)^T$$

Using (26.36) we obtain

$$E_h=\mathbf{G}^T\mathbf{W}^{-1}\mathbf{G}$$

In general G is a constraint function (here we limit ourselves to $G=|\mathbf{F}|-1$) and we want to weaken the corresponding constraint equality $G(\mathbf{x})=0$ at the element level. For this we choose a space $T(K)$ with less degrees of freedom than the space in which the original problem was to be computed (here $Q_2(K)$ for instance); hence the term 'approximate constraint' method for this procedure. Going back to incompressible materials ($G=|\mathbf{F}|-1$) we introduce the penalty function on the approximate constraint at the element level. On element K the deformation energy considered is

$$U=\int_K W(I_1, I_2)\,\mathrm{d}K+\frac{1}{\varepsilon}\int_K (|\mathbf{F}|-1)_h^2\,\mathrm{d}K$$

and replacing G by $|\mathbf{F}|-1$ in our abstract analysis gives

$$\int_K (|\mathbf{F}|-1)_h^2\,\mathrm{d}K=\left\{\int_K (|\mathbf{F}|-1)\psi_1\,\mathrm{d}x, \ldots, \int_K (|\mathbf{F}|-1)\psi_s\,\mathrm{d}K\right\}\mathbf{W}^{-1}$$
$$\times\{(|\mathbf{F}|-1)\psi_1\,\mathrm{d}K, \ldots, (|\mathbf{F}|-1)\psi_s\,\mathrm{d}K\}^T \quad (26.37)$$

We should note that if we take the displacements in $Q_1(K)$ and for $T(K)$ the space of constants over K then W^{-1} reduces to $[\text{meas}\,(K)]^{-1}$ and we obtain (26.25) (the Q_1-Q_0 element). If K is a straight-sided nine-node sub-parametric element and $T(K)=Q_1(K)$, then $\mathbf{W}^{-1}$ is the diagonal matrix with $w_{ii}^{-1}=C_i^{-1}$ and (26.37) is nothing but (26.32)! However, for the case of a curved-sided element we still have to choose $T(K)$.

26.5.2 Global coordinate constraint approximation

Now that we have a rigorous method for defining approximate constraints the focus is on the choice of the space $T(K)$. Our choice is again motivated by considerations of ease of programming and the relation of the method to the reduced integration technique. Let us consider the nine-node isoparametric element and define

$$T(K)=\langle 1, r_1, r_2, r_1r_2\rangle$$

where the r_i are the element natural coordinates. These coordinates are defined on the reference element $\hat{K}$ and (26.35) yields

$$w_{ij}=\int_K \psi_i(\mathbf{r})\psi_j(\mathbf{r})\det \mathbf{J}_K(\mathbf{r})\,\mathrm{d}\mathbf{r}$$

where the ψ_i ($i=1, \ldots, 4$) are four basic functions of $T(K)=Q_1(K)$. If K is a curved element the expression we have to integrate is of high order and we need to use 3×3 Gauss integration.

For (26.37) we have to compute

$$\int_K (|\mathbf{F}|-1)\psi_i(\mathbf{r})\det \mathbf{J}_K(\mathbf{r})\,\mathrm{d}\mathbf{r}$$

which by the definition of $|\mathbf{F}|$ includes rational functions, and we shall use here too the 3×3 quadrature rule. Hence, *a priori* the choice of a polynomial space in the natural coordinates does not bring any simplification in the computation of the penalty function.

Before studying other choices for $T(K)$, we should point out that there arises a natural question in the computation of the two integrals above: what effect does the application of an approximate quadrature rule (e.g. 2×2 Gauss integration) have on the accuracy of our computations? For the weight matrix this is equivalent to a 'lumping' procedure $w_{ii} = \text{meas}(K) \det \mathbf{J}_K(g_i)$. This warrants a separate study as to the effect on the inverse matrix $\mathbf{W}^{-1}$.

Referring to the study of Sani *et al.* [20], we know that the preceding choice of $T(K)$ does not always satisfy the Brezzi–Babuska condition and gives rise to checkerboard pressure modes. These modes do not appear if we do not include $r_1 r_2$ in our space of approximation, i.e. if we take:

$$T(K) = \langle 1, r_1, r_2 \rangle$$

However, we then do not use a natural basis such as the one defined by the values at the Gaussian nodes. Since the original problem is set in the global coordinate system (x_i) a natural choice would be

$$T(K) = \langle 1, x_1, x_2 \rangle$$

Such a procedure is commonly employed in assumed stress hybrid finite element analysis, in which the constraints are expressed in global coordinates [5]; this was also suggested by Nagtegaal, Parks, and Rice [6].

A study of this approximate constraint method for incompressible fluids [21] shows that there is no loss of accuracy when compared to $T(K) = Q_1(\hat{K})$. This approach can directly be extended to axisymmetric elements and to three-dimensional problems.

In the following we apply the formulation to three-dimensional second-order elements together with

$$T(K) = \langle 1, x_1, x_2, x_3 \rangle$$

26.5.3 First and second variation of the approximate constraint penalty function

What follows is actually valid for any constraint function $G(u)$, and not only $|\mathbf{F}| - 1$, as well as for any choice of $T(K)$. The derivation of the first variation (residual) and the second variation (Hessian) corresponding to $\int_\Omega W(I_1, I_2)\, \mathrm{d}\Omega$ is classic (see, for example, [2]), and we shall give the details for the terms derived from the penalty function only:

$$\frac{1}{\varepsilon} \int (|\mathbf{F}| - 1)_h^2 \, \mathrm{d}K$$

Given $T(K)$ and a basis $\psi_i \in T(K)$, $i = 1, \ldots, s$, and noting that $|\mathbf{F}|$ is a function of the nodal degrees of freedom $\mathbf{u}=(u_1,\ldots,u_k)$ we define

$$b_i(\mathbf{u})=\int_K (|\mathbf{F}|-1)\psi_i \, \mathrm{d}K \qquad i=1,\ldots,s \tag{26.38}$$
$$\mathbf{B}(\mathbf{u})=(b_i(\mathbf{u}))^T$$

Hence, our penalty function is on element K:

$$\frac{1}{\varepsilon}\mathbf{B}(\mathbf{u})^T W^{-1}\mathbf{B}(\mathbf{u})$$

We also introduce the vector:

$$\frac{\partial b_i}{\partial \mathbf{u}}(\mathbf{u})=\int_K \frac{\partial(|\mathbf{F}|-1)}{\partial \mathbf{u}}\psi_i \, \mathrm{d}K \qquad \text{for } i=1,\ldots,s \tag{26.39}$$

and denote by $(\partial\mathbf{B}/\partial\mathbf{u})(\mathbf{u})$ the corresponding $s\times k$ matrix. Let

$$\frac{\partial^2 b_i}{\partial \mathbf{u}\,\partial \mathbf{u}}(\mathbf{u})=\int_K \frac{\partial^2(|\mathbf{F}|-1)}{\partial \mathbf{u}\,\partial \mathbf{u}}\psi_i \, \mathrm{d}K \qquad \text{for } i=1,\ldots,s \tag{26.40}$$

and $(\partial^2\mathbf{B}/\partial\mathbf{u}\,\partial\mathbf{u})(\mathbf{u})$ be the corresponding $(s\times k\times k)$ third-order tensor.

The first variation of the penalty function is

$$\frac{\partial}{\partial \mathbf{u}}\left(\frac{1}{\varepsilon}\int_K (|\mathbf{F}|-1)_h^2 \, \mathrm{d}K\right)=\frac{2}{\varepsilon}\mathbf{B}^T(\mathbf{u})\mathbf{W}^{-1}\frac{\partial \mathbf{B}}{\partial \mathbf{u}}(\mathbf{u}) \tag{26.41}$$

and the second variation is

$$\frac{\partial^2}{\partial \mathbf{u}\,\partial \mathbf{u}}\left(\frac{1}{\varepsilon}\int_K (|\mathbf{F}|-1)_h^2 \, \mathrm{d}K\right)=\frac{2}{\varepsilon}\left\{\left[\frac{\partial \mathbf{B}}{\partial \mathbf{u}}(\mathbf{u})\right]^T\mathbf{W}^{-1}\frac{\partial \mathbf{B}}{\partial \mathbf{u}}(\mathbf{u})+\mathbf{B}^T(\mathbf{u})\mathbf{W}^{-1}\frac{\partial^2 \mathbf{B}}{\partial \mathbf{u}\,\partial \mathbf{u}}(\mathbf{u})\right\} \tag{26.42}$$

For the practical implementation of the element, it is convenient to separate in (26.42) the first from the second derivatives. Define:

$$\mathbf{RK}_1(\mathbf{u})=\left[\frac{\partial \mathbf{B}}{\partial \mathbf{u}}(\mathbf{u})\right]^T\mathbf{W}^{-1}\frac{\partial \mathbf{B}}{\partial \mathbf{u}}(\mathbf{u}) \tag{26.43}$$

$$\mathbf{RK}_2(\mathbf{u})=\mathbf{B}(\mathbf{u})^T\mathbf{W}^{-1}\frac{\partial^2 \mathbf{B}}{\partial \mathbf{u}\,\partial \mathbf{u}}(\mathbf{u}) \tag{26.44}$$

For the brick element and our choice of $T(K)$ we first have to compute the 4×4 weight matrix:

$$\mathbf{W}=\int_K \langle 1, x_1, x_2, x_3\rangle^T\langle 1, x_1, x_2, x_3\rangle \, \mathrm{d}K \tag{26.5}$$

which can be done on the reference element $\hat{K}$ by means of a higher order Gaussian rule. The expression in (26.45) involves only the shape functions of K, not their derivatives, and it must be computed only once, so that it is not a costly procedure to compute it exactly. Let us recall that by definition

$$|\mathbf{F}|-1=\det\begin{bmatrix} 1+\dfrac{\partial u_1}{\partial x_1}, & \dfrac{\partial u_1}{\partial x_2}, & \dfrac{\partial u_2}{\partial x_3} \\ \dfrac{\partial u_2}{\partial x_1}, & 1+\dfrac{\partial u_2}{\partial x_2}, & \dfrac{\partial u_2}{\partial x_3} \\ \dfrac{\partial u_3}{\partial x_1}, & \dfrac{\partial u_3}{\partial x_2}, & 1+\dfrac{\partial u_3}{\partial x_3} \end{bmatrix}-1$$

and by (26.38) we have to compute

$$b_i(\mathbf{u})=\int_{\hat{K}}(|\mathbf{F}|-1)\psi_i(\mathbf{r})\det\mathbf{J}_K(\mathbf{r})\,\mathrm{d}\mathbf{r};\qquad \psi_i(\mathbf{r})=1,\,x_1(\mathbf{r}),\,x_2(\mathbf{r}),\,x_3(\mathbf{r}) \tag{26.46}$$

and by (26.39)

$$\frac{\partial b_i}{\partial \mathbf{u}}(\mathbf{u})=\int_{\hat{K}}\frac{\partial(|\mathbf{F}|-1)}{\partial \mathbf{u}}\psi_i\det\mathbf{J}_K\,\mathrm{d}\mathbf{r} \tag{26.47}$$

which yields $\mathbf{RK}_1$. We similarly obtain $\mathbf{RK}_2$ defined in (26.44).

26.5.4 Practical implementation

It is well known that for the three-dimensional element with m nodes used in the total Lagrangian formulation the main cost of the element generation lies in computing a term of the form

$$\mathbf{A}_K=\int_K\{\mathbf{E}_i(\mathbf{u})\}^T\mathbf{C}\{\mathbf{E}_i(\mathbf{u})\}\,\mathrm{d}K \tag{26.48}$$

which implies that at each integration point we need to compute a tensor product $\{3m\times 3m\times 6\}\times\{6\times 6\}\times\{6\times 3m\times 3m\}$.

Considering our approximate constraint method, on the other hand, the third-order tensor product in (26.42) is calculated only once on the element level. Once all numerical integrations to evaluate the third-order tensor have been carried out, the final product $\{3m\times 3m\times 4\}\{4\times 4\}\{4\times 3m\times 3m\}$ is computed once only. Hence, the added cost due to the approximate constraint is negligible. Furthermore, we can obtain a very effective code by noting that $|\mathbf{F}|$ is a trilinear alternate form in (u_1, u_2, u_3).

As has been shown in the linear constraint case by Engelman *et al.* [21] (see also Zienkiewicz, Taylor, and Baynham [22]) the 27-node element is most effective. The serendipity-type elements may yield 'checkerboard nodes' [20], but proved adequate to illustrate our approach (see the example in Subsection 26.5.5).

Whether we are using directly the total Lagrangian formulation or a variable matrix method related to it, such as the BFGS method [23], a main difficulty in the penalty function approach is to find a good initial guess $\mathbf{u}_0$. Consider the simple bilinear case with $T(K) = Q_0$. The second variation of the penalty function is

$$\frac{1}{\varepsilon}(\mathbf{RK}_1(\mathbf{u}) + \mathbf{RK}_2(\mathbf{u}))$$

$$= \frac{1}{\varepsilon}\left[\text{meas}\,(K)^{-1}\left(\int_K \frac{\partial}{\partial \mathbf{u}}(|\mathbf{F}| - 1)\,\mathrm{d}K\right)^T\left(\int_K \frac{\partial}{\partial \mathbf{u}}(|\mathbf{F}| - 1)\,\mathrm{d}K\right)\right.$$

$$\left. + \text{meas}\,(K)^{-1}\left(\int_K (|\mathbf{F}| - 1)\,\mathrm{d}x\right)\int_K \frac{\partial^2}{\partial \mathbf{u}\,\partial \mathbf{u}}(|\mathbf{F}| - 1)\,\mathrm{d}K\right] \quad (26.49)$$

and we note that $\mathbf{RK}_1(\mathbf{u})$ is a positive definite matrix for any $\mathbf{u}$, but this does not hold for $\mathbf{RK}_2(\mathbf{u})$. Moreover, $\mathbf{RK}_2(\mathbf{u})$ includes the quantity $\int_K (|\mathbf{F}| - 1)\,\mathrm{d}x/\text{meas}\,(K)$ which represents the relative volume change of element K. This quantity can have quite arbitrary values. However, at the solution $\int_K (|\mathbf{F}| - 1)\,\mathrm{d}K/(\varepsilon\,\text{meas}\,(K))$ is actually the hydrostatic pressure so that the term is bounded near convergence.

The above considerations suggest an algorithm for solution in which the Hessian is replaced by a stiffness matrix that includes all the second variation terms except $\mathbf{RK}_2(\mathbf{u})$. This solution procedure corresponds to a quasi-Newton method, and converges slower than the original Newton–Raphson method, when this one is started with a proper choice for $\mathbf{u}_0$. Thus, an effective algorithm would start the solution without $\mathbf{RK}_2(\mathbf{u})$, and then check in each iteration and for each element if $(|F(\mathbf{u})| - 1)_h$ is of order ε. Whenever this condition is met $\mathbf{RK}_2(\mathbf{u}_n)$ is included in the total stiffness matrix.

26.5.5 A simple example

To illustrate the discussion of this paper we implemented our method for a 21-node brick element together with a P_1 approximate constraint. Such an element may give rise to 'checkerboard' modes [20], but not in the simple example considered here.

Figure 26.2 shows the example considered: the analysis of an infinitely long hollow cylinder with imposed displacements at the interior surface. Three different finite element meshes were used as shown in the figure. Table 26.1 gives the results obtained in the solution of these models.

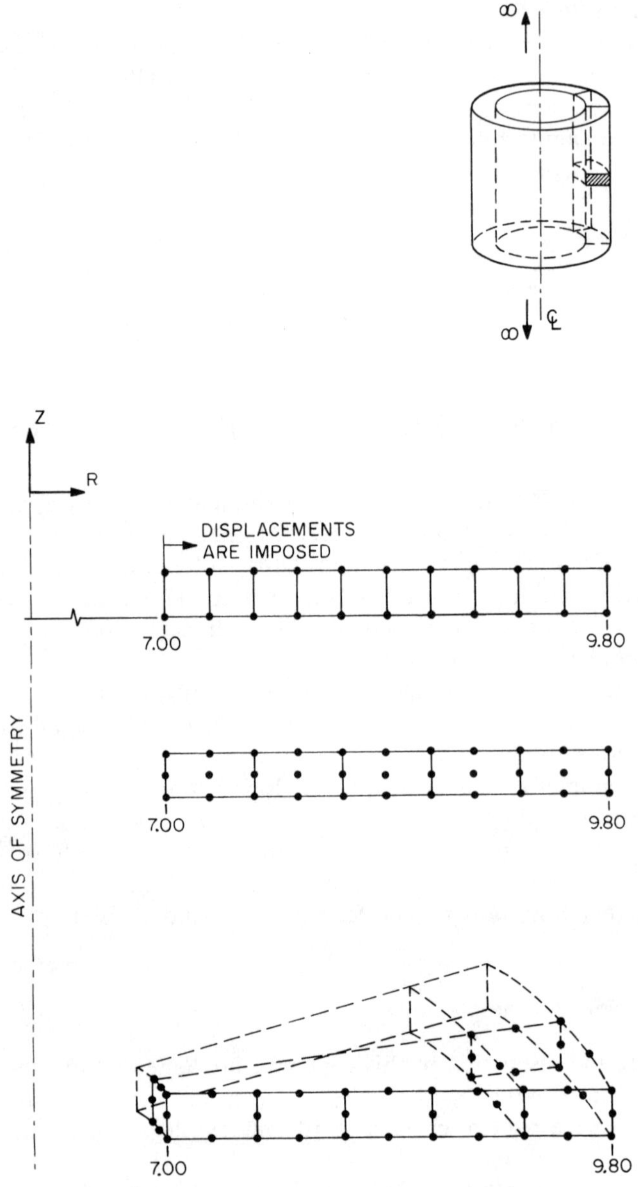

Figure 26.2 Analysis of an infinitely long cylinder

Table 26.1 Analysis of infinitely long cylinder

Initial	Displacements		
Nodal coordinates	4 node/P_0	9 node/Q_1	21 node/P_1
7.0	5.6	5.6	5.6
7.28	5.4839	5.4839	5.4840
7.56	5.3723	5.3723	5.3733
7.84	5.2650	5.2650	5.2661
8.12	5.1619	5.1619	5.1640
8.40	5.0627	5.0627	5.0650
8.68	4.9673	4.9673	4.9709
8.96	4.8756	4.8756	4.8794
9.24	4.7874	4.7874	4.7926
9.52	4.7025	4.7025	4.7081
9.80	4.6208	4.6208	4.6281

26.6 CONCLUDING REMARKS

A finite element formulation for incompressible or nearly incompressible large deformation non-linear elastic analysis has been presented. This formulation emerged from a detailed study of the selective reduced integration schemes used in kinematically linear analysis. The study has shown that the usual reduced integration schemes cannot be applied directly to higher order distorted elements or elements subjected to large strains.

In the finite element formulation presented we do not (necessarily) employ reduced integration but include the incompressible condition through an 'approximate constraint procedure'. This method appears directly applicable to low- and high-order elements, distorted elements, and large deformation conditions. It also seems to be effective computationally, in particular when used with a quasi-Newton solution of the non-linear equations. We are now pursuing more detailed analyses of the method and numerical experiences in order to gain further insights into the technique.

Finally, we should note that the procedure of using an 'approximate constraint' penalty function—presented here for the incompressibility constraint—is quite general and shows much potential for the effective treatment of other constraints that arise in structural (shell) analysis and in fluid mechanics [2].

ACKNOWLEDGMENT

The support of this work by the United States–Israel Binational Foundation is gratefully acknowledged. Discussions with Professor R. L. Sani and Dr. M. Engelman helped to clarify several points presented here.

REFERENCES

1. J. T. Oden, *Finite Elements of Nonlinear Continua*, McGraw-Hill, New York, 1972.
2. K. J. Bathe, *Finite Element Procedures in Engineering Analysis*, Prentice-Hall (in press).
3. T. J. R. Hughes, 'Generalization of selective integration procedures to anisotropic and nonlinear media', *Int. J. Num. Math. in Eng.*, **15,** 1413–1418 (1980).
4. D. S. Malkus, 'Finite elements with penalties in nonlinear elasticity', *Int. J. Num. Meth. in Eng.*, **16,** 121–136 (1980).
5. T. H. H. Pian and S. W. Lee, 'Notes on the finite elements for nearly incompressible materials', *AIAA J.*, **14,** 824–826 (1976).
6. J. C. Nagtegaal, D. M. Parks, and J. R. Rice, 'On numerically accurate finite element solutions in the fully plastic range', *J. Comp. Meth. Appl. Mech. Eng.*, **4,** 153–157 (1974).
7. M. Fortin, 'Resolution des equations des fluides incompressibles par la methode des elements finis', in *Proc. Third Int. Conf. Numerical Methods in Fluid Mechanics* (Eds. M. Cabannes and R. Temam), Springer-Verlag, 1973.
8. M. Crouzeix and P. A. Raviart, 'Conforming and non-conforming finite element methods for solving the stationary Stokes equation', *RAIRO, R.3*, **7,** 33–76 (1973).
9. P. Le Tallec, 'Numerical analysis of equilibrium problems in incompressible nonlinear elasticity', TICOM Report 80-3, Austin, 1980.
10. J. T. Oden, 'A theory of penalty methods for finite element approximation of highly nonlinear problems in continuum mechanics', *J. Comp. Struct.*, **8,** 445–449 (1978).
11. M. Bercovier, These de Doctorat d'Etat, Rouen, 1976.
12. D. S. Malkus, 'A finite element displacement model valid for any value of the compressibility', *Int. J. Solids and Structures*, **12,** 731–738 (1976).
13. M. Bercovier, 'Perturbation of mixed variational problems. Application to mixed finite element methods', *RAIRO*, (*Numerical Analysis*), **12,** 211–236 (1978).
14. M. Bercovier and M. Engelman, 'A finite element for the numerical solution of viscous incompressible flows, *J. Comp. Phys.*, **30,** 181–201 (1979).
15. T. J. R. Hughes, W. K. Liu and A. Brooks, 'Review of finite element analysis of incompressible viscous flows by the penalty function formulations', *J. Comp. Phys.*, **30,** 1–60 (1979).
16. G. Tavarra and G. Rimondi, 'Stress and strain calculation of rubber goods', *Com. Int. Rubber Conf.*, Kiev, 1978.
17. D. S. Malkus and T. J. R. Hughes, 'Mixed finite element methods—reduced and selective integration techniques: A unification of concepts', *J. Comp. Meth. Appl. Mech. Eng.*, **15,** 63–81 (1978).
18. P. G. Ciarlet and P. A. Raviart, 'The combined effect of curved boundaries and numerical integration in isoparametric finite element methods', in *The Mathematical Foundations of the Finite Element Method with Applications to Partial Differential Equations* (Ed. A. K. Aziz), pp. 409–474, Academic Press, New York, 1972.
19. E. Jankovich, F. Leblanc, M. Durand, and M. Bercovier, 'A finite element for the analysis of rubber parts, experimental and analytical assessment', *J. Computers and Structures* (in press).
20. R. L. Sani, P. M. Gresho, R. L. Lee, and D. F. Griffiths, 'The cause and cure (?) of the spurious pressures generated by certain FEM solutions of the incompressible Navier–Stokes equations', *Int. J. Num. Meth. Fluids*, **1** (1981).

21. M. Engelman, R. L. Sani, P. M. Gresho, and M. Bercovier, *Consistent versus Reduced Integration Formulations for The Penalty FEM Using Several Old and a New Element*, Submitted for publication.
22. O. C. Zienkiewicz, R. L. Taylor, and J. M. W. Baynham, 'Mixed and irreducible formulations in finite element analysis', Chapter 21 in this book.
23. K. J. Bathe and A. P. Cimento, 'Some practical procedures for the solution of nonlinear finite element equations', *J. Comp. Meth. Appl. Mech. Eng.*, **22,** 59–85 (1980).

Hybrid and Mixed Finite Element Methods
Edited by S. N. Atluri, R. H. Gallagher, and O. C. Zienkiewicz

Chapter 27

Four Theorems On the Limit Analysis In Solid Mechanics

Hsueh Dah-Wei

27.1 NOMENCLATURE

The summation convention of Cartesian tensor notation will be used in this paper, and let:

$x_i (i=1, 2, 3)$ = three-dimensional Cartesian coordinates
$\sigma_{ij} (i, j=1, 2, 3)$ = stress tensor
$\varepsilon_{ij} (i, j=1, 2, 3)$ = strain rate tensor
$v_i (i=1, 2, 3)$ = velocity components
$X_i (i=1, 2, 3)$ = body force components
$T_i (i=1, 2, 3)$ = basic surface loads
ν = multiplier load factor; thus νT_i = surface loads and ν_{ext} is to be sought in limit analysis (ν_{ext} means the extreme value of ν)
$n_i (i=1, 2, 3)$ = outward-drawn unit normal vector to a surface element
$f(\sigma_{ij}) - \sigma_T^2 = 0$ = yield condition of material; $f(\sigma_{ij})$ is a quadratic form of stress components and σ_T may be taken as varying from point to point for the non-homogeneous material
subscript p = plastic
subscript r = rigid

27.2 INTRODUCTION

Since the establishment of the well-known theorems of upper and lower bounds, considerable advance has been made in the limit analysis as a branch of applied plasticity in solid mechanics. Although thus far numerous results have been obtained, yet further progress seems to be very difficult in tackling more complicated problems. Progress is restricted by the fact that the limit theorems cannot in these cases give upper and lower bounds with a sufficient degree of approximation. Moreover, it is especially difficult to apply the lower bound theorems to most problems satisfactorily.

There exist two ways of attack for the limit analysis. One is by applying the bound theorems represented by Prager, Hodge, Drucker, and Greenberg; the other is by applying the variational principle such as in [1] and [2]. Many variational principles were mentioned in the well-known book of Washizu [3]. Chien [4] and Pian and Tong [5] all developed variational principles and contributed in many applications to finite element method.

In this paper, four strict theorems on the limit analysis for materials exhibiting rigid perfect plasticity are established and all negligences in [1] are corrected. On the basis of the theory of conditional extreme value in variational calculus, it is proved that the multiplier which is utilized in this paper is unique-reasonable. These terms can be applied to the limit analysis dealing with the non-homogeneous and the anisotropic materials. The illustrative solutions at the end of this paper point out that these theorems appear to be useful in obtaining the approximate solutions such as for strength problems in solid mechanics and for plastic-forming processes of metals by utilizing a generalized theorem established in here.

27.3 THEOREMS FOR THE LIMIT ANALYSIS IN SOLID MECHANICS

A rigid/perfect plastic body under the action of loads, when it is at the limit state, must satisfy the following equations and conditions:

(a) Equilibrium equations over the whole volume of the body:

$$\sigma_{ij,j} + X_i = 0 \tag{27.1}$$

(b) Flow law and rigid condition:

$$\varepsilon_{ij} = \lambda \frac{\partial f}{\partial \sigma_{ij}} \qquad (\text{in } V_p) \tag{27.2a}$$

$$\varepsilon_{ij} = 0 \qquad (\text{in } V_r) \tag{27.2b}$$

where λ is a positive scalar factor and ε_{ij} are related to v_i by

$$\varepsilon_{ij} = \tfrac{1}{2}(v_{i,j} + v_{j,i}) \tag{27.3}$$

(c) Boundary condition on S_σ:

$$\sigma_{ij} n_j = \nu T_i \tag{27.4}$$

(d) Boundary condition on S_v:

$$v_i = 0 \tag{27.5}$$

(e) Yield condition:

$$f=\sigma_T^2 \qquad (\text{in } V_p) \tag{27.6a}$$

$$f\leqslant\sigma_T^2 \qquad (\text{in } V_r) \tag{27.6b}$$

The limit analysis of structures consists of solving exactly or approximately the value of ν in equations (27.1) to (27.6), which yield a unique solution according to the uniqueness theorem.

For the sake of clarity and conciseness, the equilibrium conditions and the continuous conditions on the surface of the rigid/plastic boundary and on the surface of discontinuity which may occur in the plastic region are not taken into account. According to the work by Shi [6] or according to [7], even if these cases were to be taken into account, one may obtain the consequence by adding corresponding terms to it.

Theorem 27.1 *The limit state of structure is attained when the multiplier load factor ν takes the stationary value of the following expression for arbitrary variations of σ_{ij} and v_i:*

$$\nu=\text{ext}\frac{\int_V(\sigma_{ij}\varepsilon_{ij}-X_iv_i)\,dV-\int_{V_p}(\sigma_{ij}\varepsilon_{ij}/2\sigma_T^2)(f-\sigma_T^2)\,dV-\int_{S_v}\sigma_{ij}n_jv_i\,dS}{\int_{S_\sigma}T_iv_i\,dS} \tag{27.7}$$

where ε_{ij} is related to v_i by the expression (27.3), *σ_{ij} and v_i are arbitrary independent functions with the only restriction that $\sigma_{ij}\varepsilon_{ij}>0$, and $f(\sigma_{ij})$ has its maximum value in V_p.*

Proof With the arbitrary variations $\delta\sigma_{ij}$, δv_i, and $\delta\varepsilon_{ij}$ and noting that $V=V_r+V_p$, $S=S_{\sigma r}+S_{vr}+S_{\sigma p}+S_{vp}$, we transform equation (27.7) to the following equation applying the Gauss theorem:

$$\begin{aligned}
&-\int_{V_r}(\sigma_{ij,j}+X_i)\,\delta v_i\,dV+\int_{V_p}\left[\left(\frac{f-\sigma_T^2}{2\sigma_T^2}\sigma_{ij}\right)_{,j}-\sigma_{ij,j}-X_i\right]\delta v_i\,dV\\
&+\int_{V_r}\varepsilon_{ij}\,\delta\sigma_{ij}\,dV+\int_{V_p}\left(\varepsilon_{ij}-\frac{\sigma_{mk}\varepsilon_{mk}}{2\sigma_T^2}\frac{\partial f}{\partial\sigma_{ij}}-\frac{f-\sigma_T^2}{2\sigma_T^2}\varepsilon_{ij}\right)\delta\sigma_{ij}\,dV\\
&-\int_{S_v}v_in_j\,\delta\sigma_{ij}\,dS+\int_{S_{\sigma p}}\left(\sigma_{ij}n_j-\nu T_i-\frac{f-\sigma_T^2}{2\sigma_T^2}\sigma_{ij}n_j\right)\delta v_i\,dS\\
&+\int_{S_{\sigma r}}(\sigma_{ij}n_j-\nu T_i)\,\delta v_i\,dS-\int_{S_{vp}}\frac{f\ \ \sigma_T^2}{2\sigma_T^2}\sigma_{ij}n_j\,\delta v_i\,dS-\int_{S_{rp}}\frac{f-\sigma_T^2}{2\sigma_T^2}\sigma_{ij}n_j\,\delta v_i\,dS=0
\end{aligned}$$

Since the variations $\delta\sigma_{ij}$ and δv_i are arbitrary we must have

$$\sigma_{ij,j} + X_i = 0 \qquad (\text{in } V_r) \tag{27.8a}$$

$$\left(\frac{f-\sigma_T^2}{2\sigma_T^2}\sigma_{ij}\right)_{,j} - \sigma_{ij,j} - X_i = 0 \qquad (\text{in } V_p) \tag{27.8b}$$

$$\varepsilon_{ij} = 0 \qquad (\text{in } V_r) \tag{27.8c}$$

$$\varepsilon_{ij} - \frac{\sigma_{mk}\varepsilon_{mk}}{2\sigma_T^2}\frac{\partial f}{\partial \sigma_{ij}} - \frac{f-\sigma_T^2}{2\sigma_T^2}\varepsilon_{ij} = 0 \qquad (\text{in } V_p) \tag{27.8d}$$

$$v_i = 0 \qquad (\text{on } S_v) \tag{27.8e}$$

$$\sigma_{ij}n_j - \nu T_i - \frac{f-\sigma_T^2}{2\sigma_T^2}\sigma_{ij}n_j = 0 \qquad (\text{on } S_{\sigma p}) \tag{27.8f}$$

$$\sigma_{ij}n_j - \nu T_i = 0 \qquad (\text{on } S_{\sigma r}) \tag{27.8g}$$

$$\frac{f-\sigma_T^2}{2\sigma_T^2}\sigma_{ij}n_j = 0 \qquad (\text{on } S_{vp}) \tag{27.8h}$$

$$\frac{f-\sigma_T^2}{2\sigma_T^2}\sigma_{ij}n_j = 0 \qquad (\text{on } S_{rp}) \tag{27.8i}$$

Multiplying ε_q in (27.8d) by σ_{ij} and noting that $(\partial f/\partial\sigma_{ij})\sigma_{ij} = 2f$, we have

$$\sigma_{ij}\varepsilon_{ij}\left(1 - \frac{1}{2\sigma_T^2}\frac{\partial f}{\partial\sigma_{mk}}\sigma_{mk} - \frac{f-\sigma_T^2}{2\sigma_T^2}\right) = \left(1 - \frac{2f}{2\sigma_T^2} - \frac{f-\sigma_T^2}{2\sigma_T^2}\right)\sigma_{ij}\varepsilon_{ij} = 0$$

Thus, we obtain $f = \sigma_T^2$ in V_p. This equation fulfills the actual yield condition in the plastic region while it follows that $f(\sigma_{ij}) \leqslant \sigma_T^2$ in the rigid region. Substituting $f = \sigma_T^2$ into equations (27.8b, d, f, h and i), we obtain

$$\sigma_{ij,j} + X_i = 0 \qquad (\text{in } V) \tag{27.9a}$$

$$\varepsilon_{ij} = 0 \qquad (\text{in } V_r) \tag{27.9b}$$

$$\varepsilon_{ij} - \frac{\sigma_{mk}\varepsilon_{mk}}{2\sigma_T^2}\frac{\partial f}{\partial\sigma_{ij}} = 0 \qquad (\text{in } V_p) \tag{27.9c}$$

$$v_i = 0 \qquad (\text{on } S_v) \tag{27.9d}$$

$$\sigma_{ij}n_j - \nu T_i = 0 \qquad (\text{on } S_\sigma) \tag{27.9e}$$

The equation (27.9c) itself is the flow law [7]. Equations (27.8h, i) are identical. Equations (27.9) with $f = \sigma_T^2$ in V_p and $f \leqslant \sigma_T^2$ in V_r are the same equations and conditions as (27.1) to (27.6). Therefore, we can conclude that Theorem 27.1 is equivalent mathematically to the whole set of equations and conditions which must be satisfied by the limit analysis. The above proof can be extended without any difficulty to the cases in which there are several rigid regions and plastic regions at the limit state.

If the conditions on S_v are $v_i = \bar{v}_i$, where $\bar{v}_i$ are given as in plastic-forming processes, formula (27.7) should be

$$\nu = \text{ext} \frac{\int_V (\sigma_{ij}\varepsilon_{ij} - X_i v_i)\,\mathrm{d}V - \int_{V_p} (\sigma_{ij}\varepsilon_{ij}/2\sigma_T^2)(f - \sigma_T^2)\,\mathrm{d}V - \int_{S_v} \sigma_{ij} n_j (v_i - \bar{v}_i)\,\mathrm{d}S}{\int_{S_\sigma} T_i v_i\,\mathrm{d}S} \tag{27.10}$$

This generalized form can easily be proved.

We can also prove the following:

Theorem 27.2: Converse theorem *To the precise solution σ_{ij} and v_i of a rigid/perfect plastic body which is set at the limit state, the variation of the multiplier factor $\delta\nu$ equals zero.*

In order that one can apprehend and utilize these theorems as proposed in this paper in a better way, an additional explanation about the source of the multiplier seems to be necessary. In fact, the limit state of a rigid/perfect plastic body is described by the functional

$$F = \int_V \sigma_{ij}\,\tfrac{1}{2}(v_{i,j} + v_{j,i})\,\mathrm{d}V - \int_V X_i v_i\,\mathrm{d}V - \nu \int_{S_\sigma} T_i v_i\,\mathrm{d}S \tag{27.11}$$

and it will undergo the extreme value under the following conditions:

$$\tfrac{1}{2}(v_{i,j} + v_{j,i}) = 0 \qquad (\text{in } V_r) \tag{27.12}$$

$$f - \sigma_T^2 = 0 \qquad (\text{in } V_p) \tag{27.13}$$

$$v_i = 0 \qquad (\text{on } S_v) \tag{27.14}$$

and

$$\frac{\partial f}{\partial \sigma_{ij}}\,\sigma_{ij} = 2f \tag{27.15}$$

It is the problem of conditional extreme value and may be changed by the Lagrange multiplier λ, μ_{ij}, and γ_i to the extreme value problem of the functional

$$F_1 = \int_V \sigma_{ij}\,\tfrac{1}{2}(v_{i,j} + v_{j,i})\,\mathrm{d}V - \int_V X_i v_i\,\mathrm{d}V - \nu \int_{S_\sigma} T_i v_i\,\mathrm{d}S + \int_{S_v} \gamma_i v_i\,\mathrm{d}S$$
$$+ \int_{V_p} \lambda(f - \sigma_T^2)\,\mathrm{d}V + \int_{V_r} \tfrac{1}{2}(v_{i,j} + v_{j,i})\mu_{ij}\,\mathrm{d}V \tag{27.16}$$

where the variation of λ, γ_i, μ_{ij}, σ_{ij}, and v_i are arbitrary. If conditions (27.12) to (27.14) are satisfied in V_r, V_p, and on S_v respectively, then equation (27.16) will reduce to equation (27.11). Taking the variation of equation

(27.16), we obtain

$$\delta F_1 = \int_V \tfrac{1}{2}(v_{i,j} + v_{j,i})\, \delta\sigma_{ij}\, \mathrm{d}V + \int_V \sigma_{ij}\, \tfrac{1}{2}(\delta v_{ij} + \delta v_{j,i})\, \mathrm{d}V$$
$$- \int_V X_i\, \delta v_i\, \mathrm{d}V - \delta\nu \int_{S_\sigma} T_i v_i\, \mathrm{d}S - \nu \int_{S_\sigma} T_i\, \delta v_i\, \mathrm{d}S + \int_{S_v} \delta\gamma_i v_i\, \mathrm{d}S$$
$$+ \int_{S_v} \gamma_i\, \delta v_i\, \mathrm{d}S + \int_{V_p} \delta\lambda (f - \sigma_T^2)\, \mathrm{d}V + \int_{V_p} \lambda \frac{\partial f}{\partial \sigma_{ij}}\, \delta\sigma_{ij}\, \mathrm{d}V$$
$$+ \int_{V_r} \tfrac{1}{2}(v_{i,j} + v_{j,i})\, \delta\mu_{ij}\, \mathrm{d}V + \int_{V_r} \tfrac{1}{2}(\delta v_{i,j} + \delta v_{j,i})\mu_{ij}\, \mathrm{d}V \tag{27.17}$$

When $\delta F_1 = 0$ and $\delta\nu = 0$ we obtain $\nu = \nu_{\mathrm{ext}}$. From the Gauss theorem,

$$\int_{V_r} \sigma_{ij}\, \tfrac{1}{2}(\delta v_{i,j} + \delta v_{j,i})\, \mathrm{d}V = -\int_{V_r} \sigma_{ij,j}\, \delta v_i\, \mathrm{d}V + \int_{S_{rp}} \sigma_{ij} n_{jrp}\, \delta v_i\, \mathrm{d}S$$
$$+ \int_{S_{\sigma r} + S_{vr}} \sigma_{ij} n_j\, \delta v_i\, \mathrm{d}S$$

$$\int_{V_p} \sigma_{ij}\, \tfrac{1}{2}(\delta v_{i,j} + \delta v_{j,i})\, \mathrm{d}V = -\int_{V_p} \sigma_{ij,j}\, \delta v_i\, \mathrm{d}V + \int_{S_{pr}} \sigma_{ij} n_{jpr}\, \delta v_i\, \mathrm{d}S$$
$$+ \int_{S_{\sigma p} + S_{vp}} \sigma_{ij} n_j\, \delta v_i\, \mathrm{d}S$$

$$\int_{V_r} \tfrac{1}{2}(\delta v_{i,j} + \delta v_{j,i})\mu_{ij}\, \mathrm{d}V = -\int_{V_r} \mu_{ij,j}\, \delta v_i\, \mathrm{d}V + \int_{S_{\sigma r} + S_{vr}} \mu_{ij} n_j\, \delta v_i\, \mathrm{d}S$$
$$+ \int_{S_{rp}} \mu_{ij} n_{jrp}\, \delta v_i\, \mathrm{d}S$$

Equation (27.17) can be calculated and rearranged as follows:

$$\nu_{\mathrm{ext}} \int_{S_\sigma} T_i\, \delta v_i\, \mathrm{d}S = -\int_{V_p} (\sigma_{ij,j} + X_i)\, \delta v_i\, \mathrm{d}V - \int_{V_r} [(\sigma_{ij} + \mu_{ij})_{,j} + X_i]\, \delta v_i\, \mathrm{d}V$$
$$+ \int_{S_v} v_i\, \delta\gamma_i\, \mathrm{d}S + \int_{S_{vp}} (\sigma_{ij} n_j + \gamma_i)\, \delta v_i\, \mathrm{d}S$$
$$+ \int_{S_{vr}} [(\sigma_{ij} + \mu_{ij}) n_j + \gamma_i]\, \delta v_i\, \mathrm{d}S$$
$$+ \int_{V_p} \left[\tfrac{1}{2}(v_{i,j} + v_{j,i}) + \lambda \frac{\partial f}{\partial \sigma_{ij}}\right] \delta\sigma_{ij}\, \mathrm{d}V$$
$$+ \int_{V_p} (f - \sigma_T^2)\, \delta\lambda\, \mathrm{d}V$$

$$+\int_{V_r} \tfrac{1}{2}(v_{i,j}+v_{j,i})(\delta\sigma_{ij}+\delta\mu_{ij})\,\mathrm{d}V$$

$$+\int_{S_{pr}} [\sigma_{ij}n_{jpr}+(\sigma_{ij}+\mu_{ij})n_{jrp}]\,\delta v_i\,\mathrm{d}S$$

$$+\int_{S_{\sigma p}} \sigma_{ij}n_j\,\delta v_i\,\mathrm{d}S+\int_{S_{\sigma r}} [(\sigma_{ij}+\mu_{ij})n_j]\,\delta v_i\,\mathrm{d}S$$

Since the variations $\delta\lambda$, $\delta\gamma_i$, $\delta\mu_{ij}$, $\delta\sigma_{ij}$, and δv_i are arbitrary, we obtain the following equations:

(a) In V_p:

$$\sigma_{ij,j}+X_i=0 \tag{27.18a}$$

$$\tfrac{1}{2}(v_{i,j}+v_{j,i})+\lambda\frac{\partial f}{\partial\sigma_{ij}}=0 \tag{27.18b}$$

$$f-\sigma_T^2=0 \tag{27.18c}$$

(b) In V_r:

$$(\sigma_{ij}+\mu_{ij})_{,j}+X_i=0 \tag{27.19a}$$

$$\tfrac{1}{2}(v_{i,j}+v_{j,i})=0 \tag{27.19b}$$

(c) On the surface of the rigid plastic region boundary S_{rp}:

$$\sigma_{ij}n_{jpr}+(\sigma_{ij}+\mu_{ij})n_{jrp}=0 \tag{27.20}$$

(d) On the surface of the rigid region S_{vr} and $S_{\sigma r}$:

$$(\sigma_{ij}+\mu_{ij})n_j+\gamma_i=0 \quad\text{and}\quad v_i=0 \quad (\text{on } S_{vr}) \tag{27.21a}$$

$$(\sigma_{ij}+\mu_{ij})n_j-\nu_{ext}T_i=0 \quad (\text{on } S_{\sigma r}) \tag{27.21b}$$

(e) On the surface of the plastic region S_{vp} and $S_{\sigma r}$:

$$\sigma_{ij}n_j+\gamma_i=0 \quad\text{and}\quad v_i=0 \quad (\text{on } S_{vp}) \tag{27.22a}$$

$$\sigma_{ij}n_j-\nu_{ext}T_i=0 \quad (\text{on } S_{\sigma p}) \tag{27.22b}$$

From the above results we may absorb the Lagrange multiplier μ_{ij} into σ_{ij} in the rigid region. Thus we can take

$$\mu_{ij}=0 \tag{27.23}$$

Using formula (27.23) we can combine equations (27.21a) and (27.22a) and get (on the surface of $S_v=S_{vr}+S_{vp}$):

$$\gamma_i=-\sigma_{ij}n_j \tag{27.24}$$

and $v_i=0$ on the surface S_v, which is itself equation (27.14).

On the surface of the rigid plastic boundary S_{pr} we have

$$-n_{jpr}=n_{jrp}$$

Thus, equation (27.20) becomes an identical equation. Noting equation (27.23), from (27.18a) and (27.19a) we obtain the equilibrium equation; from (27.19b), we obtain the rigid condition in the rigid region; from (27.18e), we obtain the yield condition in the plastic region. Multiplying (27.18b) by σ_{ij}, we have

$$\tfrac{1}{2}(v_{i,j}+v_{j,i})\sigma_{ij} = -\lambda \frac{\partial f}{\partial \sigma_{ij}} \sigma_{ij} = -2\lambda f = -2\lambda \sigma_T^2$$

that is,

$$\lambda = -\frac{1}{4\sigma_T^2}(v_{i,j}+v_{j,i})\sigma_{ij} \tag{27.25}$$

Thus the three Lagrange multipliers are determined as above. Substituting them in equation (27.16) we have

$$F_1 = \int_V \sigma_{ij}\, \tfrac{1}{2}(v_{i,j}+v_{j,i})\, dV - \int_V X_i v_i \, dV - \nu \int_{S_\sigma} T_i v_i \, dS$$
$$- \int_{V_p} \frac{1}{4\sigma_T^2}(v_{i,j}+v_{j,i})\sigma_{ij}(f-\sigma_T^2)\, dV - \int_{S_v} \sigma_{ij} n_j v_i \, dS$$

Thus, the unique-reasonableness of the multiplier proposed in this paper is proved.

An important practical advantage of this variational theorem is that it provides much freedom to select the function σ_{ij} and v_i in applying, for example, Ritz approximation procedures. The more rationally the functions σ_{ij} and v_i are selected, the more accurate can the answers be expected. Thus, if possible, one is recommended to assign σ_{ij} to satisfy the equilibrium equation and v_i to satisfy the boundary restraint condition. When σ_{ij} and v_i are selected in such a way, we can establish the following:

Theorem 27.3 *If a stress field σ^*_{ij} is selected to satisfy the equilibrium requirements as in the lower bound theorem and if a velocity field v_i^0 is selected to satisfy the mechanism requirements in the upper bound theorem, then the limit load deduced from Theorem 27.1 will lie between the lower and upper bounds given by the bound theorems.*

Proof If σ^*_{ij} and v_i^0 are selected as in Theorem 27.3, Theorem 27.1 yields

$$\nu = \frac{\displaystyle\int_V (\sigma^*_{ij}\varepsilon^0_{ij} - X_i v_i^0)\, dV - \int_{V_p} \frac{\sigma^*_{ij}\varepsilon^0_{ij}}{2\sigma_T^2}(f^* - \sigma_T^2)\, dV}{\displaystyle\int_{S_\sigma} T_i v_i^0 \, dS}$$

$$= \frac{\displaystyle\int_V \sigma^*_{ij}\varepsilon^0_{ij} \frac{3\sigma_T^2 - f^*}{2\sigma_T^2}\, dV + \int_{V_r} \frac{\sigma^*_{ij}\varepsilon^0_{ij}(f^* - \sigma_T^2)\, dV}{2\sigma_T^2} - \int_V X_i v_i^0 \, dV}{\displaystyle\int_{S_\sigma} T_i v_i^0 \, dS} \tag{27.26}$$

where $f^* \equiv f(\sigma^*_{ij})$.

With the same σ^*_{ij} and v^0_i, the upper and lower bound theorems give respectively

$$\nu_{\text{upper}} = \frac{\int_V \sigma^0_{ij}\varepsilon^0_{ij}\,dV - \int_V X_i v^0_i\,dV}{\int_{S_\sigma} T_i v^0_i\,dS} \tag{27.27a}$$

$$\nu_{\text{lower}} = \frac{\int_V \sigma^*_{ij}\varepsilon^0_{ij}\,dV - \int_V X_i v^0_i\,dV}{\int_{S_\sigma} T_i v^0_i\,dS} \tag{27.27b}$$

where σ^0_{ij} in (27.27) is related to ε^0_{ij} according to the flow law and the yield condition.

Thus we have [8]

$$\sigma^0_{ij}\varepsilon^0_{ij} \geqslant \frac{3\sigma^2_T - f^*}{2\sigma^2_T}\sigma^*_{ij}\varepsilon^0_{ij} \tag{27.28}$$

and $\sigma^*_{ij}\varepsilon^0_{ij} > 0$; hence we have

$$\nu_{\text{lower}} \leqslant \nu \leqslant \nu_{\text{upper}}$$

Theorem 27.4 *Suppose*

(a) *$\tilde{\sigma}_{ij}$ is a stress field in equilibrium with the basic surface loads T_i (the body force X_i assumed the be zero).*

(b) $\sigma_{ij} = \beta\tilde{\sigma}_{ij}$
where

$$\beta = \left[\frac{1}{3}\,\frac{\int_V \tilde{\sigma}_{ij}\varepsilon_{ij}\,dV + \frac{1}{2}\int_{V_p} \tilde{\sigma}_{ij}\varepsilon_{ij}\,dV}{\int_{V_p} (\tilde{\sigma}_{ij}\varepsilon_{ij}\tilde{f}/2\sigma^2_T)\,dV}\right]^{1/2} \tag{27.29}$$

is a variational parameter and $\tilde{f} = f(\tilde{\sigma}_{ij})$.

(c) *v_i is a kinematically admissible velocity field.*

(d) *$\tilde{\sigma}_{ij}\varepsilon_{ij} > 0$, ε_{ij} is related to v_i by equation (27.3), then the multiplier load factor ν determined by*

$$\nu = \beta\,\frac{\frac{2}{3}\int_V \tilde{\sigma}_{ij}\varepsilon_{ij}\,dV + \frac{1}{3}\int_{V_p} \tilde{\sigma}_{ij}\varepsilon_{ij}\,dV}{\int_{S_\sigma} T_i v_i\,dS} \tag{27.30}$$

should be not lower than that obtained from the lower bound theorem with the stress field $\tilde{\sigma}_{ij}$ and not higher than that obtained from the upper bound theorem with the velocity field v_i,

$$\nu_{\text{lower}} \leqslant \nu \leqslant \nu_{\text{upper}}$$

Proof Formula (27.7) in this case becomes

$$\nu = \text{ext}\,\frac{\int_V \beta\tilde{\sigma}_{ij}\varepsilon_{ij}\,dV - \int_{V_p}[\beta\tilde{\sigma}_{ij}\varepsilon_{ij}(\beta^2\tilde{f} - \sigma_T^2)/2\sigma_T^2]\,dV}{\int_{S_\sigma} T_i v_i\,dS} \tag{27.31}$$

If $F(\beta)$ represents the numerator in equation (27.31), from $\partial F/\partial\beta = 0$, we obtain (27.30). Substituting (27.30) into (27.31), we obtain (27.29).

The inequality $\nu \leqslant \nu_{\text{upper}}$ has already been established by Theorem 27.3, so we will prove $\nu_{\text{lower}} \leqslant \nu$. If $\nu_{\text{lower}}\tilde{\sigma}_{ij}$ is the solution given by the lower bound theorem, owing to $\nu_{\text{lower}}^2\tilde{f} \leqslant \sigma_T^2$, we obtain

$$\nu_{\text{lower}}\tilde{\sigma}_{ij}\varepsilon_{ij}\frac{3\sigma_T^2 - \nu_{\text{lower}}^2\tilde{f}}{2\sigma_T^2} \geqslant \nu_{\text{lower}}\tilde{\sigma}_{ij}\varepsilon_{ij} \tag{27.32}$$

Then

$$\begin{aligned}
\nu &= \frac{\int_V \beta\tilde{\sigma}_{ij}\varepsilon_{ij}\,dV - \int_{V_p}(\beta\tilde{\sigma}_{ij}\varepsilon_{ij}/2\sigma_T^2)(\beta^2\tilde{f} - \sigma_T^2)\,dV}{\int_{S_\sigma} T_i v_i\,dS} \\
&= \frac{\int_{V_p}\frac{1}{2\sigma_T^2}\beta\tilde{\sigma}_{ij}\varepsilon_{ij}(3\sigma_T^2 - \beta^2\tilde{f})\,dV + \int_{V_r}\beta\tilde{\sigma}_{ij}\varepsilon_{ij}\,dV}{\int_{S_\sigma} T_i v_i\,dS} \\
&\geqslant \frac{\int_{V_p}\frac{1}{2\sigma_T^2}\nu_{\text{lower}}\tilde{\sigma}_{ij}\varepsilon_{ij}(3\sigma_T^2 - \nu_{\text{lower}}^2\tilde{f})\,dV + \int_{V_r}\nu_{\text{lower}}\tilde{\sigma}_{ij}\varepsilon_{ij}\,dV}{\int_{S_\sigma} T_i v_i\,dS} \\
&\geqslant \frac{\int_{V_p}\nu_{\text{lower}}\tilde{\sigma}_{ij}\varepsilon_{ij}\,dV + \int_{V_r}\nu_{\text{lower}}\tilde{\sigma}_{ij}\varepsilon_{ij}\,dV}{\int_{S_\sigma} T_i v_i\,dS} \\
&= \frac{\int_V \nu_{\text{lower}}\tilde{\sigma}_{ij}\varepsilon_{ij}\,dV}{\int_{S_\sigma} T_i v_i\,dS} = \nu_{\text{lower}}
\end{aligned} \tag{27.33}$$

Thus Theorem 27.4 is proved.

Obviously, if we select the function v_i which satisfies $v_i = \bar{v}_i$ on S_v, Theorems 27.3 and 27.4 still hold.

Theorem 27.1 as shown above can be extended to cases with finite displacement. The necessity of doing so is from the consideration that all those engaged in mechanics have long understood that the effect due to geometric changes must be taken into consideration when carrying out the limit analysis which is often of great practical value. Yet there seems to be no available analytical method that is generally applicable.

Suppose there is a rigid perfect plastic body under the action of loading, when at the limit state it must satisfy the following equations and conditions:

(a) The equilibrium equations within the entire volume of the body, that is:

$$[(\delta_{ki} + v_{k,i})\sigma_{ij}]_{,j} + X_k = 0 \tag{27.34}$$

(b) The flow law and rigid condition:

In the plastic region V_p:
$$\varepsilon_{ij} = \lambda \frac{\partial f}{\partial \sigma_{ij}} \tag{27.35a}$$

In the rigid region V_r:
$$\tfrac{1}{2}(v_{i,j} + v_{j,i} + v_{k,i}v_{k,j}) = 0 \tag{27.35b}$$

(c) The relation between the strain rate tensor and the velocity components, that is:
In V_p:

$$\varepsilon_{ij} = \tfrac{1}{2}(v_{i,j} + v_{j,i} + v_{k,i}v_{k,j}) \tag{27.36}$$

(d) The boundary conditions on S_σ where the traction is given:

$$(\delta_{ik} + v_{i,k})\sigma_{kj}n_j = \nu T_i \tag{27.37}$$

(e) The boundary conditions on S_v where the velocity component is given:

$$v_i = \bar{v}_i \tag{27.38}$$

(f) The yield conditions:

In V_p:
$$f = \sigma_T^2 \tag{27.39a}$$

In V_r:
$$f \leq \sigma_T^2 \tag{27.39b}$$

where δ_{ij} is the Kronecker symbol.

A theorem on the limit analysis in finite displacement can be stated as follows:

Theorem 27.5 *When the limit state is reached the multiplier load factor* ν

takes the stationary value as is given in the following expression:

$$\nu = \text{sta} \frac{\displaystyle\int_V [\tfrac{1}{2}(v_{i,j} + v_{j,i} + v_{k,i}v_{k,j})\sigma_{ij} - X_i v_i]\,dV - \int_{V_p} (\sigma_{ij}\varepsilon_{ij}/2\sigma_T^2)(f - \sigma_T^2)\,dV - \int_{S_v} (\delta_{ik} + v_{i,k})\sigma_{kj}n_j(v_i - \bar{v}_i)\,dS}{\displaystyle\int_{S_\sigma} T_i v_i\,dS} \tag{27.40}$$

where σ_{ij}, ε_{ij}, and v_i are arbitrary independent functions with the only restriction that for $\sigma_{ij}\varepsilon_{ij} > 0$, the maximum value of $f(\sigma_{ij})$ lies in the plastic region V_p.

Proof We may transform equation (27.40) by using the Gauss theorem:

$$\begin{aligned}
&-\int_V \{[(\delta_{ki} + v_{k,i})\sigma_{ij}]_{,j} + X_k\}\,\delta v_k\,dV \\
&-\int_{V_p} [\varepsilon_{ij} - \tfrac{1}{2}(v_{i,j} + v_{j,i} + v_{k,i}v_{k,j})]\,\delta\sigma_{ij}\,dV \\
&+\int_{V_p} \left(\varepsilon_{ij} - \frac{\sigma_{mk}\varepsilon_{mk}}{2\sigma_T^2}\frac{\partial f}{\partial\sigma_{ij}} - \frac{f - \sigma_T^2}{2\sigma_T^2}\varepsilon_{ij}\right)\delta\sigma_{ij}\,dV \\
&+\int_{V_r} \tfrac{1}{2}(v_{i,j} + v_{j,i} + v_{k,i}v_{k,j})\,\delta\sigma_{ij}\,dV \\
&+\int_{S_\sigma} [(\delta_{ik} + v_{i,k})\sigma_{kj}n_j - \nu T_i]\,\delta v_i\,dS \\
&-\int_{V_p} \frac{f - \sigma_T^2}{2\sigma_T^2}\sigma_{ij}\,\delta\varepsilon_{ij}\,dV \\
&-\int_{S_v} (v_i - \bar{v}_i)\,\delta[(\delta_{ik} + v_{i,k})\sigma_{kj}n_j]\,dS = 0
\end{aligned}$$

Since the variation $\delta\sigma_{ij}$, $\delta\varepsilon_{ij}$, and δv_i are arbitrary, we must have

$$[(\delta_{ki} + v_{k,i})\sigma_{ij}]_{,j} + X_k = 0 \qquad (\text{in } V) \tag{27.41a}$$

$$\varepsilon_{ij} - \tfrac{1}{2}(v_{i,j} + v_{j,i} + v_{k,i}v_{k,j}) = 0 \qquad (\text{in } V_p) \tag{27.41b}$$

$$\varepsilon_{ij} - \frac{\sigma_{mk}\varepsilon_{mk}}{2\sigma_T^2}\frac{\partial f}{\partial\sigma_{ij}} - \frac{f - \sigma_T^2}{2\sigma_T^2}\varepsilon_{ij} = 0 \qquad (\text{in } V_p) \tag{27.41c}$$

$$\tfrac{1}{2}(v_{i,j} + v_{j,i} + v_{k,i}v_{k,j}) = 0 \qquad (\text{in } V_r) \tag{27.41d}$$

$$(\delta_{ik} + v_{i,k})\sigma_{kj}n_j = \nu T_i \qquad (\text{on } S_\sigma) \tag{27.41e}$$

$$v_i = \bar{v}_i \qquad (\text{on } S_v) \tag{27.41f}$$

$$\frac{f - \sigma_T^2}{2\sigma_T^2}\sigma_{ij} = 0 \qquad (\text{in } V_p) \tag{27.41g}$$

Multiplying equation (27.41c) by σ_{ij} and making use of the equality $(\partial f/\partial\sigma_{ij})\sigma_{ij}=2f$, we get

$$\sigma_{ij}\varepsilon_{ij}\left(1-\frac{1}{2\sigma_T^2}\frac{\partial f}{\partial\sigma_{mk}}\sigma_{mk}-\frac{f-\sigma_T^2}{2\sigma_T^2}\right)=\left(1-\frac{2f}{2\sigma_T^2}-\frac{f-\sigma_T^2}{2\sigma_T^2}\right)\sigma_{ij}\varepsilon_{ij}=0$$

Thus we obtain $f=\sigma_T^2$ in V_p. This equation fulfils the actual yield condition in the plastic region; it follows that $f(\sigma_{ij})\leqslant\sigma_T^2$ in the rigid region. Substituting $f=\sigma_T^2$ into (27.41c), we obtain an equation which is itself the flow law. Substituting $f=\sigma_T^2$ into (27.41g), we obtain a useless identical equation and other equations numbered (27.41a, b, d, e, and f) which are in fact equations (27.34), (27.35b), (27.36), (27.37), and (27.38) respectively. So we may now conclude that Theorem 27.5 given here is equivalent mathematically to the whole set of equations and conditions which the limit analysis must satisfy. The above proof can be extended without any difficulty to the cases with several rigid regions and plastic regions at the limit state.

Note that σ_{ij}, ε_{ij}, and v_i in (27.40) are arbitary independent functions. It is not necessary for them to satisfy the equilibrium, flow law, and the relations between the strain tensor and the velocity components respectively. If a restriction is placed on the relation between ε_{ij} and v_i, as is shown in equation (27.36), we then obtain

$$\nu=\text{sta}\;\frac{\displaystyle\int_V(\sigma_{ij}\varepsilon_{ij}-X_iv_i)\,dV-\int_{V_p}(\sigma_{ij}\varepsilon_{ij}/2\sigma_T^2)(f-\sigma_T^2)\,dV-\int_{S_v}(\delta_{ik}+v_{i,k})\sigma_{kj}n_j(v_i-\bar{v}_i)\,dS}{\displaystyle\int_{S_\sigma}T_iv_i\,dS}\tag{27.42}$$

If it is further restricted to cases with small deformation, equation (27.42) will then reduce to equation (27.7).

The following converse theorem can be proved without difficulty:

Theorem 27.6: Converse theorem *The precise solution of σ_{ij}, ε_{ij}, and v_i in finite deformation of a rigid/perfect plastic body which is at the limit state will make the variation of the multiplier factor $\delta\nu$ zero.*

For the purpose of making apprehension and utilization of Theorem 27.5 better, an additional explanation seems to be necessary. In fact, at the limit state, a rigid/perfect plastic body should be described by the following functional:*

$$F=\int_V\left[\sigma_{ij}\,\tfrac{1}{2}(v_{i,j}+v_{j,i}+v_{k,i}v_{k,j})-X_iv_i\right]dV-\nu\int_{S_\sigma}T_iv_i\,dS\tag{27.43}$$

* This proof can be extended to (27.40) without any difficulty.

which is to take the stationary value under the following conditions:

$$\tfrac{1}{2}(v_{i,j}+v_{j,i}+v_{k,i}v_{k,j})=0 \qquad (\text{in } V_r)$$
$$f-\sigma_T^2=0 \qquad (\text{in } V_p)$$
$$v_i=\bar{v}_i \qquad (\text{on } S_v)$$

It is a problem of conditional stationary value and may be changed into one functional by taking the Lagrange multiplier λ, μ_{ij}, γ_i and adding three integral terms to the stationary value problem of the functional (27.43):

$$F_1=\int_V [\sigma_{ij}\,\tfrac{1}{2}(v_{i,j}+v_{j,i}+v_{k,i}v_{k,j})-X_iv_i]\,dV-\nu\int_{S_\sigma} T_iv_i\,dS$$
$$+\int_{S_v}\gamma_i(v_i-\bar{v}_i)\,dS+\int_{V_p}\lambda(f-\sigma_T^2)\,dV$$
$$+\tfrac{1}{2}\int_{V_r}(v_{i,j}+v_{j,i}+v_{k,i}v_{k,j})\mu_{ij}\,dV$$

where the variation of λ, γ_i, μ_{ij}, σ_{ij}, and v_i are arbitary. Taking the variation of the above equation, noting that when $\delta F_1=0$ and $\delta\nu=0$, $\nu=\nu_{sta}$ and finally using the Gauss theorem, we can obtain

$$\mu_{ij}=0$$
$$\lambda=-\frac{1}{4\sigma_T^2}(v_{i,j}+v_{j,i}+v_{k,i}v_{k,j})\sigma_{ij}$$
$$\gamma_k=-(\delta_{ki}+v_{k,i})\sigma_{ij}n_j$$

So far we have completely proved the theorem. From the differences in values calculated from equations (27.7) and (27.40) (or equation 27.42), the effect from the change of geometric factor on limit loading can thus be determined.

27.4 ILLUSTRATIVE SOLUTIONS

The variational theorems described in the previous section will now be illustrated by application to certain specific problems.

Example 27.1 A problem of steady plastic flow of a frictionless extrusion of a sheet when the reduction of thickness is 50 per cent is under consideration. Figure 27.1 shows the die used in this extrusion process and a tentatively assumed field of plastic stress. The sheet filling the entire width of the well-lubricated die is pushed in from the left while it leaves the die towards the right with its thickness reduced to half the original value. The following solution was given by Hill [7].

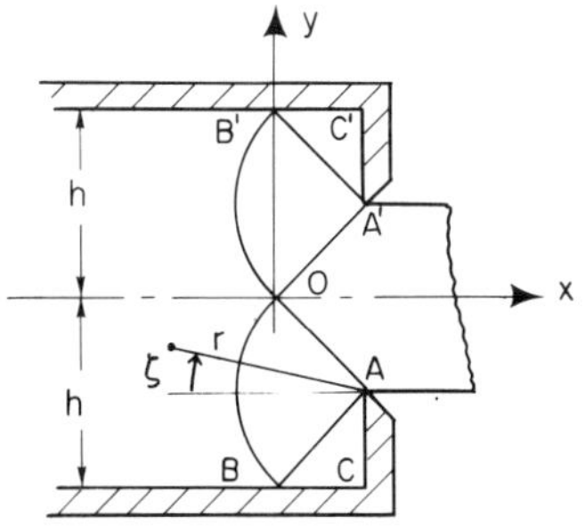

Figure 27.1 Die used in the extrusion process and assumed field of plastic stress

In the centred fans OAB and OA′B′

$$\sigma_r = \sigma_\theta = 2k\left(\frac{\pi}{4} - \tfrac{1}{2} - \theta\right), \qquad \tau_{r\theta} = k$$
$$v_r = -\sin\theta, \qquad v_\theta = -\left(\frac{\sqrt{2}}{2} + \cos\theta\right) \tag{27.44}$$

According to the above, the strain rate field is

$$\varepsilon_r = 0, \qquad \varepsilon_\theta = 0, \qquad \gamma_{r\theta} = \frac{\sqrt{2}}{2r} \tag{27.45}$$

Substituting (27.4) and (27.25) into Theorem 27.1 (formula 27.10) and noting that the boundaries of these two fans are lines of discontinuity of the velocity field, we obtain the average extrusion pressure p:

$$p = \frac{1}{\sqrt{2}}\int_{\pi/4}^{3\pi/4} \frac{\sqrt{2}}{2} k\, \mathrm{d}\theta + \int_{-\pi/4}^{\pi/4} \left(\frac{\sqrt{2}}{2} + \sin\varphi - \sin\varphi\right)\frac{k}{\sqrt{2}}\, \mathrm{d}\varphi$$
$$+ \frac{1}{h}\int_0^{h/\sqrt{2}} \frac{k}{\sqrt{2}}\, \mathrm{d}r - \frac{k}{h}\int_0^{h/2}\left(\frac{\sqrt{2}}{2} - \sqrt{2}\right)\mathrm{d}r = k\left(1 + \frac{\pi}{2}\right)$$

This result coincides with Hill's solution. It is a universal rule that when we substitute a precise solution σ_{ij} and v_i into Theorem 27.1, we will obtain the precise multiplier. Example 27.2 below also possesses this result. Now, throughout Example 27.2, we will point out the degree of preciseness of Theorem 27.1 in applications for practical problems.

Example 27.2 The indentation of a semi-infinite body by a flat rigid punch under conditions of plane strain are shown in Figure 27.2. We assume the

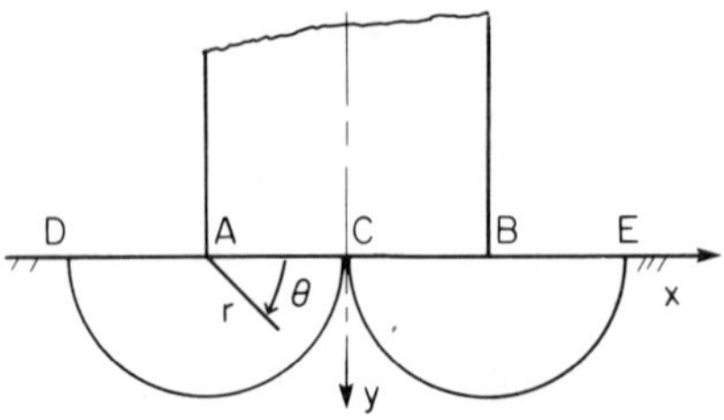

Figure 27.2 Indentation of a semi-infinite body by a flat rigid punch

underside of the punch to be well lubricated and that the stress field is

$$\sigma_r = \beta(2+\cos\theta)$$
$$\sigma_\theta = \beta(2+2\cos\theta)$$
$$\tau_{r\theta} = \beta\sin\theta$$

For simplicity, let us take the downward velocity of the punch as the unit of velocity and assume that $v_r = 0$, $v_\theta =$ constant in a semicircular plastic region. Of course, these selections are so far away from the precise solution of this problem. However, when we substitute these into Theorem 27.1, we will notice that, during the incipient plastic flow, the lubricated punch exerts a uniformly distributed pressure $5.96k$. The precise solution, according to Hill, is $5.14k$.

Example 27.3 A simply supported circular plate is subjected to uniformly distributed loading, the dead weight of the plate being neglected, that is $X_i = 0$. It is known that the carrying capacity under the Mises yield condition is about

$$p = \frac{6.5M_T}{a^2}$$

where a is the radius of the plate, $M_T = \sigma_T h^2/4$ is the plastic limit bending moment per unit width of cross-section of the plate, and (r, θ) are polar coordinates. We assume that

$$M_r = 0, \qquad M_\theta = \beta r^2/a^2, \qquad w = 1 - r/a \tag{27.46}$$

Therefore, equation (27.7) will take the following form:

$$p = \text{ext}\ \frac{\displaystyle\int_0^a \beta(r^2/a^2)(1/ra)r\,dr - (1/2M_T^2)\int_{\sqrt{(M_T/\beta a)}}^a \beta(r^2/a^3)[\beta^2(r^4/a^4) - M_T^2]\,dr}{\displaystyle\int_0^a (1-r/a)r\,dr} \tag{27.47}$$

where according to the yield condition

$$M_r^2+M_\theta^2-M_rM_\theta=\frac{\beta^2r^4}{a^4}\leqslant M_T^2$$

We obtain the surface of the rigid/perfect plastic boundary $r_0=a\sqrt{(M_T/\beta)}$ and $r<r_0$ is the rigid region of the plate under the choice of (27.46) which obviously is so far away from the practice that even the plastic region and the rigid region are exchanged with each other. From (27.47), we obtain $p=5.64M_T/a^2$ while, from [1], it is only $p=3.78M_T/a^2$.

It is worth noting, by the three examples shown above, that theorems in this paper are very useful tools for establishing solutions such as for strength problems in solid mechanics and for problems of plastic-forming processes of metals.

REFERENCES

1. L. H. Tsien and W. S. Tsoon, 'A generalized variational principle for the limit analysis in solid mechanics', *Scientia Sinica,* **13,** No. 11 (1964).
2. T. Mura and S. L. Lee, 'Application of variational principle of limit analysis', *Quart. Appl. Math.,* **21,** No. 3 (1963).
3. K. Washizu, *Variational Methods in Elasticity and Plasticity,* Pergamon Press, London, 1968.
4. W. Z. Chien, 'Studies on generalized variational principles in elasticity and their applications in finite element calculations', *Mechanics and Its Applications* (A Quarterly Journal in China), **1,** No. 1–2 (1979).
5. T. H. H. Pian and P. Tong, 'Finite element methods in continuum mechanics', in *Advances in Applied Mechanics* (Ed. C. S. Yin), Vol. 12, Academic Press, 1972.
6. P. M. Shi, 'On the variational principles of elasto-plastic body and their applications', *Acta Mechanica Sinica,* **1,** No. 3 (1957).
7. W. Prager and P. G. Hodge, *Theory of Perfectly Plastic Solids,* John Wiley & Sons, 1951.
8. Wang *et al.* 'Discussion on "The limit analysis in solid mechanics and a suggested generalized variational principle"', *Acta Mechanica Sinica,* **8,** No. 1 (1965).

Hybrid and Mixed Finite Element Methods
Edited by S. N. Atluri, R. H. Gallagher, and O. C. Zienkiewicz

Chapter 28

Mixed Models and Reduced Selective Integration Displacement Models for Vibration Analysis of Shells

Ahmed K. Noor and Jeanne M. Peters

28.1 INTRODUCTION

Although the finite element analysis for curved shell structures now spans over twenty years, the establishment of reliable and efficient finite element models for the dynamic analysis of shells continues to be the subject of intense research effort at the present. Review of the many contributions on the subject is given in a monograph [1] and a number of survey papers (see, for example, [2]).

At the present time there are two general approaches for developing shear-flexible finite element models for shells. The first approach is based on the use of three-dimensional isoparametric elements in conjunction with a continuum theory. This approach has been termed the 'degenerated shell element approach' since the three-dimensional theory is reduced, or degenerated, to a shell theory simultaneously with the finite element discretization. It was originally proposed by Ahmad, Irons, and Zienkiewicz [3] for the static analysis of moderately thick shells, and has been subsequently extended to the dynamic analysis of shells (see [4, 5]). The second approach employs two-dimensional elements used with independent interpolation functions for displacements and rotations for the discretization of two-dimensional shear deformation shell theory [6, 7].

While most of the reported studies on dynamics of shells to date have been based on Hamilton's principle—an extension of the principle of minimum potential energy (displacement or stiffness models; see, for example, [8, 9])—hybrid elements [10 to 12] and mixed elements [13, 14] have also been developed. In spite of the fact that early application of mixed models to vibration analysis did not meet with much success [15], their later use [13, 14, 16, 17] has demonstrated some of their advantages over displacement models. These include simplicity of development of elemental matrices and uniform accuracy and convergence for a wide range of geometric characteristics of the shell.

More recently, mixed models with discontinuous stress resultants at interelement boundaries were shown to have a superior performance to that of mixed models with continuous stress resultants for the non-linear analysis of shells [18]. These mixed models were also shown to be equivalent to reduced/selective integration displacement models, wherein a lower order integration formula is used for the transverse shear and extensional strain energy terms. However, some of the mixed models with discontinuous stress resultants, and their equivalent reduced/selective integration displacement models, exhibit zero-energy spurious modes or mechanisms (see [18]). The present study focuses on the question of whether the mixed shell models presented in [18] realize the full potential of mixed models in the vibration analysis of shells. Specifically, the objectives of this paper are: (a) to present two sets of simple mixed models, based on the Hu–Washizu and Hellinger–Reissner variational principles for the vibration analysis of shells; (b) to identify the conditions for the equivalence between the two sets of mixed models and between the mixed models and displacement models; and (c) to discuss the merits of using mixed models.

28.2 MATHEMATICAL FORMULATION

The analytical formulation is based on a form of the shallow shell theory with the effects of transverse shear deformation and rotary inertia included. Transverse shear effects are treated in an average sense. Lines normal to the undeformed middle surface are assumed to remain straight but not necessarily normal to the deformed middle surface. Two families of finite element models are considered in the present study; namely mixed models and displacement (or stiffness) models based on the use of reduced/selective integration. The characteristics of these models are discussed subsquently.

28.2.1 Mixed finite element models

Two sets of mixed finite element models are considered herein. The first set is based on a generalized Hu–Washizu variational principle and the second is based on a generalized Hellinger–Reissner variational principle (see [19, 20] and Appendix A). Henceforth, the two sets of models will be referred to as Hu–Washizu and Hellinger–Reissner mixed models respectively.

28.2.1.1 Hu–Washizu mixed models

The fundamental unknowns used in the development of these models consist of (see Appendix A): the five generalized displacements u_α, w, and ϕ_α; the eight stress resultants $N_{\alpha\beta}$, $M_{\alpha\beta}$, and Q_α; and the corresponding eight strain components of the middle surface $\varepsilon_{\alpha\beta}$, $\kappa_{\alpha\beta}$, and $2\varepsilon_{\alpha 3}$ (α, $\beta = 1, 2$) (see Figure 28.1 for the sign convention).

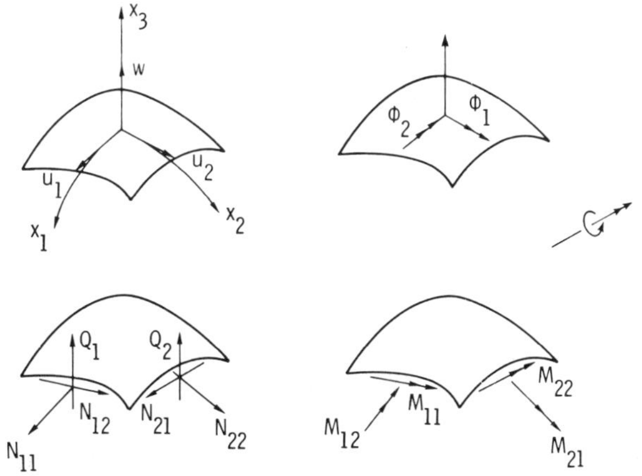

Figure 28.1 Sign convention for stress resultants and generalized displacements

The degree of the polynomial interpolation (or shape) functions used in approximating the strain components $\varepsilon_{\alpha\beta}$, $\kappa_{\alpha\beta}$, and $2\varepsilon_{\alpha 3}$ is the same as that used for approximating the corresponding stress resultants $N_{\alpha\beta}$, $M_{\alpha\beta}$, and Q_α, and differs from the degree of interpolation functions used in approximating the generalized displacements u_α, w, and ϕ_α. Moreover, the continuity of the stress resultants and strain components is not imposed at the interelement boundaries and, therefore, both the strain parameters and stress resultant parameters are eliminated on the element level.

The finite element equations for the individual elements can be cast in the following form:

$$K_{IJ}\phi_J + R_{IJ}H_J = 0 \tag{28.1}$$

$$(R^t)_{IJ}\phi_J + S_{Ij}X_j = 0 \tag{28.2}$$

$$(S^t)_{iJ}H_J - \omega^2 \mathcal{M}_{ij}X_j = 0 \tag{28.3}$$

where X_i, ϕ_J, and H_J are nodal displacements, strain and stress resultant parameters; the K, R, and S terms are 'generalized' stiffness coefficients; ω is the circular frequency of vibration of the shell; $\mathcal{M}_{ij}$ are consistent mass coefficients; and the superscript t denotes transposition. The range of the uppercase Latin indices is 1 to $8s$, where s is the number of stress resultant (or strain) nodes, or the number of parameters used in approximating each of the stress resultants (or strain components); and the range of the lowercase Latin indices is 1 to $5m$ (where m is the number of displacement nodes). A repeated index in the same term denotes summation over the full range of that index.

If the stress resultant and strain parameters are eliminated from equations (28.1), (28.2), and (28.3), one obtains the following equations in the nodal displacements X_j:

$$\mathcal{K}_{ij}X_j = \omega^2 \mathcal{M}_{ij}X_j \tag{28.4}$$

where $\mathcal{K}_{ij}$ are the linear coefficients defined as follows:

$$\mathcal{K}_{ij} = (S^t)_{iJ}(R^{-1})_{JI}K_{IL}(R^{-t})_{LK}S_{Kj} \tag{28.5}$$

and $(R^{-1})_{IJ}$, $(R^{-t})_{IJ}$ denote the inverse and transpose inverse of the array R_{IJ}.

28.2.1.2 Hellinger–Reissner mixed models

These are similar to the Hu–Washizu mixed models except that the fundamental unknowns consist of the generalized displacements and stress resultants only (see Appendix A). The finite element equations for the individual elements take the form:

$$-F_{IJ}H_J + S_{Ij}X_j = 0 \tag{28.6}$$

$$(S^t)_{iJ}H_J - \omega^2 \mathcal{M}_{ij}X_j = 0 \tag{28.7}$$

where the F terms are linear flexibility coefficients. The other terms are defined previously in connection with equations (28.1) to (28.3).

If the stress resultant parameters are eliminated from equations (28.6) and (28.7) on the element level, the resulting equations have the same form as equation (28.4), with the stiffness coefficients $\mathcal{K}_{ij}$ defined as follows:

$$\mathcal{K}_{ij} = (S^t)_{iI}(F^{-1})_{IJ}S_{Jj} \tag{28.8}$$

and $(F^{-})_{IJ}$ denotes the inverse of the array F_{IJ}.

For the shells with constant geometric and material characteristics the

Table 28.1 Integral types appearing in the expressions of the characteristic arrays for mixed models (equations 28.1 to 28.3, 28.6 and 28.7)

Array	Integral
K	$\int \bar{\bar{\mathcal{N}}}^i \bar{\bar{\mathcal{N}}}^i \, d\Omega$
R	$\int \bar{\bar{\mathcal{N}}}^i \bar{\mathcal{N}}^i \, d\Omega$
F	$\int \bar{\mathcal{N}}^i \bar{\mathcal{N}}^i \, d\Omega$
S	$\int \bar{\mathcal{N}}^i \mathcal{N}^{i'} \, d\Omega$
	$\int \mathcal{N}^i \, \partial_\alpha \mathcal{N}^{i'} \, d\Omega$
$\mathcal{M}$	$\int \mathcal{N}^{i'} \mathcal{N}^{i'} \, d\Omega$
$\mathcal{G}$	$\int \bar{\mathcal{N}}^i \, \partial_\alpha \mathcal{N}^{i'} \, \partial_\beta \mathcal{N}^{k'} \, d\Omega$

Note: $\bar{\bar{\mathcal{N}}}^i$ $(i=1$ to $s)$, $\bar{\mathcal{N}}^i$, and $\mathcal{N}^{i'}$ $(i', j' = 1$ to $m)$ are interpolation functions for strain components, stress resultants, and generalized displacements respectively; $\partial_\alpha \equiv \partial / \partial x_\alpha$ $(\alpha = 1, 2)$.

components of the arrays, K, R, S, and $\mathcal{M}$, are multiples of the integrals listed in Table 28.1.

28.2.2 Reduced selective integration displacement models

The reduced integration technique refers to using a lower order quadrature formula than the 'normal' one in evaluating the stiffness coefficients of a displacement model. This 'underintegration' of the stiffness characteristics was shown to overcome the 'locking phenomenon' observed when low-order interpolation functions are used in modelling thin shells. The procedure was first suggested by Zienkiewicz, Taylor, and Too [21], and was later applied to a number of beam, plate, and shell problems. However, the use of low-order quadrature formulae was shown to diminish the rank of the elemental matrices and may occasionally lead to undesirable zero-energy spurious modes (or mechanisms) in plane strain, plate, and shell problems (see [22 to 24]).

As a means to overcome the rank deficiency problems of reduced integration while maintaining its attributes, the reduced/selective integration concept was introduced in which the total potential energy is decomposed into components. A lower order quadrature formula is used in evaluating the constrained components, and the 'normal' integration is used for the remaining components. A generalization of the selective integration procedure has been given in [25] for cases where the segregation of the constrained energy components is not possible (e.g. in the presence of bending–extensional coupling or for a general type of anisotropy in the shell).

The selective integration concept was shown to work well for linear problems of thin isotropic plates where the constrained energy component is the transverse shear strain energy. However, for inextensional and nearly inextensional shell deformations low-order integration is required for the extensional energy as well as the transverse shear strain energy to overcome the 'membrane locking' phenomenon observed when low-order interpolation functions are used [26]. The use of such low-order formulas for both the extensional and transverse shear strain energies brings back the problem of zero-energy spurious modes.

The introduction of reduced and selective integration techniques was viewed by many analysts as mere tricks rather than methods. However, heuristic justification of these techniques on the basis of constraints was given in [22, 23, 27]. The establishment of the equivalence between some mixed models and reduced/selective integration displacement models gave credibility to the latter models and enhanced their development. However, as will be mentioned in succeeding sections, the mixed models have a number of computational advantages over their 'equivalent' reduced/selective integration displacement models.

28.3 COMMENTS ON THE FORMULATION AND FINITE ELEMENT MODELS DEVELOPED

The following comments regarding the formulation and the finite element models developed seem to be in order:

1. In evaluating the consistent mass coefficients the full (normal) Gauss–Legendre integration formula is used. The resulting mass matrix is not diagonal. Diagonal mass matrices can be obtained by using special lumping schemes (see, for example, [28 and 29]) or by using a quadrature formula in which the sampling points coincide with the nodal points (e.g. Lobatto integration formula). The effectiveness of using consistent and diagonalized mass matrices in plate vibration problems has been discussed in [29].

2. In the presence of initial stresses (or prestress) each of equations (28.3), (28.4), and (28.7) is augmented by the following term:

$$\mathcal{G}_{ijL} X_j H_L^0$$

where H_L^0 are the values of the parameters defining the prestress and $\mathcal{G}_{ijL}$ are geometric stiffness coefficients which are multiples of the integral types listed in Table 28.1.

3. Analytic expressions for the zero-energy rigid body and spurious modes are given in [18] for rectangular shallow shell mixed element models based on the serendipity and Lagrangian interpolation functions. The serendipity elements have the advantage over Lagrangian elements that their spurious modes are non-transmittable when a mesh of elements is used, regardless of the number of constraints or boundary conditions applied. However, they have the disadvantage of leading to overstiff solutions (locking phenomenon) when applied to very thin plates and shells (see [29]).

A number of approaches have been proposed to suppress the zero-energy spurious modes (see, for example, [30, 31, 32]). This topic is not discussed in the present paper.

28.4 EQUIVALENCE BETWEEN DIFFERENT MODELS

Two finite element models are said to be *equivalent* if their governing finite element equations, when expressed in terms of a common set of nodal variables and/or parameters, are *identical*. The other nodal variables and/or parameters not contained in the common set must be local to the individual elements and do not affect the assembly process. The two models are said to be *nearly equivalent* if their finite element equations, expressed in terms of a common set of variables, are *almost identical* (e.g. if the corresponding numerical coefficients in both sets of equations differ by a small percentage). In this section the equivalence between the Hu–Washizu and Hellinger–Reissner mixed models is established; then the equivalence between these two mixed models and some displacement models is discussed.

28.4.1 Equivalent mixed models

The Hu–Washizu mixed models are said to be equivalent to the Hellinger–Reissner mixed models, in the sense described above, if, after elimination of the strain parameters on the element level, the resulting finite element equations for the two models are identical (i.e. equations 28.6 and 28.7 are identical to the corresponding equations of the Hu–Washizu model). The condition for the equivalence is

$$(R^{t})_{IK}(K^{-1})_{KL}R_{LJ} = F_{IJ} \tag{28.9}$$

where $(K^{-1})_{KL}$ denotes the inverse of the array K_{KL}.

For shallow shells with constant material properties within each element, condition (28.9) is satisfied if:

(a) the same number of nodes (or parameters) and interpolation functions are used for approximating each of the generalized displacements u_α, w, ϕ_α and the stress resultants $N_{\alpha\beta}$, $M_{\alpha\beta}$, Q_α in both models and
(b) the number of parameters and the degree of interpolation polynomials used in approximating each of the strain components are the same as the number of parameters and degree of approximating polynomial for the corresponding stress resultant.

Both conditions are satisfied for the mixed models considered in the present study. Therefore, the Hu–Washizu and Hellinger–Reissner mixed models considered herein are equivalent. Note that the Hu–Washizu mixed models become *nearly equivalent* to the Hellinger–Reissner mixed models for shells with variable material characteristics if different approximations are made for the shell characteristics in the two models (e.g. shell stiffnesses are approximated in the Hu–Washizu mixed models while shell flexibilities are approximated in the Hellinger–Reissner mixed models).

28.4.2 Equivalent mixed and displacement models

A mixed finite element model is said to be equivalent to a displacement model, in the sense described above, if, after elimination of the stress resultant (and strain) parameters on the element level, the resulting finite element equations for the two models are identical (i.e. equations 28.4 for the mixed model are identical with the corresponding equations of the displacement model). The models are said to be *nearly equivalent* if their finite element equations are *almost identical.*

For the case of shallow shells with constant geometric and material properties within each element, the following two groups of equivalent mixed and displacement quadrilateral elements can be identified.

28.4.2.1 Group 1: equivalent mixed models and standard displacement models

The conditions for the equivalence are:

(a) The same number of nodes and interpolation functions are used for approximating the generalized displacements u_α, w, ϕ_α $(\alpha = 1, 2)$ in both the displacement and mixed models.
(b) The polynomial interpolation functions for the stress resultants $N_{\alpha\beta}$, Q_α, $M_{\alpha\beta}$ (and the strain components $\varepsilon_{\alpha\beta}$, $\kappa_{\alpha\beta}$, $2\varepsilon_{\alpha 3}$; $\alpha, \beta = 1, 2$) in the mixed models are of the same degree as those of the generalized displacements and are discontinuous at interelement boundaries.
(c) The Gauss–Legendre quadrature formula selected for the numerical evaluation of all terms in the displacement and mixed models is such that exact values for the integrals are obtained for mixed models with straight sides.

28.4.2.2 Group 2: equivalent mixed models and reduced/selective integration displacement models

The conditions for the equivalence are:

(a) The same number of nodes and Lagrangian interpolation functions are used for approximating the generalized displacements u_α, w, ϕ_α in both the displacement and mixed models.
(b) The polynomial interpolation functions for the stress resultants $N_{\alpha\beta}$, Q_α, $M_{\alpha\beta}$ (and the strain components $\varepsilon_{\alpha\beta}$, $\kappa_{\alpha\beta}$, $2\varepsilon_{\alpha 3}$) are of a lower degree than those of the generalized displacements and the stress resultants are discontinuous at interelement boundaries.
(c) The Gauss–Legendre quadrature formula selected for the numerical evaluation of all the terms of the mixed model is such that exact values for the integrals are obtained for straight-sided elements.
(d) A low-order Gauss–Legendre quadrature formula is used for evaluating the transverse shear and extensional energy terms in the displacement models. The number of quadrature points in the low-order formula is the same as that used for the exact evaluation of the integrals in the mixed models with straight sides.

'Exact' integration of the bending terms can be obtained by using the same low-order Gauss–Legendre formula. Therefore, the use of uniform reduced integration for the displacement models is equivalent to the use of reduced/selective integration for the transverse shear and extensional energy terms only.

Henceforth, the mixed and displacement models of Group 1 will be denoted by MDm-n and DEm respectively and the corresponding models

Table 28.2 Examples of equivalent quadrilateral mixed and displacement models for shallow shells

Number of displacement nodes, m	Number of quadrature points, n	Number of parameters per stress resultant, s	Designation	
			Mixed	Displacement
4	1	1	MD4-1	DR4-1
	4	4	MD4-4	DE4
8	4	4	MD8-4	DR8-4
	9	8	MD8-9	DE8
9	4	4	MD9-4	DR9-4
	9	9	MD9-9	DE9
12	9	9	MD12-9	DR12-9
	16	12	MD12-16	DE12
16	9	9	MD16-9	DR16-9
	16	16	MD16-16	DE16

of Group 2 will be designated MDm-n and DRm-n, where m is the number of displacement nodes in the element and n is the *total* number of quadrature points used in the Gauss–Legendre reduced integration formula.

The aforementioned equivalences are based on comparing the *exact analytic expressions* for the stiffness coefficients of the mixed and displacement models. Numerical experiments have demonstrated the superior performance of the models of Group 2 over those of Group 1. However, some of the Group 2 models contain spurious zero-energy modes.

Examples of equivalent quadrilateral mixed and displacement shallow shell elements are given in Table 28.2. The following comments on the equivalence seem to be in order:

(a) The equivalence listed in Table 28.2 holds for both straight-sided as well as curved-sided elements. It also holds for shells with variable geometric and material characteristics *provided the same approximation is used for the geometric and material properties in both the mixed and their equivalent displacement models.*
(b) The numbers of quadrature points n listed in Table 28.2 are sufficient for the *exact* evaluation of the integrals appearing in the K, R, F, and S terms (see Table 28.1) for mixed models with straight sides and constant geometric and material parameters. For elements with curved sides or with variable geometric and material properties, the number of quadrature points n given in the table lead to only *approximate values for the integrals.*
(c) The finite element equations for the individual elements (equations 28.4) *remain unchanged when the numbers of stress resultant (and strain) parameters* listed in Table 28.2 *are increased, provided the numbers of quadrature points n are not changed.*

(d) The quadrilateral mixed and displacement models listed in Table 28.2 become *nearly equivalent* when:
   (i) The number of quadrature points n listed in the table are increased (e.g. in order to evaluate the integrals associated with the K, R, F, and S terms exactly for mixed models with curved sides).
   (ii) Different approximations for the shell properties are made in mixed and displacement models (e.g. shell flexibilities are approximated in the Hellinger–Reissner mixed models while shell stiffnesses are approximated in the displacement models).

28.5 ADVANTAGES AND POTENTIAL OF MIXED MODELS

In spite of the equivalence discussed in the preceding section, the Hu–Washizu and Hellinger–Reissner mixed models combine the following advantages over their equivalent reduced/selective integration displacement models:

(a) The development of mixed models is simpler and more straightforward than the development of reduced/selective integration displacement models. This is particularly true for cases where the decomposition of energy into constrained and unconstrained components cannot be easily made (e.g. in the presence of bending–extensional coupling).

(b) The integrals forming the characteristic arrays K, R, F, S, and $\mathcal{M}$ of the mixed shell models (equations 28.1 to 28.3 and 28.6 and 28.7) can be evaluated *exactly, even when the element edges (or faces) are curved* (see [13]). In contrast, the integrals in the characteristic arrays of the displacement models can be evaluated exactly only for simple element geometries (e.g. triangular, rectangular, or parallelogram elements with straight sides). This unique advantage of the mixed models makes them eminently suited for use with reanalysis techniques in automated optimum design systems (particularly when the design variables include nodal coordinates, which is the case for configuration optimization, since these design variables appear explicitly in the expressions of the element characteristic arrays).

(c) Evaluation of the element characteristic arrays K_{IJ}, R_{IJ}, and S_{Ij} (or F_{IJ} and S_{Ij}) for the mixed models and their combination to form the array $\mathcal{K}_{ij}$ (equations 28.5 and 28.8) involves fewer arithmetic operations than the formulation of the corresponding array in displacement models. This is particularly true if the theory of group representation is used to take advantage of the similarities in algebraic form of the various integrals appearing in the characteristic arrays of the mixed models (see [13]).

Moreover, the computational effort required in evaluating $\mathcal{K}_{ij}$ can be reduced by selecting the interpolation functions for the stress resultants and strain components to be orthogonal polynomials, thereby simplifying the generation of $(R^{-1})_{IJ}$ and $(F^{-1})_{IJ}$.

The following two remarks concerning mixed models with discontinuous stress resultant fields seem to be in order:

(a) The elimination of the stress resultant parameters on the element level (condensation process) is similar to that used in hybrid models in which the stress resultant field is described within the element and the independent displacement field is defined on the element boundaries. However, the present elements differ from hybrid models in the fact that both stress resultant (strains) and generalized displacement fields are described within the elements.

(b) If the stress resultants (and strains) are eliminated on the element level the present elements can be viewed by a user as displacement models and can be easily combined with other types of displacement elements to model a shell structure.

28.6 NUMERICAL STUDIES

In order to test and evaluate the performance of the mixed models with discontinuous stress resultants (and strains) developed herein, a large number of vibration problems of shallow shells have been solved using these models. Comparison was made with analytic and converged solutions as well as with solutions obtained by other mixed and displacement models to assess the accuracy of the different models. The results of three problem sets are presented herein. These problems are: (a) a simply supported laminated orthotropic shallow spherical shell with a square platform; (b) a curved fan blade; and (c) a pear-shaped cylindrical shell. The material and geometric characteristics of the three structures are given in Figures 28.2, 28.5, and 28.8.

For the first structure only doubly symmetric vibration modes are considered and, therefore, only one-quarter of the shell is analysed. For the curved fan blade both symmetric and antisymmetric vibration modes are considered. For the pear-shaped cylinder both doubly symmetric modes as well as modes which are symmetric in the axial direction and antisymmetric in the circumferential direction are considered.

In each case the structure was analysed using: (a) standard displacement models based on normal integration, DEm; (b) displacement models based on reduced/selective integration, DRm-n; (c) displacement models based on selective integration, with reduced integration applied to transverse shear terms only, DSm-n; (d) mixed models with discontinuous strains and stress

Table 28.3 Characteristics of the mixed and displacement finite element models used in the numerical studies

(a) Mixed models

Stress resultant continuity at element boundaries	Number of displacement nodes, m	Number of parameters per stress resultant, r	Number of quadrature points, n	Designation
Continuous	4	4	4	MC4
	9	9	9	MC9
Discontinuous	4	1	4	MD4-1
	9	4	9	MD9-4

(b) Displacement models

Number of displacement nodes, m	Number of quadrature points used in evaluating: Bending energy	Extensional energy	Shear energy	Designation
	4	4	4	DE4
4	4	4	1	DS4-1
	1	1	1	DR4-1
	9	9	9	DE9
9	9	9	4	DS9-4
	4	4	4	DR9-4

resultants at element boundaries, MDm-n; and (e) Hellinger–Reissner mixed models with continuous stress resultants at element boundaries (see [13]), MCm. The characteristics of the different models are summarized in Table 28.3.

Exact analytic solutions are obtained for the spherical shell problem. Finite element solutions using classical shell theory (with transverse shear deformation neglected) along with experimental results are given in [8, 33] for the fan blade problem. Finite element solutions using the degenerate shell element approach (with transverse shear deformation included) are presented in [4, 5] for the same problem. Finite difference solutions for the pear-shaped cylinder are presented in [34].

The eigenvalues and associated eigenvectors are obtained using the subspace iteration technique with Sturm sequence checks (see [35]). For the solutions presented herein the displacement boundary conditions were enforced but the stress resultant boundary conditions were not applied. However, for the MCm models, the symmetry (and antisymmetry) conditions for both displacements and stress resultants along the centre lines were enforced. The boundary conditions used in the three structures prevented the development of mechanisms (zero-energy spurious modes) with any of the displacement or mixed models. Also, because of the equivalence between the DRm-n and MDm-n models, the solutions obtained using these models were *identical.*

28.6.1 Laminated orthotropic shallow spherical shells

The first problem considered is that of the free vibrations of the eight-layered shallow spherical shells with simply supported edges shown in Figure 28.2. The problem was selected because an exact, analytic solution can be obtained and used as a standard for assessing the accuracy of various mixed and displacement models.

An indication of the accuracy and rate of convergence of the first four doubly symmetric vibration frequencies for shells with $h/L = 0.01$ and 0.001 obtained by the different displacement and mixed four-noded and nine-noded elements is given in Figures 28.3 and 28.4 respectively. Note that the modes are classified according to the numbers of half-waves parallel to the x_1 and x_2 axes. Also, the numerical rates of convergence of the fundamental frequency predicted by the different models are given in Table 28.4. An examination of Figures 28.3 and 28.4 and Table 28.4 reveals:

(a) As expected, the four-noded element DE4 considerably over-estimates the stiffness (and the vibration frequencies) and exhibits the locking phenomenon, especially for the thinner shells. For shells with $h/L =$ 0.001 the frequencies associated with the second and higher modes obtained by the DE4 models are far removed from the exact frequencies and could not be shown in Figure 28.3.

(b) For the same number of degrees of freedom, the accuracy and rate of convergence of the frequencies obtained by the nine-noded DE9

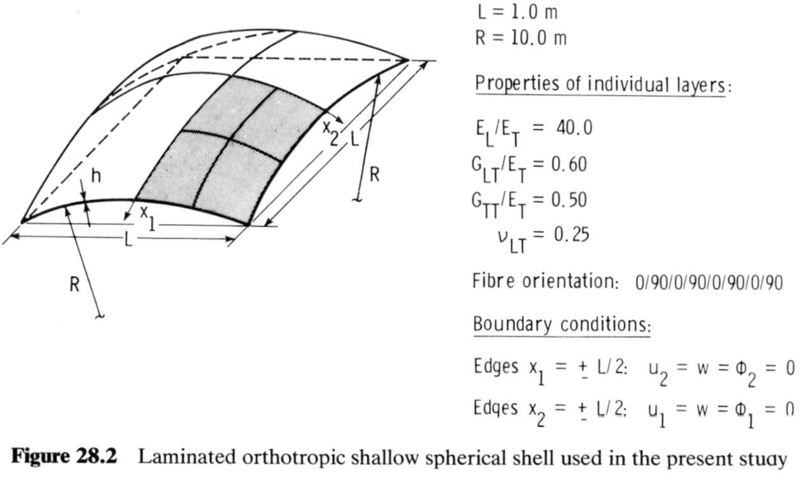

Figure 28.2 Laminated orthotropic shallow spherical shell used in the present study

Mode	$\omega_{exact}\sqrt{E_T/(\rho L^2)}$	
	h/L = 0.01	h/L = 0.001
1	0.2382	0.1516
2	1.1600	0.2245
3	1.1606	0.2566
4	1.6481	0.2569

Model	Designation
DE4	•
DS4-1	□
MC4	+
MD4-1	▲

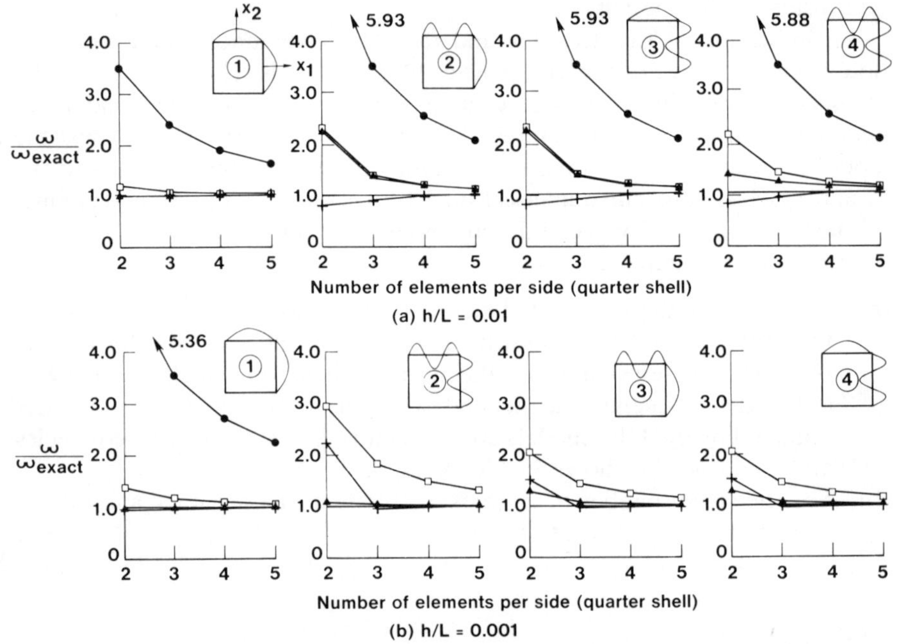

Figure 28.3 Accuracy and convergence of vibration frequencies obtained by four-noded mixed and displacement models for the shallow spherical shell shown in Figure 28.2

models are considerably higher than those obtained by the corresponding DE4 models. This is particularly true for the thinner shells (with $h/L = 0.001$) (compare Figures 28.3 and 28.4).

(c) For a given grid size, the frequencies obtained by the Hellinger–Reissner mixed models with continuous stress resultants MC4 and MC9 are more accurate than those obtained by the corresponding displacement models DE4, DS4-1 and DE9, DS9-4. For shells with $h/L = 0.01$, the higher frequencies obtained by using the MC4 and MC9 models were even more accurate than those obtained by using the corresponding MD4-1 and MD9-4 models. The situation is reversed for the thinner shells with $h/L = 0.001$. Note that, for the same grid, the MC4

Mode	$\omega_{exact}\sqrt{E_T/(\rho L^2)}$	
	h/L = 0.01	h/L = 0.001
1	0.2382	0.1516
2	1.1600	0.2245
3	1.1606	0.2566
4	1.6481	0.2569

Model	Designation
DE9	○
DS9-4	□
MC9	+
MD9-4	▲

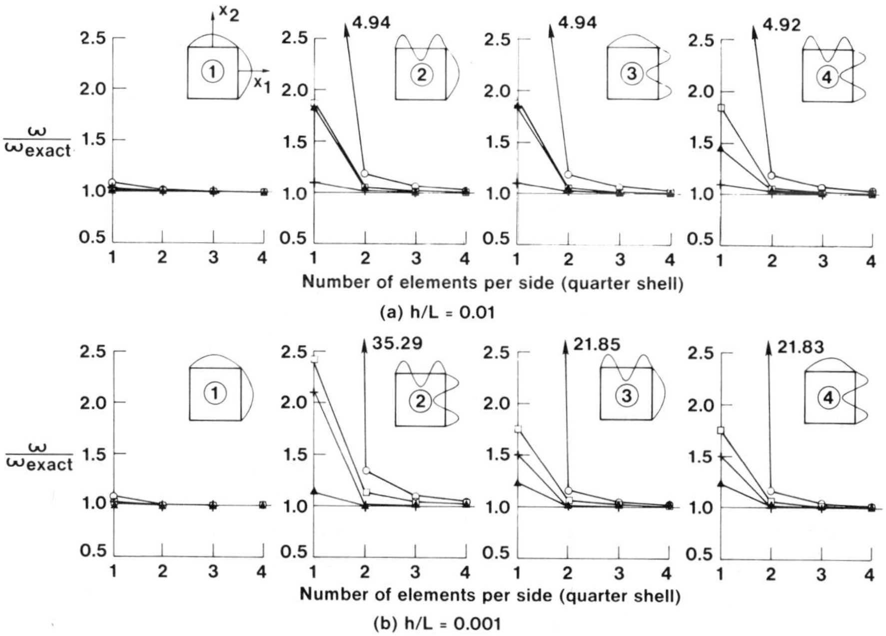

Figure 28.4 Accuracy and convergence of vibration frequencies obtained by nine-noded mixed and displacement models in the shallow spherical shell shown in Figure 28.2

and MC9 models have considerably more degrees of freedom, on the system level, than either of the corresponding mixed models with discontinuous stress resultants MD4-1 and MD9-4 or the displacement models DE4, DS4-1 and DE9, DS9-4.

(d) For a given number of degrees of freedom the mixed models with discontinuous stress resultants MD4-1 and MD9-4 are more accurate and their convergence rates are less sensitive to variations in h/L than all the corresponding mixed and displacement models (see Table 28.4).

(e) The use of selective/reduced integration for the transverse shear terms only, DS4-1 and DS9-4 models, alleviates the locking phenomenon and

Table 28.4 Numerical rate of convergence of the fundamental frequency predicted by different models for the laminated orthotropic shallow spherical shells shown in Figure 28.2. Error in fundamental frequency $= cN^{-\beta}$ (c = constant, N = total number of finite elements on side of the shell)

Model		Values of β	
		Shells with $h/L = 0.01$	Shells with $h/L = 0.001$
Four-noded elements	DE4	$1.44 \rightarrow 1.60$	$1.32 \rightarrow 1.43$
	DS4-1	$1.96 \rightarrow 1.98$	$1.86 \rightarrow 1.94$
	MC4	$3.97 \rightarrow 4.02$	$3.16 \rightarrow 3.54$
	MD4-1	$2.03 \rightarrow 2.11$	$2.02 \rightarrow 2.07$
Nine-noded elements	DE9	$2.77 \rightarrow 3.01$	$3.65 \rightarrow 4.09$
	DS9-4	$3.81 \rightarrow 3.97$	$3.00 \rightarrow 3.96$
	MC9	$4.01 \rightarrow 4.02$	$3.22 \rightarrow 3.64$
	MD9-4	4.04	$3.92 \rightarrow 4.01$

improves the performance of the element over that of the standard displacement models DE4 and DE9. However, the accuracy of such models is still considerably less than that of the MD4-1 and MD9-4 models (or their equivalent reduced integration models DR4-1 and DR9-4).

28.6.2 Curved fan blade

The second problem considered is that of the cylindrical fan blade shown in Figure 28.5. The standard for comparison (converged solution) was taken to be the solution obtained using MD16-9 elements and an 8×4 grid in half the shell. The ratios of the extensional and bending strain energies to the total strain energy of the shell, U_{ext}/U and U_{bend}/U, for the first seven

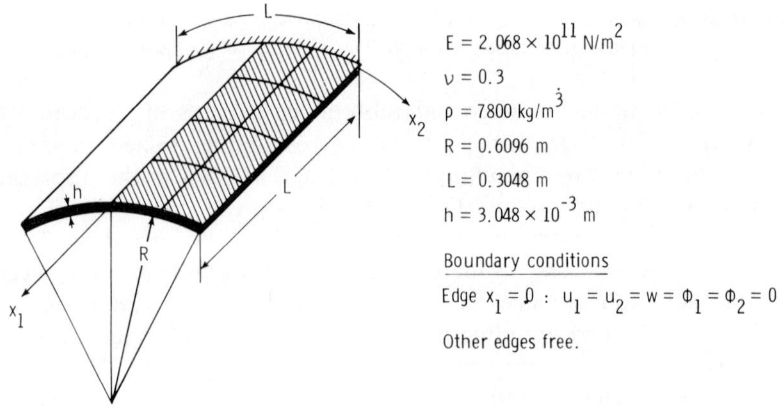

Figure 28.5 Curved fan blade used in the present study

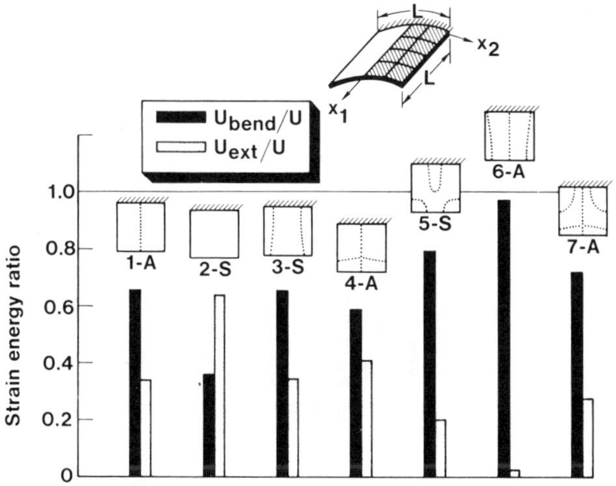

Figure 28.6 Strain energy ratios for the different vibration modes of the curved fan blade shown in Figure 28.5

vibration modes are shown in Figure 28.6. For all the seven modes the transverse shear strain energy was found to be negligible (less than 0.4 per cent).

The first twelve vibration frequencies obtained by the various displacement and mixed models along with the converged solutions and the experimental results reported in [8] are listed in Table 28.5. Also, the accuracy and convergence of the first seven frequencies obtained by the three nine-noded elements DS9-4, MC9, and MD9-4 are shown in Figure 28.7. The results

Table 28.5 Comparison of minimum vibration frequencies in hertz obtained by different displacement and mixed models for the curved fan blade shown in Figure 28.5

	Four-noded elements and 8×4 grid				Nine-noded elements and 4×2 grid					
Mode	DE4	DS4-1	MC4	MD4-1	DE9	DS9-4	MC9	MD9-4	Converged solution	[8]
1-A	296	198	88.5	83.5	88.9	87.6	87.6	85.8	86.0	85.6
2-S	272	167	140	138	144	141	140	138	138	134.5
3-S	1,448	424	253	245	266	258	253	248	249	259
4-A	1,413	430	354	342	370	352	350	344	342	351
5-S	1,487	505	397	393	420	398	394	385	386	395
6-A	1,788	789	528	573	637	575	535	544	531	531
7-A	3,997	876	744	763	859	764	748	742	727	743
8-A	2,049	809	742	733	783	746	742	734	730	751
9-S	4,053	902	786	796	890	795	786	777	774	790
10-A	4,011	1,014	811	839	922	846	818	818	804	809
11-S	4,257	1,306	949	1,192	1,337	1,137	1,016	1,094	1,001	997
12-S	4,306	1,407	1,158	1,227	1,526	1,254	1,233	1,243	1,207	1,216

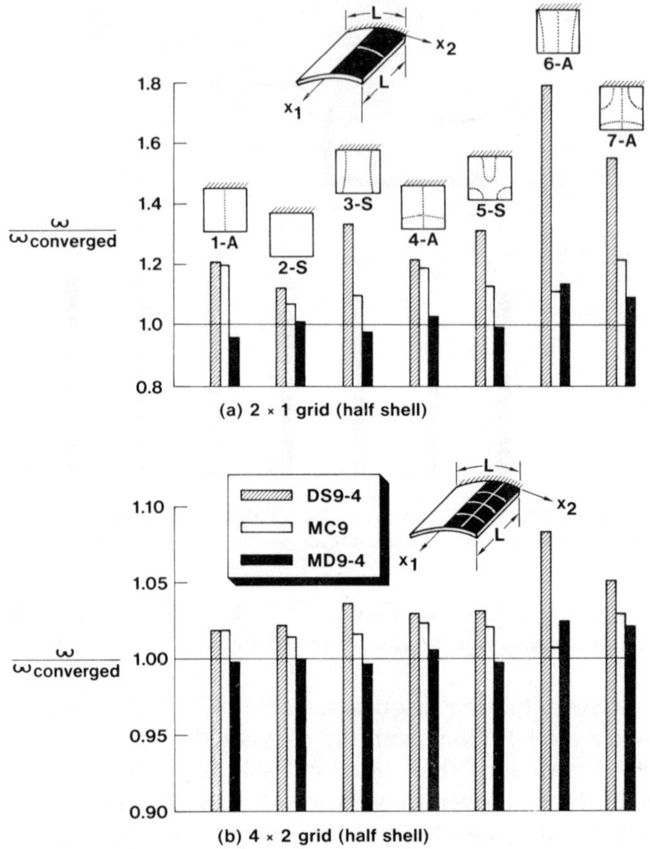

Figure 28.7 Accuracy and convergence of vibration frequencies obtained by different nine-noded elements for the curved fan blade shown in Figure 28.5

can be summarized as follows:

(a) As the ratio of the extensional energy to the total strain energy decreases, the improvement in accuracy of the frequencies obtained by the displacement model due to reduced/selective integration of the transverse shear terms only (DS9-4 model) decreases. As an example of this, U_{ext}/U for the sixth vibration mode was 0.025, and the error in the associated frequency obtained by the DS9-4 model and 2×1 grid in half the shell was 79.1 per cent. (see Figures 28.6 and 28.7). On the other hand, the use of reduced integration for both the transverse shear and extensional energy terms (DR9-4 model or, equivalently, MD9-4 model) leads to considerable improvement in accuracy. The maximum error in the first seven frequencies obtained by the MD9-4 model and a

2×1 grid in half the shell was 13.7 per cent. The maximum error reduced to 2.45 per cent. when a 4×2 grid was used.

(b) For each of the two grids considered the predictions of the mixed models with continuous stress resultants, MC4 and MC9, are more accurate than those of the corresponding displacement models but, in general, less accurate than those of the mixed models with discontinuous stress resultants, MD4-1 and MD9-4 (see Figures 28.7 and Table 28.5).

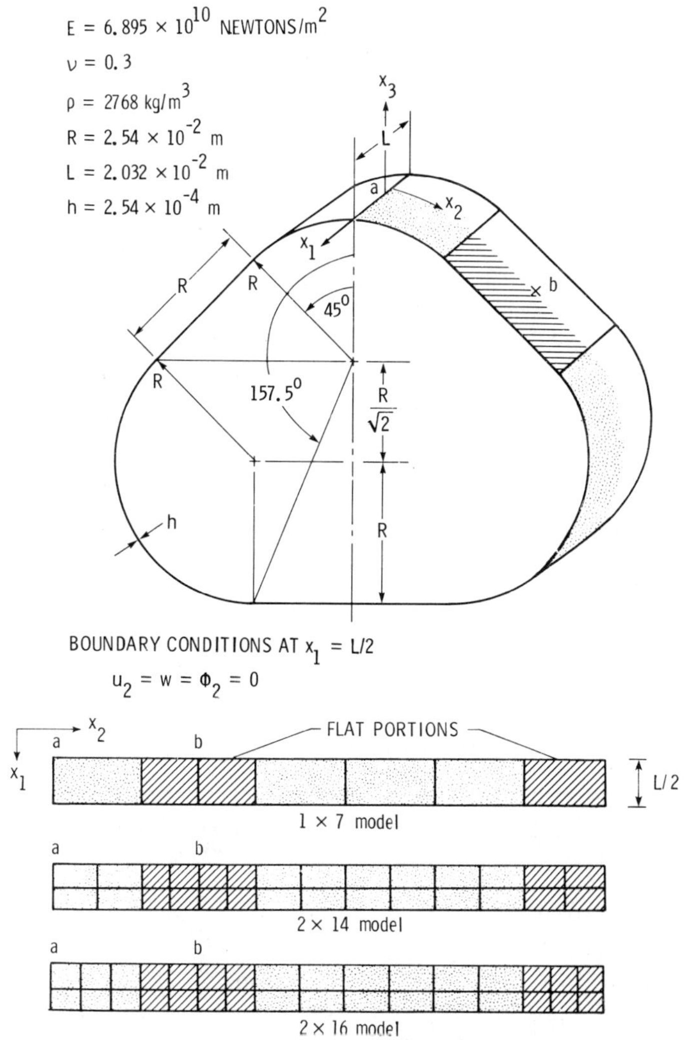

Figure 28.8 Pear-Shaped cylinder and finite element models used in the present study

Table 28.6 Comparison of minimum vibration frequencies in hertz obtained by different displacement and mixed models for the pear-shaped cylinder shown in Figure 28.8

(a) Doubly symmetric modes

	Four-noded elements and 2×14 grid				Nine-noded elements and 1×7 grid					
	DE4	DS4-1	MC4	MD4-1	DE9	DS9-4	MC9	MD9-4	solution	[35]
ω_1	20,032	2,340	1,824	2,004	3,970	2,158	1,872	1,916	1,903	1,899
ω_2	21,111	3,268	2,114	2,439	5,167	2,951	2,211	2,324	2,289	2,275
ω_3	28,802	8,716	2,904	4,735	7,748	6,537	3,743	4,411	4,253	4,252
ω_4	32,751	13,368	4,039	5,969	19,422	10,218	4,429	5,805	4,823	4,823
(b) Symmetric–antisymmetric modes										
ω_1	20,953	3,167	2,067	2,335	5,323	2,857	2,193	2,257	2,291	2,276
ω_2	26,264	5,420	2,706	3,400	6,215	4,209	2,976	3,110	3,073	3,065
ω_3	27,301	8,405	3,756	4,370	8,204	6,415	4,149	4,068	4,260	4,254
ω_4	34,329	16,319	3,832	7,420	19,746	12,579	5,176	7,271	6,985	7,021

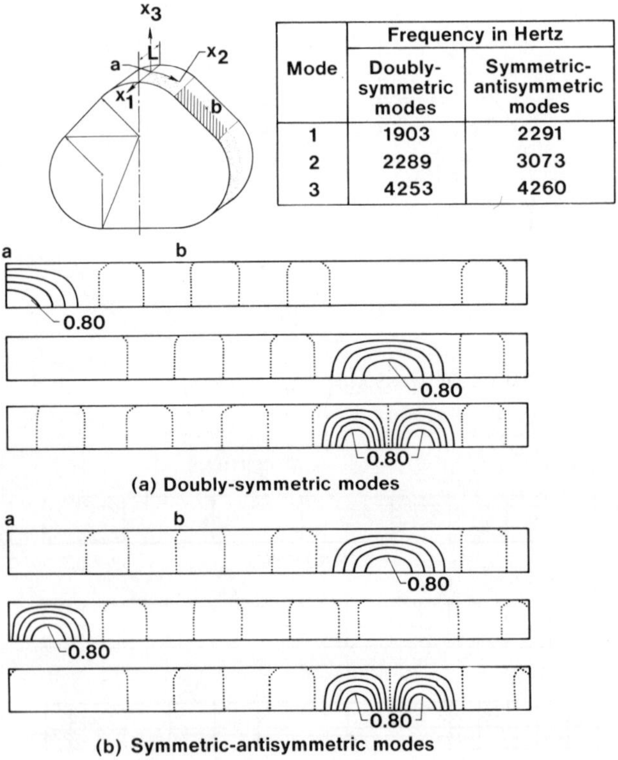

Figure 28.9 Vibration mode shapes, w/w_{max}, for the pear-shaped cylinder shown in Figure 28.8. Dotted lines denote nodal lines

28.6.3 Pear-Shaped Cylinder

As a final example consider the free vibrations of the pear-shaped cylindrical shell shown in Figure 28.8. The problem was selected to check the accuracy of various models when applied to shells with variable, discontinuous curvature. Both the doubly symmetric and the symmetric–antisymmetric modes were analysed. Symmetric–antisymmetric modes refer to the modes which are symmetric with respect to the plane $x_1 = L/2$ and antisymmetric with respect to the two planes $x_2 = \text{constant}$. The standard of comparison was taken to be the solutions obtained using the MD16-9 model and a 2×16 grid in the shell quarter.

The first four vibration frequencies obtained by the various displacement and mixed models along with the converged solutions and the finite difference results reported in [34] are listed in Table 28.6. The contour plots for the normalized transverse displacement w associated with the three lowest

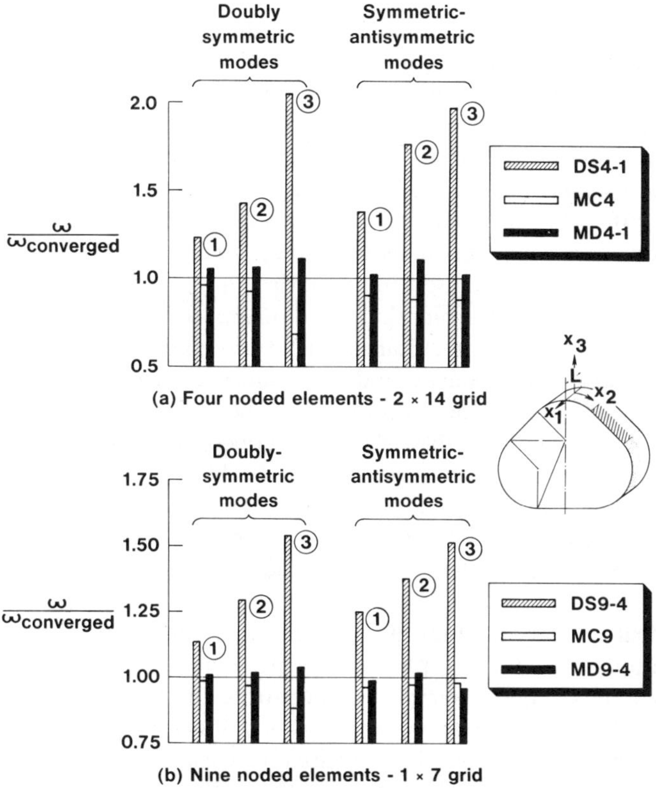

Figure 28.10 Accuracy of vibration frequencies obtained by different mixed and displacement models for the pear-shaped cylinder shown in Figure 28.8

frequencies in each of the doubly symmetric and symmetric–antisymmetric modes are given in Figure 28.9. Also an indication of the accuracy of the four-noded elements DS4-1, MC4, and MD4-1 and the nine-noded elements DS9-4, MC9, and MD9-4 is given in Figure 28.10.

As in the previous problems, for the same number of degrees of freedom, the predictions of the MD4-1 and MD9-4 models are considerably more accurate than those of the corresponding displacement models DE4, DS4-1 and DE9, DS9-4, and are generally more accurate than the predictions of the MC4 and MC9 models (see Figure 28.10).

28.7 CONCLUDING REMARKS

Two sets of simple mixed models are developed for the vibration analysis of shells. The analytical formulation is based on a form of the shallow shell theory with the effects of transverse shear deformation and rotary inertia included. The fundamental unknowns in the first model consist of eight stress resultants, eight strain components, and five generalized displacements of the shell, and the element characteristic arrays are obtained by using a generalized Hu–Washizu variational principle. In the second model the fundamental unknowns consist of eight stress resultants and five generalized displacements, and a generalized Hellinger–Reissner principle is used for developing the element characteristic arrays. The degree of the polynomial interpolation (or shape) functions used in approximating the stress resultants is the same as the degree used in approximating the strain components and is, in general, different from that used in approximating the generalized displacements. The stress resultants (and strain components) are discontinuous at the element boundaries and are eliminated on the element level.

The equivalence and 'near-equivalence' between the two sets of mixed models is established and then the equivalence between the mixed models and displacement models based on reduced/selective integration of both the transverse shear and extensional energy terms is discussed. Also, other equivalent mixed and standard displacement models are identified. The advantages of mixed models over equivalent displacement models are outlined.

Numerical results are presented for the free vibation problems of laminated orthotropic shallow spherical shells, curved fan blades, and pear-shaped cylinders. These examples demonstrate the high accuracy and effectiveness of the mixed models developed and show that the performance of these models is considerably less sensitive to variations in the geometry and material characteristics of the structure than other mixed and displacement models reported in the literature.

ACKNOWLEDGEMENT

The present work is partially supported by a National Science Foundation Grant No. PFR-7916263 and by a NASA Langley Research Center Grant No. NGR 09-010-078.

APPENDIX A MIXED VARIATIONAL PRINCIPLES USED IN PRESENT STUDY

After elimination of time and making the usual assumptions of harmonic vibrations, the functionals for the mixed variational principles used in the present study are given by:

Generalized Hu–Washizu principle

$$\pi(\varepsilon_{\alpha\beta}, \kappa_{\alpha\beta}, 2\varepsilon_{\alpha 3}, N_{\alpha\beta}, M_{\alpha\beta}, Q_\alpha, u_\alpha, w, \phi_\alpha) = V - U - T \qquad \text{(A.1)}$$

where

$$U = \frac{1}{2}\int_\Omega [C_{\alpha\beta\gamma\rho}\varepsilon_{\alpha\beta}\varepsilon_{\gamma\rho} + 2F_{\alpha\beta\gamma\rho}\varepsilon_{\alpha\beta}\kappa_{\gamma\rho} + D_{\alpha\beta\gamma\rho}\kappa_{\alpha\beta}\kappa_{\gamma\rho} + C_{\alpha 3\beta 3}(2\varepsilon_{\alpha 3})(2\varepsilon_{\beta 3})]\, d\Omega \qquad \text{(A.2)}$$

$$V = \int_\Omega [N_{\alpha\beta}(\varepsilon_{\alpha\beta} - \partial_\alpha u_\beta - k_{\alpha\beta} w) + M_{\alpha\beta}(\kappa_{\alpha\beta} - \partial_\alpha \phi_\beta) + Q_\alpha(2\varepsilon_{\alpha 3} - \phi_\alpha - \partial_\alpha w)]\, d\Omega \qquad \text{(A.3)}$$

and

$$T = \tfrac{1}{2}\omega^2 \int_\Omega [m_0(u_\alpha u_\alpha + ww) + 2m_1 u_\alpha \phi_\alpha + m_2 \phi_\alpha \phi_\alpha]\, d\Omega \qquad \text{(A.4)}$$

Generalized Hellinger–Reissner principle

$$\bar{\pi}(N_{\alpha\beta}, M_{\alpha\beta}, Q_\alpha, u_\alpha, w, \phi_\alpha) = \bar{V} - \bar{U} - T \qquad \text{(A.5)}$$

where

$$\bar{V} = \int_\Omega [N_{\alpha\beta}(\partial_\alpha u_\beta + k_{\alpha\beta} w) + M_{\alpha\beta}\partial_\alpha \phi_\beta + Q_\alpha(\phi_\alpha + \partial_\alpha w)]\, d\Omega \qquad \text{(A.6)}$$

and

$$\bar{U} = \frac{1}{2}\int_\Omega [A_{\alpha\beta\gamma\rho}N_{\alpha\beta}N_{\gamma\rho} + 2B_{\alpha\beta\gamma\rho}N_{\alpha\beta}M_{\gamma\rho} + G_{\alpha\beta\gamma\rho}M_{\alpha\beta}M_{\gamma\rho} + A_{\alpha 3\beta 3}Q_\alpha Q_\alpha]\, d\Omega \qquad \text{(A.7)}$$

In equations (A.1) to (A.7) $\varepsilon_{\alpha\beta}$, $\kappa_{\alpha\beta}$, and $2\varepsilon_{\alpha 3}$ are the extensional bending and transverse shear strains of the middle surface of the shell; $C_{\alpha\beta\gamma\rho}$, $D_{\alpha\beta\gamma\rho}$, $C_{\alpha 3\beta 3}$, and $F_{\alpha\beta\gamma\rho}$ are extensional stiffnesses, bending stiffnesses, transverse shear stiffnesses, and stiffness interaction coefficient of the shell; $A_{\alpha\beta\gamma\rho}$, $G_{\alpha\beta\gamma\rho}$, $A_{\alpha 3\beta 3}$, and $B_{\alpha\beta\gamma\rho}$ are shell compliance coefficients (inverse of shell stiffnesses, see [36]); $k_{\alpha\beta}$ are the curvature components and twist of the shell middle surface; m_0, m_1, and m_2 are density parameters of the shell (see [36]); ω is the circular frequency of vibration; Ω is the shell domain; $\partial_\alpha \equiv \partial/\partial x_\alpha$; Greek indices take the values 1, 2; and a repeated index within the same term denotes summation over its full range.

APPENDIX B NOTATION

$A_{\alpha\beta\gamma\rho}$, $A_{\alpha 3\beta 3}$, $B_{\alpha\beta\gamma\rho}$, $G_{\alpha\beta\gamma\rho}$	Shell compliance coefficients, inverse of shell stiffnesses
$C_{\alpha\beta\gamma\rho}$	Extensional stiffness coefficients
$C_{\alpha 3\beta 3}$	Transverse shear stiffnesses of the shell
$D_{\alpha\beta\gamma\rho}$	Bending stiffnesses of the shell
E	Elastic modulus of isotropic materials
E_L, E_T	Elastic modulus in direction of fibres and normal to it respectively
$F_{\alpha\beta\gamma\rho}$	Stiffness interaction coefficients of the shell
F_{IJ}	Flexibility coefficients of the shell element
G_{LT}, G_{TT}	Shear moduli in plane of fibres and normal to it respectively
H_J	Nodal stress resultants or stress resultant parameters
h	Thickness of the shell
$\mathcal{K}_{ij}$	Stiffness coefficient of the shell element
k_{IJ}	Generalized stiffness coefficients of the shell element defined in equation (28.1)
$k_{\alpha\beta}$	Curvatures and twist of the shell middle surface
$M_{\alpha\beta}$	Bending stress resultants
$\mathcal{M}_{ij}$	Consistent mass coefficients of the shell element
m	Number of displacement nodes in the element
m_0, m_1, m_2	Density parameters of the shell
$N_{\alpha\beta}$	Extensional (inplane) stress resultants
$\mathcal{N}^{i'}$	Shape functions for generalized displacements
$\bar{\mathcal{N}}^i$	Shape (or interpolation) functions for stress resultants
$\bar{\bar{\mathcal{N}}}^i$	Shape (or interpolation) functions for strain components
n	Total number of quadrature points in the Gauss–Legendre integration formula
Q_α	Transverse shear stress resultants
R	Radius of curvature of the middle surface of the shell
R_{IJ}, S_{Ij}	Generalized stiffness coefficients of the shell element defined in equations (28.1) and (28.2)

s	Number of parameters used in approximating each of the stress resultants (and strain components)
U	Total strain energy of the shell
U_{ext}, U_{bend}	Extensional and bending strain energies respectively
$\bar{U}$	Complementary energy of the shell defined in equation (A.7)
u_α, w	Displacement components in the coordinate directions
X_i	Nodal displacement parameters
x_α, x_3	Orthogonal curvilinear coordinate system
$\varepsilon_{\alpha\beta}$	Extensional strains of the shell middle surface
$2\varepsilon_{\alpha 3}$	Transverse shear strains of the shell
$\kappa_{\alpha\beta}$	Curvature changes and twist of the shell middle surface
ν	Poisson's ratio for isotropic materials
ν_{LT}	Major Poisson ratio of the individual layers
π, $\bar{\pi}$	Functionals defined in equations (A.1) and (A.5)
ρ	Density of the material of the shell
ϕ_J	Strain parameters
ϕ_α	Rotation components
Ω	Shell domain
ω	Circular frequency of vibration of the shell
∂_α	$\equiv \partial/\partial x_\alpha$

Range of indices

Uppercase Latin indices:	1 to $8s$
Lowercase Latin indices:	(i, j) 1 to $5m$
	(i', j') 1 to m
Lowercase script indices:	1 to s
Greek indices:	1, 2

Finite element model notation

DEm	Displacement model based on normal integration and having m displacement nodes
DRm-n	Displacement model based on reduced integration; number of displacement nodes $= m$ and total number of quadrature points in reduced integration formula $= n$
DSm-n	Same as DRm-n but reduced integration is used selectively for transverse shear terms only
MCm	Hellinger–Reissner mixed model with continuous stress resultants at element boundaries and m nodes (for both stress resultants and displacements)
MDm-n	Mixed model (of the Hu–Washizu or Hellinger–Reissner type) with discontinuous stress resultants (and strain components) at

element boundaries and m displacement nodes; total number of quadrature points in the Gauss–Legendre formula equals n.

Full (normal), reduced, and selective integration

If the $n_1 \times n_1$ Gauss–Legendre formula is used to integrate $\mathcal{K}_{ij}$ exactly for parallelogram elements, then (where $n_1 = \sqrt{n}$):

(a) Full (normal) integration uses $n_1 \times n_1$ quadrature points.
(b) Reduced integration uses $(n_1 - 1) \times (n_1 - 1)$ quadrature points.
(c) Selective integration uses $n_1 \times n_1$ points for the bending and extensional strain terms and $(n_1 - 1) \times (n_1 - 1)$ for the transverse shear terms.

REFERENCES

1. D. G. Ashwell and R. H. Gallagher (Eds.), *Finite Elements for Thin Shells and Curved Members*, John Wiley and Sons, 1976.
2. R. H. Gallagher, 'Shell elements', *Proc. World Congr. on Finite Element Methods in Structural Mechanics*, Vol. 1, pp. E1–E35, Bournemouth, England, 1975.
3. S. Ahmad, B. M. Irons, and O. C. Zienkiewicz, 'Analysis of thick and thin shell structures by curved finite elements', *Int. J. Num. Meth. in Eng.*, **2,** No. 3, 419–451 (1970).
4. S. Ahmad, R. G. Anderson, and O. C. Zienkiewicz, 'Vibration of thick curved shells with particular reference to turbine blades', *J. Strain Analysis*, **5,** No. 3, 200–206 (1970).
5. D. Hofmesster and D. A. Evensen, 'Vibration problems using isoparametric shell elements', *Int. J. Num. Meth. in Eng.*, **5,** No. 1, 142–145 (1972).
6. S. W. Key and Z. E. Beisinger, 'The analysis of thin shells with transverse shear strains by the finite element method', *Proc. Second Conf. Matrix Methods in Structural Mechanics*, pp. 667–710 AFFDL-TR-68-150, U.S. Air Force, December 1969).
7. K. P. Walker, 'Vibrations of cambered helicoidal fan blades', *J. Sound and Vibration*, **59,** No. 1, 35–57 (1978).
8. M. D. Olson and G. M. Lindberg, 'Dynamic analysis of shallow shells with a doubly-curved triangular finite element', *J. Sound and Vibration*, **19,** No. 3, 299–318 (1971).
9. R. H. MacNeal, 'A simple quadrilateral shell element', *Computers and Structures*, **8,** 175–183 (1978).
10. S. T. Mau and E. A. Witmer, 'Static, vibration and thermal stress analysis of laminated plates and shells by the hybrid stress finite element method with transverse shear deformation effects included', Army Materials and Mechanics Research Center Report ASRL TR-169-2, October 1972.
11. E. F. Crawley and S. W. Lee, 'The natural mode shapes and frequencies of graphite epoxy cantilevered plates and shells', Technical Memorandum FBR-78-110, Air Force Flight Dynamics Laboratory, Wright-Patterson Air Force Base, Ohio, August 1978.
12. R. L. Spilker, O. Orringer, E. A. Witmer, S. Verbiese, and S. E. French, 'Use of the hybrid stress finite element model for the static and dynamic analysis of multilayer composite plates and shells, ASRL-TR-181-2, Army Materials and Mechanics Research Center, Watertown, Massachusetts, September 1976.

13. A. K. Noor and C. M. Andersen, 'Mixed isoparametric finite element models of laminated composite shells', *Computer Methods in Applied Mechanics and Engineering*, **11,** 255–280 (1977).
14. W. Altman and M. N. Bismark-Nasr, 'Vibration of thin cylindrical shells based on a mixed finite element formulation', *Computers and Structures*, **8,** 217–221 (1978).
15. R. D. Cook, 'Eigenvalue problems with a mixed plate element', *AIAA J.*, **7,** No. 5, 982 (1969).
16. D. Talaslidis and W. Wunderlich, 'Static and dynamic analysis of Kirchhoff shells based on a mixed finite element formulation', *Computers and Structures*, **10,** 239–249 (1979).
17. S. Nemat-Nasser and C. O. Horgan, 'Variational methods for eigenvalue problems with discontinuous coefficients', in *Mechanics Today*, Vol. 5, pp. 365–376, Pergamon Press, 1980.
18. A. K. Noor and C. M. Andersen, 'Mixed models and reduced/selective integration displacement models for nonlinear shell analysis', Special ASME/AMD Publication in *Nonlinear Finite Element Analysis of Plates and Shells* (Ed. T. J. R. Hughes), pp. 119–146, AMD-48, November 1981.
19. K. Washizu, *Variational Methods in Elasticity and Plasticity*, 2nd ed., Pergamon Press, Oxford, 1975.
20. E. Hofbauer, 'Die Herleitung von Schalengleichungen uber ein erweitertes Variations-prinzip bei Berucksichtigung der Querschubverzerrungen', *Acta Mechanica*, **26,** 315–320, 1977.
21. O. C. Zienkiewicz, R. L. Taylor, and J. M. Too, 'Reduced integration technique in general analysis of plates and shells', *Int. J. Num. Meth. in Eng.*, **3,** 275–290 (1971).
22. E. D. L. Pugh, E. Hinton, and O. C. Zienkiewicz, 'A study of quadrilateral plate bending elements with "reduced" integration', *Int. J. Num. Meth. in Eng*, **12,** 1059–1079 (1978).
23. T. J. R. Hughes, M. Cohen, and M. Haroun, 'Reduced and selective integration techniques in finite element analysis of plates', *Nuclear Engineering and Design*, **46,** No. 1, 203–222 (1978).
24. N. Bicanic and E. Hinton, 'Spurious modes in two-dimensional isoparametric elements', *Int. J. Num. Meth. in Eng.*, **14,** 1545–1557 (1979).
25. T. J. R. Hughes, 'Generalization of selective integration procedures to anisotropic and nonlinear media', *Int. J. Num. Meth. in Eng.*, **15,** 1413–1418 (1980).
26. H. Stolarski and T. Belytschko, 'Reduced integration for shallow-shell facet elements', in *New Concept in Finite Element Analysis*, pp. 179–193, AMD-V. 44, ASME, 1981.
27. O. C. Zienkiewicz and E. Hinton, 'Reduced integration, function smoothing and nonconformity in finite element analysis (with special reference to thick plates)', *J. Franklin Institute*, **302,** Nos. 5/6, 443–461 (November/December 1976).
28. R. D. Krieg and S. W. Key, 'Transient shell response by numerical time integration', *Int. J. Num. Meth. in Eng.*, **7,** 273–286 (1973).
29. E. Hinton and N. Bicanic, 'A comparison of Lagrangian and serendipity Mindlin plate elements for free vibration analysis', *Computers and Structures*, **10,** 483–493 (1979).
30. D. Kosloff and G. A. Frazier, 'Treatment of hour-glass patterns in low order finite element codes', *Int. J. Num. Anal. Meth. in Geomech.*, **2,** 57–72 (1978).
31. E. Onate, E. Hinton, and N. Glover, 'Techniques for improving the performance of Ahmad shell elements', Report C/R/313/78, Department of Civil Engineering, University of Wales, Swansea, 1978.

32. H. Parisch, 'A critical survey of the nine-noded degenerated shell element with special emphasis on thin shell application and reduced integration', *Comp. Meth. in Appl. Mech. and Eng.*, **20,** 323–350 (1979).
33. G. Sander and P. Beckers, 'Delinquent finite elements for shell idealization', *Proc. World Congr. on Finite Element Methods in Structural Mechanics*, Bournemouth, Dorset, England, 12–17 Oct. 1975.
34. R. F. Hartung and R. E. Ball, 'A comparison of several computer solutions to three structural shell analysis problems', Technical Report AFFDL-TR-73-15, Air Force Flight Dynamics Laboratory, Wright-Patterson Air Force Base, Ohio, April 1973.
35. K. J. Bathe and E. L. Wilson, *Numerical Methods in Finite Element Analysis*, Prentice-Hall, 1976.
36. A. K. Noor and M. D. Mathers, 'Shear-flexible finite element models of laminated composite plates and shells', NASA TN-D-8044, December 1975.

Hybrid and Mixed Finite Element Methods
Edited by S. N. Atluri, R. H. Gallagher, and O. C. Zienkiewicz

Chapter 29

Reflections and Remarks on Hybrid and Mixed Finite Element Methods

Theodore H. H. Pian

I am overwhelmed by this gathering here to discuss issues related to hybrid and mixed finite element methods. I am sure that anyone who witnessed the early developments of finite element methods would never have dreamed that within a span of less than twenty years the research and development in this field can be so widely expanded and so firmly crystallized as indicated by the papers presented at this conference.

I would like to add my contribution to this conference by presenting some historical accounts of hybrid and mixed elements that I have been closely associated with and by giving some remarks on future developments of such methods.

There is no question that the development of finite element methods should be indebted to earlier pioneering works in matrix structural analysis. I can give many names and point out their influences. Today, I only want to acknowledge Dr. Ray Bisplinghoff. I am very proud and privileged for the opportunity to work under Ray's tutorage since the formation of the Aeroelastic and Structures Research Laboratory at M.I.T. nearly thirty-five years ago. It is obvious that the wide expansion in finite element development in solid mechanics is due to the firm establishment of variational principles. For this we should pay our tribute to Professor Eric Reissner from whom I studied the theories of elasticity, plates, and shells and learned the wide applications of variational methods in mechanics. I also wish to acknowledge the benefit I derived from my association with my very good friend, Professor Kyuichiro Washizu, since 1953. The first version of his monumental work on variational methods in elasticity and plasticity was a report to the Office of Naval Research published in 1955 by the M.I.T. Aeroelastic and Structures Research Laboratory. That work has always been used as a guide in many of my further research works and in my teaching of graduate subjects. We are sorry that Professor Washizu is unable to come to this conference on account of his new appointment at Osaka University. I think that all of us would want to send our best wishes and to congratulate him on becoming an Honorary Professor after his completion of a distinguished career at the University of Tokyo.

I would also like to acknowledge the presence, at this conference, of Professor W. Z. Chien of Tsinghua University. He began to guide his students thirty years ago in the study of variational methods in mechanics. One of the products of his team's efforts was the Hu variational functional of the Hu–Washizu principle in solid mechanics.

The first research project on finite element methods which I supervised was a study of matrix analysis methods for inelastic structures sponsored by the Air Force Flight Dynamics Laboratory. Active in that project was John Percy, Bill Loden, and the late D. R. Navaratna. Looking back, I realize how crude we were in our analyses then, and how refined and systematic we are now in solving the same elastic–plastic and creep problems. The experimental results obtained by shear-lag specimens in that programme were used by many other researchers for correlation with their finite element solutions.

The year 1964 was an extremely productive year in the development of finite element methods. The late Professor Fraeijs de Veubeke first published his equilibrium model for upper bound solutions in that year. Bob Jones published his modified potential energy principle for finite element analysis by introducing the interelement continuity as the condition of constraint.

In the Fall term of 1963, I gave a course on variational and matrix methods in structural mechanics. In one lecture in January 1964, I discussed the Reissner principle and gave an illustration of the derivation of the element stiffness matrix by a mixed formulation using compatible field and independent stresses. In such a case the stress parameters can be eliminated at the element level. I had then a project on basic research in structural mechanics sponsored by AFOSR. I asked D. R. Navaratna to calculate the stiffness matrices of a plane stress rectangular element using different numbers of stress terms. The stresses we used all satisfied the homogeneous equilibrium equations. After the job was done, I suddenly realized that I would have obtained the same results using the complementary energy principle and I would only need to interpolate the boundary displacements and hence for a problem of C^1 continuity I could avoid the difficult task of constructing shape functions. I did not tell this story in the paper that appeared in the July 1964 issue of the *AIAA Journal* and in it I only described the derivation of the element stiffness matrix by the complementary energy principle, although the example solutions were actually obtained by the mixed formulation.

Another note I should add here is that prior to the publication of that paper, I sent Dick Gallagher a copy of my manuscript for his comments. He wrote a technical comment for the *AIAA Journal* in which he pointed out that the element stiffness matrix which I obtained using five stress terms should be exactly the same as that obtained by Turner, Clough, Martin, and Topp in their 1956 historical paper. Indeed, ten years later, Frorer, Nilson, and Samuelsson showed that for rectangular elements the expressions for stiffness matrices obtained by the two different methods are exactly the same and in fact are also identical to that by Wilson's incompatible displacement model.

The first conference on matrix methods in structural mechanics held at the Wright-Patterson Air Force Base is a milestone in finite element methods. That meeting was not only the first time that compatible elements for plate-bending elements were presented but also a meeting of most researchers on hybrid and mixed models at that time. Professor Fraeijs de Veubeke pointed out the possibility of kinematic deformation modes in his equilibrium element. Bob Jones was there, although he only reported his shell analysis by the assumed displacement method. Leonard Herrmann presented, for the first time, a mixed model in finite element analysis. I reported some results obtained by elements with stress-free boundaries and by plate elements derived by my assumed stress model. I remember Professor Besseling made a comment to me that my formulation was really an application of Reissner's mixed variational principle.

I was at Cal Tech for seven months in 1965–66 and gave a series of special lectures on finite element methods in structural mechanics to faculty members and graduate students in the Firestone Flight Science Laboratory. One result of that effort was to induce Pin Tong to add an appendix to his Ph.D. thesis which is on the problem of liquid sloshing in flexible shells by a variational method. What he added was a formulation of his problem by the finite element method. But the most important consequence of my sabbatical year at Cal Tech was to bring Pin Tong to M.I.T. after his graduation. We worked together for the following seven years in different aspects of finite element methods. One of the first projects he did was to use my plate-bending element by the hybrid model to solve a low-speed viscous flow problem.

The most important step in the development of the assumed stress hybrid element is the recognition by the two of us that the variational basis of the element is a modified complementary energy principle obtained by the Lagrange multiplier method. In the Fall of 1967 when I taught a course on finite element methods, I first coined the term 'hybrid' to signify elements which maintain either equilibrium or compatibility in the element and then to satisfy compatibility or equilibrium respectively along the interelement boundary.

Although the motivation for the early developments in mixed and hybrid elements was to avoid the difficulty in C^1 continuity problem, such elements are also used in problems of C^0 continuity because they provide, in general, better accuracy in both displacements and stresses. For C° problems, both the complementary energy formulation and the mixed formulation can be used for deriving the hybrid element. It turns out that the mixed formulation is easier to implement and requires shorter computing efforts. S. W. Lee has found that for an eight-node isoparametric solid element the assumed stress formulation requires only a few percentage higher computing time in comparison to the conventional assumed displacement method. Thus, I had started with a mixed formulation and then changed to the hybrid formulation and we then switched back to the mixed formulation again in many of our finite element developments. For example, when the mixed variational principle is used it would be justified to obtain the nodal force vector and even the mass and geometric stiffness matrices for hybrid stress elements based on independently assumed shape functions in the same way as the conventional assumed displacement method. Furthermore, for either total or incremental solutions of geometrically non-linear problems by the assumed stress formulation the resulting variational functionals usually contain both stresses and displacements in the element as field variables. Such elements are really mixed/hybrid elements.

One of the methods for avoiding the difficult task of constructing shape functions for plate and shell elements is to take transverse shear into account in the formulation. Then there arrives the difficulty of locking phenomena when the ratio of the thickness and dimension of the element becomes small. Before returning to Taiwan University to head its Civil Engineering Department, Sheng-Taur Mau was with me for two years working on laminated elements by the hybrid model taking transverse shear into account. We found that the method was very effective. Bob Spilker was also working on this project. He has found later on that for assumed stress hybrid plate elements with transversed shear effects, there is no locking difficulty when the thickness becomes very small. Similarly, S. W. Lee and I had shown earlier that there was no locking difficulty for elasticity solutions for nearly incompressible materials.

Application of assumed stress hybrid elements to shell analyses is not convenient because of the complicated equilibrium equations that should be satisfied. Michio Tanaka was able to obtain finite element solutions for thin conical shells by triangular elements.

One of the most important applications of hybrid models is the treatment of stress and strains singularity. C. H. Luk did his thesis under the guidance of Pin Tong and myself on plane elastic crack problems by the hybrid stress method. He included in the assumed stresses the singular stress terms with the stress intensity factor as undetermined parameters. In such formulation those element boundaries that meet at the crack tip will contain singularities and hence will require special treatments in numerical integrations. Pin Tong realized later on that for plane crack problems complete series solutions at the crack tip are available and every assumed stress term can be made to satisfy both equilibrium and compatibility equations. In such a case the complementary energy can be expressed in terms of only boundary integrals. An element can then be formed with an embedded crack with all boundaries free of singularities. Kazumasa Moriya used both approaches in his doctoral thesis work, i.e. the first approach for three-dimensional crack problems and the second approach for bending of thin plates with through cracks. Mike Pustejovsky obtained solutions by the second approach to correlate his test results on fatigue crack propagation under the mixed mode condition.

Realizing that other schemes for relaxing the interelement continuity conditions are available, Pin Tong suggested a new hybrid displacement model by the independently assumed boundary displacements which can be made the same for neighbouring elements. This finite element model again yields element stiffness matrices. Satya Atluri developed a quadrilateral shell element using this hybrid displacement model. Professors Kikuchi and Ando also extended Tong's method to plate and shell analyses. Atluri and his colleagues at the University of Washington and at Georgia Tech have perhaps made most use of this finite element model, not for plates and shells but for plane and three-dimensional crack problems. They have extended their solutions to non-linear material problems and geometrically non-linear problems including such applications as the investigations of crack closure under fatigue loading.

At M.I.T. we have also studied non-linear solutions by assumed stress hybrid elements. Bob Spilker, working on elastic–plastic analyses, Denny Pirotin and Peter Boland, on large deflection of shells, Tom Scharnhorst, on rubber-like material, and H. C. Wang, on creep of material under large strains, all had the choice of different equilibrium conditions to be satisfied by their assumed stress increments. Using the hybrid stress method, S. W. Lee analysed non-linear creep problems and Kenji Kubomura analysed contact problems. We have found that because of its accuracy in stress evaluation, a hybrid stress element is more efficient than the assumed displacement elements for elastic–plastic and creep problems.

Bruce Irons has always been most enthusiastic about the assumed stress hybrid method but he also attributed the fact that the method has only limited adaptation to its complexity in formulation and implementation. He wishes that the formulation of hybrid elements could be simplified to a degree comparable to the use of shape function routine in the assumed displacement method. However, we know that there exist many alternatives in the choice of stresses and displacements in the mixed/hybrid element formulations. Indeed, if there is any drawback in these methods it is the demand of a thorough background in mechanics and sound judgements from the developers of mixed/hybrid elements. I think that a mixed/hybrid element will always require more effort in formulation and experimentation than that for the conventional assumed displacement method. However, such high initial development costs are worth paying for when an element is fully certified for its reliability and efficiency, and a way can be cleared for its adaptation in canned commercial computer programs.

Yes, the rapid development of finite element methods into some of the most useful engineering tools is due to the ease of formulation and implementation of the conventional assumed displacement method. However, for making progress, the seeking of more efficient methods of analysis is always an important mission. An important purpose for us to get together here in Atlanta is for the accomplishment of this mission.

Hybrid and Mixed Finite Element Methods
Edited by S. N. Atluri, R. H. Gallagher, and O. C. Zienkiewicz

Chapter 30

List of Publications by Theodore H. H. Pian

30.1 BOOKS PUBLISHED

1. *Statics of Deformable Solids* by R. L. Bisplinghoff, J. W. Mar, and T. H. H. Pian, Addison & Wesley, 1965.
2. *Finite Element Methods in Structural Dynamics* (Editor), AIAA Selected Reprint Series/Vol. 17, 1975.

30.2 TECHNICAL ARTICLES PUBLISHED

1. 'Analytical study of transmission of load from skin to stiffeners and rings of pressurized cabin structures', NACA Tech. Note 993, October 1945.
2. 'Studies of transient stresses in airplane model wing during drop tests', *Proc. Soc. Exp. Stress Analysis,* **6,** No. 1, 115–122 (1948) (by O. N. Neal, R. L. Bisplinghoff, and T. H. H. Pian).
3. 'Methods in transient stress analysis', *J. of the Aero. Sci.*, **7,** No. 5, 259–270 (May 1950) (by R. L. Bisplinghoff, G. Isakson, and T. H. H. Pian).
4. 'A mechanical analyzer for computing transient stresses in airplane structures', *J. of Appl. Mech.*, **17,** No. 3, 310–314 (September 1950) (by R. L. Bisplinghoff, T. H. H. Pian, and L. I. Levy).
5. 'Analytical and experimental studies of dynamic loads in airplane structures during landing', *J. of the Aero. Sci.*, **17,** No. 12, 765–775 (December 1950) (by T. H. H. Pian and H. I. Flomenhoft).
6. 'Structural damping in a simple built-up beam', *Proc. First U.S. National Congress of Applied Mechanics*, pp. 97–102, 1951 (by T. H. H. Pian and F. C. Hallowell).
7. 'Prediction of stresses in a structure under an arbitrary dynamic loading', *Proc. Soc. Exp. Stress Analysis,* **9,** No. 2 pp. 1–12 (1952) (by T H. H. Pian and J. N. Sidall).
8. 'A study of gust entry of sweptback wings', *Proc. Second U.S. National Congress of Applied Mechanics*, pp. 755–762, 1954 (by T. H. H. Pian and H. Ashley).
9. On the vibrations of thermally buckled bars and plates', *Proc. Ninth Int. Congr. of Applied Mechanics*, Vol. 7, pp. 307–318, Brussels, Belgium, 1956 (by R. L. Bisplinghoff and T. H. H. Pian).
10. 'Transient responses of continuous structures using assumed time functions', *Proc. Ninth Int. Congr. of Applied Mechanics*, Vol. 7, pp. 350–359, Brussels, Belgium, 1956 (by T. H. H. Pian and T. F. O'Brien).
11. 'Structural damping of a simple built-up beam with riveted joints in bending', *J. of Appl. Mech.*, **24,** No. 1, 35–38 (March 1957).
12. 'On the variational theorem for creep', *J. Aero. Sci.*, **24,** 846–847 (1957).

13. 'Creep buckling of curved beam under lateral loading', *Proc. Third U.S. National Congress of Applied Mechanics*, pp. 649–654, 1958.
14. 'Structural damping', Chap. 5 of *Random Vibration* (Ed. S. H. Crandall), Technology Press, 1959.
15. 'Reduction of strain rosettes in the plastic range', *J. Aero-Space Sci.*, **26,** 842–843 (1959).
16. 'Dynamic response of thin shell structures', *Proc. Second Symp. on Naval Structural Mechanics* (ONR Structural Mechanics Series, 'Plasticity', edited by E. H. Lee and P. S. Symonds), pp. 443–444, 1962.
17. 'Lateral response of columns of elastic-plastic behavior', *Proc. Fourth U.S. National Congress of Applied Mechanics*, pp. 1383–1390, 1962.
18. 'Impulsive loading of rigid-plastic curved beams', *Proc. Fourth U.S. National Congress of Applied Mechanics*, pp. 1039–1046, 1962 (by M. M. Chen, P. T. Hsu, and T. H. H. Pian).
19. 'Dynamic deformation and buckling of spherical shells under blast and impact loading', Collected Papers on Instability of Shell Structures, NASA TN D-1510, pp. 607–622, 1962 (by E. A. Witmer, T. H. H. Pian, and H. A. Balmer).
20. 'Large dynamic deformations of beams, rings, plates and shells', *AIAA J.*, **1,** 1848–1857 (1963) (by E. A. Witmer, H. A. Balmer, J. W. Leech, and T. H. H. Pian).
21. 'Derivation of element stiffness matrices', *AIAA J.*, **2,** 576–577 (1964).
22. 'Derivation of element stiffness matrices by assumed stress distributions', *AIAA J.*, **2,** 1333–1336 (1964).
23. 'Large elastic, plastic, and creep deflections of curved beams and axisymmetric shells', *AIAA J.*, **2,** 1613–1620 (1964) (by J. S. Stricklin, P. S. Hsu, and T. H. H. Pian).
24. 'Plane stress yield condition of oblique coordinate systems', *J. Appl. Mech.*, **31,** 145–146 (1964).
25. 'Note on the instability of circular rings confined to a rigid boundary', *J. Appl. Mech.*, **31,** 559–562 (1964) (by P. T. Hsu, J. Elkon, and T. H. H. Pian).
26. 'Application of matrix displacement method to linear elastic analysis of shells of revolution', *AIAA J.*, **3,** No. 11, 2138–2145 (1965) (by J. H. Percy, T. H. H. Pian, S. Klein, and D. R. Navaratna).
27. 'Dynamic buckling of a circular ring constrained in a rigid circular surface', *Proc. Int. Conf. Dynamic Stability of Structures*, pp. 285–297, Pergamon Press, 1966 (by T. H. H. Pian, H. A. Balmer, and L. L. Bucciarelli Jr.).
28. 'Element stiffness-matrices for boundary compatibility and for prescribed boundary stresses', *Proc. Conf. Matrix Methods in Structural Mechanics*, pp. 457–477, AFFDL-TR-66-80, November 1966.
29. 'Static and dynamic analysis of re-entry vehicle shell structures by the finite-element method', *Proc. DASA ABM Blast Vulnerability Conf.*, 19–21 October 1965, pp. 346–364, 1966.
30. 'Improvements on the analysis of shells of revolution by the matrix displacement method', *AIAA J.*, **4,** No. 11, 2069–2072 (November 1966) (by J. A. Stricklin, D. R. Navaratna, and T. H. H. Pian).
31. 'Application of the smooth-surface interpolation to the finite-element analysis', *AIAA J.*, **5,** No. 1, 187–189 (January 1967) (by A. L. Deak and T. H. H. Pian).
32. 'Buckling of radially constrained circular ring under distributed loading', *Int. J. Solids and Structures*, **3,** 715–730 (1967) (by T. H. H. Pian and L. L. Bucciarelli Jr.).
33. 'The convergence of finite element method in solving linear elastic problems', *Int. J. Solids and Structures*, **3,** 865–879 (1967) (by P. Tong and T. H. H. Pian).

34. 'Effect of initial imperfections on the instability of a ring confined in an imperfect rigid boundary', *J. Appl. Mech.*, December 1967 (by L. L. Bucciarelli Jr. and T. H. H. Pian).
35. 'Stability analysis of shells of revolution by the finite-element method', *AIAA J.*, **6,** No. 2, 355–361 (February 1968) (by D. R. Navaratna, T. H. H. Pian, and E. A. Witmer).
36. 'Numerical calculations technique for large elastic-plastic transient deformations of thin shells, *AIAA J.*, **6,** No. 12, 2352–2359 (December 1968) (by J. W. Leech, E. A. Witmer, and T. H. H. Pian).
37. 'Basis of finite element methods for solid continua', *Int. J. Num. Meth. in Eng.*, **1,** 3–28 (1968) (by T. H. H. Pian and P. Tong).
38. 'A variational principle and the convergence of a finite-element method based on assumed stress distribution', *Int. J. Solids and Structures*, **5,** 463–472 (1969) (by P. Tong and T. H. H. Pian).
39. 'Rationalization in deriving element stiffness matrix by assumed stress approach', *Proc. Second Conf. on Matrix Methods in Structural Mechanics*, pp. 441–469, AFFDL-TR-68-150, Wright-Patterson Air Force Base, 1969 (by T. H. H. Pian and P. Tong).
40. 'Buckling of radially constrained thin spherical shell under edge load', *Int. J. Non-Linear Mechanics*, **5,** 113–130 (1970) (by P. Tong and T. H. H. Pian).
41. 'Finite element stiffness methods by different variational principles in elasticity', *SIAM-AMS Proc. on Numerical Solutions of Field Problems in Continuum Physics* (Eds. G. Birkhoff and R. S. Varga) pp. 253–271, Am. Math. Soc., 1970.
42. 'Bounds to the influence coefficients by the assumed stress method', *Int. J. Solids and Structures*, **6,** 1429–1432 (1970) (by P. Tong and T. H. H. Pian).
43. 'Formulations of finite element methods for solid continua', *Recent Advances in Matrix Methods of Structural Analysis and Design*, (Eds. R. H. Gallagher, Y. Yamada, and J. T. Oden) pp. 49–83, The University of Alabama Press, 1971.
44. 'Variational formulations of numerical methods in solid continua', in *Computer-Aided Engineering* (Ed. G. M. L. Gladwell), pp. 421–448, Solid Mechanics Division, University of Waterloo, Waterloo, Ontario, 1971.
45. 'Variational formulation of finite-displacement analysis', in *High Speed Computing of Elastic Structures* (Ed. B. Fraeijs de Veubeke), pp. 43–63, Universite' de Liege, 1971 (by T. H. H. Pian and P. Tong).
46. 'Mode shapes and frequencies by finite element method using consistent and lumped masses', *Computers and Structures*, **1,** 623–628 (1971) (by P. Tong, T. H. H. Pian, and L. L. Bucciarelli).
47. 'Finite element solutions for laminated thick plates', *J. Comp. Materials*, **6,** 304–311 (April 1972) (by S. T. Mau, P. Tong, and T. H. H. Pian).
48. 'Some recent studies in assumed stress hybrid models', in *Advances in Computational Mechanics and Design* (Eds. J. T. Oden, R. W. Clough, and Y. Yamamoto), pp. 87–106, UAH Press, University of Alabama in Huntsville, 1972 (by T. H. H. Pian and S. T. Mau).
49. 'Numerical formulation of finite element methods in linear-elastic analysis of general shells', *J. Structural Mechanics*, **1,** 1–41 (1972) (by S. Atluri and T. H. H. Pian).
50. 'Finite element methods in continuum mechanics', in *Advances in Applied Mechanics* (Ed. C. S. Yih), Vol. 12, Academic Press, 1972 (by T. H. H. Pian and P. Tong).
51. 'Finite element analysis of shells of revolution by two doubly curved quadrilateral elements', *J. Structural Mechanics*, **1,** No. 3 (1972) (by S. Atluri and T. H. H. Pian).

52. 'Finite element formulation by variational principles with relaxed continuity requirements', in *The Mathematical Foundations of the Finite Element Method with Applications to Partial Differential Equations* (Ed. A. K. Aziz), pp. 671–687, Academic Press, 1972.
53. 'On the convergence of the finite element method for problems with singularity', *Int. J. Solids and Structures,* **9,** 313–321 (1973) (by P. Tong and T. H. H. Pian).
54. 'Finite element methods by variational principles with relaxed continuity requirement', in *Variational Methods in Engineering* (Eds C. A. Brebbia and H. Tottenham) pp. 3/1–3/24, Southampton University Press, 1973.
55. 'Vibration analysis of laminated plates and shells of a hybrid stress element', *AIAA J.*, **11,** 1450–1452 (1973) (by S. T. Mau, T. H. H. Pian, and P. Tong).
56. 'Hybrid models', in *Numerical and Computer Methods in Structural Mechanics,* (Eds. S. J. Fenves *et al.*), pp. 50–78, Academic Press, 1973.
57. 'A hybrid-element approach to crack problems in plane elasticity', *Int. J. Num. Meth. in Eng.*, **7,** 297–308 (1973) (by P. Tong, T. H. H. Pian, and S. Lasry).
58. 'Postbuckling analysis of shells of revolution by the finite element method', in *Thin-Shell Structures* (Eds. Y. C. Fung and E. E. Sechler), pp. 435–452, Prentice-Hall, 1974 (by P. Tong and T. H. H. Pian).
59. 'Laminated thick plate and shell analysis by the assumed stress hybrid model', *Proc. Army Symp. on Solid Mechanics,* 1972, pp. 25–35, AMMRC MS 73-2, September 1973, (by S. T. Mau, T. H. H. Pian, and P. Tong).
60. 'Elastic crack analysis by a finite element hybrid method', *Proc. Third Conf. on Matrix Methods in Structural Mechanics,* pp. 661–682, AFFDL-TR-71-160, December 1973 (by T. H. H. Pian, P. Tong, and C. H. Luk).
61. 'Derivation of geometric stiffness and mass matrices for finite element hybrid models', *Int. J. Solids and Structures,* **10,** 919–932 (1974) (by P. Tong, S. T. Mau, and T. H. H. Pian).
62. 'Elastic–plastic analysis by assumed stress hybrid model', *Proc. 1974 Int. Conf. on Finite Element Methods in Engineering* (Ed. V. A. Pulmano and A. P. Kabaila), pp. 419–434, University of New South Wales, Australia, 1974 (by T. H. H. Pian, P. Tong, C. H. Luk, and R. L. Spilker).
63. 'Nonlinear creep analysis by assumed stress finite element method', *AIAA J.*, **12,** 1756–1758 (1974).
64. 'Elastic–plastic creep analyses by assumed stress finite elements', *Trans. Third Int. Conf. on Structural Mechanics in Reactor Technology*, Commission of the European Communities, Brussels, September 1975, Vol. 5, Paper No. M2/1, pp. 1–9 (by T. H. H. Pian, R. L. Spilker, and S. W. Lee).
65. 'Crack elements', *Proc. World Congress on Finite Element Methods in Structural Mechanics* (Ed. J. Robinson), pp. F.1–F.39, Robinson and Associates, Dorset, England, 1975.
66. 'Application of finite element method to mixed-mode fracture', *Recent Advances in Engineering Science,* **6,** 255–263 (1976) (by P. Tong and T. H. H. Pian).
67. 'Nonlinear analysis by assumed stress hybrid models', Invited lecture, *Proc. 24th Japan National Congress of Applied Mechanics,* Tokyo, November 1974, pp. 1–10, University of Tokyo Press, 1976.
68. 'Notes on finite elements for nearly incompressible materials', *AIAA J.*, **14,** 824–826 (1976) (by T. H. H. Pian and S. W. Lee).
69. 'Variational principles for incremental finite element methods', Vol. 302, Nos. 5/6, pp. 473–488, J. Franklin Institute.
70. 'Large deflection analysis of thin elastic structures by the assumed stress hybrid

finite element method', *Computers and Structures*, **7,** 1–12 (1977) (by P. L. Boland and T. H. H. Pian).

71. 'Fracture mechanics analysis with hybrid "crack" finite elements', *Case Studies in Fracture Mechanics*, AMMRC MS 77-5 pp. 2.4.1–2.4.16, 1977 (by P. Tong, T. H. H. Pian, and O. Orringer).
72. 'Formulations of large deflection shell analysis by assumed stress finite element method', in *Formulations and Computational Algorithms in Finite Element Analysis* (Ed. K. J. Bathe, J. T. Oden, and W. Wunderlich), pp. 241–264, M.I.T. Press, 1977 (by T. H. H. Pian and P. L. Boland).
73. 'Creep and viscoplastic analysis by assumed stress hybrid finite elements', *Proc. Int. Conf. on Finite Elements on Nonlinear Solid and Structural Mechanics*, Geilo, Norway, 29 Aug.–1 Sept. 1977, Paper No. F06.1 (by T. H. H. Pian and S. W. Lee).
74. 'Hybrid finite element procedures for analyzing through flaws in plates in bending', *Trans. Fourth Int. Conf. on Structural Mechanics in Reactors Technology*, Paper No. M2/4, San Francisco, California, 15–19 Aug. 1977 (by H. C. Rhee, S. N. Atluri, K. Moriya, and T. H. H. Pian).
75. 'Three dimensional crack element by assumed stress hybrid model', *Proc. Fourteenth Annual Meeting of the Society of Engineering Science*, Bethlehem, Pennsylvania, 14–16 Nov. 1977, pp. 913–917 (by T. H. H. Pian and K. Moriya).
76. 'Improvement of plate and shell finite elements by mixed formulations', *AIAA J.*, **16,** No. 1, 29–34 (January 1978) (by S. W. Lee and T. H. H. Pian).
77. 'Three dimensional fracture analysis by assumed stress hybrid elements', *Numerical Methods in Fracture Mechanics, Proc. First Int. Conf. Held at the University College Swansea*, 9–13 January 1978, pp. 363–373 (by (T. H. H. Pian and K. Moriya).
78. 'Variational and finite element methods in structural analysis', *Collection of Papers of the Twentieth Israel Annual Conf. on Aviation and Astronautics*, Tel-Aviv and Haifa, 22–23 Feb. 1978, pp. 120–128.
79. 'Finite element analysis of rubber-like materials by a mixed model', *Int. J. Num. Meth. in Eng.*, **12,** 665–676 (1978) (by T. Scharnhorst and T. H. H. Pian).
80. Book Review: *Structural Mechanics Software Series, Vol. 1*, Mechanics Research Communications, Vol. 4, No. 6, 1977.
81. 'A historical note about "Hybrid Elements"', *Int. J. Num. Meth. in Eng.*, **12,** 891–892 (1978).
82. 'A study of axisymmetric solid of revolution elements based on the assumed-stress hybrid model', *Computers and Structures*, **9,** 273–279 (1978) (by R. L. Spilker and T. H. H. Pian).
83. 'Variational and finite element methods in structural analysis', *Israel J. of Technology*, **16,** No. 1–2, 23–33 (1978); also published in *RCA Review*, **39,** No. 4, 648–664 (1978).
84. 'Hybrid-stress models for elastic–plastic analysis by the initial-stress approach', *Int. J. Num. Meth. in Eng.*, **14,** 359–378 (1979) (by R. L. Spilker and T. H. H. Pian).
85. 'Formulation of contact problems by assumed stress hybrid elements', in *Nonlinear Finite Element Analysis in Structural Mechanics* (Eds. W. Wunderlich, E. Stein, and K. J. Bathe), pp. 49–59, Springer-Verlag, 1981 (by T. H. H. Pian and K. Kubomura).
86. 'Large deformations of rigid-plastic shallow spherical shells', *Int. J. Mech. Sci.*, **23,** 69–76 (1981) (by K. Kondo and T. H. H. Pian).

87. 'Calculation of interlaminar stress concentration in composite laminates', *J. Composite Materials*, **15,** 225–239 (1981) (by P. Bar-Yoseph and T. H. H. Pian).
88. 'Large deformations of rigid-plastic circular plates', *Int. J. Solids and Structures*, **17,** 1043–1055 (1981) (by K. Kondo and T. H. H. Pian).
89. 'Large deformations of rigid-plastic beams', *J. Structural Mechanics*, **9,** 139–159 (1981) (by K. Kondo and T. H. H. Pian).
90. 'Large deformations of liquid-plastic polygonal plates', *J. Structural Mechanics*, **9,** 271–293 (1981) (by K. Kondo and T. H. H. Pian).

Author Index

Ahmad, S., 537, 548, 562
Aizawa, T., 345, 346, 359
Akay, H. U., 232, 240
Altmann, W., 233, 240
Altman, W., 537, 563
Al-Yassin, Z., 16
Andersen, C. M., 233, 240, 537, 538, 542, 547, 548, 563
Anderson, R. G., 537, 548, 562
Ando, Y., 316, 345, 346, 358, 359, 364, 379
Apostal, M. C., 345, 357
Argyris, J. H., 91, 92, 453, 464
Arnold, D. N., 448, 451
Ashwell, D. G., 88, 92, 537, 562
Atluri, S. N., 51, 57, 58, 59, 60, 62, 63, 64, 65, 66, 67, 68, 69, 70, 71, 243, 250, 315, 323, 345, 358
Aziz, A. K., 141, 440, 471, 486

Babuska, I., 19, 48, 59, 70, 141, 238, 241, 244, 251, 433, 436, 438, 440, 441, 442, 450, 471, 486
Ball, R. E., 548, 557, 563
Barnard, A. J., 118
Barony, S. Y., 236, 240
Barsoum, R. S., 325, 343, 345, 358
Bathe, K. J., 137, 327, 343, 497, 510, 513, 515, 516, 517, 548, 564
Batoz, J. L., 117, 137
Baynham, J. M. W., 405, 512, 517
Bazant, Z. P., 325, 340, 343
Beckers, P., 267, 279, 548, 563
Beisinger, Z. E., 537, 562
Belytschko, T., 117, 132, 134, 541, 563
Benthem, J. P., 325, 340, 343
Bercovier, M., 141, 142, 467, 476, 485, 487, 495, 499, 501, 502, 503, 504, 512, 516, 517
Beresford, P. L., 187
Besseling, J. F., 253, 258, 261, 262, 266
Bicanic, N., 541, 542, 563, 563
Bismark-Nasr, M. N., 537, 563
Bogner, F. K., 232, 240
Boland, P., 68, 71
Boland, P. L., 308, 323
Bratianu, C., 59, 60, 62, 70
Brezzi, F., 59, 70, 141, 267, 268, 269, 271, 273, 276, 280, 471, 486, 487, 495
Bron, J., 232, 240
Brooks, A., 59, 70, 499, 516
Budiansky, B., 236, 237, 241
Bushnell, D., 236, 241
Byskov, E., 345, 358

Campbell, D. M., 3, 16
Cantin, G., 88, 92
Canuto, C., 267, 280
Carey, G. F., 487, 495
Carnoy, E., 315, 323
Chan, A. S. L., 236, 240
Charlwood, R. G., 453, 465
Chen, D. P., 59, 70
Chen, Wanji, 173, 176, 183, 188
Cheung, Y. K., 32, 49, 155, 164
Chien, W. Z., 318, 383, 404, 520, 535
Chou, S. C., 118
Chryssafi, S., 289, 307, 308, 322
Ciarlet, P. G., 139, 143, 147, 268, 269, 270, 271, 280, 440, 451, 451, 503, 516
Cimento, A. P., 517
Clough, R. W., 419, 430, 88, 92
Cohen, M., 117, 541, 563
Conner, J., 233, 240
Cook, R. D., 537, 563
Cook, T. S., 325, 336, 337, 343
Cosserat, E., 1, 16
Cosserat, F., 1, 16
Cowper, G. R., 32, 49, 78, 88, 91, 92

Crawley, E. F., 537, 562
Crouzeix, M., 499, 516
Cusens, A. R., 164

De Jong, J. E., 494, 496
De Veubeke, B. F., 58, 59, 69, 78, 92, 189, 267, 270, 279, 308, 315, 323, 453, 465
Devore, R., 439, 450
Dhalla, A. K., 453, 465
Dhatt, G., 232, 240
Dong, S. B., 361, 379
Dorr, M. R., 440, 451
Douma, T., 325, 340, 343
Dunham, R. S., 19, 48, 243, 250
Durand, M., 504, 516

Edwards, G., 78, 91
Eisenhart, L. P., 84, 92
Elias, Z. M., 453, 465
Engelman, M., 499, 503, 512, 516, 517
England, A. H., 335, 343
Erdogan, F., 325, 336, 337, 338, 339, 340, 343
Eringen, A. C., 1, 16
Estenssord, L. F., 325, 340, 343
Evensen, D. A., 537, 548, 562
Ezawa, Y., 326, 327, 343

Falk, S., 231, 240
Farris, R. J., 12, 17
Finch, J. R., 236, 237, 241
Fix, G. J., 419, 430, 487, 495
Flanagan, D. P., 134
Fleury, C., 282, 288
Flugge, W., 88, 92
Fonder, G. A., 88
Fortin, M., 516
Fox, R. C., 232, 240
Francis, J. G. F., 331, 343
Fransson, B., 115, 162, 163, 164, 166
Frazier, G. A., 542, 563
French, S. E., 537, 562
Fujitani, Y., 325, 340, 343
Fukuchi, N., 69, 71
Fung, Y. C., 489, 495
Fu, F. C. L., 244, 251

Gallagher, R. H., 243, 250, 345, 357, 453, 464, 465, 537, 562
Gathe, K. L., 117
Gibert, Ph., 289, 322
Giencke, E., 189, 190, 198
Glover, N., 542, 563
Godbole, P. N., 410, 429
Godier, J. N., 462
Goldberg, J. E., 236, 240
Goldman, N. L., 346, 356, 357, 358
Golub, G. H., 247, 251
Goodier, J. N., 32, 49, 465
Gordon, W. J., 454, 459, 465
Gould, P. L., 236, 240
Gray, W. G., 414, 430
Gresho, P. M., 510, 512, 516, 516, 517
Griffiths, D. F., 510, 513, 516
Gruzdev, A., 1, 16

Hall, C. A., 459, 465
Hantung, R. F., 548
Hapel, K. H., 203
Harbord, R., 232, 233, 240
Haroun, M., 117, 541, 563
Harts, B. J., 465
Hartung, R. F., 557, 564
Hartz, B. J., 453
Hein, V. L., 325, 338, 339, 340, 343
Heinrich, J. C., 420, 430
Hellen, T. K., 345, 358
Hellinger, E., 53, 58, 69, 243, 250
Hellon, K., 19, 48
Henshell, R. D., 73, 76, 91, 325, 343, 345, 358
Herrmann, L. R., 2, 3, 6, 7, 9, 10, 12, 15, 16, 17, 48, 98, 116, 232, 240, 243, 250, 381, 404, 420, 430, 490, 496
Hilton, P. D., 345, 358
Hinton, E., 117, 419, 430, 541, 542, 563
Hodge, P. G., 521, 522, 535
Hofbaver, E., 538, 563
Hoffmeister, D., 537, 548, 562
Horgan, C. O., 537, 563
Horrigmoe, G., 59, 70
Ho, L. W., 117, 137
Hughes, T. J. R., 59, 70, 117, 122, 132, 410, 417, 418, 420, 429, 430, 476, 476, 486, 486, 487, 488, 489, 492, 494, 495, 497, 499, 503, 516, 541, 541, 563
Hutchinson, J. R., 12, 17
Hutchinson, J. W., 345, 346, 350, 353, 354, 355, 356, 357, 358, 359
Hu, H. C., 53, 54, 69, 243, 250

Iguti, F., 233, 240
Iida, S., 345, 357
Irons, B. M., 537, 562
Irwin, G. R., 325, 343
Ishida, M., 352, 353, 359

Jain, P. C., 420, 430
Jankovich, E., 504, 516
Jenning, L., 247, 251
Jiang, H., 182
Jin, W., 185
Johnson, C., 487, 495
Jordan, S., 345, 357

Kaaoknukulchai, H., 117, 132
Kabaila, A. P., 453, 465
Kalnins, A., 91, 92, 236, 241
Kania, E., 124, 127
Kantorovich, L. V., 155
Karrholm, G., 155
Katz, I. N., 440, 441, 450
Kavanagh, K. T., 132
Kawai, T., 93, 102, 105, 109, 116
Keer, L. M., 361, 379
Kellogg, R. B., 440, 451
Kelsey, S., 453, 464
Key, S. W., 132, 488, 490, 494, 495, 537, 542, 562, 563
Kikuchi, F., 137, 140, 146
Kikuchi, N., 59, 70, 467, 471, 472, 475, 476, 478, 480, 485, 486, 487
Kim, S. J., 477, 486
Knops, R. J., 290, 323
Kobayashi, A. S., 345, 357, 358
Kohler, M., 203
Kohn, R., 281, 284, 298
Kohn, W., 243, 250
Koiter, W. T., 290, 323
Kondou, K., 93, 94, 102, 105, 116
Kondrat'ev, V. A., 139
Kosko, E., 78, 92
Kosloff, D., 542
Krieg, R. D., 542, 563
Kromnansl, J. A., 243, 250
Kross, D., 137
Krylov, V. I., 155

Lachance, L., 232, 233, 240
Ladkany, S. G., 118
Lang, K. W., 244, 251
Lasry, S. J., 325, 343
Lawther, R., 453, 465
Leblanc, F., 504, 516
Leckie, F. A., 346, 354, 356, 359
Lee, E. H., 243, 247, 250, 251
Lee, J. K., 19, 48, 59, 70, 238, 241
Lee, K. N., 244, 251, 407, 429
Lee, R. L., 510, 513, 516, 516
Lee, S. L., 520, 535
Lee, S. W., 62, 70, 184, 498, 507, 510, 516, 537, 562
Letallec, P., 499, 500, 504, 516
Levy, J. F., 410, 420, 429
Liebowicz, H., 325, 342
Lindberg, G. M., 32, 49, 78, 88, 91, 92, 537, 548, 553, 562
Lin, K. Y., 325, 336, 337, 338, 343
Liu, W. K., 132, 133, 499, 516
Liu, Yingxi, 173, 176, 183, 185
Lockner, N., 91, 92
Look, R. D., 118, 132
Loo, Y. C., 164
Love, A. E., 84, 92
Lui, W. K., 59, 70
Luk, C. H., 345, 358, 361, 364, 379
Lu, Hexiang, 13, 176
Lynch, D. R., 414, 430

Macneal, R. H., 537, 562
Maiden, D., 345, 357
Makoju, J. O., 82, 92
Malkus, D. S., 118, 122, 417, 418, 430, 487, 495, 498, 499, 503, 504, 505, 516
Mang, H. A., 243, 250
Marcal, P. V., 345, 357
Marguerre, K., 205
Marini, L. D., 267, 268, 270, 280
Mar, J. W., 325, 336, 337, 338, 343
Maskeri, S. M., 122, 124, 127
Mathers, M. D., 560, 563
Matsuo, A., 105, 116
Mau, S. T., 118, 537, 562
Mesiere, Y., 306, 323
Midlin, R. D., 13
Millard, A., 453, 465
Miller, A., 438, 450
Minagawa, S., 244, 247, 248, 251
Mindlin, R. D., 1, 2, 13, 16, 84, 92, 117
Mirza, F. A., 19, 20, 21, 48, 49
Mirza, R. A., 20
Mitzutani, A., 139

Miyazaki, N., 361, 364, 379
Morely, L. S. D., 88, 92, 453, 465
Mosolov, P. P., 288
Mriza, F. A., 21
Mukadai, Y., 105, 116
Munir, N. I., 118, 121, 124, 127
Murakawa, H., 51, 62, 64, 65, 66, 67, 68, 69, 70, 71, 315, 323
Mura, T., 535

Nadai, A., 283, 288
Nagtegaal, J. C., 42, 494, 496, 498, 502, 510, 516
Nakagaki, M., 345, 358
Nakazawa, S., 414, 430
Naylor, D. J., 420, 430
Needleman, A., 346, 356, 357, 359
Nellinger, E., 220, 239
Nemat-Nasser, S., 243, 244, 247, 248, 250, 251, 361, 379, 407, 429, 537, 563
Newton, R. E., 415, 416, 430
Nishiguchi, I., 326, 327, 343
Nishioka, T., 345, 346, 358, 361, 364, 379
Niyazaki, N., 345, 346, 358
Noor, A. K., 233, 240, 537, 538, 542, 546, 548, 563, 564
Novozhilov, V., 84, 86, 92

Oden, J. T., 19, 48, 59, 70, 137, 189, 238, 241, 243, 250, 421, 430, 431, 467, 471, 472, 472, 473, 475, 476, 477, 478, 480, 485, 486, 487, 487, 495, 497, 499, 516
Ogura, N., 345, 346, 358, 361, 364, 379
Ohayon, R., 289, 323
Okabe, M., 326, 327, 343
Okumura, H., 325, 332, 340, 343
Olson, E. T., 487, 495
Olson, M. D., 20, 22, 26, 28, 32, 47, 49, 78, 88, 91, 92, 537, 548, 553, 562
Onate, E., 420, 430, 542, 563
Orringer, O., 118, 537, 562
Orris, R. M., 247
Orris, T. M., 251
Osborn, J. E., 244, 251
Osborne, J., 411, 430, 442, 541

Pagh, E. D. L., 541
Panks, D. M., 507
Paris, P. C., 325, 343
Parisch, H., 542, 564
Parks, D. M., 345, 357, 421, 430, 498, 502, 510, 516
Pawsey, S. F., 419, 430
Pedersen, P., 282, 288
Petyt, M., 247, 251
Pian, T. H. H., 36, 49, 53, 57, 58, 59, 62, 68, 69, 70, 71, 74, 75, 91, 117, 118, 173, 184, 215, 239, 243, 249, 250, 267, 270, 279, 280, 308, 323, 325, 343, 345, 358, 361, 364, 379, 381, 389, 404, 405, 429, 431, 498, 510, 516, 520
Pironneau, O., 141, 142
Pister, K. S., 19, 48, 243, 250, 490, 496
Pitkaranta, J., 440, 451, 487, 495
Poceski, A., 232, 240
Prager, W., 243, 248, 250, 251, 521, 535
Prange, G., 220, 239
Prato, C. A., 233, 240
Prokopov, V. K., 1, 16
Pugh, D. E. L., 117, 563

Quarteroni, A., 267, 279

Rabier, P., 279, 280
Ranaweera, M. P., 346, 354, 356, 359
Rappaz, J., 267, 273, 280
Raviart, P. A., 19, 48, 267, 273, 280, 307, 323, 499, 503, 507, 516
Raviart, R. A., 499
Reddy, J. N., 19, 48, 238, 241, 243, 250
Reed, K. W., 69, 71
Reissner, E., 53, 58, 69, 220, 239, 243, 250
Rensch, H. J., 236, 241
Rhee, H. C., 58, 59, 69
Rheinbolt, W., 433, 438, 438, 450
Rice, J. R., 345, 350, 358, 421, 430, 498, 502, 507, 510, 516
Richmondi, G., 516
Rigby, F. N., 73, 82, 87, 92
Rimondi, G., 500
Rosengren, G. F., 345, 350, 358
Rubinstine, R., 64, 69, 70, 71
Rybicki, E. F., 453, 465

Sabir, A. B., 88, 92
Samuelsson, A., 155
Sanders, G., 267, 279, 315, 323, 548, 564
Sani, R. L., 510, 510, 512, 513, 516, 517
Savin, G. N., 32, 49

Schafer, H., 232, 240
Schamber, R. A., 16, 2
Schapery, R. A., 12, 17
Scherer, K., 439, 450
Schmit, L. A., 232, 240, 282, 288, 453, 465
Schoenstadt, A., 414, 430
Sen, S. K., 236, 240
Shames, I. H., 1, 16
Shaw, K. G., 325, 343, 345, 358
Shih, C. F., 345, 346, 353, 354, 355, 356, 357, 358, 359
Shiina, S., 102, 116
Shi, G., 183
Shi, P. M., 521, 535
Sih, G. C., 325, 342, 343
Simon, B., 345, 357
Slater, J. M., 411, 430
Sodhi, D., 464
Song, S. W., 59, 70
Song, Y. J., 467, 471, 472, 472, 475, 476, 478, 480, 485
Sperlich, J., 203
Spilker, R. L., 117, 118, 121, 122, 124, 127, 537, 562
Stagg, K. G., 32, 49
Stolarski, H., 541, 563
Strang, G., 281, 284, 288
Stummel, F., 147
Sullivan, C. J., 78, 80, 92
Sumi, Y., 317, 345, 358, 361, 379
Sundararajan, C., 464, 465
Szabo, B. A., 440, 441, 450, 451

Tabarrok, B., 464, 465
Tahiani, C., 232, 233, 240
Takeuchi, N., 116, 93
Talaslidis, D., 230, 231, 232, 233, 238, 240, 241, 537, 563
Tang, Limin, 173, 176, 183, 185, 188
Tanrikulu, M., 248, 251
Tao, Z., 185
Tavarra, G., 500, 516
Taylor, R. L., 84, 92, 117, 132, 187, 405, 410, 411, 419, 420, 429, 430, 453, 465, 490, 496, 512, 517, 541, 563
Tezduyar, T. E., 117
Thomas, J. M., 48
Tierstin, H. F., 1, 16
Timoshenko, S. P., 32, 49, 76, 91, 126, 462, 465
Ting, T. W., 283, 288
Toi, Y., 109, 116
Tokuda, N., 345, 358, 361, 366, 371, 379
Tong, P., 36, 49, 51, 57, 58, 59, 62, 69, 70, 75, 91, 117, 118, 243, 249, 250, 270, 280, 325, 343, 345, 358, 361, 364, 379, 381, 389, 404, 520, 520, 535
Too, J., 84, 92, 117, 419
Too, S. M., 541
Tottenham, H., 236, 240
Toukda, N., 378
Toupin, R., 253, 266
Too, J. M., 430
Trbojevic, V. M., 236, 240
Triebel, H., 439, 450
Truesdell, C., 253, 266
Tsay, C. S., 132, 133
Tseng, J., 19, 26, 28, 47, 49
Tsien, L. H., 520, 535
Tsoon, W. S., 520, 535
Turner, C. E., 353, 359

Valid, R., 289, 290, 295, 298, 300, 303, 308, 311, 313, 315, 322, 323
Verbiese, S., 537, 562
Visser, W., 232, 233, 240
Vogelius, M., 433, 450

Wachspress, E. L., 187
Walker, K. P., 537, 562
Wang, L., 527, 535
Washizu, K., 53, 54, 69, 94, 116, 139, 140, 219, 239, 243, 250, 347, 359, 520, 535, 538, 563
Watanabe, M., 94, 102, 105, 110, 115, 116
Watkins, D. S., 465
Watwood, V. B., 453, 465
Webster, J. J., 73
Weinberger, H. F., 246, 251
Wempner, G. A., 137
Westergaard, H. M., 367, 368, 370, 370, 372, 379
Wilkes, E. W., 290, 323
Williams, M. L., 325, 335, 336, 342, 366, 379
Williams, R. T., 414, 430
Will, D., 233, 240
Wilson, E. L., 187, 327, 343, 458, 563
Witmer, E. A., 118, 537, 537, 562
Woinowsky-Krieger, S., 126, 76, 91

Wolf, J. P., 59, 70
Worm, S., 206, 208
Wunderlich, W., 155, 215, 216, 217, 220, 224, 225, 228, 230, 231, 232, 233, 236, 239, 240, 241, 537, 563

Yagawa, G., 345, 346, 358, 359, 361, 364, 379
Yamada, M., 244, 247, 248, 251
Yamada, Y., 325, 326, 327, 332, 340, 343
Yamamoto, Y., 345, 358, 361, 366, 371, 378, 379
Yang, W. H., 247, 251
Ying, L-A., 59, 62, 70

Zienkiewicz, O. C., 9, 17, 32, 49, 84, 92, 117, 189, 243, 250, 258, 266, 347, 348, 359, 405, 410, 414, 419, 420, 429, 430, 453, 464, 465, 512, 517, 537, 541, 541, 548, 562, 563

Due